ABOUT THE AUTHORS

William A. Sheppard was born in Hamilton, Ontario, and became a U.S. citizen in 1965. He received the degrees of B.Sc. in 1950 and M.Sc. in 1951 from McMaster University in Hamilton and Ph.D. in 1954 from Massachusetts Institute of Technology. During 1954 and 1955 he was a Postdoctoral Fellow at the Hickrill Research Foundation. After a year teaching at Yale University, Dr. Sheppard joined the research staff of the Central Research Department, duPont, where he is now a research supervisor. During the past year he spent a semester as Visiting Professor at the State University of New York at Buffalo.

Clay M. Sharts received the B.S. degree from the University of California, Berkeley, in 1952, and the Ph.D. from the California Institute of Technology in 1959. He was a Research Chemist in the Explosives Department, duPont, from 1958 to 1962. Since 1962 he has been in the Chemistry Department of San Diego State College, where he is currently Professor of Chemistry.

ORGANIC FLUORINE CHEMISTRY

FRONTIERS IN CHEMISTRY

Ronald Breslow and Martin Karplus, Editors

T. A. Bak Contributions to the Theory of Chemical Kinetics

C. J. Ballhausen and H. B. Gray Molecular Orbital Theory

P. D. Bartlett Nonclassical Ions

R. A. Bernheim Optical Pumping

M. Bersohn and J. C. Baird Electron Paramagnetic Resonance

T. C. Bruice and S. J. Benkovic Bioorganic Mechanisms: Volumes I and II

D. P. Craig and S. H. Walmsley Excitons in Molecular Crystals

M. C. Hamilton and J. A. Ibers Hydrogen Bonding in Solids

T. L. Hill Thermodynamics of Small Systems: Parts 1 and 2

G. H. Langford and H. B. Gray Ligand Substitution Processes

R. G. Parr The Quantum Theory of Molecular Electronic Structure

J. H. Richards and J. B. Hendrickson The Biosynthesis of Steroids, Terpenes, and Acetogenins

W. A. Sheppard and C. M. Sharts Organic Fluorine Chemistry

R. Stewart Oxidation Mechanisms

N. J. Turro Molecular Photochemistry

K. B. Wiberg Computer Programming for Chemists

ORGANIC FLUORINE CHEMISTRY

WILLIAM A. SHEPPARD

Central Research Department, E. I. du Pont de Nemours and Company

CLAY M. SHARTS

San Diego State College

W. A. BENJAMIN, INC.

New York

1969

ORGANIC FLUORINE CHEMISTRY

Standard Book Number: 8053-8790-0 (Cloth)
Library of Congress Catalog Card Number 69-17691
Manufactured in the United States of America
12345MR 32109

The manuscript was put into production on August 8, 1968;
this volume was published on May 1, 1969

W. A. BENJAMIN, INC.
New York, New York 10016

Foreword

Organic fluorine chemistry is only in its adolescence compared to most other fields of chemistry. Swartz did yeoman pioneering work in the early part of the century, but because of hazards and difficulties in handling highly reactive and corrosive fluorinating agents, the field was left relatively undeveloped until the advent of Freons® in the late 1930's. The requirements of World War II for materials with unique and superior properties gave great impetus to research in fluorine chemistry. In particular, the Manhattan Project led to remarkable technological developments and provided many research workers with the background needed to start productive research programs. However, organic fluorine chemistry still remained separated from the main stream of organic chemistry, probably because of the gross differences in properties and reactions of fluorinated organic derivatives relative to hydrocarbon analogs. Research was further hindered by a lack of commercial availability of many fluorinated reagents. Progress was made mainly by industrial research chemists, particularly in the Du Pont Laboratories, and by devotees in a limited number of universities.

In the last decade the whole picture has changed. Organic chemists now recognize the changes in chemical reactivity and properties caused by introduction of fluorine or fluorinated groups into organic compounds. A variety of fluorinated reagents and intermediates is now readily available from chemical suppliers (see Appendix C), so that syntheses of fluorinated compounds by the average chemist are now practical. Consequently, many chemists use fluorine or fluorinated substituents as labels or probes, as controlling substituents for mechanism studies, or as activators of some chemical or biological property. In particular, the enhancement of biological effects by fluorine substitution has been pursued through research programs in many pharmaceutical firms.

The impact of modern instrumentation on organic fluorine chemistry cannot be overstated. Gas chromatography has permitted previously impossible separations. With F^{19} nmr spectroscopy as an analytical and research tool, the isolation and identification of new fluorine compounds are almost as routine as for classical organic compounds.

Probably the greatest impetus to fluorine research has been the ever increasing discovery of practical applications for fluorine compounds. Consequently, industrial laboratories have developed strong research programs and supported academic work in a variety of fluorine fields. Also, the government has funded major programs, particularly on nitrogen-fluorine and oxygen-fluorine chemistry and on stable fluorinated materials. Many of today's fluorine chemists entered the field via government contract work.

As evidence of the growing importance and interest in fluorine chemistry, an international symposium has been held on alternate years since 1961. A recently established Fluorine Division of the American Chemical Society is now flourishing, with regular programs at national meetings (the authors can be contacted for information on membership).

WILLIAM A. SHEPPARD

CLAY M. SHARTS

Wilmington, Delaware
San Diego, California
October 1968

Preface

In the late 1950's, we realized that no good general text on organic fluorine chemistry was available, particularly one that emphasized modern synthetic developments. With the encouragement of Prof. J. D. Roberts and of colleagues at the Du Pont Company, we undertook to prepare a general review. However, while we were still in the process of literature searching, both the excellent volume by Hudlicky and the Houben-Weyl compilation appeared (see Appendix A, Ref. A1 and A2). In addition, numerous specialized reviews (see Appendixes A and B) were published. In 1964 one of us (CMS) taught a graduate-level, special-topics course on fluorine chemistry. As a result of this experience, we reevaluated the state of the literature and realized the urgent need for a review in organic fluorine chemistry different from any available. We concluded that a book was needed that could introduce the field of fluorine chemistry to the average organic chemist and permit him to acquire quickly the background necessary to use fluorine chemistry in his research. Our objective was to present a critical interpretation of the field, using the mechanistic approach of modern organic chemistry. Also, we wanted to evaluate new synthetic methods to introduce fluorine or fluorinated units selectively and to compare these methods to older classical techniques. We have not prepared a comprehensive text for experts in fluorine chemistry, but rather a critical review that will aid all organic chemists to use organic fluorine compounds in their research. We have provided judgments as to what fluorine literature is useful and important and how to obtain the primary reference material needed for a particular problem. The book should also serve as an introduction to organic fluorine chemistry for graduate students and as the basis for a special-topics course. For the convenience of students and instructors, and the entertainment of certain readers, we have included a selection of problems.

By way of amplification, we point out again that this is a critical, not a comprehensive review of fluorine chemistry. Complete coverage of the literature would require several volumes. We have used the following organization. After a brief introduction to point out the anomalous nature of organic fluorine compounds (Chapter 1), we delve into the mechanistic aspects and discuss organic fluorine chemistry in terms of modern physicial organic chemistry

(Chapters 2 and 3). Chapter 4 is a brief summary of "classical methods" of fluorination to set the stage for Chapters 5 and 6, which review exhaustively and evaluate critically modern methods for selective introduction of fluorine (mainly one to three atoms) and fluorinated units. Chapters 7 and 8 are intended to review briefly the chemical properties of and summarize synthetic approaches to fluorinated functional groups. In Chapter 9 we have included a very important discussion on safety and toxicity, plus an indication of the importance of organic fluorine chemistry in biological applications. In Appendixes A and B we have compiled and evaluated critically the review literature of fluorine chemistry. Appendix C gives some information on the suppliers of fluorinating agents and fluorinated intermediates, and on available commercial literature. As already pointed out, we have selected the literature critically to best illustrate our discussion and give representative examples. We apologize to any readers whom we have slighted by not including their work, but in a review of this type complete coverage is impossible. We have concentrated on the literature of the last decade. (March 1968 was our cutoff date, but we include an addendum of significant publications appearing during the remainder of 1968.)

W. A. S.
C. M. S.

Acknowledgments

We express our sincere thanks to the many people who helped make this book possible: first to our respective families, who made many sacrifices so that we could complete the time-consuming task of preparing the manuscript, and who gave much-needed help in sorting references and proofreading; second, to the Du Pont Company, which generously made available a variety of needed assistance such as library facilities, secretarial help, and reproduction facilities; third, to the extremely important secretarial staff, particularly Mrs. Deena Cochran, Miss Lori Sullivan, Mrs. Janice Martin, Miss Nancy Mulligan, and Mrs. Barbara Abeyta, who contributed much in addition to superb typing skills; fourth, to Miss Mamie Reitano, Miss Jenny Chou, Mr. Ed Wonchoba, and all the staff of the Lavoisier Library at the Du Pont Experimental Station for their valuable assistance in literature searching, proofreading, and reference checking. Finally, we thank our many colleagues and friends (many of whom are much more knowledgeable in fluorine chemistry than are we), both at Du Pont and at academic institutions, for advice and encouragement. In particular, we thank Drs. T. L. Cairns, B. C. McKusick, V. A. Engelhardt, C. G. Krespan, H. E. Simmons, G. A. Boswell, S. Andreades, W. J. Middleton, D. C. England, J. F. Harris, O. W. Webster, W. C. Ripka, W. Mahler, W. H. Sharkey, C. W. Tullock, R. D. Richardson, C. S. Cleaver, and J. W. Clayton, and Professors M. F. Hawthorne, J. D. Roberts, A. Streitwieser, D. E. Applequist, R. W. Taft, J. P. Freeman, W. T. Miller, P. Tarrant, and H. E. O'Neal.

W. A. S.
C. M. S.

Contents

List of Tables

Chapter One
Introduction: The Anomalous Nature of Fluorine

Fluorine is different from the other halogens either when in the elemental state or bonded in chemical compounds. However, few chemists can clearly state even semiquantitative differences between fluorine-containing compounds and corresponding halogen-containing compounds. In this chapter we have outlined some experimental results that illustrate the difference in chemical and physical properties of organic fluorine compounds. In Chapters 2 and 3 these results will be elaborated and discussed in terms of modern organic chemistry.

This book is oriented toward applications of fluorine chemistry to organic chemistry. Consequently, emphasis is on the C–F bond. A main objective is to outline the growing importance of fluorine in the field of organic chemistry by showing the value of fluorine substitution in mechanistic studies, problems of chemical bonding, synthesis of new and unusual classes of compounds, and biological studies. In particular, we have summarized and critically evaluated new synthetic methods so that a chemist can quickly assimilate the necessary background material, consult pertinent references to primary literature, and accomplish a synthesis involving selective introduction of fluorine or a fluorinated substituent into an organic molecule.

A simple example of the unusual effect of fluorine substitution is the contrast of properties of fluorinated methanes to those of the other halogenated methanes. In Fig. 1-1, the boiling points of CH_4, CH_3F, CH_2F_2, CHF_3, and CF_4 are compared with the corresponding chloromethanes and bromomethanes. The maximum boiling point for the fluorinated methanes occurs for methylene fluoride while the chlorinated and brominated series have the expected continuous rise in boiling points with increasing halogen substitution.

Another, and more fundamental, experimental observation on fluorinated methanes is the completely unique change in bond length and bond strength in

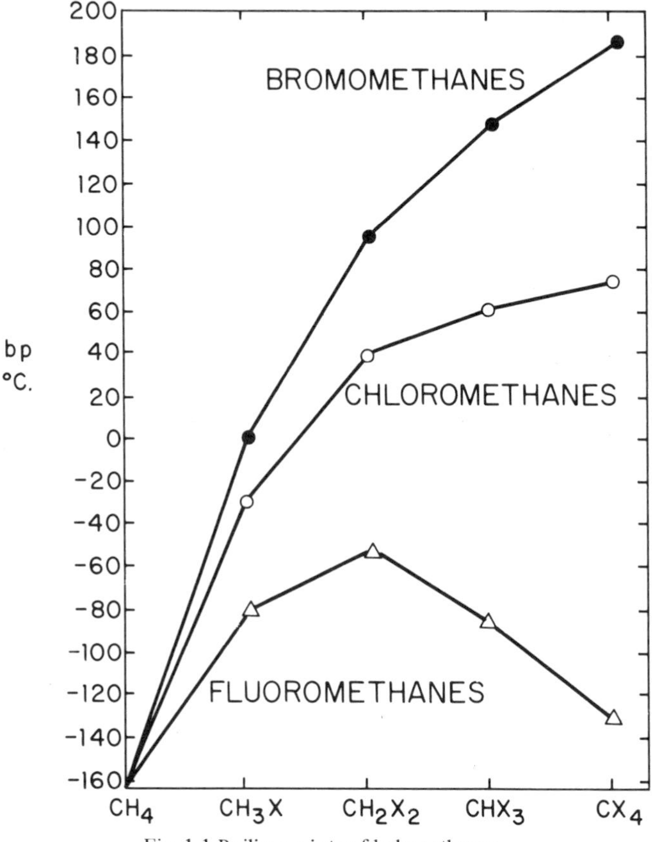

Fig. 1-1 Boiling points of halomethanes.

proceeding from methane to tetrafluoromethane. In Table 1-1, the bond lengths and bond strengths are tabulated for the series methane to tetrafluoromethane as well as for the corresponding chlorinated and brominated derivatives. As fluorine replaces hydrogen the C–F bond shortens and simultaneously the bond strength increases. This shortening and strengthening of the C–F bond with increasing substitution is in striking contrast to the chlorinated and brominated methanes, in which essentially constant bond lengths and bond strengths are observed for all substituted methanes. The same shortening and strengthening effect is found also for chlorofluoromethanes, as shown in Table 1-2. These are general phenomena characteristic of the C–F bond in almost all fluorinated organic compounds.

Fluorine is the most electronegative element. When bonded to other atoms, fluorine polarizes the bond, drawing electrons to it. The electron-attracting inductive effect of fluorine is clearly shown by the enhanced acidity of acetic acids when substituted by fluorine. All fluorinated acetic acids are stronger than

Table 1-1

Bond Lengths and Energies in Halomethanes[a]

	X = F		X = Cl		X = Br	
	r_o(C–F), Å	B (C–F), kcal	r_o(C–Cl), Å	B (C–Cl), kcal	r_o(C–Br), Å	B (C–Br), kcal
CH_3X	1.385	107.0	1.782	78.0	1.939	66.6
CH_2X_2	1.358	109.6	1.772	77.9	1.934	66.4
CHX_3	1.332	114.6	1.767	78.3	1.930	66.1
CX_4	1.317	116	1.766	78.2	1.942	66.6

[a] Based on data summarized by C. R. Patrick, *Advan. Fluorine Chem.*, **2,** 1 (1961), and from G. Glockler, *J. Phys. Chem.*, **63,** 828 (1959).

Table 1-2

Bond Lengths and Strengths in Chlorofluoromethanes[a]

Compound	r_o(C–F), Å	r_o(C–Cl), Å	B (C–F), kcal	B (C–Cl), kcal
CCl_4	–	1.765	–	78.3
CCl_3F	1.348	1.770	112.1	77.7
CCl_2F_2	1.340	1.777	112.9	77.1
$CClF_3$	1.328	1.785	114.6	76.5
CF_4	1.317	–	116.0	–

[a] Based on data summarized by G. Glockler, *J. Phys. Chem.*, **63,** 828 (1959).

the corresponding chlorinated acetic acids, confirming experimentally the relative electronegativities of fluorine and chlorine (see Table 1-3).

One of the powerful tools for studying the electron-withdrawing or -donating power of a substituent is to determine the substituent constant, σ, based on the classical method of Hammett.[1] The experimental values are frequently separated into inductive and resonance constants, σ_I and σ_R,[2] which are used qualitatively to evaluate and to compare contributions from inductive and resonance effects. A series of typical substituent parameters is given in Table 1-4.

Note that the value for the sigma inductive constant, σ_I, indicates that fluorine is marginally a better electron-withdrawing agent[3] than chlorine, bromine,

[1] L. P. Hammett, *Physical Organic Chemistry,* McGraw-Hill, New York, 1940, p. 78. For a short review on correlation between structure and reactivity in aromatic systems, see J. D. Roberts and M. C. Caserio, *Basic Principles of Organic Chemistry,* W. A. Benjamin, New York, 1964, p. 954.

[2] R. W. Taft, Jr., in *Steric Effects in Organic Chemistry* (M. S. Newman, ed.), Wiley, New York, 1956, pp. 556-675.

[3] A positive value for σ_I and σ_R indicates electron withdrawal from the ring; a negative value means electron donation.

or iodine (in that order). Other groups such as CN, NO_2, and $\overset{+}{N}(CH_3)_3$ are better electron-withdrawing groups than fluorine by significant amounts and the σ_I values are significantly larger, as expected from electronegativity considerations. What is unusual and surprising is the apparent ability of fluorine to donate electrons to the benzene ring by a resonance effect. In fact, fluorine, a first-row element, appears significantly better than the other halogens in its ability to

Table 1-3

Dissociation Constants of Halosubstituted Acetic Acids[a]

	K_a[b]		K_a[b]
CH_3CO_2H	1.77×10^{-5}		
FCH_2CO_2H	2.20×10^{-3}	$ClCH_2CO_2H$	1.40×10^{-3}
F_2CHCO_2H	5.7×10^{-2}	Cl_2CHCO_2H	3.32×10^{-2}
F_3CCO_2H	5.9×10^{-1}[c]	Cl_3CCO_2H	0.20×10^{-1}

[a] Based on data summarized by A. M. Lovelace, D. A. Rausch, and W. Postelnek, *Aliphatic Fluorine Compounds,* Reinhold, New York, 1958, p. 202; and *Handbook of Chemistry and Physics,* Chemical Rubber Publishing Co., Cleveland, 1964.

[b] In water at 25°C.

[c] Complete dissociation in aqueous solution; see A. L. Henne and C. J. Fox, *J. Am. Chem. Soc.,* **73,** 2323 (1951).

Table 1-4

Substituent Constants[a]

	σ_m	σ_p	σ_I	σ_R
NH_2	– 0.16	– 0.66	+ 0.10	– 0.76
OCH_3	– 0.12	– 0.27	+ 0.25	– 0.51
CH_3	– 0.07	– 0.17	– 0.05	– 0.11
H	0.00	0.00	0.00	0.00
F	+ 0.34	+ 0.06	+ 0.52	– 0.44
Cl	+ 0.37	+ 0.23	+ 0.47	– 0.24
Br	+ 0.39	+ 0.23	+ 0.45	– 0.22
I	+ 0.35	+ 0.28	+ 0.38	– 0.11
CF_3	+ 0.43	+ 0.54	+ 0.41	+ 0.09
CN	+ 0.68	+ 0.63	+ 0.58	+ 0.10
NO_2	+ 0.71	+ 0.78	+ 0.63	+ 0.16
$\overset{+}{N}(CH_3)_3$	+ 0.88	+ 0.82	+ 0.86	0

[a] Data based on values summarized by H. H. Jaffé, *Chem. Rev.,* **53,** 191 (1953); R. W. Taft, Jr. and I. C. Lewis, *J. Am. Chem. Soc.,* **81,** 5343 (1959); R. W. Taft, *J. Phys. Chem.,* **64,** 1805 (1960); and R. W. Taft, E. Price, I. R. Fox, I. C. Lewis, K. K. Andersen, and G. T. Davis, *J. Am. Chem. Soc.,* **85,** 709, 3146 (1963).

donate electrons through resonance. In electrophilic substitution also fluorobenzene reacts faster than chloro- or bromobenzene, although all halobenzenes are less reactive than benzene.[4] The σ_R value of -0.44 predicts that fluorine is a strong ortho-para director. Although fluorine appears similar to the rest of the halogens in directing ortho-para, it is unique in giving powerful predominance of para relative to ortho substitution. Table 1-5 gives the amounts of ortho and para isomers formed for various types of substitution reactions of fluorobenzene, chlorobenzene, and bromobenzene. The striking fact that para isomers are much more strongly favored for fluorobenzene in comparison with chloro- and bromobenzene is not easily explained, especially since this result is contrary to predictions based on steric factors.

Anomalies in fluorinated organic compounds are nowhere more prominent than in reactivities and reactions of fluoroalkenes. Organic chemists are accustomed to viewing a C=C bond as relatively invariant from compound to compound. Alkenes in which the unsaturated carbons are substituted by fluorine behave in a markedly different manner from other alkenes. This chemical difference is illustrated by data on the heats of hydrobromination, bromination, and chlorination of ethylene and various fluorinated ethylenes given in Table 1-6.

In terms of reactivity, fluorinated ethylenes are polymerized easily by radical catalysts, in contrast to the considerable difficulty encountered in polymerizing ethylene. Fluorinated ethylenes undergo rather facile addition to themselves and to activated olefins and dienes to give cyclobutane derivatives, a type of addition virtually unknown to normal olefins.[5] Many reagents add to

$$2CF_2{=}CF_2 \longrightarrow \begin{array}{ccc} F_2 & & F_2 \\ & \square & \\ F_2 & & F_2 \end{array}$$

fluorinated olefins that do not add to normal olefins. For example ethanol adds to trifluorochloroethylene to give an ether as follows:

$$CF_2{=}CFCl + C_2H_5OH \xrightarrow{C_2H_5O^-} C_2H_5O{-}CF_2{-}CFClH$$

[4] L. M. Stock and H. C. Brown, in *Advances in Physical Organic Chemistry* (V. Gold, ed.), Vol. 1, Academic, New York, 1963, pp. 35-154.

The relative rates of attack by electrophilic reagents, using benzene as a standard of 1.0, range from 0.1 and 0.7 for fluorobenzene and from 0.01 to 0.2 for chloro- and bromobenzene.

[5] J. D. Roberts and C. M. Sharts, *Org. Reactions,* **12**, 1 (1962).

Table 1-5

Directive Effects in Selected Electrophilic Substitution Reactions on Halobenzenes[a]

Type of substitution	% Isomer distribution and rate relative to benzene											
	Fluorobenzene				Chlorobenzene				Bromobenzene			
	ortho	meta	para	k_F/k_B	ortho	meta	para	k_{Cl}/k_B	ortho	meta	para	k_{Br}/k_B
Chlorination, Cl_2, $HOAc\text{-}H_2O$, 25°C	10.9	–	89.1	0.10	32.4	–	67.6	0.10	38.6	–	61.4	0.072
Nitration, NO_2BF_4 $C_4H_8SO_2$, 25°C	8.5	–	91.5	0.45	22.1	0.7	76.6	0.14	25.7	1.1	73.2	0.12
Mercuration, $Hg(OAc)_2$ HOAc, 90°C	33.7	6.9	59.4	0.53	30.1	21.5	48.5	0.090	28.2	25.8	46.0	0.080
Benzylation, $C_6H_5CH_2Cl$ $AlCl_3$, $MeNO_2$, 25°C	14.7	0.2	85.1	0.46	33.0	0.6	66.4	0.24	32.5	0.7	66.8	0.18
Bromination, Br_2[b] Fe	1.8	–	98.2	–	–	–	–	–	–	–	–	–

[a] Unless stated otherwise based on data summarized by L. M. Stock and H. C. Brown, Ref. 4.

[b] G. Olah, A. Pavlath, and G. Varsanyi, *J. Chem. Soc.*, 1823 (1957).

Table 1-6
Heats of Addition to Fluorinated Ethylenes[a]

	Heat of addition (kcal/mole) for additive		
Compound	HBr	Br_2	Cl_2
$F_2C{=}CF_2$	– 32.99	– 38.48	– 57.32
$F_2C{=}CFCl$	– 26.07	– 31.61	– 48.82
$F_2C{=}CCl_2$	– 22.05	–	– 41.08
$H_2C{=}CH_2$	– 16.8	– 23.8	– 41.5

[a] Based on data summarized by C. R. Patrick, *Advan. Fluorine Chem.*, **2**, 1 (1961).

A similar type of unusual reactivity is also found for fluorinated ketones. The influence and interaction of fluorine on and with a π system, particularly in regard to the unusual nature of fluoroalkenes and fluoroketones, is an important subject for further discussion.

When displacement reactions in organic chemistry are considered, the discussions usually center on the behavior of chloride, bromide, and iodide as nucleophilic attacking agents or as leaving groups with the usual order $I^- > Br^- > Cl^- \gg F^-$, and with little being said about the fluoride ion other than that it is a poor nucleophile. In most typical displacement reactions (solvolytic), fluoride is indeed a much poorer leaving group than the other halides. In Table 1-7 some data on displacement reactions are presented. Clearly in the solvolysis and displacement reactions listed in Table 1-7A a chloride ion is a much better leaving group than a fluoride ion. A reversal occurs in some reactions given in Table 1-7B where displacement is from an aromatic group. These observations require discussion invoking heat of solvation and limited polarizability of fluoride ions.

Many other anomalies of fluorine are described and discussed in later chapters. Only some striking examples have been selected for this introduction to show that organic fluorine compounds are indeed unique and offer a major challenge to modern chemists. We have tried to develop and provide reasonable explanations but many unanswered questions remain. Organic fluorine chemistry is a fertile and challenging field for mechanistic studies. We strive in this volume to provide road signs for those interested in employing the unique properties of fluorine to help provide definitive information on problems of chemical structure, bonding, and reactivity. The use of organic fluorine compounds in biochemistry is an infant field and although we are mainly concerned with organic fluorine chemistry, we feel compelled to include a moderate discussion on the use of fluorine in molecular biological studies and in medicinal and agricultural applications.

Table 1-7A

Fluorine-Chlorine Ratios for Reactions of Alkyl Halides[a]

Halide	Reagent	Solvent	Temp.(°C)	F/Cl ratio [b]
$(C_6H_5)_3CX$	H_2O	85% aq. acetone	25	1.0×10^{-6}
$(CH_3)_3CX$	H_2O/EtOH	80% aq. EtOH	25	1.1×10^{-5}
$C_6H_5CH_2X$	H_2O	10% aq. acetone	50	3.2×10^{-3}
CH_3X	H_2O	H_2O	100	2.9×10^{-2}
CH_3X	I^-	H_2O	25	1.2×10^{-3}

Table 1-7B

Fluorine-Chlorine Ratios for Reactions of Activated Aryl Halides (X = F or Cl)[a]

Substituted benzene	Reagent	Solvent	Temp. (°C)	F/Cl ratio[b]
1-X-2-NO_2	OMe^-	MeOH	100	3.2×10^2
1-X-2-NO_2	OEt^-	EtOH	90	1.0×10^3
1-X-4-NO_2	Piperidine	EtOH	90	2.1×10^2
1-X-2,4-$(NO_2)_2$	OMe^-	MeOH	0	8.8×10^2
1-X-2,4-$(NO_2)_2$	$C_6H_5NH_2$	EtOH	50	6.2×10^1
1-X-2,4-$(NO_2)_2$	C_6H_5NHMe	EtOH	50	7.3×10^{-1}
1-X-2-N_2^+	SCN^-	95% aq. t-BuOH	20	1.7×10^{-1}

[a] Based on data summarized by R. E. Parker, *Advan. Fluorine Chem.* **3**, 63 (1963).

[b] F/Cl ratio is the ratio of the rate constant for reaction of the fluorinated compound to the rate constant for reaction of the chlorinated compound. The Parker article cited in footnote *a* should be consulted for additional and more detailed information.

Chapter Two The Nature of Fluorine and Fluoride Ion

Fluorine is unique in that no other element has greater oxidizing power. The bonds formed by fluorine are among the strongest known, particularly to carbon, where it forms a stronger single bond than any other element (see Table 2-1). Paradoxically fluorine forms very weak bonds to other elements such as nitrogen, oxygen, and xenon. As a general rule the strength of the X–F bond decreases as the electronegativity of X increases.[1]

Table 2-1

Selected Bond Strengths Involving Fluorine or Carbon[a]

Bond	*E*, kcal	Bond	*E*, kcal
C–F	107-121[b]	C=O	178
C–Cl	81	C=N	147
C–Br	68	C=C	145.8
C–I	57	C≡N	212.6
C–O	85.5	C≡C	199.6
C–N	72.8	H–F	135
C–S	65	H–Cl	103.1
C–H	98.7	H–Br	86.5
		N–F	65
		S–F	68

[a] T. L. Cottrell, *The Strengths of Chemical Bonds,* 2nd Ed., Butterworths Scientific Publications, London, 1958.

[b] Authors estimated range; energy varied with bonding environment. Cottrell cited 106 kcal.

[1] W. L. Jolly, *Inorg. Chem.*, **3**, 459 (1964).

In this chapter, we have pointed out some important facts about fluorine both as the element and ion. These facts are essential in gaining an understanding of fluorine bonding (particularly to carbon) and reactions. Our discussion is brief, but we have provided a guide to the most important references for further reading.

2-1 FLUORINE AND FLUORINATION BY FLUORINE

Elemental fluorine is a pale yellow to yellow-green gas, bp 85.02°K, produced only by electrolysis because no chemical oxidizing agent is sufficiently powerful (Table 2-2). Details of fluorine production have been extensively reviewed.[2] Fluorine is available commercially[3] on industrial or laboratory scale and often is prepared in situ or for immediate use with commercially available electrolytic cells that are designed for laboratory use.[3, 4] Complete information on handling of fluorine is available from fluorine suppliers[5] and in recent reviews[2]

Table 2-2

Standard Electrode Potentials[a]

Reaction	E_0(volts)[b]
$Na^+ + e^- = Na$	– 2.714
$2H^+ + 2e^- = H_2$	0.000
$I_2 + 2e^- = 2I^-$	0.536
$Br_2 + 2e^- = 2Br^-$	1.07
$Cl_2 + 2e^- = 2Cl^-$	1.36
$O_3 + 2H^+ + 2e^- = O_2 + H_2O$	2.07
$F_2 + 2e = F^-$	2.65

[a] Based on data summarized by B. H. Mahan, *University Chemistry*, Addison-Wesley, Reading, Mass., 1965, p. 240.

[b] Values referred to the hydrogen ion-hydrogen couple as zero are for unit activities and a temperature of 25°C.

[2] (a) G. H. Cady, in *Fluorine Chemistry* (J. H. Simons, ed.), Vol. I, Academic, New York, 1950, p. 293. (b) M. Hudlicky, *Chemistry of Organic Fluorine Compounds*, Macmillan, New York, 1962, p. 44. (c) A. J. Rudge, *The Manufacture and Use of Fluorine and Its Compounds*, Oxford Univ. Press, London, 1962, pp. 19-45.

[3] General Chemical Division, Allied Chemical Corporation, P.O. Box 70, Morristown, New Jersey.

[4] Harshaw Chemical Division of the Kewanee Oil Company, Barclay Building, 1 Belmont Avenue, Bala Cynwyd, Pennsylvania.

[5] General Chemical Division of Allied Chemical Corporation, Product Data Sheet, PD-TA-85412, entitled *Fluorine*.

(see also Section 9-1). A word of caution: *Since fluorine in the elemental state is highly reactive with almost all compounds and materials, it must be handled carefully according to directions.*

To evaluate the role of elemental fluorine in organic fluorine chemistry, the following points are important (some are closely interrelated).

1. High oxidation potential of fluorine (Table 2-2).
2. High ionization energy of fluorine (Table 2-3).
3. High electron affinity of fluorine (Table 2-3).
4. Low polarizability of fluorine.
5. Low dissociation energy of molecular fluorine (Table 2-4).
6. High electronegativity of covalently bonded fluorine (Table 2-3).
7. High strength of the C–F bond (Table 2-1).

Elemental fluorine is highly reactive on an absolute scale. Relatively, fluorine has much greater reactivity than oxygen, nitrogen, and chlorine, elements which are good oxidizing agents and have high electronegativities. The primary reason for the unique high reactivity of fluorine must be its very low energy of dissociation (37 kcal/mole). Molecular oxygen and nitrogen are stabilized by

Table 2-3

Selected Ionization Energies, Electron Affinities, and Electronegativities[a]

	Ionization energy[b]	Electron affinity[c]	Electronegativity[d]
F	401.5	83.5	4.0
Cl	300[e]	87.3	3.0
Br	272.9	82.0	2.8
I	242.2[e]	75.7	2.5
O	313.8	–	3.5
H	315.0	0	2.1
Li	125.8	0	1.0

[a] L. Pauling, *The Nature of the Chemical Bond,* 3rd Ed., Cornell Univ. Press, Ithaca, New York, 1960, p. 95.

[b] Values in kcal/mole for reaction $X \rightarrow X^+ + e^-$; first ionization energy, see reference in above footnote *a*, p. 57.

[c] Values in kcal for $X + e^- \rightarrow X^-$

[d] Pauling scale as defined in reference given in above footnote *a*, p. 88.

[e] More recent spectral work changes these values slightly; R. E. Huffman, J. C. Larrabee, and Y. Tanaka, *J. Chem. Phys.*, 47, 856 (1967).

Table 2-4

Dissociation Energies of Selected Elements ($X_2 \rightarrow 2X\cdot$)

Element	Dissociation energy, kcal[a]
F_2	37
Cl_2	58
Br_2	46
N_2	225
O_2	118

[a] T. L. Cottrell, *The Strengths of Chemical Bonds,* 2nd Ed., Butterworths Scientific Publications, London, 1958.

multiple bonding but even singly bonded chlorine has a greater dissociation energy (Table 2-4). The low dissociation energy of fluorine assures that a kinetically significant number of fluorine atoms are available for reaction even at room temperature. The fluorine atom, resulting when fluorine dissociates, is an extraordinary electron-seeking species that forms extremely stable bonds with carbon and hydrogen. The energy released in making H–F and C–F bonds supplies energy for massive fluorine dissociation. Consequently elemental fluorine reacts violently with organic materials in extremely exothermic reactions and shows little selectivity. The lack of selectivity of fluorine relative to chlorine, bromine, and other radicals is shown by data in Table 2-5. The use of elemental fluorine as a fluorinating agent is hence restricted to special techniques such as use of very low temperatures and high dilutions.

These arguments are best elaborated by comparing the mechanism for chlorination and fluorination of hydrocarbons. A radical chain mechanism is generally accepted for chlorination of an alkane.[6] Note that the relatively modest energies for chain propagation steps are not sufficient to cause dissociation of molecular chlorine.

The proposed mechanism of fluorination[7] is somewhat similar, but an important difference is the alternative pathway 1b for initiation. Step 1a is a classical mechanism that is widely cited in the literature. Recently Miller[7a] has correctly advocated step 1b. For low-temperature fluorination ($<-40°C$),

[6] (a) J. D. Roberts and M. C. Caserio, *Basic Principles of Organic Chemistry,* W. A. Benjamin, New York, 1964, p. 89. (b) C. Walling, *Free Radicals in Solution,* Wiley, New York, 1957, p. 352.

[7] (a) W. T. Miller, Jr., S. D. Koch, Jr., and F. W. McLafferty, *J. Am. Chem. Soc.,* **78,** 4992 (1956). W. T. Miller, Jr. and S .D. Koch, *J. Am. Chem. Soc.,* **79,** 3084 (1957). (b) J. M. Tedder, *Advan. Fluorine Chem.,* **2,** 104 (1961). (c) P. Robson, V. C. R. McLoughlin, J. B. Hynes, and L. A. Bigelow, *J. Am. Chem. Soc.,* **83,** 5010 (1961) and earlier papers. For the most recent paper in series see B. C. Bishop, J. B. Hynes, and L. A. Bigelow, *J. Am. Chem. Soc.,* **85,** 1606 (1963). (d) G. A. Kapralova, L. Yu. Rusin, A. M. Chaikin, and A. E. Shilov, *Proc. Acad. Sci. USSR, Chem. Sect.* (English transl.), **150,** 505 (1963).

Table 2-5A
Relative Selectivities of Elemental Halides with Hydrocarbons at −81° in Liquid Phase[a]

	CH_3	CH_2	CH
F·	1	1.3	2.5
Cl·	1	4.6	10.3

Table 2-5B
Relative Selectivities of Radical X· for Primary, Secondary, and Tertiary Hydrogen Atoms in Alkanes at 27°[b]

X·	CH_3	CH_2	CH
F	1	1.2	1.4
Cl	1	3.9	5.1
CD_3	1	35	–
Br	1	82	1600

Table 2-5C
Relative Selectivities in Halogenation of n-Butyl Fluoride in the Gas Phase[c]

Halogen	Temp. (°C)	Relative selectivities at each position			
		FCH_2—	CH_2—	CH_2—	CH_3
F·	20	< 0.3	0.8	1.0	1
Cl·	35	0.8	1.6	3.7	1
Br·	146	10	9	82	1

[a] P. C. Anson and J. M. Tedder, *J. Chem. Soc.*, 4390 (1957).
[b] P. C. Anson, P. S. Fredricks (in part), and J. M. Teddeı, *J. Chem. Soc.*, 918 (1959).
[c] P. S. Fredricks and J. M. Tedder, *J. Chem. Soc.*, 144 (1960).

initiation by a fluorine molecule (step 1b) seems certain. And, energetically on any basis, step 1b is favorable (a similar step for Cl_2 requires 54 kcal). Which of the two initiating steps is chosen is not critical to understanding why fluorine is so much more reactive, but the large energy of the chain-propagating steps is very significant. In contrast to chlorination, in which the energy for the initiation

		ΔH, kcal/mole
Initiation	1. $Cl_2 \xrightarrow[\text{heat, } 300^\circ C]{h\nu \text{ or}} 2Cl\cdot$	+58
Propagation	2. $RH + Cl\cdot \longrightarrow R\cdot + HCl$	−3
	3. $R\cdot + Cl_2 \longrightarrow RCl + Cl\cdot$	−23
Termination	4. $R\cdot + Cl\cdot \longrightarrow RCl$	−80
	5. $Cl\cdot + Cl\cdot \longrightarrow Cl_2$	−58
	6. $R\cdot + R\cdot \longrightarrow R{-}R$	−83

		ΔH, kcal/mole
Initiation	1a. $F_2 \longrightarrow 2F\cdot$	37.5
	1b. $F_2 + RH \longrightarrow R\cdot + HF + F\cdot$	4.1
Propagation	2. $RH + F\cdot \longrightarrow R\cdot + HF$	-34.0
	3. $R\cdot + F_2 \longrightarrow RF + F\cdot$	-68.0
Termination	4. $R\cdot + F\cdot \longrightarrow RF$	-107 to -112
	5. $F\cdot + F\cdot \longrightarrow F_2$	-37.5
	6. $R\cdot \longrightarrow$ fragmentation	

step is greater than the energy for the chain-propagation steps, fluorination by molecular fluorine involves chain-propagation steps which evolve energy of the same order or greater than the energy for the initiation steps. As a result, fluorination of organic compounds by molecular fluorine is an exceedingly difficult reaction to control.

An equally serious difficulty in the use of molecular fluorine is the high energy (107–121 kcal/mole) for formation of the C–F bond. This energy exceeds the energy of the C–C bond (83 kcal/mole) and leads to fragmentation of the carbon skeleton of organic compounds.

This simple discussion explains why elemental fluorine has limited utility as a fluorinating agent in organic fluorine chemistry. Mechanism studies on addition of F_2 to olefins[8] also provide evidence in support of this discussion, as does the recent synthetic work of Merritt.[8c] More extensive excellent discussions on the theoretical aspects of this problem are available.[7] New experimental developments using low temperatures[7a, 8a] and high dilution techniques[7c] look promising for achieving some selectivity in fluorination of organic compounds with elemental fluorine. High dilution techniques for gas phase reactions were originally developed by Bigelow and co-workers,[7c] and their apparatus, called a jet reactor, is now available commercially.[9]

2-2 THE FLUORIDE ION

To understand the chemical stability (or reactivity) of fluorine bonded to carbon, we must consider the properties of the fluoride ion as compared to other ions, particularly halides. Probably the most important characteristics are charge, size, polarizability, heat of hydration, and thermodynamic properties in solution. The numerical values for some of these properties are given in Table 2-6. Note

[8] (a) W. T. Miller, Jr., J. O. Stoffer, G. Fuller, and A. C. Currie, *J. Am. Chem. Soc.*, **86**, 51 (1964). (b) A. S. Rodgers, *J. Phys. Chem.*, **69**, 254 (1965); *ibid.*, **67**, 2799 (1963). (c) R. F. Merritt and T. E. Stevens, *J. Am. Chem. Soc.*, **88**, 1822 (1966); R. F. Merritt and F. A. Johnson, *J. Org. Chem.*, **31**, 1859 (1966).

[9] Peninsular ChemResearch, Inc., Box 14318, Gainesville, Florida 32603.

Table 2-6

Selected Physical Properties of the Fluoride Ion and Halide Ions

Ion	Radius,[a] Å	Polarizability,[b] Å^3	Heat of hydration,[c] kcal/mole	State	Heat of formation,[d] kcal/mole	Free energy of formation,[d] kcal/mole	Entropy of ion,[d] cal/mole °C
F^-	1.36	0.86 (0.99)	123	gaseous	– 65.6	–	–
				aqueous	– 78.66	– 66.08	– 2.3
Cl^-	1.81	3.05	89	gaseous	– 58.3	–	–
				aqueous	– 40.023	– 31.350	13.2
Br^-	1.96	4.17	81	gaseous	– 55.3	–	–
				aqueous	– 28.9	– 24.574	19.29
I^-	2.16	6.28	72	gaseous	– 50.2	–	–
				aqueous	– 13.37	– 12.35	26.14

[a] Calculated radii, T. Moeller, *Inorganic Chemistry,* Wiley, New York, 1952, pp. 138-141.

[b] Data from Ref. 10a, p. 82; and G. Glockler, in *Fluorine Chemistry,* (J. H. Simons, ed.), Vol. I, Academic, 1950, p. 323.

[c] W. M. Latimer, *The Oxidation States of the Elements and Their Potentials in Aqueous Solutions,* 2nd Ed., Prentice Hall, New York, 1952, p. 22.

[d] Reference (*c*), p. 52, 54.

that the fluoride ion is much smaller than the other halide ions; the volume-charge density on fluorine—that is, charge per unit volume—is therefore significantly larger than for the other halide ions. As a result, the fluoride ion will more strongly affect centers of positive charge. A direct consequence of the small size of the fluoride ion and high resultant charge density is that the fluoride ion (as well as appropriately bonded fluorine) forms stronger hydrogen bonds than other ions. The relatively high heat of hydration for the fluoride ion is also expected because of its small size, and the inner hydration sphere of the fluoride ion contains five tightly bound water molecules compared with three for the chloride, two for the bromide, and three to four for the iodide ion.[10] Because the fluoride ion is so tightly bound by water and has such a high heat of hydration, it is extremely poor as a nucleophile in a solvent system containing water (polarizability will be considered later). The thermodynamic data on bond energies and heats of solvation can be used to predict that aqueous fluoride ions cannot substitute for chlorides in an alkyl chloride:

$RCl\ +$	$F^-\,(aq.) \not\longrightarrow$	$RF\ +$	$Cl^-\,(aq.)$	
D_{C-Cl}	$H_{solv.}$	D_{C-F}	$H_{solv.}$	ΔH
-81	-123	-106	-89	$+9$ kcal

However, these calculations ignore entropy effects which would work in the opposite direction since a hydrated fluoride ion has a lower entropy than a hydrated chloride. In any case, in anhydrous systems the replacement of chloride by fluoride is strongly favored thermodynamically and, indeed, the use of anhydrous metal fluorides (such as SbF_3) is the classical method for replacing chlorine by fluorine in organic compounds. The important point to be remembered from this discussion is that fluoride ions almost always must be anhydrous to achieve fluorination by replacement reactions.

In Table 1-7, data from kinetic studies show that fluorine in an alkyl fluoride is much more difficult to replace than chlorine in an alkyl chloride; for example, I^- replaces fluoride in methyl fluoride more slowly than chloride in methyl chloride by a factor of 1.2×10^{-3}. This large difference to a considerable extent results from the much greater amount of energy needed to break a C—F bond; but this is not the whole story since data in Table 1-7B show that some fluoride replacements can be faster than chloride replacements. These cases are all for aryl halides and formation of an intermediate in which little C—F bond breaking has occured is thought to be rate-determining. A good discussion of the

[10](a) *Gmelins Handbuch der Anorganischen Chemie,* 8th. Ed., Vol. 3, System No. 5 suppl., Verlag Chemie, GmbH, Weinheim/Bergstrasse, 1959, p. 206. (b) H. Remy, *Treatise on Inorganic Chemistry,* Vol. 1, Elsevier, New York, 1956, p. 75.

mechanism of these displacements has been given by Parker,[11] and we will discuss this problem in Section 3-3.

In summary, replacement of fluoride by another ion is

1. hindered by the low polarizability of F^-,
2. hindered by the high bond strength of the C–F bond.
3. helped by the presence of water, which provides a driving force by hydrating F^-,
4. helped by fluoride ion acceptors (to be discussed shortly).

Replacement of another ion by a fluoride is

1. hindered by the low polarizability of F^-,
2. helped by the high bond strength of the C–F bond,
3. hindered by the presence of water, which binds more tightly to fluoride than to other halides.

Another important point is that fluoride is a powerful base in many fluoride systems. Many metallic fluorides behave as acids toward fluoride ions and as a result bind them strongly (fluoride ion acceptors):

Acid		Base		Salt
HF	+	NaF	→	$Na^+HF_2^-$
HF	+	BF_3	→	$H^+BF_4^-$
SbF_5	+	SbF_5	→	$SbF_4^+SbF_6^-$

Finally, the exceptionally strong hydrogen bonding ability of fluoride is most clearly seen in hydrofluoric acid. Hydrofluoric acid exists as a polymer, $(HF)_n$, which can be in either linear or cyclic form in the liquid and gas phase. The details of the structure have been discussed extensively but the main reason for the polymer structure must be hydrogen bonding.[12] No doubt this strong tendency to hydrogen bond extends even to partially fluorinated hydrocarbons where fluorine is covalently bonded to carbon but still has a high electron density. Thus the maximum in boiling point for methylene fluoride in the methane series, as shown in Figure 1-1, probably results from intermolecular association by hydrogen bonding.

[11] R. E. Parker, *Advan. Fluorine Chem.* 3, 63 (1963).

[12] (a) Ref. 10a, pp. 165-172. (b) G. Glockler, in *Fluorine Chemistry* (J. H. Simons, ed.), Vol. 1, Academic, New York, 1963, p. 339.

Chapter Three The Carbon-Fluorine Bond

This chapter is devoted to a critical discussion of the chemical and physical properties of organic fluorine compounds in terms of modern organic chemistry. Our approach is to evaluate the nature of the C—F bond and its effect on the bonding, geometry, and structure in the rest of the molecule. This picture is then used to explain, as far as possible, the physical and chemical properties of fluorinated organic compounds.

3-1 BONDING CHARACTERISTICS

Before considering the detailed bonding in fluorinated methanes, we should first review the nature of the C—F bond in simple, general terms based on electronegativity. Fluorine is the most electronegative element (see Table 2-3). Stated more explicitly, when bonded to another atom by a single electron-pair bond (σ bond) fluorine is the element which will exert the greatest attraction for the electron pair. Thus, the bond will be strongly polarized because of the unequal sharing of the electron pair.

$$\mathrm{C} \qquad :\mathrm{F}$$

$$+ \longrightarrow -$$

The electron distribution is best described by a quantum mechanical approach, but for a physical picture, the C—F bond can be represented as a mixture of various parts of covalent and ionic bonding.

$$\mathrm{C}:\mathrm{F} \longleftrightarrow \mathrm{C}^{+} \quad :\mathrm{F}^{-}$$

Based on Pauling's electronegativities, the percent ionic character in carbon-halogen bonds can be calculated (Table 3-1).

Table 3-1

Calculated Ionic Character of Carbon Halide Bonds

Bond	Electronegativity diff.[a]	%Ionic character[a]
CF	1.5	43
CCl	0.5	12
CBr	0.3	2
CI	0.0	0

[a] Calculated as described in L. Pauling, *Nature of the Chemical Bond,* 3rd Ed., Cornell Univ. Press, Ithaca, N.Y., 1960, pp. 88-98.

Glockler[1] presents calculations for percent ionic resonance energy ($CH_3{}^+X^-$ contribution) in the methyl halides. These figures of 67%, 50%, 42%, and 40%, for X as F, Cl, Br, and I respectively, although based on old data, still show the greater amount of ionic character contributed by fluorine.

Perhaps a more realistic way to gain an insight into the relative amounts of ionic character in the carbon-halogen bonds is to consider the gas-phase dipole moments of the methyl halides in relation to the bond length of the carbon-halogen bond (Table 3-2). Remember that in this treatment of methyl halides we

Table 3-2

Calculated Charge Separation in Methyl Halides

	Dipole moment Debye	Bond length Å	Charge separation, Dipole moment/bond length
CH_3F	1.79	1.38	1.30
CH_3Cl	1.87	1.77	1.05
CH_3Br	1.80	1.93	0.93
CH_3I	1.65	2.18	0.76

are considering $CH_3{}^+X^-$ and not C^+X^-. The charge separation in the C–F bond is very significantly the greatest for the series.

For the hydrogen halides, calculations of ionic character based on dipole moments are given in Table 3-3.

[1] G. Glockler, in *Fluorine Chemistry* (J. H. Simons, ed.), Vol. 1, Academic, New York, 1950, p. 359.

Table 3-3

Ionic Character of Hydrogen Halides

Hydrogen halide	%Ionic Character[a]
HF	45
HCl	17
HBr	12
HI	5

[a] From reference cited in footnote *a*, Table 3-1.

The only logical conclusion that can be drawn from the above facts and discussion is that a carbon substituted by fluorine has lost a considerable amount of electron density, much more than if substituted by other halogens. This carbon has a significant positive character and will itself be more electronegative than carbon bonded to other elements. The C–F bond may be physically viewed as no longer truly covalent but as containing a substantial ionic contribution.

Another viewpoint is that fluorine exerts a much greater inductive effect through its σ bond to carbon than do other halogens and most other simple groups except nitro, cyano, and sulfonyl. Probably the best experimental evidence for the powerful inductive effect of fluorine is the comparison of the acid strengths of the mono-, di-, and trihalogen substituted halogen acids given in Table 1-3. Although the ability of both fluorine and chlorine to accommodate a negative charge leads to increased stability of the haloacetate ion compared to acetate ion, clearly fluorine is a more powerful electron-withdrawing group than chlorine and the other halogens.

3-1A. ON SATURATED CARBON

In the preceding paragraphs, we have given a simple picture of the C–F bond as highly polar, emphasizing the ideas of large ionic contribution and the positive character of the carbon.

Quantitatively, correlations of thermodynamic data to give group contributions for calculation of heats of formation, entropies, and heat capacities of fluorocarbons have been made by Bryant[2] and by Rogers and Benson[3]. Using the information in these papers a chemist can easily estimate $\Delta H_f^\circ 298$, ΔS° 298 and C_p° for organic fluorine systems of potential interest. Unfortunately we only have space to urge readers to consult these important and fundamental references.

Peters[4] has qualitatively evaluated the bonding in fluorinated methanes by a molecular orbital treatment that estimates the charge induced on a carbon

[2] W. M. D. Bryant, *J. Polym. Sci.*, **56**, 277 (1962).

[3] (a) A. S. Rogers and S. W. Benson, *Chem. Rev.*, in press. (b) J. R. Lacher and H. A. Skinner, *J. Chem. Soc.* (A), 1034 (1968).

[4] D. Peters, *J. Chem. Phys.*, **38**, 561 (1963).

bonded to one or more fluorines. As a means of discussing the problem, assume that each fluorine bonded to carbon has the ability to withdraw to it 0.15 e^- of charge, inducing on the carbon a corresponding + 0.15 units of charge. The normally sp^3 hybridized carbon can lose 25% or more electron density and will undoubtedly lose 2p charge density. After attachment of one fluorine to the carbon, the carbon can be described as $sp^{2.85}$ hybridized. For n attached fluorines, each withdrawing a constant increment (δ) of charge, the condition of carbon can be described as $sp^{(3-ns)}$. If δ is 0.15 e^-, as postulated, then the carbon in CF_4 will be described as $sp^{2.4}$. As the number of attached fluorines increases, the percentage of s character in the atomic orbital from carbon will increase and the resulting bonds will shorten. Thus the C–F bond in CF_4 is 1.32 Å compared to the C–F bond length from an sp^2 hybridized carbon ($CF_2{=}CF_2$) of 1.30 Å.

Peters has indicated a refinement to the argument in which a saturation effect is presumed. Instead of each fluorine withdrawing a constant amount of charge, one fluorine may withdraw 0.30 e^-, two fluorines may each withdraw 0.20 e^-, three fluorines may each withdraw 0.17 e^-, and four may each withdraw 0.15 e^-. Then the condition of the carbon atom would be $sp^{2.7}$ in CH_3F, $sp^{2.6}$ in CH_2F_2, $sp^{2.5}$ in CHF_3, $sp^{2.4}$ in CF_4.

Essentially Peters argues for an alteration of the hybridization of the bonding carbon in methanes caused by removal of electron density by fluorine. Bond shortening is ascribed to increasing s character in the atomic orbitals from carbon. As an increasing number of fluorines are bonded to carbon, the positive charge density at carbon increases the positive character. Hence the amount of ionic character in C–F bonds of fluorinated methanes increases with increasing fluorine substitution. Increasing ionic attraction in the C–F bonds should shorten the C–F bond lengths as well as strengthen the bonds.

Another approach to the problem of bond shortening and bond strengthening with increasing fluorine substitution in the fluorinated methanes was suggested by Brockway[5] and subsequently discussed by Pauling,[6] Hine, and others. Hine[7] has recently expressed this argument in detail and assigned the name "double bond–no bond resonance" to the following type of resonance:

```
    F               F              F⁻
    |               |
F—C—F  ⟷  F⁻  C=F⁺  ⟷  F—C—F  ⟷   etc.
    |               |              ‖
    F               F              F₊
```

[5] L. O. Brockway, *J. Phys. Chem.*, **41**, 185, 747 (1937).

[6] L. Pauling, *The Nature of the Chemical Bond*, 3rd Ed., Cornell Univ. Press, Ithaca, N.Y., 1960, pp. 314-315.

[7] J. Hine, *J. Am. Chem. Soc.*, **85**, 3239 (1963); see also J. Hine, L. G. Mahone, and C. L. Liotta, *J. Am. Chem. Soc.*, **89**, 5911 (1967).

For tetrafluoromethane, twelve resonance forms involving double bond–no bond resonance can be written. Only six such resonance forms can be written for fluoroform and two for methylene fluoride; no resonance forms of the double bond–no bond type are available to methyl fluoride. Hine has suggested that each C—F bond is increased in stability by 3.2 kcal/mole for each resonance form available to it. A value of 3.2 kcal/mole resonance energy for each resonance form agrees reasonably closely with the extra stability of CH_2F_2, CHF_3, CF_4 of 5.2, 22.8, and 36 kcal/mole (see Fig. 3-1).

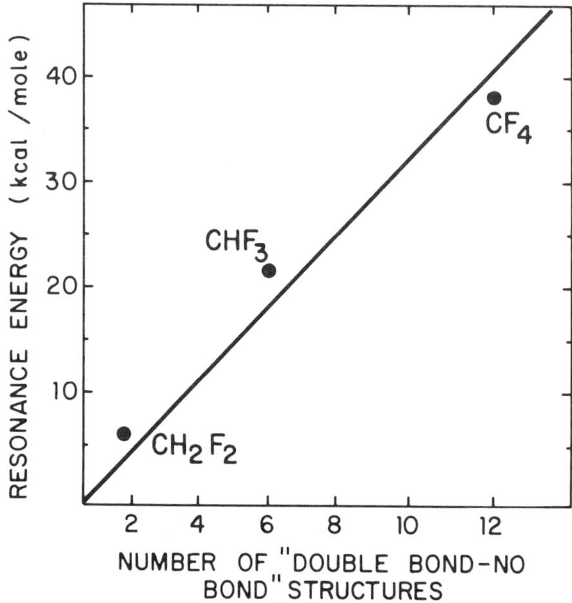

Fig. 3-1 Plot of number of double bond–no bond structures versus resonance energy of fluoromethanes.

The attractive feature of the double bond–no bond resonance explanation is the intuitively easy understanding of bond shortening in the fluorinated methanes as fluorines accumulate on carbon. Increasing double bond character results in decreasing bond length. In discussing the fluorinated methanes, Pauling has estimated 8%, 15%, and 19% double bond character in the C—F bonds of CH_2F_2, CHF_3, and CF_4 respectively. For CF_4, a 19% double bond character corresponds to an electric charge of +0.96 on the carbon atom in which Pauling chooses to write the 12 resonance forms for CF_4 in a modified form from Hine.

$$\begin{array}{c} F \\ | \\ F\text{—}C\text{—}F \\ | \\ F \end{array} \longleftrightarrow \begin{array}{c} \quad F^- \\ F^- \quad C^+\text{=}F^+ \\ | \\ F \end{array} \longleftrightarrow \text{etc.}$$

Neither of these explanations is satisfactory. The increase in positive charge on carbon (or in other words the ionic nature of the C–F bond) is a physically rational approach that explains many other properties of fluorine. However, Chesick[8] has obtained refined kinetic and thermodynamic data on the thermal conversion of decafluoro-1,2-dimethylcyclobutene to decafluoro-2,3-dimethylbutadiene-1,3 and compared it to data for isomerization of hexafluorocyclobutene

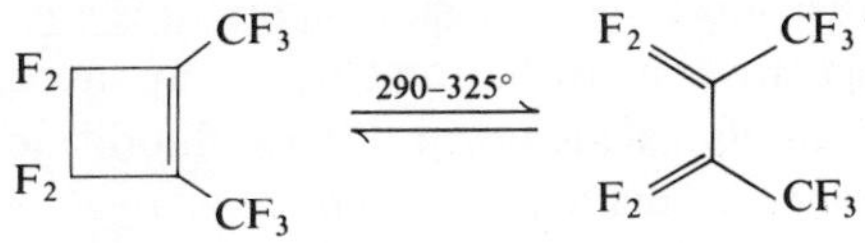

F_2, CF_3, F_2, CF_3 — 290–325° ⇌ — F_2, CF_3, F_2, CF_3

to hexafluorobutadiene-1,3.[9] A significant difference of 11.3 kcal/mole for $\Delta H°$ in the two systems is interpreted against Peters' view that changes in C–F bond strengths are attributable to changes in the hybridization of carbon.

The double bond–no bond resonance theory is a useful working rule to predict stability of fluorocarbanions and other fluorinated systems[10] but Streitwieser[11] argues against fluorine hyperconjugation from results of acidity studies. The fluorinated bicyclic compound **1**, first examined by Tatlow,[12] is more acidic than nonafluoro-t-butane **2**:

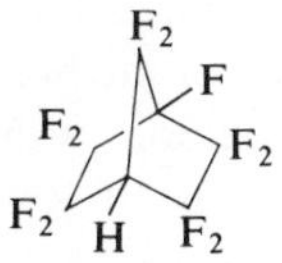

	1	**2**
	F_2, F, F_2, F_2, F_2, H, F_2	$(CF_3)_3C—H$
Relative rates for isotopic exchange:	5×10^9	10^9

[8]J. P. Chesick, *J. Am. Chem. Soc.*, **88**, 4800 (1966).

[9]E. W. Schlag and W. B. Peatman, *J. Am. Chem. Soc.*, **86**, 1676 (1964).

[10]S. Andreades, *J. Am. Chem. Soc.*, **86**, 2003 (1964), has used this approach effectively in studies on perfluorocarbanions.

[11](a) A. Streitwieser and D. Holtz, *J. Am. Chem. Soc.*, **89**, 692 (1967). (b) A. Streitwieser, A. P. Marchand, and A. H. Pudjaatmaka, *ibid.*, 693. (c) *Chem. Eng. News,* **45**, Feb. 6, 54 (1967).

[12]S. F. Campbell, R. Stephens, and J. C. Tatlow, *Tetrahedron*, **21**, 2997 (1965).

If hyperconjugation did operate, the incipient double bond could not form at the highly strained bridgehead position and **2** should be more acidic than **1**.[13] Streitwieser explains the acidity results simply by the inductive effect of fluorine. The high electron density in filled p-orbitals of the α-fluorines to some extent offsets the normally strong inductive withdrawal. As a consequence α-F does not stabilize an anion as well as a trifluoromethyl group. This conclusion was confirmed with a series of proton-exchange experiments using 9-substituted-9-tritiated fluorenes and methanolic methoxide.[11b] The trifluoromethyl substituent falls on the line for other substituents suggesting that it is normal; if negative hyperconjugation is operative the trifluoromethyl group should promote anion formation. We present additional arguments against fluoride ion hyperconjugation in the following section on aromatic compounds.

Pople has undertaken quantitative molecular orbital calculations on fluorinated compounds.[14] He uses an approximate self-consistent molecular orbital theory (complete neglect of differential overlaps or CNDO) to calculate charge distributions and to calculate electronic dipole moments that agree reasonably well with experimental values. These calculations propose that fluorine affects both the σ and π electrons of the molecule. In the σ framework the charge density alternates, and withdrawal is more effective at odd atom positions from the fluorine. However the effect drops off very rapidly. The unshared p-electrons of fluorine are calculated to feed charge density back to the organic system, also with charge alteration (even when no π system is directly involved). The overall charge density is the sum of the effect on both σ and π framework (see Fig. 3-2); the induced charges alternate in a decaying manner so that the β position is normally negative. The approach has obvious limitations but should lead to an adequate understanding of the bonding in C—F bonds.

Another interesting anomaly in fluorocarbons is that the FCF angle is often near 108° instead of the normal tetrahedral angle of 109.3° (Table 3-4). Actually Cram has recently pointed out that the bond angle ANA in compounds such as $:NA_3$ decreases as the substituent A becomes more electronegative (109° in trimethylamine and 102.5° in NF_3).[15] In this discussion related to character of carbanions, he argues that the σ bond increases in p character with increasing electron-withdrawing power of substituents, causing decrease in bond angle. This idea is easily rationalized. The Iπ repulsion energy in the system is minimized if the unshared electron pair contracts; such a contraction means more s-electron contribution in the unshared pair and consequently more p-electron density in the σ bonds. This argument also disagrees with Peters' explanation[4] discussed

[13] Streitwieser pointed out one weakness in this argument: Compound **1** may have enhanced acidity because of increased s character in the C—H bond derived from ring strain effects in norbornane structure.

[14] J. A. Pople and M. Gordon, *J. Am. Chem. Soc.*, 89, 4253 (1967).

[15] D. J. Cram, *Fundamentals of Carbanion Chemistry,* Academic, New York, 1965, p. 56.

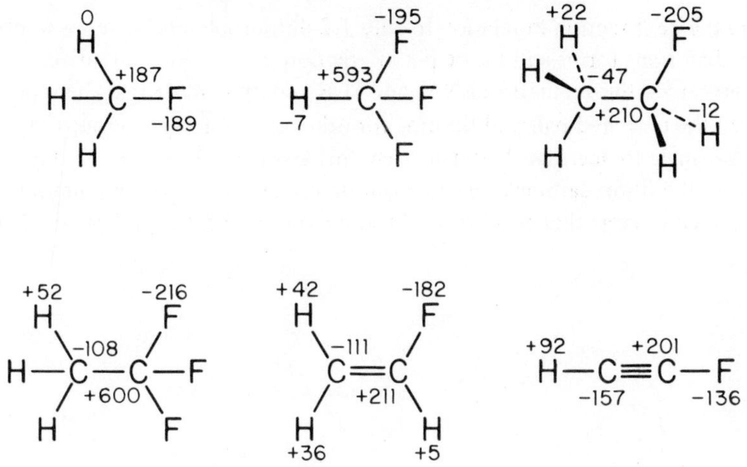

Fig. 3-2 Electron distribution in fluorocarbons (10^{-3} electron unit).

Table 3-4

Bond Angles in Fluorocarbons[a]

Compound	FCF angle	Other angle
CH_2F_2	108.3	HCH 111.9
CHF_3	108.8	
$CClF_3$	108.6	
$CHCl_3$	–	ClCCl 110.4
ClF_3	108.3	

[a] *Tables of Interatomic Distances and Configuration in Molecules and Ions,* Special Publication No. 11, The Chemical Society, Burlington House, London, 1958.

above. However, the bond angle in tris-(trifluoromethyl)amine, $(CF_3)_3N$, is 108.5°.[16] At first glance this result raises a serious question about the argument of redistribution of p-electron density; but, by consulting the organic chemists' explainers guide,[17] we account for this unexpected, nearly tetrahedral, angle by steric repulsion between the trifluoromethyl groups.

As another possibility we suggest that the interaction of the p-electrons of the two fluorines gives some weak homo-type bond between the fluorines (which can be described in quantum mechanical terms) that has sufficient importance (although no doubt relatively small compared to classical inductive effects) to cause a decrease in the angle. This p-p interaction idea will be discussed more

[16] R. L. Livingston and G. Vaugh, *J. Am. Chem. Soc.*, 78, 4866 (1956).

[17] E. M. Arnett, *J. Chem. Ed.*, in press.

fully in the next section in relation to why 1,2-difluoroolefins are more stable in the cis than trans forms and to the p-π interaction theory. An alternative explanation for the geometry of NA_3 and $^-CR_3$ compounds is that the repulsion between the unshared pair and fluorine (or other electronegative group) causes the ANA angle to decrease. Unfortunately this argument does not logically extend to the fluorocarbons. The conjugation effects in the tricyanomethide ion are sufficiently large that this ion, as the ammonium salt, is almost planar and

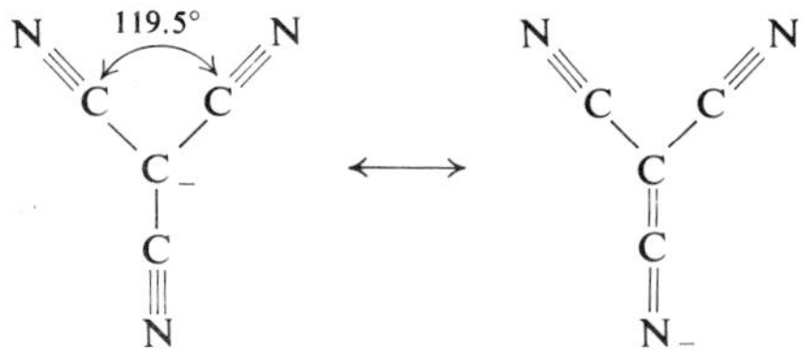

tricyanomethide ion

trigonal.[18] Although cyano groups strongly withdraw electrons inductively, the large conjugative interaction to enhance electron withdrawal can override all other effects, so that fluorine cannot be directly compared to cyano groups when considering stabilization of an unshared pair.

Related to this discussion on anions (see also pp. 28-29), an α-chlorine is reported to stabilize a carbanion better than an α-fluorine.[19] Thus the acidity of haloforms decreases in order $Br_3CH > Cl_3CH > F_3CH$ so that the corresponding anions appear to increase in stability in the series $^-CF_3 < ^-CCl_3 < ^-CBr_3$. Similar conclusions were drawn from studies on rates of S_E1 decarboxylations of trihaloacetate ions, $CX_3CO_2^-$,[20] where the rate increased in order from fluoro to bromo with a marked difference between trifluoro- and trichloroacetates (relative rates for acetates, X as F:Cl:Br::1:10^4:10^6). The obvious prediction that the more electronegative fluorine should best stabilize the trihalomethide anion does not work. One rationalization is that d-orbitals on chlorine and bromine are extremely important in delocalizing the negative charge, but bromine is not expected to be better than chlorine (3d of Cl can better overlap with 2p than 4d of Br). Another explanation is that relief of repulsive forces between the halogens becomes much more important with increasing size of the halogen, but the weak point in this

[18] C. Bugg, R. Desiderato, and R. L. Sass, *J. Am. Chem. Soc.*, **86**, 3157 (1964).

[19] J. Hine and N. W. Burske, *J. Am. Chem. Soc.*, **78**, 3337 (1956). Additional evidence was obtained from infrared studies, R. E. Glick, *Chem. Ind.* (London), 413 (1956).

[20] (a) R. E. Glick, *Chem. Ind.* (London), 716 (1955). (b) K. R. Bower, B. Gray, and T. L. Konkol, *J. Am. Chem. Soc.*, **88**, 1681 (1966).

argument is that carbanions appear to favor a geometry where the angle decreases toward 90° rather than going toward a planar configuration.[15] But, remember, hydrogen bonding is always important and may well be the controlling factor under the conditions of these experiments; more definitive experimental studies are needed.

3-1B. IN UNSATURATED SYSTEMS

In this section we will discuss the bond of fluorine to an sp^2 hybridized carbon, both in olefinic and aromatic systems. This section will supplement and expand the preceding discussion on fluorinated methanes and will conclude with comments on the effect of simple fluorinated substituents on a π system, particularly the aromatic ring.

Table 3-5

Carbon-Fluorine Bond Lengths and Strengths in Selected Compounds[a]

Compound	Length, Å	Strength, kcal/mole
CH_3F	1.39	107
CF_4	1.32	121
$H_2C{=}CHF$	1.348[b]	–
$H_2C{=}CF_2$	1.323[c]	–
$F_2C{=}CF_2$	1.31	116[d]
C_6H_5F	1.305	117[e]
C_6F_6	1.327[f]	145
$HC{\equiv}CF$	1.279[g]	114

[a] Unless indicated otherwise, data are from general sources, such as T. L. Cottrell, *The Strengths of Chemical Bonds,* 2nd Ed., Butterworths, Washington, D.C., 1958. For a recent comparative tabulation of C–F bond distances see J. L. Hencher and S. H. Bauer, *J. Am. Chem. Soc.,* **89,** 5527 (1967).

[b] V. W. Laurie, *J. Chem. Phys.,* **34,** 291 (1961).

[c] V. W. Laurie and D. T. Pence, *J. Chem. Phys.,* **38,** 2693 (1963).

[d] C. R. Patrick, *Tetrahedron,* **4,** 26 (1958).

[e] Bond dissociation energy estimated by method of L. A. Errede, *J. Phys. Chem.,* **64,** 1031 (1960).

[f] A. Almenningen, O. Bastiansen, R. Seip, and H. M. Seip, *Acta Chem. Scand.,* **18,** 2115 (1964).

[g] J. K. Tyler and J. Sheridan, *Proc. Chem. Soc.,* **119** (1960).

The bond length of the C–F bond in fluorinated alkenes and aromatics is shorter than the normal C–F bond but of the same order as a C–F in CF_4 (see Table 3-5). A value of 1.30 to 1.32 Å seems reasonable to assign to fluorine bonded to sp^2 hybridized carbons and closely corresponds to the 1.32 Å bond length of CF_4. These data add support to Peters' argument[4] that the carbon of CF_4 is hybridized to $sp^{2.1 \text{ to } 2.4}$. Although the data are extremely limited, the bond strength for aryl and olefinic C–F bonds appears to be comparable in magnitude to that in CF_4. A low order of reactivity for aryl fluoride is thus expected and in general found. A single fluorine in an aromatic ring is much less reactive than a single fluorine on an aliphatic carbon. This question is discussed more fully in the next section.

When hydrogen on an sp^2 carbon is replaced by fluorine, the strong electron withdrawal by fluorine removes electron density from the π system as well as the σ framework. Consequently fluorinated olefins are much less susceptible to electrophilic attack (normal for hydrocarbon olefins) and much more susceptible to nucleophilic attack. This question is discussed fully in a recent review[21] and will be considered again in the last section of this chapter, which deals with chemical reactivity. As examples, alcohols or amines add to fluoroolefins in high yields at moderate conditions of temperature and pressure, and with base catalysis.[22] In contrast, drastic conditions are needed to add these reagents to hydrogen substituted ethylenes.

$$\text{t-BuOH} + F_2C{=}CF_2 \xrightarrow[\text{DMF, exothermic}]{\text{t-BuONa}} \text{t-BuOCF}_2\text{CF}_2\text{H}$$

$$(C_2H_5)_2NH + F_2C{=}CFCl \xrightarrow[\text{or solvent}]{\text{no catalyst}} (C_2H_5)_2NCF_2CFClH$$

The olefinic carbon has lost a major amount of electron density due to electron withdrawal by fluorine and is very susceptible to attack by a nucleophile. A similar situation exists for cyanocarbons where the strong electron-withdrawing effect of the cyano groups makes tetracyanoethylene very susceptible to attack by a nucleophile.[23]

Thermodynamic calculations also substantiate the arguments that substitution of fluorine on the carbons of a C=C bond significantly weakens the double bond. O'Neal and Benson[24] have shown the C–C π bond strength in tetrafluoroethylene is only 40.0 kcal/mole (20 kcal/mole weaker than in ethylene).

$$CF_2H{-}CF_2H \rightleftharpoons \cdot CF_2{-}CF_2\cdot + 2H \qquad 2\Delta H^\circ(C{-}H) = 2(103.5) = 207$$

$$\Delta H_f^\circ = -218.0 \qquad \Delta H_f^\circ(R:) = ? \qquad \Delta H = 104.2$$

Then $\Delta H_f^\circ(R:) = -115$ kcal/mole

$$CF_2{=}CF_2 \rightleftharpoons \cdot CF_2{-}CF_2\cdot$$

$$\Delta H_f^\circ = -155 \qquad \Delta H_f^\circ = -115$$

Then $\Delta H^\circ = +40.0$ kcal/mole

[21] R. D. Chambers and R. H. Mobbs, *Advan. Fluorine Chem.*, **4**, 50 (1965).

[22] D. C. England, L. R. Melby, M. A. Dietrich, and R. V. Lindsey, Jr., *J. Am. Chem. Soc.*, **82**, 5116 (1960).

[23] T. L. Cairns and B. C. McKusick, *Angew. Chem.*, **73**, 520 (1961).

[24] H. E. O'Neal and S. W. Benson, *J. Phys. Chem.*, **72**, 1866 (1968).

Thus only 40 kcal/mole is required to convert tetrafluoroethylene to the diradical compared to 60 kcal/mole for ethylene.

In unsymmetrical fluorohaloolefins the nucleophilic attack is almost always on the CF_2 groups because

1. the largest induced positive charge is on carbon bearing two fluorines;
2. steric repulsion is least with very small fluorines;
3. chlorine or other halogen in the β-position can stabilize an α-negative charge by d-orbital resonance (see discussion on p. 26).

The relative reactivities of the fluoroolefins to nucleophilic attack (always on terminal CF_2) are

$$CF_2{=}CF_2 < CF_2{=}CFCF_3 < CF_2{=}C(CF_3)_2$$

One explanation[14] is that a vinyl fluorine can partly counteract its strong inductive electron withdrawal by return of electron density from the nonbonding p-electron to the π system (called p-π interaction). In spite of high electronegativity and low polarizability of fluorine, the resonance electron return is suggested to be very important for fluorine, even better than for chlorine or other halides which are much more polarizable. A plausible reason is that fluorine and carbon are almost the same size and the C–F bond is short, so that overlap is better than for the other halogens.

The acidities of the trihaloacrylic acids, $CF_2{=}CFCO_2H$ pK_a 1.80, and $CCl_2{=}CClCO_2H$ pK_a 1.21[25] ($CH_2{=}CHCO_2H$, pK_a 4.25) are cited as experimental evidence in support of better return of π-electron density from fluorine than chlorine. However, the unexpected decreased acidity for the trifluoro acid could arise from intramolecular hydrogen bonding:

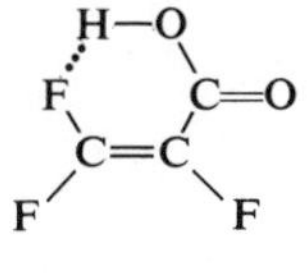

```
  H—O
 :    \
F      C=O
 \    /
  C=C
 /    \
F      F
```

The inductive and resonance effects of trihalovinyl groups need to be measured more precisely under a variety of conditions.

Finally, the fluoride-ion no-bond resonance or negative hyperconjugation theory can be invoked to explain the direction of nucleophilic attack:

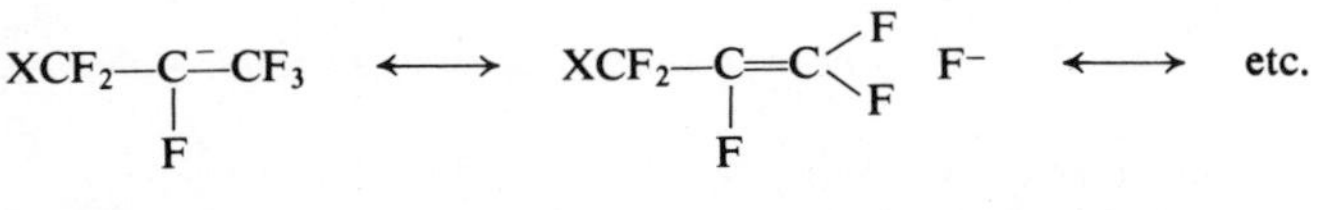

$$XCF_2{-}\overset{-}{C}(F){-}CF_3 \longleftrightarrow XCF_2{-}C(F){=}CF_2 \;\; F^- \longleftrightarrow \text{etc.}$$

[25] A. L. Henne and C. J. Fox, *J. Am. Chem. Soc.*, **76**, 479 (1954).

We reemphasize our view expressed earlier in this chapter that fluoride-ion no-bond resonance is only a working rule and is useful in this case to predict direction in addition of nucleophiles to fluoroolefins, but has no physical significance.

In Chapter 1 the high reactivity of fluoroolefins to addition of various reagents was shown by heats of addition (see Table 1-6). A weaker C=C bond can be considered to mean more diradical character: $\dot{C}F_2$–$\dot{C}F_2$. The weakening of the C=C bond would certainly occur if we had p-π interaction as discussed above. An alternative explanation is strong repulsion of π electrons by the p-electrons of fluorine. Murrell[26] has promoted the idea of π-electron repulsion for aromatic systems to explain the electronic spectra of halobenzenes, and we will elaborate on this idea when we deal with aromatic systems.

Fluoroolefins much prefer to give cycloadducts **1** rather than normal Diels-Alder products **2**.[27] Bartlett[28] has studied the mechanism of this cycloaddition

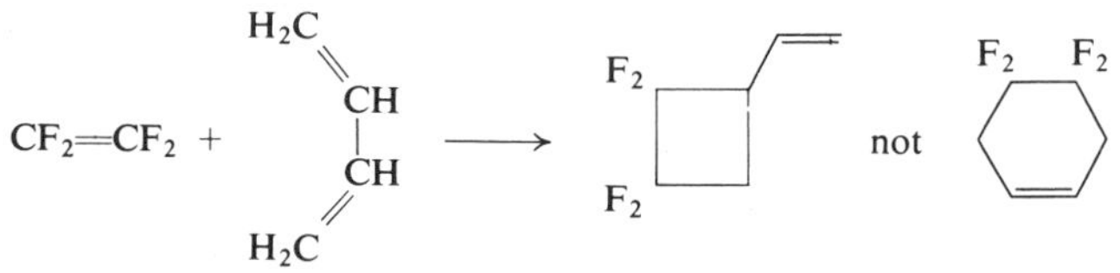

$CF_2{=}CF_2$ + $H_2C{=}CH{-}CH{=}CH_2$ ⟶ **1** not **2**

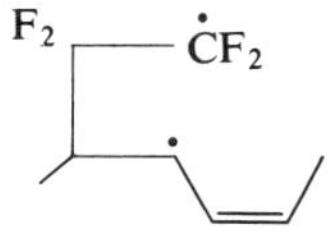

3 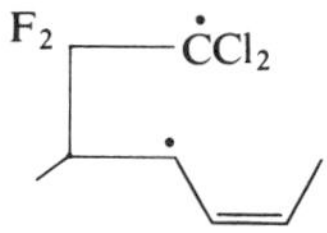 **4**

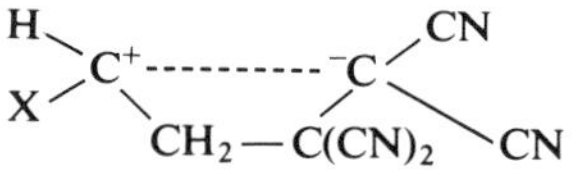

5

[26] J. N. Murrell, *The Theory of the Electronic Spectra of Organic Molecules,* Wiley, New York, 1963, pp. 189-237.

[27] J. D. Roberts and C. M. Sharts, *Org. Reactions,* **12,** 1 (1962).

[28] (a) P. D. Bartlett and L. K. Montgomery, *J. Am. Chem. Soc.,* **86,** 628 (1964). (b) P. D. Bartlett, *Science,* **159,** 833 (1968).

reaction and has proposed a diradical intermediate **3**. In the case of unsymmetrical fluorochlorocoolefins, the adduct is always that from the more stable diradical such as **4**. However, tetracyanoethylene and related cyanoolefins also give cycloaddition reactions[29] but evidence here favors an ionic intermediate **5**. For both fluoro- and cyanoolefins the sp^2-hybridized carbon is of low electron density, but the cyano groups can much better stabilize the ionic form by resonance. Release of repulsive energy between the geminal fluorines or cyanos of the olefin in going to a transition state may be much better for the four-membered ring cycloadduct than a six-membered Diels-Alder product.

Another unusual observation is that 1-fluoro-2-haloolefins are more stable in the cis than the trans form.[30] This phenomenon is apparently general for a series of 1,2-disubstituted olefins where the substituents contain unshared electron pairs (see Table 3-6) and has been suggested to result from resonance contributions by ionic forms

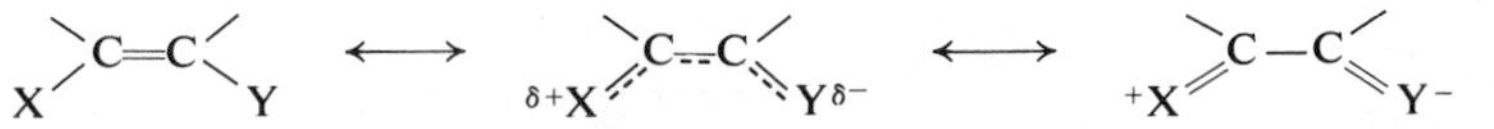

where the interaction between the positive and negative dipoles is energetically more favorable in the cis rather than the trans forms. If Y has readily accessible vacant d-orbitals, such interaction should become more important; however, steric repulsion and decrease in electronegativity will decrease the importance of this ionic contribution in progressing from fluorine to iodine. Data on a series of such olefins are given in Table 3-6. We suggest as an alternative explanation p-p interaction between the unshared electron pairs of the 1 and 2 substituents. Such an interaction must give some type of nonbonding overlap.

Additional strong evidence for p-p interaction is the conformation dependence in 1,2-difluoroethane. In the gas phase the gauche form has the same

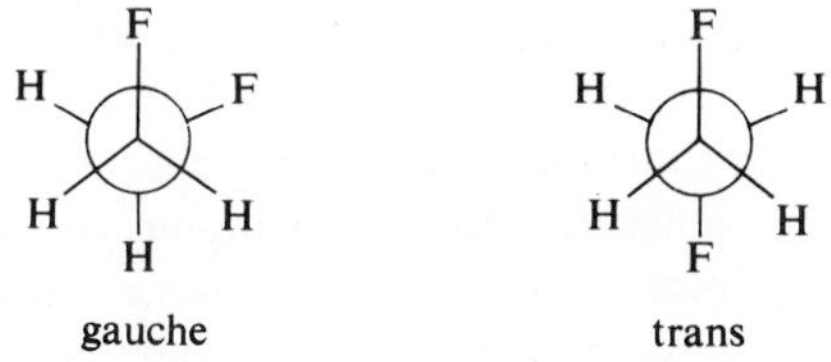

gauche trans

[29] S. Proskow, H. E. Simmons, and T. L. Cairns, *J. Am. Chem. Soc.*, **88**, 5254 (1966).

[30] (a) N. C. Craig and E. A. Entemann, *J. Am. Chem. Soc.*, **83**, 3047 (1961). (b) A. Demiel, *J. Org. Chem.*, 27, 3500 (1962). (c) H. G. Viehe, J. Dale, and E. Franchimont, *Chem. Ber.*, 97, 244 (1964).

Table 3-6A

Stabilities of 1,2-Disubstituted Olefins

HXC=CHY X	Y	% Cis at equilibrium (equilibrium temp. °C)	Reference
F	F	63 (200)	*a*
F	Cl	70 (200)	*a*
F	Br	70 (200)	*a*
F	I	67 (250)	*b*
Cl	Cl	61 (245)	*c*
Br	Br	50 (225)	*d*

[a] H. G. Viehe, *Chem. Ber.*, **93**, 1697 (1960).
[b] Ref. 30b.
[c] R. E. Wood and D. P. Stevenson, *J. Am. Chem. Soc.*, **63**, 1650 (1941). K. S. Pitzer and J. L. Hollenberg, *J. Am. Chem. Soc.*, **76**, 1493 (1954).
[d] H. G. Viehe and E. Franchimont, *Chem. Ber.*, **96**, 3153 (1963); and **97**, 602 (1964).

Table 3-6B

Comparison of Enthalpies and Entropies of Isomerization of Dihaloethylenes[a]

cis-HXC=CXH → *trans*-HCX=CXH

X	Temp. range (°K)	$\Delta H°$, cal/mole	$\Delta S°$, eu/mole	$\Delta H_0°$, cal/mole
F	480-760	928	0.13	–
Cl	458-548	720	0.26	445 ± 20
	573-623	480		
Br	417-451	130	0.60	– 100 ± 160
I	403-432	– 1550	2.4	– 1700 ± 1000

[a] Data assembled from Ref. 30a.

energy as the trans but on basis of Coulombic repulsion between the carbon fluorine dipoles the trans is calculated to be more stable by at least 0.8 kcal.[31] In this system a p-p interaction is suggested to give 0.8 kcal stabilization for the gauche form.

Another type of experimental evidence for p-p interaction comes from nmr measurements. Fluorine-fluorine spin-spin coupling through space[32] probably results from p-p interaction; the cis fluorines are just within the 2.7 Å, that appears to be the maximum distance to allow coupling and is approximately the sum of the van der Waals radii. The importance of through-space coupling

[31] P. Klaboe and J. R. Nielsen, *J. Chem. Phys.*, **33**, 1764 (1960).

[32] (a) S. Ng and C. H. Sederholm, *J. Chem. Phys.*, **40**, 2090 (1964). (b) H. S. Gutowsky and V. D. Mochel, *J. Chem. Phys.*, **39**, 1195 (1963).

has been questioned,[33] but in recent publications[34] several groups provide data and arguments that through-space interaction is definitely significant at 2 to 2.5 Å.

A related phenomenon is the decrease in FCF angle below 120° in terminal CF groups of olefins. For example, the FCF angle is 109.1° in $H_2C{=}CF_2$[35a] and 114° in $F_2C{=}CF_2$.[35b] The angle decrease in fluorocarbons discussed in the previous section must be related. Again p-p interaction is proposed as the explanation, and the energy needed to distort the angle about 10° is estimated to be only about 5 kcal.[36] The increase in the FCF angle by over 4° in $F_2C{=}CF_2$ compared to $H_2C{=}CF_2$ is expected if 1,2-interactions are also important. However, repulsion between p-electrons of fluorine and the π system could also contribute to distortion of this angle.

The high electronegativity of fluorine and the observed electron-withdrawing power in aliphatic compounds would lead to the prediction that fluorine as a substituent on an aromatic ring should be deactivating (rate retarding) and meta directing toward electrophilic substitution. However, in fluorobenzene, electrophilic substitution occurs at the ortho and para position, and the rate of substitution is only slightly less than the rate for benzene (Table 1-5). If partial rate factors are used, electrophilic substitution in the para position of fluorobenzene appears faster than at a single position of benzene. The classical explanation is resonance, as represented by contributing forms shown, and is proposed for any element with unshared pairs, which includes oxygen, nitrogen,

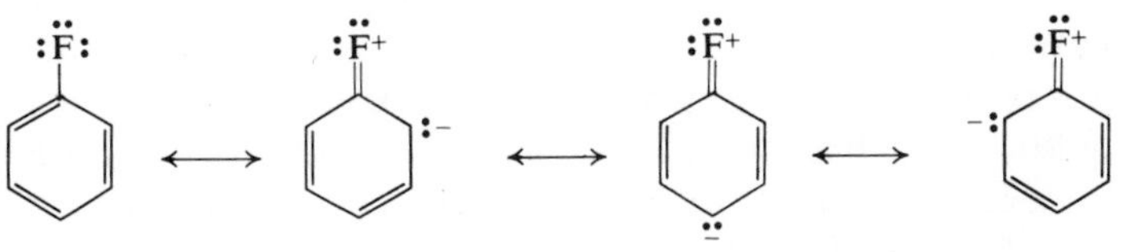

:F: ⟷ :F+ ⟷ :F+ ⟷ :F+

and chlorine. In this way the directive effects are conventionally explained; indeed the σ_R value, which is considered as a semiquantitative measure of electron donation or withdrawal by resonance, is a large negative value for fluorine (see Table 1-4), clearly indicating strong electron donation by resonance.

[33] K. L. Servis and J. D. Roberts, *J. Am. Chem. Soc.*, **87**, 1339 (1965).

[34] (a) P. C. Myhre, J. W. Edmonds, and J. D. Kruger, *J. Am. Chem. Soc.*, **88**, 2459 (1966). (b) J. P. N. Brewer, H. Heaney, and B. A. Marples, *Chem. Commun.*, **27** (1967). (c) J. Jonas, L. Borowski and H. S. Gutowsky, *J. Chem. Phys.*, **47**, 2441 (1967).

[35] (a) V. W. Laurie and D. T. Pence, *J. Chem. Phys.*, **38**, 2693 (1963). (b) I. L. Karle and J. Karle, *J. Chem. Phys.*, **18**, 963 (1950).

[36] Strain energy for cyclopropane (49° distortion) is 27 kcal/mole, K. Wiberg, G. M. Lampman, R. P. Ciula, D. S. Connor, P. Schertler, and J. Lavanish, *Tetrahedron*, **21**, 2749 (1965).

However, use of the resonance picture shown above is inadequate (and actually very misleading) to explain the striking directivity preference for para substitution over ortho substitution shown in Table 1-5. The striking preference for para directivity in fluorobenzene (89% para, 11% ortho) relative to chlorobenzene (68% para, 32% ortho) suggests some unusual effect that cannot be steric. To evaluate this effect, several points of difference between a C–F bond and a C–Cl, or other carbon halogen bond, should be considered.

1. An aryl C–F bond is very short; it is shorter than a C=C bond.
2. A C–F bond is more ionic.
3. Fluorine is a first-row element and consequently does not have accessible d-orbitals and cannot be easily polarized.
4. Fluorine is the most electronegative element.

The shortness of the C–F bond and the similar size of the orbitals containing p and π electrons means that p-π interaction should be at a maximum. The powerful inductive effect of the fluorine, which is very close to the aromatic ring because of the short C–F bond, is felt most strongly at the ortho position and drops off rapidly becoming much smaller for the para position. The net result is that the return of electron density by resonance can effectively cancel the small inductive effect at the distant para position but it is not large enough to decrease significantly the powerful electron withdrawal in the ortho position. The conventional resonance picture shown above is misleading; fluorine is *never* going to have a positive charge. The great electronegativity of fluorine permits it to withdraw electrons from the σ framework and accumulate electron density on fluorine. This accumulation of charge is only partly fed back to the π system by resonance. The surprising fact is that the feed-back must be much better than for chlorine or other halogens which do not withdraw electrons as strongly and are more polarizable. This improved feed-back must result from improved orbital overlap because of the short C–F bond and similar size of the carbon and fluorine p-orbitals. Figure 3-3 is a crude physical picture showing the effect of fluorine on an aromatic ring.

Murrell[26] recently proposed that fluorine is different from the other halogens and does not return any electron density to the π system by resonance, but instead strongly repels the π electrons. This idea, mentioned earlier in the discussion on fluoroolefins, is not new; but Murrell has developed it to explain the electronic spectra of halobenzenes. The high electron density, accumulated on fluorine by inductive withdrawal through the σ framework, has a strong repulsive effect on the π-electron density and concentrates charge density in the para position. Again the short C–F bond and small size of the fluorine atom makes the effect unusually strong for fluorine, even stronger than indicated because of the more polarizable character of the other halogens. This argument

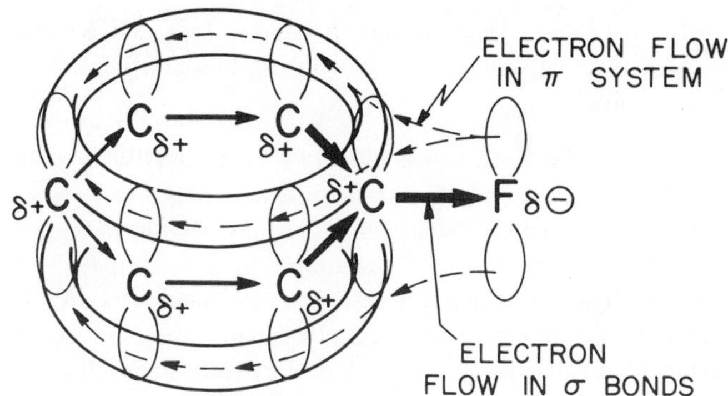

Fig. 3-3 Diagram showing accepted patterns of electron flow in σ and π framework.

is revolutionary to organic chemists, but it does have merit and must be evaluated carefully both theoretically and experimentally.

The electronic character of fluorinated substituents is readily measured using aromatic systems as probes. Thus the gross inductive and resonance effects of substituents such as XF_n and $X(R_f)_n$ are easily compared by measuring effects on a group Y:

In these series, X can be carbon or heteroatoms or groups such as N, P, O, S, SO_2. The effect measured on Y can be pK_a (Y is CO_2H, OH, NH_2, $N(CH_3)_2$), rate of ester hydrolysis (Y = CO_2R) or F^{19} nmr chemical shift (Y = F). Such data are readily compared to accumulated knowledge on many other substituents by use of Hammett[37] or Taft[38] parameters. In addition, directive effects of groups can be studied using orientation in electrophilic substitution on the aromatic ring. A series of fluorinated substituents have been examined in detail[39] and substituent

[37] H. H. Jaffé, *Chem. Rev.*, **53**, 191 (1953).

[38] R W. Taft, Jr., in *Steric Effects in Organic Chemistry* (M. S. Newman, ed.), Wiley, New York, 1956, pp. 556-675.

[39] (a) J. D. Roberts, R. L. Webb, and E. A. McElhill, *J. Am. Chem. Soc.*, **72**, 408 (1950). (b) W. A. Sheppard, *ibid.*, **84**, 3072 (1962). (c) *Ibid.*, **85**, 1314 (1963). (d) *Ibid.*, **87**, 2410 (1965). (e) F. S. Fawcett and W. A. Sheppard, *ibid.*, **87**, 4341 (1965). (f) W. A. Sheppard, *Trans. N.Y. Acad. Sci.*, Series II, **29**, 700 (1967). (g) J. W. Rakshys, R. W. Taft, and W. A. Sheppard, *J. Am. Chem. Soc.*, **90**, 5236 (1968). (h) C. L. Liotta and D. F. Smith, Jr., *Chem. Commun.*, 416 (1968).

parameters with orientation effects are summarized in Table 3-7. Several interesting observations and conclusions can be drawn from this data; some of the more significant are as follows:

1. Heteroatoms, like N, O, S, lose most of their basicity when substituted by fluorine or fluoroalkyl groups. The unshared pairs of electrons are available for resonance only in extremely demanding situations, such as stabilization of a transition state; thus oxygen of the OR_f group is now like a halogen.

2. If d-orbitals are energetically accessible in the heteroatom X, the strong inductive effect of a fluorine or fluoroalkyl substituent promotes use of some empty d-orbitals to enhance electron withdrawal by resonance.

3. The inductive effects for fluorine and fluoroalkyl groups are approximately equivalent.

4. The resonance effects (+R) remain significant for a series of fluoroalkyl groups.

The origin of the significant +R resonance effect for a CF_3 group[40] does not fit into the classical picture of conjugative interaction normally recognized for groups such as nitro or cyano. Two mechanisms have been proposed:

1. Fluoride ion hyperconjugation or no-bond resonance (which was described in the previous section in discussion of properties of fluorocarbons) represented as

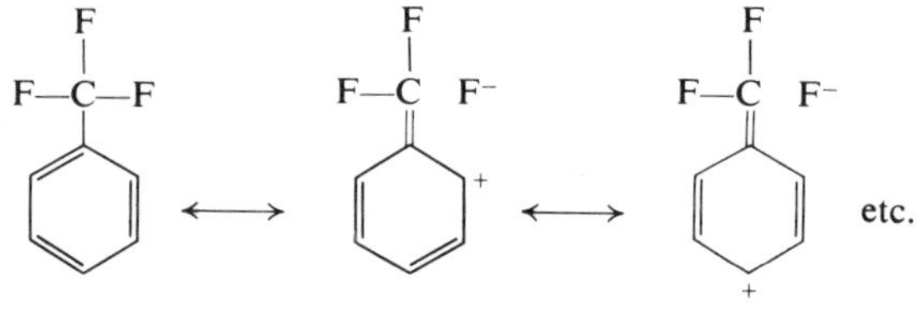

F F F
F—C—F F—C F⁻ F—C F⁻
⟷ + ⟷ etc.
+

2. π-Inductive effects resulting from strong inductive withdrawal of electrons from the neighboring position on the ring and transmitted by normal resonance mechanism to the ortho and para positions:

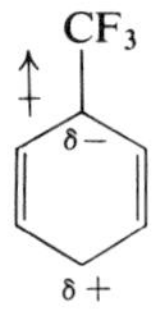

CF_3
$\delta-$
$\delta+$

[40] A −R effect is reported for CF_3 (σ_m 0.53, σ_p 0.48) in reaction of the benzoic acids with diazomethane.[39a] The mechanism of this reaction must be such that the inductive (or field effect) of the trifluoromethyl group has a dominant influence.

Table 3-7

Substituent Effects for Fluorinated Groups[a]

Substituent	σ_m	σ_p	σ_I	σ_R
A. *Ortho-para directing*				
F	0.34 (.37)	0.06 (.02)	0.45 (.47)	– 0.40 (–.45)
$N(CF_3)_2$	0.40 (.47)	0.53 (.53)	0.29 (.44)	0.23 (– .06)
OCF_3	0.38 (.47)	0.35 (.27)	0.39 (.50)	– 0.04 (–.23)
OCF_2CF_3	– (.48)	– (.28)	– (.52)	– (–.25)
SCF_3	0.40 (.46)	0.50 (.64)	0.31 (.40)	0.17 (.22)
B. *Meta directing*				
CF_3	0.42 (.49)	0.53 (.65)	0.33 (.44)	0.18 (.18)
CF_2CF_3[b]	– (.52)	– (.69)	–	–
$CF(CF_3)_2$	0.37 (.52)	0.53 (.67)	0.25 (.48)	0.26 (.17)
$C(OH)(CF_3)_2$[b]	– (.35)	– (.48)	– (.31)	– (.15)
SF_5	0.61 (.63)	0.68 (.86)	0.55 (.56)	0.11 (.27)
SO_2CF_3	0.79 (1.00)	0.93 (1.65)	0.69 (.84)	0.22 (.73)
C. *Reactive groups (orientation not measured)*				
SF_3	–	–	0.60	0.20
$P(CF_3)_2$	–	–	0.50	0.19
PF_2	–	–	0.38	0.21
PF_4	–	–	0.45	0.35

[a] Data from Ref. 39 (see also Table 6-6). Main values are from pK_a measurements on benzoic acids. Values in parentheses are from pK_a of anilinium ions. For reactive groups, data are from F^{19} chemical shift measurements on aryl fluorides.

[b] Data from pK_a of N,N-dimethylanilinium ions.

Neither of these mechanisms can satisfactorily explain all of the data.[41] As a new approach to this problem, fluorine p-π interaction has been proposed.[39d] The interaction of the p-electrons of fluorines with the π system of the aromatic ring causes significant return of electron density to the ring, partly counteracting the normal strong inductive withdrawal. This electron density is returned more effectively to the meta than the para position so that the para position appears more strongly deactivated. In Fig. 3-4 the mode of overlap is illustrated and distances shown to prove that both C-1 and C-2 of the ring are within van der Waals distance of 3.05 Å (1.70 for C and 1.35 for F) needed for an interaction. A classical resonance form can be drawn to represent this interaction:

F
F—C—F^+
H
H

[41] For a more detailed discussion of this question see Ref. 39d.

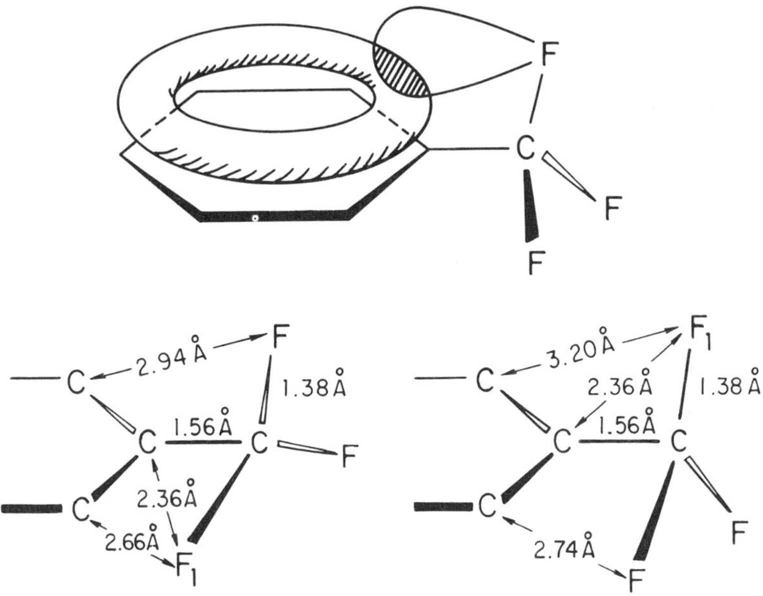

F_1 IN PLANE OF RING $\quad$ F_1 PERPENDICULAR TO PLANE OF RING

Fig. 3-4 Diagrams showing atomic distances (calculated for tetrahedral angles) and orbital overlap in benzotrifluoride.

Molecular orbital calculations also support the feed-back, through the π system mechanism[42]; the net charge density in benzotrifluoride is calculated to alternate so that the ortho and para positions appear more deficient in charge density than the meta position.

Additional data in favor of the p-π interaction mechanism has accumulated[39d, e] particularly from esr studies.[43] However, other workers[44, 45] have presented arguments and data against p-π interaction. They propose that the strong field

[42] Unpublished calculations by Dr. H. E. Simmons in collaboration with W. A. Sheppard. Much more extensive calculations have been carried out by Dr. J. A. Pople (private communication) using the CNDO method (Ref. 14).

[43] (a) P. J. Scheidler and J. R. Bolton, *J. Am. Chem. Soc.*, **88**, 371 (1966). (b) R. J. Lontz, *J. Chem. Phys.*, **45**, 1339 (1966). (c) E. T. Strom, *J. Am. Chem. Soc.*, **88**, 2065 (1966). (d) E. T. Strom, A. L. Bluhm, and J. Weinstein, *J. Org. Chem.*, **32**, 3853 (1967). (e) E. G. Janzen and J. L. Gerlock, *J. Am. Chem. Soc.*, **89**, 4902 (1967). (f) J. C. Danner and A. H. Maki, *J. Am. Chem. Soc.*, **88**, 4297 (1966). (g) The p-π interaction mechanism can be extended to organometallic systems. F. A. Cotton and R. M. Wing, *J. Organometal. Chem.*, **9**, 511 (1967) argue for π interaction between a trifluoromethyl group and manganese pentacarbonyl.

[44] M. J. S. Dewar and A. P. Marchand, *J. Am. Chem. Soc.*, **88**, 354 (1966).

[45] F. W. Baker, R. C. Parish and L. M. Stock, *J. Am. Chem. Soc.*, **89**, 5677 (1967).

effect of a trifluoromethyl group is the important factor and that polar effects are generally propagated more efficiently from the para than from the meta position with $\sigma_{p-x}/\sigma_{m-x} = 1.2$.[46]

However, an important piece of data that does not fit this theory is that the tricyanomethyl group, $C(CN)_3$ (which is inductively a much stronger electron-withdrawing group than a trifluoromethyl), does not show a +R effect![47]

The p-π interaction idea is no doubt of only secondary importance, for fluoroalkyl groups and inductive or field effects are certainly the dominant factor. A π-inductive effect, as described above, is considered to be another second-order type effect which probably also contributes to the +R character of fluoroalkyl groups. Again, Murrell's π-repulsion theory[26] needs to be considered as a possible alternative explanation.

As pointed out earlier in this section, p-π interaction is of first-order importance when fluorine is attached directly to an aromatic system. This p-π interaction in fluorobenzenes is of major significance in F^{19} nmr chemical shift measurements to determine substituent parameters and mechanism of transmission of substituent effects.[48, 49] The F^{19} shifts of fluorobenzenes cannot be correlated directly with Hammett σ parameters.[50] Taft[48] devised a convenient recipe to calculate substituent parameters from these F^{19} measurements. However, an argument has developed about the interpretation of these shifts for study of transmission of substituent effects.[49] We feel that this problem is easily understood using a simple qualitative picture which takes into account p-π interaction. The F^{19} chemical shift is a composite of several effects of which two are probably dominant. First, the electronic character of the substituent influences the chemical shift in a regular manner that can be correlated with Hammett substituent parameters. When the fluorine is isolated from the π system containing the substituent, we see this simple direct correlation. For example, in arylsulfur pentafluorides the apex fluorine is insulated from the π system by the SF_4 unit and the chemical shift for this fluorine, which ranges over 6 ppm from *p*-amino

[46](a) O. Exner, *Tetrahedron Letters*, 815 (1963). (b) M. J. S. Dewar, *Hyperconjugation*, Ronald Press, New York, 1962, p. 159.

[47]W. A. Sheppard and R. M. Henderson, *J. Am. Chem. Soc.*, **89**, 4446 (1967).

[48](a) R. W. Taft, Jr., *J. Phys. Chem.*, **64**, 1805 (1960). (b) R. W. Taft, E. Price, I. R. Fox, I. C. Lewis, K. K. Andersen, and G. T. Davis, *J. Am. Chem. Soc.*, **85**, 709, 3146 (1963).

[49](a) M. J. S. Dewar and A. P. Marchand, *J. Am. Chem. Soc.*, **88**, 3318 (1966). (b) W. Adcock and M. J. S. Dewar, *J. Am. Chem. Soc.*, **89**, 379 (1967).

[50](a) For an authoritative discussion of the factors effecting nmr chemical shifts see J. W. Emsley, J. Feeney, and L. H. Sutcliffe, *High Resolution Nuclear Magnetic Spectroscopy*, Pergamon Press, New York, 1966. (b) M. J. S. Dewar and T. G. Squires, *J. Am. Chem. Soc.*, **90**, 210 (1968), and G. L. Anderson and L. M. Stock, *ibid.*, 212, have examined F^{19} nmr shifts in aliphatic systems and report no apparent correlation of shift with electronic character of the substituent. However, in preliminary studies P. E. Peterson and W. A. Sheppard have found correlations when trifluoroacetic acid is used as solvent.

to *p*-nitro substituents, correlates perfectly with the standard Hammett σ values from ionization of benzoic acids.[39d, 51] Also, Mallory[52] has noted that in a series of fluorophenanthrenes, fluorostilbenes and 1-fluorophenyl-2-arylethanes, where the fluorine is in a different ring from the substituent, the F^{19} chemical shift can be correlated with the normal σ values. However, when the substituent can interact directly through the π system with the p-electrons of fluorine the F^{19} chemical shift is effected in a different way, and the simple correlation breaks down. Also the ring current is altered by substituents and could produce further variation in the F^{19} chemical shift. However, ring current effects are considered to be minor (0.5 ppm maximum change in shift) relative to the electronic effects (20 ppm range).

In any case, the F^{19} nmr chemical shift measurements in fluorobenzenes are a sensitive probe to determine substituent effects, particularly to evaluate contributions by the π system in transmission of electronic effects.

As a summary to this discussion on the effect of fluorine in unsaturated systems, the most important factor is strong inductive withdrawal of electrons, which is chiefly from the σ framework but must be felt throughout the molecular system. Superimposed on this inductive withdrawal is an interaction of fluorine with the π system. One unanswered question is: Does fluorine feed electron density back through the p-electrons (the accepted resonance picture which we prefer) or does it strongly repel π electrons? Perhaps both mechanisms operate to extents that depend on the overlap, which takes into account factors such as size of π orbitals, distance, angle, and mode of interaction. Another point is p-p interaction. At certain distances two fluorines (or any other atoms with unshared p-electrons) appear to have sufficient nonbonded interaction (dispersion forces) to overcome the usual repulsion. More experimental data are needed and theoretical studies[53] must be developed.

3-2 EFFECT ON PROPERTIES OF FLUOROCARBONS

Organic chemists learn organic chemistry in terms of molecules in which carbon-hydrogen and carbon-carbon are the familiar bonds. Fluorocarbons with C–F bonds appear then to be abnormal; for example, the boiling points of straight-chain perfluorocarbons compared to corresponding paraffin hydrocarbons (Fig. 3-5) appear anomalously low for their relatively higher molecular weight. The

[51] D. R. Eaton and W. A. Sheppard, *J. Am. Chem. Soc.*, **85**, 1310 (1963).

[52] F. W. Mallory, 3rd Middle Atlantic Regional meeting of the American Chemical Society, Philadelphia, Pa., Feb. 1968, and private communications.

[53] A recent example is the calculations by G. de Alti, V. Galasso, and G. Costa, *Spectrochimica Acta*, **21**, 649 (1965) on perhaloethylenes.

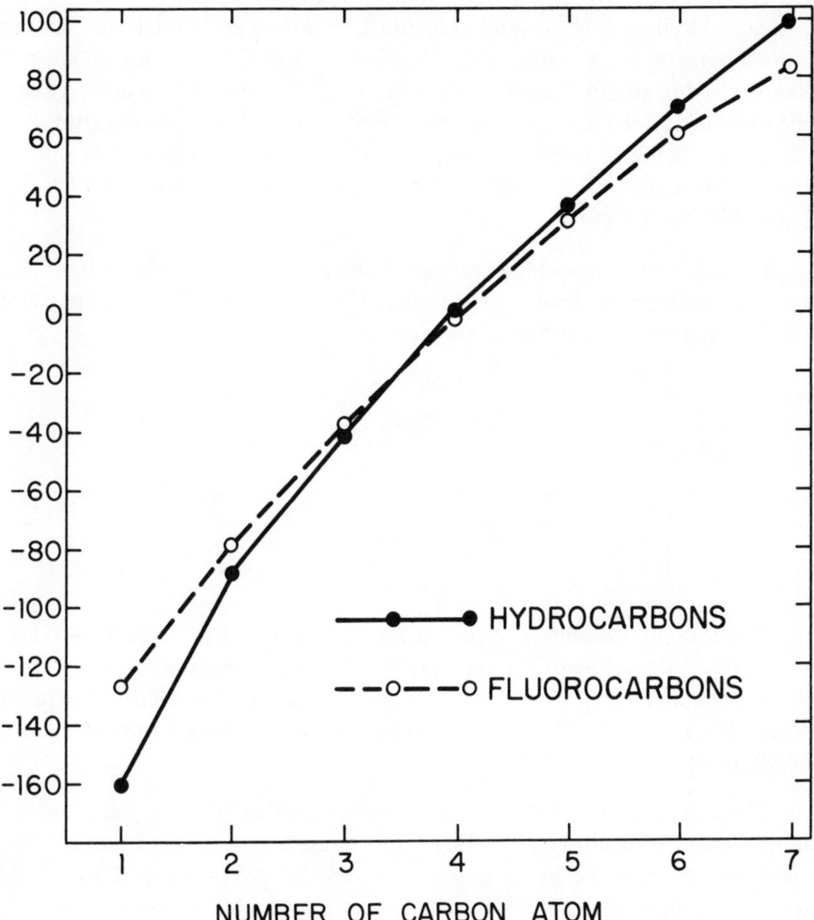

Fig. 3-5 Boiling points of straight-chain hydrocarbons and perfluorocarbons.

alternative view, as developed by Reed,[54] is that fluorocarbons are normal and hydrocarbons are unusual. Reed elaborates the "interpenetration" theory of Simons and Dunlap[55] to explain why hydrocarbons and not fluorocarbons are anomalous, by contrasting the physical properties of the two classes under eighteen different headings, such as vapor pressure, critical constants, liquid compressibility, and surface tension. The "interpenetration" theory proposes that the small size of the hydrogen atom permits attractive interactions between carbon-carbon and carbon-hydrogen centers of separate molecules in addition to

[54] T. M. Reed, III, in *Fluorine Chemistry* (J. H. Simons, ed.), Vol 5, Academic, New York, 1964, pp. 133-236.

[55] J. H. Simons and R. D. Dunlap, *J. Chem. Phys.*, **18**, 335 (1950).

the expected hydrogen-hydrogen interactions. In fluorocarbons only the fluorine-fluorine interactions are significant so that forces between molecules are much higher for hydrocarbons than fluorocarbons. Unfortunately, we do not have the space to review all the data and arguments that bear on this discussion (Ref. 54 is highly recommended as the comprehensive treatise on this subject), but we would like to comment on some properties of fluorocarbons that are readily explained by this proposal:

1. The effect of branching in the molecular structure of hydrocarbons is greater and in the opposite direction to the effect in fluorocarbons. As an example, the boiling points of the pentanes are given in Table 3-8.

Table 3-8

Boiling points (°C) of the pentanes

	X = H	X = F
$CX_3(CX_2)_3CX_3$	36.1	29.5
$(CX_3)_2{-}CXCX_2CX_3$	27.9	29.3
$(CX_3)_4C$	9.5	28.5–29.5

2. The compressibility of hydrocarbons is less that than of fluorocarbons.

3. Densities of liquids composed of branched-chain isomers in the hydrocarbon series are less than for the straight-chain structure. The densities of liquid fluorocarbons are greater for the branched-chain isomers than for the straight-chain isomers.

The key point to remember from this discussion is that fluorocarbons (compared to hydrocarbons) have low forces of attraction between molecules. In fact, fluorocarbons behave as nearly ideal liquids.[56, 57] However, for partially fluorinated hydrocarbons, intermolecular hydrogen bonding with fluorine becomes very important. The maximum boiling point for difluoromethane in the methene series was pointed out in Fig. 1-1 and discussed in Section 2-2 in terms of hydrogen bonding. Such intermolecular hydrogen bonding promoted by the high polarization in the molecule is probably at a maximum for difluoromethane.

The thermal stability of fluorocarbons is generally considerably better than that of the corresponding hydrocarbons.[56b] Thus CF_4 decomposes only slowly at

[56] For further discussion and descriptive material on physical properties of fluorocarbons, see, in addition to Ref. 54, (a) M. Hudlicky, *Chemistry of Organic Fluorine Compounds,* Macmillan, New York, 1962, pp. 288-313. (b) T. J. Brice, in *Fluorine Chemistry,* (J. H. Simons, ed.), Vol. 1, Academic, New York, 1950, pp. 436-459.

[57] T. W. Bates and W. H. Stockmayer, *Macromolecules,* 1, 17 (1968) have made a significant contribution to understanding the physical properties of perfluoroalkanes by study of conformational energies. Their conclusions have particular value to studies of polytetrafluoroethylene.

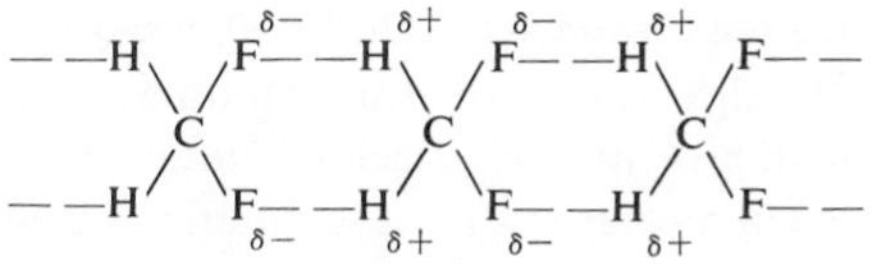

```
          δ−      δ+   δ−      δ+
— —H    F— —H    F— —H    F— —
    \  /     \  /     \  /
     C        C        C
    /  \     /  \     /  \
— —H    F— —H    F— —H    F— —
        δ−   δ+    δ−    δ+
```

temperatures of the carbon arc; long-chain perfluorocarbons do not undergo homogeneous decomposition until near-red heat temperatures to give CF_4 and carbon.[58] The mode of decomposition of fluorocarbons and hydrocarbons differs. Commonly, dehydrogenation is one of the important ways for a hydrocarbon to decompose. Fluorocarbons usually cleave at higher temperature at the weaker C—C bond leaving the strong C—F bond intact.

This discussion of physical properties has been of necessity superficial. A number of properties, such as dielectric character and surface tension, are of considerable importance in industrial applications for fluorocarbons, and we can only refer the reader to reviews for elaboration on the importance of physical properties to utility.[59] We must now turn to the important subject of chemical behavior of fluorocarbons.

3-3 CHEMICAL REACTIVITY

In Sections 2-2 and 3-1B, we pointed out some aspects about the chemical behavior of fluoride ion, fluorinated molecules, and the C—F bond. We now plan to elaborate on the earlier discussion and present a unified picture.

3-3A. SATURATED FLUOROCARBONS

Thermodynamically fluorocarbons are calculated to be unstable to hydrolysis; for example, the free energy for hydrolysis of carbon tetrafluoride is calculated[60] to be exergonic by 72.8 kcal/mole

$$\begin{array}{llll} CF_4(g) + 2\,H_2O(g) & \longrightarrow 4\,HF(g) + CO_2(g) & & -72.8 \\ -162.5 \quad 2(-57.80) & \quad 4(-64.2) & -94.05 & \text{kcal/mole} \end{array}$$

and the advantage gained by hydration of HF would increase the value substantially. But perfluorocarbons are essentially inert to hydrolysis until heated to a temperature of near 500°C. This inertness is readily explained kinetically by a high activation energy for hydrolysis. The very-difficult-to-polarize fluorine

[58] Ref. 56b, p. 432.

[59] (a) See Ref. 56a, pp. 332-357. (b) A. J. Rudge, *The Manufacture and Use of Fluorine*, Oxford Univ. Press, London, 1962, p. 65.

[60] (a) National Bureau of Standards, Circular No. 500, "Selected Values of Chemical Thermodynamic Properties," 1952. (b) J. D. Cox, H. A. Gundry, and A. J. Head, *Trans. Faraday Soc.*, **61**, 1594 (1965) report a value of −41.4 kcal/mole for the standard enthalpy to hydrolysis of CF_4 to CO_2 and HF(20 H_2O), at 25°C.

atoms act as a shield for the carbon skeleton. In other words, the short C–F bond combined with the significant size of the fluorine atom and the tight binding of the C–F bond prevents any attacking group from penetrating close enough to carbon at low temperatures to begin interaction. Chlorocarbons are much more easily hydrolyzed because chlorine, with the longer weaker bond to carbon, is more easily polarized than fluorine.

Table 3-9

Thermal Stabilities of Polymers; Degradation in Vacuo[a]

	Decomposition		
Polymer	Temperature (°C)	Main Products	Activation energy kcal/mole
$(CF_2-CF_2)_n$	450-510	$CF_2{=}CF_2$	80
$(CH_2-CH_2)_n$	380-410	Some $CH_2{=}CH_2$ chiefly various hydrocarbons	70
$(CF_2-CFCl)_n$	360-390	Monomer and propene	66
$(CF_2-CFH)_n$	400-475	HF and about 10 mole % hydrofluorocarbons	–
$(CF_2-CH_2)_n$	380-530	HF and about 50 mole % hydrofluorocarbons	48
$(CH_2-CHF)_n$	370-480	HF and about 30-40 mole % hydrocarbons	–
$(CH_2-CHCl)_n$	240-300	HCl	30

[a] Based on data summarized by C. R. Patrick, *Advan. Fluorine Chem.*, **2**, 1 (1961).

When alkanes are only partially fluorinated, the above arguments no longer apply. In a molecule such as 2,2,3,3-tetrafluorobutane, $CH_3CF_2CF_2CH_3$, another mode of decomposition is possible; the extremely stable and thermodynamically favored hydrogen fluoride is easily eliminated. Not only is such a molecule thermally unstable to elimination of HF, it is very labile to attack by base at the relatively acidic α-hydrogen. To illustrate these points, some data on thermal stabilities of polymers are given in Table 3-9. Perfluorinated compounds are the most stable. The destabilizing influence of α-hydrogen favoring hydrogen fluoride elimination is clearly shown.

In Section 2-2 we pointed out comparisons on the kinetics of fluoride and chloride displacements. A comprehensive review on the mechanisms of fluoride displacements is now available,[61] and we will draw on data compiled in this review to illustrate the important aspects of our discussion on mechanism.

A simple approach is to assume that in halide displacement, bond breaking occurs to a significant extent in the rate-determining step. Then since the fluorine-to-carbon bond is much stronger than the other carbon-halide bonds, the energy

[61] R. E. Parker, *Advan. Fluorine Chem.*, **3**, 63 (1963).

of activation for bond breaking will be higher, and consequently the rate for fluoride displacement would be slower. This simple approach, which ignores many other factors, is surprisingly effective since generally aliphatic fluorides undergo displacement reactions much more slowly than do the corresponding chlorides (see example in Table 1-7A).

But such a simple bond strength approach ignores other factors. One argument is that carbon bonded to fluorine will have a greater positive charge than carbon bonded to any other halogen and should therefore be much more susceptible to attack by a nucleophile. However, fluorine is not nearly as easily polarized as other halogens and in the transition state cannot electrically deform as the nucleophile approaches. Because of the short C–F bond, the repulsion by fluorine is magnified to the extent that much greater energy is needed to bring the nucleophile within reactive distance to carbon. This discussion presumes an S_N2 mechanism (bimolecular displacement). For reactions that go by an S_N1 mechanism (unimolecular), only the breaking of the carbon-halogen is important in the rate-determining step, and bond strengths are the best criteria to predict relative reactivities of the halides.

Another factor ignored in the above discussion is the solvation of ions. We pointed out in Section 2-2 that solvation is very important and concluded that the fluoride ion could not replace chloride in aliphatic chlorides in an aqueous system because the much greater hydration energy of fluoride completely overcame the advantage of forming the more stable C–F bond. However in non-aqueous systems, fluoride readily replaces chloride. Turning the argument around, the high energy of hydration of fluoride should promote fluoride displacement by chloride in aqueous systems if factors such as repulsion of attacking nucleophile, as discussed above, were not so important.

A good example of the effect of solvation is shown by reaction of aliphatic alcohols with hydrogen halides. The replacement of hydroxyl group goes readily for aqueous HCl, HBr, and HI but not for aqueous HF. In addition to the unfavorable hydration or solvation energy of fluoride ion, the high bond strength of HF is unfavorable for displacements. In fact, the HF bond and the HF···H hydrogen bond are so strong that even with the very high hydration energy of fluoride ion, hydrogen fluoride is still a weak acid in water.

The importance of hydrogen bonding is clearly shown in the solvolysis of benzyl halides. The hydrolysis of benzyl fluoride is acid catalyzed but that of benzyl chloride is not.[62] Hydrogen bonding to fluorine in the transition state by hydronium ions is the best explanation:

$$\overset{\delta+}{H_2O}\text{---}\underset{\displaystyle C_6H_5}{\underset{|}{CH_2}}\text{---}F\text{---}H\text{---}\overset{\delta+}{OH_2}$$

[62]C. G. Swain and R. E. T. Spalding, *J. Am. Chem. Soc.*, **82**, 6104 (1960).

The large dipole moment of the carbon fluorine bond facilitates such hydrogen bonding.

One of the examples where saturated alkyl fluorides appear to be more reactive than the corresponding chlorides is in Friedel-Crafts alkylation reaction.[63] The most likely explanation is electrophilic catalysis by the Lewis acid such as $AlCl_3$

$$ArH + R{-}X + AlCl_3 \longrightarrow \underset{\displaystyle H}{\overset{\delta+}{Ar}}{-}{-}{-}R{-}{-}{-}X{-}{-}{-}\overset{\delta-}{AlCl_3}$$

and the energy gained from forming an Al–F bond over breaking a C–F is better than that from forming Al–Cl over breaking C–Cl.

The important factors in chemical reactivity, particularly displacements for fluorines bonded to carbon in saturated systems (sp^3 hybridized carbon), are summarized as follows:

1. Fluorine will be displaced by either an S_N1 or S_N2 mechanism at a slower rate than other halogen ions.

2. The much stronger C–F bond (stronger than other carbon halogen bonds) is an important reason for the slower rate because bond breaking is usually significant in the rate determining step.

3. The low polarizability of fluorine is kinetically effective in preventing attack by a nucleophile on the carbon to which fluorine is bonded and is also important in retarding hydrolysis.

4. Solvent (or other electrophilic agent) can play an important role because of strong hydrogen bonding and high solvation of fluoride. Change in solvent from aqueous to nonaqueous will significantly change relative rates.

3-3B. UNSATURATED SYSTEMS

In Section 3-1B we pointed out the high reactivity of fluoroolefins toward nucleophilic reagents and discussed at some length the reason for this high reactivity. Since this material has already been discussed and is reviewed comprehensively by Chambers and Mobbs,[21] we summarize with the general equation shown in (Chart 3-1), where R_f and R'_f are usually highly fluorinated groups and

[63] (a) N. O. Calloway, *J. Am. Chem. Soc.*, **59**, 1474 (1937). (b) J. Bernstein, J. S. Roth, and W. T. Miller, Jr., *J. Am. Chem. Soc.*, **70**, 2310 (1948). (c) J. H. Simons and G. C. Bassler, *J. Am. Chem. Soc.*, **63**, 880 (1941).

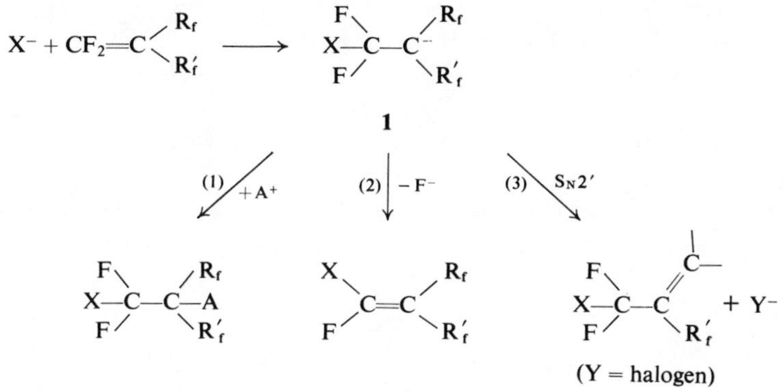

$$X^- + CF_2{=}C(R_f)(R_f') \longrightarrow X{-}CF_2{-}C^-(R_f)(R_f')$$

1

(1) $+A^+$ → $X{-}CF_2{-}C(R_f)(A)(R_f')$

(2) $-F^-$ → $(X)(F)C{=}C(R_f)(R_f')$

(3) S_N2' → $X{-}CF_2{-}C(R_f'){=}C< \; + \; Y^-$ (Y = halogen)

CHART 3-1

X^- is a nucleophile. The carbanion species **1** is formed as an intermediate and, depending on the character of R_f and R_f', can convert to products by three different courses: (1) addition of proton or cationic group is most common for fluoroolefins; (2) loss of fluoride to give a new olefin is more common with higher molecular weight and cyclic fluoroolefins; (3) this mechanism cannot be easily distinguished from a S_N2' displacement and is common for polyfluorohalogen olefins.

Greater reactivity of fluorine bonded to an sp^2-hybridized carbon is also evident in fluoroaromatic compounds. In contrast to benzene and even hexachlorobenzene, highly fluorinated benzenes are extremely labile to nucleophilic attack. Also monofluorobenzenes activated by electron-withdrawing groups are much more susceptible to nucleophilic displacement than the corresponding chloro derivatives. Some data were given in Table 1-7B. Examples of such displacements are very common in current literature,[64] and the examples in Chart 3-2 have been chosen as typical.[65]

For highly fluorinated aromatics, the strong electron-withdrawing effect decreases the charge density in both the σ and π framework, promoting attack by a nucleophile. For the *ortho*- and *para*-fluoronitrobenzenes, the rate of substitution is 10 to 1000 times faster than for the corresponding chloro compounds. This enhanced rate of fluoro compared to chloro in these activated aromatics is in sharp contrast to the aliphatic displacements discussed earlier. Obviously we have a change of mechanism, and now no carbon halogen bond breaking is

[64] For a review of such displacements see A. E. Pavlath and A. J. Leffler, *Aromatic Fluorine Compounds,* Reinhold, New York, 1962, p. 55.

[65] (a) D. J. Alsop, J. Burdon, and J. C. Tatlow, *J. Chem. Soc.,* 1801 (1962). (b) For recent studies on nucleophilic substitution in perfluoroaromatic systems, see J. Burdon, W. B. Hollyhead, and J. C. Tatlow, *J. Chem. Soc.,* 5152 (1965); and J. Burdon, *Tetrahedron,* **21**, 3373 (1965).

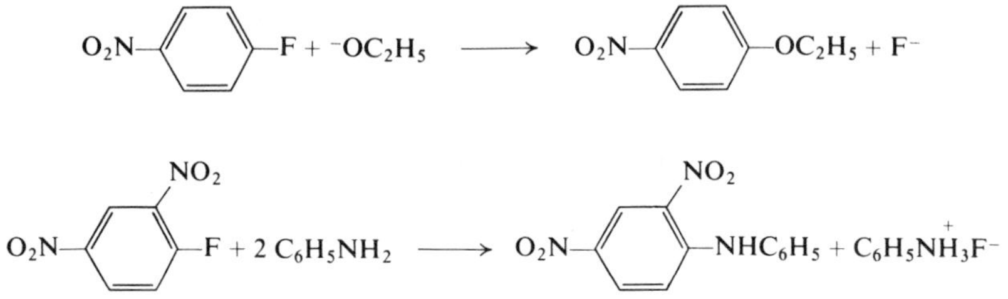

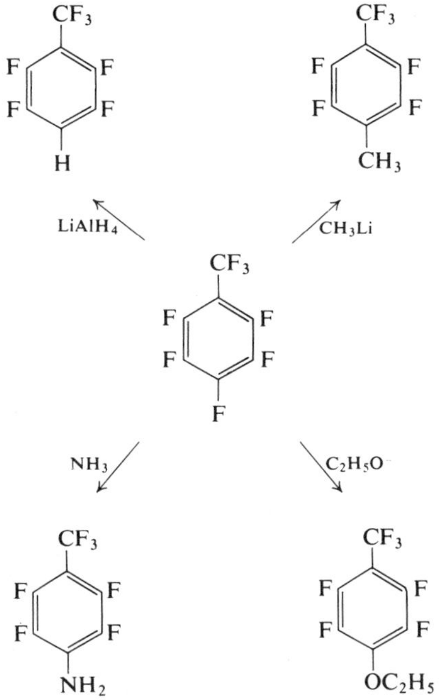

$$O_2N\text{—}C_6H_4\text{—}F + {}^-OC_2H_5 \longrightarrow O_2N\text{—}C_6H_4\text{—}OC_2H_5 + F^-$$

$$(NO_2)\,O_2N\text{—}C_6H_3\text{—}F + 2\,C_6H_5NH_2 \longrightarrow (NO_2)\,O_2N\text{—}C_6H_3\text{—}NHC_6H_5 + C_6H_5\overset{+}{N}H_3F^-$$

$CF_3C_6F_4H$ ← $LiAlH_4$ — $CF_3C_6F_5$ — CH_3Li → $CF_3C_6F_4CH_3$

$CF_3C_6F_4NH_2$ ← NH_3 — $CF_3C_6F_5$ — $C_2H_5O^-$ → $CF_3C_6F_4OC_2H_5$

CHART 3-2

occurring in the rate-controlling step. Intermediates of the type shown are strongly supported by experimental evidence provided by Bunnett[66] and Parker.[67] This subject is thoroughly reviewed by Parker.[61]

To summarize, unsaturated fluorinated systems are extremely susceptible to attack by a nucleophile. Usually a stable intermediate is formed, and, since no carbon fluorine bond breaking occurs in the rate-determining step, fluorine is often displaced faster than other halogens. Again hydrogen bonding can assist in removal of fluoride ion.

[66] J. F. Bunnett and J. J. Randall, *J. Am. Chem. Soc.*, **80**, 6020 (1958).
[67] J. Miller and A. J. Parker, *J. Am. Chem. Soc.*, **83**, 117 (1961).

$$CH_3O^- + \text{(p-fluoronitrobenzene)} \longrightarrow \left[\text{(F, } OCH_3 \text{ adduct, } NO_2^-) \longleftrightarrow \text{(F, } OCH_3 \text{ adduct, } =NO_2^-) \right] \xrightarrow{-F^-} \text{(p-nitroanisole)}$$

PROBLEMS

1. For ethylamine $K_b = 4.5 \times 10^{-4}$. Consider the data of Table 1-3 and predict K_b for $FCH_2CH_2NH_2$, $F_2CHCH_2NH_2$, and $F_3CCH_2NH_2$.

2. Briefly and concisely explain:
(a) why F_2 is extremely reactive toward most reagents;
(b) why fluoride ion is usually considered a poor nucleophile;
(c) why fluorine is produced by electrolysis of fused KHF_2;
(d) the results presented in Table 2-5; why is fluorine so nonselective?
(e) why Miller's mechanism for fluorination is reasonable (see page 12; see Ref. 2-7a);
(f) why CH_3CH_2F may be easily hydrolyzed to ethanol.

3. Cite the experimental evidence for greater ionic character in a C–F bond than in a C–Cl bond.

4. Explain why the acidities of haloforms decrease in the order $HCBr_3 > HCCl_3 > HCF_3$. Is this result consistent with the fact that fluorine is more electronegative than chlorine or bromine?

5. (a) Consider the unusual reactions of $CF_2{=}CF_2$ presented in Chapter 3 along with other data such as that in Tables 1-2 and 1-6. What do you consider as the best explanation for the vast differences in the chemistry of $CF_2{=}CF_2$ compared to $CH_2{=}CH_2$?

(b) Predict the reactivity of $CF_2{=}CHF$, $CF_2{=}CH_2$, $CFH{=}CFH$ and $CFH{=}CH_2$ in 2 + 2 (cyclobutane formation) and 2 + 4 (Diels-Alder) cycloadditions, in cationic-, anionic-, and free radical-initiated polymerizations, and toward nucleophilic attack. Compare with respective chloro analogs.

6. Predict the product(s) formed in the thermal reaction under pressure of 1,1-dichloro-2,2-difluoroethylene with each of the compounds listed.
(a) Acrylonitrile.
(b) Butadiene.
(c) Styrene.
(d) *cis-cis*-2,4-Hexadiene.
(e) Ketene.

7. Evaluate the data on substitution of fluorobenzene in Table 1-5 and the arguments by the authors in Chapter 3. Critically support or refute the authors' viewpoint. If possible, postulate a better explanation for the data of Table 1-5.

8. (a) Predict the position of nitration and the relative rate of nitration compared to fluorobenzene in the following compounds:

(1) 4-Nitrofluorobenzene.
(2) 1,3-Difluorobenzene.
(3) 3-Chlorofluorobenzene.
(4) 4-Chlorofluorobenzene.

(b) Explain your predictions.

9. (a) Predict substituent parameters (σ_m and σ_p) and directivity toward orientation of a substituent in electrophilic substitution (e.g., bromination) for the following substituents on benzene:

(1) $-(CF_2)_3CF_3$
(2) $-OSF_5$
(3) $-SOCF_3$ (sulfoxide)
(4) $-SO_2F$
(5) $-CH_2CF_3$
(6) $-N(COF)_2$

(b) Explain your predictions.

10. Compare the inductive and resonance effects of $-F$ and $-CF_3$ as substituents on an aromatic ring. How do these compare with $-Cl$ and $-CCl_3$ substituents? Support or refute the authors' explanation of the effects. If possible, postulate a better explanation.

11. Estimate the boiling points of the following compounds. Use the boiling point of the corresponding hydrocarbon as a starting point.

(a) Tetrafluoromethane.
(b) Perfluorobutane.
(c) Hexafluorobenzene.
(d) Pentafluorobenzene.
(e) 1,2,4,5-Tetrafluorobenzene.
(f) Trifluoromethylbenzene.
(g) 1,4-Bis(trifluoromethyl)benzene.
(h) Hexakis(trifluoromethyl)benzene.
(i) 2,2,3,3,4,4-Hexafluoropentane.
(j) 1,1,1,5,5,5-Hexafluoropentane.

12. Support or refute the statement: "Fluorocarbons have normal physical properties while hydrocarbons are highly unusual molecules with abnormal physical properties." Cite several pieces of evidence for your view. (Chapter 2 of Ref. A3-5 in Appendix A is a key source of data.)

13. Put the compounds listed below in the order of *decreasing* stability to (a) heat and (b) alcoholic sodium hydroxide solution.

(1) $CF_3CH_2CH_2CH_2CH_2CH_3$
(2) $CF_2HCH_2CH_2CH_2CH_2CH_3$
(3) $CFH_2CH_2CH_2CH_2CH_2CH_3$
(4) $CF_3CHFCH_2CH_2CH_2CH_3$
(5) $CF_3CH_2CHFCH_2CH_2CH_3$
(6) $CH_3CH_2CF_2CF_2CH_2CH_3$
(7) $CH_3CH_2CHFCHFCH_2CH_3$
(8) $CHF_2CF_2CF_2CF_2CF_2CHF_2$

14. Perfluoroaromatic compounds are highly reactive to nucleophilic reagents. Predict product(s) and, where possible, relative rates in each of the following reactions. Justify predictions.

(a) C_6F_5Cl, C_6F_5H, and C_6Cl_5H with sodium methoxide in methanol.
(b) Octafluoronaphthalene with sodium methoxide in methanol.
(c) $C_6F_5N(CH_3)_2$ and $C_6F_5NH_2$ with ammonia.
(d) $C_6F_5OCH_3$ and $C_6F_5NO_2$ with dimethylamine or with sodium methoxide.
(e) C_6F_5N (pentafluoropyridine) and $C_4F_4N_2$ (tetrafluoropyrimidine) with hydroxide ion or with methoxide.
(f) C_6F_5OH with potassium hydroxide in t-butanol or with sodium methoxide in methanol (reflux).

Chapter Four Classical Methods of Fluorination

The majority of organic fluorine compounds have been prepared by what we consider as "classical methods of fluorination." These classical methods were developed mainly prior to 1950 and were of great importance to industry and

Outline for Methods of Fluorination

Section	Fluorinating agent or method	Mode of action
4-1	Fluorine (F_2)	Addition of F–F to C=C. Substitution of H and X in C–H.
4-2	Halogen fluorides (ClF, ClF_3, BrF_3, IF_5)	Addition of X–F to C=C. Substitution of H and X in C–H and C–X.
4-3	Hydrogen fluoride (in addition reactions)	Addition of H–F to C=C. Substitution of X in C–X.
4-4	Oxidizing metallic fluorides (particularly CoF_3 and AgF_2)	Addition of F–F to C=C. Substitution of H and X in C–H and C–X.
4-5	Hydrogen fluoride and/or antimony fluorides (in substitution of halogen or oxygenated functions)	Substitution of X in C–X. Addition of HF to C=C.
4-6	Groups I and II halides (nonoxidizing in substitution of halogens or oxygenated functions)	Substitution of X in C–X
4-7	The Schiemann reaction	Substitution of NH_2 in aromatic compounds.
4-8	Electrochemical fluorination	Addition of F–F to C=C or C=X. Substitution of H or X in C–H and C–X.
4-9	Jet fluorination (F_2 special technique)	Addition of F–F to C=C or C=X. Substitution of H in C–H.

the Manhattan Project in leading to useful and needed organic fluorine compounds.[1]

In this chapter we will briefly review the classical methods of fluorination (see Outline). Representative data are selected for tables to illustrate the capabilities of each method. Organization here is by technique in contrast with Chapters 5 and 6 and later chapters which emphasize the result—how to achieve a particular kind of fluorine substitution. Because of the large volume of literature, including extensive reviews, our coverage is only a critical survey particularly discussing limitations, side reactions, and mechanism.

4-1 ELEMENTAL FLUORINE WITH ORGANIC COMPOUNDS

$$F_2 + \gt C{=}C\lt \longrightarrow \gt C(F){-}C(F)\lt$$

$$\gt CH \longrightarrow \gt CF$$

$$\gt CX \longrightarrow \gt CF$$

The direct reaction of fluorine with organic compounds usually gives extensive degradation and fragmentation of the starting materials and not simple reaction products. With special techniques (to be discussed later in this chapter and in Chapter 5) fluorine can be selective. The basic problem, as discussed in Chapter 2, is that the energy given up when H–F and C–F bonds are formed is greater than the energy required to break other molecular bonds such as C–C, C–Cl, and C–H. Unless an effective means of removing the excess energy or "taming" fluorine is used, extensive skeletal fragmentation of the organic molecule will occur, and can occur explosively! Classically, fluorine is tamed with a fluorine carrier an inorganic reagent which, when fluorinated to a higher oxidation state, can oxidatively fluorinate an organic compound. Practically, CoF_2 to CoF_3 and AgF to AgF_2 have been used most extensively and will be discussed in Section 4-4.

Bockemüller[2] was the first to control the direct reaction of fluorine with an alkane or to add fluorine selectively to a C=C bond by passing fluorine diluted with carbon dioxide into dichlorodifluoromethane solutions of organic compounds. He succeeded in substituting a single fluorine into cyclohexane, butyric

[1] The following references give a good background on the development of "classical" methods: (a) J. M. Hamilton, Jr., *Advan. Fluorine Chem.*, **3**, 117 (1963). (b) T. J. Brice in *Fluorine Chemistry* (J. H. Simons, ed.), Vol. 1, Academic, New York, 1950, pp. 423-462.

[2] (a) W. Bockemüller, *Ann.*, **506**, 20 (1933). (b) W. Bockemüller, in *Newer Methods of Preparative Organic Chemistry*, Interscience, New York, 1948, pp. 230-232.

Table 4-1

Selective Reaction of Fluorine with Alkanes and Alkenes

Alkane or alkene	Conditions	Product(s) (% yield)	Ref.
Substitution of Hydrogen by Fluorine			
Hexane	$-78°$, $N_2:F_2$ at least 4:1, 1.05g/hr F_2	$C_6H_{13}F$, mostly 2-fluorohexane	3
Cyclohexane	$-80°$, CCl_2F_2 solvent, $CO_2:F_2$ –1:1	Fluorocyclohexane	2
1-Chlorobutane	Large excess chlorobutane, gas phase, 21°	Excess chlorobutane recovered. Mixture of chloro-fluorobutanes: 52.3% 1,2- plus 1,3- isomers; 47.7% 1,4-	9
1-Fluorobutane	Large excess fluorobutane, gas phase, 21°	Excess fluorobutane recovered. Mixture of difluoro-butanes: No 1,1-isomer, 23.8% 1,2-, 45.7% 1,3-, 45.7% 1,4-	9
Propanoyl fluoride	$-78°$, $N_2:F_2$ at least 4:1	3-Fluoropropanoyl fluoride	3
Methyl propionate	$-78°$, $N_2:F_2$ at least 4:1	Methyl 3-fluoropropionate	3
Butyric acid	0°, CCl_4 solvent	3- and 4-fluorobutyric acid	2
Isobutyric acid		3-Fluoro-2-methylpropionic acid	2
Butyric acetic anhydride		3- and 4-fluorobutyric acid	2
Hexadecane	15°, CCl_4 solvent, $CO_2:F_2$ – 1:1	From 10 g $C_{16}H_{34}$ get 11.7 g mixed fluoro compounds	2
Addition of Fluorine to Carbon-Carbon Double Bonds			
1,1-Dichloro-2,2-difluoroethylene	$-110°$	1,1-Dichlorotetrafluoroethane (21) 2,2,3,3-Tetrachlorohexafluorobutane (79)	6
1,2-Dichloro-1,2-difluoroethylene	-110 and $-133°$, CCl_2F_2 solvent	1,2-Dichlorotetrafluoroethane (27 and 62) 1,2,3,4-Tetrachlorohexafluorobutane (62 and 31)	6, 7
1-Chloro-1,2,2-trifluoroethylene	-110 and $-150°$	Chloropentafluoroethane (29 and 26) 2,3-Dichlorooctafluorobutane (26 and 64)	6, 7
2,3-Dichlorohexafluoro-2-butene	$-77°$, CCl_3F solvent	2,3-Dichlorooctofluorobutane (41)	7
Octafluoro-2-butene	$-75°$, $CClF_3$ solvent	Perfluorobutane (88) Perfluoro-3,4-dimethylhexane (7)	6
2-Butenoic (crotonic) acid		2,3-Difluorobutyric acid	2
Hexadecene		$C_{16}H_{32}F_2$ (small yield)	2
trans-9-Octadecenoic (elaidic) acid		Difluoro derivative	2
cis-9-Octadecenoic (oleic) acid		$C_{18}H_{34}O_2F_2$	2
Indene	$-78°$, CCl_3F solvent, "4A Molecular Sieve"	*trans*-1,2-Difluoroindan (32)	5
2-Methylindene	$-78°$, CCl_3F solvent, "4A Molecular Sieve"	1,2-Difluoro-2-methylindan (*cis*-, 15; *trans*-, 28)	5
Acenaphthylene	$-78°$, CCl_3F solvent, "4A Molecular Sieve"	1,2-Difluoroacenaphthene (*cis*-, 35; *trans*-, 11)	5
1-Methylacenaphthylene	$-105°$, CCl_2F_2 solvent	*cis*-1,2-Difluoro-1-methylacenaphthene (20)	5
Δ^4-Cholesten-3-one	$-78°$, CCl_3F solvent,	*cis*-4,5-Difluorocholesten-3-one (60–70)	4
Cholesteryl chloride	Not specified; presumably $-78°$, CCl_3F	3-β-Chloro-*cis*-5,6-difluorocholestene (–)	4
Hexachlorobenzene	25°, $ClCF_2CFCl_2$ solvent	90% yield of mixture $C_6Cl_xF_{12\text{-}x}$ where $x = 5,6,7$ with average composition $C_6F_6Cl_6$: 5% of $x = 4$; 25% of $x = 5$; 37% of $x = 6$; 23% of $x = 7$; dehalogenation with Fe at 330° gives C_6F_6, C_6F_5Cl and $C_6F_4Cl_2$	10

acid, butanoyl chloride, butyric acetic anhydride, and isobutyric acid. More recently, Tedder[3] directly substituted single fluorine atoms into *n*-hexane, propanoyl fluoride, and methyl propionate by fluorinating the cooled liquid hydrocarbon with nitrogen-diluted fluorine. Bockemüller also added fluorine across the C=C bonds of hexadecene, oleic acid, elaidic acid, and crotonoic acid.[2] Very recently, Bockemüller's early studies have been "rediscovered" and extended to a variety of alkenes, conjugated alkenes, acenaphthylenes, and steroids.[4, 5] (See Table 4-1 and Section 5-1A.) Miller[6, 7] and his colleagues have elucidated the mechanism of addition of fluorine to the C=C bond of halogenated alkenes at low temperatures. To dissipate heat and reduce side reactions, Miller used a rotating metal gauze stirrer at $-77°$ to $-155°C$; he found dimerization–addition of the alkene (see Table 4-1 for some of the experimental results)[6, 7] and proposed a general mechanism which we illustrate with the specific example in Chart 4-1. In this specific case the rates of steps 3, 4, and 5 are such that equal

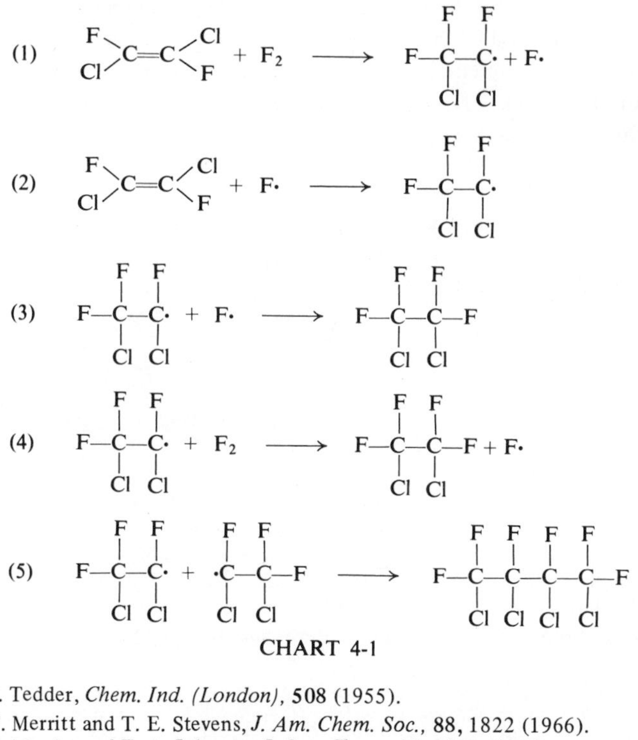

$$\text{(1)}\quad \mathrm{FClC{=}CClF} + F_2 \longrightarrow \mathrm{F{-}CFCl{-}CFCl\cdot} + F\cdot$$

$$\text{(2)}\quad \mathrm{FClC{=}CClF} + F\cdot \longrightarrow \mathrm{F{-}CFCl{-}CFCl\cdot}$$

$$\text{(3)}\quad \mathrm{F{-}CFCl{-}CFCl\cdot} + F\cdot \longrightarrow \mathrm{F{-}CFCl{-}CFCl{-}F}$$

$$\text{(4)}\quad \mathrm{F{-}CFCl{-}CFCl\cdot} + F_2 \longrightarrow \mathrm{F{-}CFCl{-}CFCl{-}F} + F\cdot$$

$$\text{(5)}\quad \mathrm{F{-}CFCl{-}CFCl\cdot} + \mathrm{\cdot CFCl{-}CFCl{-}F} \longrightarrow \mathrm{F{-}CFCl{-}CFCl{-}CFCl{-}CFCl{-}F}$$

CHART 4-1

[3] J. M. Tedder, *Chem. Ind. (London)*, **508** (1955).
[4] R. F. Merritt and T. E. Stevens, *J. Am. Chem. Soc.*, **88**, 1822 (1966).
[5] R. F. Merritt and F. A. Johnson, *J. Org. Chem.*, **31**, 1859 (1966).
[6] W. T. Miller, Jr., J. O. Stoffer, G. Fuller, and A. C. Currie, *J. Am. Chem. Soc.*, **86**, 51 (1964).
[7] W. T. Miller, Jr., and S. D. Koch, Jr., *J. Am. Chem. Soc.*, **79**, 3084 (1957).

amounts of starting 1,2-dichloro-1,2-difluoroethylene are converted to the addition product 1,2-dichlorotetrafluoroethane and to the dimerization-addition product 1,2,3,4-tetrachlorohexafluorobutane.

Fundamental studies by Tedder, Anson, and Fredricks on the direct reaction of fluorine with alkanes and alkyl halides in the gas phase have shown that fluorine shows much less selectivity in substituting for hydrogen than does chlorine or bromine.[8, 9] (See Table 2-5.)

A useful application of fluorine addition-substitution is reaction of hexachlorobenzene in a 1,1,2-trifluorotrichloroethane slurry with fluorine to form a mixture of chlorofluorocyclohexanes[10] of average composition $C_6F_6Cl_6$, which can be dehalogenated over iron to hexafluorobenzene and other products.

Selective reactions of fluorine with alkanes, alkenes, and aromatics are summarized in Table 4-1.

In the preceding discussion we have considered only the reactions of fluorine with C=C bonds and with the C–H bonds in substituted hydrocarbons. But fluorine does react with a variety of hetero single, double, and triple bonds such as C–S, C=O, and C≡N. Because of space limitation we refer the reader to Section 5-1 and recommend reviews by Tedder,[11] Bigelow,[12] and Lovelace, Rausch, and Postelnek[14] for additional information on these reactions as well as classical fluorinations.

4-2 HALOGEN FLUORIDES WITH ORGANIC COMPOUNDS

$$XF_n + \gt C{=}C\lt \longrightarrow \gt C(F){-}C(X)\lt \qquad (n = 1, 3, 5, 7)$$

$$\gt CH \longrightarrow \gt CF \text{ or } \gt CX$$

$$\gt CX \longrightarrow \gt CF$$

[8]P. C. Anson, P. S. Fredricks, and J. M. Tedder, *J. Chem. Soc.*, 918 (1959).

[9](a) P. S. Fredricks and J. M. Tedder, *J. Chem. Soc.*, 144 (1960). (b) P. S. Fredricks and J. M. Tedder, *J. Chem. Soc.*, 3520 (1961).

[10]G. M. Brooke, R. D. Chambers, J. Heyes, and W. K. R. Musgrave, *J. Chem. Soc.*, 729 (1964).

[11]J. M. Tedder, *Advan. Fluorine Chem.*, **2**, 104 (1961).

[12]L. A. Bigelow in *Fluorine Chemistry* (J. H. Simons, ed.), Vol. 1, Academic, New York, 1961, pp. 373-399.

[13]*Methoden der Organischen Chemie* (Houben-Weyl), Vol. 5, Part 3, Georg Thieme Verlag, Stuttgart, 1962, pp. 8-14.

[14]A. M. Lovelace, D. A. Rausch, and W. Postelnek, *Aliphatic Fluorine Compounds*, Am. Chem. Soc. Monograph No. 138, Reinhold, New York, 1958.

The inclusion of halogen fluorides as classical reagents is debatable since the majority of research with halogen fluorides has been carried out in fairly recent years. But two review articles by Musgrave[15] and in Houben-Weyl[16] have already appeared on this technique.

Chlorine trifluoride and chlorine monofluoride are the most reactive halogen fluorides, but the former, studied chiefly, is less reactive with organic compounds than fluorine. Addition of Cl–F occurs to unsaturated linkages including aromatic rings, but fluorine or chlorine may be substituted in C–H and carbon-halogen bonds in aliphatic and aryl compounds. Fragmentation side reactions are easier to avoid than with fluorine but are still a problem. A hazard in using chlorine trifluoride is that it is easily condensed (bp 11.30°C); contact of the liquid with organic compounds can lead to violent explosions.

A recent classical use of chlorine trifluoride is in controlled addition reactions to aromatic compounds to give saturated polychlorofluorocycloalkanes and cycloalkenes.[17, 18] The mixture of saturated chlorofluorocyclic products is

$$\text{(benzene)} + ClF_3 \longrightarrow$$

$$C_6F_{11}Cl,\ C_6F_{10}Cl_2,\ C_6F_9Cl_3,\ C_6F_8Cl_4,\ + \text{hydrochlorofluoro compounds}$$

$$C_6F_{(12-n)}Cl_n \longrightarrow C_6F_6 \text{ (hexafluorobenzene)} + C_6F_5Cl + C_6F_4Cl_2 \quad \text{(where n is 1 to 3)}$$

dehalogenated to perfluoroaromatic compounds. Some data are given on the reaction of chlorine trifluoride with benzene and pyridine in Table 4-2.

Bromine trifluoride is less reactive than chlorine trifluoride but adds to unsaturated aliphatic and aryl compounds or causes substitution for hydrogen or halogen in aliphatic or aryl C–H or carbon-halogen bonds. McBee, Lindgren, and Ligett[19] reacted bromine trifluoride with hexachlorobenzene to give a variety of products (see Table 4-2) some of which were dehalogenated to hexafluorobenzene. Bromine trifluoride is also capable of "gentle" reactions as shown by Stevens[20a, b] in a series of studies. With acetone, 2,2-difluoropropane was obtained in 90% yield; with propionitrile, 1,1,1-trifluoropropane was obtained in 86% yield (see Section 5-4 for further discussion).

[15] W. K. R. Musgrave, *Advan. Fluorine Chem.*, **1**, 1 (1960).

[16] See Ref. 13, pp. 72-79.

[17] W. K. R. Musgrave, *2nd Intern. Fluorine Symp. Estes Park, Colorado, August, 1962,* Reprints of papers, pp. 20-30.

[18] R. D. Chambers, J. Heyes, and W. K. R. Musgrave, *Tetrahedron,* **19,** 891 (1963).

[19] E. T. McBee, V. V. Lindgren, and W. B. Ligett, *Ind. Eng. Chem.*, **39,** 378 (1947).

[20] (a) T. E. Stevens, *J. Org. Chem.*, **26,** 1627 (1961). (b) T. E. Stevens, U.S. Patent 3,068,299 (1962); *Chem. Abstr.*, **58,** 10076g (1963). (c) R. A. Davis and E. R. Larsen, *J. Org. Chem.*, **32,** 3478 (1967).

Table 4-2

Selected Reactions of Halogen Fluorides with Organic Compounds

Organic compound	Halogen fluoride and conditions	Product(s) (% yield)	Ref.
CF_3CH_2Br	BrF_3, 0° to 60°, bromine solvent	CF_3CH_2F (76)	20c
$CF_2=CF_2$	$C_2F_4:IF_5:I_2$/5:1:2 mole, 10 hr, 25°	CF_3CF_2I (86)	22
$CF_2=CF_2$	$C_2F_4:BrF_3:Br_2$/3:1:1 mole, 2 hr, 25°	CF_3CF_2Br (–)	22
$CF_3CF=CF_2$	$C_3F_8:IF_5:I_2$/5:1:2 mole, 24 hr, 150°	CF_3CFICF_3 (90)	22
Perfluorocyclohexene	$C_6F_{10}:BrF_3:Br_2$/3:1:1 mole, 26 hr, 265°	Perfluorocyclohexyl bromide (78)	22
$CF_2=CFCl$	$C_2F_3Cl:IF_5:I_2$/5:1:2 mole, 4 hr, 25°	CF_2ICF_2Cl (45) + CF_3CFClI (37)	22
$CF_2=CH_2$	$C_2H_2F_2:IF_5:I_2$/5:1:2 mole, 15 hr, 103°	CF_3CH_2I (86)	22
$ClCH=CHCl$	$C_2H_2Cl_2:ClF_3$/17.3:7.8 mole, 24.5 hr, 55°	$CHClFCHCl_2$ (24), $CHClFCHClF$ (40), unidentified residue (36)	23
$Cl_2C=CHCl$	$C_2HCl_3:ClF_3$/1.8:1.0 mole, 8 hr	$CHClFCCl_2F$ (20), $CHCl_2CCl_2F$ (45)	23
Benzene	ClF_3 diluted with varying amounts of N_2, 200-350°	Complex mixtures depending on reaction conditions. Among the products are $C_6F_{11}Cl$, $C_6F_{10}Cl_2$, $C_6F_9Cl_3$, $C_6F_8Cl_4$, C_6H_5Cl, C_6H_4ClF, unidentified chlorofluorohydrocyclohexanes. See original reference for more detail	24
Hexachlorobenzene	$C_6Cl_6:BrF_3$/1.0:3.0 mole, complex reaction cycle of temperature and rate of addition of reactants used, from 0° up to 150° for 12 hr	53% yield of fluorohalocarbon product containing perfluorocyclohexadiene, perfluorobenzene, chloroheptafluorocyclohexadiene, dichlorooctafluorocyclohexene, dichlorohexafluorocyclohexadiene, trichloroheptafluorocyclohexene, trichloropentafluorocyclohexadiene and other unidentified products	19
Hexachlorobenzene	$C_6Cl_6:ClF_3$/0.35:0.60 mole, 240°, 3 hr	$C_6F_7Cl_3$(2), $C_6F_6Cl_4$ (10), $C_6F_4Cl_4$ (4), $C_6F_5Cl_5$(30), $C_6F_4Cl_6$(35), C_6Cl_6(10)	18
Hexachlorobenzene	ClF_3 used in large molar excess, conditions varied from 250° to 100°	$C_6F_9Cl_3$ (trace), $C_6F_8Cl_4$(10), $C_6F_7Cl_5$(25), $C_6F_6Cl_6$(15), $C_6F_5Cl_7$(2)	18

[23] J. Muray, British Patent 878,585 (1961); *Chem. Abstr.*, **57**, 11018*d* (1962).

[24] R. E. Banks, P. Johncock, R. H. Mobbs, and W. K. R. Musgrave, *Ind. Eng. Chem., Process Des. Develop.*, **1**, 262 (1962).

In recent work, bromine trifluoride in bromine solvent was used to fluorine selectively CF_3CH_2Br, CF_3CHBr_2, $CBrF_2CH_2Br$, $CBrF_2CHBr_2$, CF_3CH_2Br, $CBrF_2CBr_3$, and $CBr_2FCHBrF$.[20c] In this case, hydrogen was not substituted.

$$CF_3CHBr_2 \xrightarrow[Br_2\ \text{solvent}]{BrF_3,\ <35^\circ} CF_3CHBrF\ (91.7\%) + CF_3CHF_2\ (5.1\%)$$

Iodine pentafluoride is the mildest reagent of the halogen fluorides, but, again, depending on reaction conditions it can add to unsaturated bonds as well as replace hydrogen or halogen in C–H or carbon-halogen bonds. Stevens[21] reported a mild reaction where isothiocyanates were converted to thiobis-N-(trifluoromethyl)amines (see Section 5-1).

Mixtures of 1 mole of bromine trifluoride with 1 mole of bromine, or 1 mole of iodine pentafluoride with 2 moles of iodine have been shown to be equivalent, respectively, to 3 moles of bromine fluoride or 5 moles of iodine fluoride.[22] Using the above reagents, bromine fluoride and iodine fluoride were added to C=C bonds (see Section 5-2G for more extensive data).

$$3CF_2{=}CFCl + 2I_2 + IF_5 \longrightarrow CF_2ClCF_2I\ (43\%) + CF_3CFClI\ (39\%)$$

$$\text{(perfluorocyclohexene: } F_2, F_2, F_2, F_2, F, F) + Br_2 + BrF_3 \longrightarrow \text{(1-bromo-1-fluoro ring: } F_2, F_2, F_2, F_2, F_2, F, Br) \quad (78\%)$$

4-3 HYDROGEN FLUORIDE AS AN ADDITION REAGENT

Hydrogen fluoride is the most extensively used industrial fluorinating agent and is the basic chemical of the fluorochemical industry.[1, 25] In research laboratories, hydrogen fluoride has been extensively employed as catalyst, solvent, and fluorinating agent. In this chapter we will consider briefly only the important classical use of hydrogen fluoride: addition to C=C and C≡C bonds, and substitution of halogen atoms by fluorine. We will not consider additions to other systems such as the ring openings of epoxides to give α,β-hydroxy fluorides (see

[21](a) T. E. Stevens, *J. Org. Chem.*, **26**, 3451 (1961). (b) T. E. Stevens, U. S. Patent 3,025,324 (1962); *Chem. Abstr.*, **57**, 7107h (1962).

[22](a) R. D. Chambers, W. K. R. Musgrave, and J. Savory, *J. Chem. Soc.*, 3779 (1961). (b) W. K. R. Musgrave, J. Perrett, and J. Savory, British Patent 885,007 (1961); *Chem. Abstr.*, **57**, 11018i (1962).

[25]H. G. Bryce in *Fluorine Chemistry* (J. H. Simons, ed.), Vol. 5, Academic, New York, 1964, pp. 295-498.

Section 5-2A). Houben-Weyl gives much more complete and detailed coverage of hydrogen fluoride as a fluorinating agent.[26]

4-3A. ADDITION OF HYDROGEN FLUORIDE TO ALKENES

1. To Hydrocarbon Alkenes

$$HF + \gt C{=}C\lt \longrightarrow \gt C(H){-}C(F)\lt$$

Hydrocarbon alkenes substituted at positions other than on the C=C bond add hydrogen fluoride readily at temperatures lower than room temperature. Alkenes heavily substituted by halogen or halogen-like groups react with difficulty, if at all. Hydrocarbon alkenes, ethylene excepted, usually add hydrogen fluoride at $-78°$ to $0°C$. The mechanism is proton addition to give an intermediate carbonium ion which then abstracts a fluoride ion from hydrogen fluoride.

$$R_1R_2C{=}CR_3R_4 + (HF)_x \longrightarrow R_1R_2CH\overset{+}{C}R_3R_4 + H_{x-1}F_x^-$$

$$\xrightarrow{F^- \text{ from } (HF)_x} R_1R_2CHCR_3R_4F$$

Addition to unsymmetrical alkenes follows Markownikoff's rule (see examples in Table 4-3). Water must be rigorously excluded from reaction mixtures as it will compete too successfully with hydrogen fluoride for the carbonium ion, and, at higher levels of concentration, will hydrate hydrogen fluoride reducing its strength as an acid. Low temperature is necessary to prevent the side reactions of polymerization or rearrangement of the alkene or condensation with alkyl halide product, which are promoted by the powerful catalytic activity and proton-donating ability of hydrogen fluoride.[27]

1-Alkenes and 1,2-dialkylethylenes add hydrogen fluoride most easily and with the highest yields. Alkenes that can form tertiary carbonium ions tend to give extensive rearrangement and polymerization. Bicyclic alkenes frequently undergo rearrangements. Selected examples in Table 4-3 illustrate these points and outline the scope of additions of hydrogen fluoride to hydrocarbon alkenes.

[26] Ref. 13, pp. 95-143.

[27] J. H. Simons in *Fluorine Chemistry* (J. H. Simons, ed.), Vol. 1, Academic, New York, 1950, pp. 261-292.

Table 4-3

Addition of Hydrogen Fluoride to Alkenes and Cycloalkenes

Alkene or cycloalkene	Reaction conditions[a]	Product(s) (% yield)	Ref.
Ethylene	90°, 5 hr, 20-25 atm	1-Fluoroethane (81)	28
Propylene	– 45° to 0°, 2-5 hr	2-Fluoropropane (51-62)	28
Propylene	Gaseous phase, activated carbon catalyst	2-Fluoropropane (98)	29
Cyclopropane	25°, 0.25 hr	1-Fluoropropane (80)	28
Allene	– 76° to – 69°, 1 hr	2,2-Difluoropropane (50)	30
1-Butene	Not given	2-Fluorobutane (–)	31
Cyclobutene	0°, 10 hr, CCl_4 solvent	Fluorocyclobutane (10% conversion, 98% yield)	32
2-Methyl-2-butene	Not given	2-Fluoro-2-methylbutane (–)	33
Cyclohexene	– 78° to 0°	Fluorocyclohexane (80, 70)	28, 33
α-Pinene	– 50°, 6 hr, CCl_4 solvent	Limonene hydrofluoride (90)	34
Camphene	– 80°, 10 hr pet. ether (30-50°) solvent	Camphene hydrofluoride (88)	35
	– 80°, excess HF (1:2.5), pet. ether solvent	Camphene hydrofluoride (55) Isobornyl fluoride (35)	35
Bornylene	Series of experimental conditions – 70° to – 5°, 0.5 to 6 hr, pet. ether solvent	Overall 70-80% yield of mixture of varying composition, depending on conditions. Isotricyclene (0-9), epibornyl fluorides (31-48), isobornyl fluoride (22-31), camphene hydrofluoride (19-52)	36
5-Pregnen-3β-ol-20-one	– 80 to – 50, 6 hr, CH_2Cl_2 tetrahydrofuran solvent	5α-Fluoropregnan-3β-ol-20-one	37
5α-Pregnen-3β, 17α-diol-20-one diacetate	– 80 to – 50, 6 hr, CH_2Cl_2 tetrahydrofuran solvent	5α-Fluoropregnan-17α-ol-3,20-dione acetate	37
17α,21-Dihydroxy-4,9(11)-pregnadiene-3,20-dione 21-acetate	– 2°, 4 hr, 74% HF in pyridine	9α-Fluoro-17α,21-dihydroxy-4-pregnene-3,20-dione 21-acetate	38

[a]Unless otherwise specified, the reaction is carried out in the liquid phase.

[28] A. V. Grosse and C. B. Linn, *J. Org. Chem.*, **3**, 26 (1938).
[29] R. F. Sweeney and C. Woolf, U.S. Patent 2,917,559 (1959); *Chem. Abstr.*, **54**, 10858i (1960).
[30] P. R. Austin, D. D. Coffman, H. H. Hoehn, and M. S. Raasch, *J. Am. Chem. Soc.*, **75**, 4834 (1953).
[31] A. V. Grosse, R. C. Wackher, and C. B. Linn, *J. Phys. Chem.*, **44**, 275 (1940).
[32] M. Hanack and H. Eggensperger, *Chem. Ber.*, **96**, 1341 (1963).
[33] S. M. McElvain and J. W. Langston, *J. Am. Chem. Soc.*, **66**, 1759 (1944).
[34] M. Hanack, *Chem. Ber.*, **93**, 844 (1960).
[35] M. Hanack and W. Keberle, *Chem. Ber.*, **94**, 62 (1961).
[36] M. Hanack and R. Hahnle, *Chem. Ber.*, **95**, 191 (1962).
[37] A. Bowers, U.S. Patent 3,071,602 (1963); *Chem. Abstr.*, **61**, 4434d (1964).
[38] C. G. Bergstrom, R. T. Nicholson, and R. M. Dodson, *J. Org. Chem.*, **28**, 2633 (1963).

2. To Halogenated Alkenes

$$HF + X_2C{=}CX_2 \longrightarrow X_2C(H){-}CX_2(F)$$

$$=\overset{|}{C}{-}X \longrightarrow =\overset{|}{C}{-}F$$

The addition of hydrogen fluoride to halogenated alkenes has been an extremely important reaction historically since it has been used extensively in research directed toward synthesis of Freon[39] chemicals and intermediates for research in the Manhattan Project. The ease with which hydrogen fluoride adds to a halogenated alkene varies inversely with the number of halogens or pseudohalogens (particularly $-CCl_3$, $-CF_3$) substituted on the double bond. The greater the halogen substitution the more difficult the addition. To add hydrogen fluoride to tetrachloroethylene requires a catalyst and elevated temperature, whereas hexachloropropene will not add hydrogen fluoride under any known conditions. In the preceding and following discussions, fluorine is excluded when the term halogen is used.

Henne and Plueddeman[40] have correlated the difficulty of adding hydrogen fluoride to halogenated alkenes with structure. We have followed their correlation in organizing the data of Table 4-4 to illustrate the effect of increasing halogen substitution in the halogenated alkenes.

For the mechanism of this addition we propose a concerted process where charge separation develops in the transition state. The reaction could be termolecular or cyclic with respective transition states as shown. Electron withdrawal

$$\left[H{-}\overset{\delta-}{F}\cdots\overset{\delta+}{C}{=}\overset{\delta-}{C}\cdots\overset{\delta+}{H}{-}F \right] \quad \text{or} \quad \left[\begin{matrix} \overset{\delta+}{H} & - & \overset{\delta-}{F} \\ \vdots & & \vdots \\ \underset{\delta-}{C} & \text{----} & \underset{\delta+}{C} \end{matrix} \right]$$

by halogen or pseudohalogen makes the electron pair of the C=C bond less available for attack or protonation by hydrogen fluoride. Although the polymeric form of liquid hydrogen fluoride probably causes steric problems in solution work (favors the termolecular transition state?), monomeric hydrogen fluoride, a relatively small molecule that is the reactant in high-temperature gas phase addition, probably does not have any significant steric effect (favors a cyclic four-center transition state?). An anionic mechanism is ruled out because of orientation in addition. For example, if hydrogen fluoride added by an anionic mechanism to 2,2-dichloroethylene, the product expected would be

[39] Freon, a registered trademark of the E. I. du Pont de Nemours Co., Inc.

[40] A. L. Henne and E. P. Plueddeman, *J. Am. Chem. Soc.*, **65**, 1271 (1943).

Table 4-4

Addition of Hydrogen Fluoride to Halogenated Alkenes

Halogenated alkene	Reaction conditions	Product(s) (% yield)	Ref.
Addition to CH_2=CXR and RCH=CXR′			
CH_3CH=$CClCH_3$	1:1-Alkene:HF, – 78° to 23°	$CH_3CClFCH_2CH_3$ (50.8), $CH_3CF_2CH_2CH_3$ (13.4), and $CH_3CCl_2CH_2CH_3$ (11.4)	41
CH_3CH=$CClCH_3$	1:2.5-Alkene:HF, – 78° to 23°	$CH_3CClFCH_2CH_3$ (2.8), $CH_3CF_2CH_2CH_3$ (62.5), and $CH_3CCl_2CH_2CH_3$ (1.8)	41
CH_2=$CClCH_2CH_2CH_3$	1:2-Alkene:HF, – 78° to 43°, 43° for 2 hr	$CH_3CF_2CH_2CH_2CH_3$ (64.1)	42
CH_3CH=$CClCH_2CH_3$	– 80° to 60°, held at 60° until HCl evolution ceased	$CH_3CH_2CF_2CH_2CH_3$ (67-73)	43
1-Chlorocyclohexene	Ether solvent, HF bubbled in at 0° over 1.5 hr	1-Chloro-1-fluorocyclohexane (96)	44
C_6H_5CCl=CH_2	Olefin added to HF-ether solution at 0° over 1 hr	$C_6H_5CF_2CH_3$ (~25)	45
Addition to CH_2=CHX and RCH=CHX			
CH_2=CHCl	–, 4-5 hr, 10-15 atm	CH_3CHClF (68)	46
CH_2=CHF	0°, FSO_3H-HF solvent	CH_3CHF_2 (85)	47
CH_3CH=CHCl	100°, 4 hr	$CH_3CH_2CHF_2$ (trace), CH_3CH_2CHFCl (10), CH_3CHFCH_2Cl (20), $CH_3CHClCH_2Cl$ (20), tar	40
CH_3CH_2CH=CHCl	65°, 2 hr	$C_3H_7CHF_2$ (trace), C_3H_7CHClF (10), tar	40
Addition to RR′C=CHX			
$(CH_3)_2C$=CHCl	0° or – 23°, fast reaction	$(CH_3)_2CHCHClF$ (65)	40
Addition to CH_2=CX_2, RCH=CX_2, and RR′C=CX_2			
CH_2=CCl_2	65°, 3 hr	CH_3CFCl_2 (50)	40
$(CH_3)_2CHCH$=CCl_2	110°, 12 hr	$(CH_3)_2CHCH_2CF_3$(2.3), $(CH_3)_2CHCH_2CF_2Cl$ (60.3) and $(CH_3)_2CHCH_2CFCl_2$ (9.8)	48
CH_3CH_2CH=CClF	65°, 8 hr	$CH_3CH_2CH_2CClF_2$ (–)	49
CH_3CH_2CH=CF_2	– 80° to room temp. very fast	$CH_3CH_2CH_2CF_3$ (96-99)	49
$(CH_3)_2C$=CCl_2	100°, 1 hr	$(CH_3)_2CHCCl_2F$ (35)	40

[41] E. T. McBee and W. R. Hausch, *Ind. Eng. Chem.*, **39**, 418 (1947).
[42] M. W. Renoll, *J. Am. Chem. Soc.*, **64**, 1115 (1942).
[43] A. L. Henne and E. G. DeWitt, *J. Am. Chem. Soc.*, **70**, 1548 (1948).
[44] H. Hopff and G. Valkanas, *Helv. Chim. Acta*, **46**, 1818 (1963).
[45] J. A. Sedlak, G. C. Gleckler, and K. Matsuda, U.S. Patent 3,200,159 (1965); *Chem. Abstr.*, **63**, 17964*g* (1965).
[46] M. Otto, H. Theobald, R. Melan, German Patent 859,887 (1952); *Chem. Abstr.*, **47**, 11219*c* (1953).
[47] J. D. Calfee and F. H. Bratton, U.S. Patent 2,462,359 (1949); *Chem. Abstr.*, **43**, 3834*i* (1949).
[48] P. Tarrant, J. Attaway, and A. M. Lovelace, *J. Am. Chem. Soc.*, **76**, 2343 (1954).
[49] A. L. Henne and J. B. Hinkamp, *J. Am. Chem. Soc.*, **67**, 1197 (1945).

1,1-dichloro-2-fluoroethane and not the observed 1,1-dichloro-1-fluoroethane (see Table 4-4). Anion $FCH_2\bar{C}Cl_2$ is the one expected, not $\bar{C}H_2CFCl_2$, on the

$$[F{-}H\cdots\overset{\delta-}{C}H_2\overset{\delta+}{\text{=}}CCl_2\cdots F{-}H] \longrightarrow CH_3CCl_2F$$

concerted "*cationic*" (upper path)

$$H_2C{=}CCl_2 \quad (HF)_x$$

anionic (lower path, crossed out)

$$[FCH_2\bar{C}Cl_2] \longrightarrow FCH_2CHCl_2$$

basis of correlation of anion stabilities and the rules for addition to C=C bonds formulated by Park, Lacher, and Dick.[50]

Many tri- and tetrahalogen-substituted alkenes add hydrogen fluoride only in the presence of a catalyst. Henne and Arnold[51, 52] found that hydrogen fluoride could be added to tri- and tetrahalogen-substituted alkenes if a boron trifluoride catalyst was used and an appropriate temperature from 60°–180°C was selected. Other investigators have used catalysts such as aluminum fluoride, tin(IV) chloride, and antimony(V) chloride. The catalyzed addition of hydrogen fluoride to halogenated alkenes has significant industrial importance.

We believe that catalysts for the addition of hydrogen fluoride to halogenated alkenes act by abstraction of fluoride from hydrogen fluoride to form a new acid. The strength of the new acid is greater than that of hydrogen fluoride and, therefore, more able to attack the weak Lewis base, the halogenated alkene. All the effective catalysts for hydrogen fluoride addition form very stable fluoride complex ions in anhydrous hydrogen fluoride:

$$HF + BF_3 \longrightarrow H^+BF_4^-$$
$$HF + AlF_3 \longrightarrow H^+AlF_4^-$$
$$HF + SbCl_5 \longrightarrow H^+SbCl_5F^-$$
$$2HF + SnCl_4 \longrightarrow H_2SnCl_4F_2$$

Addition of hydrogen fluoride to halogenated alkenes differs markedly from addition to hydrocarbon alkenes in that the side reactions of polymerization and rearrangements rarely occur. But a serious side reaction for halogenated alkenes is substitution; in Section 4-4, we will discuss the utility of hydrogen fluoride as an agent for substituting fluorine for chlorine or bromine, but some examples given in Table 4-4 in which fluorine substitutes for chlorine in a side reaction are additions of hydrogen fluoride to 2-chloro-2-butene, 1-chloropropene, 1,1-dichloro-3-methyl-2-butene, and 1-chloro-1-phenylethylene (see also Table 4-6).

[50] J. D. Park, J. R. Lacher, and J. R. Dick, *J. Org. Chem.*, **31**, 1116 (1966).

[51] A. L. Henne and R. C. Arnold, *J. Am. Chem. Soc.*, **70**, 758 (1948).

[52] See Ref. 13, p. 107, Table 15.

Until now we have avoided discussion of alkenes bearing fluorine on the C=C bond. Fluoroalkenes differ from halogenated alkenes in that they undergo facile nucleophilic attack, particularly at terminal difluoromethylene groups. Miller, Fried, and Goldwhite[53] used potassium fluoride in formamide to add hydrogen fluoride to fluoroalkenes by an anionic mechanism.

$$F^- + CF_2{=}CClCF_3 \longrightarrow CF_3{-}\bar{C}ClCF_3$$

$$CF_3{-}\bar{C}ClCF_3 + HCONH_2 \longrightarrow HCONH^- + CF_3CHClCF_3$$

Other additions reported by this method are to chlorotrifluoroethylene (72%), perfluoropropene (60%), perfluorobutene (35%).

The addition of hydrogen fluoride by an initial anionic attack by fluoride ion is similar to base-catalyzed additions of aliphatic and aromatic alcohols, amines, and thiols to fluorinated alkenes. This class of reactions of fluoroolefins is illustrated by the first-discovered examples of addition of methanol to fluorochloroalkenes[54] and is reviewed by Miller and Fainberg,[55] in Houben-Weyl,[56] and by Park *et al.*[50] (see Section 6-1C).

$$OCH_3^- + CF_2{=}CCl_2 \longrightarrow CH_3OCF_2\bar{C}Cl_2$$

$$CH_3OCF_2\bar{C}Cl_2 + CH_3OH \longrightarrow CH_3OCF_2CCl_2H$$

4-3B. ADDITION OF HYDROGEN FLUORIDE TO ALKYNES

$$HF + RC{\equiv}CR \longrightarrow \left[\begin{matrix} RC{=}CR \\ |\quad| \\ H\quad F \end{matrix}\right] \longrightarrow \begin{matrix} RC{-}CR \\ |\quad| \\ H_2\ F_2 \end{matrix}$$

The addition of hydrogen fluoride to alkynes is a reaction that occurs readily at low temperatures to give geminally substituted difluoroalkanes.

$$R{-}C{\equiv}CH \longrightarrow RCF_2CH_3$$

This addition is a clean, high-yield classical fluorination reaction, but now of limited utility because of better new methods for synthesis of *gem*-difluoroalkanes. For example, 2,2-difluorohexane is best prepared from 2-hexanone and sulfur tetrafluoride rather than from 1-hexyne and hydrogen fluoride (see Section 5-4B.1).

In Table 4-5 selected examples of addition of hydrogen fluoride to alkynes are presented. Markownikoff addition is observed. As expected, internal alkynes

[53] W. T. Miller, Jr., J. H. Fried, and H. Goldwhite, *J. Am. Chem. Soc.*, **82**, 3091 (1960).
[54] W. T. Miller, Jr., E. W. Fager, and P. H. Griswold, *J. Am. Chem. Soc.*, **70**, 431 (1948).
[55] W. T. Miller, Jr., and A. H. Fainberg, *J. Am. Chem. Soc.*, **79**, 4164 (1957).
[56] See Ref. 13, pp. 280–306.

give both possible isomers. For example, 2-pentyne adds hydrogen fluoride to give 2,2- and 3,3-difluoropentane.

$$CH_3C{\equiv}CCH_2CH_3 \xrightarrow{2HF,\ <0^\circ} CH_3CF_2CH_2CH_2CH_3 + CH_3CH_2CF_2CH_2CH_3$$

Note that 2 moles of hydrogen fluoride add to alkynes at a temperature of 0°C or lower. Acetylene is the exception. The expected product 1,1-difluoroethane from addition of 2 moles is difficult to obtain although the reaction proceeds easily to vinyl fluoride. Recent patent literature is full of examples of studies by industrial firms in which a variety of catalysts is used to obtain either or both of the possible products (for examples, see selected Refs. 57–60). Houben–Weyl[61] has a good review.

4-4 FLUORINATION BY HIGH-VALENCY OXIDATIVE METALLIC FLUORIDES

$$MF_x + \rangle C{=}C\langle \longrightarrow \rangle C(F){-}C(F)\langle$$

$$\rangle CH \longrightarrow \rangle CF$$

$$\rangle CX \longrightarrow \rangle CF$$

Many high-valency metallic fluorides have the capability of oxidative fluorination of alkanes and alkenes as well as the ability to substitute fluorine for other appropriate groups such as halogen. Specifically we will discuss the reactions of cobalt(III) fluoride (CoF_3), silver(II) fluoride (AgF_2), and lead(IV) fluoride (PbF_4) with straight-chain and cyclic alkanes and alkenes, aromatic and

[57] C. M. Brock, C. Hundred, R. F. Drees, and J. F. Smith, U.S. Patent 3,190,930 (1965); *Chem. Abstr.*, **63**, 6859h (1965).

[58] F. J. Christoph, Jr., and G. Teufer, French Patent 1,383,927 (1965); *Chem. Abstr.*, **62**, 11686d (1965); also U.S. Patents 3,178,483 and 3,178,484 (1965).

[59] W. H. Snavely, Jr., Belgian Patent 612,914 (1962); *Chem. Abstr.*, **58**, 1344h (1963); also U.S. Patent 3,129,709 (1964).

[60] T. E. Hedge, D. F. Covley, R. L. Urbanowski, C. E. Enter, R. Steinkoegnig, French Patent 1,396,709 (1965); *Chem. Abstr.*, **62**, 10334h (1965).

[61] See Ref. 13, pp. 108-113.

Table 4-5

Addition of Hydrogen Fluoride to Alkynes

Alkyne	Conditions	Product(s) (% yield)	Ref.
Acetylene	25°, 8–12 atm, 72 hr	15% conversion to 1,1-difluoroethane and vinyl fluoride in 65:35 ratio	62
1-Propyne	– 70°, alkyne added to HF	2,2-Difluoropropane (61)	62
1-Butyne	– 70°, alkyne added to HF over 1 to 1.5 hr	2,2-Difluorobutane (46)	62
2-Butyne	– 70°	2,2-Difluorobutane (over 65)	62
1-Pentyne	– 78°	2,2-Difluoropentane (over 65)	62
2-Pentyne	– 50°	2,2- and 3,3-Difluoropentane in ratio of 85:15 (about 80% overall)	63
1-Hexyne	– 50° to 0°, 5 moles HF:1 mole ether or acetone:1 mole alkyne	2,2-Difluorohexane (80)	63
3-Hexyne	– 70°	3,3-Difluorohexane (76)	63
1-Octyne	0°, 5 moles HF:1 mole ether or acetone:1 mole alkyne	2,2-Difluorooctane (80-90)	63
1-Chloro-4-decyne	0°, HF in ether, 1.7 hr	4,4- and 5,5-Difluoro-1-chlorodecane	64
Phenylacetylene	0°, 2 hr, 4 HF:1 ether:0.25 alkyne	1,1-Difluoro-2-phenylethane (18)	65
Phenylacetylene	Gas phase, HgO/C catalyst, 150° Flow reactor	1,1-Difluoro-2-phenylethane (43)	65

[62] A. V. Grosse and C. B. Linn, *J. Am. Chem. Soc.*, **64**, 2289 (1942).

[63] A. L. Henne and E. P. Plueddeman, *J. Am. Chem. Soc.*, **65**, 587 (1943).

[64] M. S. Newman, M. W. Renoll, and I. Auerbach, *J. Am. Chem. Soc.*, **70**, 1023 (1948).

[65] K. Matsuda, J. A. Sedlak, J. S. Noland, and G. C. Gleckler, *J. Org. Chem.*, **27**, 4015 (1962).

heterocyclic hydrocarbons, and the substituted derivatives of these classes of organic compounds. Reagents with fluorinating power similar to the above which will not be discussed at length are manganese(III) fluoride (MnF_3), cerium(IV) fluoride (CeF_4), bismuth(V) fluoride (BiF_5) and uranium(VI) fluoride (UF_6). A major effort in this area was made during the war years.[66-84] Again, several excellent reviews[66-69, 84] are available, but we recommend the authoritative article by Stacey and Tatlow[67] as the most valuable.

The high-valency fluorides under discussion differ from other fluorides [lead(IV) fluoride excepted] in that elemental fluorine is required for their preparation. To obtain CoF_3, AgF_2, MnF_3, BiF_5, and UF_6, the lower oxidation states of the appropriate element must be oxidized by fluorine. Lead(IV) fluoride may also be obtained by direct fluorination but is more frequently generated in situ from lead(IV) oxide or lead tetraacetate and liquid hydrogen fluoride (see Section 5-1B).[66, 67]

The important fluorination reactions of high-valency metallic fluoride are summarized[67] and illustrated by typical examples in Chart 4-2. In the last example, substitution of halogen occurs accompanied by oxidative substitution of hydrogen. When aromatic hydrocarbons, alkenes, or halogenated hydrocarbons

[66] See Ref. 13, pp. 53-64.

[67] M. Stacey and J. C. Tatlow, *Advan. Fluorine Chem.*, **1**, 166 (1960).

[68] T. J. Brice in *Fluorine Chemistry* (J. H. Simons, ed.), Vol. 1, Academic, New York, 1950, pp. 423-462.

[69] G. H. Cady, A. V. Grosse, E. J. Barber, L. L. Burger, and Z. D. Sheldon, *Ind. Eng. Chem.*, **39**, 290 (1947).

[70] R. D. Fowler, W. B. Burford, III, J. M. Hamilton, Jr., R. G. Sweet, C. E. Weber, J. S. Kasper, and I. Litant, *ibid.*, 292.

[71] (a) E. T. McBee, H. B. Hass, P. E. Weimer, G. M. Rothrock, W. E. Burt, R. M. Robb, and A. R. Van Dyken, *ibid.*, 298. (b) E. T. McBee, B. W. Hotten, L. R. Evans, A. A. Alberts, Z. D. Welch, W. B. Ligett, R. C. Schreyer, and K. W. Krantz, *ibid.*, 310.

[72] J. H. Babcock, W. S. Beanblossom, and B. H. Wojcik, *ibid.*, 314.

[73] W. B. Burford, III, R. D. Fowler, J. M. Hamilton, Jr., H. C. Anderson, C. E. Weber, and R. G. Sweet, *ibid.*, 319.

[74] R. G. Benner, A. F. Benning, F. B. Downing, C. F. Irwin, K. C. Johnson, A. L. Linch, H. M. Parmelee, and W. V. Wirth, *ibid.*, 329.

[75] W. T. Miller, Jr., A. L. Dittman, R. L. Ehrenfeld, and M. Prober, *ibid.*, 333.

[76] E. A. Belmore, W. M. Ewalt, and B. H. Wojcik, *ibid.*, 338.

[77] R. D. Fowler, H. C. Anderson, J. M. Hamilton, Jr., W. B. Burford, III, A. Spadetti, S. B. Bitterlich, and I. Litant, *ibid.*, 343.

[78] M. Couper, F. B. Downing, R. N. Lulek, M. A. Perkins, F. B. Stilmar, and W. S. Struve, *ibid.*, 346.

[79] F. B. Stilmar, W. S. Struve, and W. V. Wirth, *ibid.*, 348.

[80] C. F. Irwin, R. G. Benner, A. F. Benning, F. B. Downing, H. M. Permelee, and W. V. Wirth, *ibid.*, 350.

[81] W. S. Struve, A. F. Benning, F. B. Downing, R. N. Lulek, and W. V. Wirth, *ibid.*, 352.

[82] E. T. McBee and L. D. Bechtol, *ibid.*, 380.

[83] C. Slesser and S. R. Schram, *Preparation, Properties, and Technology of Fluorine and Organic Fluoro Compounds,* McGraw-Hill, New York, 1951.

[84] (a) T. P. Waalkes, U.S. Patent 2,466,189 (1949); *Chem. Abstr.*, **43**, 5031f (1949); (b) A. L. Henne and T. P. Waalkes, *J. Am. Chem. Soc.*, **67**, 1639 (1945).

Oxidative Substitution of Hydrogen:

$$-\overset{|}{\underset{|}{C}}-H + 2MF_n \longrightarrow -\overset{|}{\underset{|}{C}}-F + 2MF_{(n-1)} + HF$$

$$C_7H_{16} + 32CoF_3 \xrightarrow{79\%} C_7F_{16} + 32CoF_2 + 16HF \qquad (70)$$

Oxidative Addition to Unsaturated Bonds:

$$>C{=}C< + 2MF_n \longrightarrow -\overset{F}{\underset{|}{\overset{|}{C}}}-\overset{F}{\underset{|}{\overset{|}{C}}}- + 2MF_{(n-1)}$$

$$CF_3CCl{=}CCl_2 + PbF_4 \xrightarrow{59\%} CF_3CClFCCl_2F + PbF_2 \qquad (84)$$

Substitution of Halogen:

$$-\overset{|}{\underset{|}{C}}-Cl + MF_n \longrightarrow -\overset{|}{\underset{|}{C}}-F + MF_{(n-1)} + \tfrac{1}{2}X_2$$

$$C_7H_2Cl_2F_{10} + 7AgF_2 \longrightarrow C_7ClF_{15} + 7AgF + \tfrac{1}{2}Cl_2 + 2HF \qquad (73)$$

CHART 4-2

are reacted with high-valency fluorides, more than one of the fundamental reactions can be expected as illustrated in the following two examples:

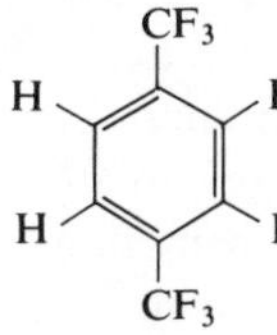

$$\text{(bis(trifluoromethyl)benzene)} + 14CoF_3 \xrightarrow[300\text{–}350°]{97\%} C_8F_{16} + 14CoF_2 + HF \qquad (72)$$

(used meta and para mixture) (perfluorodimethylcyclohexane)

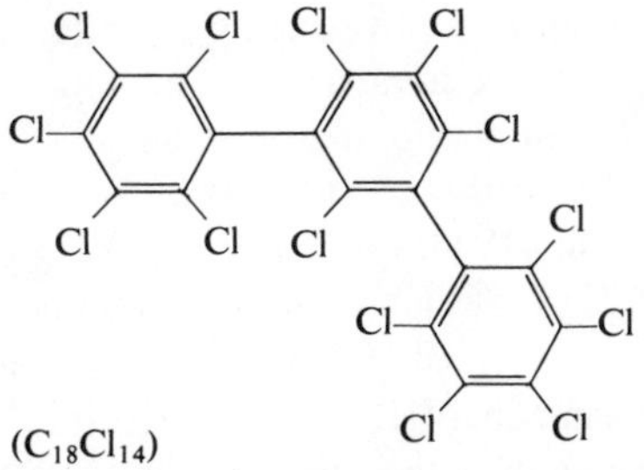

$$(C_{18}Cl_{14}) \xrightarrow[\substack{230\text{–}240° \\ 10\ hr}]{AgF_2} C_{18}Cl_5F_{27} \qquad (81)$$

For conversion of aromatic and aliphatic hydrocarbons to the corresponding saturated perfluorinated derivatives, cobalt(III) fluoride is preferred because of superior handling properties. Fluid bed reactions are possible, in contrast to most other fluoride systems, and refluorination of cobalt fluoride is nearly quantitative. Consequently, a highly efficient cyclic process can be used to great advantage.

In recent years the Birmingham group has made extensive use of cobalt(III) fluoride to prepare intermediates for conversion to perfluororoaromatic compounds.[85] For example, the known fluorination[70] of *p*-xylene was adapted to a cyclic process.[67, 85] The perfluoro-1,4-dimethylcyclohexane was then

$$26CoF_3 + p\text{-}C_6H_4(CH_3)_2 \xrightarrow{300\text{–}350°} 26CoF_2 + C_6F_{10}(CF_3)_2 + 10HF$$

$$26CoF_2 \xrightarrow{13F_2,\ 200\text{–}300°} 26CoF_3$$

dehalogenated over Fe at elevated temperature to perfluoro-*p*-xylene. Under certain conditions polyfluoroaromatic compounds may be converted to polyfluorocycloalkenes.[85b]

Silver(II) fluoride appears to be as efficient as cobalt(III) fluoride for fluorinating hydrocarbons. However, it has been used only infrequently, mainly because mixtures of silver(I) and silver(II) fluorides form an eutectic,[71, 72] and regeneration of silver(II) fluoride from the mixed silver(I) fluorides obtained in a fluorination reaction is not efficient. For laboratory scale exploratory work, silver(II) fluoride is convenient to handle and is preferred for many fluorinations. Silver(II) fluoride is often preferred over CoF_3 for replacing chlorine in perchlorofluorocarbon compounds[71] and is particularly effective in completely fluorinating fused-ring or anellated aromatic hydrocarbons.[82]

In Table 4-6, a large number of representative examples of fluorination by cobalt(III) fluoride and silver(II) fluoride are summarized. The importance of cobalt(III) fluoride fluorinations in recent developments of perfluoroaromatic chemistry cannot be overstated (see Sections 6-2C and 8-9).

Manganese(III) fluoride and cerium(IV) fluoride have not been used extensively as fluorinating agents.[66, 67, 77] Available data indicate that these agents are similar in fluorinating ability to cobalt(III) and silver(II) fluorides. They appear to be milder in action than cobalt(III) and silver(II) fluorides and should be considered for use with compounds that undergo extensive cleavage or disproportionation with the latter reagents.

Lead(IV) fluoride is not usually used directly in fluorination reactions but is generated in situ as required, usually from lead dioxide or lead tetraacetate. An old example is the addition of fluorine to tetrachloroethylene.

[85] (a) B. Gething, C. R. Patrick, and J. C. Tatlow, *J. Chem. Soc.*, 1574 (1961). (b) J. Riera and R. Stephens, *Tetrahedron*, **22**, 2555 (1966).

$$CCl_2{=}CCl_2 + 4HF + PbO_2 \longrightarrow CFCl_2CFCl_2 + 2H_2O + PbF_2 \quad (84b)$$

A more complete discussion and recent examples of the use of PbF_4 are given in Section 5-1B.

4-5 HYDROGEN FLUORIDE AND/OR ANTIMONY FLUORIDES IN SUBSTITUTION OF HALOGEN OR OXYGENATED FUNCTIONS

$$\gt CX \longrightarrow \gt CF$$

By far the most widely used method of introducing fluorine into organic compounds is substitution of fluorine for halogen or for oxygen-bonded groups such as a *p*-toluenesulfonate group. Industrially, and historically, hydrogen fluoride, with or without a catalyst, is the most important reagent for this purpose. Antimony(III) and antimony(V) mixed halogen fluorides or fluorides are generally more powerful reagents than hydrogen fluoride for halogen exchange and have been used widely not only directly for substitution but also as the catalysts with hydrogen fluoride.

The potential utility of many other metal fluorides for halogen exchange was investigated in early studies, and mercury(II) fluoride and silver(I) fluoride were found to complement effectively antimony(III and V) fluorides. In early work, almost any kind of halogen could be selectively substituted by proper use of HF, SbF_3, SbF_xCl_y $(x + y = 5)$, AgF, and HgF_2. Alkali metal fluorides were also used but received extensive investigation only in the last 20 years after polar organic solvents such as dimethylformamide, dimethylsulfoxide, and tetramethylene sulfone became readily available and were shown to make potassium fluoride an active and selective fluorinating agent (see Section 5-4A.1)

Fluorination by halogen exchange or substitution of an oxygen-bonded group has been reviewed extensively and comprehensively[95-101]; we particularly recommend the first three reviews. The scope and limitations of the various reagents for effecting substitution of halogen or oxygen-bonded groups are outlined and illustrated by selected examples in Tables 4-7, 4-8, and 4-9.

[95] See Ref. 13, pp. 119-196.

[96] A. K. Barbour, L. J. Belf, and M. W. Buxton, *Advan. Fluorine Chem.*, **3**, 181 (1963).

[97] M. Hudlicky, *Chemistry of Organic Fluorine Compounds*, MacMillan, New York, 1962, pp. 87-112.

[98] Ref. 14, pp. 3-17.

[99] P. Tarrant in *Fluorine Chemistry* (J. H. Simons, ed.), Vol. 2, Academic, New York, 1954, pp. 213-320.

[100] J. D. Park in *Fluorine Chemistry*, (J. H. Simons, ed.), Vol. 1, Academic, New York, 1950, pp. 523-552.

[101] A. L. Henne, *Org. Reactions*, **2**, 49 (1944).

Table 4-6

Fluorination by Cobalt(III) Fluoride and Silver(II) Fluoride

Compound fluorinated	Reagents and conditions	Product(s) (% yield)	Ref.
n-Pentane	CoF_3; one pass C_5H_{12} at 200-300°, N_2 diluent	C_5F_{12} (58)	71a
n-Pentane	CoF_3, C_5H_{12} at 0.3 atm, N_2 at 0.2 l./min. Two passes over CoF_3 bed, 175-325° and 300-325°	C_5F_{12} (39); 3 isomers of C_5HF_{11} (29)	86
2-Methylbutane (isopentane)	CoF_3, C_5H_{12} at 0.3 atm, N_2 at 0.2 l./min. Two passes C_5H_{12} over CoF_3 bed, 175-325° and 300-325°	Iso-C_5F_{12} (49), iso-C_5HF_{11} (7.4)	86
Cyclopentane	CoF_3, one pass C_5H_{10} at 200-300°, N_2 diluent	Cyclo-C_5F_{10}	71a
Cyclopentane	CoF_3, one pass C_5H_{10} at 170-200°, N_2 diluent	Cyclo-C_5F_{10} (11.5), cyclo-C_5HF_9 (11.4), cyclo-1H:2H–$C_5H_2F_8$ (10.3), cyclo-1H:3H–$C_5H_2F_8$ (17.5), cyclo-1H:4H/2H–$C_5H_3F_7$ (12.5), plus several other polyfluorocyclopentanes	87
Cyclopentadiene	CoF_3, 120 g C_5H_6 at 1 cc/min, 190-250°	Cyclo-C_5F_{10} (10), cyclo-C_5HF_9 (22) cyclo-1H:2H–$C_5H_2F_8$ (9), cyclo-1H:3H–$C_5H_2F_8$ (12), cyclo-1H:2H:4H–$C_5H_3F_7$ (8) plus 4 more polyfluorocyclopentanes	88
Cyclopentene	Same as for cyclopentadiene	Nearly identical to cyclopentadiene	88
n-Heptane	CoF_3; conditions were extensively studied. Graduated temperatures and 2 passes used: pass 1 at 175-200° up to 275-300°; pass 2 at 300°	*n*-C_7F_{16} (79)	71a, 74
$C_7H_xCl_4F_8$ (average composition of polychlorofluoroheptane)	AgF_2, 175-325°	C_7F_{16} (8), C_7ClF_{15} (35), $C_7Cl_2F_{14}$ (35), $C_7Cl_3F_{13}$ (20) plus low and high boilers	71b
Benzene	CoF_3 (see extensive details in the reference)	Product composition depends on conditions; $C_6H_xF_{12-x}$, x = 1-4 (see reference)	67
1,2-Dichlorobenzene	CoF_3; conditions not given, probably 300°	C_6F_{12} (7), $C_6F_{11}Cl$ (28), $C_6F_{10}Cl_2$ (39), $C_6F_9Cl_3$ (21)	89
1,3-Dichlorobenzene	CoF_3; conditions not given, probably 300°	C_6F_{12} (12), $C_6F_{11}Cl$ (36), $C_6F_{10}Cl_2$ (25), $C_6F_9Cl_3$ (7)	89
1,4-Dichlorobenzene	CoF_3; conditions not given, probably 300°	C_6F_{12} (22), $C_6F_{11}Cl$ (35), $C_6F_{10}Cl_2$ (27), $C_6F_9Cl_3$ (9)	89
Hexachlorobenzene	CoF_3; no details	Mixture $C_6F_{11}Cl$, $C_6F_{10}Cl_2$, . . . , $C_6F_6Cl_6$	90
1,2,3,4-Tetrafluorobenzene	CoF_3; 100°; N_2 atm	plus small amounts of 5 other products	85b

o-, *m*-, or *p*-Xylene	CoF_3; 175-200° up to 275-300°	1,2-, 1,3-, or 1,4-$(CF_3)_2$—C_6F_{10} (42, 50, or 42 resp.)	71, 74
o-Xylene or *p*-Xylene	CoF_3, 300-350°, CF_3—C_6F_{10}—CF_3 not isolated, products dehalogenated over Fe at 450-500°	*o*-$C_6F_4(CF_3)_2$ or *p*-$C_6F_4(CF_3)_2$ in good yield	85a
Bis(trifluoromethyl)benzene (mixture of 1,3- and 1,4-)	CoF_3, 2 passes, 175-200° up to 275-300° and 300-325°	Mixture of 1,3- and 1,4-CF_3—$C_6F_{10}CF_3$(88)	71, 74
Naphthalene	AgF_2, temp. programmed for 225-330°	Perfluorodecalin, $C_{10}F_{18}$ (56)	82
Indane, fluorene, acenaphthalene, phenanthrene, anthracene, pyrene	CoF_3, elevated temp. Perfluorocarbons treated at elevated temperature over metal gauze to dehalogenate	Perfluoroaromatics corresponding to starting hydrocarbon aromatic: perfluoroindane, perfluoroacenaphthene, etc.	91
Perchloro-*m*-terphenyl	AgF_2. 230-240°, 10 hr	Mixed chlorofluoro-saturated terphenyls such as $C_{18}Cl_5F_{27}$	79
$C_6H_5(CH_2)_xC_6H_5$ where x = 1-6	CoF_3, 300-350°	$C_6F_{11}(CF_2)_xC_6F_{11}$ where x = 1-6, x = 1 (45), 2 (66), 3 (33), 4 (28), 5 (24), 6 (8)	92a
Diphenyl ether	CoF_3, 200-300°	$C_6F_{11}OC_6F_{11}$ (15)	71a
Tetrahydrofuran	CoF_3, stirred; 100-120°	Mixture of $C_4H_2F_6O$ isomers (15) Mixed $C_4H_2F_6O$ $\xrightarrow[\text{fused}]{\text{KOH}}$ tetrafluorofuran (F at 2,3,4,5)	92b
1- and 3-Methoxyperfluorocyclohexene (ratio of 1-/3- of 7/3, resp.)	At 120° for 5 hr with excess CoF_3 At 195° for 3 hr with excess CoF_3	$C_6F_{11}OCH_3$(21.0), $C_6F_{11}OCH_2F$(54.5), $C_6F_{11}OCHF_2$(24.5); $C_6F_{11}OCH_2F$ (5.2), $C_6F_{11}OCHF_2$ (82.5), $C_6F_{11}OCF_3$ (12.2)	93
Cyanogen	AgF_2; 105-115°	CF_2—CF_2 / N═N (ring)	94
Tetrabromo-2,3-diaza-1,3-butadiene	AgF_2; 100°; gaseous starting material passed over AgF_2	$CF_3N{=}NCF_3$ (92)	94b

[86] E. J. Barber, L. L. Burger, and G. H. Cady, *J. Am. Chem. Soc.*, **73**, 4241 (1951).
[87] A. Bergomi, J. Burdon, T. M. Hodgins, R. Stephens, and J. C. Tatlow, *Tetrahedron*, **22**, 43 (1966).
[88] R. J. Heitzman, C. R. Patrick, R. Stephens, and J. C. Tatlow, *J. Chem. Soc.*, 281 (1963).
[89] P. L. Coe, B. T. Croll, and C. R. Patrick, *Tetrahedron*, **20**, 2097 (1964).
[90] P. Johncock, W. K. R. Musgrave, J. Feeney, and L. H. Sutcliffe, *Chem. Ind. (London)*, 1314 (1959).
[91] D. Harrison, M. Stacey, R. Stephens, and J. C. Tatlow, *Tetrahedron*, **19**, 1893 (1963).
[92] (a) A. K. Barbour, G. B. Barlow, and J. C. Tatlow, *Appl. Chem.*, **2**, 127 (1952). (b) J. Burdon, J. C. Tatlow, and D. F. Thomas, *Chem. Commun.*, 48 (1966).
[93] A. B. Clayton, R. Stephens, and J. C. Tatlow, *J. Chem. Soc.*, 7370 (1965).
[94] (a) H. J. Emeleus and G. L. Hurst, *J. Chem. Soc.*, 3276 (1962). (b) R. A. Mitsch and P. H. Ogden, *J. Org. Chem.*, **31**, 3833 (1966).

4-5A. HYDROGEN FLUORIDE IN UNCATALYZED SUBSTITUTIONS

Hydrogen fluoride is only a mild fluorinating agent for substitution of halogen by fluorine. (Note side reactions of addition in Section 4-3.) Only in compounds where all halogens are activated by a neighboring group can complete substitution occur, such as in benzotrichloride[102]

$$C_6H_5CCl_3 + 3HF \xrightarrow[95\%]{0^\circ} C_6H_5CF_3 + 3HCl \qquad (102)$$

or perchloropropene.[48]

$$CCl_2{=}CClCCl_3 + 3HF \longrightarrow CCl_2{=}CClCF_3 + 3HCl$$

In general, hydrogen fluoride will attack only trihalomethyl groups and leave other halogens in the compound unsubstituted. The degree of substitution will depend on neighboring substituents and other conditions.

$$3HF + CH_3CCl_3 \xrightarrow[\substack{80\% \\ 350\text{–}400\text{ atm}}]{225^\circ,\ 18\text{ hr}} CH_3CF_3 + 3HCl \qquad (103)$$

$$HF + CH_2ClCCl_3 \xrightarrow[80\%]{225^\circ} CH_2ClCCl_2F \qquad (104)$$

Additional examples of substitution of halogen using hydrogen fluoride are shown in Table 4-7.

4-5B. ANTIMONY(III) AND ANTIMONY(V) FLUORIDES IN SUBSTITUTIONS

The fluorinating power of antimony fluorides varies rather continuously from relatively mild antimony(III) fluoride through mixtures of antimony(III) fluoride and antimony(V) chloride to the very powerful antimony(V) fluoride. By proper choice of the antimony fluoride, a moderately high degree of selectivity can be achieved in substitution of fluorine for various types of halogen.

Antimony(III) fluoride is only slightly more active than hydrogen fluoride and behaves similarly when mixed with hydrogen fluoride. It is often preferred for laboratory scale preparations because it is a solid and, consequently, easier to handle than hydrogen fluoride; but also it is unreactive with C=C bonds, whereas hydrogen fluoride will frequently add (note Section 4-3).

[102] J. H. Simons and C. J. Lewis, *J. Am. Chem. Soc.*, **60**, 492 (1938).

[103] E. T. McBee, H. B. Hass, W. A. Bitterbender, W. E. Weesner, W. G. Toland, Jr., W. R. Hausch, and L. W. Frost, *Ind. Eng. Chem.*, **39**, 409 (1947).

[104] J. H. Brown and W. B. Whalley, *J. Soc. Chem. Ind. (London)*, **67**, 331 (1948).

Table 4-7

Fluorination by Halogen Substitution with Hydrogen Fluoride and Antimony(III) Fluoride

Compound to be fluorinated	Reagent and reaction conditions	Product(s) (% yield)	Ref.
CH_3CCl_3	$C_2H_3Cl_3$:HF/7.5:18, 144°, 1.75 hr	CH_3CFCl_2 (34), CH_3CF_2Cl (26)	104
CCl_4	HF, 230-240°, 67 atm	$CFCl_3$ (63)	105
CH_3CCl_3	$C_2H_3Cl_3$:HF/1:6; 225°, 16-20 hr, 350-400 atm	CH_3CF_3 (80)	103
CH_2ClCCl_3	$C_2H_2Cl_4$:HF/1:1, 225°; 0.33 hr, 50 atm	CH_2ClCCl_2F (80)	104
$CH_3CCl(CH_3)CH_2CCl_3$	$C_4H_8Cl_4$:HF/1.1:2.0, 130°, 4 hr	$CH_3CCl(CH_3)CH_2CF_3$ (40). Author suggested reaction proceeds through $CH_3C(CH_3){=}CHCCl_3$ intermediate	48
$CHCl_2CCl_3$	C_2HCl_5:HF/1.0:1.0; 225°, 0.25 hr, 50 atm	$CHCl_2CCl_2F$ (76)	104
$CH_3CH_2CCl_3$	HF, 100-119°, 14 hr	$CH_3CH_2CFCl_2$ (12.7), $CH_3CH_2CF_2Cl$ (67.5), $CH_3CH_2CF_3$ (11.3)	109
4-Cl–$C_6H_4CCl_3$	HF, 110°, 10-14 atm, 1-2 hr	4-Cl–$C_6H_4CF_3$	106
1,3,5-$(Cl_3C)_3$–C_6H_3	HF, 200°, 20 hr, 133 atm	1,3,5-$(CF_3)_3$–C_6H_3 (49)	107
4-Cl–$C_6H_4OCCl_3$	HF, 120-160°, 40-50 atm, 3 hr	4-Cl–$C_6H_4OCF_3$ (83)	108
$CCl_2{=}CClCCl_3$	SbF_3, 150°, iron autoclave	$CCl_2{=}CClCF_3$ (43), $CCl_2{=}CClCClF_2$ (28), $CCl_2{=}CClCCl_2F$ (13)	110a
Tetrachlorocyclopropene	SbF_3, 92°	1,2,3-Trichloro-3-fluorocyclopropene (16) 1,2-Dichloro-3,3-difluorocyclopropene (36)	110b
Tetrabromocyclopropene	SbF_3, 109°	1,2-Dibromo-3,3-difluorocyclopropene (51)	110b
$ClCH{=}C(CH_3)CCl_3$	SbF_3, conditions not given but probably heated to start and distilled product as formed	$ClCH{=}C(CH_3)CF_3$ (–)	111
$C_6H_5CCl_3$	Excess SbF_3 added portion by portion over 0.6 hr to refluxing reactant	$C_6H_5CF_3$ (85)	112
2-CN–$C_6H_4CCl_3$	SbF_3, heat, distill product	2-CN–$C_6H_4CF_3$ (77)	113
4-NO_2–$C_6H_4CBr_3$	$C_7H_4Br_3NO_2$:HF/1.0:1.2; heat; distill product	4-NO_2–$C_6H_4CF_3$ (90)	114a
1,1-Dichloro-1-silacyclobutane	SbF_3; *m*-xylene; 0°	1,1-Difluoro-1-silacyclobutane	114b

[105] J. H. Brown and W. B. Whalley, British Patent 576,189 (1946); *Chem. Abstr.*, **42**, 1603*d* (1948).

[106] P. Osswald, F. Müller, and F. Steinhäuser, German Patent 575,593 (1933); *Chem. Abstr.*, **27**, 4813*b* (1933).

[107] E. T. McBee and R. E. Leech, *Ind. Eng. Chem.*, **39**, 393 (1947).

[108] British Patent 765,527 (1955); *Chem. Abstr.*, **51**, 14803*f* (1957).

[109] E. T. McBee, H. B. Hass, R. M. Thomas, W. C. Toland, Jr., and A. Truchan, *J. Am. Chem. Soc.*, **69**, 944 (1947).

[110] (a) A. L. Henne, A. M. Whaley, and J. K. Stevenson, *J. Am. Chem. Soc.*, **63**, 3478 (1941). (b) S. W. Tobey and R. West, *J. Am. Chem. Soc.*, **88**, 2481 (1966).

[111] A. M. Whaley and H. W. Davis, *J. Am. Chem. Soc.*, **70**, 1026 (1948).

[112] E. Pouterman and A. Girardet, *Helv. Chim. Acta*, **30**, 107 (1947).

[113] L. M. Yagupolsky and N. I. Manko, *J. Gen. Chem.* USSR, (English trans.), **23**, 1033 (1953).

[114] (a) R. G. Jones, *J. Am. Chem. Soc.*, **69**, 2346 (1947). (b) J. Laane, *J. Am. Chem. Soc.*, **89**, 1144 (1967).

Antimony(III) fluoride is effective in substituting halogens on carbon activated by an aromatic ring (benzylic), a C=C bond (allylic), or other appropriate group (ether, for example); in contrast to hydrogen fluoride, it will give substitution in a dialkyldichloromethane. A few examples are given below, and others are in Table 4-7.

$$CCl_3\text{-}C_6H_4\text{-}CCl_3 + \text{excess } SbF_3 \xrightarrow{\text{heat}} F_3C\text{-}C_6H_4\text{-}CF_3 \quad (60\%) \qquad (115)$$

$$(H)(Cl)C{=}C(Cl){-}CCl_3 + \text{excess } SbF_3 \xrightarrow{150^\circ} (H)(Cl)C{=}C(Cl){-}CF_3 \qquad (111)$$

$$CH_3SCCl_3 + \text{excess } SbF_3 \xrightarrow{140^\circ} CH_3SCF_3 + CH_3SCF_2Cl \qquad (116)$$

$$CH_3{-}C(CH_3){-}O{-}C(Cl){-}Cl \text{ (epoxide)} + \text{excess } SbF_3 \longrightarrow CH_3{-}C(CH_3){-}O{-}C(F){-}F \text{ (epoxide)} \qquad (117)$$

$$CH_3CCl_2CH_3 + \text{excess } SbF_3 \xrightarrow{0^\circ} CH_3CF_2CH_3 \qquad (118)$$

Mixtures of antimony(III) fluoride and antimony(V) chloride or other antimony(V) compounds are very useful fluorinating reagents in the laboratory for halogen substitution. The actual reagent can be prepared directly or in situ from appropriate mixtures such as $SbCl_5/HF$, $SbCl_5/SbF_3$, or SbF_3Br_2. In all three cases, mixtures of Sb(III) and Sb(V) fluorides are present. By controlling the ratio Sb(V)/Sb(III), the fluorinating power of the reagent can be varied from weak SbF_3 to powerful SbF_5. Hudlicky[97] has presented the following order of strengths:

$$SbF_3 < (SbF_3 + SbCl_3) < (SbF_3 + SbCl_5) < SbF_3Cl_2 < SbF_5$$

The pattern of activity for halogen substitution in using Sb(III)–Sb(V) mixtures was first recognized by Henne[101] and is outlined below.

1. Activated halogens are most easily replaced, particularly in $-CCl_3$ groups (allylic, benzylic, etc.).

[115] S. D. Ross, M. Markarian, and M. Schwarz, *J. Am. Chem. Soc.*, **75**, 4967 (1953).
[116] W. E. Truce, G. H. Birum, and E. T. McBee, *J. Am. Chem. Soc.*, **74**, 3594 (1952).
[117] A. M. Whaley, U.S. Patent 2,451,185 (1948); *Chem. Abstr.* **43**, 674b (1949).
[118] (a) A. L. Henne, M. W. Renoll, and H. M. Leicester, *J. Am. Chem. Soc.*, **61**, 938 (1939). (b) A. L. Henne and M. W. Renoll, *J. Am. Chem. Soc.*, **59**, 2434 (1937).

2. The $-CCl_3$ groups are highly reactive and are converted stepwise into $-CCl_2F$, $CClF_2$, and with difficulty to $-CF_3$.

3. The $>CCl_2$ groups substituted by two alkyl groups are highly reactive and converted stepwise into $>CClF$ and $>CF_2$.

4. The $-CHCl_2$ groups react less readily and are converted stepwise to $-CHClF$ and with great difficulty to $-CHF_2$.

5. The $-CH_2Cl$ and $>CHCl$ groups are extremely difficult to attack and only rarely give $-CH_2F$ and $>CHF$.

6. The presence of fluorine on a neighboring carbon reduces the reactivity of a halogen (chlorine in $-CF_2CCl_3$ less reactive than $-CH_2CCl_3$).

7. The greater the hydrogen content of the molecule, the greater the chance for side reactions and decomposition.

8. Alkenes frequently undergo allylic shifts.

Again, examples were selected for Table 4-8 to illustrate this fluorination technique, and reviews[95, 96] should be consulted for a more detailed coverage. Unfortunately, side reactions involving eliminations and addition of hydrogen fluoride are a major limitation of this classical method.

4-5C. HYDROGEN FLUORIDE IN CATALYZED SUBSTITUTIONS

For industrial fluorinations, hydrogen fluoride in the presence of antimony(V) compounds is the most important fluorinating agent. Industrial reactions may be carried out in continuous processes and partially fluorinated material recycled along with unused hydrogen fluoride, making the operation highly economical. The industrial process for producing the refrigerant Freon 12 (difluorodichloromethane) is a good example for illustrating the importance of this method.[135]

Carbon tetrachloride at temperatures below 200°C reacts only very slowly with hydrogen fluoride and below 150° essentially not at all. At higher temperatures, reaction occurs.[104]

$$CCl_4 \xrightarrow[230\text{–}240°,\ 70\ \text{atm}]{2\ \text{moles HF}} \underset{63\%}{CFCl_3} \qquad (104)$$

$$CCl_4 \xrightarrow[460°,\ 70\ \text{atm}]{HF} \underset{54\%}{CFCl_3} + \underset{31\%}{CF_2Cl_2} \qquad (104)$$

But in the presence of antimony(V) chloride, reaction occurs easily at 100°C and 30–40 atm pressure to give a 90% yield of difluorodichloromethane.

$$CCl_4 \xrightarrow[100°,\ 30\text{–}40\ \text{atm}]{HF\text{–}SbCl_5} CF_2Cl_2$$

[135] See Ref. 1a, pp. 146-157.

Table 4-8

Fluorination by Halogen Substitution with Mixed Antimony (III) and (V) Fluorides

Compound to be fluorinated	Reagent and reaction conditions	Product(s) (% yield)	Ref.
CBr_4	$CBr_4:SbF_3:Br_2/5.0:7.5:1.5$ mole, 180-220°, autoclave, 4.5-5.6 atm, 24 hr	CF_3Br (67), CF_2Br_2 (3)	119
CBr_4	$CBr_4:SbF_3:Br_2/1.0:0.5:0.05$ mole, 120-130°, 1 hr	$CFBr_3$ (65-75), CF_2Br_2 (5-10)	120
CCl_4	100° autoclave	CF_2Cl_2 (90)	121
$CHCl_3$	100°, 58 atm	CHF_2Cl (–)	122
CH_3CCl_3	$SbF_3:SbF_3Cl_2/9:1$; product formed depends on amount fluorinating agent, reaction at room temp.	CH_3CFCl_2 (85-90) or CH_3CF_2Cl (85-90)	123
CCl_3CCl_3	SbF_3Cl_2; product formed depends on amount SbF_3Cl_2, temp. and pressure	Stepwise formation of products observed: $CCl_3CFCl_2 \rightarrow CFCl_2CFCl_2 \rightarrow CFCl_2CF_2Cl \rightarrow CF_2ClCF_2Cl$	124
$CCl_3CCl_2CCl_3$	$C_3Cl_8:SbF_3:SbCl_5:SbF_3Cl_2/3.0:1.0:0.13:0.12$, 140°, 14 hr; $C_3Cl_8:SbF_3:SbCl_5/4.7:3.1:0.59$ mole, 100°, 8 hr; $C_3Cl_8:SbF_3:SbCl_5:SbF_3Cl_2/3.8:2.8:0.50:1.3$; 100° 14 hr, then 140°, 2 hr	$CCl_3CCl_2CCl_2F$ (50) + C_3Cl_8; $CCl_2FCCl_2CCl_2F$ (48) + other products; $CCl_2FCCl_2CClF_2$ (22) + other products Authors concluded fluorination was stepwise: $C_3Cl_8 \rightarrow C_2Cl_5CCl_2F \rightarrow CFCl_2CCl_2CFCl_2 \rightarrow CFCl_2CCl_2CF_2Cl$	125
$CH_3CH_2CCl_3$	$SbF_3 + SbF_3Cl_2$; heat to distillation temp. To obtain $CH_3CH_2CF_3$ in 85% yield it is necessary to start with $CH_3CH_2CCl_2F$ (from $CH_3CH{=}CCl_2$ and HF)	$CH_3CH{=}CCl_2$ (50), $CH_3CH_2CF_3$ (5-10), $CH_3CH_2CFCl_2 + CH_3CH_2CF_2Cl$ (10)	126
$CH_3CCl_2CH_3$	$SbF_3:Br_2/20:1$, 0°-70°	$CH_3CF_2CH_3$ (85), $CH_3CFClCH_3$ (10-15)	118
$CH_3CF_2CCl_3$	SbF_3Cl_2, high temp.	$CH_3CF_2CFCl_2$ (50), $CH_3CF_2CF_2Cl$ (trace)	118b
$CH_2BrCBr_2CH_2Br$	$C_3H_4Br_4:SbF_3:Br_2/3.0:4.0:5.0$ mole, 150°	$CH_2BrCF_2CH_2Br$ (45)	109
$CF_2{=}CFCF_2Cl$	$C_3F_5Cl:SbF_3:Cl_2/0.16:0.46:0.18$ mole, 125°, 8 hr	Allylically rearranged starting material. No fluorination. $CF_3CF{=}CFCl$	127
$CHCl_2CHClCHCl_2$	$C_3HCl_5:SbF_3/2.35:3.00$ mole, 140°, distill [No $SbCl_5$ catalyst]; however,	$CHCl{=}CClCF_3$ (26), $CHCl{=}CClCF_2Cl$ (21), $CCl_2{=}CClCHF_2$ (24). See reference for explanation of product mixture	111
	$C_3HCl_5:SbF_3:SbCl_5/2.35:3.00:0.05$; heat until product distills	$CHCl{=}CClCF_3$ (92). The difference caused by Sb(V) catalyzed isomerization of $CCl_2{=}CClCHCl_2$ to $CHCl{=}CClCCl_3$, which is then converted completely to trifluoromethyl derivative	

$CH_2=CHCCl_3$	$SbF_3:SbF_3Cl_2/9:1$, − 10 to 60°, H_2O rigorously excluded	$CH_2=CHCF_3$ (51). Mandatory to exclude H_2O to prevent formation of allylically rearranged side products	128
$CCl_3CF_2CCl_2CCl_3$	SbF_3Cl_2, customary procedures: presumably 150-200° for several hr	$CCl_3CF_2CCl_2CF_2Cl$ (40), $CFCl_2CF_2CCl_2CF_2Cl$ (23), $CF_2ClCF_2CCl_2CF_2Cl$ (6), polymer (23)	127
$CFCl_2CFClCFClCFCl_2$	$SbF_xCl_{y\ (x+y=5)}$ 250°, 10 hr	$CF_2ClCFClCFClCF_2Cl$ (86.6)	129
$C_3F_7CCl_3$	$C_4F_7Cl_3:SbF_3:SbCl_3/1.0:1.0:0.56$ mole, 210°, 12 hr, autoclave	$C_3F_7CFCl_2$ (77). Illustrates clearly the effect of neighboring $-CF_2-$	130
$CCl_2=CClCCl=CCl_2$	$C_4Cl_6:SbF_3:SbF_3Cl_2/4.0:1.7:6.76$ mole, 155°, 2 hr, 6 atm	$[CCl_3CCl=CClCCl_3] \rightarrow CF_3CCl=CClCF_3$ (95)	131
$\overline{CCl=CClCCl_2CCl_2CCl_2}$	$C_5Cl_8:SbF_3:SbF_3Cl_2/1.0:2.8:0.4$, 250°, 1 hr, 6.7 atm	$\overline{CCl=CClCF_2CF_2CF_2}$ (72); proceeds through complex mechanism involving eliminations, additions, substitutions, and rearrangements	132
$\overline{CF_2CCl_2OCCl_2CF_2}$	$C_4OF_4Cl_4:SbF_3Cl_2/1.0:2.6$, 155°, 24 hr, 15 atm	$\overline{CF_2CF_2OCF_2CF_2}$ (67)	133
$CCl_3C(O)CCl_3$	Hexachloroacetone:$SbF_3:SbCl_5/1:2:1$, 10°, 105-110°, 2 hr, distill	$CFCl_2C(O)CF_2Cl$ (62), $CF_2ClC(O)CF_2Cl$ (29), $CF_2ClC(O)CF_3$ (4)	134
$CH_2ClSCCl_3$	$C_2H_2Cl_4S:SbF_3:SbCl_5/2.38:4.76:0.10$, 80°, distill	CH_2ClSCF_3 (66)	116

[119] H. Waterman, U.S. Patent 2,531,372 (1950); *Chem. Abstr.*, **45**, 1705*d* (1951).
[120] J. M. Birchall and R. N. Haszeldine, *J. Chem. Soc.*, 13 (1959).
[121] T. Midgley, Jr. and A. L. Henne, *Ind. Eng. Chem.*, **22**, 542 (1930).
[122] H. S. Booth and E. M. Bixby, *Ind. Eng. Chem.*, **24**, 637 (1932).
[123] A. L. Henne and M. W. Renoll, *J. Am. Chem. Soc.*, **58**, 889 (1936).
[124] H. S. Booth, W. L. Mong, and P. E. Burchfield, *Ind. Eng. Chem.*, **24**, 328 (1932).
E. G. Locke, W. R. Brode, and A. L. Henne, *J. Am. Chem. Soc.*, **56**, 1726 (1934); U.S. Patent 2,007,198 (1931).
[125] A. L. Henne and E. C. Ladd, *J. Am. Chem. Soc.*, **60**, 2491 (1938).
[126] A. L. Henne and A. M. Whaley, *J. Am. Chem. Soc.*, **64**, 1157 (1942).
[127] A. L. Henne and T. H. Newby, *J. Am. Chem. Soc.*, **70**, 130 (1948).
[128] R. N. Haszeldine, *J. Chem. Soc.*, 3371 (1953).
[129] R. P. Ruh, R. A. Davis, and K. A. Allswede, U.S. Patent 2,777,004 (1957); *Chem. Abstr.*, **51**, 11370*h* (1957).
[130] E. T. McBee, D. H. Campbell, and C. W. Roberts, *J. Am. Chem. Soc.*, **77**, 3149 (1955).
[131] A. L. Henne and P. Trott, *J. Am. Chem. Soc.*, **69**, 1820 (1947).
[132] K. A. Latif, *J. Indian Chem. Soc.*, **30**, 524 (1953); *Chem. Abstr.*, **49**, 883*i* (1955).
[133] A. L. Henne and S. B. Richter, *J. Am. Chem. Soc.*, **74**, 5420 (1952).
[134] C. B. Miller and C. Woolf, U.S. Patent 2,853,524 (1958); *Chem. Abstr.*, **53**, 5133*h*, 4137*h* (1959).

Mixtures of antimony(III) and antimony(V) fluorides can be used at the same temperature to give 90% yield of difluorodichloromethane.[121] For industrial operation, antimony(V)-catalyzed fluorination of carbon tetrachloride by hydrogen fluoride can be carried out over a wide range of temperatures, pressures, and concentrations to obtain either $CFCl_3$ or CF_2Cl_2.

Formally, antimony(V)-catalyzed substitution by hydrogen fluoride can be viewed as a stepwise reaction.

$$SbCl_5 + xHF \longrightarrow SbCl_{(5-x)}F_x + xHCl$$

(x is probably 3)

$$SbCl_{(5-x)}F_x + CCl_4 \longrightarrow SbCl_{(7-x)}F_{(x-2)} + CF_2Cl_2 + 2HCl$$

$$SbCl_{(7-x)}F_{(x-2)} + 2HF \longrightarrow SbCl_{(5-x)}F_x + 2HCl$$

But why are antimony(V) fluorides the most effective general compounds for fluorination and why is antimony(V) such a powerful catalyst for hydrogen fluoride? The most likely and reasonable explanation is the ability of antimony to use d-orbitals to form hexacoordinate halide complexes of high stability. The complex ions SbF_6^- and $SbCl_6^-$ are known and can form incipiently in transition states of fluorination (Chart 4-3). Whether such a transition state collapses by

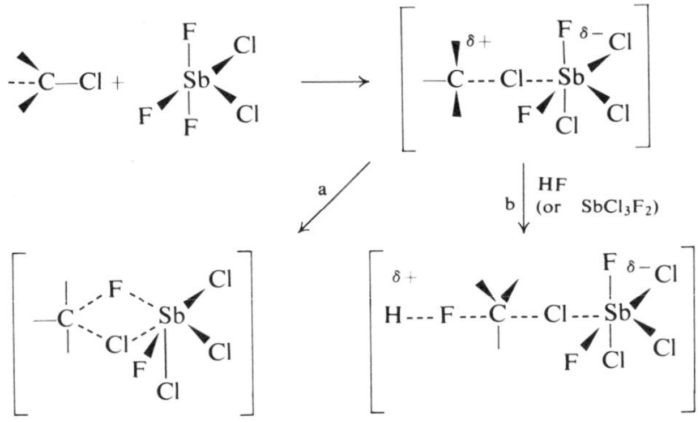

CHART 4-3

intramolecular fluorine transfer (path a) or whether a fluoride is transferred from hydrogen fluoride (path b) or a second molecule of catalyst (path b) is not known. But in any case, the apparent requirement of an intermediate complex suggests other potentially useful catalysts such as Al(III), As(V), P(V), Hg(I), Hg(II), and Ag(I). In actual practice, Al(III) has been demonstrated to have catalytic ability as have many other metals,[136a] but Sb(V) remains the most useful and most effective catalyst and Sb(V) fluorides the most effective fluorinating agents. In 1935, Booth and Swinehart[136b] actually suggested a mechanism similar to that proposed above. In the light of current understanding of coordination chemistry, collapse of the transition state by intramolecular transfer of fluoride seems the more likely pathway to fluorination although, in the presence of large excess of hydrogen fluoride, intermolecular reaction is also likely. The fluorinating activity of hydrogen fluoride in the presence of an adequate amount of Sb(V) compounds is very similar to that of mixtures of Sb(III) and Sb(V) fluorides so that the transition states must have the same form. The differences are in side reactions. The Sb(III)–Sb(V) fluorides do frequently cause rearrangements of starting materials so that products obtained in fluorinations are not always those predicted (see Table 4-8). Hydrogen fluoride in the presence of Sb(V) compounds will add to C=C bonds in addition to substitution.

We wish to emphasize the importance of the HF–Sb(V) system, particularly in industrial fluorinations. However, the system behaves sufficiently like the Sb(III)–Sb(V) system so that we chose not to present a table to illustrate its applications and refer the reader to the cited reviews.

4-6 GROUPS I AND II FLUORIDES FOR SUBSTITUTION FLUORINATION

Group I and II fluorides, of which potassium fluoride is the most important, are used effectively to replace one or two halogens or an oxygenated group, such as an ester, with fluorine. Classically, silver(I) and mercury(II) fluorides were used for controlled fluorination of partially halogenated alkyl chlorides, bromides, and iodides and supplemented the action of antimony fluorides on polyhalogenated compounds (Section 4-5). They have been replaced to a great extent in the last decade by potassium fluoride in polar nonaqueous solvents (see Section 5-4A.1). Other alkali metal fluorides are of very limited utility (sodium fluoride is less reactive and cesium fluoride is not as readily available), and none of the alkaline earth fluorides are effective in replacement-type fluorinations.

4-6A. SILVER(I) AND MERCURY(II) FLUORIDES

Many partially fluorinated compounds were first prepared by reaction of silver(I) fluoride with an appropriate halide, for examples: fluoroform, methylene

[136] (a) See Ref. 13, p. 125. (b) H. S. Booth and C. F. Swinehart, *J. Am. Chem. Soc.*, 57, 1333 (1935).

fluoride, allyl fluoride, pentyl fluoride, octyl fluoride, 2-fluoroethyl acetate, and methyl fluoroacetate.[137] Today the major value of silver fluoride is to replace isolated halogen atoms, particularly iodine and bromine, by fluorine in complex molecules that are susceptible to rearrangements or side reactions. As examples, silver(I) fluoride converted 16β-bromo-17α,20:20,21-bis(methylenedioxy)-Δ^4-pregnene-3-one into 16β-fluoro-17α,20:20,21-bis(methylenedioxy)-Δ^4-pregnene-3-one with retention of configuration[138] and α-1-bromo-2,3,4,6-tetraacetyl-*d*-galactose into β-1-fluoro-2,3,4,6-tetraacetyl-*d*-galactose.[139]

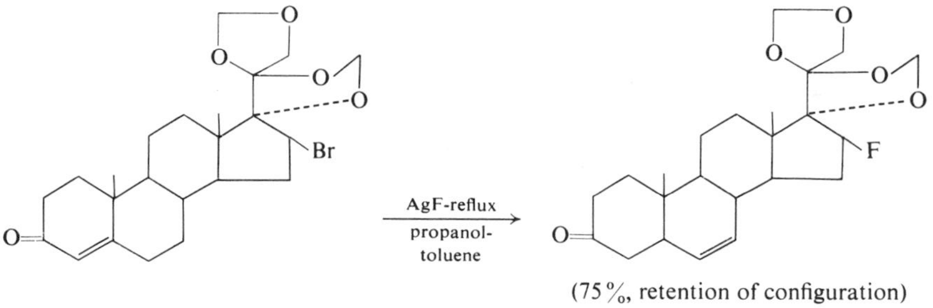

AgF-reflux / propanol-toluene

(75 %, retention of configuration)

(138)

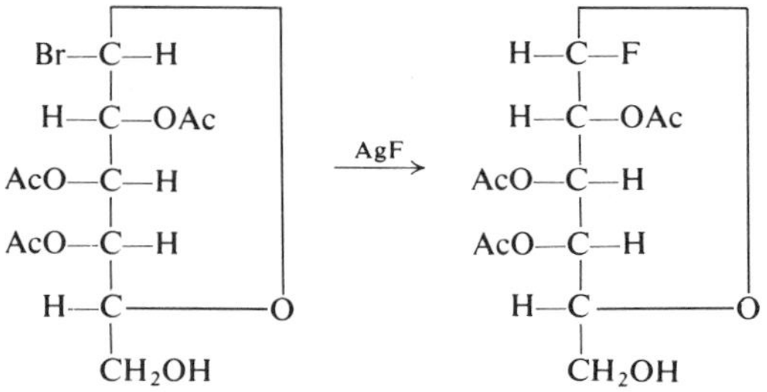

Br—C—H, H—C—OAc, AcO—C—H, AcO—C—H, H—C—O, CH_2OH $\xrightarrow{AgF}$ H—C—F, H—C—OAc, AcO—C—H, AcO—C—H, H—C—O, CH_2OH (139)

The major problems with silver(I) fluoride are

1. side reaction of elimination of hydrogen halide;

2. complexes with other silver halides to give double salts ($AgF + AgX \rightarrow AgF{\cdot}AgX$) so that a twofold excess of AgF is required;

3. high cost particularly since fluoride is only a small percentage weight of reagent.

[137] See Ref. 13, p. 206.

[138] W. T. Moreland, D. P. Cameron, R. G. Berg, and C. E. Maxwell, III, *J. Am. Chem. Soc.*, 84, 2966 (1962).

[139] F. Micheel, A. Klemer, and G. Baum, *Chem. Ber.*, 88, 475 (1955).

Mercury(II) fluoride is usually superior to silver(I) fluoride and is also preferable to mercury(I) fluoride which has been discussed elsewhere,[140] but has been replaced to a great extent by potassium fluoride.

Mercury(II) fluoride gives replacement of isolated halogen atoms, particularly iodine and bromine, by fluorine; antimony fluoride is useless in this type of fluorination. This activity of mercury(II) fluoride probably arises from the better

$$CH_3CH_2Br \xrightarrow{HgF_2,\ 0^\circ} CH_3CH_2F \ (\sim 100\%) \qquad (141)$$

$$BrCH_2CHBrCl \xrightarrow{HgF_2} BrCH_2CHClF \qquad (142), (143)$$

electrophilic character of mercury compared to antimony toward halogens causing a change in reaction mechanism.

The limitations of mercury(II) fluoride are

1. deactivation by water or systems that can produce water (hydrocarbon or chlorocarbons are preferred solvents);
2. deactivation by functional groups such as alcohols, esters, and ketones;
3. method of preparation from elemental fluorine or in situ from mercury(II) oxide and hydrogen fluoride.

Some examples in Table 4-9 illustrate the scope and limitation of mercury(II) fluoride as a fluorinating agent.

4-6B. POTASSIUM FLUORIDE AND OTHER ALKALI METAL FLUORIDES

Potassium fluoride is the most useful alkali metal fluoride for substituting fluorine for halogen or an oxygen-bonded function. Cesium and probably rubidium fluorides appear more active than potassium fluoride but are too costly for common use. Sodium fluoride, the cheapest, is very ineffective for most fluorinations (solubility too low), and lithium fluoride is a very poor fluorinating

[140] See Ref. 13, p. 196.

[141] A. L. Henne and T. Midgley, Jr., *J. Am. Chem. Soc.*, **58**, 882 (1936).

[142] A. L. Henne and M. W. Renoll, *J. Am. Chem. Soc.*, **58**, 887 (1936).

[143] J. B. Dickey, E. B. Towne, M. S. Bloom, G. J. Taylor, H. M. Hill, R. A. Corbitt, M. A. McCall, and W. H. Moore, *Ind. Eng. Chem.*, **46**, 2213 (1954).

Table 4-9

Silver(I) Fluoride and Mercury(II) Fluoride as Fluorinating Agents for Organic Halides

Halogenated compound	Fluorinating agent and conditions	Product(s) (% yield)	Ref.
1-Bromopentane	AgF, conditions not specified; HgF claimed to be superior to AgF	1-Fluoropentane (–), pentene (–), 1-fluorodecane (–)	144
Methyl 4-bromo-2-butenoate	AgF, gentle heating to reflux	Methyl 4-fluoro-2-butenoate (17)	145
Ethyl 7-bromoheptanoate	AgF, 60-80°	Ethyl 7-fluoroheptanoate (34.1)	146
21-Iodoprogesterone	AgF, CH_3CN containing some H_2O, 30-40°	21-Fluoroprogesterone (63)	147
16β-Bromo-17α,20:20,21-bis(methylene-dioxy)-Δ^4-pregnene-3-one	AgF, 2-propanol and toluene, reflux	16β-Fluoro-17α,20:20,21-bis(methylenedioxy)-Δ^4-pregnene-3-one (74.6)	138, 148
6-Methyl-11β,17α-dihydroxy-21-iodo-4-pregnene-3,20-dione	AgF, CH_3CN with trace H_2O, 40°, 1 hr	6-Methyl-11β,17α-dihydroxy-21-fluoro-4-pregnene-3,20-dione (28)	149a
6-Methyl-11β,17α-dihydroxy-21-iodo-1,4-pregnadiene-3,20-dione	AgF, CH_3CN with trace H_2O, 50°, 2 hr	6-Methyl-11β,17α-dihydroxy-21-fluoro-1,4-pregnadiene-3,20-dione (12)	149b
2,4,6-Trichloropyrimidine	AgF, triperfluorobutylamine, 90°	2,4,6-Trifluoropyrimidine (76)	150
4,6-Dichloro-5-nitropyrimidine	AgF, 170-180°, 0.75 hr	4,6-Difluoro-5-nitropyrimidine (~ 75)	151
2-Chloro-9-methylpurine	AgF, xylene, reflux, 1.25 hr	2-Fluoro-9-methylpurine (35)	152
Bromoethane	HgF_2, 0°, collect product in – 78° trap	Fluoroethane ("quantitative")	141
1,1,2-Tribromoethane	HgF_2 (amount determines product formed)	1-Fluoro-1,2-dibromoethane or 1,1-difluoro-2-bromoethane (80-90%)	141
1-Chloro-1,2-dibromoethane	HgF_2, heat until product distills	1-Fluoro-1-chloro-2-bromoethane (100)	142
1,1-Dibromoethane	HgF_2, ice-filled reflux condenser	1,1-Difluoroethane (> 90)	142
1,1,3-Tribromopropane	HgF_2, warm as required, distill product	1,1-Difluoro-3-bromopropane (37)	143
1,2-Dibromo-2-chloropropane	HgF_2, reflux, distill product	2,2-Difluoro-1-bromopropane (48.8)	143
Ethyl(chloromethyl)ether	HgF_2, 10-25°	Ethyl(fluoromethyl)ether (33)	153
n-Butyl(1,2,2,2-tetrachloroethyl)ether	HgF_2, 10-25°	*n*-Butyl(1-fluoro-2,2,2-trichloroethyl)ether (39)	153

1,1,1,2-Tetrachloropropane	HgO-HF, 100°	1,1,1-Trifluoro-2-chloropropane (80)	126
1,1-Dichloroheptane	HgO-HF, 0° or lower	1,1-Difluoroheptane (> 80)	114a, 154
2,2-Dibromoethyl acetate	HgO-HF, 0° or lower	2,2-Difluoroethyl acetate (50-70)	154
Dichlorodiphenylmethane	HgO-HF, 0° or lower	Difluorodiphenylmethane (50-70)	154
2,3-Difluorohexachloropropane	HgO-HF, autoclave, 175°, 40 hr	1,2,3-Trichloropentafluoropropane (70)	155
1,2,3,3-Tetrachlorotetrafluoropropane	HgO-HF, autoclave, 175°, 24 hr	1,2,3-Trichloropentafluoropropane (65)	155
1,4-Difluorooctachlorobutane	HgO-HF, autoclave, 100-110°	1,1,4,4-Tetrafluorohexachlorobutane (52)	155
		Heptachloro-1,1,4-trifluorobutane (37)	
DDT, bis(4-chlorophenyl)trichloromethylmethane	HgO-HF, room temp., 2-3 hr	1,2-di(4-chlorophenyl)-1,1,2-trifluoroethane (52)	156
1,2-Diphenyltetrachloroethane	HgO-HF, 15-20°, 16 hr	1,2-Diphenyltetrafluoroethane (56)	157

[144] F. Swarts, *Bull. Soc. Chim. Belg.*, **30**, 302 (1921); *Chem. Abstr.*, **16**, 3062 (1922).

[145] F. L. M. Pattison and B. C. Saunders, *J. Chem. Soc.*, 2745 (1949).

[146] F. L. M. Pattison, S. B. D. Hunt, and J. B. Stothers, *J. Org. Chem.*, **21**, 883 (1956).

[147] P. Tannhauser, R. J. Pratt, and E. V. Jensen, *J. Am. Chem. Soc.*, **78**, 2658 (1956).

[148] W. T. Moreland, R. G. Berg, and D. P. Cameron, *J. Am. Chem. Soc.*, **82**, 504 (1960).

[149] (a) F. H. Lincoln, Jr., W. P. Schneider, and G. B. Spero, U.S. Patent 2,867,631 (1959); *Chem. Abstr.*, **53**, 16220*b* (1959). (b) F. H. Lincoln, Jr., and W. P. Schneider, U.S. Patent 2,862,936 (1958); *Chem. Abstr.*, **53**, 10309*e* (1959).

[150] H. Schroeder, *J. Am. Chem. Soc.*, **82**, 4115 (1960).

[151] A. G. Beaman and R. K. Robins, *J. Medicinal and Pharm. Chem.*, **5**, 1067 (1962).

[152] A. G. Beaman and R. K. Robins, *J. Org. Chem.*, **28**, 2310 (1963).

[153] C. T. Mason and C. C. Allain, *J. Am. Chem. Soc.*, **78**, 1682 (1956).

[154] A. L. Henne, *J. Am. Chem. Soc.*, **60**, 1569 (1938).

[155] W. T. Miller, U.S. Patent 2,668,182 (1954); *Chem. Abstr.*, **49**, 2478*e* (1955).

[156] S. Cohen, A. Kaluszyner, and R. Mechoulam, *J. Am. Chem. Soc.*, **79**, 5979 (1957).

[157] L. V. Johnson, F. Smith, M. Stacey, and J. C. Tatlow, *J. Chem. Soc.*, 4710 (1952).

agent because it has such a strongly bonded ionic crystal lattice. (See Houben-Weyl[158] for a more extensive discussion of the alkali metals, including ammonium fluoride.) The general utility of potassium fluoride will be discussed here, but recent advances, particularly in use of polar solvents, will be reviewed in Section 5-4A.1.

Potassium fluoride substitutes fluorine for halogen in the following classes of compounds:

1. Carboxylic acid halides.
2. Aromatic- and alkyl-sulfonyl halides.
3. α-Halo compounds such as α-halo esters, amides, nitriles.
4. Primary alkyl halides.
5. ω-Halo compounds such as ω-halo alcohols, esters, and nitriles.
6. Aromatic halides substituted in ortho or para positions by an electron-withdrawing group, especially nitro.
7. Perchlorinated aromatic compounds, especially hexachlorobenzene.

Dry potassium fluoride, without any solvent, will fluorinate the first two classes above. More effective fluorinations are usually achieved at lower temperatures in the presence of an appropriate solvent. The above classes of compounds can usually be fluorinated by potassium fluoride in one of the solvents listed below:

1. Di- and polyalcohols, especially glycol and diethylene glycol.
2. Amides, usually formamide, dimethylformamide, acetamide, *N*-methylacetamide, *N*-methylpyrollidone.
3. Dimethylsulfoxide.
4. Aromatic hydrocarbons; xylene and nitrobenzene often used.

Polar solvents are probably effective because they dissolve sufficient potassium fluoride ion pairs (K^+F^-) to permit reaction in solution rather than at the surface of solid potassium fluoride which is easily rendered inert by coverage with potassium halide. Solvent effects are discussed further in Section 5-4A.1.

Since 1950, dry potassium fluoride in a polar solvent has been used extensively to replace oxygen-bonded groups, particularly *p*-toluenesulfonate and methylsulfonate ester groups. Alcohols may be converted by mild reactions indirectly into fluorides, avoiding relatively vigorous reaction conditions required for direct preparation of halides.

$$RCH_2OH \xrightarrow{ArSO_2Cl} RCH_2OSO_2Ar \xrightarrow[\text{solvent}]{KF} RCH_2F$$

[158]See Ref. 13, p. 145.

Again a series of examples is given in Table 4-10 to illustrate the use of potassium fluoride as a fluorinating agent.

Comments and examples on scope, limitations, or special advantages in use of potassium fluoride are given below.

1. Sulfonyl halides are often run with no solvent, but an aqueous system (such as 70% aqueous KF) is also used because the rate of hydrolysis is much slower than rate of fluorination.

$$CH_3SO_2Cl \xrightarrow[\text{steam distill}]{70\%\text{ aq. KF}} CH_3SO_2F \quad (65\%) \qquad (159)$$

2. Carboxylic acid fluorides are readily prepared using potassium fluoride, potassium acid fluoride, or potassium fluorosulfinate. An acid can be converted directly to an acid fluoride by using a less volatile acid chloride with the potassium fluoride fluorinating agent.[160]

$$CH_3(CH_2)_4C(=O){-}OH + C_6H_5C(=O){-}Cl + KHF_2 \xrightarrow[\text{(2) distill}]{\text{(1) 100°, 1 hr}} \underset{67\%}{CH_3(CH_2)_4C(=O){-}F}$$

3. Halogen atoms on a carbon alpha to an activating group (such as an ester, nitrile, or amide) are more difficult to fluorinate than those in carboxylic acid or sulfonyl halides, but easier than those isolated in aliphatic chains or bonded to aromatic rings. A solvent is not necessary but preferred because a lower fluorination temperature can be used.

$$\underset{(C_6H_{11} = \text{cyclohexyl})}{BrCH_2COOC_6H_{11}} \xrightarrow[\text{200°, 48 hr}]{KF} \underset{74\%}{FCH_2COOC_6H_{11}} \qquad (161)$$

$$ClCH_2C(=O){-}NH_2 \xrightarrow[\text{reflux}]{\text{KF-xylene}} \underset{75\%}{FCH_2C(=O){-}NH_2} \qquad (162)$$

[159] L. Z. Soborovskii, B. M. Gladshtein, M. I. Kiseleva, and V. N. Chernetskii, *Zhur. Obshchei Khim.*, **28**, 1866 (1958); *Chem. Abstr.*, **53**, 1110i (1959); *J. Gen. Chem. (USSR)* (English trans.), **28**, 1909 (1958).

[160] G. Olah, S. Kuhn, and S. Beke, *Chem. Ber.*, **89**, 862 (1956).

[161] E. Gryszkiewicz-Trochimowski, O. Gryszkiewicz-Trochimowski, and R. Levy, *Bull. Soc. Chim. France*, 462 (1953); *Chem. Abstr.*, **48**, 3895b (1954).

[162] C. Chapman and M. A. Phillips, British Patent 757,610 (1956); *Chem. Abstr.*, **51**, 11648g (1957).

Table 4-10

Potassium Fluoride as a Fluorinating Agent

Reactant	Conditions	Product(s) (% yield)	Ref.
Sulfonyl and Carboxylic Acid Halides			
Pentanoyl chloride	KHF_2, benzoyl chloride, 1 hr, 100°, distill	Pentanoyl fluoride (66.5)	160
Acetic acid	KF, benzoyl chloride, 3 hr, 100°, distill	Acetyl fluoride (77)	172a
Methanesulfonyl chloride	Sat. aq. KF, steam distill	Methanesulfonyl fluoride (65)	159
β-Nitroethanesulfonyl chloride	Sat. aq. KF, steam distill	β-Nitroethanesulfonyl fluoride (59)	159
1,2-Dimethylbenzene-4-sulfonyl chloride	73% aq. KF, 1 hr reflux, distill	1,2-Dimethylbenzene-4-sulfonyl fluoride (54)	174
6-Purinethiol	KF, HF, Cl_2 (oxidation by Cl_2 followed by displacement)	6-Purinesulfonyl fluoride (90)	175
Phenyl chlorosulfinate	NaF, acetonitrile	Phenyl fluorosulfinate (85)	176
Halogenated Alkanes and Derivatives			
2-Chloroacetamide	KF, tetrachloroethylene, 100°, then reflux	2-Fluoroacetamide (75)	162
Cyclohexyl 2-chloroacetate	KF, no solvent, 100-120°, 20 hr	Cyclohexyl 2-fluoroacetate (30-35)	161
Cyclohexyl 2-bromoacetate	KF, no solvent, 195-205°, 48 hr	Cyclohexyl 2-fluoroacetate	161
2-Chloroacet-*p*-chloroanilide	KF, bis(2-hydroxyethyl) ether, 130°	2-fluoroacet-*p*-chloroanilide (53)	172b
n-Heptyl chloride	KF, diethylene glycol	*n*-Heptyl fluoride (39)	164
6-Bromohexene-1	KF, diethylene glycol, 90°, 2 hr	6-Fluorohexene-1 (40)	164
1,5-Dibromopentane	KF, glycol, 100-110°, 6 hr	5-Bromo-1-fluoropentane (31.4) and 1,5-difluoropentane (25)	177
1,7-Dichloroheptane	KF, diethylene glycol, 125°, 4 hr, 40-50 mm	7-Chloro-1-fluoroheptane (48.5)	178
1,4-Dichlorobutane	KF, glycol, 170° distill product	1,4-Difluorobutane (44.2)	179
1,18-Dichlorooctadecane	KF, diethylene glycol, 175°	1,18-Difluorooctadecane (40)	164
Allyl bromide	KF, glycol, 180°	Allyl fluoride (30)	180
11-Bromoundecene-1	KF, diethylene glycol	11-Fluoroundecene-1 (46)	164
4-Chlorobutyronitrile	KF, diethylene glycol, 180°	4-Fluorobutyronitrile (63)	181
7-Bromoheptanonitrile	KF, diethylene glycol, 115-125°, 12 mm	7-Fluoroheptanonitrile (58.3)	182
5-Chloropentanol	KF, diethylene glycol, 160°; note ring closure competes with fluorination. With 4-chlorobutanol, only tetrahydrofuran is formed	5-Fluoropentanol (30.2) plus tetrahydropyran	183
10-Chlorodecanol	KF, diethylene glycol, 120-125°	10-Fluorodecanol (64)	183
Methyl 8-Bromooctanoate	KF, diethylene glycol, 150°	Methyl 8-fluorooctanoate (36)	183
Activated Aromatic Halides			
4-Chloronitrobenzene	KF, dimethylsulfoxide, 185-190°, 14 hr	4-Fluoronitrobenzene (72)	166
3-Trifluoromethyl-4-chloronitrobenzene	KF, dimethylformamide, 160°, 4 hr	3-Trifluoromethyl-4-fluoronitrobenzene	166
4-Chloro-2,5-difluoronitrobenzene	KF, dimethylsulfoxide, 175-180°, 2 hr	2,4,5-Trifluoronitrobenzene (45)	166

4-Chloro-1,3-dinitrobenzene	KF, no solvent, 190-200°, 7 hr	4-Fluoro-1,3-dinitrobenzene (92)	184
2-Chloropyridine	KF, dimethylsulfoxide, heated 21 days at 200-210°	2-Fluoropyridine (50)	185
See Table 5-4 for additional examples			
Polychloroaromatic and Heteroaromatic Compounds			
See Table 5-4			
Sulfonic Acid Esters			
Ethyl *p*-toluenesulfonate	KF, no solvent, 190-250°, 9 hr	Fluoroethane (85.6)	168
Vinyl *p*-toluenesulfonate	KF, diethylene glycol, 140-180°, 7.5hr, 20-50 mm	Fluoroethylene (78.9)	168
n-Heptyl *p*-toluenesulfonate	KF, diethylene glycol, 150-180°, 3 hr, 40-50 mm	1-Fluoroheptane (55.1)	168
2-Ethoxyethyl methylsulfonate	KF, diethylene glycol, 100°, 300 mm	2-Ethoxy-1-fluoroethane	186
Bis(2-chloroethyl) sulfate	KF, no solvent, 175°	1-Chloro-2-fluoroethane (60)	187
9α-Fluorohydrocortisone-21-methylsulfonate	KF, dimethylsulfoxide, 110°	21,9α-Difluoro-Δ^4-pregnene-11β,17α-diol-3,20-dione	170
17α,21-Dihydroxy-4-pregnene-3,20-dione 21-*p*-toluenesulfonate	KF, dimethylsulfoxide, 100°, 18 hr	17α-Hydroxy-21-fluoro-4-pregnene-3,20-dione (10)	188
17,20;20,21-Bismethylenedioxy-6α-hydroxymethyl-5α-pregnane-3β, 11β-diol 3-acetate-21-*p*-toluenesulfonate	KF, diethylene glycol, 205-215°, 1 hr, followed by basic hydrolysis of acetate	17,20;20,21-Bismethylenedioxy-6α-fluoromethyl-5α-pregnane-3β,11β-diol (61)	189
6-O-Methanesulfonyl-1:2:3:4-di-O-isopropylidene-D-galactose	KF, glycol, heat for 75 min	6-Deoxy-6-fluoro-1:2-3:4-di-O-isopropylidene-D-galactopyranose	169
2-Tosylvaleronitrile	KF, diethylene glycol, distill product	2-Fluorovaleronitrile (49)	171

[173] A. N. Nesmejanow and E. J. Kahn, *Chem. Ber.*, **67**, 370 (1934).
[174] W. Davies and J. H. Dick, *J. Chem. Soc.*, 2104 (1931).
[175] A. G. Beaman and R. K. Robins, *J. Am. Chem. Soc.*, **83**, 4038 (1961).
[176] H. A. Pacini and A. E. Pavlath, *J. Chem. Soc.*, 5741 (1965).
[177] F. W. Hoffman, *J. Org. Chem.*, **15**, 425 (1950).
[178] F. L. M. Pattison and W. C. Howell, *J. Org. Chem.*, **21**, 748 (1956).
[179] F. W. Hoffman, *J. Org. Chem.*, **14**, 105 (1949).
[180] E. T. McBee, C. G. Hsu, O. R. Pierce, and C. W. Roberts, *J. Am. Chem. Soc.*, **77**, 915 (1955).
[181] E. Ott, G. Piller, and H. J. Schmidt, *Helv. Chim. Acta*, **39**, 682 (1956).
[182] F. L. M. Pattison, W. J. Cott, W. C. Howell, and R. W. White, *J. Am. Chem. Soc.*, **78**, 3484 (1956).
[183] F. L. M. Pattison, W. C. Howell, A. J. McNamara, J. C. Schneider, and J. F. Walker, *J. Org. Chem.*, **21**, 739 (1956).
[184] N. N. Vorozhtsov, Jr., and G. G. Yokobson, *Zhur. Obshchei. Khim.*, **27**, 1672 (1957); *Chem. Abstr.*, **52**, 2777g (1958); *J. Gen. Chem.*, USSR, (English trans.), **27**, 1741 (1957).
[185] G. C. Finger, L. D. Starr, D. R. Dickerson, H. S. Gutowsky, and J. Hamer, *J. Org. Chem.*, **28**, 1666 (1963).
[186] F. L. M. Pattison and J. E. Millington, *Can. J. Chem.*, **34**, 757 (1956).
[187] V. V. Razumovskii and A. E. Fridenberg, *J. Gen. Chem.*, USSR (English trans.), **19**, 83 (1949).
[188] J. E. Herz and J. Fried, U.S. Patent 3,050,536 (1962); *Chem. Abstr.* **57**, 16708*d* (1962).
[189] P. F. Beal, R. W. Jackson, and J. E. Pike, *J. Org. Chem.*, **27**, 1752 (1962).

4. Primary monohalides unactivated in an aliphatic system can be replaced using potassium fluoride in a polar solvent, but secondary and tertiary halogen usually only eliminate because potassium fluoride in a nonaqueous solvent is a powerful base toward hydrogen halides.[163]

$$\text{—CH}_2\text{—}\underset{}{\overset{\text{X}}{\overset{|}{\text{CH}}}}\text{—}\overset{\text{H}}{\overset{|}{\text{CH}}}\text{—CH}_2\text{—} + \text{KF} \xrightarrow[\text{solvent}]{\text{nonaqueous}} \text{—CH}_2\text{CH{=}CH—CH}_2\text{—} + \text{KF·HX}$$

$$\text{CH}_3(\text{CH}_2)_4\text{CH}_2\text{Br} \xrightarrow[\text{diethylene glycol}]{\text{KF, 180°}} \underset{50\%}{\text{CH}_3(\text{CH}_2)_4\text{CH}_2\text{F}} \qquad (164)$$

5. Halogen atoms on aromatic rings may be readily replaced using potassium fluoride if there is an electron-withdrawing group in an ortho or para position. A nitro group is particularly effective because it stabilizes the negative charge in

[Structure: bracketed Meisenheimer-type intermediate, ring carbon bearing Cl and F, para $N(O)_2$ with delocalized charge]

the transition state or intermediate. The mechanism is probably analogous to that of nucleophilic aromatic substitutions. (See Section 3-3B and Ref. 165.)

$$p\text{-ClC}_6\text{H}_4\text{NO}_2 \xrightarrow[\text{185–190°, 14 hr}]{\text{KF-dimethylsulfoxide}} \underset{(72\%)}{p\text{-FC}_6\text{H}_4\text{NO}_2} \qquad (166)$$

6. Polychloroaromatic compounds are best run with anhydrous potassium fluoride at high temperatures (over 400°C; see Section 5-4A.1).

$$\text{C}_5\text{Cl}_5\text{N (pentachloropyridine)} \xrightarrow[\text{autoclave, 400–500°}]{\text{anhydrous KF}} \text{up to 90\%} \ \text{C}_5\text{F}_5\text{N (pentafluoropyridine)} \qquad (167)$$

[163] For a review on preparation of aliphatic monofluorides, see (a) F. L. M. Pattison, *Toxic Aliphatic Fluorine Compounds,* Elsevier, Amsterdam and New York, 1959. (b) F. L. M. Pattison, R. L. Buchanan, and F. H. Dean, *Can. J. Chem.*, **43**, 1700 (1965).

[164] F. L. M. Pattison and J. J. Norman, *J. Am. Chem. Soc.*, **79**, 2311 (1957).

[165] R. E. Parker, *Advan. Fluorine Chem.*, **3**, 75 (1963).

[166] G. C. Finger and C. W. Kruse, *J. Am. Chem. Soc.*, **78**, 6034 (1956).

[167] R. D. Chambers, J. Hutchinson, and W. K. R. Musgrave, *Proc. Chem. Soc.*, 83 (1964).

7. Oxygen-bonded groups, such as sulfonate esters that are good leaving groups, are easily replaced by potassium fluoride in polar solvent. This method is particularly valuable for molecules containing sensitive functional groups and allows conversion of an alcohol function to a fluoride through a sulfonate ester. The value of this route is illustrated by the examples shown in Chart 4-4.

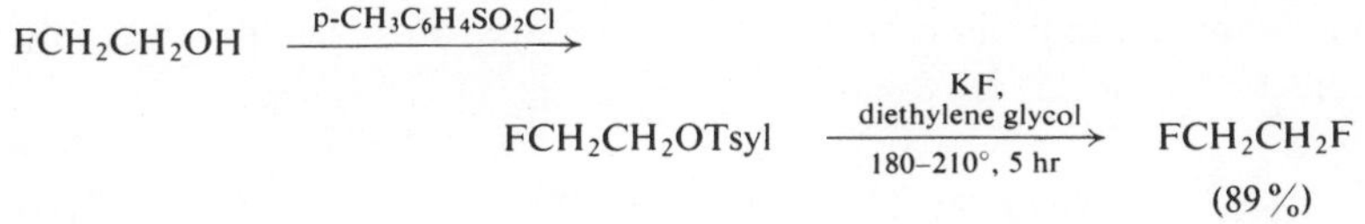

$$FCH_2CH_2OH \xrightarrow{p\text{-}CH_3C_6H_4SO_2Cl} FCH_2CH_2OTsyl \xrightarrow[180\text{–}210°,\ 5\ hr]{KF,\ diethylene\ glycol} \underset{(89\%)}{FCH_2CH_2F} \quad (168)$$

CH_3SO_3—O—H_2C (furanoside, OCH$_3$ / H; 2,3-O-isopropylidene, H_3C C CH_3) $\xrightarrow[(2)\ H_2O,\ H^+]{(1)\ KF,\ glycol}$ FCH_2 (furanose, OH / H; HO, OH) (60%) (169)

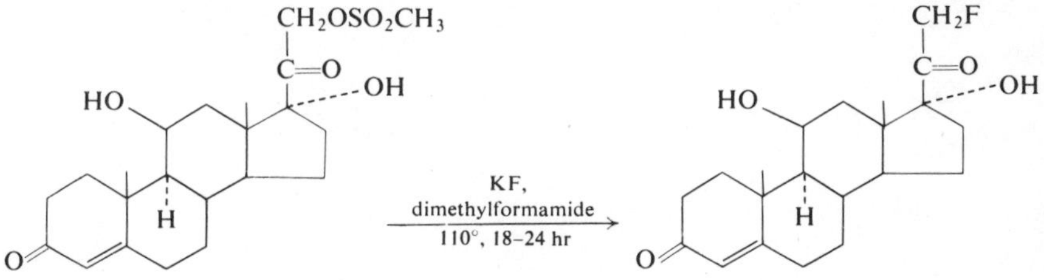

Steroid with $CH_2OSO_2CH_3$, C=O, OH, HO, H, O $\xrightarrow[110°,\ 18\text{–}24\ hr]{KF,\ dimethylformamide}$ steroid with CH_2F, C=O, OH, HO, H, O (170)

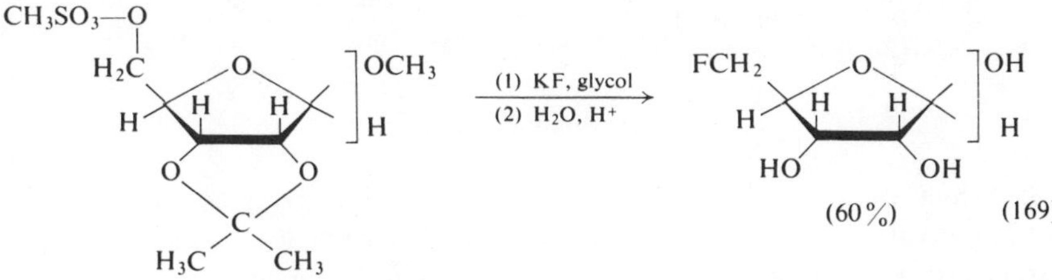

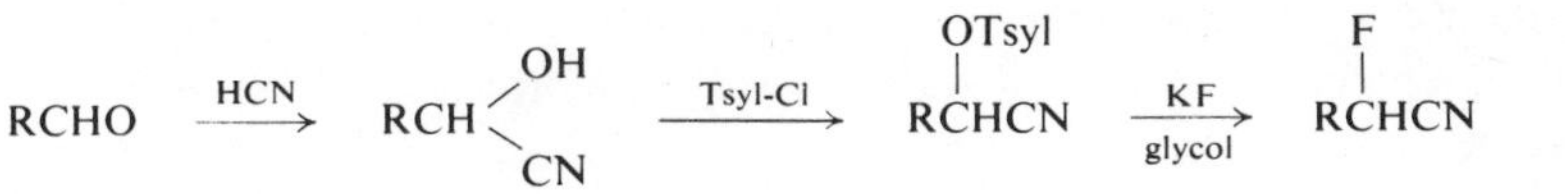

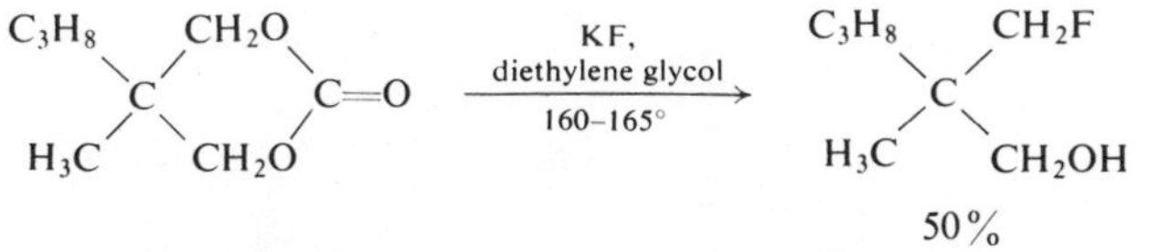

$$RCHO \xrightarrow{HCN} RCH\begin{matrix}OH\\CN\end{matrix} \xrightarrow{Tsyl\text{-}Cl} \underset{RCHCN}{\overset{OTsyl}{|}} \xrightarrow[glycol]{KF} \underset{RCHCN}{\overset{F}{|}} \quad (171)$$

$$\begin{matrix}C_3H_8 & & CH_2O\\ & C & & C{=}O\\ H_3C & & CH_2O\end{matrix} \xrightarrow[160\text{–}165°]{KF,\ diethylene\ glycol} \begin{matrix}C_3H_8 & & CH_2F\\ & C & \\ H_3C & & CH_2OH\end{matrix} \quad (172)$$

50%

CHART 4-4

[168] W. F. Edgell and L. Parts, *J. Am. Chem. Soc.*, 77, 4899 (1955).

[169] N. F. Taylor and P. W. Kent, *J. Chem. Soc.*, 872 (1958).

[170] J. E. Herz, J. Fried, P. Grabowich, and E. F. Sabo, *J. Am. Chem. Soc.*, 78, 4812 (1956).

[171] E. D. Bergmann and I. Shahak, *Bull. Res. Council Israel*, **10A**, 91 (1961).

[172] (a) I. Shahak and E. D. Bergmann, *Chem. Commun.*, 122 (1965). (b) I. Shahak and E. D. Bergmann, *J. Chem. Soc. (C)*, 319 (1967).

4-7 THE SCHIEMANN REACTION

$$ArNH_2 \longrightarrow [ArN_2^+BF_4^-] \longrightarrow ArF$$

One or two fluorines are usually introduced into an aromatic ring by the Schiemann-type reaction. An aniline or other aromatic primary amine is diazotized and the diazonium tetrafluoroborate salt precipitated, dried, and pyrolyzed to the aromatic fluoride.

$$\text{2-}CH_3C_6H_4NH_2 \xrightarrow[(3)\ NaBF_4,\ 0^\circ]{(1)\ HCl\ soln.\ (2)\ NaNO_2} \text{2-}CH_3C_6H_4N_2^+\ BF_4^-\ (81\%) \xrightarrow[104\text{-}106^\circ]{heat\ with\ sand} \text{2-}CH_3C_6H_4F\ (84\%) \quad (190)$$

The Schiemann reaction was reviewed recently and thoroughly by Suschitzky,[190] by Pavlath and Leffler,[191] and in Houben-Weyl[192] (see these reviews for typical examples). Some modifications and recent developments in use of the reaction are given in Section 5-4A.4.

4-8 ELECTROCHEMICAL FLUORINATION

$$HF \longrightarrow [F\cdot] \xrightarrow[RX]{RH} \text{perfluoroorganic}$$

Electrochemical fluorination was developed by Simons and is used industrially to prepare perfluorinated organic compounds and derivatives. An organic compound is dissolved in liquid hydrogen fluoride at 0°C to give a conducting solution through which current is passed and perfluorinated compounds are generated.

$$H_2C\langle^{CH_2-CH_2}_{CH_2-CH_2}\rangle N-CH_3 \xrightarrow[electric\ current]{HF,\ 0^\circ} F_2C\langle^{CF_2-CF_2}_{CF_2-CF_2}\rangle N-CF_3\ (39\%) \quad (193a)$$

[190] H. Suschitzky, *Advan. Fluorine Chem.*, **4**, 1 (1965).

[191] A. E. Pavlath and A. J. Leffler, *Aromatic Fluorine Compounds*, Am. Chem. Soc. Monograph No. 155, Reinhold, New York, 1962, pp. 12-16, 42-45.

[192] See Ref. 13, pp. 213-247.

[193] (a) J. Burdon and J. C. Tatlow, *Advan. Fluorine Chem.*, **1**, 129 (1960). (b) Ref. 13, pp. 38-50.

Electrochemical fluorination has been recently reviewed.[193, 194] Although this method is of considerable importance, most organic chemists will never have occasion to use the technique.

4-9 JET FLUORINATION

The technique of jet fluorination was developed by Bigelow and is a method of controlling gas phase fluorination of organic compounds in moving gas streams by good heat transfer. The technique has been used in preparation of many novel fluorine compounds, but usually complex mixtures are formed. The organic chemist will not find jet fluorination a useful tool.

The fluorination of tetrafluorosuccinonitrile can serve as an example of the type of product mixture obtained:[195] carbon tetrafluoride, nitrogen trifluoride, perfluoroethane, perfluoromethylamine, perfluoropropane, perfluorodimethylamine, perfluorobutane, N-trifluoromethylperfluoropropylidine imine, perfluoropyrollidine, perfluoromethylpropylamine, perfluoro-n-butylamine.

Additional information on jet fluorination is given in reviews.[11, 196]

PROBLEMS

1. Discuss the reaction of fluorine with cyclohexane to give cyclohexyl fluoride at -78°. Explain quantitatively in terms of the energies of a series of steps why this is possible at -78° but not at 25°.

$$\text{H, H} + F_2 \xrightarrow{-78^\circ} \text{F, H} + HF$$

2. Give the reagents and experimental details necessary to accomplish the conversions listed below. Give full structural formulas for starting materials, intermediates, and products.

(a) Perfluorobutane from octafluoro-2-butene.

(b) 1,2-Difluoroacenaphthene from acenaphthalene.

(c) *cis*-4,5-Difluorocholesten-3-one from Δ^4-cholesten-3-one.

(d) Hexafluorobenzene from hexachlorobenzene.

(e) Perfluoroethyl iodide from tetrafluoroethylene.

[194] S. Nagase, *Fluorine Chem. Rev.*, **1**, 77 (1967).

[195] L. A. Bigelow, J. B. Hynes, and B. C. Bishop, *2nd Intern. Symp. Fluorine Chem., Estes Park, Colorado, July, 1960,* Preprints, pp. 138-161.

[196] Ref. 13, pp. 24-38.

3. Explain why the rotating metal gauze reactor used by W. T. Miller, Jr. (Refs. 4-6, 4-7) is effective in the reactions of fluorine with CFCl=CFCl and CF_2=CFCl. Consider the rates of competing steps in the reactions and explain the product compositions found for CFCl=CFCl at −110° and −133°C and for CF_2=CFCl at −110° and −150° (consult Table 4-1).

4. Write the structure(s) of the major product or products expected from addition of anhydrous hydrogen fluoride to each of the following alkenes:

(a) Propene, 25°, activated C catalyst [see R. F. Sweeny and C. Woolf, U.S. Patent 2,917,559 (1959); *Chem. Abstr.*, **54**, 10858 (1960)].
(b) Allene, −70°C, 1 hr.
(c) Cyclobutene, 0°C, 10 hr, CCl_4 solvent.
(d) α-Pinene, −50°C, 6 hr, CCl_4 solvent.
(e) 17α, 21-Dihydroxy-4,9(11)-pregnadiene-3,20-dione 21-acetate, −2°, 4 hr, 74% HF in pyridine.

5. Addition of hydrogen fluoride to some alkenes can give high yields of the expected monofluoroalkanes whereas with other alkenes products are obtained which result from extensive rearrangements. Using the necessary mechanisms, compare and explain in detail the differences between the addition of anhydrous hydrogen fluoride to the following alkenes:

(a) Propene at −45° to 0°C (Ref. 4-28).
(b) Cyclohexene at −78° to 0°C (Refs. 4-28 and 4-33).
(c) Bornylene at −70° to 5°C (Ref. 4-36).
(d) Cyclopropylcarbinol at −15°C [Ref. 4-32; see also Mazur, White, Semenow, Lee, Silver, and Roberts, *J. Am. Chem. Soc.*, **81**, 4390 (1959).]

6. Write the structure(s) of the major product or products expected from the addition of anhydrous hydrogen fluoride to each of the listed alkenes.

(a) 1-Chlorocyclohexene, 0°, 1.5 hr, ether solvent.
(b) 2-Chloro-2-butene, 1:1 ratio butene/HF, −78° to 23°C.
(c) 2-Chloro-2-butene, 1:2.5 ratio butene/HF, −78° to 23°C. Explain the difference in the product composition obtained in parts (b) and (c).
(d) Chloroethene, 4-5 hr, 10–15 atm.
(e) 1,1-Dichloro-1-butene, 16 hr, 65°C (see Refs. 4-40 and 4-45).
(f) 1,2-Dichloro-1-propene (see Ref. 4-40).

7. Compare the additions of HF to 1-butyne, 2-butyne, 2-pentyne, and 3-hexyne. What are the products expected in each case and what conditions should be used?

8. Outline superior syntheses for the compounds listed below using "classical methods" to introduce fluorine. Organic starting materials may contain only the elements C, H, and O. Any inorganic reagent may be used. Give reagents and conditions for every step of each synthesis.

(a) 2,2-Difluorohexane.
(b) 3,3-Difluorohexane.
(c) 4-Fluorocyclohexane-*cis*-1,2-dicarboxylic acid.
(d) 4,5-Difluorocyclohexane-*cis*-1,2-dicarboxylic acid.

9. For the reactions indicated below, write formulas or structures for the product or products formed. Where more than one product is formed, indicate the principal product(s) and indicate relative amounts of other products.

(a) (indene)–CH_3 + F_2 $\xrightarrow[\text{4A Molecular Sieve}]{-78^\circ,\ CCl_3F}$

(b) (benzene) + excess ClF_3 $\xrightarrow[\text{programmed heating}]{N_2,\ 200\text{–}350^\circ}$

(c) (phenyl)–NCS + IF_5 $\xrightarrow{\text{pyridine}}$ (See Ref. 4-21.)

(d) $CF_2{=}CFCl$ $\xrightarrow[25^\circ,\ 4\text{ hr}]{1\text{ part } IF_5\text{–}2\text{ parts } I_2}$

(e) $CH_3\overset{O}{\overset{\|}{C}}CH_3$ + BrF_3 $\xrightarrow[\text{slow addn. } BrF_3]{\text{low temp.}}$ (See Ref. 4-20.)

10. When various fluoroalkenes are dissolved in dimethylformamide at the appropriate temperature and treated with anhydrous potassium fluoride, the following conversions are observed:

$$CF_2{=}CClCF_3 \longrightarrow CF_3CHClCF_3$$
$$CF_2{=}CFCl \longrightarrow CF_3CHFCl \quad (72\%)$$
$$CF_2{=}CFCF_3 \longrightarrow CF_3CHFCF_3 \quad (60\%)$$

Explain the results; show the detailed mechanism. (See Ref. 4-52 and Section 6-1C.)

11. Explain why cobalt(III) fluoride is the most important oxidizing high-valency metallic fluoride for fluorinating hydrocarbons. Illustrate its

utility by citing several syntheses of perfluorinated alkanes and alicyclic compounds. Why is it superior to silver(II) fluoride? How is CoF_3 used in the synthesis of perfluoro-*p*-xylene? In answering this question the following references are suggested: 4–67; 4–69 to 4–82 found in *Ind. Eng. Chem.*, **39**, (1947); 4–85; and 4–91.

12. Use standard organic reactions and the appropriate "classical" fluorinating agents to accomplish the conversions stated below. Show structures of starting materials, intermediates, and products. Use optimum procedures.

(a) Toluene → trifluoromethylbenzene.
(b) Propane → 1,1,2-trichloro-3,3,3-trifluoro-1-propene.
(c) Acetone → 2,2-difluoropropane.
(d) 2,3-Dibromo-1-propene → 1,3-dibromo-2,2-difluoropropane.
(e) Methane → difluorodichloromethane.
(f) Hexachloro-1,3-butadiene → 2,3-dichloro-1,1,1,4,4,4-hexafluoro-2-butene.

13. An order of fluorinating power for antimony fluorides and a pattern of activity of chlorine toward substitution by fluorine is given in this chapter (p. 76). Using these correlations predict the order in which chlorine will be replaced in the listed compounds; the reagent of *minimum* strength required to replace each chlorine; and the ultimate product when SbF_5 is used at high temperature.

(a) CCl_3CCl_3
(b) $CCl_3CCl_2CCl_3$
(c) $ClCH_2CCl_2CH_2Cl$
(d) Cl_2CHCCl_3
(e) $CHCl_2CHClCHCl_2$
(f) $CCl_2{=}CClCCl_3$
(g) $ClCH{=}C(CH_3)CCl_3$
(h) $ClCH_2CH_2CCl_3$
(i) $Cl_2CFCFClCFClCFCl_2$
(j) $ClCH_2CH_2CCl_2CH_2CCl_3$
(k) $Cl_3C\overset{\overset{\displaystyle O}{\|}}{C}CCl_3$
(l) $CH_2{=}CHCCl_2CCl_3$

14. What are the uses and limitations of silver(I) fluoride as a fluorinating agent? Select five examples from five references to illustrate your discussion.

15. For very practical reasons the use of potassium fluoride on a methylsulfonate or *p*-toluenesulfonate is a valuable method for introducing isolated fluorine atoms into organic compounds. What are the reasons? Give five examples from five references showing the reaction.

16. Accomplish the following conversions using the starting materials given and any inorganic or organic reagents required. Show all structures and state experimental conditions needed for each step in the syntheses.

(a) Tetrahydrofuran → 1,4-difluorobutane.
(b) 21-Iodoprogesterone → 21-fluoroprogesterone.
(c) 1-Bromopropene-1 → 1,1-difluoropropane.
(d) Toluene → 1,2-diphenyltetrafluoroethane.
(e) Acetic acid → 2-fluoroethanol.
(f) Butanol → α-fluorovaleronitrile.
(g) 2-Butene → 1,2,3,4-tetrafluorobutane.
(h) 1-Heptanol → 1-fluoroheptane.
(i) Allyl bromide → 1,6-difluorohexane.
(j) Tetrahydrofuran → 5-fluorovaleronitrile.

17. Why is the Schiemann reaction the best method of introducing F into an aromatic ring? In particular, discuss the scope and limitations of the reaction and cite at least eight representative synthetic examples illustrating the scope of the reaction. Consult leading references for information (Refs. 4-190, 4-191, 4-192, and 4-97).

Chapter Five Modern Selective Methods of Fluorination

This chapter is devoted to a critical and detailed description of new selective methods for introducing fluorine, usually one to three atoms; we particularly emphasize the synthesis of organic fluorine compounds that are not highly fluorinated. In the next chapter we will discuss synthesis using fluorinated intermediates such as carbenes, carbanions, and radicals and addition of fluorinated units from reagents such as tetrafluoroethylene, hexafluoroacetone, and sulfur chloropentafluoride.

We consider Chapters 5 and 6 the most valuable in the book. Here we bring together a large amount of modern fluorine chemistry, which, because of its recent development, is not reviewed in detail in other places, or has not been considered in an organized review directed toward the utilitarian aspect of synthesis of selected organic fluorine compounds. Critical discussions of the mechanisms of reactions are included so that probable extensions and improvements can be visualized and opportunities for needed research programs should be apparent.

5-1 ADDITION OF F_2 TO UNSATURATED GROUPS

$$R_2C{=}CR_2 + F_2 \longrightarrow R_2CF{-}CFR_2$$

5-1A. FROM ELEMENTARY FLUORINE

In Chapter 2 we explained why elementary fluorine was not easily used directly to give selective fluorination of organic compounds. In Chapter 4, the classical work on elementary fluorine, electrolytic fluorination, and heavy metal fluorides (as fluorine carriers) for fluorinating organic fluorine compounds was discussed.

In using elementary fluorine, some selectivity has been obtained by certain special techniques, such as use of a jet reactor[1] (see Chapter 2 and Section 4-9) or by fluorination of very small simple molecules such as CO or COF_2 to give CF_3OOCF_3 or CF_3OF.[2] Recently, controlled addition of elementary fluorine to a variety of unsaturated organic compounds in solution has been reported. The technique is to use low temperatures (–78°C) and dilution of substrate with an inert solvent (CCl_3F).[3] Merritt and Stevens[3a] state that this procedure has been successfully applied to conjugated olefins such as indene, acenaphthylene, stilbene, and substituted phenanthrenes as well as for α,β-unsaturated acid halides; cis addition was the predominant process with stilbene, acenaphthylene, and phenanthrenes. The process is also applicable to steroidal olefins, since Δ^4-cholesten-3-one gives the *cis*-4,5-difluoride in yields of 60 to 70%.[3a]

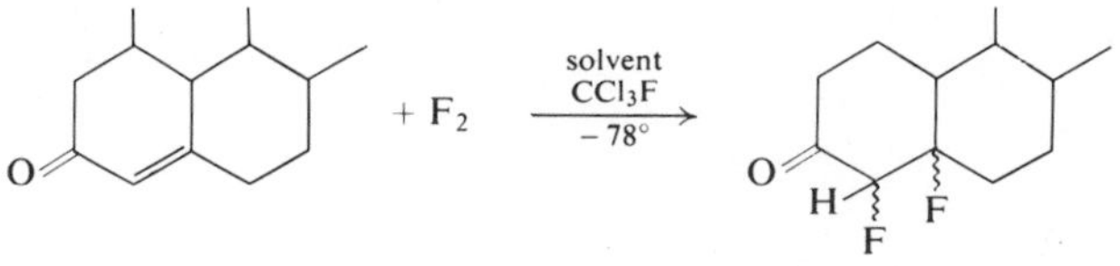

+ F_2 $\xrightarrow[-78^\circ]{\text{solvent } CCl_3F}$

O H F F

cis adduct

More recently, Merritt has reported the addition of fluorine to acetylenes:

$$C_6H_5C{\equiv}CR + 2F_2 \xrightarrow[-78^\circ]{FCCl_3} C_6H_5CF_2CF_2R \qquad (3g)$$

where R is C_6H_5, CH_3, or H.

The available evidence points strongly to a polar mechanism for these low-temperature fluorine additions. Fluorination of 1,1-diphenylethylene[3c] does not give any rearrangement products as expected if radical intermediates were

$$(C_6H_5)_2C{=}CH_2 + F_2 \longrightarrow \underset{(14\%)}{(C_6H_5)_2CFCH_2F} + \underset{(78\%)}{(C_6H_5)_2C{=}CHF} + \underset{(8\%)}{(C_6H_5)_2CFCHF_2}$$

[1] B. C. Bishop, J. B. Hynes, and L. A. Bigelow, *J. Am. Chem. Soc.*, **85**, 1606 (1963).

[2] G. H. Cady and R. S. Porter, U.S. Patents 3,230,263 and 3,230,264 (1966); *Chem. Abstr.*, **64**, 12550e,d (1966).

[3] (a) R. F. Merritt and T. E. Stevens, *J. Am. Chem. Soc.*, **88**, 1822 (1966). (b) R. F. Merritt and F. A. Johnson, *J. Org. Chem.*, **31**, 1859 (1966). (c) R. F. Merritt, *J. Org. Chem.*, **31**, 3871 (1966). (d) R. F. Merritt, *J. Am. Chem. Soc.*, **89**, 609 (1967). (e) R. F. Merritt and F. A. Johnson, *J. Org. Chem.*, **32**, 416 (1967). (f) R. F. Merritt, *ibid.*, 1633. (g) R. F. Merritt, *ibid.*, 4124.

involved (see discussion in Section 5-1B). *cis*- and *trans*-Propenylbenzene adds fluorine predominantly in a cis manner[3d] (Chart 5-1). Fluorination in methanol

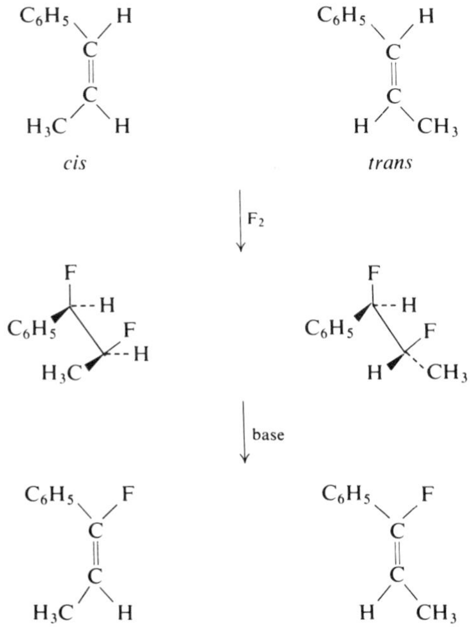

CHART 5-1

gave *dl*-erythro- and *dl*-threo-1-methoxy-1-phenyl-2-fluoropropanes as well as the vicinal difluorides which were also formed in an inert solvent. The products are similar to those formed in ionic chlorine additions, but fluorination appears to involve a polar molecular complex (**a** or **b**), rather than a bridged fluoronium

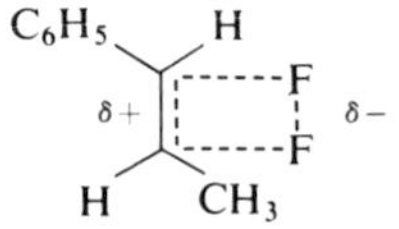

a 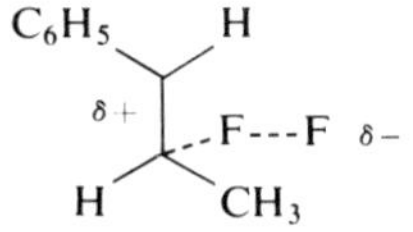 **b** 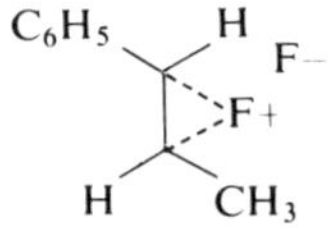 **c**

ion (**c**). The closed form (**a**) explains the high stereospecificity; the open form (**b**) explains the minor amount of trans product and the methoxy product in methanol. The high stereospecificity, even in methanol, and higher yield of vicinal dihalide in methanol compared to chlorination reactions under the same conditions argues for molecular complex (**a** or **b**) and against the fluoronium ion (**c**); "chloronium" or "bromonium" ions must be intermediates in chlorination or bromination.[4]

The polar mechanism is further strengthened by the results from fluorination of acetylenes in the presence of methanol.[3g] In addition to tetrafluoro compounds, mono- and bismethoxy compounds are obtained.

$$C_6H_5C{\equiv}CR + 2F_2 \xrightarrow{CH_3OH} \underset{\mathbf{1}}{C_6H_5CF_2CF_2R} + \underset{\mathbf{2}}{C_6H_5CF(OCH_3)CF_2R} + \underset{\mathbf{3}}{C_6H_5C(OCH_3)_2{-}CF_2R}$$

	% Composition		
	1	**2**	**3**
$R = C_6H_5$	23	57	20
$R = CH_3$	19	50	31
$R = H$	13	35	52

The controlled low-temperature fluorination has been extended to Schiff bases.[3e] The monofluoride **5** must form by F_2 addition followed by HF

$$C_6H_5CH{=}NR \xrightarrow{1.5\ F_2} \underset{\mathbf{4}\ (22\%)}{C_6H_5CF_2NFR} + \underset{\mathbf{5}}{C_6H_5CF{=}NR} + C_6H_5CH{=}\overset{+}{N}HR\ F^-$$

R is alkyl

$$\mathbf{4} + \mathbf{5} \xrightarrow{\text{excess } F_2} C_6H_5CF_2NF_2 + C_6H_5CF_3 + RF + RNF_2$$

elimination, and the trifluoride **4** forms by F_2 addition to **5**. By use of an iminochloride, the trifluoride **4** was produced as the only product in good yield.

$$C_6H_5CCl{=}NR + 2F_2 \longrightarrow \mathbf{4}\ (63\%)$$

Alkyl isocyanates do not add fluorine directly to the C=N bond at low temperatures but give instead side-chain fluorination followed by substitution on nitrogen of the carbamyl fluoride by-product.[3f] These fluorinations were run

[4] P. E. Peterson and R. J. Bopp, *J. Am. Chem. Soc.*, 89, 1283 (1967), have provided evidence for fluorine participation (via a fluoronium ion?) in reactions of 5-fluoro-1-pentyne with trifluoroacetic acid, but note that fluorine is much poorer than chlorine in such participation.

$$C_2H_5CH_2NCO + F_2 \longrightarrow C_2H_5CHFNCO + HF$$

$$C_2H_5CHFNCO \xrightarrow{F_2} C_2H_5CHFNF_2 + COF_2$$

$$C_2H_5CHFNCO \xrightarrow{C_2H_5CH_2NCO} C_2H_5CH_2NHC(O)F$$

$$C_2H_5CH_2NHC(O)F \xrightarrow{F_2} C_2H_5CH_2NFC(O)F + HF$$

in the presence of NaF to remove HF, and fluoride ion may have a catalytic effect.

Fluoride ion catalysis is another method that gives control of fluorine addition to carbonyl groups.[5]

$$CO_2 + 3F_2 \xrightarrow[\text{room temp.}]{CsF} CF_2(OF)_2 \quad (99.7\%) \qquad (5a, b)$$

$$COF_2 \xrightarrow[-78^\circ]{F_2,\ CsF} CF_3OF \quad (97\%) \qquad (5c)$$

$$CF_3COF \longrightarrow CF_3CF_2OF \quad (96\%)$$

$$(CF_3)_2CO \longrightarrow (CF_3)_2CFOF \quad (98\%)$$

$$FC(O)OOC(O)F + 2F_2 \xrightarrow[-95^\circ]{KF} FOCF_2OOCF_2OF \qquad (5d)$$

$$FC(O)OOC(O)F \xrightarrow[h\nu]{F_2,\ 25^\circ} FC(O)OF \xrightarrow[F_2]{CsF} F_2C(OF)_2 \qquad (5b)$$

$$R_fCN + 2F_2 \xrightarrow[-78^\circ]{CsF} R_fCF_2NF_2 \quad (\text{over } 90\%) \qquad (5e)$$

The reaction is limited to fluorine or perfluoroalkyl groups bonded to carbon (other groups give unstable products or are replaced, for example, $COCl_2$ gives CF_3OF and Cl_2 only). The fluorinations do not go without cesium fluoride. No

[5] (a) F. A. Hohorst and J. M. Shreeve, *J. Am. Chem. Soc.*, **89,** 1809 (1967). (b) R. L. Cauble and G. H. Cady, *ibid.*, 1962, 5161. (c) M. Lustig, A. R. Pitochelli, and J. K. Ruff, *ibid.*, 2841. (d) M. Lustig and J. K. Ruff, *Chem. Commun.*, 870 (1967). (e) J. K. Ruff, *J. Org. Chem.*, **32,** 1675 (1967).

mechanism was proposed, but ionic intermediates or transition states must again be involved.

Ionic fluorination must also be involved in some extensions of this type of reaction.

$$SOF_4 + F_2 \xrightarrow{CsF} SF_5OF \qquad (6a)$$

$$CF_3\overset{O}{\overset{\|}{C}}ONa + F_2 \longrightarrow CF_2(OF)_2 \qquad (6b)$$

$$Na_2C_2O_4 + F_2 \nearrow$$

(NaF is sometimes an effective diluent)

$$H_2N\overset{NH}{\overset{\|}{C}}SO_2H + 6F_2 \xrightarrow{NaF} F_2NCF_2NF_2 + SO_2F_2 + 4HF \qquad (6c)$$

(8–10%, formed by replacement of hydrogen, addition to imino bond, and carbon–sulfur cleavage; by-products are $F_2NSO_2F + SOF_2 + CF_2(NF_2)_2 + NF_3$)

$$(CF_3)_2C\begin{smallmatrix}OH\\ONa\end{smallmatrix} + F_2 \longrightarrow (CF_3)_2C(OF)_2 + CF_3CF(OF)_2 \qquad (6d)$$

$$(CF_3)_2C(OH)_2 + F_2 \longrightarrow (CF_3)_2CFOF \qquad (6e)$$

The fluorination of the sodium salt of hexafluoroacetone is probably an example of substitution rather than F_2 addition but is included here to provide further evidence for the ionic nature of the controlled fluorinations. The fluorination of hexafluoroacetone hydrate[6e] probably involves F_2 addition to hexafluoroacetone produced in situ.

The selective use of fluorine is an excellent advance in the fluorine field and opens up opportunities for new synthesis as well as mechanism studies. The important observation is that *ionic F_2 reactions can be controlled.* However, this method suffers from the disadvantage of special handling requirements and hazards of elementary fluorine which makes its use impractical in many laboratories.

[6] (a) J. K. Ruff and M. Lustig, *Inorg. Chem.*, **3**, 1422 (1964). (b) P. G. Thompson, *J. Am. Chem. Soc.*, **89**, 1811 (1967). (c) R. J. Koshar, D. R. Husted, and R. A. Meiklejohn, *J. Org. Chem.*, **31**, 4232 (1966). (d) P. G. Thompson and J. H. Prager, *J. Am. Chem. Soc.*, **89**, 2263 (1967). (e) J. H. Prager and P. G. Thompson, *J. Am. Chem. Soc.*, **87**, 230 (1965).

Other examples of selective addition of elementary fluorine are in preparation of α,α'-difluoroxylene (Eq. 1),[7] fluoroacetaldehyde and α,β-difluoroethyl acetate (Eq. 2),[8] and cyclic fluorocarbon peroxide (Eq. 3).[9] Again low temperatures and dilution with an inert solvent were employed, but no experimental details

$$(1)\quad CH_2{=}C_6H_4{=}CH_2 \xrightarrow{F_2} FCH_2{-}C_6H_4{-}CH_2F$$

$$(2)\quad CH_2{=}CHOC(=O)CH_3 \xrightarrow[\text{inert solvent}\; -45\text{ to }-55^{\circ}]{F_2} FCH_2CHFOC(=O)CH_3 + FCH_2C(=O)H$$

$$(3)\quad \left[CF_3C(=O){-}CH{=}C(O{-})CF_3\right]_2M \xrightarrow{F_2} CF_3CFCF_2CFCF_3 \text{ (with } O{-}O \text{ bridge between the two CF)}$$

$M = Cu(II), Ni(II)$

were available for the first two examples. In the third, the reactant is partly fluorinated, and the product is only obtained in low yield with by-products. A mixture of cis and trans isomers were obtained but, since the reaction is probably multistep involving removal of the hydrogen by elimination or enolization, no conclusion can be drawn about the details of the mechanism of fluorine addition.

5-1B. FROM A FLUORINATING AGENT

Preparation of inorganic fluorides by fluorination with elementary fluorine is a well-established technique which will not be reviewed here.[10] In many cases, fluorine is "tamed" by a heavy metal that effectively acts as a carrier for F_2; thus metal fluorides in high oxidation state and interhalogen compounds are often active, selective, fluorinating agents. Such reagents generated either in situ or prior to reaction, with or without isolation, have been effectively used for selected additions of fluorine (see Section 4-4).

[7] H. R. Davis, L. A. Errede, B. F. Landrum, U.S. Patent 3,053,909 (1962); *Chem. Abstr.*, **58**, 4466*h* (1963).

[8] A. Ya. Yakubovich, S. M. Rozenshtein, and V. A. Ginsburg, USSR Patent 162,825 (May 27, 1964); *Chem. Abstr.*, **62**, 451*g* (1965).

[9] R. L. Talbott, *J. Org. Chem.*, **30**, 1429 (1965).

[10] See R. D. W. Kemmitt and D. W. A. Short, *Advan. Fluorine Chem.*, **4**, 142 (1965); improved methods were recently described by H. W. Roesky, O. Glemser, and K. H. Hellberg, *Chem. Ber.*, **98**, 2046 (1965).

In 1931, Dimroth and Bockemüller[11] reported that 1,1-diphenylethylene adds fluorine on treatment with either lead(IV) fluoride or phenyliododifluoride in chloroform solution, but the product has recently been shown not to be the simple adduct, 1,2-difluoro-1,1-diphenylethane **(6)**, but a rearrangement product, 1,1-difluoro-1,2-diphenylethane **(7)**.[12]

$$(C_6H_5)_2C{=}CH_2 + PbF_4 \text{ or } C_6H_5IF_2 \not\longrightarrow (C_6H_5)_2CF{-}CH_2F \quad (\mathbf{6})$$

$$(C_6H_5)_2C{=}CH_2 + PbF_4 \text{ or } C_6H_5IF_2 \longrightarrow C_6H_5CF_2CH_2C_6H_5 \quad (\mathbf{7})$$

Henne and Waalkes,[13] in 1945, provided the first authentic example of selective fluorine addition to a double bond of a halogen-containing olefin by lead(IV) fluoride generated in situ from lead(IV) oxide and hydrogen fluoride. Antimony(V) fluoride,[14] cobalt(III) fluoride generated in situ from cobalt(III) oxide and hydrogen fluoride,[15] preformed cobalt(III) fluoride,[16-19] mercury(II) fluoride generated in situ from mercury(II) oxide and hydrogen fluoride,[20] chlorine trifluoride,[21] preformed manganese(III) trifluoride,[17, 19] preformed silver(II) fluoride,[19] and preformed cerium(IV) fluoride[19] all have been used for fluorination of double bonds. In some cases, yields were very good, but in general these methods as reported were of very limited synthetic application because of problems in handling the highly reactive fluorinating agents. Usually conditions were so strenuous that only olefins stabilized by extensive halogen substitution gave reasonable yields,[22] and by-product formation from hydrogen

[11] (a) O. Dimroth and W. Bockemüller, *Chem. Ber.*, **64**, 516 (1931). (b) W. Bockemüller, *ibid.*, 522.

[12] J. Bornstein, M. R. Borden, F. Nunes, and H. I. Tarlin, *J. Am. Chem. Soc.*, **85**, 1609 (1963).

[13] A. L. Henne and T. P. Waalkes, *J. Am. Chem. Soc.*, **67**, 1639 (1945); see also A. L. Henne and T. H. Newby, *J. Am. Chem. Soc.*, **70**, 130 (1948).

[14] A. L. Henne and W. J. Zimmerschied, *J. Am. Chem. Soc.*, **67**, 1235 (1945).

[15] A. F. Benning and J. D. Park, U.S. Patent 2,437,993 (1948); *Chem. Abstr.*, **42**, 4193*c* (1948).

[16] C. I. Gochenour, U.S. Patent 2,554,857 (1951); *Chem. Abstr.*, **45**, 9555f (1951).

[17] G. Fuller, M. Stacey, J. C. Tatlow, and C. R. Thomas, *Tetrahedron*, **18**, 123 (1962).

[18] A. L. Dittman and J. M. Wrightson, U.S. Patent 2,690,459 (1954); *Chem. Abstr.*, **49**, 11681*d* (1955).

[19] D. A. Rausch, R. A. Davis, and D. W. Osborne, *J. Org. Chem.*, **28**, 494 (1963).

[20] A. L. Henne and K. A. Latif, *J. Am. Chem. Soc.*, **76**, 610 (1954).

[21] J. Murray, British Patent 878,585 (1961); *Chem. Abstr.*, **57**, 11018d (1962).

[22] Lead(IV) fluoride was reported to be unsatisfactory as a fluorinating agent for hydrocarbons in the vapor phase. R. D. Fowler, H. C. Anderson, J. M. Hamilton, Jr., W. B. Burford, III, A. Spadetti, S. B. Bitterlich, and I. Litant, *Ind. Eng. Chem.*, **39**, 343 (1947).

or halogen replacement by fluorine was always a problem. Rausch and co-workers[19] evaluated the ability of many of the common heavy metal fluorides to add fluorine to olefins and discussed scope, limitation, and mechanism of these reactions.

Recently some of these methods have been modified and improved so that they are mild enough to give selective fluorine addition to double bonds of steroids. Bowers and co-workers[23] treated pregnenolone acetate (**8**) in dry methylene chloride with an excess of lead(IV) acetate and anhydrous hydrogen fluoride for 15 minutes at $-75°C$ and got a 27% yield of the difluoro derivative **9** and 63% recovered **8**. Longer reaction times or higher temperature did not

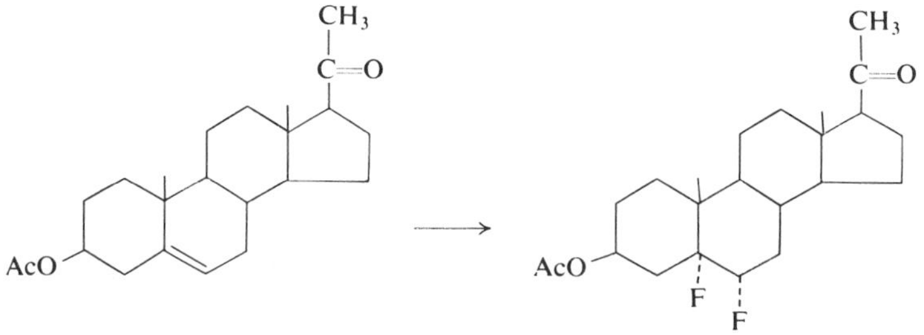

CH$_3$ C=O AcO → CH$_3$ C=O AcO F F

8 **9**

improve the yield but only led to small yield of a product from molecular rearrangement. On the basis of nmr analysis, **9** was shown to have a configuration resulting from *cis*-fluorine addition, and the authors suggest a cyclic addition mechanism.[24] They note that the ionic addition of reagents to this double bond is always trans. Bowers claims that hydrogen fluoride with osmium(IV) oxide or cobalt(III) oxide will also give the fluorine addition to steroids.[25] Examples of fluorine addition from in situ prepared PbF_4 to other unsaturated steroids have also been reported.[26] As mentioned above, the reaction of PbF_4 or phenyliododifluoride with 1,1-diphenylethylene was shown to give a molecular rearrangement product **7** and no indication of a product from direct addition.[12] Bornstein and co-workers[12] examined this reaction carefully and proposed a radical mechanism. The intermediate radical **10** from fluorine radical addition would quickly rearrange to the more stable radical **11**, which would pick up another F to give

[23] A. Bowers, P. G. Holton, E. Denot, M. C. Loza, and R. Urquiza, *J. Am. Chem. Soc.*, **84**, 1050 (1962).

[24] See E. L. Eliel, *Stereochemistry of Carbon Compounds*, McGraw-Hill, New York, 1962, p. 360.

[25] A. Bowers, U.S. Patent 3,097,199 (1963); *Chem. Abstr.*, **59**, 5238a (1963).

[26] K. Brückner and H. J. Mannhardt, German Patent 1,167,828 (1964); *Chem. Abstr.*, **61**, 4438c (1964).

$$(C_6H_5)_2C{=}CH_2 \longrightarrow \underset{\textbf{10}}{(C_6H_5)_2\underset{|}{\overset{}{C}}\!-\!\dot{C}H_2} \longrightarrow \underset{\textbf{12}}{(C_6H_5)_2\underset{F}{C}-CH_2CH_2\underset{F}{C}(C_6H_5)_2}$$

10: $(C_6H_5)_2C(F)-CH_2$

$$\downarrow$$

$$\underset{\textbf{11}}{(C_6H_5)\overset{F}{\dot{C}}-CH_2C_6H_5} \xrightarrow{F\cdot} 7$$

the product. Under certain conditions, the initial intermediate radical dimerized, and the dimer **12** was isolated.

In very recent work, Carpenter[27] reports a simplified procedure for preparation and direct use of aryliododifluorides, and presents evidence for an ionic course for the reaction. The aryliododichloride in methylene chloride is reacted with mercury(II) oxide and aqueous hydrofluoric acid, and the methylene

$$ArICl_2 + HgO + 2HF \longrightarrow ArIF_2 + H_2O + HgCl_2$$

chloride solution is used directly for fluorination. Carpenter shows that hydrogen fluoride, or some other strong acid such as trifluoroacetic acid, is necessary as a catalyst and that the methylene chloride solution contains sufficient HF to catalyze reaction, as with styrene to get 2-phenyl-1,1-difluoroethylene; he proposes an ionic mechanism with a phenonium ion intermediate:

$$RPhC{=}CH_2 + ArIF_2 \xrightarrow{H^+} RPhCFCH_2\overset{+}{I}Ar \xrightarrow{-ArI} RCF{-}CH_2 \text{ (bridged by } \overset{+}{Ph}) \xrightarrow{+F^-} RCF_2CH_2Ph$$

$$R = \text{phenyl, H}$$

From the reaction of *p*-tolyliododifluoride in chloroform with simple olefins such as cyclohexene, stilbene, and pentene-2, no well-defined products were isolated,[28] but with aromatic compounds such as pyrene, anthracene, and benzanthracene, discrete monofluorides (as well as coupling products) were claimed; however, Badger and Stephens[29] were unable to repeat this work. In

[27] W. Carpenter, *J. Org. Chem.*, **31**, 2688 (1966).

[28] B. S. Garvey, Jr., L. F. Halley, and C. F. H. Allen, *J. Am. Chem. Soc.*, **59**, 1827 (1937).

[29] G. M. Badger and J. F. Stephens, *J. Chem. Soc.*, 3637 (1956).

the reaction with a steroid isomer, phenyliododifluoride was found to add in entirety to the steroid molecule,[30] but we will discuss this work in the next section.

The fluorination of an unsaturated sugar, di-*O*-acetyl-D-arabinal, with lead(IV) acetate and hydrogen fluoride in methylene chloride at $-70°C$ also led to a rearrangement product **13** from ring contraction[31] similar to that for 1,1-diphenylacetylene. In this case, the rearrangement was suggested to occur after the initial formation of a *vicinal* difluoro adduct.

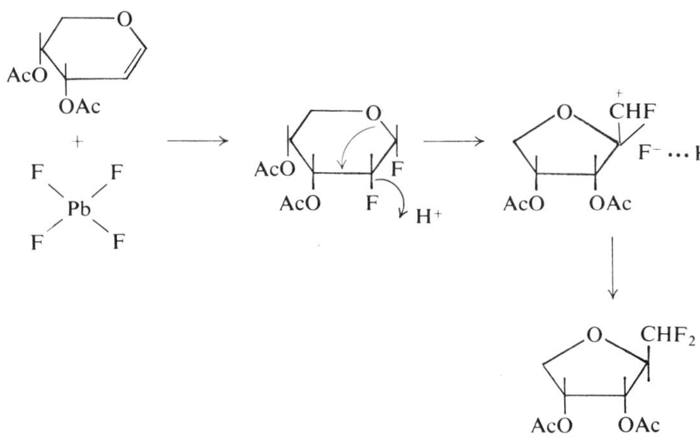

13

Another modification which avoids use of hydrogen fluoride is in situ preparation of lead(IV) fluoride from lead(IV) oxide and sulfur tetrafluoride which was used for fluorine addition to halogenated olefins.[32] The reaction is run in a pressure vessel with about fivefold excess of SF_4 to olefin and with optimum temperature of 40° to 100°C; yields varied from a few per cent to 95%. The conditions are sufficiently mild that by-products from halogen or hydrogen substitution, even for iodo olefins, were not found. Since *cis*- or *trans*-1,2-dichloroethylene both yielded a *dl*- and *meso*-1,2-dichloro-1,2-difluoroethane mixture (proper controls were run), the authors suggest that the two fluorines cannot add simultaneously and that a cyclic intermediate is not involved. They also argue against any mechanism involving prior addition of SF_4 to the olefin. This method of generating and using PbF_4 in situ should be

[30] P. G. Holton, A. D. Cross, and A. Bowers, *Steroids,* **2,** 71 (1963).

[31] (a) P. W. Kent, J. E. G. Barnett, and K. R. Wood, *Tetrahedron Letters,* No. 21, 1345 (1963). (b) P. W. Kent and J. E. G. Barnett, *Tetrahedron,* **22,** Supplement No. 7, 69 (1966).

[32] E. R. Bissell and D. B. Fields, *J. Org. Chem.,* **29,** 1591 (1964).

extended to sensitive olefins by using inert solvents and careful temperature control.

Considerably more work is needed on the addition of fluorine to olefins. We predict that conditions of solvent, temperature, reagent, and catalyst can probably be worked out so that fluorine can be added selectively, in high yield, with stereochemical control, to very sensitive olefins.

Several unusual and selective additions of fluorine to heteroatom systems should be mentioned at this point. Stevens[33] reported the reaction of IF_5 with alkyl or aryl isothiocyanates in pyridine to give an unusual sulfide **14** in

$$RNCS + IF_5 \longrightarrow \left[\begin{matrix} CF_3 \\ | \\ RN \end{matrix} \right]_2 S$$

14

yields of about 20%. This product could arise through a series of additions of F_2 or F· with C–S bond cleavage to form $RN{=}CF_2$ and SF_2; the final product is then formed by a fluoride ion catalyzed addition of SF_2 to two molecules of $RN{=}CF_2$.

Oxidative fluorination of aryl disulfides to arylsulfur trifluoride is another example of a selective fluorination of a heteroatom with no attack

$$(ArS)_2 + 6AgF_2 \xrightarrow{CCl_2FCF_2Cl} 2ArSF_3 + 6AgF$$

on an aromatic C–H system.[34] If the aromatic ring is deactivated by an electron-withdrawing group, further addition of F_2 from AgF_2 to give the arylsulfur pentafluoride[35] has also been demonstrated. Other examples of

$$ArSF_3 + 2AgF_2 \xrightarrow{120^\circ} ArSF_5 + 2AgF$$

fluorine addition to a heteroatom are arylphosphine difluoride with antimony trifluoride to give arylphosphine tetrafluoride[36] and the reaction of triphenyl phosphine with N_2F_2, N_2F_4, or SF_4 to get triphenyl difluorophosphorane.[37]

$$3C_6H_5PF_2 \xrightarrow{2SbF_3} 3C_6H_5PF_4$$

$$(C_6H_5)_3P \xrightarrow[N_2F_2 \text{ or } N_2F_4]{SF_4} (C_6H_5)_3PF_2$$

[33] (a) T. E. Stevens, *Tetrahedron Letters*, No. 17, 16 (1959). (b) T. E. Stevens, U.S. Patent 3,025,324 (1962); *Chem. Abstr.*, **57**, 7107h (1962).

[34] W. A. Sheppard, *J. Am. Chem. Soc.*, **84**, 3058 (1962).

[35] W. A. Sheppard, *J. Am. Chem. Soc.*, **84**, 3064 (1962).

[36] R. Schmutzler, *Chem. Ber.*, **98**, 552 (1965).

[37] (a) W. C. Firth, S. Frank, M. Garber, and V. P. Wystrach, *Inorg. Chem.*, **4**, 765 (1965). (b) W. C. Smith, *J. Am. Chem. Soc.*, **82**, 6176 (1960).

Many examples of preparation of inorganic fluorine compounds (containing no carbon) by selective fluorine addition are known, for example, the preparation of sulfur chloride pentafluoride from SCl_2 and a fluorinating agent,[38] but will not be reviewed in this work.

Summary

1. Selective addition of elementary F_2 to an unsaturated system can be accomplished using special techniques such as low temperature and high dilution that promote an ionic reaction, but these methods are considered to have limited utility.

2. By use of a fluorine carrier, selective addition of F_2 can be done in many unsaturated systems; PbF_4 [in situ from lead(IV) acetate and HF] appears to be the best reagent, but side reactions and rearrangements do occur. Much more work is needed to refine these methods.

5-2 ADDITION OF XF TO UNSATURATED GROUPS

$$XF + R_2C{=}CR_2 \longrightarrow \underset{X}{\underset{|}{R_2C}}{-}\underset{F}{\underset{|}{CR_2}}$$

$$XF + R_2C{=}Y \longrightarrow \overset{F}{\overset{|}{R_2C}}{-}YX$$

$$XF + R_2C\overset{O}{\frown}CR_2 \longrightarrow \underset{F}{\underset{|}{R_2C}}{-}\overset{OX}{\overset{|}{CR_2}}$$

This section covers a large area where X can be any group, fluorinated or nonfluorinated, and is organized by periodic classification of the element that forms the bond to carbon (or the heteroatom) of the unsaturated reactant.

5-2A. X IS H

The addition of hydrogen fluoride to unsaturated centers is a classical method of introducing fluorine and was reviewed in Section 4-3. Since this is an important way to introduce a single fluorine, we want to reiterate briefly the scope and limitations of the method and to point out new developments. An important consideration is that HF is highly suited to industrial operations

[38] H. L. Roberts and N. H. Ray, *J. Chem. Soc.*, 665 (1960).

because it is moderately cheap and easily handled in gas-phase flow systems over catalyst beds. It is now commonly used in conventional organic laboratories because plastic laboratory equipment, which duplicates much of the standard glassware of the organic chemist, is available from laboratory supply houses. However, pressure equipment may be required for high-temperature reactions.

The addition of HF to olefins and acetylenes often needs a catalyst or a solvent that promotes ionization. For electrophilic olefins (fluoroolefins), fluoride ion adds to give a carbanion intermediate. Nucleophilic olefins (hydrocarbon substituted) add a proton to give a stabilized carbonium ion intermediate.

$$H^+ + AF^- \xleftarrow{A} HF \xrightarrow{B} HB + F^-$$

$$RCX{=}CX_2 + F^- \longrightarrow R\bar{C}XCX_2F \xrightarrow{H^+} RC(H)XCX_2F$$

$$RCH{=}CH_2 + H^+ \longrightarrow R\overset{+}{C}HCH_3 \xrightarrow{F^- \text{ or } HF_2^-} RCH(F)CH_3$$

Of course, these are oversimplified pictures and represent the extreme cases, but they are useful to guide the choice of conditions and predict direction of addition. Examples of both types are given in Section 4-3.

Ring opening with HF addition is an important extension of addition to unsaturated systems. A recent development is the addition of HF to epoxides which has been particularly useful in steroid chemistry. The hydrogen fluoride can be used directly with an ionizing basic solvent, such as water or an ether[39] (now referred to as standard procedure), a salt such as $NaHF_2$ or amine salt,[39f, 40] or indirectly by using boron trifluoride etherate with hydrolysis in the work-up.[39f, 41] In the first two cases, the addition could go by two courses since the HF must be ionized to some extent by the solvent or base.

[39] (a) I. L. Knunyants, O. V. Kil'dasheva, and I. P. Petrov, *J. Gen. Chem. USSR (English Transl.)*, **19**, 87 and 93 (1949). (b) R. F. Hirschmann, R. Miller, J. Wood, and R. E. Jones, *J. Am. Chem. Soc.*, **78**, 4956 (1956). (c) V. Schwarz and K. Syhora, *Collection Czech. Chem. Commun.*, **28**, 637 (1963). (d) R. Deghenghi and R. Gaudry, *Can. J. Chem.*, **39**, 1553 (1961); see also H. L. Herzog, M. J. Gentles, H. M. Marshall, and E. B. Hershberg, *J. Am. Chem. Soc.*, **82**, 3691 (1960). (e) J. E. Pike, G. Slomp, and F. A. MacKellar, *J. Org. Chem.*, **28**, 2502 (1963). (f) I. G. Reshetova and A. A. Akhrem, *Bull. Acad. Sci. USSR, Div. Chem. Sci. (English Transl.)*, 68 (1965).

[40] (a) J. V. Karabinos and J. J. Hazdra, *J. Org. Chem.*, **27**, 3308 (1962). (b) G. Aranda, J. Jullien, and J. A. Martin, *Bull. Soc. Chim. France*, 1890 (1965).

[41] (a) A. Bowers, L. C. Ibáñez, and H. J. Ringold, *Tetrahedron*, **7**, 138 (1959); *J. Am. Chem. Soc.*, **81**, 5991 (1959). (b) A. A. Akhrem and I. G. Reshetova, *Bull. Acad. Sci. USSR, Div. Chem. Sci. (English Transl.)*, 102 (1963). (c) M. P. Hartshorn, D. N. Kirk, and A A. F. A. Wallis, *J. Chem. Soc.*, 5494 (1964). (d) H. B. Henbest and T. I. Wrigley, *J. Chem. Soc.*, 4765 (1957). (e) B. J. May, F. A. Nice, and G. H. Phillipps, *J. Chem. Soc. (C)*, 2210 (1966).

$$R_2C(-O-)CR_2 \xrightarrow{(a)\ H^+} R_2C(OH)-\overset{+}{C}R_2 \xrightarrow{F^-} R_2C(HO)-CR_2(F)$$

$$R_2C(-O-)CR_2 \xrightarrow{(b)\ F^-} R_2C(O^-)-CR_2(F) \xrightarrow{H^+} R_2C(HO)-CR_2(F)$$

Route a by proton opening of the ring is probable in ether solvents, but the reaction could be concerted with simultaneous attack by fluoride ion. However, the reaction with amine salts is probably by fluoride ion attack (path **b**) and would predict trans addition as is found.[40b] The BF_3–etherate system could give an $F{-}BF_2$ adduct which hydrolyzes with the aqueous base in work-up.[41d, e]

$$R_2C(-O-)CR_2 + BF_3 \longrightarrow \left[R_2C(OBF_3^-)-\overset{+}{C}R_2 \right] \xrightarrow[F^-\ \text{transfer}]{\text{inter- or intramolecular}} R_2C(OBF_2)-CR_2(F) \xrightarrow[H_2O]{OH} R_2C(OH)-CR_2(F)$$

Such an adduct could arise from a ring-opened intermediate from BF_3 attack and was suggested by side reactions such as hydride transfer,[41c] although trans opening found in some cases suggest a concerted process.[41d] Much more definitive studies are needed to define the mechanism of these useful synthetic reactions.

Hydrogen fluoride addition-opening of epoxy and larger ether rings in nucleosides[42] and sugars[43] has also been reported useful to prepare monofluoronucleosides and sugars for biological studies.

Numerous papers and over three dozen patents have appeared in recent years reporting or claiming these methods of HF addition to epoxy compounds, particularly to epoxy steroids where a large variety of potentially useful monofluoro steroids have been prepared. A representative selection of examples is given in Table 5-1 in order to show the synthetic potential of this reaction.

[42] (a) N. F. Taylor, R. F. Childs, and R. V. Brunt, *Chem. Ind. (London)*, 928 (1964). (b) S. Cohen, D. Levy, and E. D. Bergmann, *Chem. Ind. (London)*, 1802 (1964).

[43] (a) J. F. Codington, I. L. Doerr, and J. J. Fox, *J. Org. Chem.*, **29**, 558; 564 (1964). (b) J. F. Codington, I. L. Doerr, D. V. Praag, A. Bendich, and J. J. Fox, *J. Am. Chem. Soc.*, **83**, 5030 (1961).

Other miscellaneous extensions of hydrogen fluoride addition by ring opening are to a siloxane,[44] lactones,[45, 46] and cyclopropanes.[47] (See Table 5-1 for details.)

The addition of HF to heteroatom systems is of limited utility because of the instability of the products. For example, α-fluoroalcohols are in general much too unstable to isolate under normal conditions of temperature and pressure, and only unusual types such as heptafluorocyclobutanol are known.[48] Fluorinated

$$\text{(perfluorocyclobutanone)} + HF \longrightarrow \text{(heptafluorocyclobutanol: ring } F_2, F_2, F_2; \text{ C(F)(OH))}$$

tertiary amines are stable, but α-fluorinated primary and secondary amines readily lose HF. Some highly fluorinated ones, such as bis(trifluoromethyl)amine or *N*-aryl-*N*-(trifluoromethyl)amine, are readily prepared by HF addition to the azomethines but lose HF easily on heating or with an aprotic base.[49] An unusual

$$RN{=}CF_2 + HF \rightleftharpoons RNHCF_3$$
$$R = \text{Aryl or } CF_3$$

fluorinated amine, SF_5NH_2, was prepared by addition of 2 moles of HF to the S≡N bond in thiazyl trifluoride, NSF_3.[50]

5-2B. BONDING ATOM OF X IS A METAL

Silver fluoride is reported to catalyze the addition of hydrogen fluoride to fluoroolefins.[51] The reaction was suggested to go by way of addition of metal ion to the fluoroolefin followed by reaction with $H_nF_{n+1}^-$; an alternative path is

[44] E. J. P. Fear, J. Thrower, and I. M. White, *Chem. Ind. (London)*, 1877 (1961).

[45] G. A. Olah and S. J. Kuhn, *J. Org. Chem.*, **26**, 225 (1961).

[46] H. Beyer, U. Hess, and P. Bernhardt, *Chem. Ber.*, **96**, 2193 (1963).

[47] E. T. McBee, H. B. Hass, R. M. Thomas, W. G. Toland, Jr., and A. Truchan, *J. Am. Chem. Soc.*, **69**, 944 (1947).

[48] S. Andreades and D. C. England, *J. Am. Chem. Soc.*, **83**, 4670 (1961).

[49] (a) K. A. Petrov and A. A. Neimysheva, *J. Gen. Chem. USSR (English Transl.)*, **29**, 2135 (1959). See also W. A. Sheppard, *J. Am. Chem. Soc.*, **87**, 4338 (1965). (b) D. A. Barr and R. N. Haszeldine, *J. Chem. Soc.*, 2532 (1955). (c) J. A. Young, S. N. Tsoukalas, and R. D. Dresdner, *J. Am. Chem. Soc.*, **80**, 3604 (1958). (d) R. E. Banks, W. M. Cheng, and R. N. Haszeldine, *J. Chem. Soc.*, 2485 (1964).

[50] A. F. Clifford and L. C. Duncan, *Inorg. Chem.*, **5**, 692 (1966).

[51] W. T. Miller, Jr., M. B. Freedman, J. H. Fried, and H. F. Koch, *J. Am. Chem. Soc.*, **83**, 4105 (1961).

Table 5-1

Ring Opening Addition of HF to Epoxides and Other Cyclic Compounds

Cyclic reactants	Fluorinating agent and conditions	Product(s) (% yield)	Comments and ref.
Ethylene oxide	Anhydrous HF in dilute ether solution (5%) at reflux	$HOCH_2CH_2F$ (40)	A series of other aliphatic fluorohydrins also prepared from epoxides; shown to be useful intermediates. Ref. 39a
Aliphatic epoxides, includes cyclohexene, isobutylene, pinene, limonene	Anhydrous HF in ether	Fluorohydrins, plus by-products in some cases	See comparison to work in Ref. 40a. Ref. 39g
$9\beta,11\beta$-epoxy-pregnene derivatives	Anhydrous HF in tetrahydrofuran, steroid added in chloroform at $-60°$	9α-Fluoro-11β = hydroxy pregnenes (75)	Noted use of NH_4F in anhydrous HF also works. Suggest get ionization and source of F^- in THF. Ref. 39b,e; see also 39h
$16\alpha,17\alpha$-Epoxy steroids (also $5\alpha,6\alpha$ and $11\beta,12\beta$)	Aq. HF in acetone at $-60°$	16β-F-17α-Hydroxy steroid (74-98)	General procedure for adding hydrohalic acids. Ref. 39c
20-Cyano-17,20-epoxypregnenolone	48% aq. HF	17α-Fluoro-20-cyanohydrin derivative	Reaction did not go with anhydrous HF in THF. Ref. 39d
$5\alpha,6\alpha$:20ξ,21-Diepoxypregnane-$3\beta,17\beta$-diol 3-acetate	(a) HF (aq. and anhydrous) (b) KHF_2 (c) BF_3 etherate	Fluorohydrins with fluorine at 6β, or 6β and 20ξ positions	Compared relative rates of attack at two different epoxides (5,6-more reactive and gives less by-product). Studied yields relative to reagent and conditions. Ref. 39f
Cyclohexene oxide	$NaHF_2$, 150-180°	2-Fluorocyclohexanol (21)	Anhydrous HF in ether stated as unsuccessful but note Ref. 39g. Ref. 40a
C_6H_5CH—CR_2 (epoxide O bridging), R_1 on CR_2 carbon (where $R_1 = R_2 = H$ $R_1 = R_2$ = alkyl $R_1 = H, R_2$ = alkyl or aryl)	Tested a series of primary, secondary, and tertiary amines with HF. Diisopropyl amine·3HF at 105° preferred	Corresponding fluorohydrin (62-88)	Stereochemistry studied showed trans addition of HF in opening epoxide. Ref. 40b
Steroid $5\alpha,6\alpha$-epoxides	BF_3-etherate in benzene-ether, standing at room temp. Work-up by sodium carbonate wash	6-Fluoro-5-hydroxysteroids (40 to 50)	This is a follow-up of earlier reports. Ref. 41a
Steroids with epoxides at 5,6- and 20,21-positions	BF_3-etherate, ether-benzene. Stand 6 hr at room temp. Work-up with Na_2CO_3 wash	6- and 21-Fluorohydrins of steroids (60-96)	Ref. 41b
Pinene oxide	BF_3 = etherate added to cold solution of pinene oxide in ether	[structure: cyclohexene ring with OH and F-bearing isopropyl group] (18) + 6 other products	Found 50% yield of fluoroalcohol using HF in ether. Ref. 41c

16α,17α-epoxy-16β-methyl-5α-pregnan-20-one	BF_3 etherate in (a) dioxane (b) benzene	(a) Corresponding 16-methylene-D-homo-compound (no fluorine) (b) D-homo-fluorohydrin	Mechanism discussed; rearrangement products explained. Ref. 41e
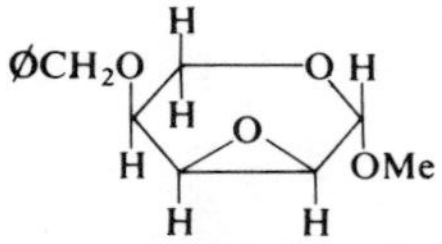 [structure: ØCH₂O, OMe] Benzyl-1-O-tosyl-β-O-arabinopyranoside	Anhydrous HF in dioxane, 120°/24 hr	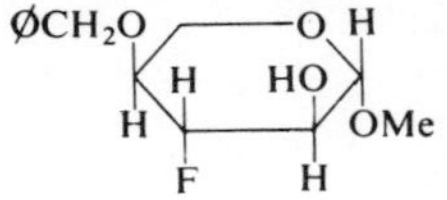 [structure: ØCH₂O, HO, F, OMe]	Ref. 42a
	Anhydrous KF in molten acetamide, 200°	Benzyl-3-desoxy-3-fluoro-β-D-xylopyranoside (50)	Ref. 42b
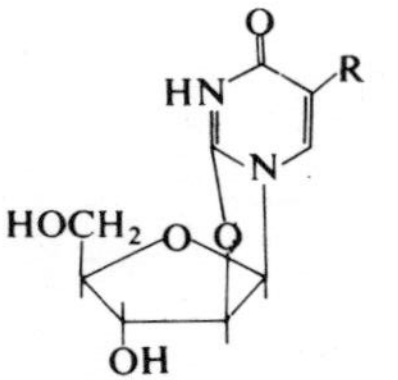 [structure: HOCH₂, HN, N, R, OH] and related nucleosides	Anhydrous HF in dioxane; heat at 120°	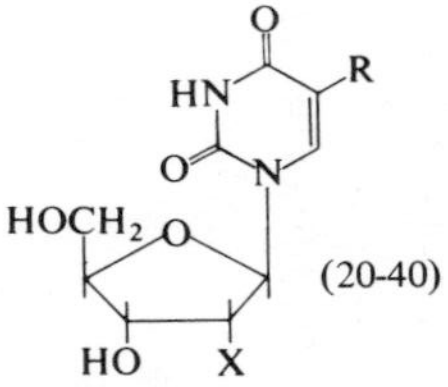 [structure: HOCH₂, HN, N, R, HO, X] (20-40)	Also with other hydrogen halides suggested mechanism is protonation on uridine. Ref. 43
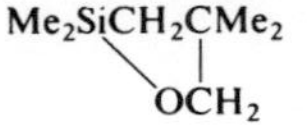 $Me_2SiCH_2CMe_2$ with OCH_2 ring	40% aq. HF, stand	$FSi(Me)_2CH_2C(Me)_2CH_2OH$ (79-84)	Ref. 44
Diketene 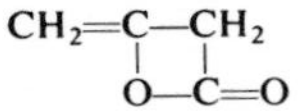 $CH_2{=}C{-}CH_2$, $O{-}C{=}O$	Anhydrous HF at below −5°	$CH_3C(=O)CH_2C(=O)F$ (65)	Useful acetylating agent. Ref. 45
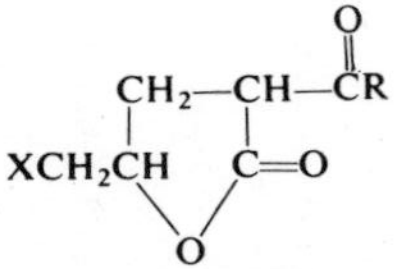 $XCH_2CH{-}CH_2{-}CH{-}C(=O)R$ with $C{=}O$, O (lactone ring)	Aq. HF	$XCH_2{-}CH(F){-}CH_2CH_2C(=O){-}R$ $R = C_6H_5$ or CH_3 X = Cl or Br	Also used HCl and HBr in place of HF. Ref. 46
1,1-Dichlorocyclopropane	Anhydrous HF, 135°, 56 atm	$CClF_2CH_2CH_3$ (25) $CF_3CH_2CH_3$ (47)	Replacement of Cl by F could occur either before or after ring opening. Ref. 47

[39](g) G. Farges and A. Kergomard, *Bull. Soc. Chim. France,* **51** (1963). (h) Several papers on the follow-up applications of HF in THF to steroid epoxides are: P. A. Diassi, J. Fried, R. M. Palmere, and E. F. Sabo, *J. Am. Chem. Soc.,* **83**, 4249 (1961); B. J. Magerlein, R. D. Birkenmeyer, and F. Kagan, *J. Am. Chem. Soc.,* **82**, 1252 (1960); C. G. Bergstrom, R. T. Nicholson, R. L. Elton, and R. M. Dodson, *J. Am. Chem. Soc.,* **81**, 4432 (1959); U. Valcavi and L. Sianesi, *Gazz. Chim. Ital.,* **93**, 309 (1963); B. J. Magerlein, F. H. Lincoln, R. D. Birkenmeyer, and F. Kagan, *J. Med. Chem.,* **7**, 748 (1964).

addition of F^- to give a stable fluorocarbanion which can associate with the silver ion; in either case the adduct $R_f^- Ag^+$ must form. A stable silver fluoride adduct[52]

$$Cl_2C{=}C(CN)_2 + 2AgF \longrightarrow [F_2C{=}C(CN)_2] \xrightarrow{AgF} F_3C\bar{C}(CN)_2Ag^+ \xrightarrow{HCl} F_3CCH(CN)_2$$

is obtained by reacting 1,1-dichloro-2,2-dicyanoethylene with AgF; the strong electron-withdrawing cyano groups must stabilize this anion and promote AgF addition so that the intermediate difluoroethylene cannot be isolated. Adducts of cesium fluoride with fluoroolefins have been postulated as intermediates for anionic polymerization of fluoroolefins,[53] addition to fluoroketones,[54] and in carbonation to perfluorocarboxylic acids.[55]

$$R_fCF{=}CF_2 + CsF \xrightarrow[\text{glycol}]{\text{triethylene}} R_f\bar{C}FCF_3 \xrightarrow{(CF_3)_2C{=}O} R_fC(F)(CF_3){-}C(OH)(CF_3)_2$$

$$R_f\bar{C}FCF_3 \xrightarrow{CO_2} R_fC(CF_3)CO_2H$$

Cesium fluoride adds to carbonyl fluoride to give a stable trifluoromethoxide salt[56]:

$$CsF + COF_2 \rightleftharpoons F_3CO^-Cs^+$$

Unfortunately this salt decomposes on attempted alkylation and has not proved useful synthetically to introduce an OCF_3 group. However, hexafluoroacetone

[52] A. D. Josey, C. L. Dickinson, K. C. Dewhirst, and B. C. McKusick, *J. Org. Chem.*, **32**, 1941 (1967).

[53] D. P. Graham, *J. Org. Chem.*, **31**, 955 (1966).

[54] D. P. Graham and V. Weinmayr, *J. Org. Chem.*, **31**, 957 (1966).

[55] D. P. Graham and W. B. McCormack, *J. Org. Chem.*, **31**, 958 (1966).

[56] D. C. Bradley, M. E. Redwood, and C. J. Willis, *Proc. Chem. Soc.*, 416 (1964).

(HFA) with potassium fluoride does form a stable salt which can be trapped and alkylated.[57]

$$(CF_3)_2C{=}O + KF \longrightarrow (CF_3)_2\overset{F}{\overset{|}{C}}O^-K^+$$

$$(CF_3)_2CFOK \xrightarrow{CH_2{-}CHCH_2Br\ (\text{epoxide})} (CF_3)_2CFOCH_2CH{-}CH_2\ (\text{epoxide})$$

(products used as polymer intermediates)

$$(CF_3)_2CFOK \xrightarrow{CH_2{=}CHCCl\ (C{=}O)} CH_2{=}CHCOCF(CF_3)_2 \quad (80\text{-}90\%)$$

A by-product in the reaction with acid halides was the corresponding acid fluoride, and evidence was provided that the fluorination went through decomposition of the ester product and was the main reaction course if the acyl halide

$$R_f{-}\overset{O}{\overset{\|}{C}}{-}O{-}C(CF_3)_2F$$

contained a strong electron-withdrawing group such as fluoralkyl or if a chlorosilane was used.[57c] Surprisingly, the cesium fluoride adduct with HFA is less stable than with fluoroolefins, since when CsF, HFA, and tetrafluoroethylene (TFE) were combined in diglyme (TFE added last), only the tertiary alcohol was produced and no other product corresponding to the formation of the intermediate anion[54]:

$$FC(CF_3)_2OCF_2CF_2^-$$

The addition of mercury(II) fluoride to fluoroolefins to give stable perfluoroalkyl mercury derivatives is a well-studied reaction.[58] Nucleophilic addition

[57](a) A. G. Pittman and D. L. Sharp, *J. Polymer Sci.*, **B3**, 379 (1965). (b) A. G. Pittman and D. L. Sharp, *Textile Res. J.*, **35**, 190 (1965). (c) A. G. Pittman and D. L. Sharp, *J. Org. Chem.*, **31**, 2316 (1966).

[58](a) P. E. Aldrich, E. G. Howard, W. J. Linn, W. J. Middleton, W. H. Sharkey, *J. Org. Chem.*, **28**, 184 (1963). (b) W. T. Miller, Jr., and M. B. Freedman, *J. Am. Chem. Soc.*, **85**, 180 (1963). (c) H. Goldwhite, R. N. Haszeldine, and R. N. Mukherjee, *J. Chem. Soc.*, 3825 (1961). (d) C. G. Krespan, *J. Org. Chem.*, **25**, 105 (1960).

$$HgF_2 + R_fC(X){=}CF_2 \longrightarrow \left(R_fC(X)(CF_3)\right)_2Hg$$

of fluoride ion was suggested as the most reasonable mechanism for addition,[58c] but Miller[51, 58b] presented strong arguments in favor of an electrophilic attack by mercury(II) ion on the π electrons of the olefinic bond (electrophilic reactions of π electrons are a previously unrecognized but important area of CF chemistry). He suggests a mechanistic representation where a cyclic mercurinium ion is formed as intermediate. (A concerted reaction is also possible.) The essential point is that while perfluoroolefins are unreactive with the strongly solvated

$$\rangle C{=}C\langle + Hg^{++} \rightleftharpoons \left[\rangle C{\cdots}Hg{\cdots}C\langle\right]^{++} + F^- \longrightarrow -\overset{|}{C}-Hg^+,\ F\overset{|}{C}-$$

$$\rangle C{=}C\langle + -\overset{|}{C}(-Hg^+)-\overset{|}{C}F- \rightleftharpoons \left[\rangle C{\cdots}Hg{\cdots}C\langle\ Hg-\overset{|}{C}-\overset{|}{C}F-\right]^{+} + F^- \longrightarrow FC-\overset{|}{C}-Hg-\overset{|}{C}-CF$$

fluoride ion in HF solution, sufficient electrophilic reactivity is developed by the addition of mercury(II) cation to the double bond to bring about reaction of fluoride ion at carbon. The addition of fluoride ion to the terminal carbon atom is favored sterically and energetically. As a follow-up to Miller's discussion, the attack of mercury(II) ion on terminal fluoroolefins was cited as evidence that difluorocarbonium ion $R-\overset{+}{C}F_2$ is more stable than the monofluorocarbonium ion $R_2-\overset{+}{C}F$.[58a] The fluoroalkyl mercury derivatives are useful intermediates to a variety of fluoroalkyl derivatives.[59, 60] Mercury(II) fluoride also adds to CS_2, splitting off sulfur,[60] to give $(CF_3S)_2Hg$ (which is a useful intermediate to supply a CF_3S group[60, 61]) and to the perfluoroazomethine $CF_3N{=}CF_2$ to give $(CF_3NCF_3)_2Hg$.[49c]

[59] (a) W. J. Middleton, E. G. Howard, and W. H. Sharkey, *J. Am. Chem. Soc.*, **83**, 2589 (1961). (b) J. A. Young and R. L. Dressler, *J. Org. Chem.*, **32**, 2237 (1967).

[60] (a) E. L. Muetterties, U.S. Patent 2,729,663 (1956); *Chem. Abstr.*, **50**, 1362c (1956). (b) E. H. Man, D. D. Coffman, and E. L. Muetterties, *J. Am. Chem. Soc.*, **81**, 3575 (1959).

[61] R. N. Haszeldine and J. M. Kidd, *J. Chem. Soc.*, 3219 (1953).

5-2C. BONDING ATOM OF X IS FROM GROUP III

Addition of BF_3 to epoxides, which on hydrolysis effectively gives HF addition, was suggested to go by way of an adduct.[41d] Complexes of boron

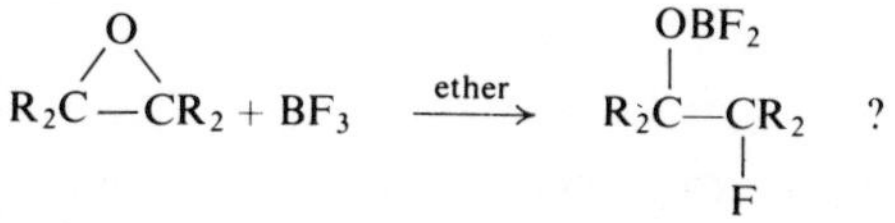

$$R_2C(-O-)CR_2 + BF_3 \xrightarrow{\text{ether}} R_2C(OBF_2)-CR_2F \quad ?$$

difluoride with β-diketones have been isolated and characterized.[62] They could form by 1,2-addition of BF_3 to the carbonyl followed by elimination of HF or condensation of BF_3 directly with the enol form.

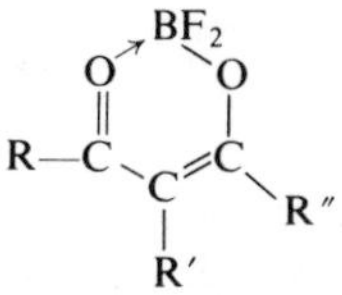

$$R-C(=O\rightarrow BF_2)-C(R')=C(-O-)R''$$

Although boron trichloride adds readily to olefins,[63] boron trifluoride is apparently too inert.

5-2D. BONDING ATOM OF X IS FROM GROUP IV

The types of reagents that have been added are COF_2, $RN{=}CF_2$, and FCH_2OH (from HF + CH_2O). The additions are believed to be fluoride ion catalyzed and, consequently, two-step: (*1*) addition of fluoride ion to give a nucleophilic anionic intermediate; and (*2*) nucleophilic displacement by this intermediate on the substrate

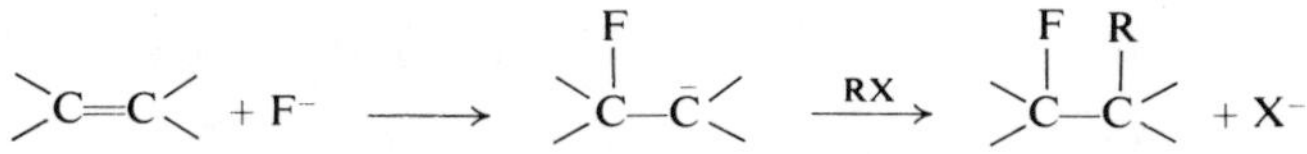

$$>C{=}C< + F^- \longrightarrow >CF-\bar{C}< \xrightarrow{RX} >CF-CR< + X^-$$

Miller[64] has provided definitive evidence for formation of fluorocarbanions from addition of fluoride ion to fluoroolefins. The fluorocarbanion intermediate can undergo a variety of reactions, and we will discuss these in detail in Section 6-1C.

[62] A. N. Sagredos, *Ann.*, **700**, 29 (1966).

[63] See discussions by E. L. Muetterties, *J. Am. Chem. Soc.*, **82**, 4163 (1960).

[64] W. T. Miller, Jr., J. H. Fried, and H. Goldwhite, *J. Am. Chem. Soc.*, **82**, 3091 (1960).

Some specific examples of nucleophilic displacements by these carbanions (which is effectively XF addition) are addition of COF_2 to fluoroolefins,[65] isocyanates,[65] or imines[66] and dimerization of difluoroazomethines.[66] In general, a source of fluoride ion must be provided in an aprotic system; cesium fluoride in acetonitrile, glyme, or dimethylformamide is probably the best available, but most of the

$$R_fCF{=}CF_2 + COF_2 \xrightarrow[\text{(from CsF)}]{F^-} R_fC(F)(CF_3){-}C(=O)F$$

$$RN{=}C{=}O + COF_2 \xrightarrow{CsF} RN(\overset{O}{\overset{\|}{C}}F)_2$$

$$RN{=}CF_2 \xrightarrow[CsF]{COF_2} RN(CF_3){-}\overset{O}{\overset{\|}{C}}F$$

$$RN{=}CF_2 \xrightarrow[RN{=}CF_2]{F^-} RN(CF_3){-}CF{=}NR$$

work has been done with potassium fluoride in formamide and with tetraethylammonium fluoride in chloroform or acetone. Since the intermediate anion must also be stabilized by electron-withdrawing groups, only fluoroolefins or electronegative heteroatom systems are usually reactive in this system. Some other examples of this type of reaction with a variety of modifications are

$$CF_3CF{=}CF_2 + HF + H\overset{O}{\overset{\|}{C}}H \longrightarrow (CF_3)_2CFCH_2OH \qquad (67)$$

$$2COF_2 + \underset{\text{(ethylene oxide)}}{CH_2{-}O{-}CH_2} \xrightarrow{\text{pyridine}} CF_3OCH_2CH_2O\overset{O}{\overset{\|}{C}}F \qquad (68)$$

In the first example the intermediate fluorocarbanion must add to formaldehyde and the adduct then protonates. In the second example, F^- from ionization of COF_2 by pyridine solvent is postulated to add to COF_2 to give CF_3O^-,

[65] F. S. Fawcett, C. W. Tullock, and D. D. Coffman, *J. Am. Chem. Soc.*, **84**, 4275 (1962).

[66] W. A. Sheppard, *J. Am. Chem. Soc.*, **87**, 4338 (1965).

[67] V. Weinmayr, U.S. Patent 2,999,884 (1961); *Chem. Abstr.*, **56**, 325i (1962).

[68] F. S. Fawcett, unpublished results; see P. E. Aldrich and W. A. Sheppard, *J. Org. Chem.* **29**, 11 (1964), Table II footnote b.

which opens the ethylene oxide; the resulting intermediate $CF_3OCH_2CH_2O^-$ now displaces fluoride from COF_2.

5-2E. BONDING ATOM OF X IS FROM GROUP V

Nitrogen trifluoride does not dissociate readily and cannot be directly added to an olefin. However, the effective addition of FNF_2 occurs by irradiation of N_2F_4 in the presence of an olefin.[69] Substitution of hydrogen by NF_2 also occurs to an equal extent and with acetylenes a rearrangement product is formed. The proposed mechanism is shown in Chart 5-2.

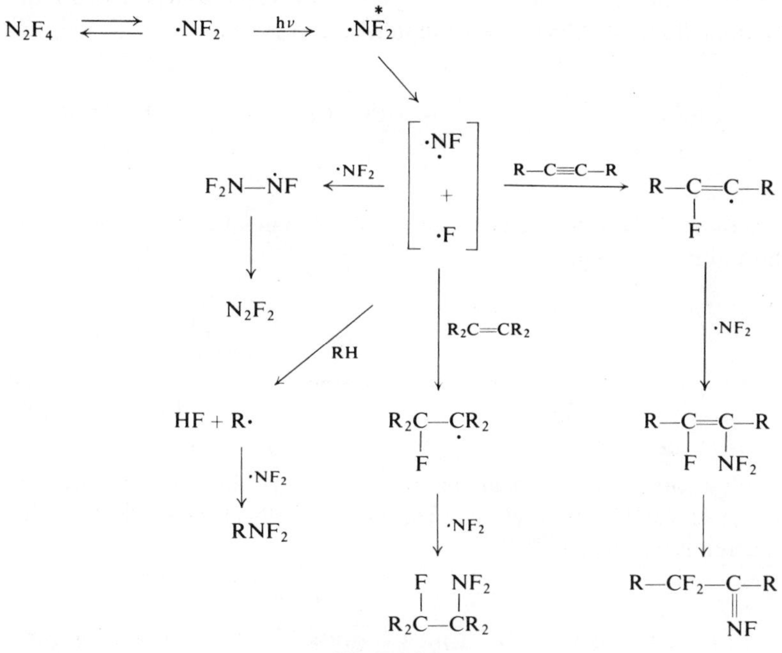

$N_2F_4 \rightleftharpoons \cdot NF_2 \xrightarrow{h\nu} \cdot NF_2^*$

$\cdot NF_2^* \rightarrow [\cdot\dot{N}F + \cdot F]$

$[\cdot\dot{N}F + \cdot F] \xrightarrow{\cdot NF_2} F_2N{-}\dot{N}F \rightarrow N_2F_2$

$[\cdot\dot{N}F + \cdot F] \xrightarrow{RH} HF + R\cdot \xrightarrow{\cdot NF_2} RNF_2$

$[\cdot\dot{N}F + \cdot F] \xrightarrow{R_2C{=}CR_2} R_2C(F){-}\dot{C}R_2 \xrightarrow{\cdot NF_2} R_2C(F){-}CR_2(NF_2)$

$[\cdot\dot{N}F + \cdot F] \xrightarrow{R{-}C{\equiv}C{-}R} R{-}C(F){=}\dot{C}{-}R \xrightarrow{\cdot NF_2} R{-}C(F){=}C(NF_2){-}R \rightarrow R{-}CF_2{-}C({=}NF){-}R$

CHART 5-2

The addition of various fluoronitrogen oxides to unsaturated centers is well established.[70a] Nitrosyl fluoride adds to fluoroolefins probably by a two-step mechanism with fluoride ion adding initially, but the reaction conditions are

[69] C. L. Bumgardner, *Tetrahedron Letters*, 3683 (1964).

[70] (a) For a review of this area, see B. L. Dyatkin, E. P. Mochalina, and I. L. Knunyants, *Russ. Chem. Rev. (English Transl.)*, **35**, 417 (1966). (b) S. Andreades, *J. Org. Chem.*, **27**, 4157, 4163, (1962). (c) B. L. Dyatkin, E. P. Mochalina, R. A. Bekker, and I. L. Knunyants, *4th Intern. Symp. Fluorine Chem., Estes Park, Colorado, 1967*, Abstracts p. 110. (d) I. L. Knunyants, B. L. Dyatkin, E. P. Mochalina, R. A. Bekker, and S. R. Sterlin, *Bull. Acad. Sci., USSR, Div. Chem. Sci. (English Transl.)*, 564 (1966). (e) B. L. Dyatkin, E. P. Mochalina, R. A. Bekker, S. R. Sterlin, and I. L. Knunyants, *Tetrahedron*, **23**, 4291 (1967).

critical and further reaction or decomposition of the nitrosofluoroalkane can occur.[70] Andreades[70b] found that tetrafluoroethylene gave chiefly an oxazetidine, but that for hexafluoropropylene a complex mixture of azomethines and nitro and nitrile derivatives are formed.

$$CF_2{=}CF_2 + NOF \longrightarrow CF_3CF_2NO \xrightarrow{CF_2=CF_2} CF_3CF_2\underset{\substack{|\\ CF_2}}{N}{-}\underset{\substack{|\\ CF_2}}{O}$$

However, the Russian workers[70a, c, e] report that FNO adds cleanly to a variety of fluoroolefins with KF as catalyst in sulfolane as solvent.

$$CF_3CF{=}CF_2 + F^- \xrightarrow[\text{sulfolane}]{\text{KF, }36^\circ} (CF_3)_2CF^- \xrightarrow{^+NO} \underset{(88\%)}{(CF_3)_2CFNO}$$

Alternatively, N_2O_4 can be used directly with KF, but without KF, only dinitro or nitro nitrites are found.

$$\text{hexafluorocyclobutene } (F_2, F_2, F, F) + KF + N_2O_4 \longrightarrow \text{1-nitroso-1-fluoro-hexafluorocyclobutane } (F_2, F_2, F_2, F, NO) \quad (51\%)$$

Fluoride ion catalysis is ineffective with olefins where the fluorocarbanion is less stable (tetrafluoroethylene or trifluorovinyl ether) and an electrophilic mechanism is postulated.[70a, d, e]

$$\overset{\delta-}{CF_2}{=}\overset{\delta+}{CF}{-}OR \xrightarrow{^+NO} [ONCF_2{-}\overset{+}{C}FOR] \xrightarrow{F^-} ONCF_2CF_2OR$$

The nitrosofluoroalkanes are also readily prepared by a two-step method when mercury(II) fluoride is added to the fluoroolefin and the intermediate mercury compound is reacted with nitrosyl chloride.[71]

$$R_fCF{=}CF_2 + HgF_2 \longrightarrow (R_f\overset{\overset{CF_3}{|}}{C}F)_2Hg \xrightarrow{NOCl} R_f\overset{\overset{CF_3}{|}}{C}FNO$$

[71] P. Tarrant and D. E. O'Connor, *J. Org. Chem.*, 29, 2012 (1964).

Recently, the reaction of NOF in CCl_4 at 0°C with a cholesteryl acetate was reported to give a nitrimine derivative which could be readily hydrolyzed to the fluoroketone.[72]

$$\text{CH}_3\text{COO-(cholesteryl)} \xrightarrow[\text{CCl}_4,\ 0^\circ]{\text{NOF}} \text{X-(5-F, 6-=N-NO}_2\text{)} \xrightarrow{\text{H}_2\text{O}} \text{X-(5-F, 6-=O)}$$

Again the initial product from NOF addition must react further.

Addition of nitrosyl fluoride, and also nitryl fluoride, to fluoroketones goes readily at low temperatures to give α-fluoronitrites and nitrates, which are reactive compounds that have a wealth of chemistry.[70]

$$R_fC(=O)R_f \xrightarrow{\text{NOF}} R_fC(ONO)(F)R_f$$

$$R_fC(=O)R_f \xrightarrow{\text{NO}_2\text{F}} R_fC(ONO_2)(F)R_f$$

The reaction mixture of nitric acid with hydrogen fluoride gives addition of the elements of $F{-}NO_2$ to olefins and provides a general route to β-fluoronitro

$$>C{=}C< + HNO_3 + HF \longrightarrow >C(F){-}C(NO_2)<$$

compounds.[73, 70a] Yields for a number of olefins, usually with one electronegative substituent, were high, but fluoroolefins were sluggish and some did not react. A carbonium ion-type intermediate, formed by addition of $\overset{+}{N}O_2$ to olefin,[70a] is reasonable, based on experimental results. This carbonium ion would easily pick up a fluoride ion.

Arsenic trichloride reacts with tetrafluoroethylene in presence of aluminum chloride to give a product corresponding to addition of $F{-}AsCl_2$.[74] The reactants

[72] G. A. Boswell, Jr., *Chem. Ind. (London)*, 1929 (1965).

[73] I. L. Knunyants, L. S. German, and I. N. Rozhkov, *Bull. Acad. Sci. USSR, Div. Chem. Sci. (English Transl.)*, 1794 (1963).

[74] A. B. Bruker, T. G. Spiridonova, and L. Z. Soborovskii, *J. Gen. Chem. USSR (English Transl.)*, **28**, 347 (1958).

are heated under pressure and the initial reaction is proposed to be fluoride exchange between the fluoroolefin and $AlCl_3$;

$$CF_2{=}CF_2 + AlCl_3 \longrightarrow CF_2{=}CCl_2 + AlF_2Cl$$

difluorodichloroethylene is, indeed, isolated as a by-product in proportionate amount. A concerted process is proposed for the second step, but other mechanisms cannot be excluded without more information.

$$CF_2{=}CF_2 + AsCl_3 + AlF_2Cl \longrightarrow \left[\begin{array}{c} CF_2 \cdots\cdots CF_2 \\ \vdots \qquad\qquad \vdots \\ F \qquad\qquad AsCl_3 \\ \vdots \\ AlFCl \end{array}\right] \longrightarrow CF_3CF_2AsCl_2 + AlFCl_2$$

5-2F. BONDING ATOM OF X IS FROM GROUP VI

Addition of RO and F from a hypofluorite to an unsaturated center is now well known but has serious limitations and requires very carefully controlled conditions at low temperatures. The hypofluorite system is a potent oxidizing agent and like N–F compounds can give explosive reactions with C–H materials. For OF_2 the types of reaction described are suggested to go by addition of F

$$R_1R_2C{=}CR_3R_4 \xrightarrow{FOF} \left[R_1R_2C(F){-}C(OF)R_3R_4\right] \longrightarrow \left[R_1R_2C(F){-}C(O\cdot)R_3R_4\right]$$

$R_1, R_2 = H$; $R_3, R_4 = C_6H_5$: $\longrightarrow C_6H_5C(=O)CHFC_6H_5$ and $C_6H_5C(=O)CH_2F$

$R_1, R_2, R_3, R_4 = CH_3$: $\longrightarrow (CH_3)_2CFC(OH)(CH_3)_2$

$$RC{\equiv}CR' + OF_2 \longrightarrow RC(=O)CF_2R' + RCF_2C(=O)R'$$

and OF to the double bond followed by further reaction (decomposition or rearrangement).[75, 76] Fluoroolefins react with OF_2 only when irradiated or heated, and the products do not contain any fluoroalkoxy compound but chiefly acid fluorides and fluorocarbons arising from fragmentation of the intermediates.[76, 77] Low-temperature addition of OF_2 to allene gave two products that are proposed to arise from an initial adduct.[78]

$$(CH_3)_2C{=}C{=}C(CH_3)_2 + OF_2 \longrightarrow \left[(CH_3)_2C{=}\underset{\displaystyle OF}{C}{-}\underset{\displaystyle F}{C}(CH_3)_2\right] \longrightarrow$$

$$(CH_3)_2\overset{F}{C}{-}\overset{O}{\overset{\|}{C}}{-}\overset{F}{C}(CH_3)_2 + (CH_3)_2\overset{F}{C}{-}\overset{O}{\overset{\|}{C}}{-}\overset{CH_2}{\overset{\|}{C}}CH_3 + HF$$

We doubt that OF_2 will have any general utility as a synthetic reagent for fluorination because it is as hazardous to handle as elementary fluorine and is not readily available.

Addition of OF_2, CF_3OF, or SF_5OF to a variety of unsaturated carbon and heteroatom systems have been described.[79–86] Examples are

$$OF_2 + 2COF_2 \xrightarrow{MF} CF_3OOOCF_3 \quad \text{(polar mechanism? since } F^- \text{ needed)} \qquad (79)$$

$$OF_2 + 2SO_2 \xrightarrow{h\nu} FSO_2OSO_2F \qquad (81)$$

$$OF_2 + SO_3 \xrightarrow{h\nu} FSO_2OOF \qquad (80)$$

$$CF_3OF + COF_2 \xrightarrow{heat} CF_3OOCF_3 \qquad (82a)$$

$$CF_3OF + CH_2{=}CH_2 \xrightarrow{h\nu} CF_3OCH_2CH_2F \qquad (82b)$$

$$CF_3OF + SF_4 \xrightarrow{h\nu} CF_3OSF_5 \qquad (84)$$

$$SF_5OF + CH_2{=}CHF \xrightarrow{h\nu} F_5SOCH_2CHF_2 \qquad (83)$$

[75] R. F. Merritt and J. K. Ruff, *J. Org. Chem.*, **30**, 328 (1965).
[76] R. A. Rhein and G. H. Cady, *Inorg. Chem.*, **3**, 1644 (1964).
[77] J. K. Ruff and R. F. Merritt, *J. Org. Chem.* **30**, 3968 (1965).
[78] R. F. Merritt, *J. Org. Chem.*, **30**, 4367 (1965).
[79] L. R. Anderson and W. B. Fox, *J. Am. Chem. Soc.*, **89**, 4313 (1967).
[80] R. Gatti, E. H. Staricco, J. E. Sicre, and H. J. Schumacher, *Angew. Chem.*, **75**, 137 (1963).
[81] G. Franz and F. Neumayr, *Inorg. Chem.*, **3**, 921 (1964).
[82] (a) R. S. Porter and G. H. Cady, *J. Am. Chem. Soc.*, **79**, 5628 (1957). (b) J. A. C. Allison and G. H. Cady, *J. Am. Chem. Soc.*, **81**, 1089 (1959).
[83] S. M. Williamson, *Inorg. Chem.*, **2**, 421 (1963).
[84] L. C. Duncan and G. H. Cady, *Inorg. Chem.*, **3**, 850 (1964).
[85] W. P. Van Meter and G. H. Cady, *J. Am. Chem. Soc.*, **82**, 6005 (1960).
[86] C. I. Merrill and G. H. Cady, *J. Am. Chem. Soc.*, **83**, 298, (1961).

Other than the first reaction[79] which is believed to be ionic, all the rest probably are radical additions where the O–F bond dissociates homolytically under influence of heat or light.

Fluorine nitrate, NO_3F, also contains an OF group and adds readily to ethylene to give $FCH_2CH_2ONO_2$ which decomposes to evolve NO_2 on heating. With tetrafluoroethylene, only COF_2 and CF_3NO_2–the expected products from decomposition of $C_2F_5ONO_2$–are found.[87]

Addition of SF_4 and CF_3SF_3 to normal olefins has not been achieved, but fluoride ion catalyzed addition to fluoroolefins gives 1 : 1 and 2 : 1 adducts.[88]

$$CF_3CF{=}CF_2 + SF_4 \xrightarrow{F^-} [(CF_3)_2CF]_2SF_2 + (CF_3)_2CFSF_3$$

Reaction of SF_4 with isocyanates or nitriles gives iminosulfur difluorides and is suggested to go through the intermediate SF_4 adduct which decomposes or rearranges.[89a]

$$RNCO + SF_4 \longrightarrow \left[\begin{array}{c} O \\ \| \\ RN{-}CF \\ | \\ F_2S{-}F \end{array}\right] \longrightarrow RN{=}SF_2 + COF_2$$

$$R{-}C{\equiv}N + SF_4 \longrightarrow \left[\begin{array}{c} F \\ | \\ R{-}C{=}N \\ \quad | \\ F{-}SF_2 \end{array}\right] \longrightarrow RCF_2N{=}SF_2$$

Sulfur hexafluoride is too stable and inert to be activated for addition to any unsaturated system. However, when S_2F_{10} is heated with certain monohalo-olefins, the addition of $F{-}SF_5$ is accomplished.[89b] The yields are very low (less than 10%) and the product structure was only inferred from mass spectrometric

$$F_5SSF_5 + XCH{=}CH_2 \xrightarrow{125\text{–}140^\circ} \begin{array}{c} XCHCH_2F \\ | \\ SF_5 \end{array} \quad \text{(where X is Cl or Br)}$$

[87] B. Tittle and G. H. Cady, *Inorg. Chem.*, **4**, 259 (1965).

[88] R. M. Rosenberg and E. L. Muetterties, *Inorg. Chem.*, **1**, 756 (1962).

[89] (a) W. C. Smith, C. W. Tullock, R. D. Smith, and V. A. Engelhardt, *J. Am. Chem. Soc.*, **82**, 551 (1960). (b) M. Tremblay, *Can. J. Chem.*, **43**, 219 (1965). (c) S. Temple, *J. Org. Chem.*, **33**, 344 (1968).

analysis. Side reactions (telomerization or oxidative degradation) are a major problem when the reaction is extended to other olefins, but adducts with allene and acetylenes were also reported. The results are easily rationalized by a free radical mechanism.

Sulfuryl fluoride adds to fluoroolefins in the presence of alkali metal fluorides (or tetramethylammonium bromide) and polar aprotic solvent to give the corresponding sulfonyl fluorides or sulfones.[89c] Sulfonyl fluorides also add to fluoroolefins; this is the best route to aryl perfluoroalkyl sulfones. The mechanism must be ionic involving initial addition of F^- to the fluoroolefin.

$$CF_2{=}CF_2 + SO_2F_2 \xrightarrow[\text{diglyme}]{\text{CsF}} F_3CCF_2SO_2CF_2CF_3 \quad (83\%)$$

$$CF_3CF_2CF_2OCF{=}CF_2 + SO_2F_2 \longrightarrow$$

$$\underset{(83\%)}{CF_3CF_2CF_2OCF(CF_3)SO_2F} \longrightarrow \underset{(23\%)}{[CF_3CF_2CF_2OC(CF_3)F]_2SO_2}$$

$$RSO_2F + CF_2{=}CF_2 \longrightarrow RSO_2CF_2CF_3$$

R = alkyl or aryl

5-2G. BONDING ATOM OF X IS HALOGEN

The addition of a positive halogen species X^+ (Cl^+, Br^+, I^+) in the presence of a fluoride ion source to give the net addition of XF trans across a double bond of an olefin has been developed into a useful and selective fluorination for a wide

$$X^+ + R_2C{=}CR_2 \longrightarrow R_2C\overset{X^+}{\cdots}CR_2 \xrightarrow{F^-} R_2C(F){-}C(X)R_2$$

X = Br or I

variety of olefinic compounds. The source of X^+ is generally a positive halogen reagent such as halogenated amides (for example, *N*-bromosuccinimide), but the free halogen can be used in certain cases (I_2 or Br_2). The fluoride ion source is generally anhydrous HF in an ether solvent, but silver fluoride in benzene has recently been claimed to have advantages.

The most extensive study of this reaction has been reported by Bowers and co-workers[90] at Syntex who demonstrated the addition for simple olefins and

[90] (a) A. Bowers, *J. Am. Chem. Soc.*, **81**, 4107 (1959). (b) A. Bowers, L. C. Ibáñez, E. Denot, and R. Becerra, *J. Am. Chem. Soc.*, **82**, 4001 (1960). (c) A. Bowers, E. Denot, and R. Becerra, *ibid.*, 4007. (d) See also H. Schmidt and H. Meinert, *Angew. Chem.*, **72**, 493 (1960); E. D. Bergmann and I. Shahak, *J. Chem. Soc.*, 1418 (1959); R. H. Andreatta and A. V. Robertson, *Australian J. Chem.*, **19**, 161 (1966). (e) The use of halogen plus silver fluoride in benzene was recently claimed to have advantages for BrF or IF addition; L. D. Hall, D. L. Jones, and J. F. Manville, *Chem. Ind. (London)*, 1787 (1967).

used it for preparation of fluorinated steroids. For example, cyclohexene and *N*-bromoacetamide added with stirring to a mixture of anhydrous hydrogen fluoride and tetrahydrofuran at −80°C and warmed to 0°C for 2 hours gave *trans*-1-fluoro-2-bromocyclohexane in good yield. Similarly cyclohexene treated with *N*-iodosuccinimide and hydrogen fluoride in diethyl ether at −80°C gave *trans*-1-fluoro-2-iodocyclohexane in 73% yield. The well-accepted cyclic iodonium or bromonium ion is proposed as the intermediate to explain the stereospecific addition.

The reaction has been extended to addition of XF to a wide variety of olefins including polyhalogenated,[91-94] unsaturated sugars,[95] alkenoic acids,[96, 97] especially as intermediates to biologically active ω-fluorinated acids[96] and for nmr studies.[97] Surprisingly, cis (not trans) addition was found for sugars,[95] and was explained either by conformational changes in the carbonium ion intermediate or by equilibrium anomerization. Another possible explanation may be steric interference caused by association of HF through hydrogen bonding with the ether of the sugar, so that the cyclic bromonium ion is not formed and fluoride

ion is hindered from trans attack. For highly fluorinated olefins the direction of addition was used as evidence for addition of fluoride ion to the olefin to give a transient carbanion which displaces on iodine.[92]

A novel application of this procedure to get geminal addition is the reaction of ethyl diazoacetate with *N*-bromosuccinimide and HF in ether to form

[91] M. Hauptschein and M. Braid, *J. Am. Chem. Soc.*, **83**, 2383 (1961).

[92] C. G. Krespan, *J. Org. Chem.*, **27**, 1813 (1962).

[93] R. D. Chambers, W. K. R. Musgrave, and J. Savory, *J. Chem. Soc.*, 3779 (1961).

[94] C. Ching-yun, J. Hsi-kwei, C. Bing-qi, and L. Meng-lan, *Acta Chim. Sinica (English Transl.)*, **32**, 23 (1966).

[95] (a) P. W. Kent, F. O. Robson, and V. A. Welch, *J. Chem. Soc.*, 3273 (1963). (b) P. W. Kent and J. E. G. Barnett, *J. Chem. Soc.*, 6196 (1964). (c) P. W. Kent and M. R. Freeman, *J. Chem. Soc. (C)*, 910 (1966); K. R. Wood, P. W. Kent, and D. Fisher, *ibid.*, 912.

[96] (a) F. H. Dean and F. L. M. Pattison, *Can. J. Chem.*, **43**, 2415 (1965). (b) F. L. M. Pattison, D. A. V. Peters, and F. H. Dean, *ibid.*, 1689. (c) F. L. M. Pattison, R. L. Buchanan, and F. H. Dean, *ibid.*, 1700. (d) F. H. Dean, J. H. Amin, and F. L. M. Pattison, *Org. Syn.*, **46**, 10 (1966), gives a detailed experimental procedure for preparation of 1-bromo-2-fluoroheptane by addition of BrF to 1-heptene.

[97] A. K. Bose, K. G. Das, and P. T. Funke, *J. Org. Chem.*, **29**, 1202 (1964).

ethyl bromofluoroacetate.[98] Commercially available chlorine monofluoride has been added directly to fluorinated nitriles to prepare compounds such as $C_2F_5NCl_2$,[99] but because ClF is almost as reactive as fluorine it will probably be limited to use on simple, highly fluorinated compounds.

Only a summary of leading references in this area is given above, and a representative selection of examples is given in Table 5-2 to illustrate the synthetic potential of this reaction for introducing a single fluorine. A large number of patents on applications of this method to steroids are not included because of space limitations and overlap with published work, but they show the importance of this method in the medicinal field. The product can, by HX elimination, serve as precursor for fluoroolefins.[96] Although this method does require handling anhydrous hydrogen fluoride, the technique is simple and requires only a polythene bottle and lead-in tubes. Since the reaction proceeds through an intermediate carbonium ion (possibly a bridged bromonium or iodonium ion), molecular rearrangements of the Wagner-Meerwin type can occur.

As discussed in Section 5-1B, aryliododifluoride had been claimed as an agent capable of adding fluorine selectively to certain olefinic bonds, but these reports have not been verified. However, phenyliododifluoride has been found to add in entirety to a steroid dienone.[30] The evidence points to a phenyliodonium salt, with addition across the 6,7-double bond; spectral studies suggest that **b** is more probable than **a**.

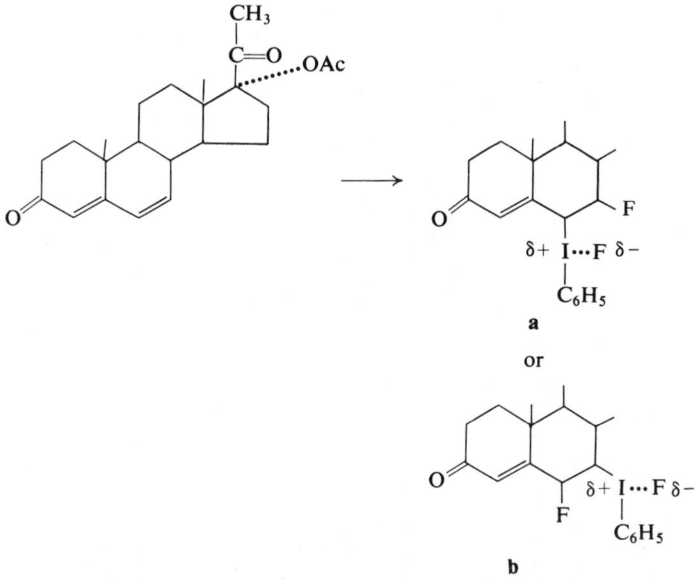

a

or

b

[98] H. MacHleidt, R. Wessendorf, and M. Klockow, *Ann.*, **667**, 47 (1963).
[99] J. B. Hynes and T. E. Austin, *Inorg. Chem.*, **5**, 488 (1966).

Table 5-2

Addition of Halogen Fluorides (XF) to Olefins and Other Unsaturated Centers

Unsaturated reactant	Conditions[a]	Product(s) (% yield)	Comments
Cyclohexene	AgF, I_2 in CH_3CN at $-80°$	*trans*-1-Bromo-2-fluorocyclohexane (60)	In an attempted repeat of this study only *N*-acetyl-2-iodocyclohexylamine was isolated. Ref. 90d
Styrene	Br_2, AgF, benzene, or acetonitrile	2-Bromo-1-fluoro-1-phenylethane (71)	Claimed advantage over HF method in that sensitive olefins not polymerized. However X replaced by F of AgF in some cases (acenaphthylene). Attempts to effect "XF" addition using HgF_2, ZnF_2, or PbF_2 in place of AgF were not successful. Ref. 90e
Methylenecyclohexane	I_2, AgF, benzene, or acetonitrile	1-(Iodomethyl)-cyclohexyl fluoride (70)	
Cyclohexene	Add NBS and olefin to HF in THF or ether at $-80°$	*trans*-1-Fluoro-2-bromo-cyclohexane (40)	Ref. 90b
Cyclohexene	NIS, HF, ether at $-80°$	*trans*-1-Fluoro-2-iodocyclohexane (73)	Ref. 90a
Steroid unsaturated in 6-, 9- or 16-position	NBA, HF, ether or methylene chloride at $-80°$	BrF adduct, e.g., 9α-bromo-11β-fluoro steroid (70-80)	Ref. 90a,b
Δ^5-Steroids	NIS, HF in CH_2Cl_2, and THF at $-80°$	6-Iodo-5-fluoro steroid	Abnormal direction of addition noted and discussed. Ref. 90c
Triacetyl D-glucal	NBS, HF in ether at $-80°$	AcOCH$_2$, O, F, OAc, Br, AcO (~ 30) + AcOCH$_2$, O, OAc, AcO, F, Br (~ 15)	Noted and discussed cis addition in contrast to trans for above. Products used as intermediates to new sugars. Ref. 95a
Di-O-acetyl D- and L-arabinals	As above, followed by cold NaOMe	O, F, Br, HO, OH (24)	Explanation offered for cis addition. Ref. 95b
Triacetyl D-galactal	NBS or NIS, HF, ether $-70°$	Two epimeric *cis*-bromo or iodofluorides (51 and 22)	Ref. 95c
$CH_2{=}CHCO_2CH_3$, also series of cinnamates, fumarates, maleates	NBA, HF, THF, and CH_2Cl_2	$CH_2FCHBrCO_2CH_3$ (50) corresponding BrF adduct (51-60)	Prepared as potential intermediates to fluorinated amino acids; noted β-F more easily lost than α-Br. Used products to study rotational isomerism by nmr. Ref. 97

$CH_2{=}CHCO_2CH_3$	NBA, HF, ether-pyridine at −20° or 1,3-dibromo-5,5-dimethylhydantoin, HF, ether-CCl_4 at −30°	$FCH_2CHBrCO_2CH_3$ (45) $BrCH_2CHFCO_2CH_3$ (trace)	See next example. Needed for nmr spin-coupling studies. Ref. 96a
$CH_3CO_2CH_2CH{=}CH_2$	NBA, HF in ether or ether–pyridine	$CH_3CO_2CH_2CHFCH_2Br$ (up to 69)	Product used to prepare $CH_2{=}CFCO_2R$, Ref. 96a
$RCH{=}CH_2$, R = alkyl, aryl	NBA, HF, ether −80°	$RCHFCH_2Br$ (70-80)	Comparison of reagents, solvents and conditions discussed. Converted products to monofluoroalkanoic acids; this route compared to perchloryl fluoride method. Ref. 96b,c
$CH_3(CH_2)_4CH{=}CH_2$	1-heptene added to mixture of NBS and HF in ether at −80°	$CH_3(CH_2)_4CHFCH_2Br$ (60-70)	Product 90% pure. Detailed experimental directions; also includes procedure for conversion to α-fluoroheptanoic acid. Ref. 96d
$R_fCF{=}CF_2$ R_f=F or fluoroalkyl (also $CCl_2{=}CF_2$)	KF, I_2, CH_3CN, 100°, autogenous pressure, 10 hr	$R_fCFI{-}CF_3$	Mechanism proposed is F^- addition to give transient fluoro carbanion which displaces on I_2. Ref. 92
$(CF_3)_2C{=}CF_2$	KF, I_2, CH_3CN, 120-130°, 8 hr, autogenous pressure	$(CF_3)_3CI$ (45-55)	Series of perfluoroalkyl iodides and bromides prepared to study inductive order of perfluoroalkyl groups. Ref. 94
$(CF_3)_2C{=}CF_2$	$I_2 + IF_5$, 130°, 60 hr	$(CF_3)_3CI$ (69)	Noted need for extended reaction time using procedure of ref. 91. Ref. 100a
$(CF_3)_2C{=}CF_2$	KF, I_2, $C_6H_5NO_2$ 70-180°, 10 hr	$(CF_3)_3CI$ (87)	Ref. 100b
$XYC{=}CF_2$ where Y = fluoroalkyl,F,Cl,H X = F,Cl,H	$2I_2 + IF_5$ 100-175° Catalyst is Al-aluminum iodide preheated; olefin then added	$CF_3{-}CXYI$ (20 to 80)	Ref. 91
$XYC{=}CF_2$ X,Y is R_f,F,H,Cl	$Br_2 + BrF_3$ $2I_2 + IF_5$	$CF_3{-}CXYBr$ $CF_3{-}CXYI$ } (45 to 99)	Also with $CCl_2{=}CCl_2$ to get $CCl_2F{-}CCl_2F$ (51%) mechanism discussed. Ref. 93
$-CCl{=}CCl-$	Cl_2, HF, 250 to 600° ZrF_4/C	$-CCl_2{-}CF_2-$	General syntheses of fluorocarbons exchange of Cl for F also occur. Ref. 100c
N_2CHCO_2Et	NBS, HF, ether, −70°	$FBrCHCO_2Et$ (50-60)	Suggest addition of Br^+ to get diazonium salt intermediate $^+N_2CHBrCO_2Et$ in which F^- displaces N_2. Ref. 97

[a] Anhydrous HF was always employed. The abbreviations are NBS, N-bromosuccinimide; NBA, N-bromoacetamide; and NIS, N-iodosuccinimide.

[100] (a) J. A. Young and T. M. Reed, *J. Org. Chem.*, **32**, 1682 (1967). (b) E. P. Mochalina, L. Dyatkin, I. V. Galakhov, and I. L. Knunyants, *Proc. Acad. Sci., USSR*, Chem. Soc. (English Transl.), **169**, 816 (1966). (c) C. F. Baranauckas and S. Gelfand, Canadian Patent 685,640 (1964).

Summary

The following synthetic aspects were reviewed in this section:

1. HF addition to olefins and acetylenes is a classical useful method of introducing a single fluorine. Opening of an epoxide with HF to give β-fluoroalcohol is a valuable, recently exploited development, particularly in the steroid area.

2. Mercury(II) fluoride addition to fluoroolefins gives a mercury derivative that is a useful intermediate for synthesis of a variety of fluoroalkyl derivatives.

3. Alkali metal additions to fluorinated olefins and ketones provide potentially useful synthetic intermediates.

4. Fluoride ion addition to fluoroolefins or fluoroketones is an effective approach to addition of a variety of electrophilic agents X, such as –COF and $-CH_2OH$.

5. Nitrogen fluoride derivatives can be added but because of multiple reaction paths or product instability have only limited utility. F–NO addition, particularly by fluoride ion catalysis to fluoroolefins, is the most promising for synthesis.

6. Addition of O–F and S–F species to a variety of unsaturated systems can be done by either radical or ionic methods. Fluoride ion catalysis is again essential for selective ionic additions.

7. Halogen fluorides are readily added to olefins by use of positive halogen in the presence of HF in an aprotic solvent and provide an excellent mild method of introducing a single fluorine with an α-halogen handle for elimination or substitution.

5-3 REPLACEMENT OF H BY F

$$RH \rightarrow RF$$

As reviewed in Chapter 4, most fluorinating agents or techniques that are sufficiently active or strenuous to replace hydrogen by fluorine are not selective and sometimes badly disrupt the whole system with extensive breaking of C–C bonds. One special technique described earlier is the jet reactor in which selectivity is attained by low-temperature operation and high dilution with an inert gas, but is limited to volatile substrates. Two significant developments have recently been made in selective replacement of H by F.

1. For active hydrogen on nitrogen, elementary fluorine in a cold aqueous solution or with an alkali metal fluoride diluent in the solid state at low temperature are used.

2. For active hydrogen on carbon, perchloryl fluoride is used.

5-3A. USE OF ELEMENTARY FLUORINE IN AQUEOUS SOLUTION

This technique was originally reported for preparation of *N,N*-difluorourea in 75% yield by bubbling fluorine diluted with an inert gas through a cold aqueous solution of urea.[101, 102] If the reaction is heated to 85°C after addition of F_2, difluoroamine is obtained.[103] This technique has been extended to fluorination of amides, urethanes and substituted ureas,[104] and amines.[105] In Chart 5-3, examples are given for the passage of fluorine diluted with nitrogen through a 5 to 16% aqueous solution of the compounds at 0° to 5°C. A variety

Amides:

$CH_3CONHCH_3 \longrightarrow CH_3NF_2$ (7%)

Urethanes:

$NH_2CO_2C_2H_5 \longrightarrow FNHCO_2C_2H_5$ (30%)

$CH_3NHCO_2C_2H_5 \longrightarrow CH_3NF_2$ (60%)

N-Substituted Ureas:

$CH_3NHCONHCH_3 \longrightarrow CH_3NFCONHCH_3$ (27%) + CH_3NF_2 (4%)

$C_2H_5NHCONHC_2H_5 \longrightarrow C_2H_5NFCONHC_2H_5$ (9%) + $C_2H_5NF_2$ (5%)

Amines[105]:

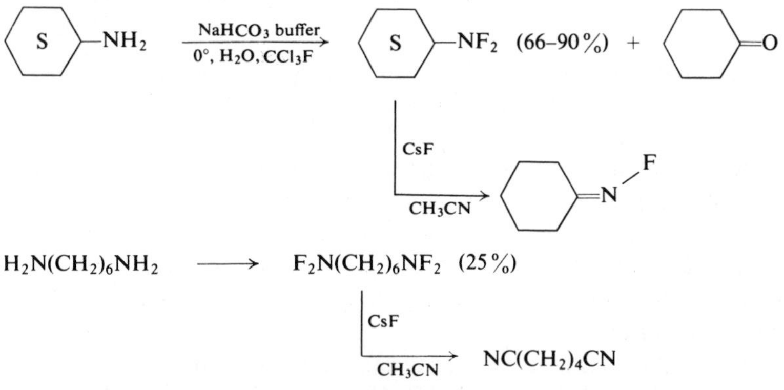

(S)—NH_2 $\xrightarrow[0°,\ H_2O,\ CCl_3F]{NaHCO_3\ buffer}$ (S)—NF_2 (66–90%) + cyclohexanone (=O)

(S)—NF_2 $\xrightarrow[CH_3CN]{CsF}$ cyclohexylidene=N—F

$H_2N(CH_2)_6NH_2 \longrightarrow F_2N(CH_2)_6NF_2$ (25%)

$F_2N(CH_2)_6NF_2$ $\xrightarrow[CH_3CN]{CsF}$ $NC(CH_2)_4CN$

CHART 5-3

[101] V. Grakauskas, *140th Meeting Am. Chem. Soc., Chicago, Illinois, 1961.* Abstract p. 23M.

[102] Direct fluorination of urea at 0°C in a flow reactor using elementary fluorine diluted with nitrogen is reported to give an unstable product (*N,N*-difluorourea?) which decomposes to difluoroamine (HNF_2) on heating to 85°C with overall yield of 15 to 30% [E. A. Lawton and J. Q. Weber, *J. Am. Chem. Soc.*, **85**, 3595 (1963)]. Isolation of *N,N*-difluorourea from the same reaction has also been reported [E. A. Lawton, E. F. C. Cain, D. F. Shefhan, and M. Warner, *J. Inorg. Nucl. Chem.*, **17**, 188 (1961)]. Unfortunately experimental details are not sufficiently detailed, probably because of the unstable and explosive character of the products, to allow direct comparison of the two methods.

[103] Belgian Patents 658,521 and 658,522 (1965); British Patent 1,018,377 assigned to Allied Chem. Corp.; *Chem. Abstr.*, **64**, 7729 (1966).

[104] R. E. Banks, R. N. Haszeldine, and J. P. Lalu, *Chem. Ind. (London)*, 1803 (1964); and *J. Chem. Soc.* (C), 1514 (1966).

[105] C. M. Sharts, *J. Org. Chem.*, **33**, 1008 (1968); U.S. Patent 3,395,318 (1967).

of other aliphatic amines is also reported. Two important techniques were emphasized: *buffered* aqueous solution and an inert organic solvent (CCl_3F) as a second phase to remove product as formed.

The reaction of fluorine with an unbuffered aqueous solution of sulfamide at 0° to 5°C gives *N,N*-difluorosulfamide.[106a] This compound is extremely

$$2F_2 + H_2NSO_2NH_2 \xrightarrow[0^\circ]{F_2/He} 2HF + NF_2SO_2NH_2 \quad \text{(stable at 0° or below)}$$

$$NF_2SO_2NH_2 \xrightarrow{OCl^-} ClNF_2$$

$$NF_2SO_2NH_2 \xrightarrow[>5^\circ]{H^+} HNF_2 + HOSO_2NH_2$$

$$NF_2SO_2NH_2 \xrightarrow{\phi_3P} \text{stable 1:1 complex}$$

unstable and reactive (see equation) and cannot be isolated at room temperature.

Cyanamide is also fluorinated in a buffered aqueous solution[106b] to give a product that is readily isomerized by fluoride ion (see Section 6-1B).

$$NH_2CN \xrightarrow[\text{phosphate-buffered aqueous solution}]{F_2,\ 5\text{–}9^\circ} \underset{(20\%)}{F_2NCN} \xrightarrow[24^\circ]{CsF} F_2C\langle{}^{N}_{N}\rangle$$

Although bond cleavage does occur in some of the systems, it can apparently be controlled and selective so that moderately high yields of unique products are obtained and attack on the hydrocarbon radicals is at a minimum. Decomposition of the initial N–F product, particularly in the aqueous media is a major problem; thus *N,N*-difluorourea readily decomposes to difluoroamine (HNF_2) on heating in water.

The method has been extended to fluorination of the aromatic-type boron cage compounds $B_{12}H_{12}^{-2}$ and $B_{12}H_6F_6^{-2}$ (from anhydrous HF on $B_{12}H_{12}^{-2}$) in water at 0°C to give $B_{12}F_{11}OH^{-2}$ and BF_4^- in almost equal amounts. [107]

The mechanisms of these aqueous fluorinations are not understood. A free radical mechanism was suggested as the only possibility in one paper.[107] A pertinent observation is that in the reaction of F_2 with F_3CCO_2H to give $F_3C(=O)OF$ (trifluoroacetyl hypofluoride) a small amount of water is needed or no product is formed.[108] These authors show that OF_2 is not responsible and

[106] (a) R. A. Wiesboeck and J. K. Ruff, *Inorg. Chem.*, **4**, 123 (1965). (b) M. D. Meyers and S. Frank, *Inorg. Chem.*, **5**, 1455 (1966).

[107] W. H. Knoth, H. C. Miller, J. C. Sauer, J. H. Balthis, Y. T. Chia, and E. L. Muetterties, *Inorg. Chem.*, **3**, 159 (1964).

[108] G. L. Gard and G. H. Cady, *Inorg. Chem.*, **4**, 594 (1965).

suggest formation of an active intermediate such as a hydrate of ·F, ·OF, or HOF. Another possibility is that the fluorination is ionic similar to that for perchloryl fluoride or N–F compounds (see next section) and involves a nucleophilic attack by an anion (or a basic nitrogen) on one fluorine with a concerted elimination of the other as fluoride ion. The water may be necessary as a weak acid to give a small amount of anion and/or to hydrate the incipient fluoride ion:

$$F{-}F \leftarrow {}^{-}N\begin{matrix} R \\ R_1 \end{matrix} \longrightarrow F^- + FN\begin{matrix} R \\ R_1 \end{matrix}$$

An alternative ionic scheme could involve hypofluorous acid (HOF) or hypofluorite anion (^-OF) as a key intermediate. Protic bases cannot be used because of the instability of the products. Ionic attack by polarizable enolic forms of ureas and urethanes on O–F compounds is also suggested as a route to difluoroureas.[104] Currently this procedure is applicable to water-soluble materials that are at least weakly acidic.

One major problem is the low stability of many of the products, particularly to aqueous base. Also pH will influence the equilibrium

$$OH^- + HOF \rightleftharpoons OF^- + H_2O$$

which could be very important if HOF or OF^- were the active species for fluorination. Careful control of pH may allow useful extensions of this method.

An extension to solid phase systems appears to have considerable potential and could overcome some of the problems of the aqueous system. A mixture of aminoiminomethanesulfinic acid and sodium fluoride can be fluorinated directly at 0°C with 10% F_2 in nitrogen to give a mixture of cleavage products:

$$H_2N\overset{\overset{NH}{\|}}{C}SO_2H + 6F_2 \longrightarrow F_2NCF_2NF_2 + SO_2F_2 + 4HF + F_2NSO_2F + \text{other products}$$

The main one of interest is bis(difluoroamino)difluoromethane in 8–10% yield.[6c] The authors suggest formation of this product by the replacement of hydrogen, addition to the imino bond, and C–S bond cleavage. This technique of fluorinating a mixture of reactant and sodium fluoride was also applied to the synthesis of $(F_2N)_2C{=}NF$ or $CF(NF_2)_3$ from guanine or guanylurea sulfate and to the fluorination of urea, thiourea, guanidine, and melamine.[109] Replacement of hydrogen in imines using metal halide catalysis has also been reported.[5e]

[109] (a) R. J. Koshar, D. R. Husted, and R. A. Meiklejohn, *4th Intern. Symp. Fluorine Chem., Estes Park, Colorado, July, 1967,* Abstracts, p. 74. (b) R. J. Koshar, D. R. Husted, and C. D. Wright, *J. Org. Chem.*, **32**, 3859 (1967). (c) R. A. Davis, J. L. Kroon, and D. A. Rausch, *J. Org. Chem.*, **32**, 1662 (1967).

$$(CF_3)_2C{=}NH + F_2 \xrightarrow[\text{or CsF}]{\text{KF}} (CF_3)_2C{=}NF + HF$$

Diluted fluorine passed over reactant (particularly salts) at low temperature in a static bed reactor has provided a route to perfluorinated O–F compounds[6d, e] (see Section 5-1A).

A final comment on limitations of these procedures–they require special equipment and techniques for safe handling of elemental fluorine and will be useful only to relatively few chemists.

5-3B. PERCHLORYL FLUORIDE

The use of perchloryl fluoride to introduce fluorine selectively in place of hydrogen in an active methylene compound, such as a malonate ester, was first reported in 1957 when the chemical was offered in experimental and developmental quantities by Pennsalt Chemicals Corp.[110] Since the original literature reports by the Pennsalt workers,[111] publications and patents in the order of a hundred have appeared. We will briefly summarize the scope and limitation of this method of fluorination and particularly point out safety problems. Although we cannot give an exhaustive review, we have critically evaluated the most useful and significant synthetic developments in Table 5-3.

Perchloryl fluoride is a surprisingly stable molecule; the fluorine and three oxygens are bonded to chlorine so that it has C_{3v} tetrahedral symmetry and essentially no electric dipole moment. It is a colorless gas, bp –46.8°C which when pure is shock and thermally stable to 500°C, but is a strong oxidizing agent and reacts *explosively* with many organic materials or other oxidizable substances.[112] Perchloryl fluoride has no unusual problems in toxicity [acute vapor toxicity (LD_{50}) of 2000 to 4000 ppm], but normal precautions to avoid exposure are recommended.

The general procedure for fluorination as reported originally[110, 111] was to pass perchloryl fluoride into a solution of an active methylene compound such as diethyl malonate in ethanol containing one equivalent of sodium ethoxide for each active hydrogen on the methylene group. Complete

[110] "Perchloryl Fluoride" dsA-1819 and dsF-1819, Pennsalt Chemicals Corp., Philadelphia, Pennsylvania, 1957.

[111] (a) C. E. Inman, E. A. Tyczkowski, R. E. Oesterling, and F. L. Scott, *Experientia*, **14**, 355 (1958). (b) C. E. Inman, R. E. Oesterling, and E. A. Tyczkowski, *J. Am. Chem. Soc.*, **80**, 6533 (1958). (c) See also C. E. Inman, R. E. Oesterling, and E. A. Tyczkowski, U.S. Patents 3,030,408 (1962) and 3,141,040 (1964); *Chem. Abstr.*, **57**, 9671d (1962) and **61**, 8240d (1964).

[112] For a discussion of explosions and safety problems with perchloryl fluoride see *Chem. Eng. News*, **38**, 62 (Jan. 25, 1960) and **37**, 60 (July 13, 1959).

fluorination of the active hydrogen was claimed; the work-up was simply to filter off the precipitate of sodium chlorate, concentrate the alcohol solution, and distill. Diethyl difluoromalonate was reported prepared in 84% yield by this procedure.

Recently the reaction of perchloryl fluoride with diethyl malonate was reinvestigated and shown to give five products in ethanol[113] (see Chart 5-4), of which the originally claimed product **1** (Chart 5-4) was only in

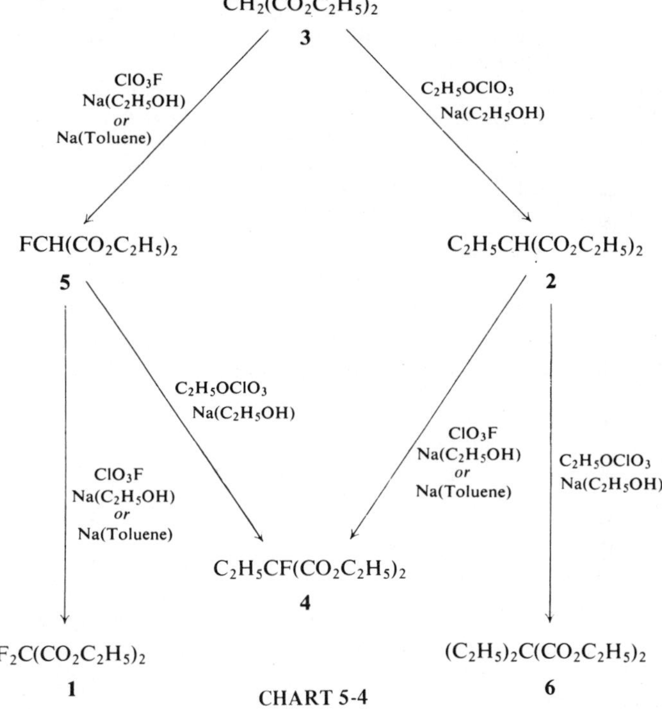

CHART 5-4

36% yield and varying amounts of **2**, **3**, **4**, and **5** were found depending on reaction conditions (see Table 5-3A). A high yield of **1** (over 95%) was obtained only from reaction of **5** with sodium in toluene. Ester **6** was obtained starting with **2** in ethanol and provides support for $EtOClO_3$ being present as a reactant. Apparently the mixtures were not detected by the analytical method used (carbon hydrogen analysis!) but a low fluorine analysis was mentioned in later publications.[111c]

In the original work two mechanisms (A and B) were proposed. Mechanism A requires direct heterolysis of perchloryl fluoride similar to

[113] H. Gershon, J. A. A. Renwick, W. K. Wynn, and R. D. Ascoli, *J. Org. Chem.*, **31**, 916 (1966).

Table 5-3A

Reactions of Perchloryl Fluoride: Composition of Fluorinated Mixture from Malonate Esters

Reacting ester[a]	Base[b]	Products, % of solvent-free mixture[a]					
		1	2	3	4	5	6
3	$Na(C_2H_5OH)$	36.1	25.2	14.4	15.3	8.7	–
3	$Na(C_2H_5OH)$	26.2	5.2	40.2	0.5	28.2	–
3	Na(toluene)	29.2	0	29.3	0	41.6	–
2	$Na(C_2H_5OH)$	–	33.3	–	51.5	–	15.2
2	Na(toluene)	–	4.0	–	96.0	–	0
5	$Na(C_2H_5OH)$	85.6	–	–	6.7	7.6	–
5	Na(toluene)	95.5[c]	–	–	–	–	–

[a]Numerals refer to malonate ester as given in Chart 5-4 (see Ref. 113).

[b]The ester, base and ClO_3F are all in the molecular proportion 1:1:1 except in the first reaction where the proportion is 1:2:2. The sodium added to solvent, malonate added, and then $FClO_3$ bubbled in at 10° to 15°C reaction temperature.

[c]Remaining 4.5% was two unidentified materials.

that found for reaction with inorganic reagents. However, the second step, which requires the fluoride to displace ClO_3^-, seems impossible since alkoxides in the alcohol solution are much more nucleophilic toward carbon than fluoride. In Mechanism B, the anion is suggested to displace

Mechanism A:

$$\mathrm{(A)(B)CH^- + FClO_3 \longrightarrow \left[(A)(B)CHClO_3\right] + F^- \xrightarrow{F^-} (A)(B)CHF + ClO_3^-}$$

$$\mathrm{(A)(B)CHF \xrightarrow{\text{base, fast}} (A)(B)CF^- \xrightarrow{FClO_3} (A)(B)CF_2 + ClO_3^-}$$

Mechanism B:

$$\mathrm{(A)(B)CH^- + FClO_3 \xrightarrow{a} \left[(A)(B)CH \cdots F \cdots ClO_3\right]^- \longrightarrow (A)(B)CHF + ClO_3^-}$$

$$\mathrm{(A)(B)CH^- + FClO_3 \xrightarrow{b} \left[(A)(B)CH \cdots \overset{\delta+}{F} \cdots \overset{\delta-}{ClO_3}\right] \longrightarrow (A)(B)CHF + ClO_3^-}$$

$$\mathrm{(A)(B)CHF \xrightarrow{\text{base, fast}} (A)(B)CF^- \xrightarrow{FClO_3} (A)(B)CF_2 + ClO_3^-}$$

on fluorine, either directly (route **b**) or by a cyclic transition state with concerted removal of ClO_3^- (route **a**). Such a displacement is reasonable because the fluorine will be at a much lower electron density than normal (perhaps the lowest in the system in regards to nucleophilic attack) and ClO_3^- is an excellent leaving group. The monofluoro ester is more acidic than the starting ester and should reform an anion in a fast step—but this anion will be a less nucleophilic species. The mixture of products actually found in this reaction[113] is to be expected and readily accomodated within the general picture given in Mechanism B if reactions involving ethanol as the perchlorate ester are included. But we will present a more rational mechanism after we review scope and limitations.

The main side reaction in use of perchloryl fluoride is oxidation and this can mean *violent explosions* under certain conditions.[112, 114] Amines are reported to give explosive reactions but recently *N*-perchlorylpiperidine was isolated as the first member of the new class of organic compounds.[115]

Piperidine in aqueous solution gives *N*-perchlorylpiperidine,[116] but 2,6-substituted-piperidines give the corresponding *N*-fluoro derivatives. These products are unstable if α-hydrogen is present. The reactions of perchloryl fluoride with other tertiary amines were also studied.

O= ⟨ring⟩ NH + ClO_3F —ether→ O= ⟨ring⟩ $NHClO_3F$

↓

O= ⟨ring⟩ N—F ←excess amine— $[$O= ⟨ring⟩ N(H)(F)$]^+$ ClO_3^-

(92%)

By-products or side reactions from oxidation have been reported with phenoxides to give quinonoid products,[117] with nitro alkenes to give vicinal

[114] S. A. Fuqua and R. M. Silverstein, *J. Org. Chem.*, **29**, 395 (1964), footnote 2.

[115] (a) D. M. Gardner, R. Helitzer, and C. J. Mackley, *J. Org. Chem.*, **29**, 3738 (1964). (b) F. L. Scott, R. E. Oesterling, E. A. Tyczkowski, and C. E. Inman, *Chem. Ind. (London)*, 528 (1960), discuss the reaction of perchloryl fluoride with nitrogeneous bases under controlled condition and report that amine hydrofluorides, hydrochlorates, and oxidation products are obtained.

[116] D. M. Gardner, R. Helitzer, and D. H. Rosenblatt, *J. Org. Chem.*, **32**, 1115 (1967).

[117] A. S. Kende and P. MacGregor, *J. Am. Chem. Soc.*, **83**, 4197 (1961).

dinitro compounds,[118] with acylamide malonates to give oxidative dimerization,[119] and with indene to give a mixture of oxyfluorination and other products.[120] Thiols are cleanly oxidized to disulfides. The side products resulting from participation of the alcohol solvent in the reaction with incomplete fluorination were discussed above,[113] and this study should be of considerable value to design experimental conditions to avoid such by-products. Another major problem is instability of products, particularly aliphatic fluorides that can readily lose HF under the basic conditions or can give ester hydrolysis and decarboxylation.[121] Aromatic systems are readily attacked by perchloryl fluoride under the catalytic action of Lewis acids to give perchloryl aromatic compounds.[122] This reaction does not present any problems under the usual fluorination conditions; note, however, that the perchlorylaromatics are dangerous compounds that can explode violently under certain conditions.

The perchloryl fluoride fluorination has been applied to a wide variety of active methylene compounds (see Table 5-3B for details) such as α-keto esters, enol ethers, enol esters, enamines, enamides, gem-dinitriles, nitroalkanes, and phenols (to dieneones). Recently preparation of α-fluorinated heterocyclics from the corresponding lithium derivatives was reported. Perchloryl fluoride has been particularly useful in the steroid field where high yields of unique products are usually obtained, and in preparing intermediates for synthesis of compounds of potential biological interest. However, because of side reactions, the perchloryl fluoride method was claimed to be inferior to less direct and tedious synthetic routes to monofluoroalkanoic acids.[96c] Selected literature examples to illustrate all of these types of fluorination are given in Table 5-3.

Some additional comments on mechanism are now appropriate. Kende and MacGregor[117] stated that the chemistry of perchloryl fluoride can be rationalized if perchloryl fluoride is regarded as an ambient electrophile. Freeman[118b] proposed that highly mesomeric carbanions (malonic ester anions and nitroalkane anions) react with perchloryl fluoride by a "bond-breaking" transition state to give C–F bonds, whereas the "more effective" charge-localized nucleophiles (alkoxides, amines, most oxime anions) react by a "bond-making" transition state to form O–Cl or N–Cl bonds. This does not explain the different behavior of perchloryl fluoride. We have considered the hypothesis that the mesomeric carbanions attack fluorine, and considerable C–F bonding develops in the transition state;

[118](a) H. Shechter and E. B. Roberson, Jr., *J. Org. Chem.*, **25**, 175 (1960). (b) J. P. Freeman, *J. Am. Chem. Soc.*, **82**, 3869 (1960).

[119]H. Gershon and A. Scala, *J. Org. Chem.*, **27**, 463 (1962).

[120](a) M. Neeman and Y. Osawa, *J. Am. Chem. Soc.* **85**, 232 (1963). The oxyfluorination reaction has been adopted to useful synthetic purposes. (b) See also M. Neeman and Y. Osawa, *Tetrahedron Letters*, 1987 (1963).

[121]H. MacHleidt and V. Hartmann, *Ann.* **679**, 9 (1964).

[122]C. E. Inman, R. E. Oesterling, and E. A. Tyczkowski, *J. Am. Chem. Soc.*, **80**, 5286 (1958).

Table 5-3B

Reactions of Perchloryl Fluoride: General Synthetic Applications

Reactants	Conditions[a]	Product(s) (%yield)	Comments
$RCH(CO_2Et)_2$ R = H or F	NaOEt or Na in toluene	$RCF(CO_2Et)_2$ $CF_2(CO_2Et)_2$	See part A; the example including α-keto-ester also given. Ref. 110, 111, 112
$EtO_2CCH_2C^{-}(K^+)(CO_2Et)(COCO_2Et)$	EtOH as solvent	$EtO_2CCH_2CF(H)CO_2Et$ (73) isolated from basic hydrolysis in workup	$EtO_2CCH_2CF(CO_2Et)(COCO_2Et)$ could not be isolated pure. Ref. 123
As above	EtOH, below 0°	$EtO_2CCH_2CF(CO_2Et)(COCO_2Et)$ (78)	
$RCH(CO_2Et)_2$ where R = CH_3, C_6H_5, $MeO_2CCH_2CH_2$, NO_2	NaH in DMF, or KOEt in dioxane	$RCF(CO_2Et)_2$ (66-78)	See Ref. 111. Product could not be isolated if reaction run over 0°. Used for synthesis of fluorolactic acid and related acids for biological studies. Ref. 124
2-Carbethoxycyclohexanone	NaOEt in EtOH	2-Fluoro-2-carbethoxy-cyclohexanone (59)	Used in synthesis of α-fluorocarboxylic acids for toxicity studies; compared with other synthetic routes. Ref. 96c
$CH_3CH_2CEt{=}CHOCH_3$	pyridine	$CH_3CH_2CEtFCO_2CH_3$ (34) + $CH_3CH_2CEtFCO_2H$ (small?)	
$RCH(CO_2Et)_2$ R is 15 different alkyl groups up to C-20	Na in toluene	$RFC(CO_2Et)_2$ (57-89)	Converted to 2-fluoro fatty acids for studies of antifungal activity relative to unfluorinated acids. Ref. 125a
$RCH(CN)(CO_2Et)$	(a) NaOEt, EtOH	(a) $RCF(CO_2Et){-}C(OEt){=}NH$ (72–90)	Animal toxicities of products studied. Ref. 125b
	(b) Na in toluene or K in DMF	(b) $RCF(CN)CO_2Et$ (35-80)	

Table 5-3B (Cont.)

Reactants	Conditions[a]	Product(s) (% yield)	Comments
$(MeO_2C)_2C(NHCOCH_3)CH_2CH(CO_2Me)_2$	Not given	$(MeO_2C)_2C(NHCOCH_3)CH_2C(CO_2Me)_2$ (46)	Converted to 4-fluoroglutamic acid for preparation of potential antimetabolites. Ref. 126
$ArCOCH_2COAr$	Pyridine at −10 to 10°	$ArCOCF_2COAr$ (69)	Noted hazard of explosion in reaction. Product used as intermediate in a synthetic scheme. Ref. 114
$RCH(CO_2Et)COR$ R is $-CH_2CH{=}CMe_2$ and similar groups	NaOEt, EtOH	$RCF(CO_2Et)COR'$ (65) + $RCHFCO_2Et$ (10)	Ketone prepared in high yield by basic hydrolysis; used in studies of fluorinated ketones and esters; stability of cyclic fluorinated ketones discussed. Ref. 127
2-Carbethoxycycloalkanones		2-Fluoro-2-carbethoxy-cycloalkanones (65-93)	
$RCH(CO_2Et)COCO_2R'$ $R = CH_3, C_2H_5, C_6H_5CH_2$	$NaOCH_3$ in CH_3OH	$RCF(CO_2Et)COCO_2R'$ (32–67)	Hydrolyzed to fluoropyruvic acid derivatives, used to study effect in carbohydrate metabolism. Ref. 128
Methyl 3-camphorcarboxylate	$NaNH_2$	Methyl 3-fluoro-3-camphor-carboxylate	Converted to 3-fluorocamphor which is, as stated, not easily prepared by other methods. Ref. 129
$RCH(CN)_2$ R = alkyl or aryl	NaH in glyme at −10° (also used NaF as base for R = Br)	$RCF(CN)_2$ (approx. 50-75)	Examples given for $R = C_2H_5$, $C_6H_5CH_2$, and C_6H_5. Ref. 130

Substrate	Conditions	Product (Yield %)	Remarks
R_2CHNO_2 (R is alkyl or alkylaryl)	NaOMe in MeOH, 0°, or NaH in THF	R_2CFNO_2 (32-82)	By-product of ketones (32-55) and vicinal dinitro compounds (1-7). Useful only for secondary; primary nitro compounds give mainly by-products. Oximic and hydroxamic acids were generally oxidized or cleaved. Scope, limitations, and mechanism discussed. Ref. 118
$(CO_2Et)_2CHNHR$	NaOEt in EtOH	$[(CO_2Et)_2C(NHR)-]_2$ (40)	No fluorinated product. Mechanism discussed. Ref. 119
Tetracycline derivatives	$NaOCH_3$ in CH_3OH or NaOH in H_2O	Fluorinated tetracycline (R₁, R₂, R₃; HO, F, OH, $CONH_2$) (53–70) *or* hemiketal form (R₁, H, R₂, R₃; HO, F, O-bridge, OH, $CONH_2$) (15–83)	Hemiketal product formed when hydroxyl group is on ring carbon where R_1 is attached. Ref. 131
Androstene and estrone derivatives with ring D 2-(hydroxymethylene)-ketone (O, =CHOH, D)	t-BuOK in t-BuOH	Ring D 2,2-difluoro ketone (O, F, F) (25–30)	Mechanisms discussed, via 2-fluoro-2-formyl ketone (O, CHO, F) → fluoroenolate (O^-, F). Ref. 132

Table 5-3B (Cont.)

Reactants	Conditions[a]	Product(s) (% yield)	Comments
Pregnedienones, testosterone derivatives, and other related steroids containing 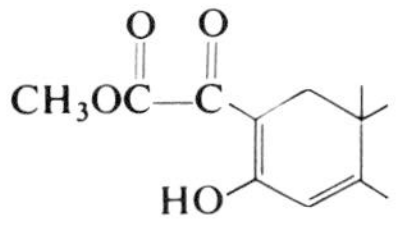 $CH_3OC(=O)-C(=O)-$ (HO) *a*	$NaOCH_3$ in CH_3OH at $-10°$ or NaOH in CH_3OH	From *a*: 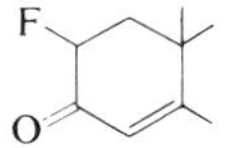 (F, O) (24)	Oxalyl group removed in workup; other examples of similar type fluorinations included. Ref. 133
$RC(=O)CH_2C(=O)-C(=O)OEt$ *b*		From *b*: depending on conditions, $RC(=O)CH_2F$ (67); $RC(=O)CHF_2$ (56)	
Enol acetates progesterones 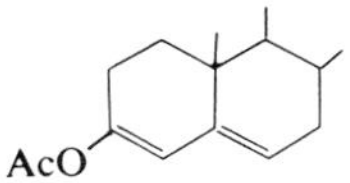 (AcO)	Dioxane	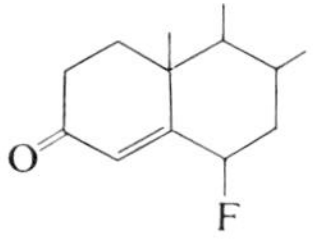 (O, F) (over 44) (Isolated by halogenation at 17 position)	Discusses preferential reaction at 6- vs. 17- position of bisenol acetate. Ref. 134
Enol ether of pregnenolones 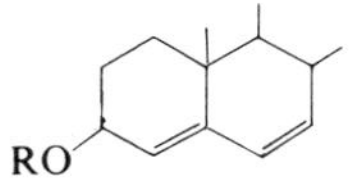 (RO)	Pyridine or aq. THF	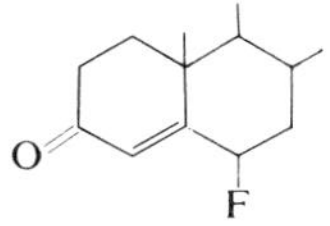 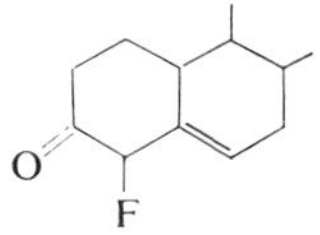 (O, F) (52–86); (O, F) (by-product in aq. THF)	Used for conversion to fluorocortisones. Some discussion of mechanism. Also applied to testosterone enol diacetate. Ref. 135
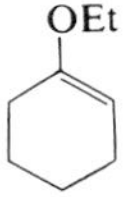 (OEt)	Pyridine 0°, aq. acid workup	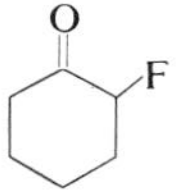 (O, F) (75–90)	Ref. 136

Enamines of cholestenone 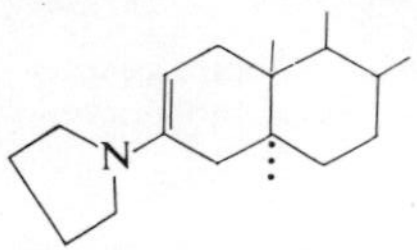	Benzene	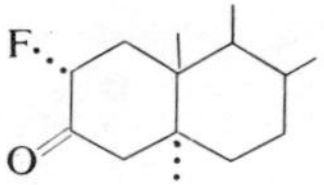 (72)	Ref. 137
Enamines of steroids 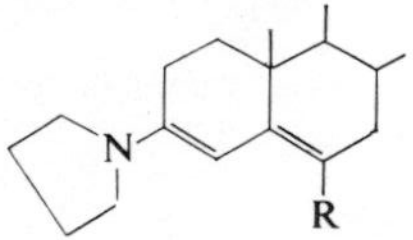 R = H or Me	Benzene, pyridine, ether, ether–pyridine 0°. Workup by basic hydrolysis	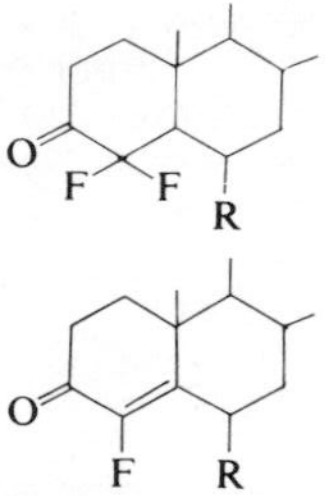 F F R; F R	Yield and relative amount of two products depends on solvent. Ref. 120b, 135a, 138
Estrone and derivatives 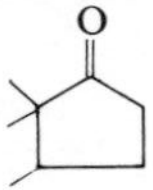	Added 4-pipecoline dehydrated in xylene; added benzene and little pyridine, then bubbled in $FClO_3$	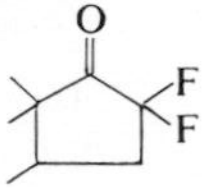 F F (15)	Enamine derivative believed formed and is considered reactive species for fluorination. Ref. 139
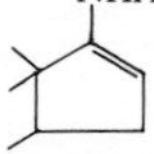 NHAc	Pyridine. Acid hydrolysis in workup	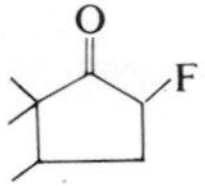 F (70)	Points out advantage that enamide in 17 position is easily prepared and reactive towards $FClO_3$. Enol acetates or other reagents very unreactive or hard to prepare. Ref. 140
Enamides of steroids 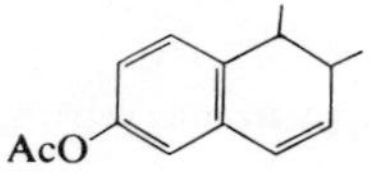 AcO	Dioxane-water?	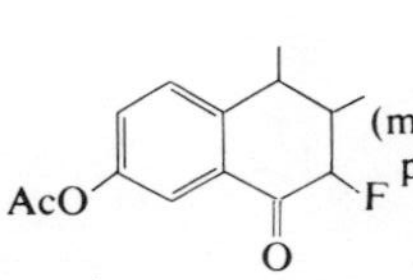 AcO F (major product) plus unidentified side products	Oxyfluorination reaction first reported for indene, suggested to go by addition of $F{-}ClO_3$, to get intermediate 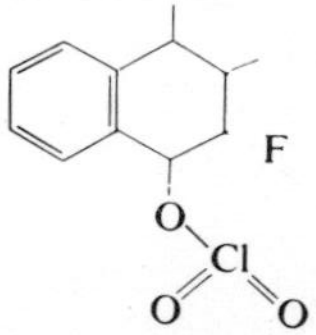 F O Cl O O Ref. 120

Table 5-3B (Cont.)

Reactants	Conditions[a]	Product(s) (% yield)	Comments
Estradiol and related phenolic steroids 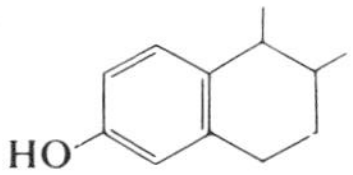	DMF or DMF-pyridine	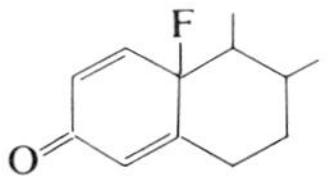 (70)	Discussed mechanism from aspect of electrophilic character of F in $FClO_3$. No attack at other positions. Ref. 141
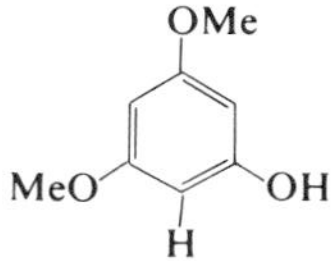	Pyridine or DMF	For X = F, equal amounts of 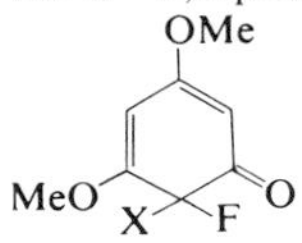 ... and 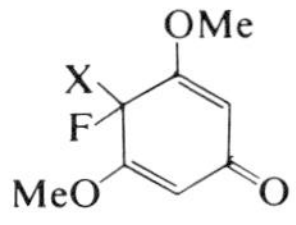 ... Also where X = Cl chlorophenol by-products are formed	Suggest *o*- and *p*-fluorophenols are precursors to product. Reactions applied in synthesis of fluorogriseofulvin analogs. Ref. 142
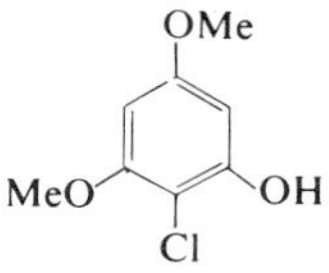 Sodium-2,6-dimethylphenoxide	(a) Hydrocarbon, dioxane (b) DMF or ethanol	(a) Dimer of 6-fluoro-2,6-dimethylcyclohexa-2,4-dien-1-one (20) (b) Dimer of 6-hydroxy-2,6-dimethylcyclohexa-2,4-dien-1-one	Discusses medium effect and mechanism for attack of anion on the ambident electrophile $FClO_3$. Ref. 117
2-Lithiothiophene or 5-Methyl-2-lithiothiophene 2-Lithiothionapthene	Ether, 0°	2-Fluorothiophene or 5-Methyl-2-fluorothiophene (44-49) 2-Fluorothionapthene (70)	Discusses mechanisms. Note, cannot be applied to other organometallics that do not delocalize charge. Phenyl lithium gives only perchlorylbenzene. Ref. 143

2,2,6,6-Tetramethylpiperidine	Bubble $FClO_3$ into 2% sodium hydroxide solution	*N*-fluoro-2,2,6,6-tetramethylpiperidine (50)	Studied reaction of $FClO_3$ with a selection of amines. N-F compounds isolated only with piperidine derivatives highly substituted in 2,6-position. Oxidation products found with other amines. Ref. 116

[a] Unless indicated otherwise $FClO_3$ bubbled into reaction mixture at about 10 to 25°.

[123] F. H. Dean and F. L. M. Pattison, *Can. J. Chem.*, **41**, 1833 (1963).

[124] L. K. Gottwald and E. Kun, *J. Org. Chem.*, **30**, 877 (1965).

[125] (a) H. Gershon and R. Parmegiani, *J. Med. Chem.*, **10**, 187 (1967). (b) H. Gershon, S. G. Schulman, and A. D. Spevack, *ibid.*, 536 (1967).

[126] V. Tolman and K. Veres, *Tetrahedron Letters*, **No. 32**, 3909 (1966).

[127] (a) H. Machleidt, *Ann.*, **667**, 24 (1963) and **676**, 66 (1964). (b) H. Machleidt and V. Hartman, *Ann.*, **679**, 9 (1964). (c) R. Tschesche and H. Machleidt, U.S. Patent 3,227,736 (1966); *Chem. Abstr.*, **64**, 9600c (1966).

[128] D. R. Grassetti, M. E. Brokke, and J. F. Murray, Jr., *J. Med. Chem.*, **9**, 149 (1966).

[129] H. Lange and M. Lipp, *Naturwissenschaften*, **47**, 397 (1960).

[130] A. D. Josey, U.S. Patent 3,114,763 (1963); *Chem. Abstr.*, **60**, 7918g (1964).

[131] H. H. Rennhard, R. K. Blackwood, and C. R. Stephens, *J. Am. Chem. Soc.*, **83**, 2775 (1961).

[132] C. H. Robinson, N. F. Bruce, and E. P. Oliveto, *J. Org. Chem.*, **28**, 975 (1963); see also C. H. Robinson, U.S. Patent 3,126,398 and 3,126,376 (1964), *Chem. Abstr.*, **60**, 14569a, 14568d (1964).

[133] (a) C. E. Holmlund, L. I. Feldman, H. M. Kissman, and M. J. Weiss, *J. Org. Chem.*, **27**, 2122 (1962). (b) H. M. Kissman, A. S. Hoffman, and M. J. Weiss, *ibid.*, **26**, 973 (1961). (c) H. M. Kissman, A. M. Small, and M. J. Weiss, *J. Am. Chem. Soc.*, **82**, 2312 (1960). (d) A. H. Nathan, B. J. Magerlein, and J. A. Hogg, *J. Org. Chem.*, **24**, 1517 (1959). (e) A. H. Nathan, J. C. Babcock, and J. A. Hogg, *J. Am. Chem. Soc.*, **82**, 1436 (1960).

[134] G. R. Allen, Jr, and N. A. Austin, *J. Org. Chem.*, **26**, 5245 (1961).

[135] (a) B. J. Magerlein, J. E. Pike, R. W. Jackson, G. E. Vandenberg, and F. Kagan, *J. Org. Chem.*, **29**, 2982 (1964). (b) B. J. Magerlein, F. H. Lincoln, R. D. Birkenmeyer, and F. Kagan, *Chem. Ind. (London)*, 2050 (1961). (c) B. M. Bloom, V. V. Bogert, and R. Pinson, Jr., *ibid.*, 1317 (1959). (d) Y. Osawa and M. Neeman, *J. Org. Chem.*, **32**, 3055 (1967).

[136] S. Nakanishi, K. Morita, and E. V. Jensen, *J. Am. Chem. Soc.*, **81**, 5259 (1959).

[137] R. B. Gabbard, E. V. Jensen, *J. Org. Chem.*, **23**, 1406 (1958).

[138] (a) S. Nakanishi, R. L. Morgan, and E. V. Jensen, *Chem. Ind. (London)*, 1136 (1960). (b) R. Joly and J. Warnant, *Bull. Soc. Chim. (France)*, 569 (1961). (c) S. Nakanishi, *Steroids*, **3**, 337 (1964).

[139] A. H. Goldkamp, *J. Med. and Pharm. Chem.*, **5**, 1176 (1962).

[140] (a) S. Nakanishi and E. V. Jensen, *J. Org. Chem.*, **27**, 702 (1962). (b) S. Nakanishi, *J. Med. Chem.*, **7**, 108 (1964).

[141] J. S. Mills, J. Barrera, E. Olivares, and H. Garcia, *J. Am. Chem. Soc.*, **82**, 5882 (1960).

[142] (a) D. Taub, *Chem. Ind. (London)*, 558 (1962). (b) D. Taub, C. H. Kuo, and N. L. Wendler, *ibid.*, 557 (1962).

[143] R. D. Schuetz, D. D. Taft, J. P. O'Brien, J. L. Shea, and H. M. Mark, *J. Org. Chem.*, **28**, 1420 (1963).

the high strength of the incipient C–F bond, then, provides a strong driving force, whereas for the heteroatoms the bond to fluorine is of high energy and does not form, but competing attack on chlorine gives only fluoride displacement.

We do not think that any of the above proposals are adequate to explain the mechanism of perchloryl fluoride reactions. We now propose a simple mechanism that accommodates all the experimental observations and should be useful in further synthetic work. We believe that the *more nucleophilic oxygen* or other heteroatom (compared to carbon) *always attacks the chlorine.* For localized nucleophiles (such as alkoxides), simple fluoride ion displacement occurs, but for the mesomeric ions an intramolecular (cyclic) transfer of fluoride

$$RO^- + FClO_3 \longrightarrow [RO\cdots ClO_3\cdots F] \longrightarrow ROClO_3 + F^-$$

$$R_2C{=}Y{-}O^- + FClO_3 \longrightarrow [\text{cyclic intermediate}] \longrightarrow R_2CF{-}Y{=}O + ClO_3^-$$

ion can occur in the intermediate to give a C–F bond. The high energy gained by forming the C–F bond provides a strong driving force for this fluoride transfer, and fluorine never has to achieve a highly energetically unfavorable state with positive charge. Formation of by-products such as ethyl perchlorate in ethanol is expected and would lead to side reactions of the type shown in Chart 5-4.

5-3C. OTHER HYDROGEN REPLACEMENT REACTIONS

Other positive fluorine-type compounds should also be capable of selective fluorination of organic substrates that have a canonical form with a free electron pair on carbon. Thus N–F compounds have been used for preparation of 2-fluoro-2-nitropropane and also difluoromalonic ester.[144]

$$Me_2\bar{C}NO_2 + F{-}N(CF_2)_5 \longrightarrow Me_2CFNO_2 + {}^-N(CF_2)_5$$

eventually converted into $(CF_2)_3(CO_2H)_2$

[144] R. E. Banks and G. E. Williamson, *Chem. Ind. (London)*, 1864 (1964).

This reaction does not have synthetic utility like that of perchloryl fluoride, but it is of interest from the point of view of mechanism. Again oxygen attack on nitrogen followed by intramolecular fluoride transfer is a good possibility.

Sulfur tetrafluoride has also been found to give hydrogen replacement in what is considered as an abnormal fluorination.[145] Low yields in fluorination of aliphatic ketones and some other carbonyl derivatives with SF_4 may well result because of a side reaction of α-fluorine substitution; conditions that suppress this side reaction will be discussed in the next section on SF_4 chemistry.

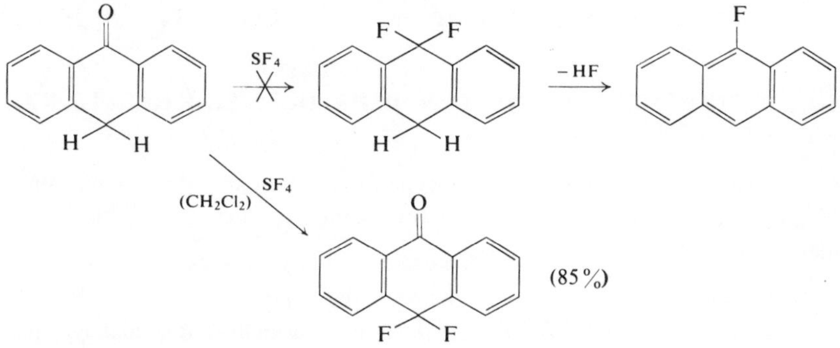

Summary

The following useful synthetic aspects were reviewed in this section:

1. Fluorine can be used selectively to convert N–H to N–F by use of cold dilute aqueous solutions or solid phase system with NaF support (also for synthesis of O–F compounds). The method is limited for general synthetic purposes because of problems in handling elementary fluorine and the hazardous properties of the products, but is a major advance in use of elementary fluorine.

2. Perchloryl fluoride is a valuable reagent for replacement of active hydrogen of a methylene compound by fluorine. This method has been applied extensively to preparation of compounds with potential biological interest.

5-4 REPLACEMENT OF X BY F

New developments discussed in this section are confined to two large areas:

1. Improved reagents, methods, and techniques for direct replacement of a substituent group X by fluorine,

$$RX \rightarrow RF$$

[145] D. E. Applequist and R. Searle, *J. Org. Chem.*, **29**, 987 (1964).

2. Direct conversion of a C=X group (mainly carbonyl) to a difluoromethylene group,

$$R_2C{=}X \longrightarrow R_2CF_2$$

$$\text{which includes } RC({=}X){-}Y \longrightarrow RCF_3$$

5-4A. REPLACEMENT OF HALOGEN OR HALOGEN-LIKE GROUPS, RX TO RF

Again this is an area of classical-type fluorination and was reviewed in Chapter 4. However, it is a useful selective method for introducing fluorine. The main topics are:

1. replacement of halogen, or halogen-like groups;
2. decomposition of fluoroformates, fluorosulfinates, or similar systems;
3. decomposition of diazonium salts (Schiemann reaction).

We will discuss new developments in each of these areas and critically evaluate each method for synthetic utility.

1. Use of Alkali Metal Fluorides

As pointed out in both Chapters 2 and 3, fluoride ion cannot be used to replace other halides in aqueous or protic solvents because of high free energy of solution. The use of anhydrous HF or SbF_3 in classical preparations of fluorocarbons from chlorocarbons is still of great importance, particularly industrially. A recent development is use of alkali metal fluorides in aprotic solvents, such as amides, that have high dielectric constants and often high boiling points. Very high-temperature reactions of stable chlorocarbons with alkali fluorides (sometimes molten salts) is another innovation that has some limited utility. Finally the advantages of several new agents, such as SF_4, α,α-difluoroamines, and AsF_3, will be discussed.

Aprotic solvents that have high dielectric constant can dissolve finite amounts of alkali metal fluorides and help stabilize ionic or highly polarized intermediates or transition states. Aqueous or protic solvents may do a better job at solvation and stabilization of ionic species, but as we discussed in Chapter 2, the high heat of hydration of fluoride ion completely outweighs all other factors. Parker[146] has studied the effect of solvation on the properties of anions in dipolar aprotic solvents. He points out several strong solvent-solute interactions; for anions, only ion-dipole and hydrogen bonding are important. These

[146] (a) A. J. Parker, *Quart. Rev. (London)*, **16**, 163 (1962). (b) J. Miller and A. J. Parker, *J. Am. Chem. Soc.*, **83**, 117 (1961). (c) A. J. Parker, *J. Chem. Soc.*, 1328 (1961).

interactions are particularly important in nucleophilic substitution reactions such as in replacement of fluoride for chloride or other groups, both in saturated (S_N2) and aromatic (S_NAr) systems. Parker showed that rate increases up to 10^5 are found for nucleophilic substitution reactions when the hydrogen-bonding capacity of the solvent ion is decreased (for example, from methanol to dimethylacetamide). He suggested that in dipolar aprotic solvents the solvation of anions reverses from what it is in protic solvents so that now solvation decreases in the series of anions:

$$\text{transition state ion} > \text{picrate}^- > I^- > SCN^- > Br^- > N_3^-, Cl^- >> F^-$$

The solvation effect is particularly striking for the smaller ions such as fluoride which are not easily solvated by large dipolar aprotic molecules. Solute–solvent interaction is one of the fundamental problems in chemistry which needs much more study[147] and obviously has great significance to practical problems such as synthesis of fluorinated compounds by substitution reactions.

The dipolar aprotic solvents that have been found most useful in fluoride ion replacements are amides (dimethylformamide and *N*-methylpyrrolidone) and sulfones (tetramethylene sulfone) that are high boiling so that high reaction temperatures can be used at atmospheric pressure. Since the time of the pioneer work in this area by Bergmann and Finger,[148] extensive studies on both aliphatic (saturated and unsaturated)[149] and aromatic[150] systems have been reported.

Lower boiling solvents, particularly acetonitrile, are also used for more active halides (often in pressure vessels) such as in the preparation of COF_2 and SF_4 from NaF with $COCl_2$ and SCl_2, respectively.[151] Seel[152] showed that sulfur

[147] Significant contributions to the problems of solvent interaction have recently been reported by R. W. Taft, G. B. Klingensmith, and S. Ehrenson, *J. Am. Chem. Soc.*, **87**, 3620 (1965), and by C. D. Ritchie and A. L. Pratt, *J. Am. Chem. Soc.*, **86**, 1571 (1964).

[148] (a) E. D. Bergmann and I. Blank, *J. Chem. Soc.*, 3786 (1953). (b) G. C. Finger and C. W. Kruse, *J. Am. Chem. Soc.*, 78, 6034 (1956). (c) G. C. Finger and L. D. Starr, *J. Am. Chem. Soc.*, **81**, 2674 (1959).

[149] (a) J. T. Maynard, *J. Org. Chem.*, **28**, 112 (1963). (b) K. Fukui, S. Yoneda, F. Taniniota, and H. Kitano, *J. Chem. Soc. Japan, Ind. Chem. Sect.*, **65**, 1179 (1962).

[150] (a) G. W. Holbrook, L. A. Loree, and O. R. Pierce, *J. Org. Chem.*, **31**, 1259 (1966). (b) G. Fuller, *J. Chem. Soc.*, 6264 (1965). (c) N. Ishikawa, *J. Chem. Soc. Japan, Pure Chem. Sect.*, **86**, 962 (1965), English abstract p. A57. (d) K. O. Christe and A. E. Pavlath, *Chem. Ber.*, **96**, 2537 (1963). (e) G. C. Finger, D. R. Dickerson, T. Adl, and T. Hodgins, *Chem. Commun.*, 430 (1965). (f) L. D. Starr and G. C. Finger, *Chem. Ind. (London)*, 1328 (1962). (g) G. C. Finger, L. D. Starr, D. R. Dickerson, H. S. Gutowsky, and J. Hamer, *J. Org. Chem.*, **28**, 1666 (1963). (h) R. D. Chambers, J. Hutchinson, and W. K. R. Musgrave, *Proc. Chem. Soc.*, 83, (1964). (i) W. T. Miller, J. H. Fried, and H. Goldwhite, *J. Am. Chem. Soc.*, **82**, 3091 (1960).

[151] (a) C. W. Tullock, F. S. Fawcett, W. C. Smith, and D. D. Coffman, *J. Am. Chem. Soc.*, **82**, 539 (1960). (b) C. W. Tullock, U.S. Patent 2,928,720 (1960); *Chem. Abstr.*, **54**, 25641d (1960). (c) C. W. Tullock and D. D. Coffman, *J. Org. Chem.*, **25**, 2016 (1960). (d) F. S. Fawcett and C. W. Tullock, U.S. Patent 3,088,975 (1963); *Chem. Abstr.*, **59**, 4821 (1963). (e) H. A. Pacini and A. E. Pavlath, *J. Chem. Soc.*, 5741 (1965).

[152] (a) F. Seel, H. Jonas, L. Riehl, and J. Langer, *Angew. Chem.*, **67**, 32 (1955). (b) F. Seel and L. Riehl, *Z. Anorg. Allgem. Chem.*, **282**, 293 (1955).

dioxide combines with potassium fluoride to give an adduct, potassium fluorosulfinate, KSO_2F, which is a very active fluorinating agent with concurrent loss of SO_2, usually in a polar aprotic solvent. This agent, which may be considered as a solvated form of KF, has been used to great advantage in preparation of phosphorus–fluorine and sulfur–fluorine compounds[153] and acid fluorides.[154]

$$RCOCl + KSO_2F \xrightarrow{100^\circ} \underset{(80\text{–}95\%)}{RCOF} + KCl + SO_2$$

An important advantage of the alkali metal fluoride–aprotic solvent systems is simplicity of the experimental procedure. The fluorinated product boils much lower than the starting chlorinated reactant and can be simply distilled from the reaction mixture as formed. Also the reagents require no special handling techniques (other than exclusion of water), and the reaction can be run in standard equipment.

For certain aliphatic systems, Maynard[149a] suggested that the reaction involves initial dechlorination and dehydrochlorination of the substrate followed by addition of the elements of hydrogen fluoride and finally replacement of all but highly hindered chlorine atoms. Consequently, products are often complex mixtures with hydrogen in place of halogen or are derived from addition or elimination reactions. Surprisingly, cyclic perchloroolefins go rapidly giving perfluoro compounds very cleanly. Also, Maynard examined nineteen polar solvents[149a]; in the aliphatic systems, *N*-methylpyrrolidone is preferred because of its stability at the required temperatures, its high boiling point, its low volatility, and its ready availability; dimethyl sulfone is probably as good but it suffers from the disadvantage of being solid at room temperature (mp 109°C!). In perhaloaromatic systems the replacement reaction was straightforward until three fluorine atoms had been introduced, but continued reaction gave not only further fluorination but also significant amounts of hydrogen-containing species.[150a] Perfluorinated products are obtained only in some specialized cases. Tetramethylene sulfone was claimed to be the preferred solvent for aromatic compounds.[150b]

Practically, potassium fluoride is the best alkali metal fluoride for replacements. Sodium fluoride is much less reactive, although it is used for more active halides and can be activated by complexing with an organic silicontrifluoride.[155a]

[153] (a) F. Seel and J. Langer, *Z. Anorg. Allgem. Chem.*, **295**, 316 (1958). (b) R. Schmutzler, *J. Inorg. Nucl. Chem.*, **25**, 335 (1963). (c) G. Allen, M. Barnard, J. Emsley, N. L. Paddock, and R. F. M. White, *Chem. Ind. (London)*, 952 (1963). (d) F. Seel and D. Golitz, *Chimia (Aarau)*, **17**, 207 (1963).

[154] F. Seel and J. Langer, *Chem. Ber.*, **91**, 2553 (1958).

[155] (a) R. Müller, C. Dathe, and H. J. Frey, *Chem. Ber.*, **99**, 1614 (1966), report that $Na_2(RSiF_5)$ with benzotrichloride gives benzotrifluoride in 80–90% yield compared to a 15% yield with NaF alone. (b) G. Schiemann, B. Cornils, and E. Schleusener, *Tetrahedron Letters*, 669 (1967), report that a high-temperature reaction of LiF with $CHBr_3$ gives only a few per cent exchange and considerable by-products; Na_3AlF_6 gives much greater amount of exchange.

Lithium fluoride gives some replacement under drastic conditions.[155b] Cesium fluoride (and also rubidium fluoride) are probably more active than potassium fluoride but not as readily available. Potassium bifluoride is also active but very corrosive in glass equipment. Micropulverization of the potassium fluoride seemed to help to decrease tar formation but did not significantly improve yields.

A series of representative examples are given in Table 5-4; for the most up-to-date and definitive discussion the work by Maynard[149a] and Pierce[150a] should be consulted first. Good mechanism studies, particularly on solvation effects, are needed here.

Some of the problems and limitations of the exchange reactions on perchlorinated compounds in a solvent are circumvented by high-temperature (400°–500°C range) exchange *without solvent,* particularly aromatic derivatives with an alkali metal fluoride. Compounds such as fluoranil,[156] pentafluoropyridine,[150h, 157] hexafluorobenzene,[158] and octafluoronaphthalene[159] were prepared by this route as well as aliphatic fluorinated compounds,[160, 161] some of which are valuable precursors to hexafluorobenzene.[158] Techniques with and without solvent have been compared; the no-solvent method has the advantage of less side products, but the disadvantage of requiring special high-temperature autoclaves or reactors which are not always conveniently set up for small-scale laboratory preparations. A number of examples of this method are given in Table 5-4; because of space limitation only the most pertinent have been selected, but this area has great potential and large numbers of patents and publications have appeared over the past decade.

2. New Reagents for Substitution

Several new or unusual reagents have recently been shown useful for selective replacement of halogens, hydroxyl, and other groups by fluorine. Specific examples to illustrate the use of each of these methods are given in Table 5-5.

a. *α-Fluorinated Amines,* such as (2-chloro-1,1,2-trifluorotriethyl)amine, are mild reagents that convert C–OH to C–F and have been used extensively in the

[156] K. Wallenfels and W. Draber, *Chem. Ber.*, **90**, 2819 (1957).

[157] (a) R. D. Chambers, J. Hutchinson, and W. K. R. Musgrave, *J. Chem. Soc.*, 3573 (1964). (b) R. E. Banks, R. N. Haszeldine, J. V. Latham, and I. M. Young, *J. Chem. Soc.*, 594 (1965).

[158] N. N. Vorozhtsov, Jr., V. E. Platonov, and G. G. Yakobson, *Bull. Acad. Sci. USSR, Div. Chem. Sci. (English Transl.)*, 1389 (1963).

[159] G. G. Yakobson, V. D. Shteingarts, and N. N. Vorozhtsov, Jr., *Bull. Acad. Sci. USSR, Div. Chem. Sci. (English Transl.)*, 1464 (1964).

[160] (a) W. Verbeek and W. Sundermeyer, *Angew. Chem. Intern. Ed. Engl.* **5**, 314 (1966). (b) G. W. Parshall, *J. Org. Chem.*, **27**, 4649 (1962).

[161] G. Urata, *J. Chem. Soc. Japan, Pure Chem. Sect.*, **83**, 936 (1962), English abstract p. A59.

Table 5-4

Replacement of Halides Using Alkali Metal Fluorides

Reactant	Conditions[a]	Product(s) (% yield)	Comments
$CCl_3CCl_2CCl_3$	Reactant added slowly to KF in NMP at 195°. Product distilled as formed	$CF_3Cl_2CF_3$ (69) $CF_3CH_2CF_3$, $CF_3CHClCF_3$, $CF_3CCl{=}CHF$, $CF_3CCl{=}CF_2$ } small amounts	Hexachlorocyclopropane and hexachloropropene gave same products but in different amounts. Gives comparison of yields for 19 dipolar aprotic solvents. Ref. 149a
$CCl_2{=}CClCCl{=}CCl_2$	As above	$CF_3CH{=}CFCF_3$ (65) (chiefly trans)	Other four carbon chlorinated species also studied. Ref. 149a
CCl_4	As above?	$CHCl_3$ (53) CHF_3 minor	Series of two carbon chlorinated species studied; fluorinated product usually in low yield. Ref. 149a
Octachlorocyclopentene	As above	Octafluorocyclopentene (72)	Series of cyclic chlorofluorolefins run. Ref. 149a
RCl R = aliphatic	As above or KF in several solvents	RF (high), RF (18-63), includes a variety of substituents such as ester, nitriles	Claimed as general useful procedure for making simple aliphatic fluorides. Comparison with protic solvents included. Ref. 149a
$C_6H_5CH(Br)CH_2NR_2$	KF, DMF 145-150°	$C_6H_5CH(F)CH_2NR_2$	Prepared for biological studies. Ref. 162a
$(Cl)(CN)C{=}C(Cl)(CN)$	Alkali metal fluoride, inert organic polar solvent, 140-250°	$NCC(F){=}C(F)CN$	Ref. 162b
$CF_3CCl{=}CHCl$	KF in $HCONH_2$	$CF_3CH{=}CHF$ (55)	Ref. 150i
$CH_2{=}C(CH_2Cl)CO_2C_2H_5$	KF in TMS, 170°	$CH_2{=}C(CH_2F)CO_2C_2H_5$ (60)	Used for polymer studies. Ref. 163
C_6Cl_6	KF in NMP, 195°	Sym. $C_6Cl_3F_3$ (23) $C_6Cl_2F_4$ (34) C_6ClF_5 (small)	Extension of study in aliphatic series. Ref. 148a

C_6Cl_6 [and also C_6Br_6, C_6Cl_5CN, $C_6Cl_4(CF_3)_2$]	KF in NMP, 200°, 4 hr	C_6F_5H (< .5) 1-H-2-Cl–C_6F_4 (< .5) C_6F_5Cl (< .5) 1,3-Cl_2–C_6F_4 (8) 1,2-Cl_2–C_6F_4 (1.4) 1,4-Cl_2–C_6F_4 (.1) 1-H-3,5-Cl_2–C_6F_3 (1.2) 1,3,5-Cl_3–C_6F_3 (49.2) 1,2,3,5-Cl_4–C_6F_2 (7) 1,2,3,4-Cl_4–C_6F_2 (35)	Overall yield, 70%, compares effects of solvent, temperature, time and substituents. Found TMS gave slightly higher product yields. Ref. 150a
C_6Cl_6 (also C_6Br_6 and C_6Cl_5X)	KF in TMS, 230-240°, 18 hr	C_6F_6 (.4) C_6ClF_5 (25) $C_6Cl_2F_4$ (24) Sym. $C_6Cl_3F_3$ (30)	Studied effect of solvent, temperature, substituents; discussed mechanism, particularly solvation effects of Parker (146). Ref. 150b
2-Chloropyridine	KF in dimethyl sulfone or TMS, 200-210°, 21 days	2-Fluoropyridine (50-59)	Ran a series of other pyridines containing up to four halogens, maximum of 2 fluorines introduced. Ref. 150g
Pentachloropyridine	KF in TMS or KF in NMP, 280°	3,5-Dichlorotrifluoropyridine (65) 3-Chlorotetrafluoropyridine (small)	Compared with autoclave method; without solvent at 500° get up to 90% of C_5F_5N. Ref. 150h
Tetrachlorophthalic anhydride	KF in DMF reflux, 5 hr	Octafluoroanthraquinone (2)	Suggested to form from tetrafluorophthalic anhydride by decarboxylation. See Ref. 164 for higher yield preparation in absence of solvent. Ref. 150d
CCl_3SCl	NaF in TMS, 170-250°	CF_3SCl and CF_3SSCF_3 in 3:2 ratio (47)	Large series of simple active halogen compounds including Cl on S and P. See few additional examples below. Ref. 151c
$(CNCl)_3$	NaF in TMS up to 248°	$(CNF)_3$ (74)	Ref. 151c
$SOCl_2$	NaF in CH_3CN, 80°	SOF_2 (77)	Several additional examples of SCl to SF included. Ref. 151c
$C_6H_5POCl_2$	NaF in TMS, 120°	$C_6H_5POF_2$ (65)	Several additional examples of PCl to PF included. Ref. 151c
SCl_2	NaF in CH_3CN, 80°	SF_4 (65-90)	Comparison made of solvent, fluorinating agent and conditions. Ref. 151a,b
$COCl_2$	NaF in CH_3CN	COF_2 (–)	60% chloride converted to fluoride. Ref. 151c,d

Table 5-4 (Cont.)

Reactant	Conditions[a]	Product(s) (% yield)	Comments
C_6H_5SOCl	NaF in CH_3CN	C_6H_5SOF (85)	Best method of series tried, including KF, NH_4F, AsF_3. Ref. 151e
$(NPCl_2)_3$ or $(NPCl_2)_4$	KSO_2F in $C_6H_5NO_2$, 100°	$(NPF_2)_3$ (64) or $(NPF_2)_4$ (75)	Many additional examples of conversion PCl to PF by KSO_2F are in literature. Ref. 153a
$RC(=O)Cl$ or R_3SiCl	KF adduct of $CF_3C(=O)CF_3$ in DMF at room temp.	$RC(=O)F$ (53-90) or R_3SiF (90)	Propose intramolecular transfer involving ester intermediate such as $RCO_2CF(CF_3)_2$. Ref. 166
2,4-Dinitrochlorobenzene	KF, CsF, or RbF at 195°, 2 hr	2,4-Dinitrofluorobenzene (51-98)	Compared fluorinating ability of alkali metal fluorides. NaF and LiF not reactive in the system. Also some comparison of solvents. Ref. 165
Hexachlorobenzene	KF, 450-500°	C_6F_6 (21) C_6F_5Cl (20) $C_6F_4Cl_2$ (14) $C_6F_3Cl_3$ (12)	Proposes stepwise replacement of fluorine; second and third go in meta. Ref. 158
Octachloronaphthalene	KF, 300-330°, 25 hr	$C_{10}F_8$ (24) $C_{10}F_7Cl$ (8)	Monochloro is chiefly 1-Cl isomer also prepared decafluorobiphenyl in same way. Ref. 159
Tetrachloro-*p*-quinone	KF, 200° or over	Tetrafluoro-*p*-quinone (25-30)	Also obtained other fluorochloroquinones. Studied reactions, temperature and fluorinating agent. Ref. 156
1,3,5-Tricyano-trichlorobenzene	KF, 250-275°	1,3,5-Tricyanotrifluoro benzene	Several other cyanofluorobenzenes also prepared. Ref. 167
Tetrachloroterephthaloyl chloride	CsF, 190°, 2 hr or KF at 230°, 30 hr	Tetrachloroterephthaloyl fluoride (72-87)	Ref. 168
Tetrachloroterephthaloyl fluoride	CsF, 220°, 26 hr	Tetrafluoroterephthaloyl fluoride (52)	KF not active for this replacement. Ref. 168
Pentachloropyridine	KF, 470-500°, 18 hr	Pentafluoropyridine (40-83) 3-Chlorotetrafluoropyridine (34-37) 3,5-Dichlorotrifluoropyridine (17-19)	Compared with reaction in solvent (see above). Ref. 157a,b

2,4,6-Trichloropyrimidine	KF, 230-300°, 1.5 hr Sb_2O_3 *added as catalyst*	2,4,6-Trifluoropyrimidine (48) $C_4N_2ClF_2$ (6)	Note use of Sb_2O_3 as catalyst; claimed to get higher yields. Ref. 169
$C_6Cl_3F_3$ (sym)	KF-KCl melt, 750-780°	C_6F_6 (12) C_6ClF_5 (35) $C_6Cl_2F_4$ (29)	Studied variation in salt-melt composition and temperatures on yield. Higher yields of C_6F_6 starting with C_6ClF_5. Claim advantage over solid-phase system of short contact time. Ref. 170
1,1,4,4-Tetrafluorotetrachlorocyclohexa-2,5-diene	KF, 400-700°	C_6F_6 (trace to 17); other chlorofluorobenzenes (up to 28) Octafluorocyclohexa-1,4-diene (1-3) and chlorofluorocyclohexadienes (up to 12)	Suggest aromatic compounds formed by fluoride ion catalyzed pyrolysis of the hexadienes. Ref. 160b
CCl(CN)=CCl(CN) (cis or trans)	KF, 250-300°	CF(CN)=CCl(CN) (62) CF(CN)=CF(CN) (14)	Studied variations in yields with conditions. Ref. 171
CH_3Cl_2 CH_2Cl_2	LiF-NaF-KF melt, 500°	CH_3F (85) CH_2F_2 (82)	Exchange to get CH_3F observed previously with CH_3I and AgF or HgF_2. CH_3F used for bromination and pyrolysis to C_6F_6. Ref. 160a
3,6-Dichlorophthalic anhydride	KF, 260-270°, 1 hr	3,6-Difluorophthalic anhydride (63)	Ref. 172
$CSCl_2$	LiF-NaF-KF melt, 560-600°	CSF_2 (61)	Ref. 173

[a] Common solvents abbreviated as: tetramethylene sulfone TMS, *N*-methylpyrrolidone NMP, dimethylformamide DMF; formula or name given for others.

[162] (a) N. B. Chapman, R. M. Scrowston, and R. Westwood, *J. Chem. Soc. (C)*, 528 (1967). (b) H. J. Cenci, French Patent 1,325,307 (1963); *Chem. Abstr.*, **59**, 13830h (1963).
[163] J. A. Powell and R. K. Graham, *J. Polymer Sci.*, Part A, Gen. Papers **3**, 3451 (1965).
[164] G. G. Yakobson, V. N. Odinokov, and N. N. Vorozhtsov, Jr., *Tetrahedron Letters*, **No. 49**, 4473 (1965).
[165] N. N. Vorozhtsov, Jr. and G. G. Yakobson, *J. Gen. Chem. USSR (English Transl.)*, **31**, 3459 (1961).
[166] A. G. Pittman and D. L. Sharp, *J. Org. Chem.*, **31**, 2316 (1966).
[167] K. Wallenfels, F. Witzler, and K. Friedrich, *Tetrahedron*, **23**, 1353 and 1845 (1967).
[168] G. G. Yakobson, V. N. Odinokov, T. D. Petrova, and N. N. Vorozhtsov, *J. Gen. Chem. USSR (Eng. Transl.)*, **34**, 2987 (1964).
[169] B. A. Ivin, V. I. Slesarev, and E. G. Sochilin, *ibid.*, 4183 (1964).
[170] H. C. Fielding, L. P. Gallimore, H. L. Roberts, and B. Tittle, *J. Chem. Soc. (C)*, 2142 (1966).
[171] K. Wallenfels and F. Witzler, *Tetrahedron*, **23**, 1359 (1967).
[172] E. D. Bergmann, M. Bentov, and A. Levy, *J. Chem. Soc.*, 1194 (1964).
[173] W. Sundermeyer and W. Meise, *Z. Anorg. Allgem. Chem.*, **317**, 334 (1962).

Table 5-5

Replacement of Substituent Group by Fluorine (by Fluorinating Agent or Decomposition)

Reactants	Fluorinating agents and conditions	Product(s) (% yield)	Comments
n-C_4H_9OH	$(ClCFHCF_2)N(C_2H_5)_2$, molar amount, alcohol added slowly, exothermic, product distilled	n-C_4H_9F (67)	Examples of other reactions also included. Ref. 174a
$CH_3CH_2\overset{O}{\overset{\Vert}{C}}OH$	As above	$CH_3CH_2\overset{O}{\overset{\Vert}{C}}F$ (40)	
$C_6H_5\overset{O}{\overset{\Vert}{C}}OH$	$(CF_2H)N(CH_3)_2 \cdot 3HF$	$C_6H_5\overset{O}{\overset{\Vert}{C}}F$ (90)	
3β-Hydroxyandrost-5-en-17-one and other steroids	$(ClCFHCF_2)N(C_2H_5)_2$ excess in CH_2Cl_2	3β-Fluoroandrost-5-en-17-one (96)	
3β-Hydroxypregn-5-en-20-one	(a) $(ClCFHCF_2)N(C_2H_5)_2$ molar equiv., dry THF, room temp., 10 min	(a) 3-β-Fluoropregn-5-en-20-one (52)	Also reported dehydration of benzamide to benzonitrile. Ref. 174c
HO, H (structure) 3β-Hydroxy-5α-androst-17-one	1.5 molar equivalents of amine in (a) dry CH_2Cl_2, 0° (b) dry THF, 0° or 25° (c) dry CH_3CN, 0°	R, H (structure) R = αF, β-$O_2CCHClF$, α-NHAc; H (structure); Yields of 4 products dependent on reaction conditions	Noted inversion of configuration in OH replacement. Also noted by-products of ester and from dehydration in some cases; a mechanism is proposed. Ref. 174b Shows course of reaction greatly dependent on reactants, solvent and temperature. Discusses results in terms of mechanism and proposes common intermediate. Ref. 174d

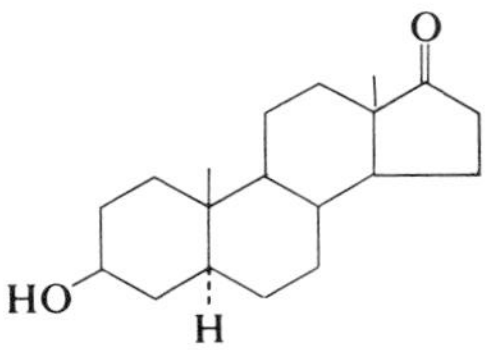

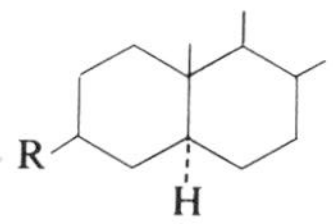

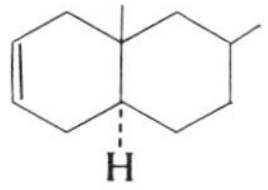

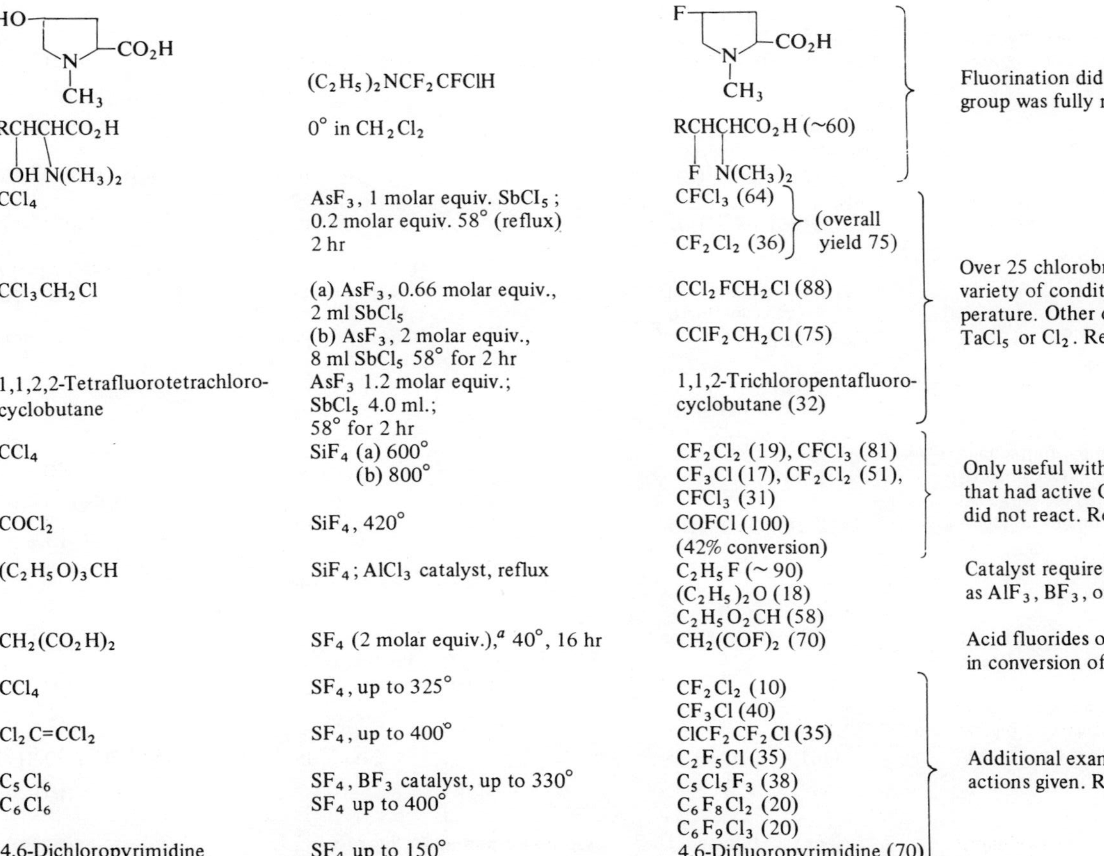

HO-pyrrolidine-CO_2H (N-CH_3)	$(C_2H_5)_2NCF_2CFClH$	F-pyrrolidine-CO_2H (N-CH_3)	Fluorination did not work unless amino group was fully methylated. Ref. 174f
$RCH(OH)CH(N(CH_3)_2)CO_2H$	0° in CH_2Cl_2	$RCH(F)CH(N(CH_3)_2)CO_2H$ (~60)	
CCl_4	AsF_3, 1 molar equiv. $SbCl_5$; 0.2 molar equiv. 58° (reflux) 2 hr	$CFCl_3$ (64), CF_2Cl_2 (36) (overall yield 75)	Over 25 chlorobromocarbons studied under variety of conditions of catalyst and temperature. Other catalysts were $NbCl_5$, $TaCl_5$ or Cl_2. Ref. 176
CCl_3CH_2Cl	(a) AsF_3, 0.66 molar equiv., 2 ml $SbCl_5$	CCl_2FCH_2Cl (88)	
	(b) AsF_3, 2 molar equiv., 8 ml $SbCl_5$ 58° for 2 hr	$CClF_2CH_2Cl$ (75)	
1,1,2,2-Tetrafluorotetrachlorocyclobutane	AsF_3 1.2 molar equiv.; $SbCl_5$ 4.0 ml.; 58° for 2 hr	1,1,2-Trichloropentafluorocyclobutane (32)	
CCl_4	SiF_4 (a) 600°	CF_2Cl_2 (19), $CFCl_3$ (81)	Only useful with very stable chlorocarbons that had active Cl. Chlorinated aromatics did not react. Ref. 177
	(b) 800°	CF_3Cl (17), CF_2Cl_2 (51), $CFCl_3$ (31)	
$COCl_2$	SiF_4, 420°	COFCl (100) (42% conversion)	
$(C_2H_5O)_3CH$	SiF_4; $AlCl_3$ catalyst, reflux	C_2H_5F (~ 90), $(C_2H_5)_2O$ (18), $C_2H_5O_2CH$ (58)	Catalyst required, but can use others such as AlF_3, BF_3, or $SnCl_4$. Ref. 178
$CH_2(CO_2H)_2$	SF_4 (2 molar equiv.),[a] 40°, 16 hr	$CH_2(COF)_2$ (70)	Acid fluorides often isolated as by-products in conversion of RCO_2H to RCF_3. Ref. 180
CCl_4	SF_4, up to 325°	CF_2Cl_2 (10), CF_3Cl (40)	Additional examples of replacement reactions given. Ref. 181
$Cl_2C{=}CCl_2$	SF_4, up to 400°	$ClCF_2CF_2Cl$ (35), C_2F_5Cl (35)	
C_5Cl_6	SF_4, BF_3 catalyst, up to 330°	$C_5Cl_5F_3$ (38)	
C_6Cl_6	SF_4 up to 400°	$C_6F_8Cl_2$ (20), $C_6F_9Cl_3$ (20)	
4,6-Dichloropyrimidine	SF_4 up to 150°	4,6-Difluoropyrimidine (70)	

Table 5-5 (Cont.)

Reactants	Fluorinating agents and conditions	Product(s) (% yield)	Comments
$C_6H_5C(OH)(CF_3)_2$	SF_4, 150°	$C_6H_5CF(CF_3)_2$ (91)	A series of α,α-bis(trifluoromethyl)benzyl alcohols converted with SF_4; HF used as solvent in some cases. Ref. 182
$C_6F_5CH(OH)CF_2X$ X = Cl or F	SF_4 (~ 2 molar equiv.); *N*-pentane solvent, 50-85° for 15 hr	$C_6F_5CHFCF_2X$ (90)	Used for synthesis of perfluorostyrene. Ref. 182b
$CHF_2C(CF_2H)(OH){-}C{\equiv}CH$	Excess SF_4, −78° to 25°, 16 hr	$CHF_2CF(CHF_2){-}C{\equiv}CH$ (32) $CHF_2C(CHF_2){=}C{=}CHF$ (26)	Reaction run on various hydroxyacetylenes derived from haloacetones. Product varied from all acetylenic fluoride to all rearranged allenic fluoride depending on substituent groups. Ref. 182d
Tropolone	SF_4, 60°, 10 hr	2-Fluorocyclohepta-2,4,6-trien-1-one (28)	Ref. 180
$(RO)_3B$ R is alkyl such as hexyl	SF_4 added to ester in ether at −70°, atm. pressure	RF (30-84)	Ref. 184
$(CH_3)_3SiCl$	SF_4, 20°	$(CH_3)_3SiF$ (90)	Examples of other conversions of Si–X → Si–F. Ref. 185
$(CH_3)_2Si(OC_2H_5)_2$	SF_4, 20°	$(CH_3)_2SiF_2$ (77) C_2H_5F (depends on $(C_2H_5O)_2SO$ condition)	
$C_2H_5OC(O)F$	(a) Pyridine (molar equiv.) 100-110° several hours (b) BF_3-etherate, 50°, product distilled as formed	C_2H_5F (63) C_2H_5F (86)	Several other aliphatic fluorides prepared in similar yields. Ref. 186

$C_6H_5OC(O)F$	Pyrolysis at 700 to 800° over Pt gauze in quartz reactor	C_6H_5F (70-90) C_6H_5OH (3-10)	$C_6H_5SC(O)F$ also works. Prepared fluoroformates from phenol and COClF. Ref. 187b
$XC_6H_4OC(O)F$ X = Br, F, CF_3, CH_3	As above	XC_6H_4F (15-57) by-products of XC_6H_4OH, XC_6H_5, C_6H_5F, $X_2C_6H_4$ usually in minor amounts	Yields depend on substituent; OCH_3 lost completely
m- or *p*-$C_6H_4(OCOF)_2$	As above	$C_6H_4F_2$ (6-15) FC_6H_4OCOF (16-90)	
n-$C_4H_9OS(O)F$	1% *N,N*-dimethylaniline, 200°	1-C_4H_9F (89) C_4H_8 (3)	2-C_4H_9F and $(C_4H_9O)_2SO$ were by-products under certain conditions. Ref. 190
n-C_4H_9OH	Excess SOF_2, with (0°) or without (100°) pyridine	$(n\text{-}C_4H_9O)_nSO$ (85)	Examining routes to alkyl fluorides from alcohols. Ref. 191
$(n\text{-}C_4H_9O)_2SO$	HF at 0°	*n*-C_4H_9F (45) *n*-C_4H_9OH (96)	
C_6H_5OH	SOF_4, 150° for 9 hr	C_6H_5F (30) $C_6H_5OSO_2F$ (15) $(C_6H_5O)_2SO_2$ (30)	Several substituted phenols also used. Ref. 193
$(C_6H_5)_2NC(OR)=\overset{+}{N}H_2\,F^-$ (or BF_4^-) R is primary alkyl group	Pyrolysis at 185° alone or with KHF_2, with and without solvent such as mesitylene, wax, or a glycol	RF (21-73)	Only olefin formed from secondary alcohols. Starting salt prepared from $(C_6H_5)_2NCN$, ROH and HF with *t*-BuOK. Ref. 197
$X\text{-}C_6H_4\text{-}N_2^+PF_6^-$	Salt prepared by diazotization using 65% hexafluorophosphonic acid; decomposed in mineral oil at 165-170°	$X\text{-}C_6H_4\text{-}F$ (60–75) (lower yields for some difficult ones where no aryl fluoride obtained by fluoroborate method)	Yields from decomposition of hexafluorophosphates always higher than from corresponding fluoroborate salts. Ref. 199

[a] All SF_4 reactions are run at autogenous pressure unless indicated otherwise.

steroid area.[174] The α-fluorinated amines are best prepared by adding a secondary amine to a fluoroolefin, for example (2-chloro-1,1,2-trifluorotriethyl)amine is prepared by bubbling trifluorochloroethylene through diethylamine.[175] They were first used for conversion of simple aliphatic alcohols and carboxylic acids to the corresponding alkyl or acyl fluorides.[174a] The conversion of a carbonyl group to CF_2 was reported for benzaldehyde but benzamide was dehydrated to benzonitrile[174c] (we will discuss these reactions in the next section on carbonyl replacements). Extensive studies have been done in the steroid area; on the basis of the accumulated evidence, a mechanism is proposed that an intermediate (**1**) is rapidly formed and decomposed to fluoro product, with or without inversion of configuration, depending on the starting alcohol. Knox and co-workers[174d, e] have shown that formation of by-products is highly dependent on the nature of the reacting alcohol (ability to form carbonium ion), solvent, and temperature.

X(C$_2$H$_5$)$_2$N—CF$_2$ ⇌ X(C$_2$H$_5$)$_2\overset{+}{N}$=C—F F$^-$ $\xrightarrow{StOH}$ X(C$_2$H$_5$)$_2$N—CF(OSt) + HF

→ X(C$_2$H$_5$)$_2\overset{+}{N}$=C—OSt F$^-$ (**1**)

1 $\xrightarrow{StO^-}$ StOSt

1 → dehydration

1 → X(C$_2$H$_5$)$_2$N—C=O + StF + olefin

1 $\xrightarrow{H_2O}$ X(O=)C—OSt

where X = —CHFCl, St = Steroid

The side reactions lead to ether or ester formation and to simple dehydration or dehydration accompanied by rearrangement. The intermediate (**1**) is proposed to be common for all products. By applying this mechanism, reasonable control and prediction of the course of the reaction are possible.

[174] (a) N. N. Yarovenko and M. A. Raksha, *J. Gen. Chem. USSR. (English Transl.)*, **29**, 2125 (1959). (b) D. E. Ayer, *Tetrahedron Letters*, 1065 (1962) and *J. Med. Chem.*, **6**, 608 (1963). (c) Z. Arnold, *Collection Czech. Chem. Commun.*, **28**, 2047 (1963). (d) L. H. Knox, E. Velarde, S. Berger, D. Cuadriello, and A. D. Cross, *Tetrahedron Letters*, 1249 (1962) and *J. Org. Chem.*, **29**, 2187 (1964). (e) L. H. Knox, E. Velarde, S. Berger, I. Delfin, R. Grezemkovsky, and A. D. Cross, *J. Org. Chem.*, **30**, 4160 (1965). (f) A. Cohen and E. D. Bergman, *Tetrahedron*, **22**, 3545 (1966).

[175] Other routes are reaction of an amide with PCl_5 followed by HF (Ref. 174c) or by SF_4 or COF_2 on an amide (see discussion of these reactions in next section on carbonyl replacements).

b. *Arsenic Trifluoride,* AsF_3, is useful for selective replacement of one to three chlorines in saturated haloalkanes[176] if a catalyst, such as antimony pentachloride, is present. At atmospheric pressure, one or two fluorine atoms can be introduced selectively into a CCl_3 group of polyhaloalkanes but mono- and dichloroalkanes give tar. The order of reactivity of groups to fluorination is in general the same as that found in classical work on antimony fluorides (see Section 4-5):

$$CCl_3 > CCl_2 > CHCl_2 \gg > CHCl > CH_2Cl$$

However, monochloro substitution (primary, secondary, and tertiary) gave only tar or dehydrohalogenation. The advantages of arsenic trifluoride are

1. liquid state;
2. readily miscible with many organic liquids, allowing a wide range of reaction conditions;
3. convenience in use in glassware at atmospheric pressure.

The degree of fluorination of selected halogenated groups was established and the influence of various reaction conditions (time, temperature, catalyst, and concentration) has been determined. No mechanism has been proposed, but the fluorination is probably through some active species, present in low concentration, which originates from the catalyst and is regenerated by arsenic trifluoride. Possible active species are $SbCl_4F$ or an ionic material such as $SbCl_4^+\ AsF_3Cl^-$.

c. *Silicon Tetrafluoride,* SiF_4, is also reported to be a new, low cost fluorinating agent for organic chlorine compounds.[177] The method was shown applicable to the preparation of fluorochloromethanes and carbonyl chloride fluoride, but is limited to reactions where the chlorinated starting material has high thermal stability. The reaction is run in the vapor phase at 500° to 800°C; however, silicon tetrafluoride was reported to react with ethyl orthoformates in the liquid phase with catalysts to give ethyl fluoride.[178] The significance of this fluorination is that silicon tetrafluoride is a waste product of fertilizer manufacture and, consequently, is a potentially low cost source of fluorine in synthesis of useful fluorocarbons. Hexafluorosilicates have also been used for fluorination[155a, 179] but appear economically unattractive because of technical problems such as corrosion and filtration.

[176] H. A. Pacini, E. G. Teach, F. H. Walker, and A. E. Pavlath, *Tetrahedron,* **22,** 1747 (1966).

[177] K. O. Christe and A. E. Pavlath, *J. Org. Chem.,* **29,** 3007 (1964).

[178] K. G. Mason, J. A. Sperry, and E. S. Stern, *J. Chem. Soc.,* 2558 (1963).

[179] (a) J. Dahmlos, U.S. Patents 2,935,531 (1960) and 3,113,157 (1963); *Chem. Abstr.,* **54,** 19481e (1960) and **56,** 3352b (1962). (b) Wasag-Chemie Aktiengesellschaft, British Patent 909,078 (1962); *Chem. Abstr.,* **57,** 9663c (1962).

d. *Sulfur Tetrafluoride,* SF_4, will be discussed below as a unique and valuable reagent for replacing a carbonyl oxygen by two fluorines, but it is also a convenient reagent for converting a carboxylic acid to an acid fluoride under very mild conditions.[180] Halogen exchange reactions of sulfur tetrafluoride were shown to occur[181] from chloroalkanes and alkenes, acyl chlorides, cyanuric chloride, and chloropyrimidines. Usually only partial substitution occurred; the degree of fluorination is a function of temperature and of type of halogen to be replaced. At the high temperatures required (200°–400°C), alkenes and aromatic compounds were converted to chlorofluoroalkanes and chlorofluorocyclohexenes, respectively.

Very acidic hydroxyl groups such as in carboxylic acids, as pointed out above (and also sulfonic acids), are easily replaced by fluoride using sulfur tetrafluoride. Normal aliphatic alcohols give chiefly ethane and very little alkyl fluoride,[180] but acidic alcohols such as α,α-bis(trifluoromethyl)benzyl alcohol give a high yield of fluoride.[182] Acetylenic alcohols derived from haloacetones are also readily fluorinated with sulfur tetrafluoride; the product is the result of

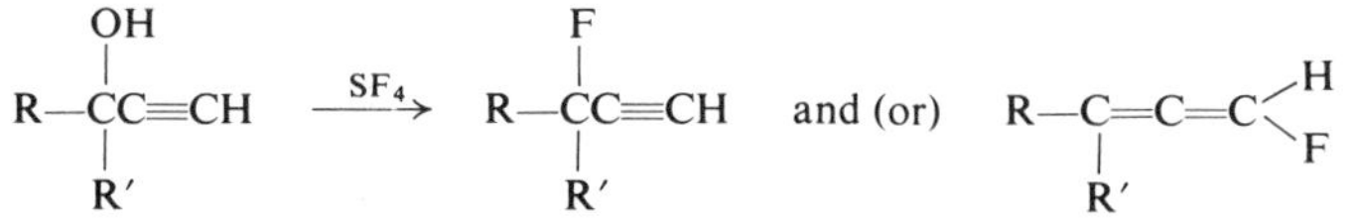

$$\text{R—C(OH)(R')—C}\equiv\text{CH} \xrightarrow{SF_4} \text{R—CF(R')—C}\equiv\text{CH} \quad \text{and (or)} \quad \text{R—C(R')}=\text{C}=\text{C}\begin{smallmatrix}\text{H}\\\text{F}\end{smallmatrix}$$

where R and R′ contain halogen (including fluorine)

direct displacement in some cases, but the allene is also formed, often as the only product.[182d] The rearranged product is suggested to come from a S_Ni decomposition of the intermediate:

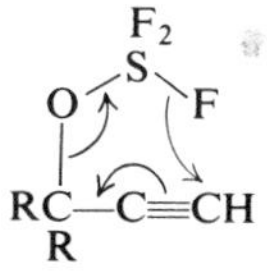

$$\text{RC(R)(O–}SF_2\text{–F)—C}\equiv\text{CH}$$

The mechanisms of these hydroxyl replacements must involve as intermediate the fluorosulfur ether, R_3COSF_3, which can decompose by a S_Ni mechanism or undergo a S_N2 displacement of OSF_3^- by F^- (from HF produced in initial step).

[180] W. R. Hasek, W. C. Smith, and V. A. Engelhardt, *J. Am. Chem. Soc.*, **82**, 543 (1960).

[181] C. W. Tullock, R. A. Carboni, R. J. Harder, W. C. Smith, and D. D. Coffman, *J. Am. Chem. Soc.*, **82**, 5107 (1960).

[182] (a) W. A. Sheppard, *J. Am. Chem. Soc.*, **87**, 2410 (1965). (b) J. M. Antanucci and L. A. Wall., *SPE Trans.*, **3**, 225 (1963). (c) E. E. Frisch, German Patent 1,207,369 (1965); *Chem. Abstr.*, **62**, 3935g (1965). (d) R. E. A. Dear and E. E. Gilbert, *J. Org. Chem.*, **33**, 819 (1968).

Inorganic hydroxy groups such as on phosphorus or arsenic are also readily replaced by fluorine using SF_4.[183]

In reaction of esters with SF_4,[180] the alkyl fluoride from the alcohol part is believed to arise from cleavage (possibly by HF?) of the α,α-difluoroether intermediate $R'CF_2OR$ from carbonyl fluorination. A similar reaction is found for boron esters, but olefin by-products are also found in some cases.[184]

$$(RO)_3B + SF_4 \xrightarrow[-70^\circ]{\text{ether}} RF$$

The fluorination of organosilicon compounds by means of SF_4 (and also SOF_2) has been described and compared with fluorination of carbon compounds.[185] The conversion of Si–X to Si–F, where X is Cl, OH, OSi, or OR goes very readily for a large series.

3. Indirect Replacements

Hydroxyl groups can be replaced indirectly by fluorine by decomposition of fluoroformates or fluorosulfinates.

$$ROH \nearrow ROC(=O)F \xrightarrow[\text{pyridine or } BF_3]{\Delta} RF + CO_2$$

$$ROH \searrow ROS(=O)F \longrightarrow RF + SO_2$$

In the original work, the fluoroformate was prepared from the chloroformate by halogen exchange and decomposed under very moderate conditions by use of pyridine or BF_3 catalysts.[186] The fluoroformates can now be prepared very easily from reaction of alcohols with COF_2 or COClF,[187] which are, in turn, prepared by inexpensive fluoride exchange reactions.[177, 151d, e] The exciting development here is that the fluoroformate decomposition can be extended to preparation of aromatic fluorides[187b] in high yields by using gas phase decomposition at 600° to 800°C. The method can be used to introduce up to three fluorines on the aromatic ring, and thermally stable substituents can be present

[183] W. C. Smith, *J. Am. Chem. Soc.*, **82**, 6176 (1960).

[184] A. Dornow and M. Siebrecht, *Chem. Ber.*, **95**, 763 (1962).

[185] R. Müller and D. Mross, *Z. Anorg. Allgem. Chem.*, **324**, 78 and 86 (1963).

[186] S. Nakanishi, T. C. Myers, and E. V. Jensen, *J. Am. Chem. Soc.*, **77**, 3099 and 5033 (1955).

[187] (a) P. E. Aldrich and W. A. Sheppard, *J. Org. Chem.*, **29**, 11 (1964). (b) K. O. Christe and A. E. Pavlath, *J. Org. Chem.*, **30**, 3170 and 4104 (1965) and *J. Org. Chem.*, **31**, 559 (1966).

on the ring. A S_Ni-type mechanism is proposed for this decomposition in contrast to the ionic mechanism suggested for the aliphatic cases. The aromatic fluoroformates cannot be induced to decompose in solution under ionic conditions

Aliphatic in solution: ionic

$$ROC(=O)F \longrightarrow \left[ROC(=O)^{+} + FX^{-} \right] \longrightarrow RF + CO_2$$

F^- or ^-FX

X = base such as BF_3 or pyridine

Aromatic in gas phase: S_Ni

$$X\text{-}C_6H_4\text{-}OC(=O)F \longrightarrow \left[X\text{-}C_6H_4 \overset{\delta+}{\cdots} \overset{\delta-}{O} \cdots C{=}O \cdots F \right] \longrightarrow X\text{-}C_6H_4\text{-}F + CO_2$$

but only disproportionate to diarylcarbonate; a free-radical mechanism where the aryl radicals have sufficient lifetime to separate from the cage was excluded because no isomerism was found. The authors favor an ion-pair intermediate, where the p-electrons of fluorine overlap with the π cloud of the aromatic ring, which can evolve CO_2 and collapse to arylfluoride by either a homolytic or heterolytic electron reorganization. But more extensive studies are needed on the mechanism of fluoroformate decomposition, particularly for the aliphatic series such as reported for chloroformates.[188] (Stereochemical studies have not even been done on the alkyl fluoroformates.)

Aliphatic chloroformates, where the aliphatic group does not readily form a carbonium ion, are converted directly to alkyl fluorides with silver fluoro-

$$\text{(1-apocamphyl)}\text{-}OCOCl \xrightarrow[C_6H_5Cl,\ 25^\circ]{AgBF_4} \underset{(52\%)}{\text{(1-apocamphyl)}\text{-}F} + \underset{(24\%)}{\text{(1-apocamphyl)}\text{-}C_6H_4Cl}$$

borate.[189] The only example was 1-apocamphane chloroformate. Phenyl chloroformate gave phenyl fluoroformate only and no evidence for a carboxylium ion, $[RO\overset{...}{-}C\overset{...}{=}O]^+$.

Decomposition of fluorosulfinate esters has recently been studied as a route to alkyl fluorides.[190] Yields were best when decomposition was carried out in

[188] E. S. Lewis and W. C. Herndon, *J. Am. Chem. Soc.*, **83**, 1955, 1959, and 1961 (1961).

[189] P. Beak, R. J. Trancik, J. B. Mooberry, and P. Y. Johnson, *J. Am. Chem. Soc.*, **88**, 4288 (1966).

[190] A. Zappel, *Chem. Ber.*, **94**, 873 (1961).

$$\mathrm{RO\overset{O}{\overset{\|}{S}}Cl} \xrightarrow{MF} \mathrm{RO\overset{O}{\overset{\|}{S}}F} \xleftarrow{?} \mathrm{ROH + SOF_2}$$

$$\mathrm{RO\overset{O}{\overset{\|}{S}}F} \xrightarrow{\text{(tertiary amine)}} \mathrm{RF + SO_2\ (+\ olefin + RO\overset{O}{\overset{\|}{S}}OR)}$$

the range of 100° to 200°C in a silver apparatus using a tertiary amine as catalyst. Considerable amount of by-product (isomeric fluoride, dialkyl sulfite ester, or olefin) was found under certain conditions but was almost negligible with the tertiary amines. The reaction of alcohols with thionyl fluoride, either with or without pyridine, was reported to give only the corresponding dialkyl sulfite and no alkyl fluoride.[191] The dialkyl sulfite was split by excess HF (anhydrous?) at 0°C to give alkyl fluoride, alcohol, and SO_2. By proper choice of conditions (solvent, temperature, and method of addition), the fluorosulfinate should be easily prepared from thionyl fluoride and alcohol; it could be subsequently converted to the alkyl fluoride using the conditions described by Zappel.[190] Also the conversion of carboxylic acids to acyl fluorides with thionyl fluoride is claimed.[192] In view of the results with COF_2 and SF_4, direct preparation of alkyl fluorides from the corresponding alcohol with SOF_2 should be possible if the right conditions are found.

Several arylfluorides have been synthesized in an unusual displacement of a phenolic group with SOF_4,[193] but arylfluorosulfonates and diarylsulfates are also formed. However, sulfuryl fluoride, SO_2F_2, in pyridine did not fluorinate a sugar and SO_2FCl introduced a Cl and not F.[194] Alkyl fluorides are formed from decomposition of alkyl fluorosulfonates in the presence of a compound such as KF which combines with the SO_3 by-product.[195]

Another method of forming alkyl fluorides by decomposition is pyrolysis of sulfonium salts that contain at least two aromatic groups bonded to sulfur; for example, triphenylsulfonium tetrafluoroborate was heated to get fluorobenzene.[196a] Another unusual route is through an alkylpseudouronium salt,[196b] as shown in the following equation:

$$\mathrm{(C_6H_5)_2NCN + HOCH_2R} \xrightarrow{t\text{-BuOK}} \mathrm{(C_6H_5)_2N\overset{NH}{\overset{\|}{C}}OCH_2R} \xrightarrow{HF}$$

$$(\mathrm{R = C_9H_{19}}) \qquad \mathrm{(C_6H_5)_2N\overset{\overset{+}{N}H_2F^-}{\overset{\|}{C}}OCH_2R} \xrightarrow{300^\circ} \mathrm{FCH_2R + (C_6H_5)_2N\overset{O}{\overset{\|}{C}}NH_2}$$

[191] K. Wiechert and R. Hoffmeister, *J. Prakt. Chem.* (4), **10**, 290 (1960).

[192] H. L. Roberts, British Patent 908,177 (1962); *Chem. Abstr.*, **58**, 5579b (1963).

[193] R. Cramer and D. D. Coffman, *J. Org. Chem.*, **26**, 4164 (1961).

[194] K. W. Buck and A. B. Foster, *J. Chem. Soc.*, 2217 (1963).

[195] A. Zappel and H. Jonas, French Patent 1,283,898 (1962).

[196] (a) Hashimoto Chem. Ind. Co., Ltd., *Japan. Publ.*, 11,362 (1964). (b) J. H. Amin, J. Newton, and F. L. M. Pattison, *Can. J. Chem.*, **43**, 3173 (1965).

Decomposition of aroylfluorides provides another new route to arylfluorides; tris(triphenylphosphine)rhodium chloride is required as catalyst.[197]

$$\mathrm{ArC(=O)F} \xrightarrow[80\text{–}120°]{[(C_6H_5)_3P]_3RhCl} \mathrm{ArF} + \mathrm{CO}$$

4. The Schiemann Reaction [198]

Several innovations have been reported in preparation of arylfluorides by decomposition of aryldiazonium salts (see Section 4-7). The most important and practical modification is use of aryldiazonium hexafluorophosphates.[199] The aryldiazonium hexafluorophosphates, $ArN_2^+PF_6^-$, are easily prepared in high purity (less soluble in water) and are stable in anhydrous form to both shock and heat, regardless of the substituent on the aromatic ring. The stability is an important safety advantage, since most diazonium salts are explosive when dry and complete absence of water is necessary to get a high yield in the decomposition to arylfluoride. Also the salts can be isolated in higher state of purity and give better yields of arylfluoride.

Another modification is use of tetrahydrofuran (THF) as solvent for preparation of the aryldiazonium tetrafluoroborate; the salt is filtered off, dried, and decomposed in an inert solvent.[200] The presence of THF during diazotization is claimed to give fluoroaromatics (in the subsequent pyrolysis step) in higher yields and of purer quality than when THF is absent. Dimethyl sulfoxide is also mentioned as being equally as effective as THF.

An "improved method" for decomposing aryldiazonium fluoroborates to arylfluorides at room temperature using copper powder or copper(I) chloride in acetone[201] is apparently not reproducible.[202]

Direct introduction of the diazonium group into an aromatic ring is possible if tertiary amino substituents are present and has been used as a route to some *p*-fluorobenzene derivatives.[203]

[197] G. A. Olah and P. Kreienbühl, *J. Org. Chem.*, **32**, 1614 (1967).

[198] For a review of this reaction including discussion of mechanisms, see A. Roe, *Org. Reactions,* **5**, 193(1949); H. Suschitzky, *Advan. Fluorine Chem.*, **4**, 1 (1965); and A. E. Pavlath and A. J. Leffler, *Aromatic Fluorine Compounds,* Am. Chem. Soc. Monograph No. 155, Reinhold, New York, pp. 12-16 and 42-45.

[199] (a) K. G. Rutherford, W. Redmond, and J. Rigamonti, *J. Org. Chem.*, **26**, 5149 (1961). (b) K. G. Rutherford and W. Redmond, *Org. Syn.*, **43**, 12 (1960).

[200] T. L. Fletcher and M. J. Namkung, *Chem. Ind. (London)*, 179 (1961).

[201] E. D. Bergmann, S. Berkovic, and R. Ikan, *J. Am. Chem. Soc.*, **78**, 6037 (1956).

[202] I. K. Barben and H. Suschitzky, *Chem. Ind. (London)*, 1039 (1957).

[203] C. Sellers and H. Suschitzky, *J. Chem. Soc.*, 6186 (1965).

Arylation by decomposition of diazonium fluoroborates has been discussed in terms of mechanisms of the decomposition reaction,[204] and two distinct mechanisms, ionic and radical, can apparently operate. No new evidence on the mechanism of pyrolysis of aryldiazonium salts to arylfluorides has been proposed since the reviews.[198]

5-4B. REPLACEMENT OF CARBONYL OXYGEN BY TWO FLUORINES

$$R_2C{=}O \longrightarrow R_2CF_2$$

A major advance in fluorination techniques was the discovery that sulfur tetrafluoride is a selective reagent for introducing fluorine into both organic and inorganic substrates, particularly for converting carbonyl to difluoromethylene and carboxyl to trifluoromethyl groups. Other reagents that can also effect these conversions have since been found but none with the versatility of sulfur tetrafluoride.

1. Sulfur Tetrafluoride

Sulfur tetrafluoride is readily prepared from sulfur dichloride and sodium fluoride in acetonitrile[151a, b] (as discussed in Section 5-4A.1) and is a stable gas, bp –40°C, which is rapidly hydrolyzed. Since it liberates HF on hydrolysis, it must be handled with caution in well-ventilated areas, taking precautions similar to those needed for handling phosgene.

An extensive survey and discussion of the scope, limitations, and mechanisms of fluorination of organic fluorine compounds was presented in the original publication by Hasek, Smith, and Engelhardt[180] and reviewed by Smith.[205] The various applications, extensions, and modifications of sulfur tetrafluoride fluorination are too numerous to compile completely in this review (we have already discussed simple halogen or hydroxyl replacements in Section 5-4A2.d), but a selection of representative examples to illustrate utility and conditions is given in Table 5-6. Smith's review[205] should be consulted for more extensive discussion.

Fluorination is usually done in stainless steel pressure vessels and the reaction is catalyzed by Lewis acids, particularly HF or BF_3, and controlled by dilution with solvents. The procedure recommended originally[180, 206] was to charge the reagent in a pressure vessel with an excess of SF_4 and a catalyst such as anhydrous HF or BF_3 (catalyst not used in some reactions) and heat for several

[204] (a) G. Valkanas and H. Hopff, *J. Org. Chem.*, **29**, 489 (1964). (b) R. Huisgen and W. D. Zahler, *Chem. Ber.*, **96**, 747 (1963).

[205] W. C. Smith, *Angew. Chem. Intern. Ed. Engl.*, **1**, 467 (1962).

[206] W. R. Hasek, *Org. Syn.*, **41**, 104 (1961).

Table 5-6

Fluorination of Carbonyl or Carboxyl Groups

Reactants	Fluorinating agents and conditions[a, b]	Product(s) (% yield)	Comments
$CH_3(CH_2)_5CO_2H$	SF_4, 100-130°, 10 hr	$CH_3(CH_2)_5CF_3$ (70-80)	Very detailed experimental directions. Ref. 206
Phthalic anhydride	SF_4, 350°, 11 hr	1,2-Bis(trifluoromethyl)-benzene (45)	Example selected for each class of compounds, many other examples reported. Ref. 180
1-Carbomethoxy-3,3,4,4-tetrafluorocyclobut-1-ene	SF_4, BF_3, 140°, 16 hr	1-(Trifluoromethyl)-3,3,4,4-tetrafluorocyclobut-1-ene (10)	
$C_6H_5CON(CH_3)_2$	SF_4, 130°, 6 hr	$C_6H_5CF_2N(CH_3)_2$ (17)	
C_6H_5CHO	SF_4, 150°, 6 hr	$C_6H_5CHF_2$ (81)	
$C_6H_5COCF_3$	SF_4, 100°, 8 hr	$C_6H_5CF_2CF_3$ (65)	
$COCl_2$	SF_4, TiF_4, 250°, 4 hr	CF_4 (90), COF_2 (9)	
Chloranil	SF_4, HF, 270°, 2.5 hr	1,1,4,4-Tetrafluorotetrachlorocyclohexa-2,5-diene (75)	
p-Benzoquinone	SF_4, 200°, 4 hr	1,2,4-Trifluorobenzene (30)	
$HO_2CC{\equiv}CCO_2H$	SF_4, 170°, 8 hr	$CF_3C{\equiv}CCF_3$ (80)	
$ArC(=O)R_f$ R_f is fluoroalkyl group	SF_4, 150°, 8 hr	$ArCF_2R_f$ (32-84)	Compounds used to study electronic effect of fluoroalkyl groups. Ref. 182a
$(CH_3)_2CHCO_2H$	SF_4, 160°, 9 hr	$(CH_3)_2CHCF_3$ (65)	Used for preparation of monomers in polymer studies. Ref. 216
2,2,3,3-Tetrafluorocyclobutane-carboxylic acid	SF_4, 130°, 16 hr	1-Trifluoromethyl-2,2,3,3-tetrafluorocyclobutane (67)	Used as intermediate in synthesis of fluorinated butadienes. Ref. 217
R_fCO_2Ar Ar = variety of substituted benzenes R_f = F or fluoroalkyl	SF_4, HF (1 to 3 molar equiv.), 175° for 8 hr	$ArOCF_2R_f$ (30-83)	Simple procedure for $ArOCF_3$ was to prepare $ArOC(=O)F$ in situ from phenol and COF_2 and react it directly with SF_4 (HF formed in first step acts as catalyst and solvent.) Ref. 207c
R_fCO_2R R = variety of electronegative substituted alkyl groups, R_f = fluoroalkyl, or F, or OR	SF_4, HF, as above	$ROCF_2R_f$ (21-54)	Includes some difunctional derivatives such as $CF_3OCH_2CH_2OCF_3$; works best if R is substituted by halogen or other electronegative groups. Ref. 207d

Tetramethylcyclobutan-1, 3-dione	SF_4, 160°, 60 hr	1,1,3,3-Tetrafluorotetramethylcyclobutane (73)	Used for pyrolysis to propenes. Ref. 219
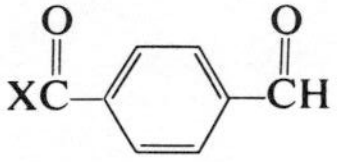 XC(=O)–C$_6$H$_4$–CH(=O) X = H or OH	SF_4, 150° for 13 hr	 HF_2C–C$_6$H$_4$–CF_2X (47–80) X = H or F	Products used for preparation of polymers. Ref. 218
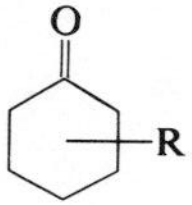 O, R	SF_4, 25 to 70°, 14 to 17 hr	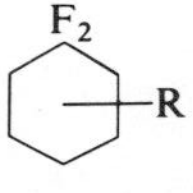 F_2, R (14–57)	Products used for conformation studies by nmr. Ref. 220
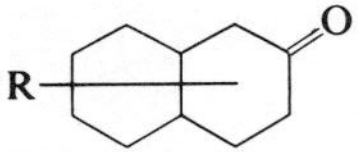 R, O	By method of Ref. 180	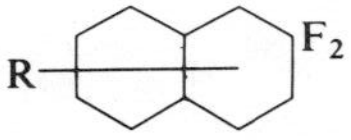 R, F_2 (15–48)	
Cholestan-3-one	SF_4 (approx. 200 molar equiv.); H_2O (approx. 20 molar equiv.); CH_2Cl_2 solvent, 16 hr, 15°	3,3-difluorocholestanone (80)	Examples for fluorination of a variety of steroids given including CO_2H to CF_3. Ref. 208
Series of steroids	SF_4, H_2O (to generate HF in situ); CH_2Cl_2 solvent, room temp.; 16 hr, similar to above example	Steroid products containing CF_3, CHF_2, and *gem*-difluoro groups at C-3, C-17 and C-20 positions (25-95)	First pointed out importance of HF as not only a catalyst but as a necessary solvent in some SF_4 fluorinations, mechanism discussed. Ref. 207a
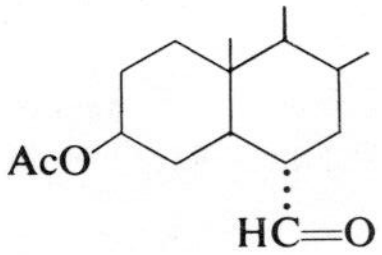 AcO, HC=O	SF_4, run as above	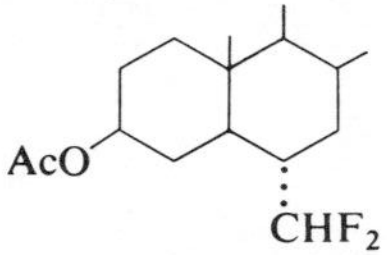 AcO, CHF_2 (60)	Used bis(methylene-dioxy) blocking group in 20,21-positions. Opened during 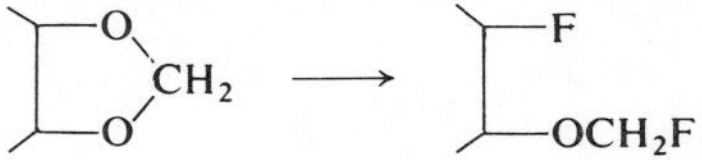 O, CH_2, O ⟶ F, OCH_2F ditions (using THF and larger excess of SF_4, HF). Ref. 221
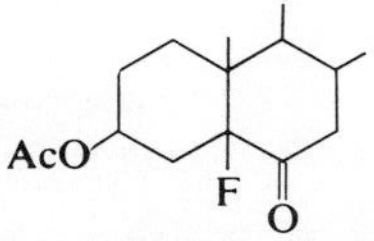 AcO, F, O Series of pregnanes and androstanes	SF_4, H_2O, solvent of CH_2Cl_2, 20°	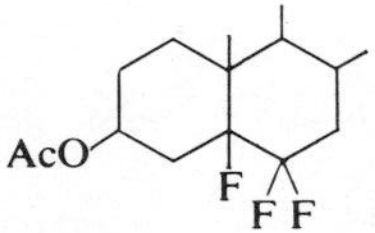 AcO, F, F F (29–82)	Product hydrolyzed with HF elimination to 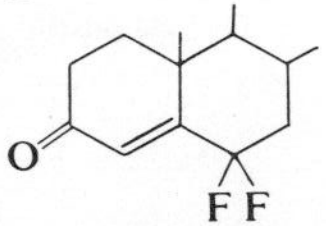 O, F F Starting fluoroketones from NOF addition reactions. Ref. 222

Table 5-6 (Cont.)

Reactants	Fluorinating agents and conditions[a, b]	Product(s) (% yield)	Comments
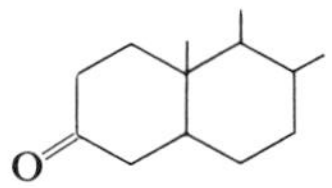 O (3-keto steroid) Series of 3 and 20-keto-steroids studied	SF_4, $CHCl_3$ solvent, generated HF in situ with ethanol	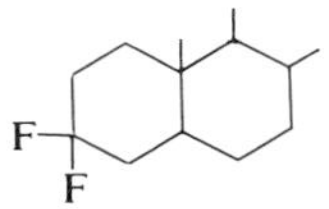 F, F (3,3-difluoro steroid) (2–37)	Procedure given in Ref. 207a appears better for steroids and other sensitive compounds. Ref. 223
$RCH(NH_2)CO_2H$ (R is for series of natural and unnatural amino acids)	SF_4, HF (5 molar equiv.), 120° for 8 hr	$RCH(NH_2)CF_3$ (2-80)	Other acids containing amino groups also fluorinated. Ref. 207b
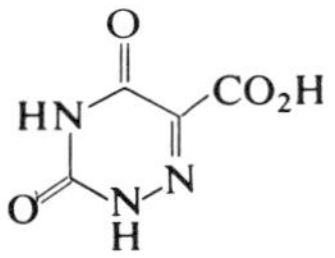 6-azauracil-5-carboxylic acid (CO_2H)	SF_4, 40 molar excess; H_2O, 4 molar excess; 50° for 24 hr	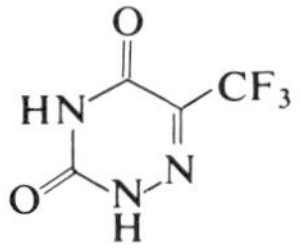 CF_3 (53)	Used for biological studies, follow-up of earlier work (Ref. 225). Ref. 224
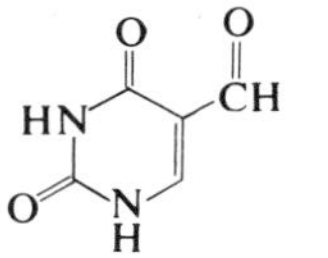 5-formyluracil (CH=O)	SF_4, 50 molar excess; H_2O, 8 molar excess; 100° for 10 hr	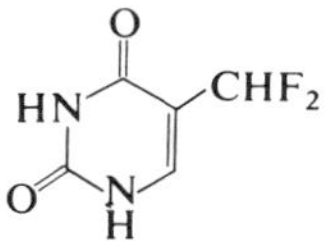 CHF_2 (60)	Used for biological studies (cancer research). Ref. 225
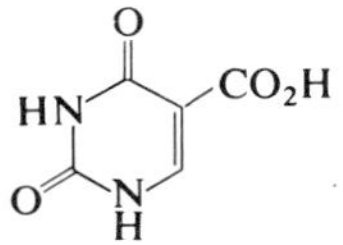 uracil-5-carboxylic acid (CO_2H)	SF_4, conditions as above	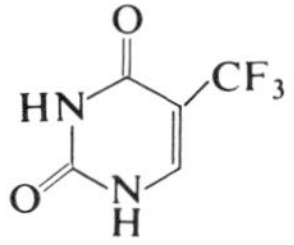 CF_3 (77)	Used in biological studies (cancer research) also has been prepared by other methods (Ref. 227). Ref. 226

ArN(COF)$_2$	SF_4, HF (5 molar equiv.); (CCl_3F solvent in some cases); 140° for 2 hr	ArN(CF_3)(COF) (22) ArN(CF_3)$_2$ (2-5)	Reaction difficult to reproduce. Went best with electron withdrawing groups on ring. Ref. 207e
ArN(CF_3)(COF)	As above	ArN(CF_3)$_2$ (2-39)	
Series of nitrogen compounds such as urea, cyanamide, tetrazoles	SF_4, 150 to 300°, 2 to 60 hr	$CF_3N{=}SF_2$ was only fluorinated product (2-60); except CF_4(12) from 5-aminotetrazole	Ref. 228
$CH_2{-}S{-}C({=}S){-}S{-}CH_2$ (cyclic, 1,3-dithiolane-2-thione)	SF_4, 110°, 8 hr	$CH_2{-}S{-}CF_2{-}S{-}CH_2$ (cyclic) (82)	Also included examples of thiuram disulfides converted to trifluoromethyl amines Ref. 210
CS_2	SF_4, AsF_3 catalyst, 475°, 2 hr	CF_4 (100)	
CS_2	SF_4, BF_3 catalyst, 180°	$(CF_3S)_2$ (small) $(CF_3S)_2S$ (28)	
$C_6H_5AsO(OH)_2$	SF_4, 70°, 10 hr	$C_6H_5AsF_4$ (45)	Additional examples of this type of fluorination given. Ref. 183
$C_6H_5POX_2$ X = F or C_6H_5	SF_4, 150°, 10 hr	$C_6H_5PF_4$ (58-62)	
$(C_6H_5)_3P{=}O$	SF_4, 150°, 6 hr	$(C_6H_5)_3PF_2$ (67)	
C_6H_5CHO	$C_6H_5SF_3$, 50 to 70°	$C_6H_5CF_2H$ (71-80) C_6H_5SOF (82-89)	Advantage over SF_4 is reaction run with standard equipment at atmospheric pressure. Very detailed directions given in Ref. 209b. Ref. 209
$CH_3(CH_2)_5CO_2H$	$C_6H_5SF_3$, 110-125°, 2 hr in Teflon reactor	$CH_3(CH_2)_5CF_3$ (28) $CH_3(CH_2)_5COF$ (44)	
C_6H_5CHO	$HCF_2N(CH_3)_2{\cdot}3HF$, 85°, 3 hr	$C_6H_5CF_2H$ (74)	This reaction mentioned as part of a general study of fluorination by replacements using α-fluorinated amines. Ref. 174c
C_6H_5CHO	COF_2, 150°, 12 hr	$C_6H_5CF_2H$ (80)	Series of examples of fluorinations with COF_2 given. Comparison with SF_4 and mechanisms discussed. Intermediate 1-fluorocyclohexyl fluoroformate isolated from cyclohexanone. Ref. 65
Cyclohexanone	(1) COF_2, pyridine, 50°, 12 hr (2) $BF_3\cdot$ etherate in hexane, reflux	1,1-Difluorocyclohexane (42 overall)	
$(CH_3)_2NCHO$	COF_2, 25°, 12 hr	$HCF_2N(CH_3)_2$ (82)	
$(CH_3)NHCHO$	COF_2, 75°, 12 hr	$HCF_2N(CH_3)COF$ (6) $HCON(CH_3)COF$ (34)	

Table 5-6 (Cont.)

Reactants	Fluorinating agents and conditions[a, b]	Product(s) (% yield)	Comments
Acetone	BrF_3 in HF at $-15°$	$CH_3CF_2CH_3$ (92)	Product further attacked by BrF_3 unless volatile enough to distill as formed. Ref. 215
Methyl ethyl ketone	As above	CH_3CF_2H, CH_3CF_3 (approx. equal amount, total about 90)	
CH_3CN	BrF_3 in HF at about $-80°$	CH_3CF_3 (81)	
$CH_3CH_2CH_2CN$	As above	6-Product mixture (29). Major fraction (66) assigned structure of expected product $CH_3CH_2CH_2CF_3$	

[a] All reactions with SF_4 or COF_2 are run in stainless steel autoclaves at autogenous pressure (unless indicated otherwise). Procedure requires charging all reactants in autoclave cooled to $-80°$, evacuating and charging SF_4 (or COF_2).

[b] Unless indicated otherwise, the fluorinating agent is used in equivalent amounts to moderate excess (usually 1 to 2 equivalents).

[216] E. B. Davidson and C. G. Overberger, *J. Org. Chem.*, **27**, 2267 (1962).

[217] R. E. Putnam and J. E. Castle, *J. Am. Chem. Soc.*, **83**, 389 (1961).

[218] S. A. Fuqua, R. M. Parkhurst, and R. M. Silverstein, *Tetrahedron*, **20**, 1625 (1964).

[219] H. G. Gilch, *J. Org. Chem.*, **30**, 4392 (1965).

[220] (a) S. L. Spassov, D. L. Griffith, E. S. Glazer, K. Nagarajan, and J. D. Roberts, *J. Am. Chem. Soc.*, **89**, 88 (1967). (b) J. T. Gerig and J. D. Roberts, *ibid.*, **88**, 2791 (1966).

[221] D. G. Martin and J. E. Pike, *J. Org. Chem.*, **27**, 4086 (1962).

[222] G. A. Boswell, Jr., *ibid.*, **31**, 991 (1966).

[223] J. Tadanier and W. Cole, *ibid.*, **26**, 2436 (1961).

[224] M. P. Mertes and S. E. Saheb, *J. Heterocyclic Chem.*, **2**, 491 (1965).

[225] M. P. Mertes and S. E. Saheb, *J. Med. Chem.*, **6**, 619 (1963).

[226] M. P. Mertes and S. E. Saheb, *J. Pharm. Sciences*, **52**, 508 (1963).

[227] C. Heidelberger, D. Parsons, and D. C. Remy, *J. Am. Chem. Soc.*, **84**, 3597 (1962).

[228] B. Marinier and J. L. Boivin, *Can. J. Chem.*, **42**, 1759 (1964).

hours at autogenous pressure. The preferred reaction temperature was 50°–100°C for aldehydes and ketones, and 100°–200°C for acids. For carbonyl groups that are highly hindered or deactivated by electron-withdrawing groups, more strenuous conditions are required. Anhydrous hydrogen fluoride can serve an important function as solvent as well as catalyst in certain cases.[207] For sensitive compounds, such as steroids, a modified procedure is recommended[207a, 208]; about 10 molar excess of SF_4 is added to a solution of the reactant in methylene chloride containing some water. The water reacts with some of the SF_4 to produce HF, and the reaction mixture is allowed to stand for 15 to 20 hours at 20°C. Steroid carbonyls are fluorinated in high yield with no significant amount of decomposition by this procedure. Large quantities of HF are needed if a free amino group is present,[207b] or in the fluorination of fluoroformates to trifluoromethyl ethers[207c,d] and of fluoroformylanilines to *N*-trifluoromethylanilines.[207e] Actually in the preparation of the trifluoromethyl ethers, the molar quantity of HF generated in the prior reaction of COF_2 with the alcohol is sufficient to serve as catalyst and solvent.

$$\mathrm{ROH + COF_2 \longrightarrow RO\overset{O}{\overset{\|}{C}}F + HF \xrightarrow[175^\circ]{SF_4} ROCF_3 + SOF_2}$$

A convenient extension is use of phenylsulfur trifluoride (or other alkyl- or arylsulfur trifluorides) in place of SF_4.[209] The phenylsulfur trifluoride is conveniently prepared in standard glass laboratory equipment from phenyl disulfide and silver(II) fluoride and can be used for fluorination at atmospheric pressure in standard equipment—a major advantage for small-scale laboratory synthesis where pressure equipment is not readily available.

Carbonyls of aldehydes and ketones are easily fluorinated to difluoromethylene groups, usually preferentially to an acid group that is in the same molecule (acid converted to acid fluoride only). Both acids and esters are converted to trifluoromethyl groups, but an acid can usually be fluorinated preferentially to an ester. In the few examples studied, thiocarbonyl compounds are also fluorinated like carbonyl.[210] Amides can be converted to α,α-difluoroamines, but this reaction is very difficult to control. Alcohols, as discussed in the previous section, can be converted to the corresponding alkyl fluorides in

[207] (a) D. G. Martin and F. Kagan, *J. Org. Chem.*, **27**, 3164 (1962). (b) M. S. Raasch, *J. Org. Chem.*, **27**, 1406 (1962). (c) W. A. Sheppard, *J. Org. Chem.*, **29**, 1 (1964). (d) P. E. Aldrich and W. A. Sheppard, *J. Org. Chem.*, **29**, 11 (1964). (e) F. S. Fawcett and W. A. Sheppard, *J. Am. Chem. Soc.*, **87**, 4341 (1965).

[208] F. Kagan and D. G. Martin, U.S. Patent 3,211,723 (1965); *Chem. Abstr.*, **60**, 641b (1964).

[209] (a) W. A. Sheppard, *J. Am. Chem. Soc.*, **84**, 3058 (1962). (b) W. A. Sheppard, *Org. Syn.*, **44**, 39 and 82 (1964).

[210] R. J. Harder and W. C. Smith, *J. Am. Chem. Soc.*, **83**, 3422 (1961).

many cases. In general, active hydrogen is not tolerated in the reaction, but free amino groups can be protected as the HF salt.[207b] As discussed in Section 5-2F, nitriles, isocyanates,[89] and some other nitrogen-containing groups[211] react by an addition rearrangement mechanism to give iminosulfur difluorides.

There are some side reactions and problems in running SF_4 fluorinations:

1. Tar formation or general decomposition often occurs because the reaction temperature is too high or because of impurities (Cl_2 or HCl) in the SF_4.[212]

2. Hydrogen substitution through the enol form of the carbonyl compound is known[145] and is probably a troublesome side reaction leading to decomposition products, particularly for aliphatic aldehydes and ketones. Use of HF or BF_3 helps reduce this problem.

3. Freidel–Crafts-type reactions are occasionally observed with an activated aromatic compound[207c] and may lead to tarry products if activated aromatic rings are present.

A simple work-up procedure is to dissolve the product in an inert solvent such as dichloromethane or trichlorofluoromethane and to add an excess of sodium fluoride (conveniently as pellets) to remove all HF (let stand for several hours or overnight), filter, and distill.

A two-step mechanism was proposed originally for the reaction.[180]

$$\gt C{=}O \xrightarrow{XF_n} -\overset{\delta+}{C}{\cdots}O{\cdots}\overset{\delta-}{XF_n} \xrightarrow{SF_4} \left[-\overset{\delta+}{C}-O-\overset{\delta-}{XF_n} \quad F-SF_3 \right]$$

$$\downarrow$$

$$\left[F-C(-O-SF_2-F) \quad F-XF_{n-1} \right] \longleftarrow -\underset{F}{\overset{|}{C}}-O-SF_3 + XF_n \quad \mathbf{1}$$

$$\downarrow$$

$$-\underset{F}{\overset{|}{C}}-F + O{=}SF_2 + XF_n$$

[211] B. Marnier and J. L. Boivin, *Can. J. Chem.*, **42,** 1759 (1964).

[212] G. E. Arth and J. Fried, U.S. Patent 3,046,094 (1962); *Chem. Abstr.*, **57,** 9460e (1962), claim a method for removing contaminants of chlorinating agents by contacting SF_4 with liquid mercury. The purified SF_4 is then useful for fluorination of ketone groups in steroids.

The carbonyl group is suggested to be polarized by coordination with the Lewis acid XF_n so that the SF_4 can react in a slow step, as shown, to give the $R_2\underset{|}{\overset{}{C}}(F)OSF_3$ intermediate. This intermediate can decompose to fluoride and SOF_2 in a rapid step with XF_n catalysts, analogous to the decomposition of alkychloroformates[186] or chlorosulfites.[213] The fluorination could also occur by the complex of SF_4 with the Lewis acid which could have an ionic form:

$$BF_3 \cdot SF_4 \rightleftharpoons BF_4^- SF_3^+ \tag{214}$$

The initial attack on the carbonyl could then be by SF_3^+ to give $+\overset{|}{\underset{|}{C}}-O-SF_3$ which picks up F^- and decomposes as before. Another possibility is that HF adds across the carbonyl to give an intermediate $F\overset{|}{\underset{|}{C}}OH$ which is rapidly converted to gem-difluorides by sulfur tetrafluoride.[207a] Much work is yet needed to understand the mechanism of this fluorination.

The reaction of arylsulfur trifluorides with benzaldehydes is particularly suited to mechanism studies, since the reaction can be run in solvents at atmospheric pressure over a reasonable temperature range, and effects of substituents on both aryl rings can be used to deduce the charge distribution in the transition state[209a]; spectral techniques should be useful to detect intermediates.

Safety problems in handling and use of sulfur fluorine compounds must be recognized.[205] As pointed out above, SF_4 is easily hydrolyzed and HF is a hydrolysis product. Also HF is the solvent, catalyst, or by-product in many of the fluorination reactions. Thus, in addition to the inherent inhalation toxicity of sulfur tetrafluoride equivalent to that of phosgene,[151] the extremely noxious hazardous properties of HF both in inhalation and in skin contact must be guarded against. All reactions with SF_4 or other sulfur fluorine compounds should be handled only in well-ventilated areas, wearing proper protective clothing and rubber gloves to avoid HF burns.

Although sulfur tetrafluoride and arylsulfur trifluorides are susceptible to rapid hydrolysis, they can be handled and manipulated in glass equipment provided that the equipment has been thoroughly dried (under vacuum or as done for Grignard reactions). Sulfur tetrafluoride is best stored in stainless steel cylinders and arylsulfur trifluorides in Teflon® containers.

[213] D. J. Cram, *J. Am. Chem. Soc.*, **75**, 332 (1953).

[214] Stable adducts of SF_4 were first described by N. Bartlett and P. L. Robinson, *Chem. Ind. (London)*, 1351 (1956), and were discussed further by A. L. Oppegard, W. C. Smith, E. L. Muetterties, and V. A. Engelhardt, *J. Am. Chem. Soc.*, **82**, 3835 (1960).

2. Other Reagents

The conversion of benzaldehyde to α,α-difluorotoluene with an α-fluorinated amine was mentioned earlier,[174c] but apparently not examined further. Fawcett, Tullock, and Coffman[65] showed that carbonyl fluoride, COF_2, has utility in fluorination of carbonyl groups to difluoromethylene; selected examples are given in Table 5-6. Carbonyl fluoride and sulfur tetrafluoride have been compared as to efficiency in oxygen replacement.[65] Carbonyl fluoride is more effective with amides, but is considerably less reactive with acids to give the CF_3 group; it is susceptible to catalysis by bases (pyridine, dimethylformamide, cesium fluoride) but gives isolable intermediates in some cases. In contrast, sulfur tetrafluoride is more reactive in forming CF_3 compounds from acids and its reactions are catalyzed by Lewis acids. A major economic advantage of carbonyl fluoride is that all the fluorine is utilized and carbon dioxide is the product, whereas with sulfur tetrafluoride in a similar reaction thionyl fluoride is the product.

The mechanism, based on accumulated data for other carbonyl fluoride reactions, is suggested to involve fluoride ion addition to (or base polarization of) the carbonyl, which displaces fluoride from carbonyl fluoride. The intermediate,

$$(1)\quad \underset{\text{(or B)}}{F^-} + \rangle C{=}O \rightleftharpoons F\overset{|}{\underset{|}{C}}O^- \xrightarrow{COF_2} F\overset{|}{\underset{|}{C}}{-}O\overset{O}{\overset{\|}{C}}F + F^-\ \text{(or B)}$$

$$(2)\quad F\overset{|}{\underset{|}{C}}{-}O\overset{O}{\overset{\|}{C}}F \xrightarrow[\text{catalyzed}]{F^-\text{ or base}} F\overset{|}{\underset{|}{C}}F + CO_2$$

α-fluoroalkylfluoroformate, is isolated in some cases but no doubt readily decomposes, particularly by base catalysis, as noted earlier for fluoroformates,[186] to a difluoromethylene product and carbon dioxide.

Bromine trifluoride has limited utility for conversion of carbonyls or ketones to difluoromethylene and nitriles to trifluoromethyl groups.[215] The reactions were carried out in a solution of hydrogen fluoride at –20°C (iodine

$$CH_3\overset{O}{\overset{\|}{C}}CH_3 + BrF_3 \xrightarrow{HF} CH_3CF_2CH_3$$

$$RCN + BrF_3 \xrightarrow{HF} RCF_3$$

pentafluoride was also used as solvent but has the disadvantage of freezing at 10°C), and the product was only isolated successfully if it was volatile enough

[215] T. E. Stevens, *J. Org. Chem.*, **26**, 1627 (1961).

to be easily swept out of the reaction mixture in an inert gas stream. Fragmentation products only were isolated from higher boiling ketones.

$$(CH_3)_2CH\overset{\overset{\displaystyle O}{\|}}{C}CH_3 + BrF_3 \longrightarrow CH_3CF_2CH_3 + CF_3CH_3$$

$$CH_3CH_2\overset{\overset{\displaystyle O}{\|}}{C}CH_3 + BrF_3 \longrightarrow CH_3CHF_2 + CF_3CH_3$$

The mechanism suggested was one similar to that proposed for sulfur tetrafluoride, where the oxygen or nitrogen coordinated with bromine trifluoride to initiate reaction. The high reactivity of bromine trifluoride to fluorination by indiscriminate attack on C–H and C–C bonds will limit the utility of this fluorination method, but the selective direct conversion of a nitrile to a trifluoromethyl group is a unique discovery. Conversion of an isothiocyanate to thiobis-(*N*-substituted)-*N*-(trifluoromethyl)amines with iodine pentafluoride[33] was discussed earlier in Section 5-1B. This is a complex reaction of a series of steps probably involving initial addition of fluorine followed by cleavage and readdition of SF_2, but is included here because the overall result is conversion of a carbon to a CF_3 group.

Table 5-6 should be consulted for selected examples to illustrate these techniques.

Summary

1. Halogen replacement by fluoride in both aliphatic and aromatic compounds is effectively done by an alkali metal fluoride in an aprotic, dipolar solvent (KF in *N*-methylpyrrolidone or tetramethylene sulfone is a particularly effective system).

2. Perfluoroaromatics can be prepared from perchloroaromatics only by use of KF at high temperature in 500°C range without a solvent.

3. Several new reagents, such as α-fluorinated amines, arsenic trifluoride, sulfur tetrafluoride, have limited utility for selective replacement of moderately activated halogens, hydroxyls, or other groups.

4. Decomposition of fluoroformates is now useful for preparation of both alkyl and aryl fluorides and has good potential because of availability of COF_2 or COClF.

5. In the Schiemann reaction for preparing fluoroaromatics, aryldiazonium hexafluorophosphates are improved reactants because of good handling stability.

6. Sulfur tetrafluoride converts C=O to CF_2 and $-CO_2H$ to CF_3–a major advance for selective synthesis of gem-difluorides and trifluoromethyl compounds.

7. Reagents such as α-fluorinated amines, COF_2 and BrF_3, have some limited utility in the synthesis of CF_2 and CF_3 groups.

PROBLEMS

1. (a) Summarize the methods of adding fluorine (F–F) to double bonds:

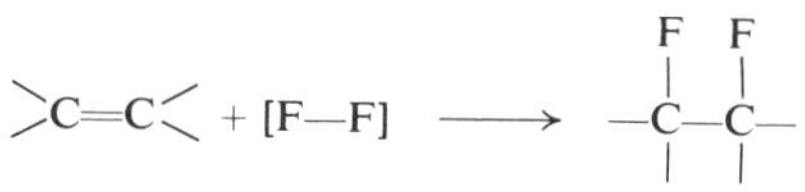

$$\gt C{=}C\lt \; + \; [F{-}F] \longrightarrow -\overset{F}{\underset{|}{\overset{|}{C}}}-\overset{F}{\underset{|}{\overset{|}{C}}}-$$

State the scope and limitations, advantages and disadvantages of each of the methods. Cite three examples *not given* in the tables or text of this book. Review Sections 4-1 and 5-1 and consult the references in answering this question.

(b) Suppose you were assigned the synthesis of 2,3-difluoro-1,2,3,4-tetrahydronaphthalene from 1,4-dihydronaphthalene and told to use one of the methods in Sections 4-1 or 5-1. Which method would you use and why?

2. Give experimental conditions to accomplish the conversions given below. Write structures for intermediates when appropriate. Use the best methods for fluorinations from Chapters 4 and 5.

(a) Urea → difluoroamine.
(b) $CH_3NHCO_2C_2H_5 \rightarrow CH_3NF_2$.
(c) Cyclohexylamine → *N*-fluorocyclohexylimine.
(d) Diethyl malonate → diethyl 2-fluoromalonate.
(e) Diethyl malonate → diethyl 2,2-difluoromalonate.
(f) Diethyl malonate → 2-fluorohexanoic acid.
(g) Δ^4-Cholesten-3-one → *cis*-4,5-difluorocholestan-3-one.

3. Addition of HF to epoxides is a useful fluorinating method. For each of the transformations listed below use this addition at some point. Also use classical methods as required, particularly those in Section 4-6B. Give structures of all starting materials, intermediates, and products. Give experimental details for each synthesis step.

(a) Propene → 2-fluoro-1-propanol.
(b) 2-Methyl-1-propene → 2-methyl-1,2-difluoropropane.
(c) 3-Methyl-3-pentanol → 3-methyl-3-fluoro-2-pentanol.
(d) $5\alpha,6\alpha$-Oxidopregnane-$3\beta,17\beta$-diol-20-one 17-acetate → 6β-fluoropregnane-$3\beta,5\alpha,17\alpha$-triol-20-one 17-acetate.

(e)  $\overbrace{CH_2{-}CHCH_2Br}^{O} + CF_3\overset{O}{\overset{\|}{C}}CF_3 \longrightarrow (CF_3)_2CFOCH_2CHFCH_2OH$

4. The authors of this work have proposed a mechanism for reactions of perchloryl fluoride with a carbanion to give fluorination. Criticize the authors' mechanism in comparison with the mechanisms advanced by others (Refs. 5-111, 5-117, 5-118b). Reach a conclusion supporting or not supporting the authors.

5. A rational mechanism for fluorination of carbonyl groups by SF_4 has been proposed (Section 5-4B).

(a) Tabulate the evidence that supports this mechanism and discuss any result which is in disagreement.

(b) Outline an experimental approach that clearly defines the rate-controlling step, nature of intermediates and transition states, and catalytic effects involved in fluorination. (Hint: consider use of reagents related to SF_4 and variations in structure of reactants and use of spectral techniques.)

6. Write structure(s) for the product or products formed in the reactions outlined below:

(a) $CCl_2{=}CCl{-}CCl{=}CCl_2 \xrightarrow[N\text{-methylpyrrolidone}]{\text{KF; }195^\circ}$

(b) 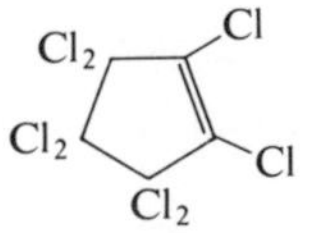 1,2-dichloro-3,3,4,4,5,5-hexachlorocyclopentene (Cl_2, Cl_2, Cl_2 on ring carbons; Cl, Cl on the double bond) $\xrightarrow[N\text{-methylpyrrolidone}]{\text{KF; }195^\circ}$

(c) $CH_2{=}C(CH_2Cl){-}COOC_2H_5 \xrightarrow[\text{tetramethylene sulfone}]{\text{KF; }170^\circ}$

(d) Hexachlorobenzene $\xrightarrow[\text{tetramethylene sulfone}]{\text{KF, }230\text{--}240^\circ\text{, 18 hr}}$

(e) 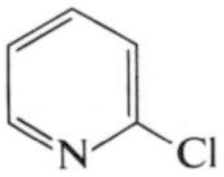 2-chloropyridine $\xrightarrow[\text{tetramethylene sulfone}]{\text{KF, }200^\circ\text{, 21 days}}$

(f) Hexachlorobenzene $\xrightarrow[\text{KF}]{450^\circ}$

(g) 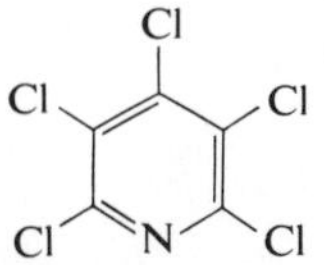 pentachloropyridine $\xrightarrow[500^\circ\text{, 18 hr}]{\text{KF}}$

7. When octachloropropane is heated with potassium fluoride at 195°C in *N*-methylpyrrolidone, the major product is 2,2-dichlorohexafluoropropane. Other products reported to be formed in small yields are $CF_3CF_2CF_3$, $CF_3CHClCF_3$, $CF_3CCl{=}CHF$, and $CF_3CCl{=}CF_2$. Suggest detailed mechanisms that explain the observations.

8. Select the appropriate, commercially available, starting materials that contain only carbon, hydrogen, and oxygen. Use an appropriate reagent from Section 5-4 to synthesize the following compounds.

(a) 1,2-Bis(trifluoromethyl)benzene.
(b) 3-Trifluoromethyl-1,1,2,2-tetrafluorocyclobutane (Hint: make $CF_2{=}CF_2$ from $CHClF_2$.).
(c) *p*-Difluorobenzene.
(d) 2-Methyl-1,1,1-trifluoropropane.
(e) Methyl trifluoromethyl ether.
(f) *p*-Chlorophenyl trifluoromethyl ether.
(g) 4,4-Difluoroheptane.
(h) 1,1-Difluorohexane.
(i) *N,N*,-Bis(trifluoromethyl)aniline.
(j) 3β-Fluoropregn-5-en-20-one.
(k) 1-Fluorobutane (by two methods).
(l) Heptafluoroisopropylbenzene.

9. Give experimental conditions to accomplish the conversions given below. Use optimum synthetic methods available from Chapters 4 and 5.

(a) $CH_3\overset{O}{\overset{\|}{C}}CH_3 \rightarrow CF_3\overset{OF}{\overset{|}{C}}FCF_3$

(b) Tetrafluoroethylene → heptafluorocyclobutanol.
(c) $CF_3CF{=}CF_2 \rightarrow (CF_3)_2CFCH_2OH$

(d) $CF_3CF{=}CF_2 \rightarrow (CF_3)_2CF\overset{O}{\overset{\|}{C}}F$

(e) $CH_2{=}CHF \rightarrow F_5SOCH_2CHF_2$
(f) $CH_2{=}CH_2 \rightarrow CF_3OCH_2CH_2F$
(g) $CF_2{=}CF_2 \rightarrow CF_3CF_2SO_2CF_2CF_3$
(h) Cyclohexene → *trans*-1-fluoro-2-iodocyclohexane.
(i) Δ^5-Pregnene-3β,17α-diol-20-one 17-acetate → 5α-bromo-6β-fluoropregnane-3β,17α-diol-20-one 17-acetate.
(j) $CH_3(CH_2)_4CH{=}CH_2 \rightarrow CH_3(CH_2)_4CF{=}CH_2$

(k) $CH_2{=}CHCH_2O\overset{O}{\overset{\|}{C}}CH_3 \rightarrow CH_2{=}CF\overset{O}{\overset{\|}{C}}OCH_3$

(l) $CH_3(CH_2)_4CH{=}CH_2 \rightarrow CH_3(CH_2)_4\overset{F}{\overset{|}{C}}H\overset{O}{\overset{\|}{C}}OH$

10. Synthesize the compounds listed below using fluorination techniques from Chapters 4 and 5. Unless otherwise specified, you may use as starting material any commercially available organic compound containing C, H, and O as listed in standard catalogs (e.g., Eastman, Aldrich). Fluorinated methanes, ethanes, or ethylenes may also be used. State all reagents and experimental conditions used.

(a) 1-Fluoropentane.
(b) 1,5-Difluoropentane.
(c) 1,2-Difluoropentane.
(d) 3,3-Difluoropentane.
(e) 2,2,4,4-Tetrafluoropentane.
(f) 2,2,3,3,4,4-Hexafluoropentane.
(g) 1,1,1,2-Tetrafluoro-4-pentene
(h) 1,1,1,4,4-Pentafluoropentane.
(i) 1,1,5,5-Tetrafluoropentane.
(j) Perfluoropentane.
(k) 1-(Trifluoromethoxy)-2-fluoropentane.
(l) 5,5,5-Trifluoropentanoic acid.
(m) 1,5-Dibromo-2-fluoropentane.
(n) 2,2,3,3,4,4-Hexafluorohexachloropentane.
(o) 1,2,3,4,5-Pentafluoropentane.

11. Synthesize the compounds listed below using fluorination techniques from Chapters 4 and 5. Unless otherwise specified, you may use as starting material any commercially available organic compound containing C, H, and O as listed in standard catalogs (e.g, Eastman, Aldrich). Fluorinated methanes, ethanes, or ethylenes may also be used. State all reagents and experimental conditions used.

(a) 1-Fluorobenzene.
(b) 1,2-Difluorobenzene.
(c) 1,2,4-Trifluorobenzene.
(d) 1,3-Difluorobenzene.
(e) 4-Fluorobenzotrifluoride.
(f) 4-(Pentafluoroethyl)aniline.
(g) 4-(Trifluoromethoxy)benzotrifluoride.
(h) Hexafluorobenzene.
(i) 2,3,4,5,6-Pentafluorotoluene.
(j) Perfluoroanisole.
(k) 1,2-Bis(trifluoromethyl)-3,6-difluorobenzene.
(l) Octafluoroanthraquinone.
(m) 1,3-Difluoro-4,6-dinitrobenzene.
(n) Pentafluorobenzene.

(o) 2-Methoxy-6-fluoronaphthalene. [See W. Adcock and M. J. S. Dewar, *J. Am. Chem. Soc.*, **89**, 386 (1967)].
(p) 1-Amino-7-fluoronaphthalene. (See Ref. in part o.)
(q) 2-Fluoro-8-nitronaphthalene. (See Ref. in part o).

12. Synthesize the compounds listed below using fluorination techniques from Chapters 4 and 5. Unless otherwise specified, you may use as starting material any commercially available organic compound containing C, H, and O as listed in standard catalogs (e.g., Eastman, Aldrich). Fluorinated methanes, ethanes, or ethylenes may also be used. State all reagents and experimental conditions used.
(a) 6β-Fluorotestosterone acetate [see Y. Osawa and M. Neeman, *J. Org. Chem.*, **32**, 3055 (1967).]
(b) 2,5-Anhydro-1-deoxy-1,1-difluoro-D-mannitol.
(c) Cholesteryl chloride → 3β-chloro-5,6-difluorocholestane.
(d) Δ^{16}-Allopregnene-3β-ol-20-one acetate → 16β-fluoro-17α-bromoallopregnane-3β-ol-11,20-dione acetate.
(e) 16α,17α-Oxido-$\Delta^{5,\,9\,(11)}$-pregnadiene-3β-ol-20-one → 6β-fluoro-16α,17α-oxido-$\Delta^{4,\,9\,(11)}$-pregnadiene-3,20-dione.
(f) 9β,11β-Epoxy-4-pregnene-17α,21-diol-3,20-dione 21-acetate → 9α-fluoro-4-pregnene-11β,17α,21-triol-3,20-dione 21-acetate.
(g) Δ^5-Pregnene-3β,17α-diol-20-one 17-acetate → 6β-fluoropregnane-3β,5α,17α-triol-20-one 17-acetate.

13. Synthesize the compounds listed below using fluorination techniques from Chapters 4 and 5. Unless otherwise specified, you may use as starting material any commercially available organic compound containing C, H, and O as listed in standard catalogs (e.g., Eastman, Aldrich). Fluorinated methanes, ethanes, or ethylenes may also be used. State all reagents and experimental conditions used.
(a) 3,3,4,4-Tetrafluoro-1-trifluoromethyl-1-cyclobutene.
(b) 2,2,4,4-Tetrafluoro-1,1,3,3-tetramethylcyclobutane.
(c) Fluorocyclobutane.
(d) Fluorocyclopentane.
(e) 2-Carbethoxy-2-fluorocyclohexanone.
(f) 1-Fluorocyclooctene.
(g) Hexafluoro-2-butyne.
(h) 1-Chloro-1,1-difluorobutane.
(i) 11-Fluoroundecene-1.
(j) 2-Fluorooctanoic acid.
(k) 2-Fluoroheptanonitrile.

Chapter Six Addition of Fluorinated Units

This chapter tells how to introduce fluorinated units in contrast to Chapters 4 and 5 which outlined methods for introducing individual fluorine atoms (usually one to three). Some overlap occurs with other sections or between sections in this chapter, but repetition or duplication is kept to the minimum needed to complete discussions.

6-1 FROM REACTIVE INTERMEDIATES

Fluorinated radicals, carbenes, carbanions, or carbonium ions are useful intermediates. We will discuss how they can be generated and used in synthesis with pertinent data on structure and mechanism. Because of the voluminous literature, particularly on radicals and carbanions, we cite only leading references to provide a critical survey.

6-1A. RADICALS

For simplicity, we will divide the description into two sections, fluorocarbon and heteroatom radicals, although both are often involved in a particular synthetic reaction.

1. Fluorocarbon

The major work in this area is on fluoroalkyl radicals (the free electron on carbon). Recently, some unsaturated fluorinated radical species, particularly fluoroaryl, have been reported.

a. *Summary of Common Methods of Generation*

1. Addition of radical species (X·) to an unsaturated fluorinated unit

$$X\cdot + CF_2{=}CF_2 \rightarrow XCF_2{-}CF_2\cdot$$

where X· is any type of radical species including alkyl, hydrogen, halogen, or oxygen.

2. Abstraction of X· from a saturated fluorocarbon

$$R\cdot + R_fX \rightarrow RX + R_f\cdot$$

where X includes H or halogen, particularly iodine.

3. Decomposition of a fluorinated species: (a) fluoroalkyl halides by heat or light

$$R_fI \rightarrow R_f\cdot + I\cdot$$

(b) fluorinated peroxides

$$R_f\overset{O}{\overset{\|}{C}}O_2H \rightarrow \cdot OH + R_f\overset{O}{\overset{\|}{C}}O\cdot \rightarrow R_f\cdot + CO_2$$

(c) ketone or aldehyde by photolysis or pyrolysis

$$R\overset{O}{\overset{\|}{C}}R_f \rightarrow R\cdot + R_f\overset{O}{\overset{\|}{C}}\cdot \rightarrow R_f\cdot + CO$$

where R = H, alkyl, or fluoroalkyl; (d) homolytic dissociation of a perfluorocarbon

$$CF_3{-}CF_3 \rightarrow 2\ CF_3\cdot$$

Fluoroalkyl radicals can give all the reactions usually associated with radicals:

1. addition (to unsaturated system)—important in polymerization;
2. chain transfer;
3. substitution;
4. coupling;
5. rearrangement;
6. fragmentation;
7. disproportionation.

Usually synthetic reactions with radicals involve more than one of the types of reactions as outlined above; we prefer to illustrate the types of reactions by discussing some radical chemistry involving fluorinated species, particularly more recent work that has good synthetic potential. Additional examples illustrating the various methods of radical generation and utility are given in Table 6-1 (see page 208).

Formation of C–C or carbon-heteroatom bonds by *free radical addition to unsaturated molecules* has been extensively reviewed recently.[1,2] These reviews cover both addition of fluorinated radical species to unsaturated systems (with free electrons on carbon as well as heteroatom species) and radical addition to fluorinated unsaturated systems. Discussion of fluorinated radical species and their synthetic utilization is also included to some extent in other reviews[3-6] but never in a comprehensive form.

One important aspect of fluorocarbon radical chemistry is in fluorination with elementary fluorine (see Sections 4-1 and 2-1) where addition of F· to olefins (particularly halogenated olefins) or hydrogen abstraction by F· generates an intermediate radical species.[7-9] This intermediate radical can undergo many of the typical reactions–abstraction, coupling, substitution, rearrangement, or fragmentation–to give products. However, we have already discussed fluorination (see above). The synthetic utility of fluorocarbon radicals generated in this way lies chiefly in F_2 addition as discussed in Sections 4-1 and 5-1, although dimeric products can be obtained in high yield under certain conditions.[7]

The most important industrial application of fluorinated radical chemistry is in polymerization or telomerization of fluorinated olefins. (Telomers are short-chain polymers, n of 2 to about 6.)

Initiation:

$$X\cdot + CF_2{=}CF_2 \rightarrow XCF_2CF_2\cdot$$

Propagation:

$$XCF_2CF_2\cdot + nCF_2{=}CF_2 \rightarrow X(CF_2CF_2)_nCF_2CF_2\cdot$$

Chain transfer:

$$X(CF_2CF_2)_nCF_2CF_2\cdot + RX \rightarrow X(CF_2CF_2)_nCF_2CF_2R + X\cdot$$

Termination:

$$2X(CF_2CF_2)_nCF_2CF_2\cdot \rightarrow X(CF_2CF_2)_{2n+2}X$$
$$X(CF_2CF_2)_nCF_2CF_2\cdot + Y\cdot \rightarrow X(CF_2CF_2)_{n+1}Y$$

[1] C. Walling and E. S. Huyser, *Org. Reactions,* **13,** 91 (1963).

[2] F. W. Stacey and J. F. Harris, Jr., *Org. Reactions,* **13,** 150 (1963).

[3] C. Walling, *Free Radicals in Solution,* Wiley, New York, 1957, Chapter 6.

[4] (a) G. Sosnosky, *Free Radical Reactions in Preparative Organic Chemistry,* Macmillan, New York, 1964, Chapter 2. (b) J. M. Tedder and J. C. Walton, *Prog. Reaction Kinetics,* **4,** 37 (1967).

[5] M. Hudlicky, *Chemistry of Organic Fluorine Compounds,* Macmillan, New York, 1962, see Chapters 5 to 7, particularly pp. 244-258.

[6] *Methoden der Organischen Chemie (Houben-Weyl),* Vol. 5, Part 3, Halogen-Verbindungen Fluor und Chlor (Muller, ed.), 4th Ed., Thieme, Stuttgart, Germany, 1962.

[7] W. T. Miller, Jr., J. O. Stoffer, G. Fuller, and A. C. Currie, *J. Am. Chem. Soc.,* **86,** 51 (1964).

[8] A. S. Rodgers, *J. Phys. Chem.,* **69,** 254 (1965).

[9] J. M. Tedder, *Advan. Fluorine Chem.,* **2,** 104 (1961).

The importance of fluorinated polymers, particularly Teflon®, has been reviewed.[10, 11] Peroxides are commonly used as initiators. Persulfate-initiated polymerization of tetrafluoroethylene in an aqueous emulsion is a preferred technique. By controlling conditions, either telomers or 1:1 adducts of radical-initiating species to the fluoroolefin can be the major products. The telomerization reaction has become particularly important for synthesis of useful fluorocarbon intermediates.

$$CH_3OH + CF_2{=}CF_2 \longrightarrow H(CF_2CF_2)_{n+m}CH_2OH + H(CF_2CF_2)_n\overset{\displaystyle OH}{\overset{|}{C}}H(CF_2CF_2)_mH \quad (12\text{–}14)$$

A peroxide or azonitrile initiator is commonly used to generate the attacking radical $\cdot CH_2OH$.

$$CF_2ClCFClI + nCF_2{=}CF_2 \xrightarrow[\text{initiator}]{\text{benzoyl peroxide}} CF_2ClCFCl(CF_2CF_2)_nI \quad (15, 16)$$

n	% Yield
1	39
2	23
≥3	25

These intermediates were coupled and dehydrohalogenated with zinc to give $CF_2{=}CF(CF_2CF_2)_{2n}CF{=}CF_2$ which were converted by standard reactions to difunctional derivatives such as acids, amides, nitriles, and diols.[15, 16]

Telomers have been prepared from a variety of fluorinated olefins, including $CF_2{=}CFCl$, $CF_3CF{=}CF_2$, $CF_2{=}CH_2$, $CH_2{=}CHF$. The initiating radical is generated from a range of reagents such as fluoroalkyl iodides, halogens or interhalogens, and alcohols by use of heat, light, or peroxides.

Actually telomers are probably formed to some extent in any radical-type addition to any fluorinated olefin. The extent of telomer formation depends on relative reactivities of the olefin and intermediate radicals in the propagation steps compared with rates of chain transfer or termination and can be influenced greatly by the conditions and relative proportions of reagents (for a discussion

[10] See Ref. 5, Chapter 10, p. 342.

[11] H. G. Bryce in *Fluorine Chemistry,* (J. H. Simons, ed.), Vol. 5, Academic, New York, 1964, p. 440.

[12] C. D. VerNooy, III, U.S. Patent 3,022,356 (1962); *Chem. Abstr.*, **57**, 3291c (1962).

[13] W. E. Hanford and J. R. Roland, U.S. Patent 2,402,137 (1946); *Chem. Abstr.*, **40**, 5585^9 (1946).

[14] R. M. Joyce, Jr., U.S. Patent 2,559,628 (1951); *Chem. Abstr.*, **46**, 3063b (1951).

[15] I. L. Knunyants, Li Dzhi-yuan, and V. V. Shokina, *Bull. Acad. Sci. USSR, Div. Chem. Sci. (English Transl.)*, 1361 (1961).

[16] I. L. Knunyants and M. P. Krasuskaya, *Bull. Acad. Sci. USSR, Div. Chem. Sci. (English Transl.)*, 173 (1963).

of this problem, see Refs. 1 and 3). Orientation of addition to unsymmetrical olefins is an important question that has recently received attention.[17, 18]

Most work on radical addition to olefins is on reactions leading to 1:1 adducts. Addition of fluorinated alkyl halides to olefins or acetylenes is particularly important and useful for synthetic and mechanistic studies. Over sixty examples of addition of fluorinated halides to olefins (often halogenated) have been reviewed.[1, 3, 4] The most common is addition of CF_3I using heat (about 200°C), light (ultraviolet of 2200 to 3000 Å), or radical source (peroxides or azo initiator) to initiate radical formation. For unsymmetrical olefins, such as $RCH{=}CH_2$, addition of the attacking radical is always on the terminal CH_2. With perfluoroolefins, such as $RCF{=}CF_2$, the steric factor is most important and the attack is usually on the CF_2 group, but other halogens or fluorinated groups influence orientation of attacking reagent so that mixtures are often obtained. The question of orientation in these additions is not simple, and discussions in reviews[1–4] should be consulted.

The valuable synthetic utility of fluoroalkyl iodides in radical reactions with olefins is best illustrated by recent work by Brace[19–27] and others.[28–33]

In the following are listed typical radical additions of iodoperfluoroalkanes to various unsaturated materials both for synthetic or mechanistic studies.

1. Addition to vinyl monomers[20].

$$CF_3CF_2CF_2I + CH_2{=}CHOCCH_3 \xrightarrow[50°,\ 7\ hr]{AVN} CF_3CF_2CF_2CH_2CHOCCH_3$$

(In both the reactant and the product, the C of $OCCH_3$ carries =O; in the product, the CH of CH_2CHO carries I.)

(96%)

(AVN = azobis-2,4-dimethylvaleronitrile; peroxide or light will not initiate)

[17] T. J. Dougherty, *J. Am. Chem. Soc.*, **86**, 460 (1964).
[18] M. Hauptschein, M. Braid, and A. H. Fainberg, *J. Am. Chem. Soc.*, **83**, 2495 (1961).
[19] N. O. Brace, *J. Am. Chem. Soc.*, **84**, 3020 (1962).
[20] N. O. Brace, *J. Org. Chem.*, **27**, 3033 (1962).
[21] N. O. Brace, *ibid.*, 4491.
[22] N. O. Brace, *ibid.*, 3027.
[23] N. O. Brace, *J. Org. Chem.*, **28**, 3093 (1963).
[24] N. O. Brace, *J. Am. Chem. Soc.*, **86**, 523 (1964).
[25] N. O. Brace, *J. Org. Chem.*, **29**, 1247 (1964).
[26] N. O. Brace, *J. Org. Chem.*, **31**, 2879 (1966).
[27] N. O. Brace, *J. Org. Chem.*, **32**, 430 (1967).
[28] G. V. D. Tiers, *J. Org. Chem.*, **27**, 2261 (1962).
[29] W. O. Godtfredsen and S. Vangedal, *Acta Chem. Scand.*, **15**, 1786 (1961).
[30] K. Abildgaard, Japanese Patent 4,622 (1964).
[31] D. J. Burton and L. J. Kehoe, *Tetrahedron Letters*, 5163 (1966).
[32] E. T. McBee, R. D. Battershell, and H. P. Braendlin, *J. Org. Chem.*, **28**, 1131 (1963).
[33] W. F. Beckert and J. U. Lowe, Jr., *J. Org. Chem.*, **32**, 1215 (1967).

2. Preparation of long-chain alkanoic and alkenoic acids with perfluoroalkyl terminal segments that have unique surface-active and wettability properties[21]:

$$CF_3(CF_2)_5CF_2I + CH_2{=}CHCH_2CO_2C_2H_5 \xrightarrow[81°,\ 14\ hr]{ABN} CF_3(CF_2)_6CH_2CHICH_2CO_2C_2H_5\ (100\%)$$

(can be converted to alkenoic acid by reaction with base to eliminate HI)

$$\xrightarrow{Zn\ in\ ethanol} CF_3(CF_2)_6(CH_2)_3CO_2C_2H_5\quad (86\%)$$

(ABN = azobisisobutyronitrile)

3. Preparation of α-amino alkanoic (and alkenoic) acids with perfluoroalkyl terminal segments for biological studies[27]—conventional amino acid syntheses were applied to the alkanoic acids above, but a more direct route was also used:

$$R_fI + CH_2{=}CH(CH_2)_3C(NHCOCH_3)(CO_2C_2H_5)—CO_2C_2H_5 \xrightarrow[80°]{ABN} R_fCH_2CHI(CH_2)_3C(NHCOCH_3)(CO_2C_2H_5)—CO_2C_2H_5\ (81\%)$$

$$\xrightarrow{(1)\ NaOH\quad (2)\ H_2—Pt\quad (3)\ \Delta,\ CO_2} R_f(CH_2)_5CH(NHCOCH_3)CO_2H$$

4. Addition to cyclic olefins—products valuable to study the factors in radical addition and conformational problems using nmr [19, 23]:

$$(CF_3)_2CFI + \text{cyclohexene} \xrightarrow[68°,\ 21\ hr]{ABN} \text{1-}(CF_3)_2CF\text{-2-iodocyclohexane}$$

(80%) trans/cis = 2.7

By-products:

1-R_f-cyclohexene; 3-R_f-cyclohexene; R_f-cyclohexane; 2,2'-di-R_f-bicyclohexyl; R_fH

5. Addition to bicyclic olefins–useful to study effect of variation in structure of olefin on mode of addition. Concludes that the strongly polar and bulky iodo- and perfluoroalkyl groups take positions in the adduct as far as possible from each other.[22]

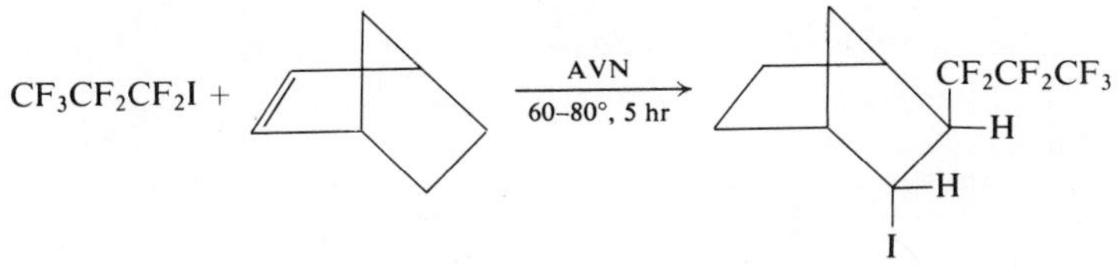

$CF_3CF_2CF_2I$ + [bicyclic olefin] $\xrightarrow[60\text{–}80°,\ 5\ hr]{AVN}$ [adduct: $CF_2CF_2CF_3$, H, H, I]

(100%)

6. Unusual cyclization to a five-membered ring in addition to formation of 1,6-heptadiene–gives insight into reaction mechanism[24, 26]:

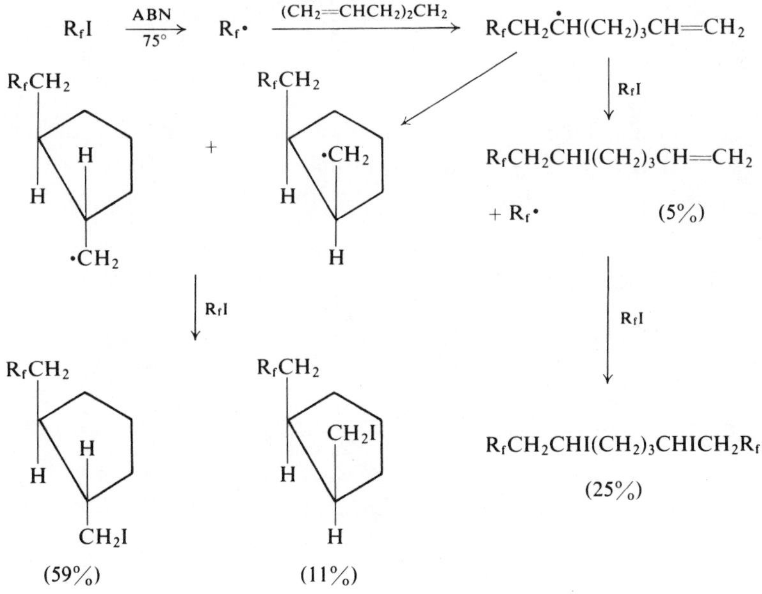

$R_fI \xrightarrow[75°]{ABN} R_f\cdot \xrightarrow{(CH_2{=}CHCH_2)_2CH_2} R_fCH_2\dot{C}H(CH_2)_3CH{=}CH_2$

$\xrightarrow{R_fI}$ $R_fCH_2CHI(CH_2)_3CH{=}CH_2 + R_f\cdot$ (5%) $\xrightarrow{R_fI}$ $R_fCH_2CHI(CH_2)_3CHICH_2R_f$ (25%)

Cyclization: [cyclopentane with R_fCH_2, H, H, $\cdot CH_2$] + [cyclopentane with R_fCH_2, H, $\cdot CH_2$, H] $\xrightarrow{R_fI}$ [cyclopentane with R_fCH_2, H, H, CH_2I] (59%) + [cyclopentane with R_fCH_2, H, CH_2I, H] (11%)

(R_f is $CF_3CF_2CF_2$—)

7. Lactone formation in addition to alkenoic acids and esters[25]:

$$R_fI + CH_2{=}CHCH_2C(CH_3)_2CO_2C_2H_5 \xrightarrow[70°]{\text{azoinitiation}}$$

$$\underbrace{R_fCH_2CHICH_2C(CH_3)_2CO_2C_2H_5}_{1} + \underbrace{R_fCH_2\underset{|}{C}HCH_2C(CH_3)_2}_{2}\ (O\text{—}C{=}O \text{ lactone ring}) + C_2H_5I$$

(1 + 2: 83%)

8. The rate of addition of R_fI to olefins generally increases with increase in size or branching of the perfluoroalkyl group (R_f). Brace carried out competitive experiments to demonstrate this and has discussed it from the point of view of dissociation energy of R_fI, the stability of $R_f\cdot$, and stability of the radical adducts.[23]

9. The mild conditions for addition of R_fI to unsaturation has been used to introduce trifluoromethyl groups into steroids.[29, 30]

O, OAc, C_2H_5O $\xrightarrow[\text{pyridine}]{CF_3I,\ h\nu}$ O, OAc, C_2H_5O, CF_3

(60%)

An unusual modification of normal radical-induced addition of perfluoroalkyl halides is by copper(I) chloride ethanolamine catalysis.[31] These reaction conditions were shown applicable to a series of fluorinated haloolefins with

$$CF_2Br_2 + CH_2{=}CH(CH_2)_5CH_3 \xrightarrow[HOCH_2CH_2NH_2]{CuCl} BrCF_2CH_2CHBr(CH_2)_5CH_3$$

reflux in *t*-BuOH for 24 hr (68%)

$$CF_3CF_2CF_2I + CH_2{=}CH(CH_2)_5CH_3 \xrightarrow[\text{conditions}]{\text{same}} CF_3CF_2CF_2CH_2CHI(CH_2)_5CH_3$$

(57%)

$$+\ CF_3CF_2CF_2CH{=}CH(CH_2)_5CH_3$$

(42%)

both external and internal double bonds. No significant amount of telomer formed. The procedure offers the advantage that only glass apparatus typical of organic reactions is required; however, reactants or products must be stable to the reagents (note olefin by-products in second example given). No mechanism is proposed but incipient radical transfers involving ligands associated with copper seem probable.

Trifluoromethyl iodide added to cyclohexene at room temperature when magnesium amalgam was present.[32] This reaction provides a very mild method to add perfluoroalkyl iodides to olefins but needs to be developed.

In a reaction of nickel carbonyl with pentafluoroiodobenzene to give decafluorobenzophenone, pentafluorobenzene radical is proposed as an intermediate.[33] The proposed mechanism is shown below.

$$C_6F_5I + Ni(CO)_4 \longrightarrow [C_6F_5CONiI(CO)_n] + (3\text{-}n)CO$$

$$[C_6F_5CONiI(CO)_n] \longrightarrow C_6F_5^{\bullet} + [NiI(CO)_{n+1}]$$

$$C_6F_5^{\bullet} + RH \longrightarrow C_6F_5H + R^{\bullet}$$

$$2C_6F_5^{\bullet} \longrightarrow C_6F_5\text{—}C_6F_5$$

$$C_6F_5^{\bullet} + [CO] \rightleftharpoons C_6F_5\text{—}CO^{\bullet}$$

$$C_6F_5^{\bullet} + [C_6F_5CONiI(CO)_n] \longrightarrow C_6F_5COC_6F_5 + [\dot{N}iI(CO)_n]$$

$$C_6F_5CO^{\bullet} + C_6F_5^{\bullet} \longrightarrow C_6F_5COC_6F_5$$

The yield of benzophenone ranges up to 34% depending on solvent and conditions; by-products are pentafluorobenzene (up to 23%) and decafluorobiphenyl (up to 40%). No benzil product was detected (in contrast to analogous reaction on iodobenzene) suggesting that the pentafluorobenzoyl radical is a very reactive and short-lived species (the fluoroacyl radical is also a very unstable species–see discussion on p. 197). Other routes to fluoroaryl radicals will be described later in this section (pp. 199-201).

The fluoroalkyl iodide radical reaction also has been adopted to a general procedure for perfluoroalkylation of aromatic compounds.[34–37] For the reaction of perfluoroalkyl iodides with a wide variety of aromatic hydrocarbons and their derivatives, Tiers claims yields of 60 to 65% for monosubstitution and, depending

$$ArH + 2R_fI \xrightarrow{250^\circ} R_fAr + R_fH + I_2$$

on conditions, up to 25% for disubstitution. Tiers carried out reactions on benzene, toluene, naphthalene, halobenzenes, perfluoroalkylbenzenes, benzonitrile, phthalic anhydride, and even fully formed dyes.

Another important utilization of fluorinated radicals, particularly from perfluoroalkyl iodides, is in synthesis of fluoroorgano metals and metalloids.[38–40]

[34] G. V. D. Tiers, *J. Am. Chem. Soc.*, **82**, 5513 (1960).

[35] Minnesota Mining and Mfg. Co., British Patent 840,725 (1960); *Chem. Abstr.*, **55**, 6496h (1961).

[36] E. S. Huyser and E. Bedard, *J. Org. Chem.*, **29**, 1588 (1964).

[37] (a) I. L. Knunyants and V. V. Shokina, Russian Patent 156,555 (1963); *Chem. Abstr.*, **60**, 6792 (1964). (b) I. L. Knunyants and V. V. Shokina, *Bull. Acad. Sci. U.S.S.R. (English Transl.)*, 68 (1967).

[38] H. C. Clark, *Advan. Fluorine Chem.*, **3**, 19 (1963).

[39] H. J. Emeleus, *Angew. Chem. Intern. Ed. Engl.*, **1**, 129 (1962).

[40] R. E. Banks and R. N. Haszeldine, *Advan. Inorg. Chem. Radiochem.*, **3**, 337 (1961).

The typical procedures are displacement of alkyl groups with perfluoroalkyl halides.

$$2MR_3 + R_fI \longrightarrow R_2MR_f + R_4MI$$

$$\xrightarrow{R_fI} RM(R_f)_2 + RI$$

$$\xrightarrow{R_fI} (R_f)_3M + RI$$

where M is P, As, Sb; R is CH_3; R_f is a fluorinated alkyl radical.

A more convenient method often useful at reasonable temperatures and pressures is direct reaction of the element in finely divided form with R_fI. As a typical example of the general scheme, the reaction with sulfur is shown.[41, 42]

$$4M + 6R_fI \longrightarrow (R_f)_3M + (R_f)_2MI + R_fMI_2 + MI_3$$

$$C_3F_7I + S \xrightarrow{220^\circ} \underset{(11\%)}{C_3F_7SC_3F_7} + \underset{(47\%)}{C_3F_7SSC_3F_7} + \underset{(19\%)}{C_3F_7\text{—}SSSC_3F_7} + I_2$$

This type of chemistry is reviewed[38–40] in description of synthesis of fluoroorganometallic and metalloid chemistry and has been extended recently as outlined in the following examples and in Table 6-1 (see page 208).

$$\underset{(\text{or } R_3As)}{CF_3I + R_3P} \longrightarrow \underset{(\text{or } R_2AsCF_3)}{R_2PCF_3} \qquad (43)$$

$$R_3SnSnR_3 + C_nF_{(2n+1)}I \xrightarrow[\text{or heat}]{h\nu} R_3SnC_nF_{(2n+1)} + R_3SnI \qquad (44, 45)$$

$$CF_2\text{=}CFI \xrightarrow[NO]{h\nu} CF_2\text{=}C\begin{matrix} F \\ NO \end{matrix} \quad \text{(interesting monomer)} \qquad (46)$$

The reaction has recently been extended to perfluoroaromatics.[47]

$$Ar_fI + M \xrightarrow[\text{several hours in sealed tube}]{120^\circ\text{–}140^\circ} (Ar_f)_nM$$

where M is Hg, S, Se, P, Sb, Ge, Sn, and Ar_f is C_6F_5.

[41] M. Hauptschein and A. V. Grosse, *J. Am. Chem. Soc.*, **73**, 5461 (1951).
[42] G. V. D. Tiers, *J. Org. Chem.*, **26**, 3515 (1961).
[43] W. R. Cullen, *Can. J. Chem.*, **40**, 426 (1962).
[44] H. C. Clark and C. J. Willis, *J. Am. Chem. Soc.*, **82**, 1888 (1960).
[45] H. D. Kaesz, J. R. Phillips, and F. G. A. Stone, *Chem. Ind. (London)*, 1409 (1959).
[46] C. E. Griffin and R. N. Haszeldine, *Proc. Chem. Soc.*, 369 (1959).
[47] S. C. Cohen, M. L. N. Reddy, and A. G. Massey, *Chem. Comm.*, 451 (1967).

A simplified method for preparation of fluoroalkyl iodides is proposed to involve a radical intermediate.[48] The procedure is to reflux a suspension of the sodium or potassium salt of the corresponding fluorocarboxylic acid with a small excess of iodine in dimethylformamide. The advantage of the procedure is use

$$R_fCO_2^-M^+ + I_2 \longrightarrow [R_fCO_2I] + MI$$

$$[R_fCO_2I] \longrightarrow R_fCO_2\cdot + I\cdot$$

$$R_fCO_2\cdot + I\cdot \longrightarrow CO_2 + R_f\cdot$$

$$R_f\cdot \xrightarrow{I_2} R_fI + I\cdot$$

of normal laboratory equipment at atmospheric pressure and high yields. Trifluoromethyl and heptafluoropropyl iodides were prepared in 70 to 80% yields, but the method is proposed to be of general utility for any fluorinated acid.

Photolysis of fluorinated ketones is a useful method of generating fluoroalkyl radicals and has also been studied extensively. Hexafluoroacetone dissociates to trifluoroacetyl and trifluoromethyl radicals in a primary process;

$$CF_3\overset{O}{\overset{\|}{C}}CF_3 \xrightarrow{h\nu} CF_3\overset{O}{\overset{\|}{C}}\cdot + \cdot CF_3$$

$$CF_3\overset{O}{\overset{\|}{C}}\cdot \longrightarrow CF_3\cdot + CO$$

$$CF_3\overset{O}{\overset{\|}{C}}\cdot \xrightarrow{Br_2} CF_3\overset{O}{\overset{\|}{C}}Br$$

$$CF_3\overset{O}{\overset{\|}{C}}\cdot \not\longrightarrow CF_3\overset{O}{\overset{\|}{C}}\overset{O}{\overset{\|}{C}}CF_3$$

$$\cdot CF_3 \xrightarrow{Br_2} CF_3Br$$

the trifluoroacetyl can be trapped but rapidly loses CO.[49a, b] Part of the preceding interpretation was questioned recently.[50] Formation of trifluoroacetyl bromide and bromotrifluoromethane is better explained by direct reaction of triplet hexafluoroacetone with bromine.

$$\underset{\text{singlet}}{CF_3\overset{O}{\overset{\|}{C}}CF_3} \longrightarrow \underset{\text{triplet}}{CF_3\overset{O}{\overset{\|}{C}}CF_3} + Br_2 \longrightarrow 2CF_3Br + CO$$

$$\underset{\text{triplet}}{CF_3\overset{O}{\overset{\|}{C}}CF_3} + Br_2 \longrightarrow CF_3Br + CF_3\overset{O}{\overset{\|}{C}}Br$$

[48] D. Paskovich, P. Gaspar, and G. S. Hammond, *J. Org. Chem.*, **32**, 833 (1967).

[49] (a) B. G. Tucker and E. Whittle, *Proc. Chem. Soc.*, 93, (1963). (b) E. A. Dawidowicz and C. R. Patrick, *J. Chem. Soc.*, 4250 (1964).

[50] J. S. McIntosh and G. B. Porter, *Trans. Faraday Soc.*, 119 (1968).

Although perhaloacetones containing a predominance of fluorine photolyze with C–C fission as the primary step, those containing a predominance of chlorine photolyze with C–Cl fission.[51] Fluoroalkyl radicals generated from fluorinated ketones have been used much more extensively for mechanistic studies (see below) than in synthetic schemes. Recently, extensive synthetic studies have been made of reactions of free radicals with hexafluoroacetone. In this way fluorinated radicals with the free electron on either carbon or oxygen are generated as intermediates.[52]

$$C_6H_{12} + CF_3C(=O)CF_3 \xrightarrow[135^\circ]{t\text{-butyl peroxide}}$$

$$C_6H_{11}C(CF_3)_2OH\ (61\%) + C_6H_{11}OCH(CF_3)_2\ (12\%) + C_6H_{10}[C(CF_3)_2OH]_2\ (17\%)$$

The photolysis of polyfluoroacyl fluorides, chlorides, and bromides gives fluoroalkyl radicals which usually react with a halide or couple but can add to fluorinated olefins.[53]

$$n\text{-}C_3F_7C(=O)Cl \xrightarrow{h\nu} \underset{(81\%)}{n\text{-}C_3F_7Cl} + \underset{(4\%)}{n\text{-}C_6F_{14}}$$

$$n\text{-}C_3F_7C(=O)Cl + CF_2{=}CF(CF_2)_2H \xrightarrow{h\nu}$$

$$\underset{(23\%)}{n\text{-}C_6F_{14}} + \underset{(26\%)}{n\text{-}C_3F_7[CF_2CF(CF_2)_2H]Cl} + Cl[CF_2CF(CF_2)_2H]_2Cl$$

[51] R. N. Haszeldine and F. Nyman, *J. Chem. Soc.*, 3015 (1961).
[52] E. G. Howard, P. B. Sargeant, and C. G. Krespan, *J. Am. Chem. Soc.*, **89**, 1422 (1967).
[53] J. F. Harris, Jr., *J. Org. Chem.*, **30**, 2182 (1965).

For perfluoroacyl fluorides, the following scheme was proposed:

$$C_3F_7\overset{\overset{\displaystyle O}{\|}}{C}F \xrightarrow{h\nu} \left[C_3F_7\overset{\overset{\displaystyle O}{\|}}{C}F\right]^* \longrightarrow C_3F_7\cdot + \cdot COF$$

$$2C_3F_7\cdot \longrightarrow n\text{-}C_6F_{14} \quad (58\%)$$

$$C_3F_7\cdot + C_3F_7\overset{\overset{\displaystyle O}{\|}}{C}F \longrightarrow C_3F_7O\dot{C}F\text{—}C_3F_7 \xrightarrow{C_3F_7\cdot} C_3F_7OCF(C_3F_7)_2$$

$$2\cdot COF \longrightarrow COF_2 + CO$$

In the presence of perfluorinated olefins, the major product is oxetane.[54]

$$n\text{-}C_3F_7\overset{\overset{\displaystyle O}{\|}}{C}F + CF_2{=}CFCF_3 \xrightarrow{h\nu} \text{oxetane ring: O–C(F)(}C_3F_7\text{-}n\text{)–C(F)(}CF_3\text{)–C}F_2\text{–}$$

For perfluoroacyl chlorides, the electronically excited acyl chloride fragments in one or both of the ways shown. The acyl radicals undergo fast dissociation. The perfluoroalkyl radicals react with a chlorine radical and to

$$C_3F_7\overset{\overset{\displaystyle O}{\|}}{C}Cl \longrightarrow C_3F_7\cdot + \cdot\overset{\overset{\displaystyle O}{\|}}{C}Cl \longrightarrow CO + \cdot Cl$$

$$C_3F_7\overset{\overset{\displaystyle O}{\|}}{C}Cl \longrightarrow C_3F_7\overset{\overset{\displaystyle O}{\|}}{C}\cdot + Cl\cdot ; \quad C_3F_7\overset{\overset{\displaystyle O}{\|}}{C}\cdot \longrightarrow C_3F_7\cdot + CO$$

$$C_3F_7\cdot + Cl\cdot \longrightarrow C_3F_7Cl$$

$$2C_3F_7\cdot \longrightarrow n\text{-}C_6F_{14}$$

a lesser extent couple. With fluorinated olefins, only a small yield of the oxetane by-product is formed and addition predominates. Perfluoroacyl bromides give alkyl bromides only on irradiation.

[54] J. F. Harris, Jr., and D. D. Coffman, *J. Am. Chem. Soc.*, **84**, 1553 (1962).

Fluoroacyl radicals have also been generated by reacting perfluoroacyl chlorides with nickel carbonyl. Depending on the conditions, these radicals are trapped or decomposed to fluoroalkyl radicals or carbenes.[55] A series of novel fluorochemicals was prepared by this method as shown.

$R_fCOCl + Ni(CO)_4$

- $\xrightarrow[C_6H_5CN]{25^\circ}$ $[R_fCO\cdot] \longrightarrow [O{=}C(R_f){-}C(R_f){=}O] \xrightarrow{R_fCO\cdot} R_fCOOC(R_f){=}C(R_f)COOCR_f$
- $\xrightarrow[ArH]{150^\circ}$ $[R_fCO\cdot] \longrightarrow [R_f\cdot] \longrightarrow R_fAr + R_fH$
- $\xrightarrow[\text{no medium}]{150^\circ}$ $[R_fCO\cdot] \longrightarrow R_fCOR_f$

  $[R_f\cdot] \longrightarrow R_f{-}R_f$

  $[R_f'CF:] \longrightarrow R_f'CF{=}CFR$

Decomposition of fluoroazoalkanes is another common method of generating fluoroalkyl radicals[56–67] that initiate polymerizations,[56] and they are used for mechanistic studies.[58–67]

$CF_3N{=}NCF_3$—

- $\xrightarrow{\text{heat or light}}$ $(CF_3)_2N{-}N(CF_3)_2$ (66%)
- $\xrightarrow[CO]{465^\circ}$ CF_3CF_3 (91%)
- $\xrightarrow[Br_2]{}$ CF_3Br
- $\xrightarrow{\text{S refluxing}}$ $CF_3S_3CF_3$
- $\xrightarrow{CF_3CF{=}CF_2}$ C_7F_{16} (33%—reported for $C_2F_5N{=}NC_2F_5$ decomposition only)
- $\xrightarrow[N_2F_4]{}$ CF_3NF_2 (55%)

[55] J. J. Drysdale and D. D. Coffman, *J. Am. Chem. Soc.*, **82**, 5111 (1960).
[56] W. J. Chambers, C. W. Tullock, and D. D. Coffman, *J. Am. Chem. Soc.*, **84**, 2337 (1962).
[57] J. A. Young and R. D. Dresdner, *J. Org. Chem.*, **28**, 833 (1963).
[58] A. P. Stefani, L. Herk, and M. Szwarc, *J. Am. Chem. Soc.*, **83**, 4732 (1961).
[59] A. P. Stefani and M. Szwarc, *J. Am. Chem. Soc.*, **84**, 3661 (1962).
[60] I. M. Whittemore, A. P. Stefani, and M. Szwarc, *ibid.*, 3799.
[61] H. Komazawa, A. P. Stefani, and M. Szwarc, *J. Am. Chem. Soc.*, **85**, 2043 (1963).
[62] M. Feld, A. P. Stefani, and M. Szwarc, *J. Am. Chem. Soc.*, **84**, 4451 (1962).
[63] P. S. Dixon and M. Szwarc, *Trans. Faraday Soc.*, **59**, 112 (1963).
[64] J. M. Pearson and M. Szwarc, *Trans. Faraday Soc.*, **60**, 553 (1964).
[65] G. E. Owen, Jr., J. M. Pearson, and M. Szwarc, *ibid.*, 564.
[66] G. E. Owen, Jr., J. M. Pearson, and M. Szwarc, *Trans. Faraday Soc.*, **61**, 1722 (1965).
[67] W. G. Alcock and E. Whittle, *Trans. Faraday Soc.*, **62**, 664 (1966).

More extensive use of perfluoroazoalkanes in synthesis is expected since a convenient and simple preparation has been reported[56]:

$$2ClCN + 2Cl_2 + 6NaF \longrightarrow \underset{(31\%)}{CF_3N{=}NCF_3} + 6NaCl$$

$$R_fCN + AgF + Cl_2 \longrightarrow R_fCF_2N{=}NCF_2R_f + [RCF{=}NCl + R_fCF_2NCl_2]$$

where R_f includes CF_3, C_3F_7, C_8F_{17}, $H(CF_2)_3$, and $H(CF_2)_5$ with yields of corresponding azoalkenes from 22 to 84%. The chloroimines and dichloroamines were found as minor by-products in some cases.

A related route to polyfluorinated aromatic radicals is by oxidation of the polyfluoroaryl hydrazines[68] giving products as shown:

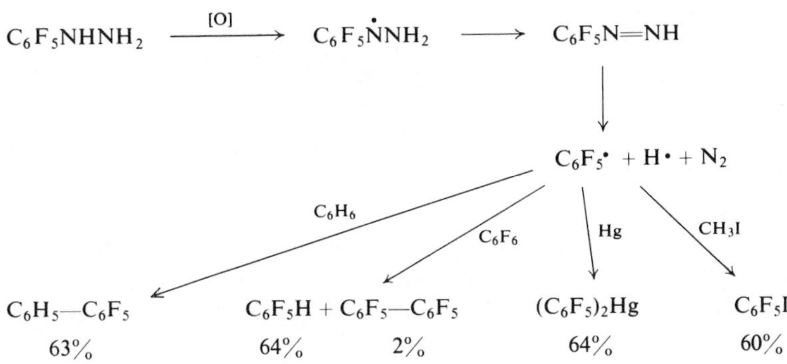

A series of oxidizing agents was studied, but selenium oxides appear best. By-products were also formed and under some conditions only mixtures were found.

Because of the high energy required to rupture C–C or C–F bonds, fluoroalkyl radicals useful for synthetic reactions are not usually generated by thermolysis or photolysis of fluorocarbons. Polytetrafluoroethylene does form radicals when irradiated,[69] however. These radicals have an appreciable lifetime because they are trapped in the polymer chains. The long-lived radicals can be reacted with a variety of reagents. The fate of these radicals is particularly important in relation to the changes in properties of the polymer.

[68] J. M. Birchall, R. N. Haszeldine, and A. R. Parkinson, *J. Chem. Soc.*, 4966 (1962).

[69] (a) Ya. S. Lebedev, Yu. D. Tsvetkov, and V. V. Voyevodskii, *Polymer Sci. USSR (English Transl. Vysokomol. Soedineniya)*, **5**, 712 (1964). (b) D. W. Ovenall, *J. Chem. Phys.*, **38**, 2448 (1963); *J. Phys. Chem. Solids*, **26**, 81 (1965). (c) S. Siegel and H. Hedgpeth, *J. Chem. Phys.*, **46**, 3904 (1967).

Pentafluorophenyl radicals are proposed as intermediates in high-temperature reactions of hexafluorobenzene[70a] and in photolysis reactions of hexafluorobenzene with a hydrogen source present.[70b] In the latter case, experimental details are meager but the product is a complex mixture:

$$C_6F_6 \xrightarrow{h\nu} C_6F_6^* \text{ (singlet or triplet?)}$$

$$\downarrow HR$$

$$C_6F_5\cdot + HF + R\cdot$$

- $\xrightarrow{RH}$ $C_6F_5H + R\cdot$
- $\xrightarrow{C_6F_6}$ $C_{12}F_{11}\cdot$ $\longrightarrow$ $C_{12}F_{10}$
- $\xrightarrow{R\cdot}$ C_6F_5R
- $\xrightarrow{C_6F_5\cdot}$ $C_{12}F_{10}$ $\xrightarrow{h\nu/RH}$ higher molecular weight product
- $\xrightarrow{R\cdot}$ RR

In the former example, Antonucci and Wall[70a] report replacement of fluorine by nonnucleophilic, or weakly nucleophilic, reagents at temperatures of 300° to 850°C. A high product yield is reported with only small amounts of by-products. For example, hexafluorobenzene with reagents such as bromine, chlorine, and tetrafluoroethylene gave as major products bromopentafluorobenzene, chloropentafluorobenzene, and octafluorotoluene. The mechanism must be free radical, but an addition–elimination process (1) is considered more probable than fluorine abstraction (2).[70c]

(1) $C_6F_6 + X\cdot \longrightarrow$ [$C_6F_6X\cdot$ (X and F on the same carbon; ring F, F, F, F, F)] $\longrightarrow$ C_6F_5X + F· (as XF or F_2)

(stabilized by resonance forms)

(2) $C_6F_6 + X\cdot \longrightarrow C_6F_5\cdot + FX$

$C_6F_5\cdot + X_2 \longrightarrow C_6F_5X + X\cdot$

[70](a) J. M. Antonucci and L. A. Wall, *J. Res. Natl. Bur. Standards,* **70A,** 473 (1966). (b) D. Bryce-Smith, B. E. Connett, A. Gilbert, and E. Kendrick, *Chem. Ind. (London),* 855 (1966). (c) Fluorine abstraction from SF_6 by methyl radicals occurs above 140°C, L. Batt and F. R. Cruickshank, *J. Phys. Chem.,* **70,** 723 (1966). Hydrogen atoms are claimed to abstract fluorine from ethyl fluoride, P. M. Scott and K. R. Jennings, *Chem. Commun.,* 700 (1967).

For the tetrafluoroethylene reaction, difluorocarbene addition to a double bond, followed by rearrangement, is also possible.

Below 100°C, the reaction of aryl radicals with hexafluorobenzene occurs by an addition–elimination mechanism as shown in (1) of the preceding reaction.[71a] At high temperatures (600°C),[71b] direct displacement, as in (2) of the preceding reaction, must also occur in order to explain the results.

Iodopentafluorobenzene could be a source of pentafluorophenyl radicals; however, iodopentafluorobenzene does not dissociate in refluxing toluene[33] (photolysis has not been reported; however, see page 193).

Fluoroalkyl radicals are also generated as intermediates in free radical addition of reagents such as halogens, hydrogen halides, and mercaptans to fluoroolefins. This area is well reviewed.[2–4] A good example is the free radical addition of trifluoromethanesulfenyl chloride to haloolefins.[72]

$$CF_3SCl + CF_2{=}CFCF_3 \xrightarrow{h\nu} \underset{(63\%)}{CF_3SCF_2\overset{\overset{\displaystyle Cl}{|}}{C}FCF_3} + \underset{(37\%)}{CF_3S\overset{\overset{\displaystyle CF_3}{|}}{C}FCF_2Cl} + \underset{(\sim 10\%)}{CF_3SCF_2\overset{\overset{\displaystyle SCF_3}{|}}{C}FCF_3} + \underset{(\sim 10\%)}{Cl(CF_2\overset{\overset{\displaystyle CF_3}{|}}{C}F)_2Cl}$$

(first two products combined: 26%)

A mixture of products is usually obtained when the fluoroolefin is unsymmetrical. From orientation studies, the mechanism proposed involves initial addition of a chlorine atom rather than a trifluoromethyl sulfur ($CF_3S\cdot$) radical. This is contrasted with radical addition of CF_3SH, where $CF_3S\cdot$ is the adding or chain-carrying species.[73] Many free radical additions to fluoroolefins give simple products and have extensive synthetic utility. A selection of recent examples is given in Table 6-1 to supplement those covered in reviews.

Fluoroalkyl radicals have been generated by several other routes as summarized in Table 6-1. In general, the synthetic utility of radicals generated by these other methods has not been demonstrated although these methods have been profitably employed in mechanism studies. Often conditions are such that synthetic opportunities are very limited.

[71] (a) P. A. Claret, G. H. Williams, and J. Coulson, *J. Chem. Soc.* (C), 341 (1968). (b) E. K. Fields and S. Meyerson, *J. Org. Chem.*, **32**, 3114 (1967).

[72] J. F. Harris, Jr., *J. Am. Chem. Soc.*, **84**, 3148 (1962).

[73] J. F. Harris, Jr., and F. W. Stacey, *J. Am. Chem. Soc.*, **83**, 840 (1961).

Diradical species are proposed as intermediates in cycloaddition reactions of fluorinated olefins,[74–77] but we will discuss this important reaction in Section 6-2A under reactions of fluoroolefins.

b. *Structure and Properties of Fluoroalkyl Radicals*

In the above discussion, we have referred several times to mechanistic studies involving fluorinated radical species. Groups under the leadership of Szwarc,[58–65] Whittle,[49a, 67, 78–83] Pritchard,[84–87] Kutschke,[88–90] and several others[4b, 91–96] have reported extensive physical chemical investigations of properties and reactions of fluoroalkyl radicals, particularly $\cdot CF_3$, generated usually by photolysis of hexafluoroazomethane or hexafluoroacetone. Some typical experimental approaches and results are outlined below.

1. *Relative and absolute rates of reaction*—for hydrogen abstraction and addition in saturated and unsaturated hydrocarbons. In dilute isooctane solution, the only products are CF_3H, C_2F_6, and N_2 from initial photolysis of $CF_3N{=}NCF_3$.

[74] (a) Paul D. Bartlett, L. K. Montgomery and B. Seidel, *J. Am. Chem. Soc.*, **86**, 616 (1964). (b) L. K. Montgomery, K. Schueller, and P. D. Bartlett, *ibid.*, 622. (c) P. D. Bartlett and L. K. Montgomery, *ibid.*, 628. (d) P. D. Bartlett, *Science*, **159**, 833 (1968). (e) P. D. Bartlett, G. E. H. Wallbillich, A. S. Wingrove, J. S. Swenton, L. K. Montgomery, and B. D. Kramer, *J. Am. Chem. Soc.*, **90**, 2049 (1968). (f) J. S. Swenton and P. D. Bartlett, *ibid.*, 2056.

[75] J. D. Park, H. V. Holler, and J. R. Lacher, *J. Org. Chem.*, **25**, 990 (1960).

[76] G. J. Janz and J. J. Stratta, *J. Org. Chem.*, **26**, 2169 (1961).

[77] P. D. Bartlett, G. E. H. Wallbillich, and L. K. Montgomery, *J. Org. Chem.*, **32**, 1290 (1967).

[78] (a) R. D. Giles and E. Whittle, *Trans. Faraday Soc.*, **62**, 128 (1966). (b) R. D. Giles, L. M. Quick, and E. Whittle, *Trans. Faraday Soc.*, **63**, 662 (1967). (c) J. W. Coomber and E. Whittle, *ibid.*, 1394.

[79] W. G. Alcock and E. Whittle, *Trans. Faraday Soc.*, **62**, 134 (1966).

[80] R. D. Giles and E. Whittle, *Trans. Faraday Soc.*, **61**, 1425 (1965).

[81] A. M. Tarr, J. W. Coomber, and E. Whittle, *ibid.*, 1182.

[82] P. J. Corbett and E. Whittle, *J. Chem. Soc.*, 3247 (1963).

[83] J. C. Amphlett and E. Whittle, *Trans. Faraday Soc.*, **63**, 80, 2695 (1967).

[84] G. O. Pritchard, J. R. Dacey, W. C. Kent, and C. R. Simonds, *Can. J. Chem.*, **44**, 171 (1966).

[85] G. O. Pritchard and J. T. Bryant, *J. Phys. Chem.*, **69**, 1085 (1965).

[86] G. O. Pritchard and R. L. Thommarson, *J. Phys. Chem.*, **69**, 1001 (1965).

[87] D. M. Tomkinson and H. O. Pritchard, *J. Phys. Chem.*, **70**, 1579 (1966).

[88] K. O. Kutschke, *Can. J. Chem.*, **42**, 1232 (1964).

[89] J. L. Holmes and K. O. Kutschke, *Trans. Faraday Soc.*, **58**, 333 (1962).

[90] S. J. W. Price and K. O. Kutschke, *Can. J. Chem.*, **38**, 2128 (1960).

[91] H. Carmichael and H. S. Johnston, *J. Chem. Phys.*, **41**, 1975 (1964).

[92] N. L. Arthur and T. N. Bell, *Chem. Comm.*, 166 (1965).

[93] D. T. Clark and J. M. Tedder, *Trans. Faraday Soc.*, **62**, 399 (1966).

[94] I. V. Berezin, *Doklady USSR, Phys. Chem. Sect. (English Transl.)*, **148**, 72 (1963).

[95] F. J. Wunderlich and R. E. Rebbert, *J. Phys. Chem.*, **67**, 1382 (1963).

[96] (a) M. G. Bellas, O. P. Strausz, and H. E. Gunning, *Can. J. Chem.*, **43**, 1022 (1965). (b) G. S. Lawrence, *Trans. Faraday Soc.*, **63**, 1155 (1967). (c) H. G. Meunier and P. I. Abell, *J. Phys. Chem.*, **71**, 1430 (1967).

$$CF_3^{\bullet} + RH \xrightarrow{k_1} CF_3H + R^{\bullet}$$

$$CF_3^{\cdot} + \begin{matrix}\text{unsaturated}\\ \text{substrate}\end{matrix} \xrightarrow{k_2} {}^{\bullet}\text{substrate—}CF_3$$

The relative ratio of products is independent of concentration. The k_2/k_1 ratio for addition to unsaturated substrates (ranges from about 10 to 10^3) is always greater for $CF_3\cdot$ than $CH_3\cdot$; however k_2/k_1 for $CF_3\cdot$ increases more rapidly in the series ethylene, propylene, and isobutene than respective methyl affinities. Tetrafluoroethylene is less reactive than ethylene with regard to $CF_3\cdot$ addition, whereas for $CH_3\cdot$ addition it is 10 times more reactive.

2. *Energies and entropies of reaction*—in a series of terminal olefins, $CH_2{=}C\lessdot$, the entropies did not change, but activation energies varied to give rate differences of 200-fold. The energy of $CF_3\cdot$ reaction for hydrogen abstraction from isobutane was 4 kcal/mole more than for addition to benzene.

3. *Substituent effects*—$CF_3\cdot$ and oxygen atoms are very similar in their addition reactions to a series of unsaturated centers substituted by various groups. In addition of $CF_3\cdot$ to aromatic compounds, the relative rate constant of $CF_3\cdot$ addition is related to the atom localization energy of the most reactive center. Thus, electron-donating groups such as CH_3 and OCH_3 enhance, whereas electron-withdrawing groups such as CF_3 or CN slow down addition. Bulky groups such as *o*-*t*-butyl decrease the rate.

4. *Media effects*—reactions of $CF_3\cdot$ are similar in the gas phase and in various solvents in the liquid phase.

5. *Deuterium isotope effects*—very small secondary deuterium isotope effects similar to values for $\cdot CH_3$.

The main conclusions about properties of fluoroalkyl radicals are

1. $CF_3\cdot$ is more reactive but more selective than $CH_3\cdot$;
2. $CF_3\cdot$ is electrophilic, whereas $CH_3\cdot$ is slightly nucleophilic;
3. $CF_3\cdot$ shows more steric effect than $CH_3\cdot$;
4. in the transition state for addition of $CF_3\cdot$ to an unsaturated system, the new C–CF_3 bond looks like a σ bond (rather than π-complex type)—the forming C–CF_3 bond is still very long and poorly formed similar to the C–CH_3 bond in $CH_3\cdot$ additions;
5. $CF_3\cdot$ coupling to give CF_3CF_3 occurs in a cage reaction;
6. $CF_3\cdot$ adds irreversibly to an unsaturated system;
7. $R_f\overset{\overset{\displaystyle O}{\|}}{C}\cdot$ has a very low activation energy for dissociation to $R_f\cdot$ + CO;
8. order of ease of H abstraction from RH is $CF_3 > C_2F_5 > C_3F_7 > CH_3$.

The question of orientation in addition to unsymmetrical olefins or other unsaturated systems was mentioned earlier.[17, 18] Some simple rules have been given.[3, 4] Recently, more sophisticated techniques of analysis and separation of reaction products have shown that many reaction products previously thought to be only a single adduct are actually mixtures.[97, 98]

The factors that influence the stabilization of the radical intermediates and rates of reaction are not simple, although steric effects are often the most important factor; many more definitive experimental studies and careful analysis of results are needed in this area.

The transient and highly reactive fluoroalkyl radicals have been stabilized by trapping at low temperatures. In this way, the chemistry and properties of these radicals were studied.[99] The radicals were generated from hexafluoroethane in the gas phase by electrical discharge, carried in a gas stream a very short distance to a cold finger, and condensed. The $CF_3\cdot$ radical was associated with a red deposit (difluorocarbene generated from octafluorocyclobutane gave a blue deposit) and was trapped with Cl_2 or $CF_2{=}CF_2$ by warming (see Chart 6-1). On standing, decomposition also occurred to give tetrafluoromethane and difluorocarbene (discussed in Section 6-1B.1b). The visible spectrum of the captive $CF_3\cdot$ radicals is quite complex. The infrared spectra of $CF_3\cdot$ in solid matrices[100] in the gas phase[101] have also been reported. The technique of freezing fluorinated radicals in a solid matrix has been used for several esr studies on radicals of fluorinated methyls,[102] fluoroalkylacetamide,[103, 104] perfluoroalkyl,[105] polytetrafluoroethylene[106–108] fluoroformyl,[109, 110] fluoronitrobenzenes,[111] and fluoro-substituted tropolonimine nickel chelates.[112–114] Normally the fluorine nuclei couple with the free electron so that some spin density appears on

[97] M. Hauptschein, R. E. Oesterling, M. Braid, E. A. Tyczkowski, and D. M. Gardner, *J. Org. Chem.*, **28**, 1281 (1963).

[98] E. R. Bissell, *J. Org. Chem.*, **29**, 252 (1964).

[99] S. V. R. Mastrangelo, *J. Am. Chem. Soc.*, **84**, 1122 (1962).

[100] D. E. Milligan, M. E. Jacox, and J. J. Comeford, *J. Chem. Phys.*, **44**, 4058 (1966).

[101] G. A. Carlson and G. C. Pimentel, *J. Chem. Phys.*, **44**, 4053 (1966).

[102] (a) R. W. Fessenden and R. H. Schuler, *J. Chem. Phys.*, **43**, 2704 (1965). (b) M. T. Rogers and L. D. Kispert, *J. Chem. Phys.*, **46**, 3193 (1967).

[103] R. J. Lontz and W. Gordy, *J. Chem. Phys.*, **37**, 1357 (1962).

[104] R. J. Lontz, *J. Chem. Phys.*, **45**, 1339 (1966).

[105] Ya. S. Lebedev, *Kinetics and Catalysis (U.S.S.R.) (English Transl.)*, **3**, 540 (1962).

[106] T. Matsugashita and K. Shinohara, *J. Chem. Phys.*, **35**, 1652 (1961).

[107] P. Barnaba, D. Cordischi, A. D. Site, and A. Mele, *J. Chem. Phys.*, **44**, 3672 (1966).

[108] Yu. D. Tsvetkov, Ya. S. Lebedev, and V. V. Voevodskii, *Polymer Sci. U.S.S.R., (English Transl. Vysokomol. Soedin.)* **2**, 45 (1961).

[109] P. J. Adrian, E. L. Cochran, and V. A. Bowers, *J. Chem. Phys.*, **43**, 462 (1965).

[110] D. E. Milligan, M. E. Jacox, A. M. Bass, J. J. Comeford, and D. E. Mann, *J. Chem. Phys.*, **42**, 3187 (1965).

[111] A. Carrington, A. Hudson, and H. C. Longuet-Higgins, *Mol. Phys.*, **9**, 377 (1965).

[112] D. R. Eaton, A. D. Josey, R. E. Benson, W. D. Phillips, and T. L. Cairns, *J. Am. Chem. Soc.*, **84**, 4100 (1962).

[113] D. R. Eaton, A. D. Josey, W. D. Phillips, and R. E. Benson, *Mol. Phys.*, **5**, 407 (1962).

[114] D. R. Eaton, A. D. Josey, and W. A. Sheppard, *J. Am. Chem. Soc.*, **85**, 2689 (1963).

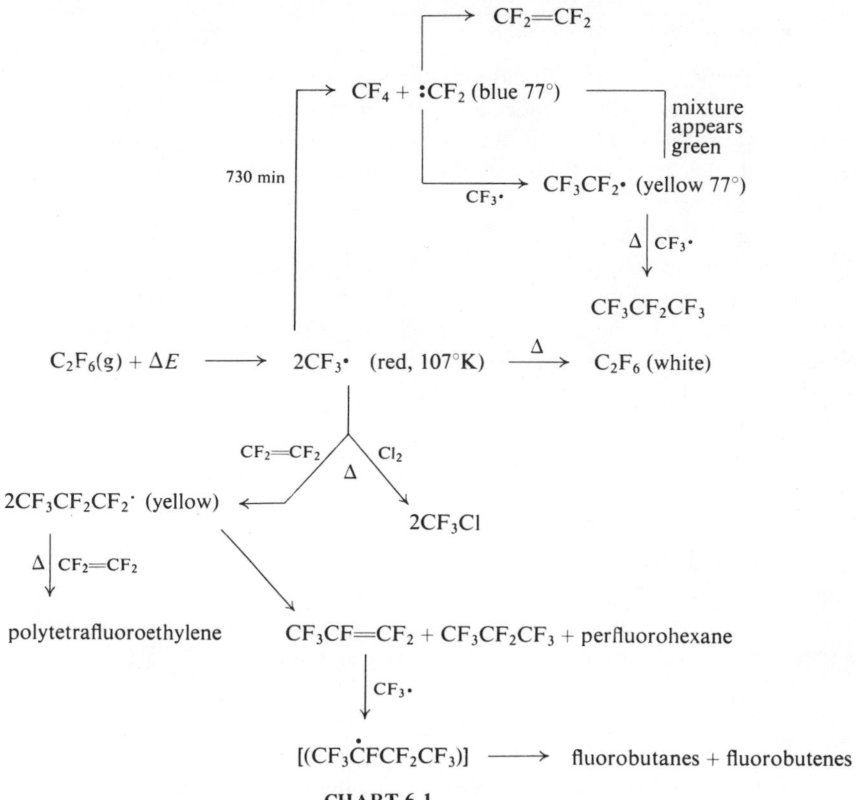

CHART 6-1

fluorine. These results have been interpreted in terms of mechanism of transmission of electronic effects. Fluorinated methyl radicals are thought to be nonplanar; in contrast, $CF_3\cdot$ is almost tetrahedral with a gradual change toward planarity as fluorine is replaced by hydrogen.[99–102]

A mass spectrometric study of several fluoroalkylazomethanes and fluoroalkyl mercury compounds led to data on vertical ionization potentials of radicals as tabulated below.[115] The results are discussed in terms of inductive and resonance effects of fluorine and trifluoromethyl substituents.

Radical	I_{vert} eV
$CF_3\dot{C}F_2$	9.98
$CF_3CF_2\dot{C}F_2$	10.06
$(CF_3)_2\dot{C}F$	10.5
$CF_3\dot{C}H_2$	10.6
$\ddot{C}F_2$	11.7

[115] I. P. Fisher, J. B. Homer, and F. P. Lossing, *J. Am. Chem. Soc.*, 87, 957 (1965).

2. Fluorinated Heteroatom Radicals

This section is devoted to radical intermediates where the free electron is on the heteroatom which is bonded either to fluorine or to fluoroalkyl groups. Few such radical species are reported for heteroatoms from Groups I to IV. The useful synthetic chemistry from the other groups is limited to a few species that have been reviewed.[116–120] Some of this chemistry is of practical interest in the rocket propellant field.[120] We have already mentioned some work on addition of fluorinated heteroatom species to unsaturated fluorinated systems. Reactions such as fluorocarbon radicals with metals and metalloids must also involve heteroatom radicals as intermediates. We will only outline here the synthetic potential of the important types using selected examples for illustration and summarize in Table 6-1 recent work not already covered in reviews.

a. *Nitrogen Species*

Because of interest in nitrogen–fluorine compounds as high-energy materials for rocket propellants,[120] the difluoroamino radical ($\cdot NF_2$) is probably the most intensively studied fluorinated heteroatom radical species. Since several comprehensive reviews on nitrogen–fluorine compounds[116-118] are available, particularly the recent review by Colburn[118] on "The difluoroamino free radical and its reaction," we will only illustrate the synthetic utility by a brief outline.

The difluoroamino radical ($\cdot NF_2$) is best prepared from dissociation of N_2F_4 ($\Delta H = 20 \pm 1$ kcal/mole which is much less than for any of the halogens), particularly because of the equilibrium

$$F_2NNF_2 \rightleftharpoons 2\cdot NF_2$$

which favors $\cdot NF_2$ at higher temperatures. But dissociation of $X{-}NF_2$ where X is H, F, Cl, or other species is also reported. A recent report by Petry summarizes the chemistry of $ClNF_2$.[121a]

The $\cdot NF_2$ radical behaves like a typical radical species: it dimerizes and couples with other free radicals, abstracts atoms and groups from other molecules, and adds to unsaturated systems; for example,

[116] (a) C. J. Hoffman and R. G. Neville, *Chem. Rev.*, **62**, 1 (1962). (b) J. K. Ruff, *Chem. Rev.*, **67**, 665 (1967).

[117] C. B. Colburn, *Adv. Fluorine Chem.*, **3**, 92 (1963).

[118] C. B. Colburn, *Chem. Britain*, **2**, 336 (1966).

[119] H. L. Roberts, *Quart. Rev. (London)*, **15**, 30 (1961).

[120] "Advanced Propellant Chemistry" in Advan. Chem. Ser. No. 54, Am. Chem. Soc., Washington, D.C., 1966.

[121] (a) R. C. Petry, *J. Am. Chem. Soc.*, 89, 4600 (1967). (b) R. C. Petry and J. P. Freeman, *J. Am. Chem. Soc.*, **83**, 3912 (1961).

Coupling:

$$(CH_3)_3CN{=}NC(CH_3)_3 \longrightarrow N_2 + (CH_3)_3C\cdot \xrightarrow{\cdot NF_2} (CH_3)_3CNF_2 \qquad (121\text{–}122)$$

$$RC(=O)H + \cdot NF_2 \longrightarrow HNF_2 + RC(=O)\cdot \xrightarrow{\cdot NF_2} RC(=O)NF_2 \qquad (121\text{–}122)$$

$$RI \xrightarrow{h\nu} I\cdot + R\cdot \xrightarrow{\cdot NF_2} RNF_2 \qquad (123)$$

An alternative mechanism for the above reactions is direct attack of radical species on N_2F_4. An $\cdot NF_2$ will be displaced and could couple with radical species or dimerize back to N_2F_4.

Mechanism studies show that when alkanes and N_2F_4 are heated, reaction occurs by hydrogen abstraction.[124, 125] Irradiation of N_2F_4 with an alkane gives an alkyldifluoroamine.[126] However, the ultraviolet irradiation of benzophenone

$$RH + \cdot NF_2 \longrightarrow HNF_2 + R\cdot \xrightarrow{\cdot NF_2} RNF_2$$

with N_2F_4 in Pyrex® gives no reaction (only benzophenone excited) until an ether is added, and then only α substitution (no β) occurs on the ether.[127]

$$\phi C(=O)\phi + N_2F_4 + \text{(tetrahydrofuran)} \longrightarrow \text{2-(difluoroamino)tetrahydrofuran (H, } NF_2\text{)}$$

A possible explanation is formation of the tetrahydrofuranyl radical through hydrogen abstraction by benzophenone triplet:

$$C_6H_5C(=O)C_6H_5 \longrightarrow [C_6H_5C(=O)C_6H_5]^* \xrightarrow{\text{tetrahydrofuran}} C_6H_5{-}\dot{C}(OH){-}C_6H_5$$

$$+ \text{(tetrahydrofuran-2-yl radical, H)} \xrightarrow{\cdot NF_2} \text{(2-H, 2-}NF_2\text{ tetrahydrofuran)}$$

The author suggests that N_2F_4 (effectively $\cdot NF_2$) will be a valuable trap to catch short-lived radicals.

[122] C. L. Bumgardner and M. Lustig, *Inorg. Chem.*, **2**, 662 (1963); see also M. Lustig, C. L. Bumgardner and J. K. Ruff, *Inorg. Chem.*, **3**, 917 (1964) and Ref. 126.

[123] J. W. Frazer, *J. Inorg. Nucl. Chem.*, **16**, 63 (1960).

[124] J. Grzechowiak, J. A. Kerr, and A. F. Trotman-Dickenson, *Chem. Comm.*, 109 (1965).

[125] J. Grzechowiak, J. A. Kerr, and A. F. Trotman-Dickenson, *J. Chem. Soc.*, 5080 (1965).

[126] C. L. Bumgardner, *Tetrahedron Letters*, 3683 (1964).

[127] M. J. Cziesla, K. F. Mueller, and O. Jones, *Tetrahedron Letters*, 813 (1966).

Table 6-1. Fluorinated

Radical species	Source of radical and method of generation[a]	Reactant and/or method of detection
A. SATURATED FLUOROCARBONS		
1. From fluoroalkyl halides		
$CF_3\cdot$	CF_3I, ABN,[a] 80°, 20 hr	Cyclohexene
$CF_3\cdot$	CF_3I, uv irradiation	Allene
$CF_3\cdot$, $CF_3CH_2CHF\cdot$, and $CF_3CHFCH_2\cdot$	CF_3I, DTBP,[a] then $CF_3\cdot + CH_2{=}CHF$	$CH_2{=}CHF$; addition at *both* $CH_2{=}$ and $CHF{=}$
$CF_3CF_2CF_2\cdot$	$CF_3CF_2CF_2I$, ABN, 70-80°, 7 hr	Norbornadiene in ethyl acetate
$CF_3CF_2CF_2\cdot$	$CF_3CF_2CF_2I$, AVN,[a] 50-80°, 7 hr	$CH_3CO_2CH{=}CH_2$
$CF_3(CF_2)_4CF_2\cdot$	$CF_3(CF_2)_5I$, ABN, 85°, 17 hr	$CH_2{=}CH(CH_2)_8CO_2C_2H_5$
$(CF_3)_2CF\cdot$	$(CF_3)_2CFI$, ABN, 60°, 7 hr	$CH_2{=}CHCH_2CH_2CH_2CH{=}CH_2$
$CF_3(CF_2)_5CF_2\cdot$	$CF_3(CF_2)_5CF_2I$, 200°, 6 hr	$CH_2{=}CHCO_2C_2H_5$ in ethyl acetate
$CF_3(CF_2)_5CF_2\cdot$	$CF_3(CF_2)_5CF_2I$, uv irradiation	1-Octene
$CF_3CF_2CF_2\cdot$ $CF_2ClCFCl\cdot$	$CF_3CF_2CF_2I$, ABN, 50-100°, 8 hr $CF_2ClCFClI$, DTBP	Methylenecyclopropane $CF_2{=}CF_2$
$CF_3(CF_2)_5CF_2\cdot$	$CF_3(CF_2)_5CF_2I$, 320°, 14 hr	Phthalic anhydride
$\cdot(CF_2)_nCFClCF_2Cl$	$I(CF_2)_nCFClCF_2Cl$ (n = 2,4,6), heat 250°, 8 hr	Benzene
$CF_3\cdot$	CF_3I, 198°	Chlorobenzene
$CF_3CF_2CF_2\cdot$ and $CF_3CF_2CF_2S\cdot$	$CF_3CF_2CF_2I$, S, 250°	Sulfur
$CF_3CF_2CF_2\cdot$ and $CF_3CF_2CF_2S\cdot$	$CF_3CF_2CF_2I$, S, 300°, 10 hr	Sulfur
$CF_3\cdot$ (?) (and $CH_3\dot{A}sCF_3$) $CF_3\cdot$	$(CH_3)_2AsI + CF_3I$ heated 1 day at 170° CF_3I, 100°	$(CH_3)_2AsI + CF_3I$ $(CH_3)_3Bi$
$BrCF_2\cdot$	CBr_2F_2, *t*-butylperbenzoate, 120°	$CF_2{=}CFH$
$BrCF_2\cdot$	CF_2Br_2, $(C_6H_5COO)_2$	$CH_2{=}CF_2$
$BrCF_2\cdot$	$BrCF_2Br$, Zn-Cu couple, 25-100° ether solvent	Cyclohexene 1-Hexene
$CF_2BrCFCl\cdot$ (?)	$CF_2BrCFClBr$, CuCl in ethanolamine, reflux in *t*-butyl alcohol	2-Octene

Radicals in Synthesis

Product(s) (yield, %)	Comments
1-Iodo-2-(trifluoromethyl)cyclohexane: $\frac{\text{trans}}{\text{cis}} = 1.8$	Series of perfluoroalkyl iodides added to cyclohexene and cyclopentene. Mechanism discussed. Products useful in conformation studies. Ref. 19, 23
$CF_3CH_2CI{=}CH_2$(–)	Studied kinetics of addition reaction. Rate law and mechanism presented. Ref. 96c
CF_3CH_2CHFI, CF_3CHFCH_2I, and higher mixed telomers $CF_3(CH_2CHF)_nI$	$CF_3\cdot$ added at *both* ends of vinyl fluoride. And propagating intermediates also added to both ends of $CH_2{=}CHF$. This nondirectivity is unusual. Ref. 17
H, I, C_3F_7, H (50), I, H, C_3F_7, H (50)	Addition of C_3F_7I to norbor nene β-pinene also studied in order to determine the effect of structure on radical additions. Ref. 22
$CH_3CO_2CHICH_2CF_2CF_2CF_3$ (96)	Similar reactions reported for $CF_3(CF_2)_5CF_2I$, $CF_3(CF_2)_9CF_2I$, and $BrCF_2CF_2I$. Ref. 20
$CF_3(CF_2)_5CH_2CHI(CH_2)_8COOC_2H_5$ (85)	Prepared 28 perfluoroalkyl segmented alkanoic acids $[R_f(CH_2)_nCOOH]$ to study surface active and wettability properties. Ref. 21
$CH_2{=}CHCH_2CH_2CH_2CHICH_2CF(CF_3)_2$ (28), $(CF_3)_2CFCH_2CHICH_2CH_2CH_2CHICH_2CF(CF_3)_2$ (6) *cis/trans*-1-(Iodomethyl)-2-(heptafluoroisopropylmethyl)-cyclopentane (66)	Mechanism discussed. Cyclization may occur during addition or by subsequent rearrangements of open chain adduct by radical mechanism. Ref. 24, 26
$CF_3(CF_2)_6CH_2CH$–CH_2–$CHCO_2C_2H_5$, ring closed by C(=O)–O (22)	Lactone formation also found in additions to 5-hexenoic acid and 2,2-dimethyl-4-pentenoic acid. Ref. 25
$CF_3(CF_2)_6CH_2CHICO_2C_2H_5$ (17) $CF_3(CF_2)_6CH_2CHIC_6H_{13}$ (87)	Also added C_3F_7I and CF_2Br_2 to other olefins. Used light and peroxides as initiators. Ref. 28
1-Iodo-1-(1,1-dihydroperfluorobutyl)-cyclopropane (96)	Ref. 191a
$CF_2ClCFCl(CF_2CF_2)_nI$; n = 1-4	Extensive chemistry reported on telomers such as synthesis of $CF_2{=}CF(CF_2CF_2)_{2n}CF{=}CF_2$. Ref. 15
Perfluoroheptylphthalic anhydride (38)	See discussion in text (page 193). Process limited to thermally stable aromatic compounds. Many examples cited. Ref. 35, 34
$C_6H_5(CF_2)_nCFClCF_2Cl$ (n = 2,4,6)	Converted to $C_6H_5(CF_2)_nCF{=}CF_2$, then to $C_6H_5(CF_2)_nCOOH$. Also synthesized $HOOC(CF_2)_n(p\text{-}C_6H_4)(CF_2)_nCOOH$ by disubstitution and further reactions. Ref. 37
(Trifluoromethyl)-chlorobenzene; o(24.9), m(11.5), p(13.7)	Similar results obtained with iodobenzene and bromobenzene. Intermediate hexadiene radical proposed. Ref. 36
$C_3F_7SSC_3F_7$ (47.0), $C_3F_7SSSC_3F_7$ (18.7)	Products suggest intermediate $C_3F_7S\cdot$ radical. $(C_3F_7)_2S$ suggested as first product formed which reacts further with S. Ref. 41
$C_3F_7SC_3F_7$ (11), $C_3F_7SSC_3F_7$ (53), $C_3F_7SSSC_3F_7$ (14.5)	Disputes Hauptschein and Grosse (41). Favors radical mechanism with both $C_3F_7S\cdot$ and $C_3F_7\cdot$ involved. Ref. 42
CF_3H, $CH_3As(CF_3)_2$, $CH_3As(CF_3)I$, and other products	Intermediate radical possible but evidence not strong. Ref. 43
$(CH_3)_2BiCF_3$ (82), $CH_3Bi(CF_3)_2$ (18)	Could form $(CF_3)_3Bi$ if large excess of CF_3I used at higher temperature. Reaction of Bi with CF_3I did not give products. Ref. 191b
$BrCF_2CHFCF_2Br$ (20), $BrCF_2CF_2CHFBr$ (9)	Results in agreement with Haszeldine, *J. Chem. Soc.*, 2300 (1957). Ref. 191c
$CF_2BrCH_2CF_2Br$ (33-55), $CF_2BrCH_2CF_2CH_2CF_2Br$ (23-29)	With CCl_3Br only monoaddition occurred: $CCl_3CH_2CF_2Br$ obtained in 60% yield. Ref. 191d
1-Bromo-2-(bromodifluoromethyl)cyclohexane(–) $CH_3(CH_2)_3CH(CF_2Br)CH_2Br$(–)	Reaction claimed for wide variety of alkenes. Ref. 192e
$CH_3CHCHBr(CH_2)_4CH_3$ with CFClCF$_2$Br on C-2; $CH_3CHBrCH(CH_2)_4CH_3$ with CFClCF$_2$Br on C-3 (86)	See discussion in text (page 192). Using this method also added following compounds to 1-octene: CF_2Br_2, CF_2BrCF_2Br, $CF_2BrCFClBr$, $CF_2ClCFCl_2$, CF_3CClBr_2, CF_3CCl_3, CF_3CBr_3, $CF_3CF_2CF_2I$. Ref. 31

Table 6-1

Radical species	Source of radical and method of generation	Reactant and/or method of detection
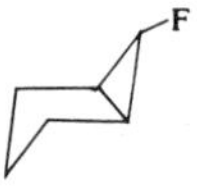 F	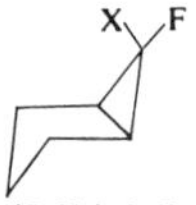 X, F (X = Cl,Br), $(C_4H_9)_3SnH$ to abstract X	$(C_4H_9)_3SnH$
2. From radical addition		
$CF_3\dot{C}FCF_2Br$ and $CF_3CFBrCF_2\cdot$	$CF_2{=}CFCF_3$, HBr, x rays	Addition of HBr
$CF_2ClCFCl\cdot$	$F_2 + CFCl{=}CFCl$, $-110°$	$2CF_2ClCFCl\cdot \rightarrow (CF_2ClCFCl-)_2$
$HOCH_2-(CF_2CF_2)_x\cdot$	CH_3OH + 1% initiator (e.g., DTBP) + $CF_2{=}CF_2$	$CH_3OH \xrightarrow{R\cdot} \dot{C}H_2OH \xrightarrow{CF_2=CF_2} HOCH_2CF_2\dot{C}F_2 \cdot \xrightarrow{CF_2=CF_2}$ etc.
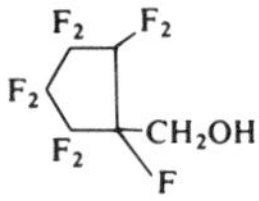 F_2, F_2, F_2, F_2, CH_2OH, F	Perfluorocyclopentene + CH_3OH + excess benzoyl peroxide; 110°, 8 hr	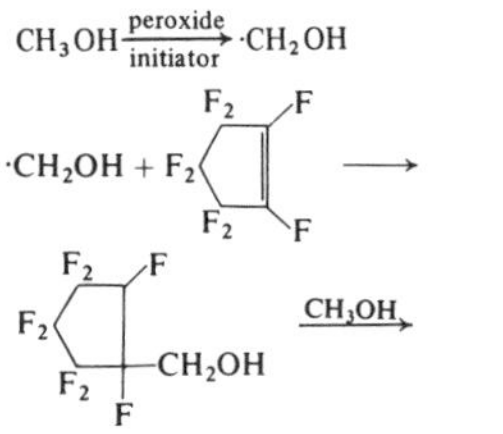 $CH_3OH \xrightarrow[\text{initiator}]{\text{peroxide}} \cdot CH_2OH$ $\cdot CH_2OH +$ (F_2, F, F_2, F_2, F) $\longrightarrow$ (F_2, F, F_2, F_2, CH_2OH, F) $\xrightarrow{CH_3OH}$
$CH_2ClCF_2\cdot$	CF_3SCl, $CF_2{=}CH_2$, irradiation by uv or x rays	Addition of CF_3SCl to olefins
$HSCF_2CF_2\cdot$	$CF_2{=}CF_2$, H_2S, x ray	Addition of H_2S to $CF_2{=}CF_2$
$H_2PCF_2CF_2\cdot$ (?)	PH_3, $CF_2{=}CF_2$, 150°, 8 hr	Thermal addition of PH_3 to fluoro-olefin
$H_2PCHFCF_2\cdot$ and $H_2PCF_2CHF\cdot$	$PH_3 + CF_2{=}CFH$, uv	Addition of PH_3 to $CF_2{=}CFH$
$CH_3C(O)CFClCFCl\cdot$	CH_3CHO, $CFCl{=}CFCl$, DTBP, 105-110°, 15 hr	Addition: CH_3CHO and $CFCl{=}CFCl$
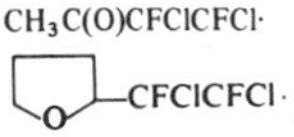 (O ring)–$CFClCFCl\cdot$	 (O ring) (THF) + $CFCl{=}CFCl$, x ray irradiation	Addition: THF + $CFCl{=}CFCl$
$CH_3C(O)CF_2CF_2\cdot$	CH_3CHO, $CF_2{=}CF_2$, photolysis Hg lamp	Addition of CH_3CHO to $CF_2{=}CF_2$, $CF_2{=}CFCl$, $CF_2{=}CFBr$
$C_6H_{11}OC(CF_3)_2$ and $C_6H_{11}C(CF_3)_2O\cdot$	$CF_3\overset{\overset{O}{\|}}{C}CF_3$, cyclohexane with DTBP, $(C_6H_5CO_2)_2$, or $C_6H_5C(CH_3)_2OOH$, or irradiated	Cyclohexane adds either to carbon or oxygen of carbonyl groups
$CH_3CO_2CH_2O\overset{CF_3}{\underset{CF_3}{C}}\cdot$ but chiefly $CH_3CO_2CH_2\overset{CF_3}{\underset{CF_3}{C}}-O\cdot$	$CF_3\overset{\overset{O}{\|}}{C}CF_3$, methyl acetate, irradiation	Methyl acetate adds mainly to carbon
3. From decomposition of other compounds		
$CF_3CF_2CF_2\cdot$	$CF_3CF_2CF_2CO\overset{-}{O}\overset{+}{M}$, I_2, dimethylformamide, reflux	Iodine
$CF_3\cdot$	CF_3NO, photolysis	Gas phase

(cont.)

Product(s) (yield, %)	Comments
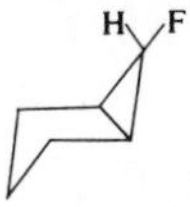 H F (bicyclic structure)	Stereochemistry preserved. Argument made for pyramidal radical. Similar reaction carried out with other isomer. Ref. 192f
$CF_3CFBrCF_2H$ and CF_3CFHCF_2Br ratio 42:58, no yields	Product ratio 38:62 when initiated by uv light. Ref. 191e
$CF_2ClCFClCFClCF_2Cl$ (79), CF_2ClCF_2Cl (21)	At −110° molecular F_2 reacts to generate $CF_2ClCFCl\cdot$ plus $F\cdot$ Ref. 7
$HOCH_2(CF_2CF_2)_nH$; n = 1-12	Modification of procedures can give $H(CF_2CF_2)_mCH(OH)(CF_2CF_2)_nH$ where $n + m < 12$. Ref. 12, 14
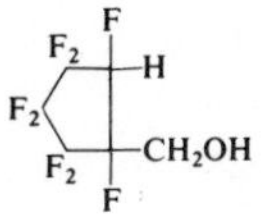 F, F_2, H, F_2, CH_2OH, F_2, F (ring structure) (80.1)	*Cis-* and *trans-* isomers obtained. No telomers were formed. Ref. 192g
$CF_3SCF_2CH_2Cl$ (40.3), $CF_3SCH_2CF_2Cl$ (11.2), CF_3SSCF_3 (12)	Unexpected initial attack by $Cl\cdot$ rather than $CF_3S\cdot$. Additions also reported for $CF_2{=}CFCF_3$, $CF_2{=}CFH$, $ClCF{=}CF_2$, $CH_2{=}CHCl$, $CH_3OCF{=}CF_2$. See text (page 201). Ref. 73
HCF_2CF_2SH (60)	Free radical addition of H_2S also made to $CF_2{=}CFCl$, $CF_2{=}CFH$, $CF_2{=}CH_2$, $CF_2{=}CFBr$, $CH_2{=}CHF$, and $CH_3OCF{=}CF_2$. Ref. 191f
$HCF_2CF_2PH_2$ (53), $(HCF_2CF_2)_2PH$ (7), $H_2PCF_2CF_2PH_2$ (9)	Product mix suggestive of radical intermediate. PH_3 also added to $CF_2{=}CFCl$, $CF_2{=}CCl_2$, $CF_2{=}CH_2$, $(CH_3)_2C{=}CF_2$, $CF_3CF{=}CF_2$. Ref. 191g
HCF_2CFHPH_2 (63), $H_2CFCF_2PH_2$ (11)	Product yields establish relative stability of competing radicals. Anionic mechanism ruled out because of expected attack at $CF_2{=}$. Ref. 191h
$CH_3C(O)CFClCFClH$ (20)	Additions of CH_3CHO, CH_3CH_2CHO, and $(CH_3)_2CHCHO$ to $CFCl{=}CFCl$ and $CF_2{=}CCl_2$ are reported. Ref. 191i
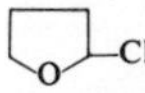 (tetrahydrofuranyl)–$CFClCFClH$ (84)	X-ray initiated additions of THF, $C_2H_5OC_2H_5$, and dioxane to $CFCl{=}CCl_2$, $CFCl{=}CFCl$ and $CF_2{=}CCl_2$ are reported. Ref. 191j
$CH_3C(O)CF_2CHF_2$ (7.2), $CH_3C(O)CF_2CFClH$ (10.0), $CH_3C(O)CF_2CFBrH$ (20.3)	Side-reaction products were oxetanes  $CH_3\dot{C}HCFXCF_2O$ (ring) Ref. 191r
$C_6H_{11}C(CF_3)_2OH$, $C_6H_{11}OCH(CF_3)_2$, $HO(CF_3)_2C{-}C_6H_{10}C(CF_3)_2OH$ where C_6H_{11} and C_6H_{10} = cyclohexyl	Product yield depended on initiator used and the temperature. Extensive discussion of mechanism is presented. Ref. 52
$CH_3CO_2CH_2C(CF_3)_2OH$ (88), $HOC(CF_3)_2CH_2CO_2CH_2C(CF_3)_2OH$ (trace)	Addition of hexafluoroacetone to acetaldehyde, benzaldehyde, methyl formate, trichlorosilane, and other compounds is also reported. Ref. 52
$CF_3CF_2CF_2I$ (70-80)	Authors claim this as a general method for synthesis of R_fI. Ref. 48
$(CF_3)_2NONO$ (96)	Proposed following mechanism $CF_3NO \rightarrow CF_3\cdot + NO\cdot$; $CF_3\cdot + CF_3NO \rightarrow (CF_3)_2NO\cdot$; $(CF_3)_2NO\cdot + NO\cdot \rightarrow (CF_3)_2NONO$ or $(CF_3)_2NO\cdot + CF_3NO \rightarrow (CF_3)_2NONO + CF_3\cdot$ Ref. 191k

Table 6-1

Radical species	Source of radical and method of generation	Reactant and/or method of detection
$CF_3\cdot$	$CF_3N{=}NCH_3$, photolysis	$CF_3N{=}NCH_3$
$CF_3\cdot$	CF_3CF_3, ^{60}Co radiation, x ray at 3 atm argon	CF_3CF_3
$CF_3\cdot$	CF_3CF_3, thermal decomposition, single-pulse shock (1300-1600°K)	Hydrogen, argon diluent
$CF_3\cdot$	CF_3COOH, photolysis, gas phase, 90°-190°	Neat, isobutane or 1-butene
$CF_3\cdot$ (?)	$CF_3C(O)CF_3 \xrightarrow{\text{pyrolysis}} CF_3\cdot$ 900°, 2 mm	Te + $CF_3\cdot$, obtain tellurium mirrors
$ClCF_2\cdot$	$ClCF_2C(O)CF_2Cl$, photolysis at 3130 Å	Benzene, hexafluorobenzene, or cyclopentane
$CF_3\cdot$	$CF_3C(O)C(O)CF_3$, photolysis	R–H such as $(CH_3)_3CH$
$n\text{-}C_7F_{15}\cdot$	$n\text{-}C_7F_{15}COF$, uv irradiation, liquid state, 5 days	Radical dimerizes
$n\text{-}C_7F_{15}\cdot$	$n\text{-}C_7F_{15}COCl$, uv irradiation, liquid state, 7 days	Recombination of radical with Cl and dimerization
$n\text{-}C_3F_7\cdot$	$n\text{-}C_3F_7COCl$, uv irradiation, liquid state, 5 days	$CF_2{=}CFC_5F_{11}$
$C_7F_{15}CO$	$C_7F_{15}COF$, irradiate uv	$CF_2{=}CFCF_3$
$C_3F_7\dot{C}O$	$C_3F_7COCl + Ni(CO)_4$, benzonitrile solvent, 25°	Radical dimerizes and reacts further with C_3F_7COF
$C_3F_7\dot{C}O$ and $C_3F_7\cdot$	$C_3F_7COCl + Ni(CO)_4$, 150°, aromatic hydrocarbon solvent	Benzene, toluene, trifluoromethylbenzene, bromobenzene
$CF_3\cdot$	$CH_3\cdot + CF_3N{=}NCF_3$, displacement at 140°, gas phase	$[(CH_3)_3C{-}O]_2$ decomposes to give $CH_3\cdot$
$CF_3\cdot$	$CF_3N{=}NCF_3$, irradiation of $CF_3N{=}NCF_3$, 180°, autogenous pressure	Tetrafluoroethylene or ethylene
$CF_3\cdot$	$CF_3N{=}NCF_3$, 450°, 30 min	N_2F_4
$CF_3\cdot$	$CF_3OSO_2OCF_3$, heat at 260°, or irradiate	Iodine
$CF_2ClCFCl\cdot$	$CF_2ClCFClSO_2Cl$, 100°, 7 days	–
$CF_3\cdot$	$CF_3H + Br_2$, thermal 360-450°, photochemical 275-360°	$CF_3H + Br_2$
$mCF_2CF_2\cdot$ $m(n\text{-}CF_2)_2CX(CF_2)_n\cdot$ X = F or CF_3	Copolymer of tetrafluoroethylene with other fluorinated olefins. Ionizing radiation at elevated temperature	Studied melt viscosity
B. Unsaturated Fluorocarbons 1. Vinyl		
$CF_2{=}CF$	$CF_2{=}CFI$, uv irradiation; ratio of alkene to $CF_2{=}CFI$ was 0.15 mole to 0.10 mole	$CH_2{=}CH_2$ $CH_2{=}CFH$ $CH_2{=}CF_2$ $CF_2{=}CFH$ $CF_2{=}CHCl$
$CF_2{=}CF\cdot$	$CF_2{=}CFI$, photolysis	Neat with NO or O_2
$CF_2{=}CF\cdot$	$CF_2{=}CFI$, photolysis	NO, nitrogen oxide
$CF_2{=}CF\cdot$ (?)	$CF_2{=}CFCl$, pyrolysis, 700°	Trifluoromethane
2. Aryl		
$C_6F_5\cdot$	$C_6F_5I{-}Ni(CO)_4$/1: 1.04 toluene, reflux	C_6F_5I and $Ni(CO)_4$
$C_6F_5\cdot$ and $C_6F_5Hg\cdot$	C_6F_5I, 200°	Mercury; No positive evidence for radical intermediate

(cont.)

Product(s) (yield, %)	Comments
$(CF_3)_2NN(CH_3)_2$ or $CF_3(CH_3)NN(CH_3)CF_3$ (25) also N_2, CH_3CF_3, CF_3CF_3, CHF_3	Mechanism discussed. Ref. 191l
CF_4, $(CF_2)_3$, C_3F_8, C_4F_{10}, C_2F_2	Physical study. Reactions are partly radical, partly ionic. Ref. 191m
CHF_3, C_2F_4	Physical studies determined primary process to be $CF_3CF_3 \rightarrow 2CF_3\cdot \rightarrow 2CF_2: + 2F\cdot$ $D(CF_3CF_3) = 93 \pm 4$ kcal/mole. Ref. 191n
CF_3CF_3 and CO_2 major products, CO, CF_3H, H_2 minor products	Primary process is $CF_3COOH \rightarrow CF_3\cdot + COOH$. Ref. 191o
$CF_3TeTeCF_3$ (red liquid)	Application of classical technique of reaction of radicals with metal mirrors. First perfluoroalkyl tellurium compound isolated but could not isolate Bi and Pb compounds. Ref. 191p
Mixtures of adducts and substitution products. No yield reported, only ratios of products	Mechanism studies; activation energy for addition to benzene and perfluorobenzene are 5.3 and 2.4 kcal/mole, respectively. Decomposition rates of adducts also studied. Ref. 191q
$CF_3CF_3 + CO$; If R–H present get $CF_3CR(OH)C(O)CF_3$ and $CF_3CH(OH)C(O)CF_3$	Authors suggest mechanism. Ref. 191s
C_7F_{15}–C_7F_{15} (58)	Similar results obtained with C_3F_7COF, $Cl(CF_2)_8COF$, and $H(CF_2)_4COF$. See text (page 197). Ref. 53
n-$C_7F_{15}Cl$ (67), *n*-$C_{14}F_{30}$ (5.1)	Similar results obtained with C_3F_7COCl and $ClCO(CF_2)_3COCl$. See text (page 197). Ref. 53
n-$C_3F_7[CF_2CF(C_5F_{11})]Cl$ (44), $Cl[CF_2CF(C_5F_{11})]_2Cl$ (27)	Similar addition occurs with $CF_2{=}CF(CF_2)_2H$. Ref. 53
Perfluoro-2-heptyl-3-methyloxetane (91)	Comprehensive paper on oxetane formation. Contents too extensive for description here. Ref. 54
$C_3F_7COOC(C_3F_7){=}C(C_3F_7)COOC_3F_7$	Similar reaction with $H(CF_2)_4COCl$ occurs in 19% yield. Ref. 55
Respective products: $C_6H_5C_3F_7$ (70-80), isomeric $CH_3C_6H_4C_3F_7$ (70-80) isomeric $CF_3C_6H_4C_3F_7$ (30) isomeric $BrC_6H_4C_3F_7$ (30-40)	Isomeric products not separated. See text (page 198) for reaction with no solvent present. Ref. 55
CF_3H, $CF_2{=}CH_2$ plus small amounts $CF_3N{=}NCH_3$ and $[CF_3(CH_3)N-]_2$	Mechanism discussed. Postulate $(CH_3)_2N$–$\dot{N}CF_3$ intermediate. Ref. 191t
Polytetrafluoroethylene or polyethylene	Other reactions illustrated in text (page 198). Ref. 56
CF_3NF_2 (65)	Reactions of $C_2F_5N{=}NC_2F_5$ to give $C_2F_5NF_2$ also reported. Ref. 57
CF_3I (87), SO_2	$(FCH_2O)_2SO_2$ and $(F_2CHO)_2SO_2$ can be used as fluoromethyl- and difluoromethyl-alkylating agents. Ref. 191u
SO_2 (18), $CF_2ClCFCl_2$ (19), recovered $CF_2ClCFClSO_2Cl$ (80)	Generated other intermediate radicals by free radical addition such as $Br\cdot$ to give $BrCF_2\dot{C}FSO_2Cl$. Ref. 191v
CF_3Br, HBr	Mechanism study leading to $D_{CF_3-H} = 109.5$ kcal/mole and $D_{CF_3-CF_3} = 93$ kcal/mole. Ref. 191w
Crosslinked polymer	Conclude crosslinking via radical species. Possible species discussed in relation to polymers. Ref. 191x
$CF_2{=}CFCH_2CH_2I$ $CF_2{=}CFCH_2CFHI$ (50) $CF_2{=}CFCH_2CF_2I$ (24) $CF_2{=}CFCFHCF_2I$ (39) $CF_2{=}CFCHClCF_2I$	Butenes were dehydrohalogenated to corresponding 1,3-butadienes: $CF_2{=}CFI$ was dimerized in 50% conversion to $CF_2{=}CFCF{=}CF_2$. Ref. 192h
Red polymer, $(C_2F_3)_n$? $(C_2F_3NO)_2$ with NO, F_2CO, CFIO, C_2F_3OI, $(FCO)_2CF_2$ with O_2	Quantum yields of products were high. Chain mechanism postulated. Ref. 191y
$CF_2{=}CFNO$ (yield not reported)	Product dimerizes at 80° to give *cis*- and *trans*- $\begin{array}{cccc} O & N & CF & CF_2 \\ \vert & \vert & \vert & \vert \\ CF_2 & CF & N & O \end{array}$ Ref. 46
$CF_3CF{=}CF_2$ (60), HCl (–)	Radical intermediate $CF_2{=}CF\cdot$ probable because CF_3H is stable to 800°C. Ref. 191z
C_6F_5–C_6F_5 (35.2), $C_6F_5COC_6F_5$ (20.4), C_6F_5H (23.2)	See discussion in text (page 193). Conditions widely varied. C_6F_5Br reacted poorly. Ref. 33
$C_6F_5HgC_6F_5$	May form intermediate $C_6F_5M\cdot$ radicals where M is Hg, S, Se, P, Sb, Ge, Sn. Reported formation of $(C_6F_5)_2S$, $(C_6F_5)_3P$, $(C_6F_5)_3Sb$, $(C_6F_5)_4Ge$, $(C_6F_5)_4Sn$. Ref. 47

Table 6-1

Radical species	Source of radical and method of generation	Reactant and/or method of detection
F F F F F F C_6H_4X (ring structure, •)	Decomposition of benzyl peroxides in refluxing hexafluorobenzene	$(XC_6H_4CO_2)_2$ substituted with X = H, CH_3, Cl, Br
F F F F F F C_6H_5 (ring structure, •) or F F F F F F C_6H_5 (ring structure, •)	(ring)—NO_2 ⟶ (ring)• + NO_2 Flow reactor, 600°	(ring)• + F(ring)F (F F F F) → radical
C. Fluorinated heteroatom radicals		
$\cdot NF_2$	N_2F_4, $h\nu$, Hanovia EH-4 lamp	Biacetyl
$\cdot NF_2$ and :NF	N_2F_4, $h\nu$ (2537 Å)	SO_2
$\cdot NF_2$ and :NF	N_2F_4, $h\nu$ (2537 Å)	(a) Methane (30 min) (b) *n*-Butane (c) *trans*-2-Butene (d) Ethylene (e) 2-Butyne
$\cdot NF_2$	N_2F_4, $h\nu \geqslant 2750$ Å, 25-40°	(a) CH_3I (b) C_2H_5I
$\cdot NF_2$	$N_2F_4 \rightleftarrows 2\cdot NF_2$, thermal > 180°, < 40 mm-Hg	Kinetic study with isobutane, cyclopentane, *n*-butane, neopentane. H abstraction
$\cdot NF_2$	$N_2F_4 \rightleftarrows 2\cdot NF_2$, thermal > 178°	Kinetics of reaction of $\cdot NF_2$ with acetone
$\cdot NF_2$	N_2F_4, benzophenone, Hanovia Hg lamp type 608-36A	Diethyl ether (or dioxane, tetrahydrofuran, diisopropyl ether)
$\cdot NF_2$	N_2F_4, thermal	Alkene, NaF $RCH{=}CH_2 \xrightarrow{N_2F_4} [RCH(NF_2)CH_2NF_2] \xrightarrow{NaF} RC(=NF)-C{\equiv}N$ Example: Allyl acetate
$\cdot NF_2$	N_2F_4, thermal	Addition to styrene or 4-chloro styrene

(cont.)

Product(s) (yield, %)	Comments
2,3,5,6-tetrafluoro-4-fluorophenyl–C_6H_4X biphenyl (up to 76)	Must be attack of $C_6H_4X\cdot$ on C_6F_6. Thought to be different than normal arylation of benzene. Ref. 71a, 192a
C_6H_5–C_6F_5 major product; 8 other products detected	Evidence for formation of $C_6F_5\cdot$ from isolation $C_6F_5-C_6F_5$ Suggested $C_6F_6 + NO_2 \rightarrow NOF + C_6F_5\cdot$ Ref. 71b
$CH_3C(=O)NF_2$ (80)	Similarly, $\cdot NF_2$ also added to radicals generated from glyoxal, benzil, azoisobutyronitrile, azoisobutane, and hexaphenylethane. Ref. 121b
FSO_2NF_2 (89), FN=NF (60)	$\frac{1}{2}N_2F_4 \rightleftharpoons \cdot NF_2 \xrightarrow{h\nu} \cdot NF_2^*$; $\cdot NF_2^* \rightarrow F\cdot + :NF$; $F\cdot \xrightarrow{\cdot NF_2} NF_3$; $F\cdot + SO_2 \rightarrow FSO_2\cdot \xrightarrow{\cdot NF_2} FSO_2NF_2$; $:NF \rightarrow N_2F_2$ Ref. 122
(a) CH_3NF_2 (50), FN=NF (50) (b) $CH_3CH_2CH_2CH_2NF_2$, $CH_3CH_2CH(NF_2)CH_3$ (ratio 1.4:1) (c) $CH_3CH{=}CHCH_2NF_2$ (~37), $CH_3CHFCH(NF_2)CH_3$ (~37) (d) $FCH_2CH_2NF_2$ (–) (e) $CH_3CF_2C(=NF)CH_3$ (–), $CH_3CF{=}C(NF_2)CH_3$ (–)	Again postulate $\cdot NF_2$ excited to form $\cdot NF_2^*$ which decomposes to $F\cdot$ and :NF. Interesting mechanism presented. Ref. 126
(a) CH_3NF_2 (up to 52) (b) $C_2H_5NF_2$ (up to 70)	This was first report of alkyldifluoroamines. Ref. 123
Not isolated	Kinetics used to find strength of N–H bond in HNF_2 to be greater than 72.5 kcal/mole. Strength of C–H bonds also determined. Ref. 124
Not isolated. Decompose at temp. of study. ? $CH_3C(=O)CH_2NF_2$?	Kinetics used to determine bond dissociation energy of C–H as 92.1 ± 1 kcal/mole. Ref. 125
$CH_3CH_2{-}O{-}CH(NF_2)CH_3$	General reaction is substitution of NF_2 for α-hydrogen. Relative reactivities of ether summarized. Ref. 127
$N{\equiv}C{-}C(=NF){-}CH_2OC(=O)CH_3$ (58)	Similar products in 30-70% yields from acrylonitrile, vinyl fluoride, styrene, propylene, methyl acrylate, allyl acetate, and vinylsulfur pentafluoride. Ref. 129b
$C_6H_5CHNF_2CH_2NF_2$ (–) or *p*-$ClC_6H_4CHNF_2CH_2NF_2$ (–)	Products dehydrofluorinated and converted into $X{-}C_6H_4{-}C(=N{-}F){-}C{\equiv}N$ for comparison with other compounds. Ref. 128b

Table 6-1

Radical species	Source of radical and method of generation	Reactant and/or method of detection
$\cdot NF_2$	N_2F_4, thermal, 75°, 4 hr	*trans*-Stilbene
$\cdot NF_2$	N_2F_4, thermal	(a) Cyclooctatetraene (75°, 2 hr) (b) 6,6-Diphenylfulvene (40°) (c) Acenaphthalene (50-55°, 1.5 hr)
$\cdot NF_2$	N_2F_4, thermal, 40°C	Kinetics of addition to anthracene, stilbene, phenanthrene, naphthalene, and 1,2-benzanthracene studied
$\cdot NF_2$	N_2F_4, thermal, 41-155°, gas phase	Kinetic study of addition of N_2F_4 to ethylene, propene, but-1-ene, *trans*-but-2-ene, *cis*-but-2-ene, iso-butene, 2-methylbut-2-ene, 2,3-di-methylbut-2-ene
$\cdot NF_2$	N_2F_4, thermal	2-Acetoxycyclopentene
$\cdot NF_2$	N_2F_4, thermal, 30°, 5 hr	Divinyl ether (0.02 mole), N_2F_4 (0.10 mole)
$\cdot NF_2$	N_2F_4, thermal, 80°, 3 hr	Diallyl ether
$\cdot NF_2$	N_2F_4, thermal, 80°, 3 hr	$C_6H_5C{\equiv}CC_6H_5$
$\cdot NF_2$	N_2F_4, thermal, 70°, 2 hr	Isopropenylacetylene
$\cdot NF_2$	N_2F_4, thermal, 188°, 3 hr	Perfluoro-2-heptyne
$\cdot NF_2$	N_2F_4, uv irradiation 25°, 1.3 hr	$CF_3N{=}NCF_3$, $C_2F_5N{=}NC_2F_5$

(cont.)

Product(s) (yield, %)	Comments
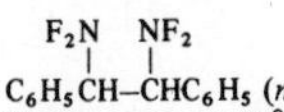 $C_6H_5CH(NF_2)-CH(NF_2)C_6H_5$ (*meso* 37%, *dl* 50%) (decomposes at > 80°)	Dehydrofluorination with $(CH_3)_3N$ gave *syn-* and *anti-* 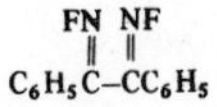 $C_6H_5C(=NF)-C(=NF)C_6H_5$ Ref. 131
(a) 3,8-Bis(difluoramino)cycloocta-1,4,6-triene (65-70), unknown stereochemistry (b) *cis-* (29) and *trans-* (7) 3,5-Bis(difluoramino)-4 (diphenylmethylene)cyclopentene-1 (c) *cis-* (30-34) and *trans-* (40-47) 1,2-Bis(difluoramino)-acenaphthene	Addition to 1,3-cyclooctadiene occurred by 1,2- and 1,4-addition. Derivative chemistry of some compounds reported. Ref. 130
Expected 1,2- or 1,4- addition as appropriate	Rate law d(adduct)$/dt = k(N_2F_4)$ (Arom) Rates measured for *trans*-stilbene and anthracene. Relative rates determined for all addends. Ref. 134a
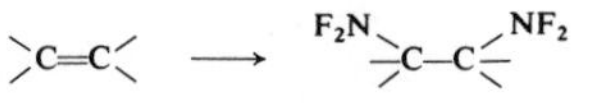 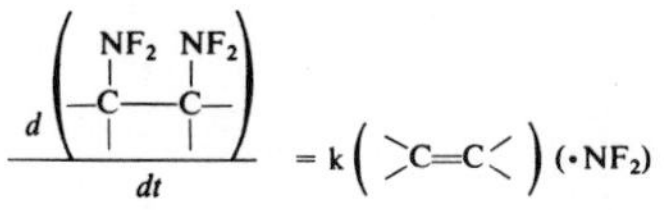 $>C=C< \longrightarrow F_2N-C-C-NF_2$ $\dfrac{d\left(-C(NF_2)-C(NF_2)-\right)}{dt} = k\left(>C=C<\right)(\cdot NF_2)$	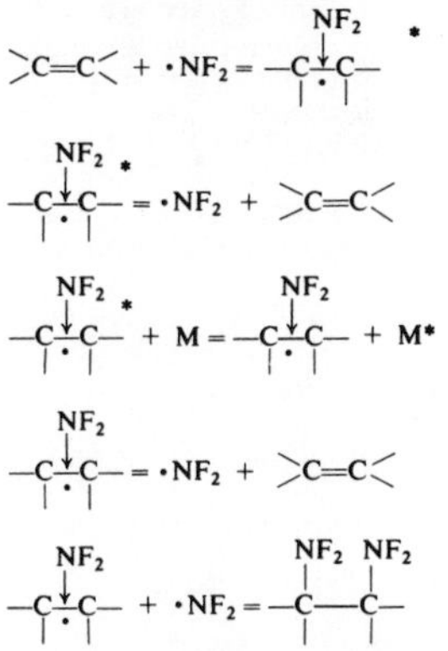 $>C=C< + \cdot NF_2 = [-C(NF_2)-\dot{C}-]^*$ $[-C(NF_2)-\dot{C}-]^* = \cdot NF_2 + >C=C<$ $[-C(NF_2)-\dot{C}-]^* + M = -C(NF_2)-\dot{C}- + M^*$ $-C(NF_2)-\dot{C}- = \cdot NF_2 + >C=C<$ $-C(NF_2)-\dot{C}- + \cdot NF_2 = -C(NF_2)-C(NF_2)-$ (M is a third body). Mechanism deduced from second-order kinetics. Ref. 134b
cis- (41) and *trans-* (59) 1,2-Bis(difluoramino)-2-acetoxycyclopentane	Product rearranged with 96-100% H_2SO_4 to 5-(difluoramino)-*N*-fluorovalerolactam. Ref. 133
$CH_2=CH-OCH(NF_2)CH_2NF_2$ (71) $F_2NCH_2CH(NF_2)OCH(NF_2)CH_2NF_2$ (13)	Hydrolysis of $CH_2=CHOCH(NF_2)CH_2NF_2$ gives $F_2NCH_2CH(NF_2)(OH)$ (58), a unique synthesis of this alcohol. Ref. 192a
$F_2NCH_2CH(NF_2)CH_2OCH_2CH(NF_2)CH_2NF_2$ (48) *cis-* (24) and *trans-* (28) 3,4-Bis(difluoraminomethyl)-tetrahydrofuran	Mechanism suggested for formation of cyclic products. Also added N_2F_4 to *N,N*-diallyl acetamide and acrylic anhydride. Ref. 192b
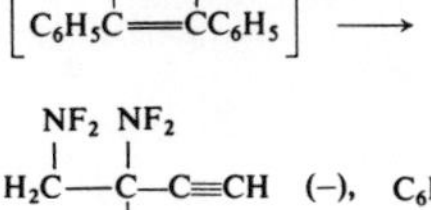 $[C_6H_5C(NF_2)=C(NF_2)C_6H_5] \longrightarrow$	Similar addition and rearrangement with dimethyl acetylenedicarboxylate. Ref. 132
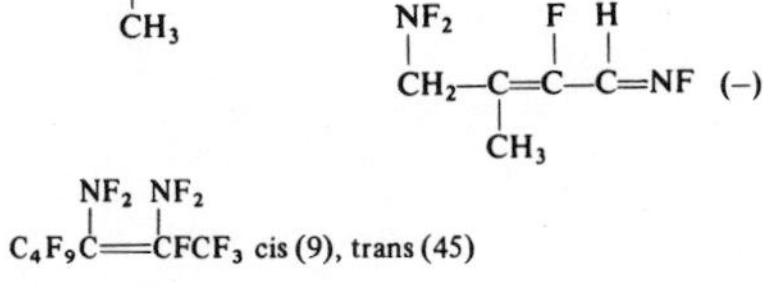 $H_2C(NF_2)-C(NF_2)(CH_3)-C\equiv CH$ (–), $C_6H_5C(=NF)-CF(NF_2)C_6H_5$ (52) $CH_2(NF_2)-C(CH_3)=C(F)-C(H)=NF$ (–)	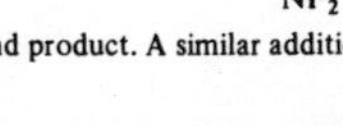 Postulate $CH_2(NF_2)-C(CH_3)=C=C(H)(NF_2)$ as initially formed intermediate rearranging to second product. A similar addition to allene is reported. Ref. 132
$C_4F_9C(NF_2)=C(NF_2)CF_3$ cis (9), trans (45)	Rearranges thermally to  $R_fC(=NF)-CF(NF_2)R_f'$. Also reported are reactions of N_2F_4 with perfluoro-2-butyne, dicyanoacetylene, diphenylacetylene, methylphenylacetylene, phenylpropyoyl fluoride, 1,1,1-trifluoropropyne, 2,4-hexadiyne, allene. Ref. 129c
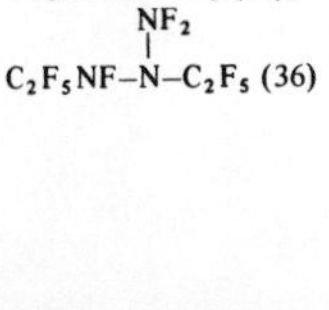 $CF_3NF-N(NF_2)-CF_3$ (50), $C_2F_5NF-N(NF_2)-C_2F_5$ (36)	First report of this highly fluorinated linear triazine system. Ref. 136

Table 6-1

Radical species	Source of radical and method of generation	Reactant and/or method of detection
$\cdot NF_2$	$ClNF_2$, thermal	Ethylene, 130°, 12 hr
		trans-Butene-2, 120°, 7 hr
$(CF_3)_2N\cdot$	$(CF_3)_2NCl$, heat, 100-150°, 6 hr	Ethylene
$(CF_3)_2N\cdot$	$(CF_3)_2NCl$, uv, 22 hr	Trifluoroethylene
$(CF_3)_2N\cdot$	$(CF_3)_2NCl$, uv irradiation	$CH_2{=}CH_2$ $HC{\equiv}CH$ $CF_2{=}CFH$ $(CF_3)_2NCH{=}CHCl$
$C_2F_5\dot{N}CF_3$	$C_2F_5NClCF_3$, heat, or uv	Trifluoroethylene
$C_4F_9\dot{N}CF_3$	$C_4F_9NClCF_3$, heat, or uv	Ethylene
$(C_3F_7)_2N\cdot$	$(C_3F_7)_2NCl$	Acetylene
$(CF_3)_2CFNCF_3$	$(CF_3)_2CFNClCF_3$, heat, or uv	Ethylene
$[CF_3(CF_2)_5]_2N\cdot$	$[CF_3(CF_2)_5]_2NCl$, heat, or uv	Ethylene
$(CF_3)_2N\cdot$	$CF_3N{=}CF_2$ + F· from AgF_2, HgF_2, CoF_3, PbF_4, MnF_3	Examined product
$(CF_3)_2N\cdot$	$(CF_3)_2NBr$, hν from Hg lamp, 5 days	Dimerization of radical
$(CF_3)_2N\cdot$	$(CF_3)_2NBr$, hν from Hg lamp	NO_2
$(CF_3)_2N\cdot$	$(CF_3)_2NBr$, thermal	(a) Ethylene (20°, 12 hr) (b) Acetylene (50°, 3 hr) (c) Tetrafluoroethylene (60°, 1 hr)
$(CF_3)_2N\cdot$	$(CF_3)_2NBr$, uv initiation	$CF_2{=}CF_2$
$(CF_3)_2N\cdot$	$(CF_3)_2NOCF_3$, hν irradiation or by pyrolysis at 775°	Dimerization and at high temperature, disproportionation
$(CF_3)_2N\cdot$	$(CF_3)_2NONO$	Disproportionation to products
$CF_3\dot{N}SF_5$	CF_3NClSF_5, 150°, 1 hr	Hexafluoropropylene
$(CF_3)_2P\cdot$	$(CF_3)_2PP(CF_3)_2$ (a) 6 days, room temp. (b) I_2 cat., 34 hr, 165°	(a) Ethylene (b) Tetrafluoroethylene
$(CF_3)_2P\cdot$ and $CF_3\dot{P}I$	CF_3I + P, heat, 195-230°	Products formed infer radical intermediates
$CF_3\dot{P}{-}P(CF_3){-}P(CF_3){-}\dot{P}CF_3$? or $CF_3\dot{P}{-}\dot{P}CF_3$	$(CF_3P)_4$, 170°, 70 hr	$CF_3C{\equiv}CCF_3$
$\cdot OF$	OF_2 (a) room temp., 30 mm (b) mixed at liquid N_2 temp., warmed to room temp.	(a) Ethylene (30 mm) (b) Tetrafluoroethylene
$\cdot OF$	OF_2, −78°, < 2 hr	Tetramethylethylene

(cont.)

Product(s) (yield, %)	Comments
$ClCH_2CH_2Cl$ (24), $ClCH_2CH_2NF_2$ (47), $F_2NCH_2CH_2NF_2$ (19), $CH_3CHClCHClCH_3$, *dl* and *meso*; $CH_3CH(NF_2)CHClCH_3$, *erythro* and *threo*; $CH_3CH(NF_2)CHNF_2CH_3$, *dl* and *meso* (95 overall for mixture)	Similar reaction for propylene. Light initiated reactions led only to dichloro compounds. Ref. 121a
$(CF_3)_2NCH_2CH_2Cl$ (50)	Dehydrochlorination gave $(CF_3)_2NCH{=}CH_2$, which was polymerized. Ref. 143
$(CF_3)_2NCHFCF_2Cl$ (42), $(CF_3)_2NCF_2CHFCl$ (7)	Dehydrochlorination gave $(CF_3)_2NCF{=}CF_2$ and $(CF_3)_2NCF{=}CFCl$ which were polymerized. $(CF_3)_2NCl$ also added to acetylene. Ref. 143
$(CF_3)_2NCH_2CH_2Cl$ (85) $(CF_3)_2NCH{=}CHCl$ (27) $(CF_3)_2NCF_2CFClH + (CF_3)_2NCFHCF_2Cl$ (50) $(CF_3)_2NCHClCHClN(CF_3)_2 + [(CF_3)_2N]_2CHCHCl_2$	Similar additions carried out with $R_fR_f'NCl$ where R_f and R_f' varied from CF_3- to $CF_3(CF_2)_5-$. Ref. 143
$(C_2F_5)(CF_3)NCHFCF_2Cl$ (–), $(C_2F_5)(CF_3)NCF_2CHFCl$ (–)	$(C_2F_5)(CF_3)N–Cl$ also added to acetylene. Ref. 143
$(C_4F_9)(CF_3)NCH_2CH_2Cl$ (–)	$(C_4F_9)(CF_3)N–Cl$ also added to trifluoroethylene. Ref. 143
$(C_3F_7)_2NCH{=}CHCl$(–)	$(C_3F_7)_2NCl$ also added to ethylene and trifluoroethylene. Ref. 143
$(CF_3)_2CFNCH_2CH_2Cl$ (–)	Dehydrochlorinated to $(CF_3)_2CFNCH{=}CH_2$. Ref. 143
$[CF_3(CF_2)_5]_2NCH_2–CH_2Cl$	Dehydrochlorinated to $[CF_3(CF_2)_5]_2NCH{=}CH_2$. Ref. 143
$(CF_3)_2NN(CF_3)_2$, $(CF_3)_2NF$ and/or $(CF_3)_2NCF{=}NCF_3$. Relative amounts depend on metal fluoride and temperature	Proposed mechanism: $CF_3N{=}CF_2 + F\cdot \rightarrow (CF_3)_2N\cdot$ $2(CF_3)_2N\cdot \rightarrow (CF_3)_2NN(CF_3)_2$ $(CF_3)_2N\cdot + F\cdot \rightarrow (CF_3)_2NF$ $(CF_3)_2N\cdot + CF_3N{=}CF_2 \rightarrow (CF_3)_2NCF_2\dot{N}CF_3 \rightarrow (CF_3)_2NCF{=}NCF_3 + F\cdot$. However, with certain metal fluorides, an ionic mechanism through F^- catalysis is considered more likely. Ref. 144a
$(CF_3)_2N–N(CF_3)_2$ (40)	Also isolated CF_3Br, CF_4, C_2F_6, $CF_3N{=}CF_2$, and $(CF_3)_2NH$ (from H_2O). Ref. 144b
$(CF_3)_2NONO$	Obtained $(CF_3)_2N–NO$ with NO and $(CF_3)_2NC(O)Br$ with CO. Ref. 144b
$(CF_3)_2NCH_2CH_2Br$ (82) $(CF_3)_2NCH{=}CHBr$ (10)	No fluorination occurred in spite of known reaction of $(CF_3)_2N\cdot$ to fluorinate and give $CF_3N{=}CF_2$. Reaction may involve ionic intermediate. Ref. 144b
$(CF_3)_2NCF_2CF_2Br$ (71) $(CF_3)_2NCF_2CF_2Br$ (96)	Similar additions with $CF_2{=}CFH$, $CF_2{=}CH_2$, $CFH{=}CFH$, $CH_2{=}CH_2$, $CF_3CF{=}CF_2$, $CF_2{=}CFCl$, $CFCl{=}CFCl$. Ref. 143b
$(CF_3)_2NN(CF_3)_2$	At high temperature, $(CF_3)_2NF$ and $CF_3N{=}CF_2$ form by disproportionation of two radicals. Ref. 145
$(CF_3)_2NF$ and $CF_3N{=}CF_2$	Products form from disproportionation of two radicals. Ref. 145
$(CF_3)(SF_5)N–CF(CF_3)CF_2Cl$ and/or $(CF_3)(SF_5)N–CF_2CFClCF_3$ (37)	Orientation of addition not proved. Ref. 146
(a) $(CF_3)_2PCH_2CH_2P(CF_3)_2$ (94.5) (b) $(CF_3)_2PCF_2CF_2P(CF_3)_2$ (60)	Addition claimed to 22 other alkenes and to acetylenes. Ref. 153 Addition to olefinic bond is a general reaction
$(CF_3)_3P$, $(CF_3)_2PI$, CF_3PI_2 formed in varying amounts depending on temperature and ratio of starting material	Authors suggest intermediate radical species. No conclusive proof provided. Ref. 154
$F_3C–C{=}C(CF_3)–P(CF_3)–P–$ (ring: $(F_3C)C{=}C(CF_3)$, $P(CF_3)–P(CF_3)$)	Nature of intermediate radical (biradical?) not known. Radical mechanism probable. Ref. 156
FCH_2CH_2F and CH_3CH_2F, CF_3CF_3 and COF_2	No discussion. Ethylene reactions frequently exploded. Ref. 161
$CH_3–C(F)(CH_3)–C(OH)(CH_3)–CH_3$ (65–70)	Available evidence suggested addition of elements $F\cdot$ and $OF\cdot$ to give $CH_3–C(F)(CH_3)–C(OF)(CH_3)–CH_3$ (65-70) as intermediate which then lost $F\cdot$ and abstracted H. Ref. 160

Table 6-1

Radical species	Source of radical and method of generation	Reactant and/or method of detection
$\cdot OF$	OF_2, −78°, 2.5 hr	1,1-Diphenylethylene
$\cdot OF$	OF_2, low temperature	$C_6H_5CH{=}CHC_6H_5$ (cis and trans)
$\cdot OF$	OF_2, −78°, 70 min	$C_6H_5C{\equiv}CCH_3$
$\cdot OF$	OF_2, irradiation with high-pressure Hg lamp	Perfluorocyclobutene
$\cdot OF$	OF_2, −78°	Tetramethylallene
$CF_3O\cdot$	CF_3OF	Ethylene
$CF_3O\cdot$	CF_3OF, thermal	$CFCl{=}CFCl$, flow reactor
$CF_3O\cdot$	CF_3OF, 250°, flow reaction	Sulfur trioxide
$CF_3O\cdot$	CF_3OF, $h\nu$ from Hg lamp, 3 days	Sulfur tetrafluoride
$CF_3O\cdot$	CF_3OF, thermal (200-225°) or irradiation (2537 Å)	Perfluoropropene
$CF_3O\cdot$	CF_3OF, uv irradiation, 24 hr	Tetrafluorohydrazine
(S)–$C(CF_3)_2$–$O\cdot$	$CF_3C(=O)CF_3$, cyclohexane with di-*tert*-butylperoxide, $(C_6H_5CO_2)_2$, or $C_6H_5C(CH_3)_2OOH$ or by irradiation	Cyclohexane adds to either carbon or oxygen of carbonyl
$F{-}S(=O)_2{-}O\cdot$	FSO_2OOSO_2F	(a) Tetrafluoroethylene (b) Perfluorocyclopentene
$SF_5O\cdot$	SF_5OOSF_5, thermal, 150°	(a) Benzene (b) Toluene (90°, 10 hr) (c) Chlorobenzene (no conditions)
$SF_5O\cdot$	SF_5OF, thermal, 95°, 12 hr	Perfluorocyclopentene
$SF_5O\cdot$	SF_5OF	(a) Vinyl fluoride (b) Vinylidene fluoride
$SF_5O\cdot$	SF_5OF, 17 hr, room temp.	Tetrafluorohydrazine
$\cdot SF_5$	SF_5Cl, thermal	(a) Ethylene (90°, 10 hr) (b) Propene (100°, 3 hr) (c) 1,3-Butadiene (100°, 2 hr) (d) Isobutene (100°, 3 hr) (e) Cyclohexene (20°, 18 hr) (f) Propyne (90°, 4 hr) (g) Vinyl chloride (150°, 6 hr)
$\cdot SF_5$	SF_5Cl, 160-170°	Acetylene
$\cdot SF_5$	SF_5Cl, irradiate	$CF_3C{\equiv}N$
$\cdot SO_2F$	SO_2FCl, thermal	Ethylene

(cont.)

Product(s) (yield, %)	Comments
$C_6H_5C(=O)CH_2F$ (66) plus trace of biphenyl and polymeric material. *No* fluorobenzene	Postulate $C_6H_5-C(OF)(C_6H_5)-CH_2F$ as intermediate which loses F· and then rearranges with loss of C_6H_5·. Ref. 160
$C_6H_5C(=O)CHFC_6H_5$ (17) plus α-fluorostilbenes (cis and trans). Majority of product was polymer	Additions to 1-butene and 1-propene were uncontrollable. Styrene polymerized. Unsaturated esters and nitriles gave only polymer. Ref. 160
$C_6H_5-C(=O)-CF_2-CH_3$ (81), $C_6H_5C(=O)F$ (18)	Similar reaction obtained with diphenylacetylene and pentyne-2. Ref. 160
$CF_3CF_2CF_2COF$ (64) plus COF_2, CF_4, CF_3CF_2COF and perfluorocyclobutane	Similar oxidations occurred with perfluoropropene and 2-perfluorobutene. Ref. 162
$CH_3C(F)(CH_3)-C(=O)-C(F)(CH_3)-CH_3$ (–), $CH_3C(F)(CH_3)-C(=O)-C(=CH_2)CH_3$ (–)	Products unstable. Structures assigned on basis of spectra. Ref. 163
$CF_3OCH_2CH_2F$	Ref. 192c
$CF_3OCFClCF_2Cl$ (–)	Converted to $CF_3OCF{=}CF_2$ for polymer studies. Ref. 167
CF_3OOSO_3F (~10), SO_2FOOSO_2F (major product)	Reaction with sulfur dioxide gave several products. Ref. 165d
CF_3OSF_5 (10), $(CF_3O)_2SF_4$ (8). Also isolated SiF_4, COF_2, SO_2F_2, SOF_2, CF_3OOCF_3	Obtained 35% yield of CF_3OSF_5 in a thermal (100°) reaction in Monel bomb, 1 week. Ref. 165c
Telomers $CF_3O(CF_2CF(CF_3))_nOCF_3$ when $n > 2$	Depending on procedure, telomer oils (n = 2,3,4, . . .) or polymer formed. Ref. 171b
CF_3ONF_2 (40)	CF_3ONF_2 is a good oxidizing agent. Ref. 166, 192d
(S)$-C(CF_3)_2-OH$, (S)$-OCH(CF_3)_2$, (S) ring bearing two $C(CF_3)_2OH$ groups	Product yield depends on initiator and temperature. Ref. 52
(a) $FSO_2OCF_2CF_2OSO_2F$ (–) (b) cyclopentane ring with F_2, F_2, F_2 and two adjacent carbons each bearing F and OSO_2F	Vacuum line techniques employed. Yields and exact conditions not specified. Ref. 174a
(a) $C_6H_5OSF_5$ (50) + SOF_4 + HF (b) $CH_3C_6H_4OSF_5$ (54) (c) $Cl(C_6H_4)OSF_5$ (10:1/para:ortho, no yield)	Appears to be a general reaction for aromatic compounds with electron withdrawing substituents. Explosions occurred with phenol, anisole, and aniline. Ref. 175b
cyclopentane ring with F_2, F_2, F_2, F_2 and one carbon bearing F and OSF_5	Addition also reported to $CF_2{=}CF_2$, $CCl_2{=}CCl_2$, $CH_2{=}CH_2$, $CH_2{=}CHCl$. Ref. 165a
(a) $F_2CHCH_2OSF_5$ (b) $F_3CCH_2OSF_5$	Vacuum line work. No yields or conditions given. Ref. 165b
SF_5ONF_2 (40)	With irradiation, 60% yield obtained. Ref. 166
(a) $ClCH_2CH_2SF_5$ (47) (b) $CH_3CHClCH_2SF_5$ (78) (c) $CH_2{=}CHCHClCH_2SF_5$ (–) (d) No reaction of SF_5Cl; polymer (e) (S) with adjacent F and Cl + (S) with adjacent SF_5 and Cl (13) (f) $CH_3CCl{=}CHSF_5$ (40) (g) $Cl_2CHCH_2SF_5$ (85)	Addition reaction is clean. Polymer formed with styrene. Addition occurred with acetylene to give $CHCl{=}CHSF_5$; no experimental details. Ref. 182
$CHCl{=}CHSF_5$ (40)	Compound dehydrohalogenated to give $HC{\equiv}CSF_5$. Reaction reported for acetylene. Ref. 184b
$CF_3CCl{=}NSF_5$ (32)	Addition of SF_5Cl reported to $C_3F_7C{\equiv}N$, cyanogen (2 moles). Ref. 146
$Cl(CH_2CH_2)_nSO_2F$: For n = 1 (8), 2 (50), 3 (28), 4 (10), 5 (4)	Radical initiators such as benzoyl peroxide were also used. Ref. 187

Table 6-1

Radical species	Source of radical and method of generation	Reaction and/or method of detection
$CF_3S\cdot$	CF_3SCl, uv irradiation	Cyclohexane
$CF_3S\cdot$	CF_3SCl, uv irradiation	$CF_2=CFCl$
$CF_3S\cdot$	$(CF_3S)_2Hg$, 225°, 14 hr	Perfluoropropene, 3000 atm

[a] DTBP = Di-*tert*-butylperoxide
AVN = Azobis-2,4-dimethylvaleronitrile
ABN = Azobisisobutyronitrile

[191] (a) B. C. Anderson, *J. Org. Chem.*, **27**, 2720 (1962). (b) T. N. Bell, B. J. Pullman, and B. O. West, *Australian J. Chem.*, **16**, 636 (1963). (c) A. T. Coscia, *J. Org. Chem.*, **26**, 2995 (1961). (d) W. Durrell, A. M. Lovelace, and R. L. Adamczak, *J. Org. Chem.*, **25**, 1661 (1960). (e) F. W. Stacey and J. F. Harris, Jr., *J. Org. Chem.*, **27**, 4089 (1962). (f) J. F. Harris, Jr., and F. W. Stacey, *J. Am. Chem. Soc.*, **85**, 749 (1963). (g) G. W. Parshall, D. C. England, and R. V. Lindsey, Jr., *J. Am. Chem. Soc.*, **81**, 4801 (1959). (h) R. Fields, H. Goldwhite, R. N. Haszeldine, and J. Kirman, *J. Chem. Soc.* (C), 2075 (1966). (i) H. Muramatsu and K. Inukai, *J. Org. Chem.*, **27**, 1572 (1962). (j) H. Muramatsu, K. Inukai, and T. Ueda, *J. Org. Chem.*, **29**, 2220 (1964). (k) R. N. Haszeldine and B. J. H. Mattinson, *J. Chem. Soc.*, 1741 (1957). (l) A. H. Dinwoodie and R. N. Haszeldine, *J. Chem. Soc.*, 2266 (1965). (m) L. Kevan and P. Hamlet, *J. Chem. Phys.*, **42**, 2255 (1965). (n) E. Tschuikow-Roux, *J. Chem. Phys.*, **43**, 2251 (1965). (o) A. M. Mearns and R. A. Back, *Can. J. Chem.*, **41**, 1197 (1963). (p) T. N. Bell, B. J. Pullman, and B. O. West, *Australian J. Chem.*, **16**, 722 (1963). (q) J. R. Majer, D. Philips, and J. C. Robb, *Trans. Faraday Soc.*, **61**, 122 (1965); *ibid.*, 110. (r) E. R. Bissell and D. B. Fields, *J. Org. Chem.*,

(cont.)

Product(s) (yield, %)	Comment
$CF_3SC_6H_{11}$ (45), $C_6H_{11}Cl$ (28), CF_3SSCF_3 (35)	$CF_3SCH_2-C_6H_4-CH_3$, mixed $C_4H_9SCF_3$, and mixed $C_4H_9SCF_3$ obtained from toluene, *n*-butane, and isobutane, respectively. Ref. 188
$CF_3SCF_2CFCl_2$ + $CF_3SCClFCF_2Cl$ in ratio of 4.3:5.7 (no yield)	Converted mixture to $CF_3SCF{=}CF_2$ and $CF_3CF_2CFCl_2$ with Zn dust. Ref. 167
$CF_3S\left(CF_2-\underset{}{\overset{CF_3}{\overset{\mid}{C}F}}\right)_xSCF_3$	Structure postulated. Ref. 189

29, 249 (1964). (s) I. M. Whittemore and M. Szwarc, *J. Phys. Chem.*, **67**, 2492 (1963). (t) L. Batt and J. M. Pearson, *Chem. Commun.*, 575 (1965). (u) G. A. Sokol'skii and M. A. Dmitriev, *J. Gen. Chem. U.S.S.R. (English Transl.)*, **31**, 2821 (1961). (v) R. E. Banks, G. M. Haslam, R. N. Haszeldine, and A. Peppin, *J. Chem. Soc.* (C), 1171 (1966). (w) P. Corbett, A. M. Tarr, and E. Whittle, *Trans. Faraday Soc.*, **59**, 1609 (1963). (x) E. R. Lovejoy, M. I. Bro, and G. H. Bowers, *J. Appl. Polymer Sci.*, **9**, 401 (1965). (y) J. Heicklen, *J. Phys. Chem.*, **70**, 618 (1966). (z) G. Pass, *J. Chem. Soc.*, 824 (1965).

[192] (a) P. A. Claret, J. Coulson, and G. H. Williams, *Chem. Ind. (London)*, 228 (1965). (b) S. F. Reed, Jr., *J. Org. Chem.*, **32**, 2894, 3675 (1967). (c) J. A. C. Allison and G. H. Cady, *J. Am. Chem. Soc.*, **81**, 1089 (1959). (d) W. P. Van Meter and G. H. Cady, *J. Am. Chem. Soc.*, **82**, 6005 (1960). (e) A. D. Ketley, U.S. Patent 3,310,589 (1967). (f) T. Ando, F. Namigata, H. Yamanaka, and W. Funasaka, *J. Am. Chem. Soc.*, **89**, 5719 (1967). (g) R. F. Stockel and M. T. Beachem, *J. Org. Chem.*, **32**, 1658 (1967). (h) J. D. Park, R. J. Seffl, and J. R. Lacher, *J. Am. Chem. Soc.*, **78**, 59 (1956).

Addition:

$$RCH{=}CHR' + N_2F_4 \xrightarrow[\text{liquid or gas phase}]{\text{heat,}} \underset{\displaystyle RCH{-}CHR'}{\overset{NF_2\ \ NF_2}{|\quad\ |}}$$

This is a general reaction for N_2F_4 with a majority of alkenes.[128–130] It was thoroughly investigated by Petry and Freeman at the Rohm and Haas Laboratories, Redstone Arsenal, beginning in 1958.[128a] Similar work was carried out in this period by Logothetis and Sausen at the du Pont Central Research Department Laboratories.[129] A selection of examples is given in Chart 6-2.

$$ClCH_2CH{=}CH_2 + N_2F_4 \xrightarrow[\text{st. steel bomb}]{120°,\ 4\ \text{hr}} ClCH_2CH(NF_2)CH_2NF_2\ (90\%) \qquad (128a)$$

$$C_4H_9CH{=}CH_2 + N_2F_4 \xrightarrow[\text{pressure tube}]{100°,\ 3\ \text{hr}} C_4H_9CH(NF_2)CH_2NF_2\ (82\%) \qquad (128a)$$

$$C_6H_5CH{=}CH_2 + N_2F_4 \xrightarrow[\text{pressure}]{\Delta} C_6H_5CH(NF_2)CH_2NF_2 \qquad (128a,b)$$

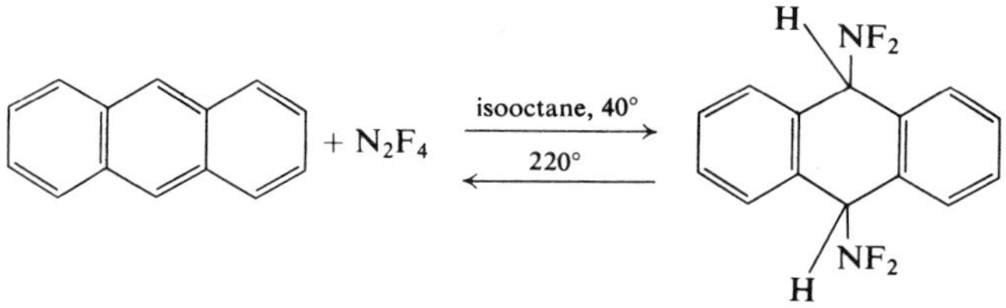

$$\text{anthracene} + N_2F_4 \underset{220°}{\overset{\text{isooctane, } 40°}{\rightleftharpoons}} \text{9,10-bis(difluoroamino)-9,10-dihydroanthracene (H, } NF_2 \text{ at C-9 and C-10)} \qquad (129a)$$

$$C_6H_5CH{=}CHC_6H_5 \longrightarrow C_6H_5CH(NF_2){-}CH(NF_2)C_6H_5 \quad meso\ 37\%,\ d,l\ 50\% \qquad (129b)$$

(decomposes over 80°)

$$\xrightarrow[CH_2Cl_2,\ 35°]{(C_2H_5)_3N} C_6H_5C({=}NF){-}C({=}NF)C_6H_5 \qquad (131)$$

CHART 6-2

[128] (a) R. C. Petry and J. P. Freeman, *J. Org. Chem.*, **32**, 4034 (1967). (b) T. E. Stevens, *ibid.*, 670.

[129] (a) A. L. Logothetis, *J. Org. Chem.*, **31**, 3686 (1966). (b) A. L. Logothetis and G. N. Sausen, *ibid.*, 3689. (c) G. N. Sausen and A. L. Logothetis, *J. Org. Chem.*, **32**, 2261 (1967).

[130] T. S. Cantrell, *ibid.*. 911.

[131] F. A. Johnson, C. Haney, and T. E. Stevens, *J. Org. Chem.*, **32**, 466 (1967).

$$\text{cyclooctatetraene} + N_2F_4 \xrightarrow{60\text{–}70^\circ} \text{5,8-bis(difluoramino)cycloocta-1,3,6-triene} \quad (65\text{–}70\%) \quad (130)$$

$$Cl\text{–}C_6H_4\text{–}CH{=}CH_2 + N_2F_4 \longrightarrow Cl\text{–}C_6H_4\text{–}CH(NF_2)CH_2NF_2 \quad (128b)$$

$$\xrightarrow[CH_2Cl_2]{\text{pyridine}} Cl\text{–}C_6H_4\text{–}C({=}NF)\text{–}CN$$

$$CH_2{=}CHCN + N_2F_4 \xrightarrow[NaF]{120^\circ,\ 3\ hr} N{\equiv}C\text{–}C({=}N\text{–}F)\text{–}C{\equiv}N \quad (66\%) \quad (129b)$$

CHART 6-2 (*cont.*)

Frequently, the bis(difluoramino) product cannot be isolated but instead a dehydrofluorinated derivative is obtained.[129b, 131] The dehydrofluorination reaction is promoted by bases such as pyridine or NaF, or catalytic amounts of HF (see examples in Chart 6-2).

$$RCH(NF_2)CH_2NF_2 \xrightarrow[\text{pyridine or other base}]{NaF} RC({=}NF)\text{–}CN$$

Addition of N_2F_4 to acetylenes does not *usually* give the expected bis-(difluoramino)ethylene derivatives. Instead, rearrangement products are obtained.[129c, 132]

$$R\text{–}C{\equiv}C\text{–}R' \xrightarrow{N_2F_4} \left[RC(NF_2){=}C(NF_2)\text{–}R' \right] \longrightarrow R\text{–}C({=}N\text{–}F)\text{–}CF(NF_2)\text{–}R'$$

[132] R. C. Petry, C. O. Parker, F. A. Johnson, T. E. Stevens, and J. P. Freeman, *J. Org. Chem.*, **32**, 1534 (1967).

$$NC{-}C{\equiv}C{-}CN \xrightarrow{N_2F_4} \left[NC{-}\underset{}{\overset{NF_2}{C}}{=}\overset{NF_2}{C}{-}CN \right] \longrightarrow NC{-}\overset{NF}{\overset{\|}{C}}{-}\underset{F}{\overset{NF_2}{C}}{-}CN \quad (33\%) \qquad (129c)$$

$$C_6H_5C{\equiv}CC_6H_5 \xrightarrow{N_2F_4} \left[C_6H_5\overset{NF_2}{C}{=}\overset{NF_2}{C}{-}C_6H_5 \right] \longrightarrow C_6H_5\overset{NF}{\overset{\|}{C}}{-}\underset{F}{\overset{NF_2}{C}}{-}C_6H_5 \qquad (132)$$

The only class of acetylenes that adds N_2F_4 to give stable bis(difluoroamino)-ethylenes is di(perfluoroalkyl)acetylenes.[132] These derivatives undergo rearrangement on gentle heating. The rearrangement of 2,3-bis(difluoroamino) hexafluoro-

$$CF_3{-}C{\equiv}C{-}CF_3 \xrightarrow{N_2F_4} \underset{\text{stable and isolable, } cis \text{ and } trans}{CF_3{-}\overset{NF_2}{C}{=}\overset{NF_2}{C}{-}CF_3} \xrightarrow{\text{heat}} CF_3{-}\underset{F}{\overset{NF_2}{C}}{-}\overset{NF}{\overset{\|}{C}}{-}CF_3 \qquad (132)$$

2-butene probably occurs via a fluorine radical mechanism because the compound initiates polymerization of perfluoroalkenes at the temperature of rearrangement. This is similar to the conversion of *cis*-difluorodiazine to *trans*-difluorodiazine, where *cis*-N_2F_2 initiates radical polymerizations.

Other interesting rearrangements of alkene adducts with N_2F_4 are known.[133]

$$\text{1-acetoxycyclopentene} + N_2F_4 \longrightarrow \text{1-acetoxy-1,2-bis(difluoroamino)cyclopentane} \xrightarrow{H_2SO_4} \text{N-fluoro-6-(difluoroamino)-2-piperidone} \qquad (134)$$

Studies have been made on the kinetics of addition of N_2F_4 to alkenes. All evidence favors initial dissociation of N_2F_4 to $\cdot NF_2$ radicals with an $\cdot NF_2$ radical attack on the double bond.[134]

$$N_2F_4 \rightleftharpoons 2NF_2\cdot$$

$$\cdot NF_2 + {>}C{=}C{<} \longrightarrow F_2N{-}\overset{|}{\underset{|}{C}}{-}\overset{|}{\underset{|}{C}}\cdot$$

$$F_2N{-}\overset{|}{\underset{|}{C}}{-}\overset{|}{\underset{|}{C}}\cdot + N_2F_4 \longrightarrow F_2N{-}\overset{|}{\underset{|}{C}}{-}\overset{|}{\underset{|}{C}}{-}NF_2 + NF_2\cdot$$

[133] T. E. Stevens and W. H. Graham, *J. Am. Chem. Soc.*, 89, 182 (1967).

[134] (a) H. Cerfontain, *J. Chem. Soc.*, 6602 (1965). (b) A. J. Dijkstra, J. A. Kerr, and A. F. Trotman-Dickenson, *J. Chem. Soc. (A)*, 582 (1966); *J. Chem. Soc. (A)*, 105 (1967); *ibid.*, 864.

The irradiation of N_2F_4 has been studied by Bumgardner[135] who suggested the following scheme:

$$N_2F_4 \rightleftharpoons \cdot NF_2 \xrightarrow{uv} \cdot NF_2^*$$

$$\downarrow$$

$$NF_3 \xleftarrow{\cdot NF_2} F\cdot + :NF \longrightarrow N_2F_2$$

In the presence of other agents, particularly unsaturated compounds, substitution and addition by either $\cdot NF_2$ or $F\cdot$ can occur to give complex products. Thus *trans*-butene-2 gives both substitution of an $\cdot NF_2$ group for a hydrogen and addition of elements of NF_3 across the double bond to give $CH_3CHFCH(NF_2)CH_3$ (see discussion in Section 5-2E, page 121).

A novel example of addition of the elements $F—NF_2$ by irradiation of N_2F_4 is the addition to hexafluoroazomethane. The addition appears to be general for perfluoroazoalkanes.[136]

$$CF_3N{=}NCF_3 + N_2F_4 \xrightarrow[2537\ \text{Å}]{h\nu} CF_3N(NF_2){-}N(F){-}CF_3$$

Spectral studies, such as electron spin resonance and infrared on $\cdot NF_2$, have been used to determine structure of $\cdot NF_2$ and related NF-containing radicals and to get bond energies.[137–142]

Nitrogen fluorine compounds and $\cdot NF_2$ certainly have broad opportunities for research in many areas, particularly for synthetic organic reactions. However, these compounds present a safety problem. Because of their high-energy content and potent oxidizing power, their use will probably be restricted to laboratories that are equipped to handle hazardous material.

Bis(trifluoromethyl)amino radicals are readily prepared by heating or irradiating $(CF_3)_2NX$; they dimerize or add to unsaturated compounds such as ethylene, tetrafluoroethylene, or acetylene as well as adding to reagents such as

[135] C. L. Bumgardner, *Tetrahedron Letters*, 3683 (1964).
[136] G. N. Sausen, *J. Org. Chem.*, **33**, 2336 (1968); *ibid.*, 2330.
[137] A. P. Modica and D. F. Hornig, *J. Chem. Phys.*, **43**, 2739 (1965).
[138] M. D. Harmony and R. J. Myers, *J. Chem. Phys.*, **37**, 636 (1962).
[139] J. B. Farmer, M. C. L. Gerry, and C. A. McDowell, *Mol. Phys.*, **8**, 253 (1964).
[140] C. B. Colburn, R. Ettinger, and F. A. Johnson, *Inorg. Chem.*, **3**, 455 (1964).
[141] P. H. Kasai and E. B. Whipple, *Mol. Phys.*, **9**, 497 (1965).
[142] M. E. Jacox and D. E. Milligan, *J. Chem. Phys.*, **46**, 184 (1967).

NO_2, NO, CO.[143, 144] The $(CF_3)_2N\cdot$ is also formed by irradiation of $(CF_3)_2NOCF_3$.[145] An unusual radical species $CF_3(SF_5)N\cdot$ is probably involved in addition of $CF_3(SF_5)NCl$ to perfluoropropylene.[146]

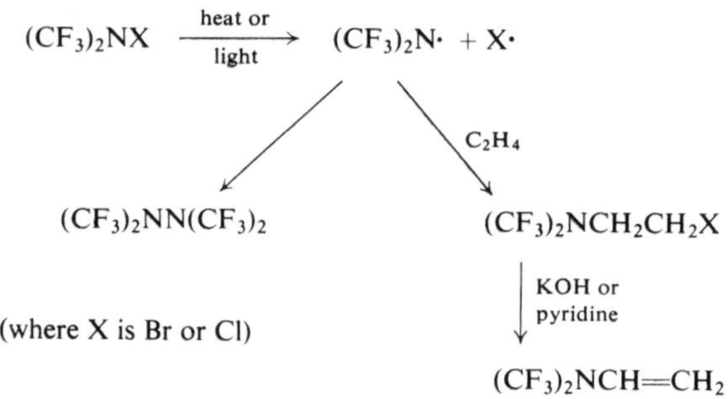

Difluoromethylenimino radical, $CF_2{=}N\cdot$, was generated by photolysis of tetrafluoro-2,3-diaza-1,3-butadiene[147]; it appears to be a relatively stable intermediate since it does not decompose to FC≡N and $CF_2{=}NF$ and is readily trapped.

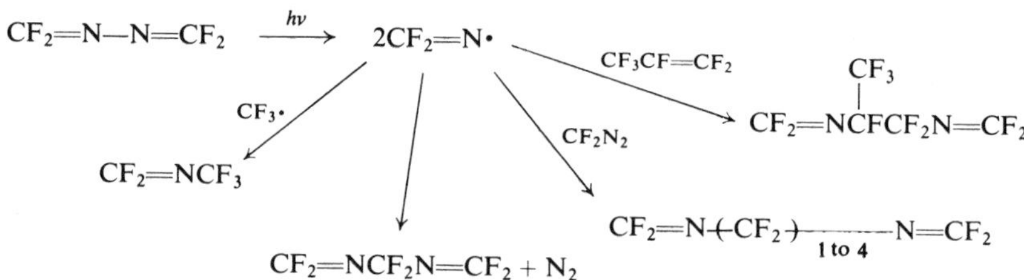

Tetrafluorosuccinimide radical is formed from *N*-bromotetrafluorosuccinimide but was shown in mechanism studies not to be the chain-carrying species in bromination reactions.[148]

[143] (a) F. S. Fawcett, U.S. Patents 3,311,599 and 3,359,319 (1967); *Chem. Abstr.*, **67**, 2751c (1967). (b) R. N. Haszeldine and A. E. Tipping, *J. Chem. Soc.*, 6141 (1965).

[144] (a) J. A. Young, W. S. Durrell and R. D. Dresdner, *J. Am. Chem. Soc.*, **81**, 1587 (1959). (b) H. J. Emeleus and B. W. Tattershall, *Z. Anorg. Allgem. Chem.*, **327**, 147 (1964).

[145] (a) A. H. Dinwoodie and R. N. Haszeldine, *J. Chem. Soc.*, 1681 (1965). (b) R. N. Haszeldine and A. E. Tipping, *J. Chem. Soc. (C)*, 1236 (1966). (c) R. N. Haszeldine and A. E. Tipping, *J. Chem. Soc. (C)*. 1241 (1967).

[146] C. W. Tullock, D. D. Coffman, and E. L. Muetterties, *J. Am. Chem. Soc.*, **86**, 357 (1964).

[147] P. H. Ogden and R. A. Mitsch, *J. Am. Chem. Soc.*, **89**, 3868 (1967).

[148] R. E. Pearson and J. C. Martin, *J. Am. Chem. Soc.*, **85**, 3142 (1963).

b. *Phosphorus and Arsenic*

The radical $\cdot PF_4$ has been identified by esr studies on irradiated hexafluorophosphate salts,[149–151] and $\cdot PF_2$ has also been observed.[150, 151]

No synthetic chemistry of these species is reported. Tetrafluorodiphosphine, F_2PPF_2, the phosphorus analog of tetrafluorohydrazine, gives products F_2PH and F_2PX on reaction with Bronsted–Lowry acids HX. One mechanism suggested $F_2P\cdot$ radicals as intermediates.[152] No reactions with organic substrates were mentioned.

The addition of tetrakis(trifluoromethyl)diphosphine to olefins and acetylenes[153, 154] probably involves the bis(trifluoromethyl)phosphine radical, $(CF_3)_2P\cdot$, but no mechanism is proposed. Preparation and chemistry of this radical[153] from $(CF_3)_2PX$ needs study.

$$(CF_3)_2P{-}P(CF_3)_2 + CH_2{=}CH_2 \longrightarrow (CF_3)_2PCH_2CH_2P(CF_3)_2$$

The reaction of tetraphenylcyclotetraphosphine,[155] $(C_6H_5P)_4$, with CF_3I to give $C_6H_5P(CF_3)_2$ must involve radical attack by $\cdot CF_3$. Some phosphorus radicals are probably involved as intermediates.

Fluoroalkyl phosphorus radical species probably are also involved in the synthesis of cyclopolyphosphine.[156]

$$\begin{matrix}(CF_3P)_4 \\ \text{or} \\ (CF_3P)_5\end{matrix} + CF_3C{\equiv}CCF_3 \xrightarrow[70\ \text{hr}]{170^\circ} \begin{matrix}F_3CC{-}PCF_3 \\ \| \quad | \\ F_3CC{-}PCF_3\end{matrix} + \begin{matrix} & & CF_3 \\ & & | \\ F_3C{-}C & \diagup & P \\ \| & & \quad \diagdown \\ & & \quad PCF_3 \\ \| & & \quad \diagup \\ F_3C{-}C & \diagdown & P \\ & & | \\ & & CF_3\end{matrix}$$

(55%) (31%)

No authenticated examples of fluorinated arsenic radicals have been reported. Again in the chemistry of $(CF_3)_2AsX$ the radical species[40, 157] $(CF_3)_2As\cdot$ is probably generated. Additional work to find this radical needs to be done. The reaction of arsenobenzene, $(C_6H_5As)_6$, with CF_3I to give products[158] such

[149] P. W. Atkins and M. C. R. Symons, *J. Chem. Soc.*, 4363 (1964).
[150] J. R. Morton, *Can. J. Phys.*, **41**, 706 (1963).
[151] R. W. Fessenden and R. H. Schuler, *J. Chem. Phys.*, **45**, 1845 (1966).
[152] K. W. Morse and R. W. Parry, *J. Am. Chem. Soc.*, **89**, 172 (1967).
[153] A. B. Burg and L. R. Grant, U.S. Patent 3,118,951 (1963); *Chem. Abstr.*, **60**, 10718B (1964).
[154] F. W. Bennett, H. J. Emeleus, and R. N. Haszeldine, *J. Chem. Soc.*, 1565 (1953).
[155] M. A. A. Beg and H. C. Clark, *Can. J. Chem.*, **39**, 564 (1961).
[156] W. Mahler, *J. Am. Chem. Soc.*, **86**, 2306 (1964).
[157] W. R. Cullen, *Can. J. Chem.*, **38**, 439 (1960).
[158] W. R. Cullen and N. K. Hota, *Can. J. Chem.*, **42**, 1123 (1964).

as $C_6H_5As(CF_3)_2$ must involve $\cdot CF_3$ and probably arsenic radical intermediates. The reactions of $(C_6F_5)_2AsCl$ with mercury[159] to give $(C_6F_5)_2AsAs(C_6F_5)_2$ or $(C_6F_5)_2AsHgCl$ could involve $(C_6F_5)_2As\cdot$ and provide a source of these radicals for reaction with unsaturated organic compounds.

c. *Oxygen*

Oxygen difluoride was discussed in Section 5-2F. It adds to olefins and acetylenes, probably by dissociation to $\cdot OF$ and $\cdot F$.[160–163] Usually the products are mixtures resulting from rearrangements of the initial adduct.

Reactions of OF_2 with methane lead to trifluoromethyl hypofluorite[164]; this explosive reaction is controlled by using reduced pressure. A multistep radical reaction is suggested.

$$CH_4 + 4OF_2 \rightarrow CF_3OF + 4HF + \tfrac{3}{2}O_2$$

Fluoroalkyl hypohalites and peroxides can give $R_fO\cdot$ radicals. Some synthetic reactions are reported, particularly by Cady and co-workers,[165] for addition of CF_3OF or SF_5OF to olefins and in reaction with inorganic reagents (see Section 5-2F).

$$CF_3OF \xrightarrow{h\nu} [CF_3OF]^* \longrightarrow F^\bullet + CF_3O^\bullet \xrightarrow{\bullet NF_2} CF_3ONF_2 \qquad (166)$$

A practical use of CF_3OF radical addition was in synthesis of trifluoromethyl trifluorovinyl ether for polymerization studies.[167, 168] The addition

$$CF_3OF + CFCl{=}CFCl \longrightarrow CF_3OCFClCF_2Cl \xrightarrow[\text{dimethyl sulfoxide}]{Zn} CF_3OCF{=}CF_2$$

was run in the gas phase at high dilution in an apparatus (jet reactor) normally used for controlled fluorination. Extensive degradation occurred if the con-

[159] M. Green and D. Kirkpatrick, *Chem. Commun.*, 57 (1967).

[160] R. F. Merritt and J. K. Ruff, *J. Org. Chem.*, **30**, 328 (1965).

[161] R. A. Rhein and G. H. Cady, *Inorg. Chem.*, **3**, 1644 (1964).

[162] J. K. Ruff and R. F. Merritt, *J. Org. Chem.*, **30**, 3968 (1965).

[163] R. F. Merritt, *ibid.*, 4367.

[164] L. B. Marantz, *J. Org. Chem.*, **30**, 4380 (1965).

[165] (a) S. M. Williamson and G. H. Cady, *Inorg. Chem.*, **1**, 673 (1962). (b) S. M. Williamson, *Inorg. Chem.*, **2**, 421 (1963). (c) L. C. Duncan and G. H. Cady, *Inorg. Chem.*, **3**, 850 (1964). (d) W. P. Van Meter and G. H. Cady, *J. Am. Chem. Soc.*, **82**, 6005 (1960). (e) C. I. Merrill and G. H. Cady, *J. Am. Chem. Soc.*, **83**, 298 (1961).

[166] W. H. Hale, Jr., and S. M. Williamson, *Inorg. Chem.*, **4**, 1342 (1965).

[167] W. S. Durrell, E. C. Stump, G. Westmoreland, and C. D. Padgett, *J. Polymer Sci.*, Part A, **3**, 4065 (1965).

[168] A more practical route to $R_fOCF{=}CF_2$ is reported by C. G. Fritz, E. P. Moore, Jr., and S. Selman, U.S. Patent 3,114,778 (1963) [*Chem. Abstr.*, **60**, 6750B (1964)] and J. F. Harris and D. I. McCane, U.S. Patent 3,180,895 (1965) [*Chem. Abstr.*, **63**, 1701E (1965)] by pyrolysis of perfluorinated 2-alkoxypropionic acid (available for perfluorinated acid fluoride with hexafluoropropylene epoxide).

ditions were not carefully controlled. Radical addition of $CF_3O\cdot$ (or $F\cdot$) is the most probable mechanism.

The photochemical addition of CF_3OOCF_3 to CO must involve $CF_3O\cdot$,[169] but again a more practical synthetic route to PFMC is reported.[170]

$$CF_3OOCF_3 + CO \longrightarrow CF_3O\overset{\overset{\displaystyle O}{\|}}{C}OCF_3 \text{ (PFMC)}$$

Trifluoromethoxy radicals must also be involved in the following sequence[171]:

$$CF_3OF + COF_2 \xrightarrow[200\text{ atm}]{275^\circ} \underset{(90\%)}{CF_3OOCF_3} \xrightarrow[h\nu \text{ or heat}]{CF_2{=}CFCF_3} CF_3O(C_3F_6)_nOCF_3$$

We considered that OF_2,R_fOF, and R_fOOR_f are good sources of $FO\cdot$ and $R_fO\cdot$, but they are not reagents that normally an organic chemist can use for synthesis; they are too hazardous to handle and too difficult to control in reactions with organic compounds.

Earlier, the reaction of free radicals with perfluoroacetone to give radicals with the free electron on oxygen or on carbon was discussed.[52] Chlorofluoroacetones were also studied. Whether the free electron resides on carbon or on oxygen is a question dependent both on reaction conditions and substitution of the ketone. The equations in Chart 6-3 illustrate the point.[52]

The first member of the fluorine–nitrogen–oxygen radical series is trifluoroamine oxide, $F_3NO\cdot$, prepared by electric discharge through an NF_3–O_2 mixture at $-196°C$.[172] It was stated to be a strong oxidizing agent to many organic and inorganic reagents and to form stable 1:1 complexes with AsF_5 and SbF_5. It was photolyzed to $F_2NO\cdot$, a relatively stable radical identified by esr.

More practical fluorine–nitrogen–oxygen radicals for the organic chemist are the fluorocarbon nitroxides. They are isolable, stable, free radicals [$(CF_3)_2NO\cdot$ is a purple gas, bp $-20°C$] prepared from fluoroalkyl nitroso compounds.[173]

[169] E. L. Varetti and P. J. Aymonino, *Chem. Commun.*, 680 (1967).

[170] B. C. Anderson and G. R. Morlock, U.S. Patent 3,226,418 (1965); *Chem. Abstr.*, **64**, 9598 (1966) report that carbonyl fluoride may be easily dimerized or trimerized (to PFMC) by heating with a basic catalyst.

[171] (a) R. S. Porter and G. H. Cady, *J. Am. Chem. Soc.*, 79, 5628 (1957). (b) H. L. Roberts, *J. Chem. Soc.*, 4538 (1964).

[172] W. B. Fox, J. S. Mackenzie, N. Vanderkooi, B. Sukornick, C. A. Wamser, J. R. Holmes, R. E. Eibeck, and B. B. Stewart, *J. Am. Chem. Soc.*, **88**, 2604 (1966).

[173] (a) W. D. Blackley and R. R. Reinhard, *J. Am. Chem. Soc.*, **87**, 802 (1965). (b) W. D. Blackley, *J. Am. Chem. Soc.*, **88**, 480 (1966). (c) S. P. Makarov, A. Ya. Yakubovich, S. S. Dubov, and A. N. Medvedev, *Proc. Acad. Sci. U.S.S.R., Chem. Sect. (English Transl.)*, **160**, 195 (1965). (d) S. P. Makarov, M. A. Englin, A. F. Videiko, V. A. Tobolin, and S. S. Dubov, *Proc. Acad. Sci. U.S.S.R., Chem. Sect. (English Transl.)*, **168**, 483 (1966). (e) R. E. Banks, R. N. Haszeldine, and M. J. Stevenson, *J. Chem. Soc.* (C), 901 (1966). (f) R. N. Haszeldine and B. J. H. Mattinson, *J. Chem. Soc.*, 1741 (1957).

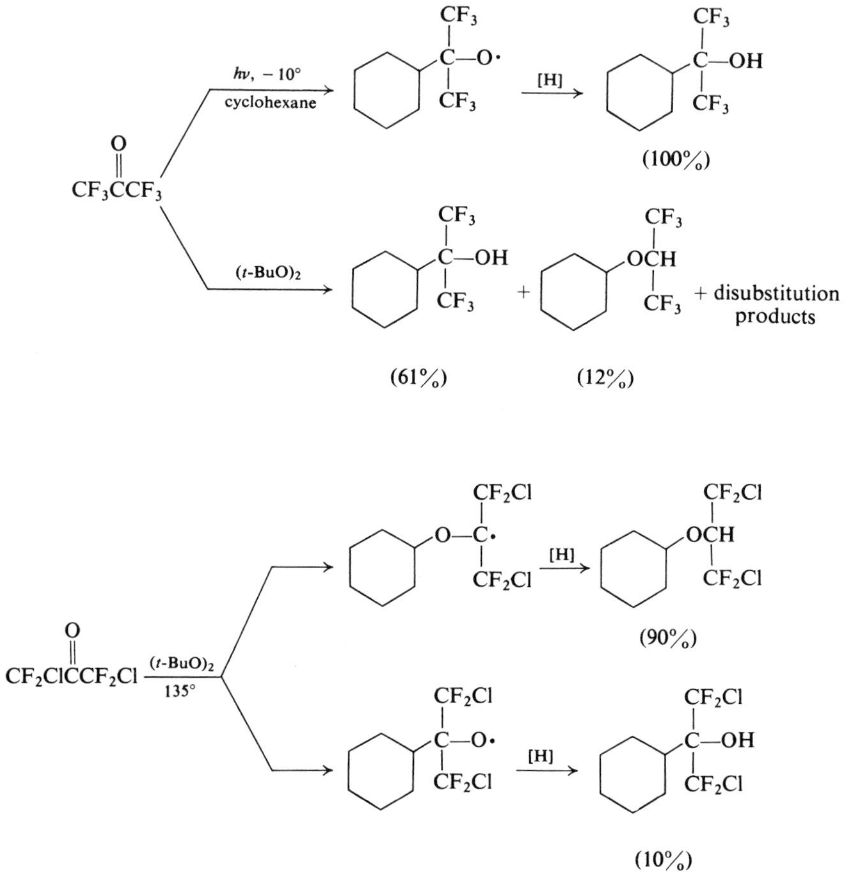

CF_3COCF_3 $\xrightarrow[\text{cyclohexane}]{h\nu,\ -10^\circ}$ $C_6H_{11}C(CF_3)_2O\cdot$ $\xrightarrow{[H]}$ $C_6H_{11}C(CF_3)_2OH$ (100%)

CF_3COCF_3 $\xrightarrow{(t\text{-BuO})_2}$ $C_6H_{11}C(CF_3)_2OH$ (61%) + $C_6H_{11}OCH(CF_3)_2$ (12%) + disubstitution products

$CF_2ClCOCF_2Cl$ $\xrightarrow[135^\circ]{(t\text{-BuO})_2}$ $C_6H_{11}O\dot{C}(CF_2Cl)_2$ $\xrightarrow{[H]}$ $C_6H_{11}OCH(CF_2Cl)_2$ (90%)

$CF_2ClCOCF_2Cl$ $\xrightarrow[135^\circ]{(t\text{-BuO})_2}$ $C_6H_{11}C(CF_2Cl)_2O\cdot$ $\xrightarrow{[H]}$ $C_6H_{11}C(CF_2Cl)_2OH$ (10%)

CHART 6-3

The esr spectra of these radicals show that some spin delocalization or spin polarization occurs to fluorine (surprisingly the F^{19} nmr spectra can be measured in spite of the paramagnetic species). These radicals are useful reagents to introduce a $(CF_3)_2NO$ group (see Chart 6-4).

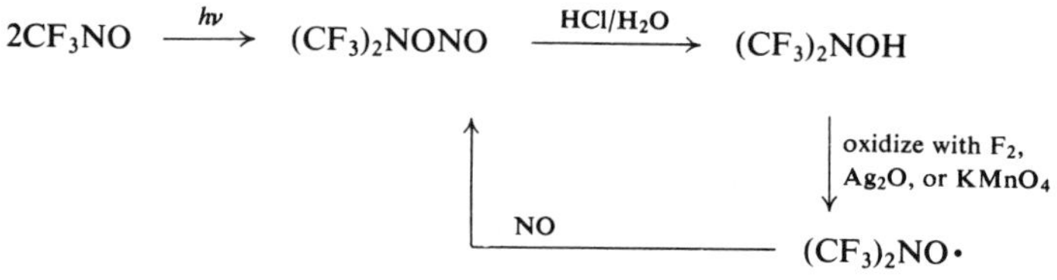

$2CF_3NO \xrightarrow{h\nu} (CF_3)_2NONO \xrightarrow{HCl/H_2O} (CF_3)_2NOH$

$(CF_3)_2NOH \xrightarrow{\text{oxidize with } F_2,\ Ag_2O, \text{ or } KMnO_4} (CF_3)_2NO\cdot$

$(CF_3)_2NO\cdot \xrightarrow{NO} (CF_3)_2NONO$

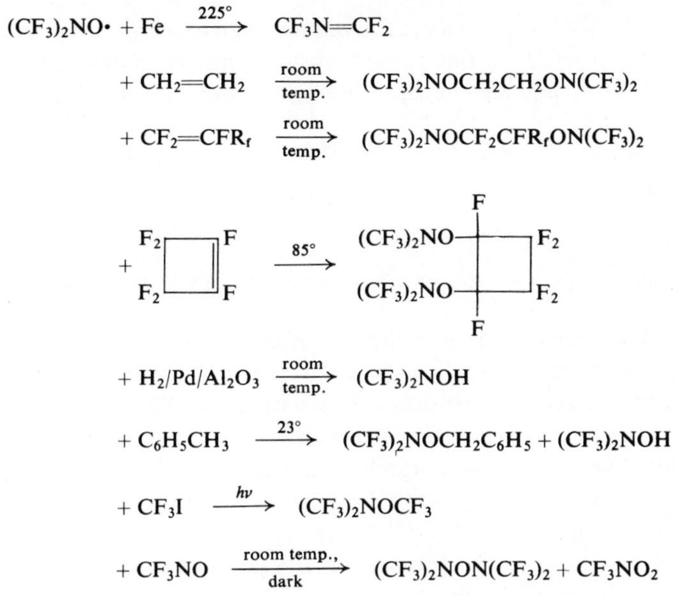

$$(CF_3)_2NO\cdot + Fe \xrightarrow{225^\circ} CF_3N{=}CF_2$$

$$+ CH_2{=}CH_2 \xrightarrow[\text{temp.}]{\text{room}} (CF_3)_2NOCH_2CH_2ON(CF_3)_2$$

$$+ CF_2{=}CFR_f \xrightarrow[\text{temp.}]{\text{room}} (CF_3)_2NOCF_2CFR_fON(CF_3)_2$$

$$+ \text{perfluorocyclobutene } (F_2, F_2, F, F) \xrightarrow{85^\circ} \text{1,2-bis}[(CF_3)_2NO]\text{-1,2-difluoro-3,3,4,4-tetrafluorocyclobutane}$$

$$+ H_2/Pd/Al_2O_3 \xrightarrow[\text{temp.}]{\text{room}} (CF_3)_2NOH$$

$$+ C_6H_5CH_3 \xrightarrow{23^\circ} (CF_3)_2NOCH_2C_6H_5 + (CF_3)_2NOH$$

$$+ CF_3I \xrightarrow{h\nu} (CF_3)_2NOCF_3$$

$$+ CF_3NO \xrightarrow[\text{dark}]{\text{room temp.,}} (CF_3)_2NON(CF_3)_2 + CF_3NO_2$$

CHART 6-4

Other fluorine–nitrogen–oxygen radical intermediates are suggested in preparation or reaction of perfluoronitroso compounds[173f]; for example,

$$CF_3I \text{ or } CF_3NO \xrightarrow{h\nu} \cdot CF_3 \xrightarrow{CF_3NO} (CF_3)_2NO\cdot \xrightarrow{\cdot CF_3} (CF_3)_2NOCF_3 \ (49\%)$$

$$(CF_3)_2NO\cdot \xrightarrow{\cdot NO} (CF_3)_2NONO$$

Several fluorine–sulfur–oxygen radicals have been reported, but these are often generated under conditions where an organic reactant would be destroyed. Peroxydisulfuryl difluoride readily dissociates to fluorosulfate free radicals and can be added to fluoroolefins, but the reaction is extremely vigorous.[174] The $SF_5O\cdot$ species has been proposed to form when SF_5OF is added to olefins[165a, b]

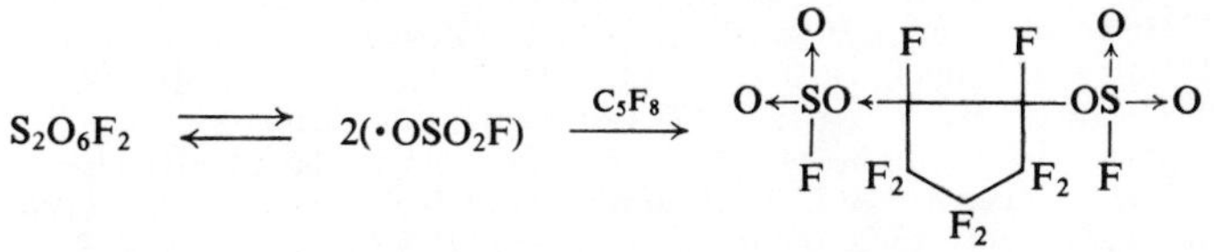

$$S_2O_6F_2 \rightleftharpoons 2(\cdot OSO_2F) \xrightarrow{C_5F_8} \text{1,2-bis}(FSO_2O)\text{-1,2-difluoro-3,3,4,4,5,5-hexafluorocyclopentane}$$

[174] (a) J. M. Shreeve and G. H. Cady, *J. Am. Chem. Soc.*, **83**, 4521 (1961). (b) F. B. Dudley and G. H. Cady, *J. Am. Chem. Soc.*, **85**, 3375 (1963).

(see Section 5-2F). The peroxide, SF_5OOSF_5, should also provide a source of $SF_5O\cdot$, but the reactions of this peroxide with organic materials are attack on an aromatic system[175] and telemorization with hexafluoropropene.[176] A free radical

$$C_6H_6 + SF_5OOSF_5 \xrightarrow{150^\circ} \underset{(50\%)}{C_6H_5OSF_5} + SOF_4 + HF$$

$$SF_5OOSF_5 + CF_3CF{=}CF_2 \xrightarrow{150^\circ} SF_5O(CF_2{-}\overset{\overset{\displaystyle CF_3}{|}}{C}F)_nOSF_5 \quad (n = 2 \text{ to } 4)$$

mechanism is expected; however, the only evidence is substitution in toluene and chlorobenzene to give chiefly para (not meta) orientation which indicates an ionic transition state for substitution. In any case, since SF_5OF and SF_5OOSF_5 oxidatively degrade most organic compounds,[175, 176] they will probably only have limited synthetic value in organic synthesis.

Several oxygen–fluorine radical species, such as $FO\cdot$, $CF_3O\cdot$, and $CF_3OO\cdot$, have been studied by esr.[120, 177–181] However, no synthetic opportunities have developed from these studies because the radicals are usually generated at low temperatures in a solid matrix.

d. *Sulfur*

The sulfur pentafluoride radical, $\cdot SF_5$, is generated from SF_5Cl or S_2F_{10} by heat or light and adds to a variety of ethylenes and acetylenes and fluorinated nitriles[119, 182–186] (see Chart 6-5). Fluoroolefins give telomers in addition to 1:1 adducts. For most reactions with olefins, SF_5Cl dissociates to the radicals $\cdot SF_5$ and $Cl\cdot$ and the initial addition is by the chlorine radicals. However, under certain conditions an ionic mechanism is proposed for addition.

[175] (a) H. L. Roberts, *J. Chem. Soc.*, 2774 (1960). (b) J. R. Case, R. Price, N. H. Ray, H. L. Roberts, and J. Wright, *J. Chem. Soc.*, 2107 (1962).

[176] (a) J. R. Case and G. Pass, *J. Chem. Soc.*, 946 (1964). (b) C. I. Merrill and G. H. Cady, *J. Am. Chem. Soc.*, **85**, 909 (1963).

[177] A. D. Kirshenbaum, *Inorg. Nucl. Chem. Letters,* **1**, 121 (1965).

[178] A. D. Kirshenbaum and A. G. Streng, *J. Am. Chem. Soc.*, **88**, 2434 (1966).

[179] F. J. Adrian, *J. Chem. Phys.*, **46**, 1543 (1967).

[180] R. W. Fessenden and R. H. Schuler, *J. Chem. Phys.*, **44**, 434 (1966).

[181] (a) F. Neumayr and N. Vanderkooi, Jr., *Inorg. Chem.*, **4**, 1234 (1965). (b) N. Vanderkooi, Jr., and W. B. Fox, *J. Chem. Phys.*, **47**, 3634 (1967).

[182] J. R. Case, N. H. Ray, and H. L. Roberts, *J. Chem. Soc.*, 2066, 2070 (1961).

[183] Imperial Chemical Industries Ltd., French Patent 1,294,361 (1962).

[184] (a) C. W. Tullock, D. D. Coffman, and E. L. Muetterties, *J. Am. Chem. Soc.*, **86**, 357 (1964). (b) F. W. Hoover and D. D. Coffman, *J. Org. Chem.*, **29**, 3567 (1964); see also D. D. Coffman, U.S. Patent 3,284,496 (1966); *Chem. Abstr.*, **66**, 55005y (1967).

[185] (a) B. Cohen and A. G. MacDiarmid, *Chem. Ind. (London)*, 1866 (1962). (b) B. Cohen and A. G. MacDiarmid, *Inorg. Chem.*, **4**, 1782 (1965).

[186] (a) H. L. Roberts, *J. Chem. Soc.*, 3183 (1962). (b) M. Tremblay, *Can. J. Chem.*, **43**, 219 (1965).

$$SF_5Cl + RCH{=}CH_2 \xrightarrow[\text{or peroxide}]{\text{heat, light}} RCHClCH_2SF_5 \xrightarrow{\text{base}} RCH{=}CHSF_5 \qquad (182)$$

$$SF_5Cl + HC{\equiv}CH \longrightarrow \underset{(35\%)}{F_5SCH{=}CHCl} + \underset{(5\text{–}10\%)}{F_5SCH{=}CHCH{=}CHCl} \qquad (182,\ 184)$$

$$SF_5Cl + C_2F_4 \longrightarrow Cl(C_2F_4)_nSF_5 \qquad (183)$$

$$SF_5Cl + CF_3CN \xrightarrow{h\nu} \underset{(32\%)}{SF_5N{=}CClCF_3} + S_2F_{10} \qquad (184)$$

$$S_2F_{10} \longrightarrow 2[\cdot SF_5] \xrightarrow{Cl_2} 2SF_5Cl$$

$$2[\cdot SF_5] \xrightarrow[125\text{–}140^\circ]{XCH{=}CH_2} XCH(SF_5)CH_2F \quad (\text{X is Cl or Br}) \qquad (185,\ 186)$$

$$2[\cdot SF_5] \xrightarrow{CF_3CF{=}CF_2} CF_3CF(SF_5){-}CF_2SF_5$$

CHART 6-5

Sulfur chloride pentafluoride is readily prepared[184]

$$SF_4 + Cl_2 + CsF \rightarrow SF_5Cl + CsCl$$

and easily converted to S_2F_{10} using hydrogen.

Unfortunately, the radical addition of S_2F_{10} to olefins usually goes in low yield and often with side reactions (note ·F addition in Chart 6-5), probably because S_2F_{10} is a potent oxidizing agent and disproportionates readily to SF_4 and SF_6. An added disadvantage in working with S_2F_{10} is its relatively high toxicity (like phosgene).

A free radical reaction of sulfuryl chlorofluoride with fluorinated olefins probably involves $\cdot SO_2F\cdot$[187]

$$ClSO_2F + nRCH{=}CH_2 \xrightarrow[\text{initiation}]{\text{radical}} Cl(CH(R)CH_2)_nSO_2F$$

The addition of CF_3SCl or CF_3SH to fluoroolefins via $CF_3S\cdot$ was discussed in the fluorocarbon radical section.[72, 73] Irradiation of CF_3SCl generates the $CF_3S\cdot$ species[188] which can substitute into alkanes.

[187] G. V. D. Tiers, U.S. Patents 3,050,555 and 3,050,556 (1962); *Chem. Abstr.*, **58**, 6696f and d (1963).

[188] J. F. Harris, Jr., *J. Org. Chem.*, **31**, 931 (1966).

$$CF_3SCl + C_6H_{12} \xrightarrow{h\nu} CF_3SC_6H_{11} + C_6H_{11}Cl + CF_3SSCF_3 + HCl$$

(cyclohexane) (45%) (28%) (35%)

Mixtures were usually obtained where more than one type of hydrogen was present. A radical chain mechanism can account for all products found.

$$CF_3SCl \xrightarrow{h\nu} CF_3S\cdot + Cl\cdot$$

$$Cl\cdot + R{-}H \longrightarrow HCl + R\cdot$$

$$R\cdot + CF_3SCl \longrightarrow RCl + CF_3S\cdot$$

$$R\cdot + CF_3SCl \longrightarrow RSCF_3 + Cl\cdot$$

$$CF_3S\cdot + R{-}H \longrightarrow CF_3SH + R\cdot$$

$$CF_3SH + CF_3SCl \longrightarrow CF_3SSCF_3 + HCl$$

The radical addition of CF_3SH to olefins was used to prepare products for verification of structure and analysis of product mixtures.

$$CF_3SH + \text{cyclohexene} \xrightarrow{h\nu} \text{cyclohexyl}{-}SCF_3$$

Bis(trifluoromethyl)disulfide reacts with hexafluoropropene to give a polymer[189]; initiation and termination by $CF_3S\cdot$ is probable.

$$CF_3SSCF_3 + n\text{-}CF_3CF{=}CF_3 \longrightarrow CF_3S(C_3F_6)_nSCF_3$$

Bis(trifluoromethyl)disulfide on irradiation is suggested to give $CF_3S\cdot$ which initiates polymerization of tetrafluorothiirane as shown.[190]

$$CF_3SSCF_3 \xrightarrow{h\nu} CF_3S\cdot + \text{tetrafluorothiirane} \longrightarrow CF_3SCF_2CF_2\cdot$$

$$CF_3SCF_2CF_2\cdot \xrightarrow{\text{tetrafluorothiirane}} CF_3S(SCF_2CF_2)_2\cdot \xrightarrow{\text{tetrafluorothiirane}} {-}(SCF_2CF_2){-}_n$$

[189] H. S. Eleuterio, U.S. Patent 2,958,685 (1960); *Chem. Abstr.*, **55**, 6041c (1961).

[190] W. R. Brasen, H. N. Cripps, C. G. Bottomley, M. W. Farlow, and C. G. Krespan, *J. Org. Chem.*, **30**, 4188 (1965).

Other radicals such as chlorine and fluoroalkylsulfenyl chlorides can also be used as photoinitiators; other radicals have been shown to attack the sulfur and open the ring to give radical intermediates.

6-1B. CARBENES

Divalent carbon species, known as carbenes, :CXY, are an important class of reactive intermediates that, when X and/or Y are halogens, are particularly valuable in synthesis of halocyclopropanes and dihalomethylene derivatives. Since several reviews[193, 194] are available, we summarize only the synthetic application of fluorinated carbenes, covering in particular recent work using selected examples supplemented with data in Table 6-2 to illustrate experimental potential.

The area of heteroatom carbenes has been reviewed very recently[196] and will only be summarized to point out the synthetic potential of these fluorinated species.

The following types of fluorinated carbenes will be discussed in this section: CF_2, CFX (where X is H, Cl, Br, OR, or other heteroatom groups), and CFR and CR_fR (where R is alkyl, aryl, fluoroalkyl, aryl, carboethoxy, and R_f is fluoroalkyl).

1. Methods of Preparation and Reactions of Fluorinated Carbenes

a. *Basic Hydrolysis of Haloforms*

$$\text{CHXYZ} \xrightarrow{\text{base}} {}^{-}\text{CXYZ} \longrightarrow \text{:CXY} + \text{Z}^{-}$$

where X, Y, and Z are halogens and at least X or Y is fluorine.

Hine provided the initial evidence based on kinetic and mechanistic studies for intermediacy of halocarbenes.[194a] The effects of different halogen on the relative rates of the two steps has been concisely summarized[194]; actually, difluorocarbene appears to form in a one-step concerted mechanism. Difluorocarbene is the most stable halocarbene and preferentially dimerizes under many conditions.

Difluoromethylation of anions or active hydrogen compounds is the most important synthetic use of haloform-generated carbenes in organic synthesis. The difluorocarbene is highly electrophilic and rapidly attacks anions; in protic systems, where hydrolyses are usually run and the alcohol or mercaptan is in large excess, attack could be directly on the oxygen or sulfur to give insertion in

[193] W. E. Parham and E. E. Schweizer, *Org. Reactions,* **13,** 55 (1963).

[194] (a) J. Hine, *Divalent Carbon,* Ronald Press, New York, 1964, Chapter 3. (b) W. Kirmse, *Carbene Chemistry,* Academic, New York, 1964, Chapter 8.

[196] O. M. Nefedov and M. N. Manakov, *Angew. Chem. Internat. Edit. Engl.,* **5,** 1021 (1966).

Table 6-2. Selected Examples of Synthesis

Carbene	Source and method of generating carbene	Reactant and/or method of detection; conditions
Basic hydrolysis of haloforms		
$:CF_2$	$CHClF_2$, xcs NaOH, aq. dioxane 68-70°	C_6H_5OH, p-$CH_3C_6H_4OH$, p-$CH_3O{-}C_6H_4OH$, p-$O_2N{-}C_6H_4{-}OH$, 2,4-$(CH_3)_2{-}C_6H_3{-}OH$, 2,4-$(Cl)_2{-}C_6H_3{-}OH$, 2-Naphthol
$:CF_2$	$CHClF_2$, xcs NaOH, aq. dioxane or similar polar solvent	XC_6H_4SH, where X is bromine or CH_3
$:CF_2$	$CHClF_2$, sodium alcoholate, alcohol solvent	$NaOCH_3$, CH_3OH, 5 hr 60° $NaOC_4H_9$, C_4H_9OH, 2 hr 80° $NaSCH_2CH_2OH$, $HSCH_2CH_2OH$ $(CH_3)_2C{=}N{-}ONa$
$:CF_2$	$CHClF_2$, sodium dialkylphosphate, hydrocarbon solvent	$NaOP(OC_2H_5)_2$, 30-35° $NaOP(O\text{-iso-}C_3H_8)_2$ $NaOP(OC_4H_9)_2$
$:CF_2$	$CHClF_2$, sodium t-butoxide, 1,2-dimethoxyethane solvent	$CH_3{-}C{-}CH_2CH(CO_2C_2H_5)_2$ (C bearing a cyclic O–O ring) $C_6H_5CH(CO_2C_2H_5)_2$ $(C_6H_5)_2\bar{C}CN$ 2-carbethoxyindole
$:CF_2$	$CHClF_2$, NaOH, tetraglyme, heat to 95°	Tetramethylethylene
$:CFCl$	$CHCl_2F$, NaOH, tetraglyme heat to 95°	Tetramethylethylene
$:CFCl$	$CHCl_2F$, ethylene oxide, $(CH_3)_4N^+Br^-$ catalyst, heat from 135° to 170°	$CH_2{=}CHOC_2H_5$
$:CFCl$	$CHCl_2F$, ethylene oxide, $(CH_3)_4N^+Br^-$ catalyst, 130-165°, 5-7 hrs	Propene 2-methylpropene Cyclohexene Cyclooctene Styrene 1,1-diphenylethene
$:CFCl$	$CHCl_2F$, ethylene oxide, $(CH_3)_4N^+Br^-$ catalyst, 130-150°, 5 hr	1,3-butadiene
Energetic decomposition of fluorocarbons		
$:CF_2$?	$CHClF_2$, pyrolysis	$:CF_2$ dimerizes
$:CBrF$?	$CHBr_2F$, pyrolysis over noble metal such as Pt in presence of H_2	$:CBrF$ loses Br
$:CBrF$?	CBr_3F, pyrolysis 530-550°, under N_2	$:CBrF$ loses Br
$:CClF$?	$CHCl_2F$, pyrolysis 650-700°	$:CClF$ loses Cl
$:CClF$	$CHCl_2F$, $CH_2{-}CH_2$ (epoxide, O bridging), $NH_4^+Br^-$, 150°	$CH_2{=}CHOC_4H_9$

and Reactions of Fluorine-Containing Carbenes

Product(s) (yield, %)	Comments
$C_6H_5OCHF_2$ (65), *p*-$CH_3C_6H_4OCHF_2$ (66), *p*-$CH_3O{-}C_6H_4{-}OCHF_2$ (53), *p*-$O_2N{-}C_6H_4{-}OCHF_2$ (9.5), 2,4-$(CH_3)_2{-}C_6H_3{-}OCHF_2$ (56), 2,4-$(Cl)_2{-}C_6H_3{-}OCHF_2$ (44), 2-(difluoromethoxy)-naphthalene (66)	Mechanism proposed: $HCClF_2 + OH^- \rightleftarrows H_2O + {:}CF_2 + Cl^-$; $ArO^- + {:}CF_2 \rightleftharpoons ArOCF_2^-$; $ArOCF_2^- + H_2O \rightleftharpoons ArOCHF_2 + OH^-$. Orthoformate esters formed as side reaction products. Dioxane required for better yields. Ref. 197
$XC_6H_4SCF_2H$ (55-80)	Ref. 198
CH_3OCF_2H (23), $CH_3CH_2CH_2CH_2OCF_2H$ (–, low), $HOCH_2CH_2SCF_2H$ (60), $(CH_3)_2C{=}NOCF_2H$ (–, low)	The $CHClF_2$ gas bubbled into alcoholate solution. Very simple experimental procedure of general utility for preparation of difluoromethyl ethers. Ref. 199
$HCF_2OP(OC_2H_5)_2$ (48.6), $HCF_2OP(O\text{-iso-}C_3H_8)_2$ (48.5), $HCF_2OP(OC_4H_9)_2$ (67.3)	With $NaOP(OCH_3)_2$ in CH_3OH an equilibrium amount of $NaOCH_3$ results in formation of HCF_2OCH_3. Ref. 200
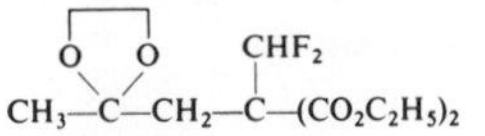 $CH_3{-}C(OCH_2CH_2O){-}CH_2{-}C(CHF_2){-}(CO_2C_2H_5)_2$ (60) [ketal ring: O–CH₂CH₂–O on C], $C_6H_5C(CHF_2)(CO_2C_2H_5)_2$ (80), $(C_6H_5)_2C(CHF_2)CN$ (excellent), *N*-difluoromethyl-2-carbethoxyindole (–)	Carbostyril gave both C- and O-alkylation yielding 2-difluoromethoxyquinoline and *N*-difluoromethyl-2-quinolone. Ref. 201
1,1-Difluorotetramethylcyclopropane (6)	Low yield of cyclopropane is expected; ${:}CF_2$ will react faster with OH^-. Ref. 202
1-Fluoro-l-chlorotetramethylcyclopropane (43)	Ref. 202
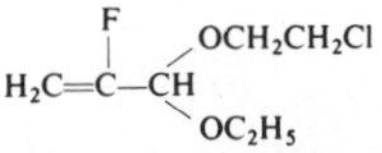 $H_2C{=}CF{-}CH(OCH_2CH_2Cl)(OC_2H_5)$ (52)	Similar reaction reported for 10 other enol ethers. Cyclopropane derivatives formed as side reaction products from cyclic enol ethers. Ref. 203a
2-Fluoro-2-chloro-1-methylcyclopropane (15), 2-Fluoro-2-chloro-1,1-dimethylcyclopropane (39), 7-Fluoro-7-chloronorcarane (45), 9-Fluoro-9-chloro-bicyclo[6.1.0] nonane (57), 2-Fluoro-2-chloro-1-phenylcyclopropane (45), 2-Fluoro-2-chloro-1,1-diphenylcyclopropane (20)	Two isomers formed in each reaction. No separations made. With $CHClF_2$ get only $F_2CHOCH_2CH_2Cl$ by insertion of ${:}CF_2$ into O–H bonds. Ref. 203b
1-Chloro-1-fluoro-2-vinylcyclopropane (32), 2,2′-Dichloro-2,2′-difluorodicyclopropane (30)	Similar additions with 1,4-pentadiene, 1,5-hexadiene, 2,3-dimethyl-1,3-butadiene, 1,5-cyclooctadiene. Ref. 203c
$CF_2{=}CF_2$	Classical synthesis of tetrafluoroethylene. Complete kinetics and mechanism study. Ref. 206
Hexafluorobenzene	Postulate ${:}CBrF$ intermediate. Ref. 207. Also see Ref. 209, 210, 211, 212, 213, 220.
Hexafluorobenzene	Postulate ${:}CBrF$ intermediate. Ref. 208
Hexafluorobenzene	Postulate ${:}CClF$ intermediate. Ref. 214. Also see Ref. 215, 216, 217.
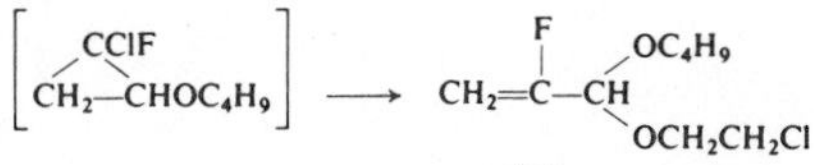 $[CH_2{-}CHOC_4H_9$ (cyclopropane with CClF)$] \longrightarrow CH_2{=}CF{-}CH(OC_4H_9)(OCH_2CH_2Cl)$ (53)	Intermediate carbene and cyclopropane postulated. Note use of ethylene oxide as base to absorb HCl. Ref. 203a

Table 6-2

Carbene	Source and method of generating carbene	Reactant and/or method of detection; conditions
$:CF_2$	$CF_2{=}CF_2$, mercury sensitized photolysis, 21-224°, $P_{CF_2=CF_2} < 60$ mm	$CF_2{=}CF_2$
$:CF_2$	$CF_2{=}CF_2$, uv irradiation	F–N=O
$:CF_2$	$CF_2Cl\overset{O}{\overset{\|}{C}}CF_2Cl$, uv irradiation	Photochemical study
$:CF_2$	Perfluorocyclopropane, thermal decomposition at 160-170°	Cyclohexene 2-Butene *cis*-2-Butene *trans*-2-Butene
$:CF_2$	1,1-Difluorotetrachlorocyclopropane, thermal decomposition at 200°	Cyclohexene *cis*-2-Butene
$:CF_2$	CF_2N_2, heat 140° for 4 hr	$CF_2{=}CFCF{=}CF_2$
$:CF_2$	Perfluorocyclopropane, thermal decomposition	$CH_2{=}CH_2$ used for detection
$(CF_3)_2C:$?	$CF_3{-}CF{=}CF_2$, heat 425° over C	S vapor
Decarboxylation of trihaloacetate salts		
$:CF_2$	$ClCF_2CO_2Na$, 30 hr, reflux in cyclohexene	Cyclohexene
$:CF_2$	$ClCF_2CO_2Na$, diglyme, 125-130°, 10-15 min	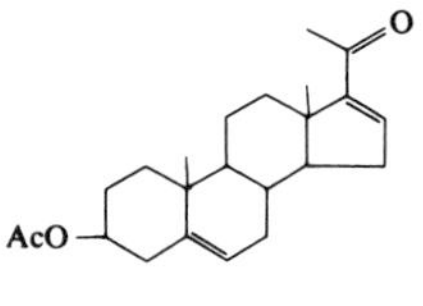
$:CF_2$	$ClCF_2CO_2Na$, diglyme under reflux	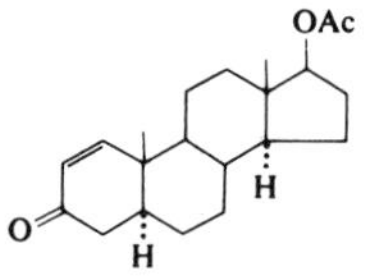
$:CF_2$	$ClCF_2CO_2Na$, diglyme under reflux	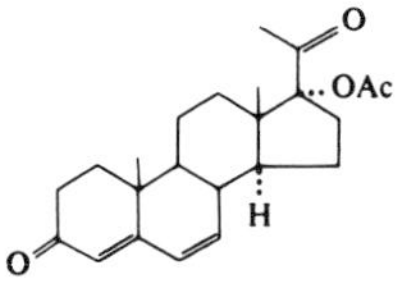

(cont.)

Product(s) (yield, %)	Comments
Perfluorocyclopropane	Primarily a photochemical mechanism study. Ref. 224
Perfluorocyclopropane + perfluoro-2-methyloxazetidine + perfluoro-2-ethyloxazetadine	$:CF_2$ inserted into an N–F bond. Ref. 225
CO, CF_2ClCF_2Cl, CF_2Cl_2, $CF_2ClCF_2CF_2Cl$	Quantum yields obtained and mechanisms postulated. Ref. 226. See also Ref. 227.
7,7-Difluoronorcarane (85-99) 2,2-Difluoro-1-ethylcyclopropane *cis*-2,3-Dimethyl-1,1-difluorocyclopropane *trans*-2,3-Dimethyl-1,1-difluorocyclopropane	Perfluorocyclopropane decomposition is of less importance than examples which follow. Ref. 228
7,7-Difluoronorcarane (67) *cis*- (86) and *trans*- (2) 1,1-Difluoro-2,3-dimethylcyclopropane	Important source of $:CF_2$ because of easy synthesis: $CCl_2{=}CF_2 + :CCl_2 \rightarrow$ 1,1-difluorotetrachlorocyclopropane Similar reactions reported for 1,1,2-trichlorotrifluorocyclopropane. Ref. 228
Perfluorovinylcyclopropane (85)	Kinetics of rearrangement of product to perfluorocyclopentene reported. Ref. 229
1,1-Difluorocyclopropane (–)	Kinetics of thermal decomposition studied. Ref. 230
$CF_3\overset{\overset{S}{\Vert}}{C}CF_3$ (–)	Alternative to carbene is formation of 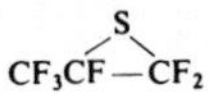 $CF_3CF{-}CF_2$ (three-membered ring bridged by S) Ref. 231
7,7-Difluoronorcarane (60-65)	Ref. 234
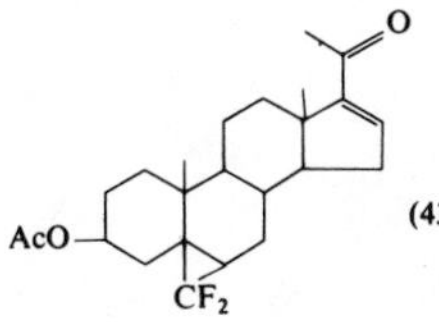 (43)	Paper reports many additions of $:CF_2$ to double bonds in steroids. Comparisons of steric requirements of $:CCl_2$ and $:CF_2$ are made. Ref. 233
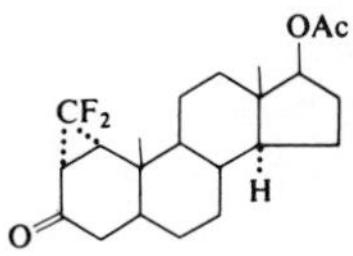 (33-42)	Other additions to steroidal enones reported. Ref. 195a
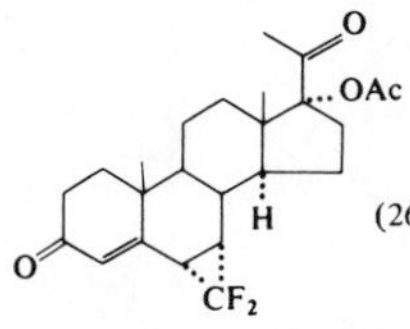 (26-74)	Additions to other dienones and trienones are reported. Ref. 195b

Table 6-2

Carbene	Source and method of generating carbene	Reactant and/or method of detection; conditions
$:CF_2$	$ClCF_2CO_2Na$, diglyme under reflux	[steroid structure: OAc, OCH_3, H–C=, O=, H]
$:CF_2$	$ClCF_2CO_2Na$, reflux in glyme for 72 hr	$(C_6H_5)_3P$, $CH_3(CH_2)_5CHO$ p-$CH_3C_6H_4CHO$, $(C_6H_5)_3P$ (2-Tetrahydrofuranyl)carboxaldehyde, $(C_6H_5)_3P$ C_6H_5CHO, $(C_6H_5)_3P$ p-F–C_6H_5CHO, $(C_6H_5)_3P$
$:CF_2$	$ClCF_2CO_2Na$, 160-180°, N-methylpyrrolidone	$(C_4H_9)_3P$, Cyclohexanone $(C_4H_9)_3P$, $C_6H_5C(=O)CH_3$
$:CF_2$	$ClCF_2CO_2Na$, N-methylpyrrolidone, heat	2-Acetylfuran
$:CF_2$	$ClCF_2CO_2Na$, diglyme, heat	$C_6H_5C(=O)CF_3$
$:CF_2$	$ClCF_2CO_2Na$, heat	$C_6H_5C(=O)CF_2CF_3$
$:CF_2$	$ClCF_2CO_2Na$, 50°, triglyme	$C_6H_5C(=O)CF_3$, $(C_6H_5)_3P$
$:CFCl$	$CFCl_2CO_2Na$, decarboxylate in presence of $(C_6H_5)_3P$	Benzaldehyde, Trifluoroacetylbenzene, Cyclopentanone
Basic cleavage of haloketones and related compounds		
$:CFCl$	$ClCF_2C(=O)CF_2Cl$, $(CH_3)_3CO^-K^+$, 0-5°	Cyclohexene α-Methylstyrene
$:CFCl$	$ClCF_2C(=O)CF_2Cl$, $(CH_3)_3CO^-K^+$ 0-10°, 2 hr	Isobutylene
$:CFCl$	$FCCl_2CO_2CH_3$, $CH_3O^-Na^+$ cyclohexene solvent, 0° to reflux	Cyclohexene

(cont.)

Product(s) (yield, %)	Comments
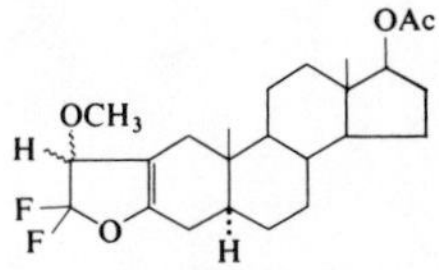 (two epimers 42, 28)	Ref. 195c
$CH_3(CH_2)_5CH{=}CF_2$ (16) $p\text{-}CH_3{-}C_6H_4{-}CH{=}CF_2$ (60) 2(2,2-Difluorovinyl)-tetrahydrofuran (75) $C_6H_5CH{=}CF_2$ (22), $p\text{-}F{-}C_6H_4CH{=}CF_2$ (65)	General reaction proceeds via the ylid. $:CF_2 + (C_6H_5)_3P \rightarrow (C_6H_5)_3P{=}CF_2$ $RCH{=}CF_2 \leftarrow$ RCHO Ref. 235
(Difluoromethylene)cyclohexane (46) $C_6H_5C(CH_3){=}CF_2$ (34.6)	Yields poorer than with aldehydes. Many side reaction products formed. Ref. 195d
2-(2,2-Difluoro-1-methylvinyl)furan	Reactions of methyl *n*-pentyl ketone, benzophenone, acetylthiophene, plus compounds in ref. 195d are reported. Ref. 236
$C_6H_5C(CF_3){=}CF_2$ (65-70)	Similar reactions for $p\text{-}R{-}C_6H_4\overset{O}{\overset{\Vert}{C}}CF_3$ where $R{=}CH_3, OCH_3$, F. Ref. 237
$C_6H_5C(CF_2CF_3){=}CF_2$ $(CF_3)(F)C{=}C(C_6H_5)(CF_3)$ $(CF_3)(F)C{=}C(CF_3)(C_6H_5)$	Product composition depends on reaction time and amount of F^- present. Ref. 238
$C_6H_5C(CF_3){=}CF_2$ (77)	Reaction with acetophenone was unsatisfactory. Ref. 239
$C_6H_5CH{=}CFCl$ (49), $C_6H_5(CF_3)C{=}CFCl$ (56), (Fluorochloromethylene)cyclopentane (9)	cis and trans isomers formed. The $(C_6H_5)_3P{=}CFCl$-ylid is an intermediate. May use $CHCl_2F$ and potassium *t*-butoxide to form :CFCl; yields usually lower. Ref. 238b
7-Fluoro-7-chloronorcarane (36) 1-Chloro-1-fluoro-2-methyl-2-phenylcyclopropane (44)	Very simple procedure for generating :CClF. Ref. 241
1-Fluoro-1-chloro-2,2-dimethylcyclopropane (35)	Product reduced with organotin hydrides to give 1,1-dimethyl-2-fluorocyclopropane. Ref. 242
7-Chloro-7-fluoronorcarane (35)	The carbene :CFCl was found as a result of side reactions in carrying out the condensation. Ref. 245 $C_6H_5\overset{O}{\overset{\Vert}{C}}CH_3 \xrightarrow{FCCl_2COOCH_3} C_6H_5\overset{O}{\overset{\Vert}{C}}CH_2{-}\overset{O}{\overset{\Vert}{C}}CCl_2F$

Table 6-2

Carbene	Source and method of generating carbene	Reactant and/or method of detection; conditions
:CFCl	$FCCl_2CO_2CH_3$, NaH, CH_3OH	Cyclohexene
:CFCl	$ClCF_2C(=O)CF_2Cl$, $NaOCH_3$	Norbornene
:CFCl	$ClCF_2C(=O)CF_2Cl$, $(CH_3)_3O^-K^+$, $-10°$, 2 hr	*Relative rate* Tetramethylethylene 31.0 Trimethylethylene 6.5 2-Methylpropene 1.0 *cis*-2-Butene 0.14 *trans*-2-Butene 0.097
$:CF_2$	$C_6H_5SO_2CHF_2$, $CH_3O^-Na^+$	Studied kinetics
$:CF_2$	$ClCF_2CO_2C_2H_5$, $(CH_3)_3CO^-K^+$, 0-25°	2,3-Dimethylindole
Organometallic reagents on perhalocarbons		
$:CF_2$	CF_2Br_2, C_4H_9Li	RCH=CHR $(C_6H_5)_3P$
$:CF_2$	CF_2Br_2, Lithium alkyl or Grignard reagent	Grignard or lithium alkyl such as C_4H_9Li or C_4H_9MgBr
$:CF_2$	CF_3Br, C_4H_9Li, ether.	Cyclohexene
Decomposition of halogenated heteroatom molecules		
$(CF_3)_2C$:	$(CF_3)_2C{=}N_2$, heat at 200°, 8 hr	Benzene
$(CF_3)_2C$:	$(CF_3)_2C{=}N_2$, heat at 150°, 8 hr	Cyclohexane
$(CF_3)_2C$:	$(CF_3)_2C{=}N_2$, photolysis	Benzene
$(CF_3)_2C$:	$(CF_3)_2C{=}N_2$ or 3,3-bis(trifluoromethyl)diazirine	*cis*- and *trans*-2-Butene
$(CF_3)_2C$:	$(CF_3)_2C{=}N_2$, pyrolysis in flow system	–
$(CF_3)_2C$:	3,3-Bis(trifluoromethyl)diazirine, pyrolysis in flow system	–
$(CF_3)_2C$:	$(CF_3)_2C{=}N_2$, 150°, 8 hr	*cis, cis*-1,5-Cyclooctadiene
$(CF_3)_2C$:	3,3-bis(trifluoromethyl)diazirine, 165°, 12 hr	Cyclohexene
$CF_3\ddot{C}H$	CF_3CHN_2, uv irradiation in gas phase	Excess liquid CF_3CHN_2 present, pressure about 2.5 atm

(cont.)

Product(s) (yield, %)	Comments
7-Chloro-7-fluoronorcarane (60)	Reduction by Bu_3SnH gave two epimers of 7-fluoronorcarane. Ref. 246
Exo epimers of 3-fluoro-3-chlorotricyclo [3.2.1.0^{2,4}] octane(12)	Products rearrange thermally to 3-chloro-4-fluorobicyclo[3.2.1] oct-2-ene. Ref. 243
1-Fluoro-1-chloro-tetramethylcyclopropane (60) 1-Fluoro-1-chloro-2,2,3-trimethylcyclopropane (60) 1-Fluoro-1-chloro-2-methylcyclopropane (–) 1-Fluoro-1-chloro-2,3-dimethylcyclopropane (35) 1-Fluoro-1-chloro-2,3-dimethylcyclopropane (30)	Syn and anti products formed. *Syn*-chloro preferred. Relative rates of reaction measured as shown in "reactant" column. Ref. 243
–	Kinetics support formation of $:CF_2$ and $C_6H_5SO_2H$. Ref. 247
2,3-Dimethyl-3-(difluoromethyl)-3H-indole	Could not obtain any cyclopropane product in a thermal decomposition of $F_2ClCO_2^-Na^+$. Ref. 245
Substituted 1,1-difluorocyclopropanes $(C_6H_5)_3P{=}CF_2$	No details of reactions given. Suggests $CF_2Br_2 + C_4H_9Li \rightarrow C_4H_9Br + CF_2LiBr \rightarrow CF_2 + LiBr$. Ref. 250
$C_3H_8CH{=}CHC_4H_9$ (62)	Mechanism suggested involving $:CF_2$. This is a general method of synthesizing odd carbon alkenes. Ref. 251
7,7-Difluoronorcarane	Proposed mechanism: $C_4H_9Li + CF_3Br \rightarrow C_4H_9Br + CF_3Li$ $CF_3Li \rightarrow :CF_2 + LiF$ Cyclohexene + $:CF_2 \rightarrow$ 7,7-difluoronorcarane. Lifetime of $:CF_2$ determined as 5×10^{-4} to 10^{-3} sec. Ref. 252
7,7-Bis(trifluoromethyl)-1,3,5-cycloheptatriene (88)	Claimed as first addition of a perfluorinated carbene to benzene. Ref. 254
$C_6H_{11}CH(CF_3)_2$ (12), $C_6H_{11}N{=}NCH(CF_3)_2$, $C_6H_{11}NHN{=}C(CF_3)_2$ (where C_6H_{11} is cyclohexyl)	Side-reaction products suggested to arise by $C_6H_{11}\cdot + (CF_3)_2C{=}N_2 \rightarrow C_6H_{11}N{=}N\dot{C}(CF_3)_2 \leftrightarrow C_6H_{11}N{-}N{=}C(CF_3)_2$. Ref. 254
7,7-Bis(trifluoromethyl)cyclohepta-1,3,5-triene (40), 2,2-Bis(trifluoromethyl)bicyclo [3.2.0] hepta-3,6-diene (55)	These initially formed products thermally rearrange to $C_6H_5CH(CF_3)_2$ and $C_6H_5C(CF_3){=}CF_2$. Ref. 254
See Table 6-3	See Table 6-3 and the text. Ref. 255
$CF_2{=}CFCF_3$ (80-92), $(CF_3)_2C{=}C(CF_3)_2$ (–)	Trapping experiments used to prove intermediary of $(CF_3)_2C$:. Ref. 255
$CF_2{=}CFCF_3$, $(CF_3)_2C{=}N{-}N{=}C(CF_3)_2$ (overall 65% yield)	Azine results from attack of $(CF_3)_2C$: on the diaziridine to give possibly 1,3-Diazatetratris(trifluoromethyl)bicyclobutane as intermediate. Ref. 255
9,9-Bis(trifluoromethyl)bicyclo[6.1.0] non-4-ene	Ref. 255
7,7-Bis(trifluoromethyl)norcarane (47), 3-cyclohex-1-enyl-2H-hexafluoropropane (44)	Also 9% 2-Cyclohex-1-enyl-2H-hexafluoropropane. Ref. 255
$CF_2{=}CHF$ (10), *cis*-$CF_3CH{=}CHCF_3$ (6), *trans*-$CF_3CH{=}CHCF_3$ (13), N_2 (74), polytrifluoromethylcarbene (7), 3,4,5-*tris*(Trifluoromethyl)-Δ^1-pyrazoline (40), 1-(Trifluoromethyl)-2,2,3-trifluorocyclopropane (1),	When liq. CF_3CHN_2 was absent and the pressure was 0.11 atm, the products were $CF_2{=}CHF$ (22%), *cis*-$CF_3CH{=}CHCF_3$ (22%) and *trans*-$CF_3CH{=}CHCF_3$ (26%). Ref. 257

Table 6-2

Carbene	Source and method of generating carbene	Reactant and/or method of detection; conditions
CF_3CH:	CF_3CHN_2, uv irradiation	*trans*-2-Butene
$CF_3\ddot{C}H$	CF_3CHN_2, uv irradiation	$CH_2{=}CH_2$ Cyclohexene
$C_2F_5CF_2\ddot{C}H$	$C_2F_5CF_2CHN_2$, uv irradiation	–
$CF_3\ddot{C}H$	CF_3CHN_2, mix −78°, and warm to room temp. then higher	Chloral $CHCl_2CO_2H$ C_6H_5SH *p*-$ClC_6H_4{-}SCl$ $HgCl_2$
$CF_3\overset{O}{\overset{\Vert}{C}}{-}\ddot{C}{-}\overset{O}{\overset{\Vert}{C}}OC_2H_5$	$CF_3\overset{O}{\overset{\Vert}{C}}{-}\overset{N_2}{\overset{\Vert}{C}}{-}\overset{O}{\overset{\Vert}{C}}{-}OC_2H_5$, uv irradiation	Cyclohexene
:CF_2	Difluorodiazirine, uv irradiation, room temp.	No other olefin $CF_2{=}CCl_2$
:CF_2	Difluorodiazirine, heat, usually to 150-160°, 3-5 hr	$CF_2{=}CFCF_3$ *cis*- and *trans*-CFCl=CFCl $CF_2{=}CFH$ $CF_2{=}CFCl$ $CF_2{=}CFCN$ $CF_2{=}CFOH_3$
:CF_2	Difluorodiazirine, uv irradiation at room temperature. Olefin present in large excess	Isobutylene 1,3-Butadiene *trans*-2-Butene *cis*-2-Butene 1,1-Difluoro-2-vinylcyclopropane
:CF_2	Difluorodiazirine, uv irradiation at 36° and 91°C	Relative rate study on addition of :CF_2 to alkenes. Competitive additions related to 2-methylbutene. Analysis by gas chromatography
:CF_2	Difluorodiazirine, heat above 160°	*cis*-2-Butene 1,3-Butadiene
:CF_2	Difluorodiazirine, uv irradiation	Cl_2 I_2 N_2O_4 $ClNO_2$

(cont.)

Product(s) (yield, %)	Comments
trans-2,3-Dimethyl-1-(trifluoromethyl)cyclopropane (23), *cis*-2,3-Dimethyl-*trans*-1-(trifluoromethyl)cyclopropane (0.4), 6,6,6-Trifluoro-2-hexene, 5,5,5-Trifluoro-3-methyl-2-pentene	A similar mixture of addition and insertion products was obtained from *cis*-2-butene. Ref. 257b
N_2 (59), (Trifluoromethyl)cyclopropane (26), *trans*$(CF_3CH=)_2$ N_2 (70), 7-(trifluoromethyl)-norcarane (46), *trans*$(CF_3CH=)_2$	Also recovered $CF_2=CFH$ in each reaction plus starting materials. Ref. 257
$C_3F_7CH=CHC_3F_7$ (47), $C_2F_5CH=CF_2$ (31)	Formation of $C_2F_5CH=CF_2$ indicates migration of C_2F_5 rather than F. Ref. 257
$CF_3\overset{O}{CH-CH}CCl_3$ (53.8), $CCl_3\overset{O}{\overset{\|}{C}}CH_2CF_3$ (31.1) $CHCl_2\overset{O}{\overset{\|}{C}}-OCH_2CF_3$ (72.5) $C_6H_5SCH_2CF_3$ (89) *p*-$ClC_6H_4-SCHClCF_3$ (77.2) $(CF_3CHCl)_2Hg$	Reaction of CF_3CHN_2 with methyl acrylate gave 86.5% 3-carboxy-methoxy-5-trifluoromethyl-1-pyrazoline. Ref. 258
7-Trifluoroacetyl-7-carbethoxynorcarane	With other substrates the carbene behaves like $CF_3\overset{O^-}{\overset{\|}{C}}=\overset{\pm}{C}-\overset{O}{\overset{\|}{C}}-OC_2H_5$ Ref. 261
$CF_2=CF_2$ (high) 1,1-Dichlorotetrafluorocyclopropane (38)	Subsequent addition to give perfluorocyclopropane does not occur at room temperature. Ref. 266a
Perfluoropropylmethylcyclopropane (30) *cis*- and *trans*-1,2-Dichlorotetrafluorocyclopropane (47) Pentafluorocyclopropane (65) Chloropentafluorocyclopropane (85) Cyanopentafluorocyclopropane (66) Methoxypentafluorocyclopropane (72)	Addition of $:CF_2$ claimed to be stereo-specific suggesting singlet carbene. Heating difluorodiazine alone at 165-180° gave $CF_2=CF_2$ (14%), perfluorocyclopropane (62%) and $CF_2=N-N=CF_2$ (24%). Ref. 266a
1,1-Difluoro-2,2-dimethylcyclopropane (71) 1,1-Difluoro-2-vinylcyclopropane (35) *trans*-1,1-Difluoro-2,3-dimethylcyclopropane (33) *cis*-1,1-Difluoro-2,3-dimethylcyclopropane (26) 2,2,2′,2′-Tetrafluorodicyclopropane (15)	Tetrafluoroethylene always formed. Author concludes reaction involves ground-state singlet difluorocarbene. Ref. 266a
Various gem-difluoro-cyclopropanes. Products not isolated	Relative rates of $:CF_2$ addition: tetramethylethylene (6.46), trimethylethylene (2.16), 1,1-dimethylethylene (1.00), *trans*-2-butene (.097), *cis*-2-butene (.082), propene (.0325), 1-butene (.0096). Ref. 266c
cis-1,1-Difluoro-2,3-dimethylcyclopropane (83) 1,1-Difluoro-2-vinylcyclopropane (71)	Yields in thermal reactions higher than in uv-initiated reactions. Ref. 266a
CF_2Cl_2 (90-95) CF_2I_2 (21) $CF_2(NO_2)_2$ CF_2Cl_2 (80) CF_2ClNO_2 (15) $CF_2(NO_2)_2$ (1-2)	Ref. 195e

Table 6-2

Carbene	Source and method of generating carbene	Reactant and/or method of detection; conditions
$:CF_2$	Difluorodiazirine, uv irradiation of liquid solution of acid dissolved in CH_2Cl_2 or $CHCl_3$	CF_3COOH C_3F_7COOH C_4H_9COOH C_6H_5COOH CF_3SO_3H
$:CF_2$	Difluorodiazirine, uv irradiation of gas phase mixture with alcohol	CH_3OH CF_3CH_2OH $(CH_3)_2CHOH$
$(CH_3O)FC:$	Methoxyfluorodiazirine, heat, 50-95°	No other reactant $CF_2{=}CF_2$
$(CN)FC:$	Cyanofluorodiazirine, heat, 100° 1.5 hr	$CF_2{=}CF_2$
$ClFC:$	Chlorofluorodiazirine, uv irradiation, 25°	Cl_2
$(NF_2)FC:$	Fluoro(difluoroamino)diazirine, heat, 75°	$CF_2{=}CF_2$ No trapping agent
$(NF_2)ClC:$	Chloro(difluoroamino)diazirine, heat, 75°	No other agent
$(CF_3)_2C:$	Bis(trifluoromethyl)diazirine, uv irradiation	Cl_2
$:CF_2$	F F CF_3–C–C–F $C_6H_5CH_2N$–N=N pyrolyze 325° over nickel balls	(a) No trapping agent (b) $(CH_3)_2C{=}C(CH_3)_2$
Splitting from hetero group other than nitrogen		
$:CF_2$	$(CF_3)_3Sb$, pyrolyze 180-220°	No trapping agent
$:CF_2$	$(CH_3)_3SnCF_3$, heat at 150° in sealed tube	No trapping agent $CF_2{=}CF_2$
$:CF_2$	$(CH_3)_3SnCF_3$, NaI, dimethoxyethane, reflux, 16 hr	Cyclohexene
$:CF_2$	CF_3GeI_3, heat at 180° for 3 days	No trapping agent
$:CF_2$	$CF_3Fe(CO)_4I$, heat ~ 180° for 96 hr	No trapping agent $CF_2{=}CF_2$ (only 100°), $CH_2{=}CH_2$ (only 100°)
$CF_3CF:?$	$(CF_3CF_2)_2Fe(CO)_4$	$CF_2{=}CF_2$

(cont.)

Product(s) (yield, %)	Comments
$CF_3COOCHF_2$ (75) $C_3F_7COOCHF_2$ (50) $C_4H_9COOCHF_2$ (88) $C_6H_5COOCHF_2$ (75) $CF_3SO_2OCHF_2$ (71)	Structures established by nmr, elemental analysis, and chemical conversion to amides and methyl esters. Ref. 267
CH_3OCHF_2 (69) $CF_3CH_2OCHF_2$ (51) $(CH_3)_2CHOCHF_2$ (83)	CH_3OCHF_2 and $(CH_3)_2CHOCHF_2$ were hydrolytically unstable readily giving HF. Ref. 267
$CH_3OCF{=}CFOCH_3$ Methoxypentafluorocyclopropane (61.5) + above product	$CF_2{=}CF_2$ addition product identical with product from $:CF_2$ plus $CH_3OCF{=}CF_2$. Ref. 269
Cyanopentafluorocyclopropane	Product identical with product from $:CF_2$ plus $CF_2{=}CFCN$. Ref. 269
$CFCl_3$	Thermal decomposition of chlorofluorodiazirine was not reported. Ref. 269
Difluoraminopentafluorocyclopropane (36.8) $F_2C{=}NF$	Reactions of fluoro(difluoramino)diazirine reported with BF_3 and $AlCl_3$. Ref. 269
$ClFC{=}NF$	Rearrangement: $Cl\ddot{C}NF_2 \rightarrow ClFC{=}NF$. Ref. 269
$CF_3CCl_2CF_3$	Best general synthesis reported of bis(trifluoromethyl)diazirine. Ref. 270
(a) $C_6H_5CH_2N{=}CFCF_3$ (65), $CF_2{=}CF_2$ (45), $C_6H_5CH_2N\langle$ (ring: $CFCF_3$, CF_2) (2), $C_6H_5CH_2NHCOCF_3$ (12) (b) 1,1-Difluorotetramethylcyclopropane (18), $C_6H_5CH_2N{=}CFCF_3$	This is not a convenient method of generating difluorocarbene. Ref. 272
$CF_2{=}CF_2$ (–), perfluorocyclopropane (–)	Speculate that $(CF_3)_3Sb \rightarrow CF_3\cdot + Sb(CF_3)_2$ $CF_3\cdot \xrightarrow{Sb} :CF_2$. Ref. 273
Perfluorocyclopropane (95), Perfluorocyclopropane (nearly quant.)	$(CH_3)_3SnCF_3 \rightarrow (CH_3)_3SnF + :CF_2$ $2\ :CF_2 \rightarrow CF_2{=}CF_2$ $:CF_2 + CF_2{=}CF_2 \rightarrow$ perfluorocyclopropane. Ref. 274
7,7-Difluoronorcarane	Gem-difluorocyclopropanes formed similarly from 18 other alkenes, generally in 50-90% yields. Ref. 278
Perfluorocyclopropane (–), $CF_2{=}CF_2$ (–), perfluorocyclobutane (–)	Carried only to 40% decomposition of CF_3GeI_3. Ref. 275
$CF_2{=}CF_2$ (high), perfluorocyclopropane (–) Perfluorocyclopropane (–), 1,1-Difluorocyclopropane (–)	Also pyrolyzed $C_3F_7Fe(CO)_4I$. Ref. 276
$CF_3CF{=}CFCF_3$, $CF_2{=}CFCF_2CF_3$	Evidence for formation of CF_3CF: is inconclusive. Ref. 276

Table 6-2

Carbene	Source and method of generating carbene	Reactant and/or method of detection; conditions
$:CF_2$	$(CF_3)_3PF_2$, heat, 120-200°	No trapping agent (120°, 24 hr) No trapping agent (200°, 10 min) I_2 (120°, 24 hr)
$:CF_2$	$(CF_3)_3PF_2$, heat 100°	$CF_3C{\equiv}CCF_3$, Perfluoro-1,2-dimethylcyclopropene(−)
:CFCl	$C_6H_5HgCCl_2F$, NaI, dimethoxyethane, 85° to reflux	Cyclooctene
$CHF_2CF:$	$CHF_2CF_2SiF_3$, heat to 150°	2-Methylpropane Tetramethylethylene
$CHF_2CF:$	$CHF_2CF_2SiF_3$, 150°, 10 hr	Various alkenes to obtain cyclopropanes for nmr studies

[195](a) C. Beard, N. H. Dyson, and J. H. Fried, *Tetrahedron,* **28**, 3281 (1966). (b) C. Beard, I. T. Harrison, L. Kirkham, and J. H. Fried, *Tetrahedron,* **28**, 3287 (1966). (c) P. Hodge, J. A. Edwards, and J. H. Fried, *Tetrahedron Letters,* 5175 (1966).

(cont.)

Product(s) (yield, %)	Comments
$CF_2{=}CF_2$ (10), perfluorocyclopropane (80), $\leftarrow CF_2 \rightarrow_x$ (10) $CF_2{=}CF_2$ (80), perfluorocyclopropane (10), $\leftarrow CF_2 \rightarrow_x$ (10) CF_2I_2 (30), $ICF_2CF_2CF_2I$ (–)	At 120° (24 hr) no reaction occurred with BF_3, CO, NF_3, N_2O, PF_3, CS_2, SO_2, or CF_3I. Ref. 277a
Perfluoro-1,2-dimethylcyclopropene (–) Perfluoro-1,3-dimethylbicyclobutane (25)	Ref. 277b
9-Fluoro-9-chlorobicyclo[6.1.0]nonane	Additions of :CFCl also reported to cyclohexene, *cis*- and *trans*-2-hexene, vinyl acetate, acrylonitrile, 2,5-dihydrofuran. Ref. 279
$(CH_3)_3CHFCF_2H$ (61), $(CH_3)_2CHCH_2CHFCHF_2$ (<1) 1-Fluoro-1-(difluoromethyl)-tetramethylcyclopropane (>75)	Insertion of this carbene is general for alkanes: ethane, propane, butane, pentane, hexane, cyclohexane, cyclopentane. Addition to alkenes occurs in 75-90% yields with ethylene, tetrafluoroethylene, *cis*- and *trans*-2-butene, and cyclohexene. Ref. 282a
R_1R_2C–CR_3R_4 ring with $>CFCHF_2$ (cyclopropane: R_1—C(R_2), R_3—C(R_4), CFCHF$_2$)	^1H and ^{19}F nmr spectra observed on a series of isomers to obtain coupling constants and chemical shifts as a function of structure. Ref. 282b

(d) S. A. Fuqua, W. G. Duncan, and R. M. Silverstein, *Tetrahedron Letters*, 521 (1965). (e) R. A. Mitsch, *J. Heterocyclic Chem.*, 1, 233 (1964).

the heteroatom–hydrogen bond, but kinetic evidence points to anion reaction. Note that difluorocarbenes preferentially attack heteroatoms (O, S, P, etc.) and

$$ArOH + CClF_2H \xrightarrow[H_2O/dioxane]{NaOH} ArOCF_2H \quad (197)$$

$$ArSH \longrightarrow ArSCF_2H \quad (55\text{–}80\%) \quad (198)$$

$$CH_3OH \longrightarrow CH_3OCF_2H \quad (23\%) \quad (199)$$

$$NaOP(OR)_2 \longrightarrow CHF_2P(O)(OR)_2 \quad (200)$$

$$(R = n\text{-Bu}, 70\%)$$

do not C-alkylate, e.g., phenols, whereas dichlorocarbene does the reverse (Reimer–Tiemann reaction). In aprotic systems, carbanions are difluoromethylated:

$$Na^+ \; R\bar{C}\text{—}(CO_2Et)_2 + CClF_2H \xrightarrow[glyme]{NaO\text{-}t\text{-}Bu} RC(CF_2H)(CO_2Et)_2 \quad (60\text{–}80\%) \quad (201)$$

$$R = \text{—}CH_2C(CH_3)\text{(–O–CH}_2\text{–CH}_2\text{–O–)} \text{ or —}C_6H_5$$

Addition of dihalocarbenes to olefins also occurs in aprotic media[202]:

$$(CH_3)_2C{=}C(CH_3)_2 + CCl_2FH + \underset{\text{pellets}}{NaOH} \xrightarrow[\text{high shear stirring}]{\text{tetraglyme, }95^\circ} \text{1-fluoro-1-chloro-2,2,3,3-tetramethylcyclopropane} \quad (54\%)$$

(Most of the report by Robinson dealt with dichlorocarbene; difluorocarbene addition was poor.) An adduct has been used in situ for preparation of other

[197] T. G. Miller and J. W. Thanassi, *J. Org. Chem.*, **25**, 2009 (1960).

[198] (a) R. Van Poucke, R. Pollet, and A. De Cat, *Tetrahedron Letters*, 403 (1965). (b) R. Pollet, R. Van Poucke, and A. De Cat, *Bull. Soc. Chim. Belges*, **75**, 40 (1966).

[199] L. Z. Soborovskii and N. F. Baina, *J. Gen. Chem. U.S.S.R., (English Transl.)*, **29**, 1113 (1959).

[200] *Ibid.*, 1115.

[201] T. Y. Shen, S. Lucas, and L. H. Sarett, *Tetrahedron Letters*, 43 (1961).

[202] G. C. Robinson, *Tetrahedron Letters*, 1749 (1965).

products[203a]; a modification of this procedure has been developed as a general synthesis of dihalocyclopropanes.[203b, c] This method was exploited mainly for

$$CH_2{=}CHOC_4H_9 + CHCl_2F + \text{(ethylene oxide)} \xrightarrow[HOCH_2CH_2Cl]{Br^-} \left[\text{1-fluoro-1-chloro-2-butoxycyclopropane } (CH_2{-}CH{-}OC_4H_9, \; C(F)(Cl))\right]$$

$$\downarrow$$

$$CH_2{=}\underset{\displaystyle F}{C}{-}CH\begin{cases}OC_4H_9\\OCH_2CH_2Cl\end{cases} \quad (53\%) \xleftarrow{\text{(ethylene oxide)}} \left[CH_2{=}\underset{\displaystyle F}{C}{-}\overset{+}{C}HOC_4H_9 \;\; Cl^-\right]$$

$$\downarrow H^+$$

$$H_2C{=}\underset{\displaystyle F}{C}\overset{\displaystyle O}{\overset{\|}{C}}H$$

dichlorocarbene but was shown to be useful for chlorofluoro- and difluorocarbene additions to a large variety of alkenes and dienes.

$$HCX_3 + H_2C\underset{O}{-}CH_2 + \;>C{=}C<\; \xrightarrow[\text{tetraethylammonium bromide (catalyst)}]{150^\circ} \text{(cyclopropane)}CX_2$$

Phenylfluorocarbene has also been generated in an aprotic system and trapped with olefins.[203d]

$$C_6H_5CHBrF + KO\text{-}t\text{-}Bu + \text{olefin} \rightarrow \text{1-phenyl-1-fluorocyclopropane} \quad (56\text{–}81\%)$$

b. *Energetic Decomposition of Fluorocarbons*

Pyrolysis or irradiation of fluorocarbons ranging from haloforms to tetrafluoroethylene or fluorinated cyclopropanes provides a good method of generating halocarbenes, particularly difluorocarbene for mechanistic, structural, and thermodynamic studies. A major synthetic utility of this method is in commercial

[203] (a) J. Buddrus, F. Nerdel, and P. Hentschel, *Tetrahedron Letters*, 5379 (1966); F. Nerdel, J. Buddrus, W. Brodowski, P. Hentschel, D. Klamann, and P. Weyerstahl, *Ann.*, **710**, 36 (1967). (b) P. Weyerstahl, D. Klamann, C. Finger, F. Nerdel, and J. Buddrus, *Chem. Ber.*, **100**, 1858 (1967). (c) P. Weyerstahl, D. Klamann, M. Fligge, C. Finger, F. Nerdel, and J. Buddrus, *Ann.*, **710**, 17 (1967). (d) R. A. Moss, *Tetrahedron*, 1961 (1968).

preparation of tetrafluoroethylene.[204, 205] A recent kinetic study of the pyrolysis of chlorodifluoromethane showed that the reaction was unimolecular and that

$$CHClF_2 \xrightarrow{>650^\circ} HCl + [:CF_2] \longrightarrow CF_2{=}CF_2 \qquad (90\text{–}95\%)$$

difluorocarbene was an intermediate.[206] Pyrolysis of $CFCl_2H$ is a potentially good route to hexafluorobenzene,[207–218] and :CClF is proposed as an intermediate.[219–223]

$$6CHFX_2 \xrightarrow{\Delta} C_6F_6 + 3X_2 + 6HX \qquad (\text{X is Cl or Br})$$

$$CFX{=}CFX \not\longrightarrow C_6F_6$$

The mechanism is not clear since

$$CHFClCHFCl \xrightarrow[\Delta]{Pt} C_6F_6 + 2HCl$$

suggesting that $CF{\equiv}CF$ is also an intermediate. Energetic decomposition of di- and tribromofluoromethane is claimed to give CF based on ultraviolet absorption spectra.[223]

[204] (a) A. J. Rudge, *Fluorine Manufacture and Uses,* Oxford Univ. Press, New York, 1962, p. 72. (b) H. G. Bryce in *Fluorine Chemistry,* (J. H. Simons, ed.), Vol. 5, Academic, New York, 1964, p. 442.

[205] J. W. Edwards and P. A. Small, *Nature,* **202,** 1329 (1964).

[206] Yu. A. Panshin, *Russian J. Phys. Chem. (English Transl.),* **40,** 1197 (1966).

[207] Imperial Smelting Corp., Belgian Patent 649,355 (1964); see French Addn. 84,081; *Chem. Abstr.,* **62,** 9057c (1965).

[208] L. A. Wall, M. Hellman, and W. J. Pummer, U.S. Patent 2,927,138 (1960); *Chem. Abstr.,* **54,** 14187e (1960).

[209] R. N. Haszeldine and J. M. Birchall, British Patent 1,014,252 (1965); *Chem. Abstr.,* **64,** 9633c (1966).

[210] Sperry Rand Corp., Japanese Patent 20,121 (1963).

[211] F. R. Callihan and C. L. Quatela, U.S. Patent 3,158,657 (1964); *Chem. Abstr.,* **62,** 3974b (1965).

[212] D. W. Cottrell and W. Hopkin, British Patent 990,157 (1965); *Chem. Abstr.,* **62,** 9057c (1965).

[213] A. K. Barbour and A. E. Pedler, British Patent 990,156 (1965); (see Ref. 217).

[214] R. A. Falk, French Patent 1,334,724 (1963); *Chem. Abstr.,* **60,** 2819h (1964).

[215] F. Nyman, British Patent 968,900 (1964); *Chem. Abstr.,* **61,** 16010f (1964).

[216] National Smelting Co. Ltd., French Patent 1,321,627 (1963); *Chem. Abstr.,* **58,** 4467B (1963).

[217] National Smelting Co. Ltd., Belgian Patent 614,161 (1962); *Chem. Abstr.,* **58,** 4467b (1963).

[218] L. A. Wall and W. J. Pummer, U.S. Patent 3,033,905 (1962); *Chem. Abstr.,* **57,** 11099d (1962).

[219] Y. Desirant, *Bull. Soc. Chim. Belges,* **67,** 676 (1958).

[220] K. O. Christe and A. E. Pavlath, *Chem. Ber.,* **97,** 2092 (1964).

[221] R. E. Banks, J. M. Birchall, R. N. Haszeldine, J. M. Simon, H. Sutcliffe, and J. H. Umfreville, *Proc. Chem. Soc.,* 281 (1962).

[222] L. A. Wall, W. J. Pummer, J. E. Fearn, and J. M. Antonucci, *J. Res. Natl. Bur. Standards,* **67a,** 481 (1963).

[223] A. J. Yarwood and J. P. Simons, *Proc. Chem. Soc.,* **62** (1962).

Mercury-sensitized irradiation of tetrafluoroethylene is a common source of difluorocarbene which has been reported to add to tetrafluoroethylene or insert in heteroatom bonds:

$$CF_2{=}CF_2 \xrightarrow[Hg]{h\nu} [:CF_2] \xrightarrow{CF_2=CF_2} \underset{CF_2-CF_2}{\overset{CF_2}{\triangle}} \quad (224)$$

$$[:CF_2] \xrightarrow[CF_2=CF_2]{NOF} \begin{array}{c} CF_3N-O \\ | \quad\ \ | \\ CF_2-CF_2 \end{array} \quad (225)$$

$$\left(\text{via } CF_3N{=}O \text{ or } \begin{array}{c} O-NF \\ | \quad\ | \\ CF_2-CF_2 \end{array} \right)$$

Mastrangelo[99] used a radio frequency discharge on perfluorocyclobutane to produce difluorocarbene that was trapped on a cold finger (see also Chart 6-1, page 205). In addition to spectral measurements, the chemical reactions given in the accompanying scheme were also reported.

$$\begin{array}{c} CF_2-CF_2 \\ | \quad\ \ | \\ CF_2-CF_2 \end{array} \xrightarrow{\Delta E} 4{:}CF_2 \ \text{(blue 95°K)} \longrightarrow 2CF_2{=}CF_2 \ \text{(white)}$$

$$4{:}CF_2 \xrightarrow{Cl_2,\ 77°K} 2CF_2Cl\cdot \ (77°K)$$

$$4{:}CF_2 \xrightarrow{15\text{–}30\ min} [\colon CF] + CF_3\cdot$$

$$2CF_2Cl\cdot \xrightarrow{\Delta} CF_2ClCF_2Cl \qquad 2CF_2Cl\cdot \xrightarrow[\Delta]{Cl_2} 2CF_2Cl_2$$

$$[\colon CF] \xrightarrow{?} [C + {:}CF_2] \qquad CF_3\cdot \xrightarrow{\text{under certain conditions}} {:}CF_2 + CF_4$$

The disproportion of fluorinated radicals to generate difluorocarbenes, as suggested by Mastrangelo,[99] has been also proposed in irradiation of dichlorotetrafluoroacetone.[226, 227]

[224] N. Cohen and J. Heicklen, *J. Chem. Phys.*, **43**, 871 (1965).
[225] S. Andreades, *Chem. Ind. (London)*, 782 (1962).
[226] R. Bowles, J. R. Majer, and J. C. Robb, *Trans. Faraday Soc.*, **58**, 1541 (1962).
[227] *Ibid.*, 2394.

$$ClCF_2C(=O)CF_2Cl \longrightarrow CO + \cdot ClCF_2 \longrightarrow \text{normal radical reactions}$$

$$\cdot ClCF_2 \longrightarrow :CF_2 + CF_2Cl_2$$

$$:CF_2 + CF_2Cl_2 \xrightarrow{\cdot CF_2Cl} CClF_2CF_2CF_2Cl$$

Thermal decomposition of chlorofluorocyclopropanes is also a useful source of difluorocarbene.[228]

$$\text{cyclo-}(CF_2CX_2CX_2) \xrightarrow{160-200^\circ} [:CF_2] + CX_2{=}CX_2 \qquad (X = Cl \text{ or } F)$$

$$[:CF_2] \xrightarrow{>C=C<} \text{1,1-difluorocyclopropane} \quad (70\text{–}95\%) \text{ stereospecific addition}$$

The decomposition involves exclusive (> 90%) ejection of difluoro- rather than chlorofluoro- or dichlorocarbene, clearly illustrating greater stabilization of carbene by fluorine than by chlorine. This method has the advantage that no acidic or basic side products are present, and the difluorocarbene is generated under relatively mild conditions in a heterogeneous system. The halogenated cyclopropanes can be prepared either by addition of dichlorocarbene to a fluoroolefin or by addition of difluorocarbene (generated by any of the methods

$$CCl_3SiF_3 \xrightarrow{140^\circ} [:CCl_2] \xrightarrow{CF_2{=}CXY} \text{cyclo-}(CF_2CXYCCl_2)$$

$$(X = Y = F, 59\%;\ X = F, Y = Cl, 87\%;\ X = Y = Cl, 66\%)$$

described in this section) to haloolefins. Decomposition of perfluoroallylcyclopropane[229] and hexafluorocyclopropane[230] also give carbenoid intermediates.

Fluorocarbenes could be intermediates in the high-temperature reaction of fluoroolefins with sulfur.[231] However, a more probable mechanism is addition of S to the fluoroolefin to form an episulfide that rearranges.

[228] J. M. Birchall, R. N. Haszeldine, and D. W. Roberts, *Chem. Commun.*, 287 (1967).
[229] R. A. Mitsch and E. W. Neuvar, *J. Phys. Chem.*, **70**, 546 (1966).
[230] B. Atkinson and D. McKeagan, *Chem. Commun.*, 189 (1966).
[231] K. V. Martin, *J. Chem. Soc.*, 2944 (1964).

$$CF_3CF{=}CF_2 \xrightarrow[425°]{\text{carbon}} CF_3\ddot{C}CF_3 \xrightarrow[\text{vapor}]{S} CF_3\overset{\overset{S}{\|}}{C}CF_3$$

$$CF_3CF_2\ddot{C}F \xrightarrow{S} CF_3CF_2\overset{\overset{S}{\|}}{C}F \quad \text{(small impurity)}$$

Monofluorocarbene was formed in the gas phase by dehydrohalogenation of excited dihalomethane molecules and subsequent reaction with ethylene to give cyclopropylfluoride.[232] The methanes were excited through hot atom substitution of tritium for hydrogen:

$$T + CH_2F_2 \longrightarrow H + CHTF_2^* \longrightarrow \begin{cases} :CHF \xrightarrow{CH_2{=}CH_2} \text{cyclo-}(CH_2CH_2CHF) \\ :CTF \xrightarrow{CH_2{=}CH_2} \text{cyclo-}(CH_2CH_2CTF) \end{cases}$$

The radioactive tracer was advantageous in product analysis.

c. *Decarboxylation of Trihaloacetate Salts*

Thermal decomposition of sodium trihaloacetates in aprotic media is an excellent synthetic method for halocarbene intermediates which are used to make halocyclopropanes (by addition to olefins) or terminal difluoroolefins (by carbonyl replacement via phosphine ylids). The olefin addition reaction[233, 234]

$$CF_2ClCO_2Na \xrightarrow[\text{diglyme}]{\Delta} CO_2 + {}^-CF_2Cl \longrightarrow Cl^- + [:CF_2]$$

$$[:CF_2] \xrightarrow{\text{cyclohexene}} \text{7,7-difluorobicyclo[4.1.0]heptane } (CF_2) \qquad (234)$$

[232] (a) Y. N. Tang and F. S. Rowland, *J. Am. Chem. Soc.*, **88**, 626 (1966). (b) Y. N. Tang and F. S. Rowland, *J. Am. Chem. Soc.*, **89**, 6420 (1967).

[233] (a) L. H. Knox, E. Verlarde, S. Berger, D. Cuadriello, P. W. Landis, and A. D. Cross, *J. Am. Chem. Soc.*, **85**, 1851 (1963). (b) C. Beard, I. T. Harrison, L. Kirkham, and J. H. Fried, *Tetrahedron Letters*, 3287 (1966). (c) T. L. Popper, F. E. Carlon, H. M. Marigliano, and M. D. Yudis, *Chem. Commun.*, 277 (1968).

[234] J. M. Birchall, G. W. Cross, and R. N. Haszeldine, *Proc. Chem. Soc.*, 81 (1960).

has been applied to synthesis of steroidal dihalocyclopropanes in 60–65% yield.[233] The relative reactivities, geometry, and steric requirements of the halocarbenes are discussed.[233, 234] The mechanism of decomposition of the chlorodifluoroacetate is concerted to generate $:CF_2$ directly.[195] Some unusual products from addition to steroidal enones have been described.[233]

Difluorocarbene is suggested as an intermediate in the one-step synthesis of 1,1-difluoroolefins from aldehydes and ketones by a modified Wittig synthesis. Difluorocarbene is trapped initially as a phosphine ylide and then reacts as a normal Wittig reagent.[235–239]

$$ClCF_2CO_2Na \longrightarrow [:CF_2] \xrightarrow{R_3P} R_3P{=}CF_2 \xrightarrow{RC(=O)R_1} \begin{matrix} R \\ R_1 \end{matrix}\!\!>C{=}CF_2$$

Methods that are generally used for preparing phosphine ylids do not work for difluoromethylene phosphine.[240] Initially the reaction was reported general for aromatic, heteroaromatic, and aliphatic aldehydes by heating aldehyde, triphenylphosphine, and sodium chlorodifluoroacetate in anhydrous glyme or diglyme. Ketones require the more reactive tributylphosphine in a polar solvent such as *N*-methylpyrrolidone.[236] The reaction can be extended to prepare β-phenyl-substituted perfluoroolefins,[237–239] but fluoride-catalyzed isomerization was found for fluoroalkyl groups other than trifluoromethyl. Recently a mixed

$$C_6H_5C(=O)CF_2R_f + (C_6H_5)_3P + CClF_2CO_2Na \longrightarrow C_6H_5C(CF_2R_f){=}CF_2 \quad (R_f = F)$$

$$\downarrow$$

$$R_fCF{=}C(CF_3)C_6H_5 \quad \text{cis and trans} \quad (R_f = CF_3, -CF_2CF_3)$$

halogen ylid was prepared from decomposition of sodium dichlorofluoroacetate or dehydrohalogenation of dichlorofluoromethane in the presence of triphenylphosphine; it reacted with carbonyl derivatives to make terminal chlorofluoroolefins.[238b]

[235] S. A. Fuqua, W. G. Duncan, and R. M. Silverstein, *J. Org. Chem.*, **30**, 1027 (1965).

[236] *Ibid.*, 2543.

[237] F. E. Herkes and D. J. Burton, *J. Org. Chem.*, 32, 1311 (1967).

[238] (a) D. J. Burton and F. E. Herkes, *Tetrahedron Letters*, 4509 (1965). (b) D. J. Burton and H. C. Krutzsch, *Tetrahedron Letters*, 71 (1968).

[239] P. M. Barna, *Chem. Ind. (London)*, 2054 (1966).

[240] Chlorofluoromethylene phosphine has been prepared, but an early report of reaction of difluoromethylene with triphenylphosphine could not be substantiated (see A. J. Speziale and K. W. Ratts, *J. Am. Chem. Soc.*, **84**, 854 (1962), and Ref. 238b).

d. *Basic Cleavage of Haloketones and Related Compounds*

This route is particularly effective for generating chlorofluorocarbene for cyclopropane formation.[241–244]

$$CFCl_2\overset{O}{\overset{\|}{C}}CFCl_2 + t\text{-}BuO^-K^+ \xrightarrow{\text{cyclohexane}} CFCl_2\overset{O}{\overset{\|}{C}}O\text{-}t\text{-}Bu + [:CFCl] + Cl^-$$

(The formation of the carbene can be either concerted or two step through $^-CFCl_2$.)

$\downarrow$ cyclohexene

7-chloro-7-fluorobicyclo[4.1.0]heptane (F, Cl)

(36%)

(241)

The stereoselectivity and reactivity of :CFCl has been compared to other halocarbenes,[244] and the products have been used for ring-opening studies.[242]

$$CFCl_2\overset{O}{\overset{\|}{C}}OCH_3 \xrightarrow[\text{or } NaOCH_3 \text{ in cyclohexene}]{NaH/CH_3OH} \text{(bicyclo[4.1.0]heptane with F, Cl)} \quad (60\%) \qquad (245, 246)$$

$$C_6H_5SO_2CHF_2 \xrightarrow[C_6H_5S^-,\ CH_3OH]{NaOCH_3} [:CF_2] \longrightarrow C_6H_5SCF_2H + CH_3OCF_2H \qquad (247)$$

Mechanistic studies[247] show free difluorocarbene is formed in two steps involving an anion intermediate; in contrast, difluorocarbene from a difluorohaloform gives the carbene by a concerted reaction without forming an anion. An unusual product has been reported in studies of halocarbenes generated by this route.

[241] B. Farah and S. Horensky, *J. Org. Chem.*, **28**, 2494 (1963).

[242] J. P. Oliver, U. V. Rao, and M. T. Emerson, *Tetrahedron Letters*, 3419 (1964).

[243] L. Ghosez, G. Slinckx, M. Glineur, P. Hoet, and P. Laroche, *Tetrahedron Letters*, 2773 (1967).

[244] R. A. Moss and R. Gerstl, *Tetrahedron*, **23**, 2549 (1967); *J. Org. Chem.*, **32**, 2268 (1967).

[245] R. A. Moore and R. Levine, *J. Org. Chem.*, **29**, 1883 (1964).

[246] (a) T. Ando, H. Yamanaka, S. Terabe, A. Horike, and W. Funasaka, *Tetrahedron Letters*, 1123 (1967). (b) T. Ando, H. Yamanaka, and W. Funasaka, *ibid.*, 2587.

[247] J. Hine and J. J. Porter, *J. Am. Chem. Soc.*, **82**, 6178 (1960).

$$CClF_2CO_2CH_3 + \text{2,3-dimethylindole} \xrightarrow{\text{KO-}t\text{-Bu}} \text{3-(difluoromethyl)-2,3-dimethyl-3H-indole} \quad (248)$$

e. *Organometallic Reagents on Perhalocarbons*

A fluorinated carbene prepared by halogen–metal interchange on a perhalocarbon can also be used in synthesis. The initial report was reaction of iodotrifluoromethane with butyllithium,[249] but dibromodifluoromethane with

$$CF_3I + CH_3Li \xrightarrow{-74°} CF_2{=}CF_2 \quad (43\%)$$

n-butyllithium was claimed to be a better route to difluorocarbene.[250–252] The difluorocarbene was trapped by addition to olefins and to triphenylphosphine

$$Br_2CF_2 + n\text{-BuLi} \longrightarrow n\text{-BuBr} + [BrCF_2Li] \xrightarrow{-\text{LiBr}} [{:}CF_2] \xrightarrow{(C_6H_5)_3P} ?$$

$$[{:}CF_2] \longrightarrow \text{7,7-difluorobicyclo[4.1.0]heptane } (F_2)$$

$$[{:}CF_2] \xrightarrow{n\text{-BuLi}} C_3H_7CH{=}CHCH_2C_3H_7 + CH_3F + CH_3Br$$

(this latter experiment could not be repeated; see Ref. 240) and reacted with excess butyllithium to give olefins. The half-life of the difluorocarbene generated by this method was determined to be about 5×10^{-4} second from experiments in a flow system.

Perfluoroalkylmagnesium halides did not give any evidence of carbene formation.[32]

f. *Decomposition of Halogenated Heteroatom Molecules*

(1) *Diazo Compounds or other Nitrogen Derivatives.* Difluorodiazomethane has never been isolated but was suggested as an intermediate in amination of nitrosotrifluoromethane[253]; it presumably gave difluorocarbene that was

[248] C. W. Rees and C. E. Smithen, *J. Chem. Soc.*, 938 (1964).
[249] O. R. Price, E. T. McBee, and G. F. Judd, *J. Am. Chem. Soc.*, **76**, 474 (1954).
[250] V. Franzen, *Angew. Chem.*, **72**, 566 (1960).
[251] V. Franzen and L. Fikentscher, *Chem. Ber.*, **95**, 1958 (1962).
[252] V. Franzen, *ibid.*, 1964.
[253] S. P. Makarov, A. J. Yakubovich, V. A. Ginsberg, A. S. Filatov, M. A. Englin, N. F. Privezentseva, and T. J. Nikoforova, *Proc. Acad. Sci. U.S.S.R., Chem. Sect. (English Transl.)*, **141**, 1130 (1961).

$$CF_3NO + NH_3 \xrightarrow{-70^\circ} [CF_3N{=}NH] \xrightarrow[-HF]{NH_3} [CF_2N_2] \xrightarrow{(C_6H_5)_3P} (C_6H_5)_3P{=}CF_2$$

trapped with triphenylphosphine. Unfortunately no experimental details are given for this reaction, but some related reactions with amines and hydrazine support the postulated mechanism.

Several fluoroalkyldiazomethanes have been prepared by different routes.

$$(R_f)_2C{=}O \xrightarrow[POCl_3]{NH_3} (R_f)_2C{=}NH \xrightarrow[P_2O_5]{NH_2NH_2} (R_f)_2C{=}NNH_2 \xrightarrow[C_6H_5CN]{Pb(OAc)_4} (R_f)_2CN_2 \qquad (254, 255)$$

$$(R_f = CF_3, C_2F_5)$$

$$CF_3CF{=}CF_2 \xrightarrow{HNO_3,\ HF} (CF_3)_2CFNO_2 \xrightarrow{H_2/Pd} (CF_3)_2CHNH_2 \xrightarrow{HNO_2} (CF_3)_2CN_2 \qquad (256)$$

$$R_fCH(R)NH_2 \xrightarrow{HNO_2} R_fC(R)N_2$$

(from reduction of an amide or oxime)

for $R_f = CF_3$ or C_3F_7, $R = H$ (257, 258)

$R_f = CF_3$, $R = CH_3$ (259)

$$(F_3C)(R)C{=}NNH_2 \xrightarrow{AgO} (F_3C)(R)C{-}N_2 \xleftarrow{NH_2Cl} (F_3C)(R)C{=}NOH \qquad (259)$$

$R = CH_3$ or C_6H_5

[254] D. M. Gale, W. J. Middleton, and C. G. Krespan, *J. Am. Chem. Soc.*, 87, 657 (1965).

[255] D. M. Gale, W. J. Middleton, and C. G. Krespan, *J. Am. Chem. Soc.*, 88, 3617 (1966).

[256] E. P. Mochalina and B. L. Dyatkin, *Bull. Acad. Sci. U.S.S.R., Div. Chem. Sci. (English Transl.)*, 899 (1965).

[257] (a) R. Fields and R. N. Haszeldine, *J. Chem. Soc.*, 1881 (1964). (b) J. H. Atherton and R. Fields, *J. Chem. Soc.* (C), 1450 (1967).

[258] B. L. Dyatkin and E. P. Mochalina, *Bull. Acad. Sci., U.S.S.R., Div. Chem. Sci. (English Transl.)*, 1136 (1964).

[259] (a) R. A. Shepard and P. L. Sciaraffa, *J. Org. Chem.*, **31**, 964 (1966). (b) R. A. Shepard and S. E. Wentworth, *J. Org. Chem.*, **32**, 3197 (1967).

These fluoroalkyldiazomethanes are very stable if highly fluorinated. Thus, bis(trifluoromethyl)diazomethane is stable to isolation and short-term storage at room temperature and, unlike most diazo compounds, is not sensitive to acids. The chemical reactions (Chart 6-6) of carbenes generated from these diazo

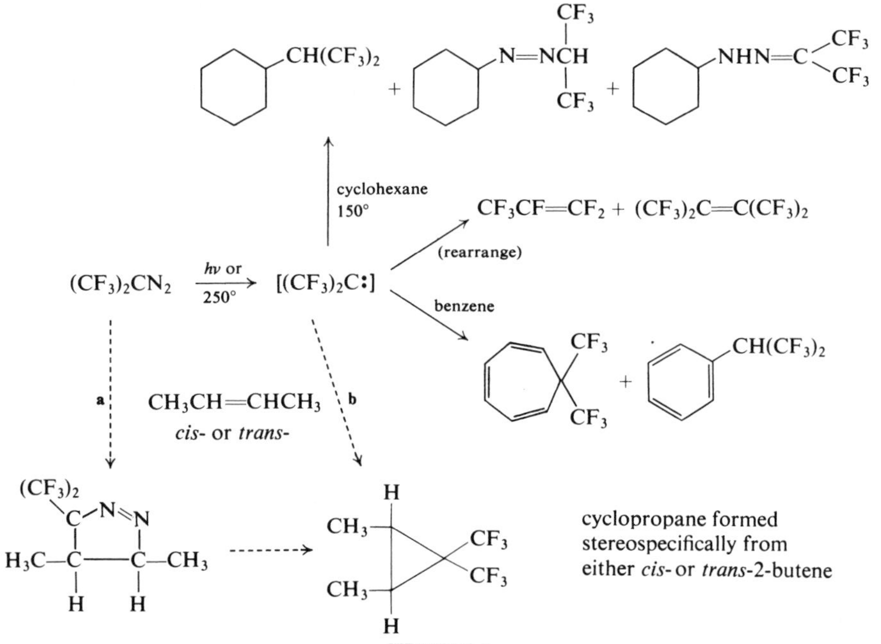
CH(CF₃)₂
N=NCH
CF₃
CF₃
NHN=C
CF₃
CF₃
cyclohexane
150°
CF₃CF=CF₂ + (CF₃)₂C=C(CF₃)₂
(rearrange)
(CF₃)₂CN₂
hν or
250°
[(CF₃)₂C:]
benzene
CF₃
CF₃
CH(CF₃)₂
a
CH₃CH=CHCH₃
cis- or trans-
b
(CF₃)₂
H₃C—C——C—CH₃
H H
H
CH₃
CF₃
CF₃
CH₃
H
cyclopropane formed
stereospecifically from
either cis- or trans-2-butene

CHART 6-6

compounds by photolysis or pyrolysis is in part unusual in that isomerization by fluorine migration does occur under certain conditions. Normal addition or insertion reactions of the carbene species are found. However, by-products arising from addition to olefins (note stereospecific addition) suggest that a pyrazoline may form with electron-rich olefins (path **a**) and decompose to give cyclopropane rather than by direct addition of the carbenes to the olefin as in path **b** (see discussion below on difluorodiazarine reactions). In addition to carbene insertion, an unusual azo product was found with cyclohexane and was proposed to arise from a radical chain reaction.[254] Apparently bis(trifluoromethyl)-carbene has different energies depending on its source of formation.[260]

[260] (a) See Ref. 255 for a full discussion. Singlet difluorocarbene appears to discriminate to some extent. (See following discussion on diazirine chemistry and section on structure.) (b) Bis(trifluoromethyl)diazomethane reacts with a variety of transition-metal complexes to give products which correspond to the insertion of the $C(CF_3)_2$ moiety into a metal-metal, metal-hydrogen, or metal-halogen bond [J. Cooke, W. R. Cullen, M. Green, and F. G. A. Stone, *Chem. Commun.*, 170 (1968)]. These reactions are exothermic at room temperature, suggesting again that the diazomethane adds to form an unstable intermediate (analogous to a pyrazoline) which subsequently loses N_2 and that free carbene is not involved.

Trifluoromethylcarbene (CF_3CH:) also readily isomerizes to $CF_2{=}CFH$ by fluorine migration as well as giving the typical carbene reactions (olefin addition and insertion products).[259] With heptafluoropropylcarbene, the pentafluoroethyl rather than fluorine migrates (based on spectral evidence only).[257] Pyrazoline

$$C_2F_5CF{=}CHF \;\not\leftarrow\; CF_3CF_2CF_2CH{:} \longrightarrow C_2F_5CH{=}CF_2$$

and polymer form in decomposition of CF_3CHN_2, probably from an excited diradical of the diazoalkane. The formation of $CF_3CH{=}CHCF_3$ in this case is suggested to result from carbene attack on the diazoethane rather than carbene dimerization.

Reactions of 2,2,2-trifluorodiazoethane with a series of reagents (olefins, aldehydes, acids, alcohols, mercaptans, sulfenyl halides, mercuric chloride) have been reported to give products (as shown) that are typical of diazo compounds.[258]

$$\underset{\diagdown O \diagup}{CF_3CH{-}CHCCl_3} \qquad CCl_3\overset{O}{\overset{\|}{C}}CH_2CF_3 \qquad CHCl_2CO_2CH_2CF_3$$

$$C_6H_5SCH_2CF_3 \qquad ClC_6H_4SCHClCF_3 \qquad (CF_3CHCl)_2Hg$$

Although ionic or diradical intermediates are probably formed in certain cases from decomposition of pyrazoline intermediates; carbenes are no doubt involved in some.

Trifluoroacetyl(ethoxycarbonyl)carbene has been prepared by irradiation. of the corresponding diazoacetic ester.[261–263] This intermediate reacts as the

$$F_3C\overset{O}{\overset{\|}{C}}{-}\overset{N_2}{\overset{\|}{C}}{-}CO_2C_2H_5 \xrightarrow{h\nu} \left[\underset{\mathbf{a}}{F_3C\overset{O}{\overset{\|}{C}}{-}\ddot{C}{-}CO_2C_2H_5} \longleftrightarrow \underset{\mathbf{b}}{F_3C{-}\overset{O^-}{\overset{|}{C}}{=}\overset{+}{C}{-}CO_2C_2H_5} \right]$$

normal carbene (form **a**) in C–H insertion and C=C addition, but adds to acetone and dehydrogenates alcohols to aldehydes by an ionic concerted mechanism involving form **b**.

Fluorinated carbenes are also produced by decomposition of fluorinated diazirines. Considerable work in this area is on difluorodiazirine, which is prepared by reductive-defluorination of bis(difluoroamino)difluoromethane,[264]

[261] F. Weygand, H. Dworschak, K. Koch, and St. Konstas, *Angew. Chem.*, **73**, 409 (1961).

[262] F. Weygand, *ibid.*, 70.

[263] F. Weygand, W. Schwenke, and H. J. Bestmann, *Angew. Chem.*, **70**, 506 (1958).

[264] (a) R. A. Mitsch, *J. Heterocyclic Chem.*, **3**, 245 (1966). (b) R. L. Rebertus, J. J. McBrady, and J. G. Gagnon, *J. Org. Chem.*, **32**, 1944 (1967).

$$C_3Cl_3N_3 \xrightarrow[125^\circ]{F_2} F_2C(NF_2)_2 \xrightarrow[\text{dicumene chromium or tetraethylammonium iodide}]{\text{dicyclopentadienyliron}} F_2C\langle{N \atop N}$$

or by isomerization of difluorocyanamide.[265]

$$NH_2CN \xrightarrow[\text{phosphate buffered aqueous solution}]{F_2,\ 5\text{–}9^\circ} \underset{(20\%)}{F_2NCN} \xrightarrow[24^\circ]{CsF} F_2C\langle{N \atop N}$$

Extensive accounts of the properties and chemistry of difluorodiazirine are available.[264–268] It is more stable than diazomethane or diazirine but can explode under certain conditions. It decomposes thermally or by ultraviolet irradiation to give difluorocarbene which has been found to give normal difluorocarbene reactions. The addition of difluorocarbenes to olefins is stereospecific (suggesting a singlet state), and the yield of cyclopropane ranges from 15 to 71% depending on substitution of olefins, temperature of reaction, and

$$\underset{(14\%)}{CF_2{=}CF_2} + \underset{(63\%)}{\text{cyclo-}(CF_2)_3} + \underset{(24\%)}{CF_2{=}N{-}N{=}CF_2}$$

$$\uparrow$$

$$F_2C\langle{N \atop N} \xrightarrow[\text{heat}]{h\nu\text{ or}} [{:}CF_2] \xrightarrow{X{-}Y} F_2CXY$$

(X = Y is Cl, I, or NO_2; X = Cl, Y = NO_2)

$$[{:}CF_2] \xrightarrow{RCOOH} RCOOCF_2H$$

$$[{:}CF_2] \xrightarrow{CH_3CH{=}CHCH_3} \text{1,1-difluoro-2,3-dimethylcyclopropane}$$

F F, H, H, H_3C, CH_3

(stereospecific addition to both cis and trans; relative rates compared to other halocarbenes)

[265] M. D. Meyers and S. Frank, *Inorg. Chem.*, **5**, 1455 (1966).

[266] (a) R. A. Mitsch, *J. Heterocyclic Chem.*, **1**, 59 and 271 (1964). (b) R. A. Mitsch, *J. Am. Chem. Soc.*, **87**, 758 (1965). (c) A. S. Rodgers and R. A. Mitsch, *J. Am. Chem. Soc.*, in press.

[267] R. A. Mitsch and J. E. Robertson, *J. Heterocyclic Chem.*, **2**, 152 (1965).

[268] J. L. Hencher and S. H. Baier, *J. Am. Chem. Soc.*, **89**, 5527 (1967).

amount of excess olefin. A number of new perfluorocyclopropanes were prepared in 30 to 85% yield by addition to fluorocarbon olefins.

A series of α-fluorodiazirines,

$$\begin{matrix}F \\ X\end{matrix}\!\!>\!C\!<\!\begin{matrix}N \\ \| \\ N\end{matrix} \qquad (X = OCH_3, Cl, CN, NF_2)$$

gave reactions similar to those for difluorodiazirine.[269] However, when difluoroaminofluorodiazirine is decomposed, a competing reaction of intramolecular fluorine migration (such as noted for the fluoroalkylcarbenes described above) is also found.

$$\begin{matrix}F \\ F_2N\end{matrix}\!\!>\!C\!<\!\begin{matrix}N \\ \| \\ N\end{matrix} \xrightarrow{\Delta} \left[\begin{matrix}F \\ F_2N\end{matrix}\!\!>\!C\!:\right] + N_2 \longrightarrow \begin{matrix}F \\ F\end{matrix}\!\!>\!C{=}NF$$

Bis(trifluoromethyl)diazarine has been synthesized by two routes.[255, 270]

$$\begin{matrix}F_3C \\ F_3C\end{matrix}\!\!>\!C\!<\!\begin{matrix}NH_2 \\ NH_2\end{matrix} \xrightarrow[NaOH]{NaOCl} \underset{\text{bp } -12^\circ}{\begin{matrix}F_3C \\ F_3C\end{matrix}\!\!>\!C\!<\!\begin{matrix}N \\ \| \\ N\end{matrix}} \xleftarrow{Pb(OAc)_4} \text{(from below)}$$

$$(CF_3)_2C{=}NH + HN_3 \longrightarrow (CF_3)_2C\!<\!\begin{matrix}N_3 \\ NH_2\end{matrix} \xrightarrow[\text{heat}]{h\nu \text{ or}} (CF_3)_2C\!<\!\begin{matrix}NH \\ | \\ NH\end{matrix}$$

This diazirine also gives the singlet carbene with typical reactions on heating or irradiation. The products from reaction of *cis*- or *trans*-2-butene with bis-(trifluoromethyl)diazomethane and bis(trifluoromethyl)diazirine on heating are different as shown in Table 6-3; the diazomethane was suggested to go by addition to the electron-rich olefin to give the pyrazoline that decomposed to diradical intermediates, whereas the diazirine gave a singlet carbene that is fairly indiscriminate but not completely so.

Perfluoro-2,3-diazabuta-1,3-diene is unusually stable relative to its hydrocarbon analog but photolyzes readily to give only diazapentadiene.[271]

[269] (a) R. A. Mitsch, E. W. Neuvar, R. J. Koshar, and D. H. Dybvig, *J. Heterocyclic Chem.*, **2**, 371 (1965). (b) R. L. Rebertus and P. E. Toren, *J. Org. Chem.*, **32**, 4045 (1967).

[270] R. B. Minasyan, E. M. Rokhlin, N. P. Gam'aryan, Ya. V. Zeifman, and I. L. Knunyants, *Bull. Acad. Sci. U.S.S.R., Div. Chem. Sci. (English Transl.)*, 746 (1965).

[271] R. A. Mitsch and P. H. Ogden, *Chem. Commun.*, 59 (1967).

Table 6-3
Reaction at 150° with 2-Butene[255]

Products (relative yields)		$(CF_3)_2C=\overset{+}{N}=\overset{-}{N}$ trans-2-Butene	$(CF_3)_2C=\overset{+}{N}=\overset{-}{N}$ cis-2-Butene	Bis(trifluoromethyl)diazirine trans-2-Butene	Bis(trifluoromethyl)diazirine cis-2-Butene
Overall yield		53	57	61	71
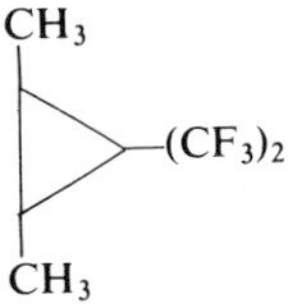 1,2-dimethyl-3,3-bis(trifluoromethyl)cyclopropane, CH_3, CH_3, $-(CF_3)_2$	cis	–	39	–	55
	trans	100	8	57	8
$(F_3C)_2C=C(CH_3)CH_2CH_3$		–	49	–	2.5
$H_3C(H)C=C(H)CH_2CH(CF_3)_2$		–	–	39	–
$H_3C(H)C=C(CH(CF_3)_2)CH_3$		–	–	4	–
$H_3C(H)C=C(CH_3)CH(CF_3)_2$		–	–	–	8
$H_3C(H)C=C(CH_2CH(CF_3)_2)H$		–	–	–	27

$$CF_2{=}\dot{N}{-}N{=}CF_2 \xrightarrow{h\nu} [CF_2{=}N\cdot] + [:CF_2] + N_2 \longrightarrow CF_2{=}NCF_2N{=}CF_2 \xrightarrow{F^-} CF_3N{=}C{=}NCF_3$$

No tetrafluoroethylene was detected, so that $CF_2=N\cdot$ is suggested to be formed more rapidly than difluorocarbene. Difluorocarbene is also generated from pyrolysis of fluorinated triazoles.[272]

[272] W. Carpenter, A. Haymaker, and D. W. Moore, *J. Org. Chem.*, **31**, 789 (1966).

$$CF_3CF{=}CF_2 + RN_3 \longrightarrow CF_3{-}\underset{|}{\overset{F}{C}}{-}CF_2 \text{ (triazoline: } RN{-}N{=}N\text{)}$$

$(R = C_6H_5CH_2)$

$$\xrightarrow{250\text{–}325^\circ,\ Ni} [{:}CF_2] + N_2 + RN{=}C(F)CF_3 \quad (65\%)$$

$$[{:}CF_2] \longrightarrow CF_2{=}CF_2\ (45\%)$$

$$[{:}CF_2] \xrightarrow{(CH_3)_2C=C(CH_3)_2} \text{1,1-difluoro-2,2,3,3-tetramethylcyclopropane } (F_2, (CH_3)_2, (CH_3)_2)\ (18\%)$$

(2) *Splitting from Heteroatom Groups Other Than Nitrogen.* A mild method of generating difluorocarbene under neutral conditions is by pyrolysis of organometallic or metalloid derivatives. Early observations that pyrolysis of tris(trifluoromethyl)arsine and stibine at 180° to 220°C gave small yields of tetrafluoroethylene and perfluorocyclopropane,[273] no doubt by way of difluorocarbene, have been utilized with trifluoromethyl derivatives of tin, germanium phosphorus, and iron carbonyl[274–276] (Chart 6-7). Note that the iron carbonyl is claimed to be the only source of difluorocarbene that adds to ethylene.[276] Also, difluoromethylene is generated from any of the series of trifluoromethylfluorophosphorane $(CF_3)_nPF_{5-n}$, where $n = 1,2,3$. Difluorocarbene is also formed in flash photolysis of trifluoromethylphosphino compounds.[277c]

A useful modification of conditions with $(CH_3)_3SnCF_3$ is to use sodium iodide in glyme; a temperature of only 80°C is required to get addition to cyclohexene.[278]

$$(CH_3)_3SnCF_3 + NaI \xrightarrow[\text{cyclohexene}]{80^\circ,\ \text{glyme}} \text{7,7-difluorobicyclo[4.1.0]heptane } (F_2)\ (73\%) + (CH_3)_3SnI\ (90\%)$$

[273] P. B. Ayscough and H. J. Emeleus, *J. Chem. Soc.*, 3381 (1954).

[274] (a) H. C. Clark and C. J. Willis, *J. Am. Chem. Soc.*, **82**, 1888 (1960). (b) H. C. Clark and C. J. Willis, *J. Am. Chem. Soc.*, **84**, 898 (1962).

[275] R. B. King, S. L. Stafford, P. M. Treichel, and F. G. A. Stone, *J. Am. Chem. Soc.*, **83**, 3604 (1961).

[276] This result has not been substantiated; difluoromethylene generated from other sources does not give any significant amount of 1,1-difluorocyclopropane in reaction with ethylene (private communication from Dr. W. Mahler, Central Research Department, E. I. du Pont de Nemours and Co.; see also Ref. 230).

[277] (a) W. Mahler, *Inorg. Chem.*, **2**, 230 (1963). (b) W. Mahler, *J. Am. Chem. Soc.*, **84**, 4600 (1962). (c) R. G. Cavell, R. C. Dobbie, and W. J. R. Tyerman, *Can. J. Chem.*, **45**, 2849 (1967).

[278] D. Seyferth, H. Dertouzos, R. Suzuki, and J. Y.-P. Mui, *J. Org. Chem.*, **32**, 2980 (1967).

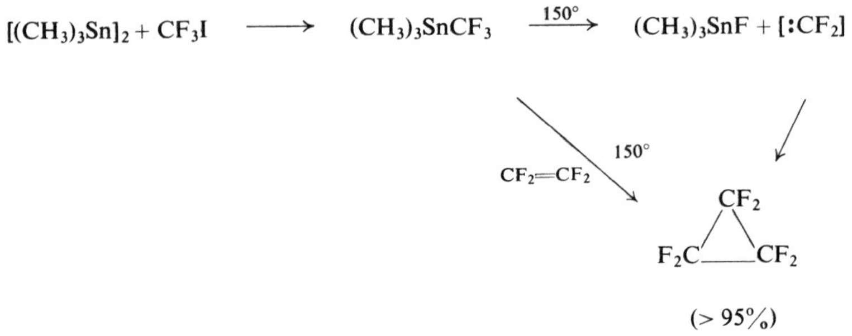

$$[(CH_3)_3Sn]_2 + CF_3I \longrightarrow (CH_3)_3SnCF_3 \xrightarrow{150°} (CH_3)_3SnF + [:CF_2]$$

$(CH_3)_3SnCF_3 \xrightarrow[150°]{CF_2{=}CF_2}$ cyclo-C_3F_6 (F_2C–CF_2–CF_2 ring) $\leftarrow [:CF_2]$

(> 95%) (274)

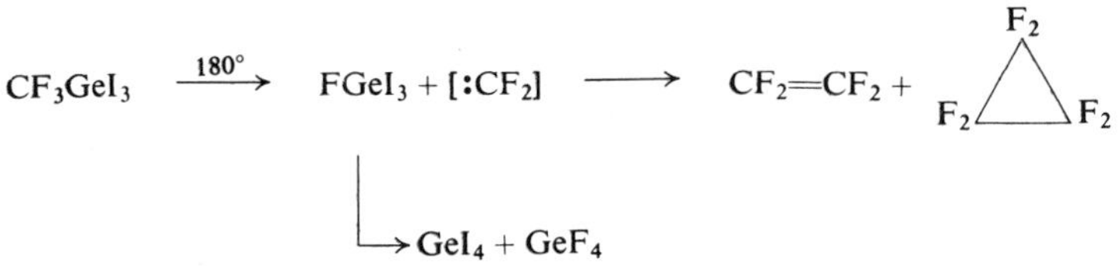

$$CF_3GeI_3 \xrightarrow{180°} FGeI_3 + [:CF_2] \longrightarrow CF_2{=}CF_2 + \text{cyclo-}(CF_2)_3 \quad (275)$$

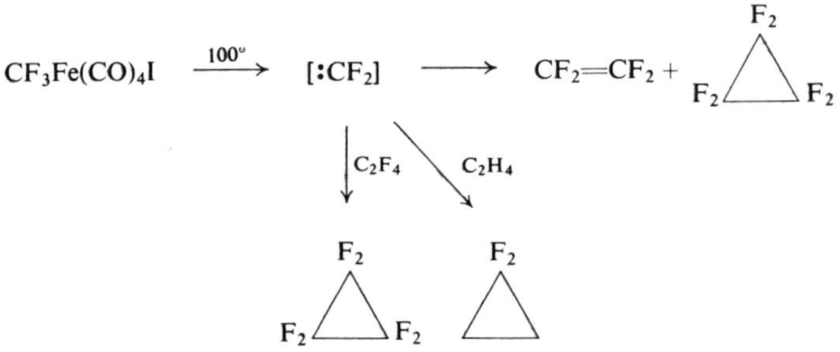

$FGeI_3 \longrightarrow GeI_4 + GeF_4$

$$CF_3Fe(CO)_4I \xrightarrow{100°} [:CF_2] \longrightarrow CF_2{=}CF_2 + \text{cyclo-}(CF_2)_3 \quad (276)$$

$[:CF_2] \xrightarrow{C_2F_4}$ cyclo-$(CF_2)_3$ (F_2, F_2, F_2)

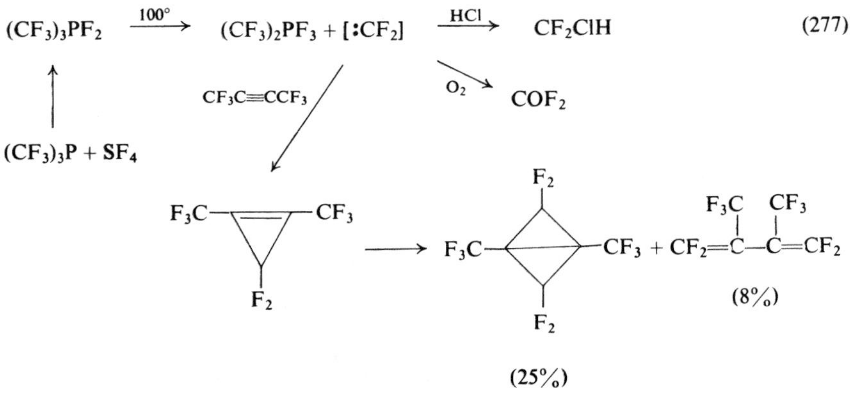

$[:CF_2] \xrightarrow{C_2H_4}$ cyclopropane with F_2

$$(CF_3)_3PF_2 \xrightarrow{100°} (CF_3)_2PF_3 + [:CF_2] \xrightarrow{HCl} CF_2ClH \quad (277)$$

$(CF_3)_3P + SF_4 \longrightarrow (CF_3)_3PF_2$

$[:CF_2] \xrightarrow{O_2} COF_2$

$[:CF_2] \xrightarrow{CF_3C{\equiv}CCF_3}$ F_3C–$C{=}C$–CF_3 cyclopropene ring with F_2

$\longrightarrow$ F_3C–bicyclo[1.1.0]butane (F_2, F_2)–CF_3 (25%) $+ CF_2{=}C(CF_3){-}C(CF_3){=}CF_2$ (8%)

CHART 6-7

This catalysis technique is useful in preparation of chlorofluorocarbene from $ArHgCCl_2F$.[279]

$$C_6H_5HgCl + CHCl_2F + t\text{-BuOK} \longrightarrow C_6H_5HgCCl_2F + t\text{-BuOH} + KCl$$

$$C_6H_5HgCCl_2F + NaI + \text{cyclohexene} \xrightarrow[85^\circ,\ 3\ hr]{CH_3OCH_2CH_2OCH_3} C_6H_5HgI + \text{7-chloro-7-fluorobicyclo[4.1.0]heptane} + NaCl \quad (70\%)$$

This route provides a source of :CFCl generated under mild, neutral conditions, and was used to prepare a selection of chlorofluorocyclopropanes.

Difluorocarbene is lost rapidly and quantitatively from trifluoromethyl-substituted boranes when an amine catalyst is added,[280] but only polymer formed and none of the usual $:CF_2$ products (C_2F_4, C_3F_6, etc.) were detected. The

$$CF_3BF_2 \xrightarrow{R_3N} BF_3 + \text{-}(CF_2)\text{-}_n$$

authors suggest that the $:CF_2$ is generated in an excited electronic state but further work is obviously needed before any conclusion can be drawn. The pyrolysis of the ammonium salt of trifluoromethylfluoroboric acid did, however, give typical difluorocarbene products.[280b]

$$CF_3BF_3^- NH_4^+ \xrightarrow{>150^\circ} NH_4^+ BF_4^- + CF_2{=}CF_2$$

The synthetic utility of this difluorocarbene source was not evaluated further.

The decomposition of chlorofluoroalkyltrichlorosilanes has been shown to go by α rather than β elimination on basis of product composition.[280–282]

$$CFCl_2CF_2SiCl_3 \xrightarrow{\alpha\text{-elimination}} SiCl_3F + CFCl_2\ddot{C}F \longrightarrow CFCl{=}CFCl \quad (80\%)$$

$$CFCl_2CF_2SiCl_3 \xrightarrow[\text{(or F migration vs. Cl)}]{\beta\text{-elimination}} SiCl_3F + CCl_2{=}CF_2 \quad (7\%)$$

[279] D. Seyferth and K. V. Darragh, *J. Organometal. Chem.*, **11**, P9 (1968).

[280] (a) T. D. Parsons, J. M. Self, and L. H. Schaad, *J. Am. Chem. Soc.*, **89**, 3446 (1967). (b) R. D. Chambers, H. C. Clark and C. J. Willis, *J. Am. Chem. Soc.*, **82**, 5298 (1960).

[281] (a) R. N. Haszeldine and J. C. Young, *Proc. Chem. Soc.*, 394 (1959). (b) W. I. Bevan, R. N. Haszeldine, and J. C. Young, *Chem. Ind. (London)*, 789 (1961). (c) D. Seyferth, J. M. Burlitch, and J. K. Heeren, *J. Org. Chem.*, **27**, 1491 (1962).

[282] (a) R. N. Haszeldine and J. G. Speight, *Chem. Commun.*, 995 (1967). (b) J. Lee, C. Parkinson, P. J. Robinson, and J. G. Speight, *J. Chem. Soc.* (B), 1125 (1967).

The decomposition of halosilanes provides a good general route to other halocarbenes.[282]

$$SiHCl_3 + C_2F_4 \xrightarrow{h\nu} CHF_2CF_2SiCl_3 \xrightarrow{SbF_3} CHF_2CF_2SiF_3 \qquad (282)$$

$$CHF_2CF_2SiF_3 \xrightarrow{150^\circ} CHF_2\ddot{C}F + SiF_4$$

$$CHF_2\ddot{C}F \xrightarrow{>CH} \;>CCHFCHF_2 + CF_2{=}CHF$$

$$CHF_2\ddot{C}F + >C{=}C< \longrightarrow \text{cyclo-}(>C\text{—}C<)CFCHF_2 \quad (75\text{–}90\%)$$

The cyclopropanes were used for nmr studies[282b]; the insertion reaction, relative to rearrangement to trifluoroethylene, was also studied.[282a]

The thermal decomposition of hexafluoropropylene-1,2-epoxide (HFPO) in the presence of thiocarbonyl compounds gives fluorinated thiiranes.[190, 283]

$$(R_f)_2C{=}S + CF_3\overset{O}{CF\text{—}CF_2} \xrightarrow[\text{over } 170^\circ]{\Delta} (R_f)_2\overset{S}{C\text{—}CF_2} + CF_3\overset{O}{\overset{\|}{C}}F$$

$(R_f = F, \text{perfluoroalkyl})$

This was the first reported example of trapping of difluorocarbene from fragmentation of HFPO (the other product is trifluoroethyl fluoride). The pyrolysis of perfluorocarbon epoxides between 125° and 200°C is claimed to be a convenient method to generate difluorocarbene for typical carbene reactions.[283b]

2. Structure and Properties of Fluorinated Carbenes

Spectral studies have been made on a number of fluorinated carbenes, particularly using the technique of trapping at low temperature–usually in a matrix or vapor phase in an inert gas. From analysis of spectral data,[284, 285] the difluorocarbene generated by a variety of techniques in the gas phase is

[283] (a) F. C. McGrew, U.S. Patent 3,136,744 (1964); *Chem. Abstr.*, **61**, 4312b (1964). (b) *Chem. Eng. News*, **45**, 18 (Aug. 7, 1967).

[284] J. P. Simons and A. J. Yarwood, *Nature*, **192**, 943 (1961).

[285] D. E. Milligan, D. E. Mann, M. E. Jacox, and R. A. Mitsch, *J. Chem. Phys.*, **41**, 1199 (1964).

believed to be a ground-state singlet species (not diradical) in nonlinear configuration (< FCF estimated 108°) with a lifetime of as long as 20 msec (in relatively good agreement with lifetime determined by chemical methods).

Triplet difluorocarbene can be generated by reaction of ground-state oxygen atoms with C_2F_4 at room temperature. It either converts to singlet or, at a high oxygen pressure, reacts to give carbonyl fluoride. Singlet CF_2 does not react with oxygen at room temperature but can dimerize or add to tetrafluoroethylene.[286]

$$CF_2{=}CF_2 + O^3 \longrightarrow CF_2{}^3 + COF_2$$

Hg-sensitized photolysis of N_2O (→ O^3)

$CF_2{}^3 \xrightarrow{O} COF_2$

$CF_2{}^3 \rightsquigarrow CF_2{}^1$

$CF_2{}^1 \xrightarrow{C_2F_4}$ CF₂—CF₂ ring with CF₂ (cyclo-C_3F_6: CF_2 / CF_2—CF_2)

$CF_2{}^1 \not\xrightarrow{O}$

$CF_2{}^1 \longrightarrow CF_2{=}CF_2$

Perfluoroalkylmethylenes generated at low temperatures in a matrix were ground-state triplets and by esr studies were concluded to be bent with the smallest angle (about 140°) for bis(trifluoromethyl)methylene.[287]

The thermochemistry of difluoromethylene has been widely discussed. Values for the heat of formation of CF_2 have ranged from -17 to -45 kcal/mole, but the most recent evidence supports a value of $-37{\pm}3$ kcal/mole.[288]

The electron structure of difluorocarbene has been discussed in terms of simple molecular orbital calculations and related to spectroscopic, thermodynamic, and chemical properties.[289] Although quantitative data are lacking, difluorocarbene is much less electrophilic than other halocarbenes. The order of reactivity towards olefin addition (or decreasing selectivity from insertion into a π bond) is

$$CF_2 < CCl_2 < CBr_2 < CH_2$$

The electron deficiency of the carbene carbon is partially relieved by π overlap with the p-electron of halogen, so that the halocarbenes are not as electrophilic as $:CH_2$. Based on simple geometry of overlap, the back-bondings of unshared p-electron into the 2p of carbon is better for the $2p_F$ as compared to $3p_{Cl}$ and $4p_{Br}$. Consequently, difluorocarbene is the least electron deficient.

The sluggish character of difluorocarbene in dimerization association or in reaction with radical species is also discussed in terms of molecular orbital calculations.

[286] (a) N. Cohen and J. Heicklen, *J. Phys. Chem.*, **70**, 3082 (1966). (b) T. Johnston and J. Heicklen, *J. Chem. Phys.*, **47**, 475 (1967).

[287] E. Wasserman, L. Barash, and W. A. Yager, *J. Am. Chem. Soc.*, **87**, 4974 (1965).

[288] R. F. Pottie, *J. Chem. Phys.*, **42**, 2607 (1965).

[289] J. P. Simons, *J. Chem. Soc.*, 5406 (1965).

3. Fluorinated Heteroatom Carbenes

A number of heteroatom compounds have an electronic structure on the heteroatom similar to that of carbene carbon. These carbene analogs have been reviewed recently by Nefedov and Manakov.[196] The fluorinated ones of interest as synthetic intermediates are limited to Groups IV and V of the periodic table.

a. *Group IV*

Silicon difluoride ($:SiF_2$) has been studied extensively. It is readily generated by passing SiF_4 over silicon powder at 1200°C[290–292] and can be trapped as a monomer at −196°C. Over −80°C, it forms a plastic polymer $(SiF_2)_n$. The half-life of gaseous SiF_2 at room temperature is ca. 150 seconds, much greater than that of $:CF_2$. Other routes to $:SiF_2$ are

$$Si_2F_6 \xrightarrow{700^\circ} [:SiF_2] + SiF_4 \tag{292}$$

$$Cl_2SiF_2 \xrightarrow[\text{ether}]{\text{Mg}} (SiF_2)_n \text{ polymer only isolated} \tag{293}$$

Silicon difluoride gives insertion and addition reactions[294, 295] (see Chart 6-8), but, unlike carbenes, shows preference for insertion in a C–F bond over insertion in C–H or addition to a π system. From spectral studies on SiF_2, the bond angle $<$ FSiF is shown to vary from about 100° to 124° depending on the energy level and state.[296]

Germanium difluoride is a stable monomeric species, prepared by heating germanium powder with hydrogen fluoride.[297] Apparently its reactions with organic substrates have not been studied.

b. *Group V*

Fluoroazene, or fluoronitrene ($:NF$), has been proposed as an intermediate in a number of reactions[299–302] (Chart 6-9).

[290] D. C. Pease, U.S. Patent 3,026,173 (1962); *Chem. Abstr.*, **57**, 3081i (1962).
[291] D. C. Pease, U.S. Patent 2,840,588 (1958); *Chem. Abstr.*, **52**, 19245c (1958).
[292] P. L. Timms, R. A. Kent, T. C. Ehlert, and J. L. Margrave, *J. Am. Chem. Soc.*, **87**, 2824 (1965).
[293] M. Schmeisser and K. P. Ehlers, *Angew. Chem. Intern. Ed. Engl.*, **3**, 700 (1964).
[294] J. C. Thompson, J. L. Margrave, and P. L. Timms, *Chem. Commun.*, 566 (1966).
[295] P. L. Timms, D. D. Stump, R. A. Kent, and J. L. Margrave, *J. Am. Chem. Soc.*, **88**, 940 (1966).
[296] V. M. Khanna, G. Besenbruch, and J. L. Margrave, *J. Chem. Phys.*, **46**, 2310 (1967).
[297] E. L. Muetterties, *Inorg. Chem.*, **1**, 342 (1962).
[298] C. L. Bumgardner, K. J. Martin, and J. P. Freeman, *J. Am. Chem. Soc.*, **85**, 97 (1963).
[299] W. H. Graham, *J. Am. Chem. Soc.*, **88**, 4677 (1966).
[300] C. L. Bumgardner and M. Lustig, *Inorg. Chem.*, **2**, 662 (1963).
[301] J. F. Haller, Ph.D. Thesis, Cornell University, 1942.
[302] (a) F. A. Johnson, *Inorg. Chem.*, **5**, 149 (1966). (b) W. le Noble and D. Skulnik, *Tetrahedron Letters*, 5217 (1967).

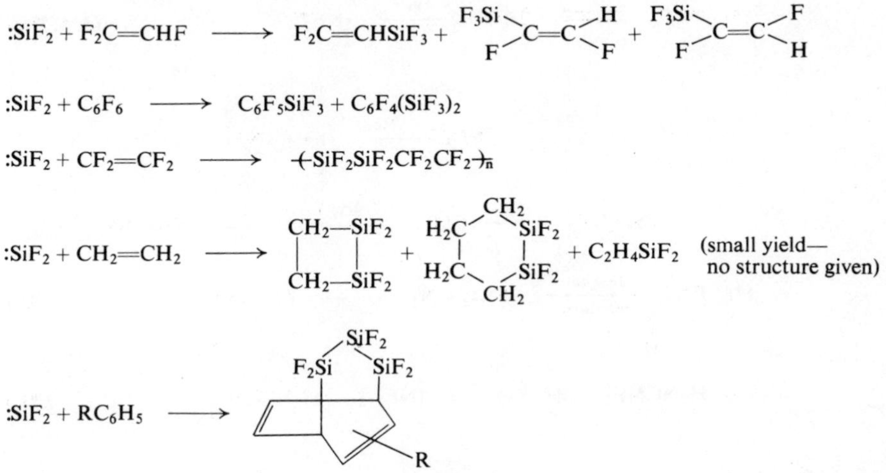

$$:SiF_2 + F_2C{=}CHF \longrightarrow F_2C{=}CHSiF_3 + (F_3Si)(F)C{=}C(H)(F) + (F_3Si)(F)C{=}C(F)(H)$$

$$:SiF_2 + C_6F_6 \longrightarrow C_6F_5SiF_3 + C_6F_4(SiF_3)_2$$

$$:SiF_2 + CF_2{=}CF_2 \longrightarrow \left(SiF_2SiF_2CF_2CF_2\right)_n$$

$$:SiF_2 + CH_2{=}CH_2 \longrightarrow \text{cyclo-}(CH_2{-}SiF_2{-}SiF_2{-}CH_2) + \text{cyclo-}(H_2C{-}CH_2{-}SiF_2{-}CH_2{-}SiF_2{-}CH_2) + C_2H_4SiF_2$$ (small yield—no structure given)

$$:SiF_2 + RC_6H_5 \longrightarrow$$ (bicyclic adduct bridged by $F_2Si{-}SiF_2{-}SiF_2$, bearing R)

CHART 6-8

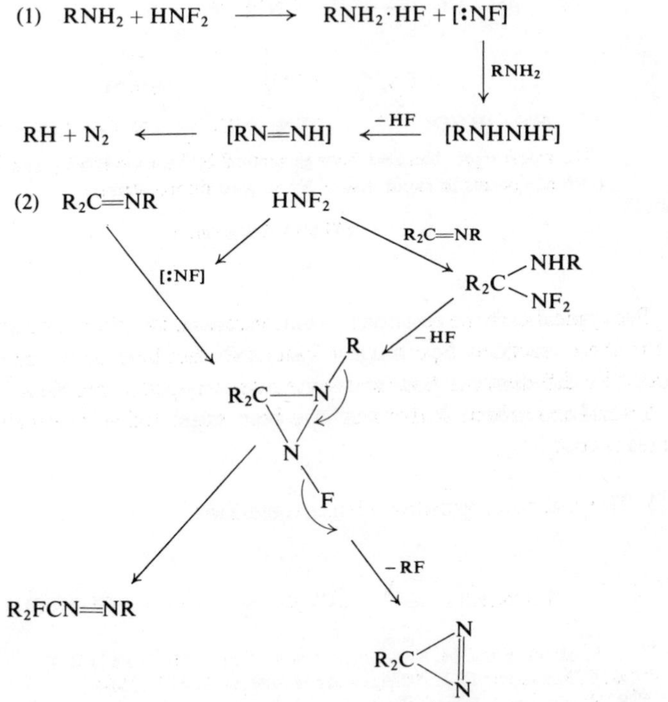

(1) $RNH_2 + HNF_2 \longrightarrow RNH_2{\cdot}HF + [{:}NF]$ (298)

$[{:}NF] \xrightarrow{RNH_2} [RNHNHF]$

$[RNHNHF] \xrightarrow{-HF} [RN{=}NH] \longrightarrow RH + N_2$

(2) $R_2C{=}NR \quad HNF_2$ (299)

$HNF_2 \longrightarrow [{:}NF]$; $R_2C{=}NR + [{:}NF] \longrightarrow$ diaziridine $R_2C(NR)(NF)$

$HNF_2 \xrightarrow{R_2C=NR} R_2C(NHR)(NF_2) \xrightarrow{-HF}$ diaziridine $R_2C(NR)(NF)$

Diaziridine $\longrightarrow R_2FCN{=}NR$

Diaziridine $\xrightarrow{-RF} R_2C\langle N{=}N$ (diazirine)

(Relative amounts of products depend on the structure of the imine)

CHART 6-9

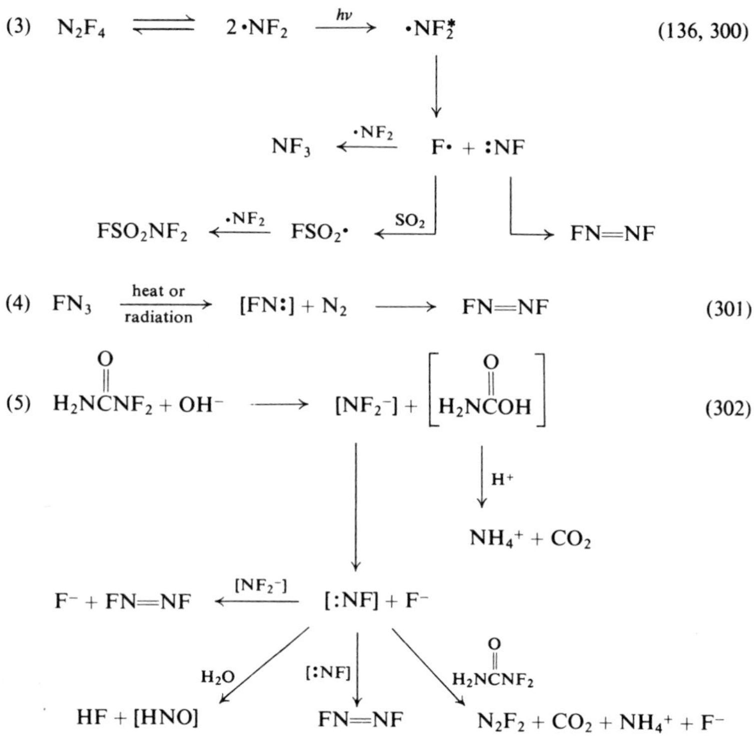

$$(3)\quad N_2F_4 \rightleftharpoons 2\cdot NF_2 \xrightarrow{h\nu} \cdot NF_2^* \qquad (136, 300)$$

$$\cdot NF_2^* \longrightarrow F\cdot + :NF$$

$$F\cdot + \cdot NF_2 \longrightarrow NF_3$$

$$F\cdot \xrightarrow{SO_2} FSO_2\cdot \xrightarrow{\cdot NF_2} FSO_2NF_2$$

$$:NF \longrightarrow FN{=}NF$$

$$(4)\quad FN_3 \xrightarrow[\text{radiation}]{\text{heat or}} [FN:] + N_2 \longrightarrow FN{=}NF \qquad (301)$$

$$(5)\quad H_2N\overset{O}{\overset{\|}{C}}NF_2 + OH^- \longrightarrow [NF_2^-] + \left[H_2N\overset{O}{\overset{\|}{C}}OH\right] \qquad (302)$$

$$\left[H_2N\overset{O}{\overset{\|}{C}}OH\right] \xrightarrow{H^+} NH_4^+ + CO_2$$

$$[NF_2^-] \longrightarrow [:NF] + F^-$$

$$[:NF] + F^- \xrightarrow{[NF_2^-]} F^- + FN{=}NF$$

$$[:NF] \xrightarrow{H_2O} HF + [HNO]$$

$$[:NF] \xrightarrow{[:NF]} FN{=}NF$$

$$[:NF] \xrightarrow{H_2N\overset{O}{\overset{\|}{C}}NF_2} N_2F_2 + CO_2 + NH_4^+ + F^-$$

(The anion F_2N^- has also been generated by base on HNF_2 and kinetic evidence points to rapid loss of F^- to give fluoronitrene)

CHART 6-9 (cont.)

The typical carbene reactions of dimerization, insertion, and addition are found in these reactions. Spectral and kinetic studies have been made on :NF generated by different methods and sometimes trapped in matrices.[303–305]

Several azocarbene derivatives have been suggested as intermediates in other reactions:

1. Thermal decomposition (Lossen reaction),

$$CF_3\overset{O}{\overset{\|}{C}}NHOH \xrightarrow{\Delta} CF_3\overset{O}{\overset{\|}{C}}{-}\ddot{\underset{\cdot\cdot}{N}}: \longrightarrow CF_3NCO \qquad (306)$$

[303] D. E. Milligan and M. E. Jacox, *J. Chem. Phys.*, **40**, 2461 (1964).
[304] M. E. Jacox and D. E. Milligan, *Appl. Optics*, **3**, 873 (1964).
[305] R. W. Diesen, *J. Chem. Phys.*, **41**, 3256 (1964).
[306] I. L. Knunyants and G. A. Sokol'skii, *Proc. Acad. Sci. U.S.S.R., Chem. Sect. (English Transl.)*, **132**, 565 (1960).

2. Hoffman reaction on perfluoroamides,

$$C_3F_7\overset{O}{\overset{\|}{C}}N^{-}\begin{matrix}Na^{+}\\ Br\end{matrix} \xrightarrow[\text{solvent}]{\text{in polar}} C_2F_5CF_2Br + NCO^{-} \tag{307}$$

$$C_3F_7\overset{O}{\overset{\|}{C}}N^{-}\begin{matrix}Na^{+}\\ Br\end{matrix} \xrightarrow[\text{anhydrous}]{\text{pyrolysis}} C_3F_7\overset{O}{\overset{\|}{C}}\ddot{N}: + NaBr \longrightarrow C_3F_7NCO\ (83\%)$$

The product formed depends on conditions.

3. Defluorination of primary perfluoroamine (reductive, using dicyclopentadienyliron or dicumenechromium),

$$R_fCF_2{-}NF_2 \longrightarrow R_f{-}\underset{F}{\underset{|}{C}}{=}NF \tag{308}$$

$$\downarrow +e^-,\ -F^-$$

$$R_fC{\equiv}N \longleftarrow R_f{-}\underset{F}{\underset{|}{C}}{=}\dot{\ddot{N}}$$

4. Fluorination of cyanogen chloride was suggested to involve trifluoromethylnitrene, CF_3N:, which dimerized or reacted with halogen.[309] This nitrene was also suggested as the intermediate when *N,N*-dihalotrifluoromethylamine was irradiated; however it was not trapped by an excess of O_2 or CO.

$$CF_3NCl_2 \text{ (or } CF_3NFCl) \xrightarrow{h\nu} CF_3N{=}NCF_3$$

5. Univalent fluoroalkyl or aryl phosphorus derivatives could be intermediates in the following reactions[310]:

[307] (a) D. A. Barr and R. N. Haszeldine, *J. Chem. Soc.*, 30 (1957). (b) D. A. Barr and R. N. Haszeldine, *Chem. Ind. (London)*, 1050 (1956).
[308] R. A. Mitsch, *J. Am. Chem. Soc.*, **87**, 328 (1965).
[309] J. B. Hynes, B. C. Bishop, and L. A. Bigelow, *Inorg. Chem.*, **6**, 417 (1967).
[310] W. Mahler and A. B. Burg, *J. Am. Chem. Soc.*, **80**, 6161 (1958).

$$CF_3PI_2 + Hg \xrightarrow{20°} \left[CF_3\overset{\displaystyle I}{\overset{|}{P}}HgI \right] \longrightarrow [CF_3P:] \longrightarrow (CF_3P)_m \quad (m = 4, 5, 6)$$

$$\left.\begin{array}{l} (CF_3)_2P\text{—}P(CF_3)_2 \xrightarrow{350°} \\ (CF_3)_2PH \xrightarrow[350°]{} \end{array}\right\} \longrightarrow [CF_3P:] \longrightarrow (CF_3P)_m$$

$$\underset{(X = Br \text{ or } I)}{C_6F_5PX_2 + Hg} \longrightarrow C_6F_5\overset{\displaystyle X}{\overset{|}{P}}HgX \longrightarrow (PC_6F_5)_5 \quad (91\%) \qquad (311)$$

6-1C. FLUORINATED ANIONS

Negatively charged species are stabilized when substituted by highly electronegative fluorine atoms. These fluorocarbanions (negative charge localized on carbon) are of major importance in synthetic organic fluorine chemistry; fluorinated anions (the negative charge localized on a heteroatom) are reviewed briefly in a second section.

1. Fluorinated Carbanions

Partially or completely fluorinated carbanions are believed to be intermediates in several types of reactions:

1. decarboxylation of alkali metal salts of fluorocarboxylic acids

$$R_fCF_2\overset{\displaystyle O}{\overset{\|}{C}}O^- \longrightarrow R_fCF_2^- + CO_2$$

2. fluorocarbon metallic compounds

$$R_fX \longrightarrow R_fM(X) \longrightarrow R_f^- + \overset{+}{M}(X)$$

3. addition of bases to fluoroolefins

$$B^- + R_fCF{=}CF_2 \longrightarrow BCF_2\text{—}\bar{C}FR_f$$

(which includes F^- as base; see Section 5-2B);

[311] (a) A. H. Cowley and R. P. Pinnell, *J. Am. Chem. Soc.*, **88**, 4533 (1966). (b) A. H. Cowley, *J. Am. Chem. Soc.*, **89**, 5990 (1967).

4. bases on monohydrofluorocarbons

$$R_fH + B^- \longrightarrow BH + R_f^-$$

5. base-promoted elimination of a leaving group X (iodide or ester)

$$B^- + R_fI \longrightarrow BI + R_f^-$$

$$B^- + R_f\overset{O}{\overset{\|}{C}}OR \longrightarrow R_f^- + B\overset{O}{\overset{\|}{C}}OR$$

Perfluorocarbanions react in several ways depending on conditions:

1. proton abstraction

$$R_f^- + HB \text{ (or } H^+) \longrightarrow R_fH + B^-$$

2. β-fluorine elimination

$$R_fCF_2CF_2^- \longrightarrow R_fCF{=}CF_2 + F^-$$

3. α-fluorine elimination

$$R_fCF_2^- \longrightarrow [R\ddot{C}F] + F^-$$

$[R\ddot{C}F]$ → reactions of carbene (see Section 6-1B)

4. Attack on, or addition to, a nucleophilic center

$$R_f^- + \overset{\delta+}{R_2C}{=}\overset{\delta-}{X} \longrightarrow R_2\underset{R_f}{\underset{|}{C}}X^-$$

$$R_2\underset{R_f}{\underset{|}{C}}X^- \xrightarrow{H^+} R_2\underset{R_f}{\underset{|}{\overset{H}{\overset{|}{C}}}}X$$

$$R_2\underset{R_f}{\underset{|}{C}}X^- \xrightarrow{-R} R\overset{X}{\overset{\|}{C}}R_f$$

In this section, we only illustrate, using selected examples, the utility of fluorocarbanions as intermediates in adding fluorinated units. The literature in this area is again much too extensive for a complete review. For background reading on carbanions, Cram[312] has a good review which includes some discussion of fluorinated carbanions.

a. *Decarboxylation of alkali metal salts of fluorocarboxylic acids* leads to products resulting from proton addition, or α or β elimination of fluoride ion. This reaction is particularly useful in preparing terminal fluoroolefins[313, 314] and in generating difluorocarbene[233–239] (see discussion on pages 257–258).

$$CF_3CF_2CF_2CO_2Na \xrightarrow{245-253^\circ} NaF + CO_2 + CF_3CF{=}CF_2 \quad (99\%) \qquad (313)$$

The rates of decarboxylation of trihaloacetates, X_3CCO_2Na, where X is Br, Cl, F, are 10^6, 10^4, and 1, respectively.[315] Since the decarboxylation was shown to be first order and gave only CX_3H and CO_2 in over 90% yield, the stability of the intermediate fluorocarbanion was inferred to be $CBr_3^- > CCl_3^- > CF_3^-$. This relatively large difference in stability is not in the order expected from electronegativity and can be explained by (*1*) d-orbital participation by Br and Cl (not energetically reasonable for F), (*2*) relief of steric repulsions for the larger bromine or chlorine atoms, or (*3*) back donation of charge from unshared p-electrons of halogen (predicted poorest for fluorine because of polarizability, which for fluorine, chlorine, and bromine is 0.86, 3.1, and 4.2 Å^3, respectively; however, the short C–F bond and better overlap of 2p with 2pπ tends to increase back donation for fluorine relative to other halogens).

The β-fluoride elimination does not necessarily involve a carbanion intermediate but could be a concerted process:

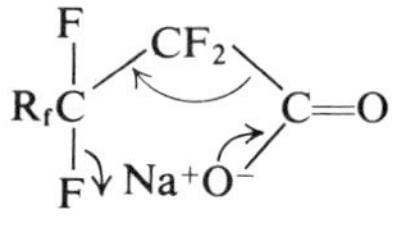

F
R_fC — CF_2 — C=O
F Na^+O^-

A simple mechanism study is needed.

b. *Fluorocarbon metallic compounds,* particularly of magnesium or lithium, have recently become useful synthetic intermediates for introducing fluorinated alkyl, alkenyl, or aryl units. Although a free carbanion is probably not formed in most reactions, carbanion character must develop during reaction so that this area merits inclusion in this section. Several good reviews on fluoroorganometallic

[312] D. J. Cram, *Fundamentals of Carbanion Chemistry,* Academic, New York, 1965.

[313] L. J. Hals, T. S. Reid, and G. H. Smith, Jr., *J. Am. Chem. Soc.,* **73,** 4054 (1951).

[314] I. Auerbach, F. H. Verhoek, and A. L. Henne, *J. Am. Chem. Soc.,* **72,** 299 (1950).

[315] B. R. Brown, *Quart. Rev. (London),* **5,** 131 (1951). See also Section 3-1A (p. 20 and Ref. 3-20).

chemistry are available,[38–40, 316, 317] but a definitive discussion devoted to the synthetic utility of these reagents is lacking.

Because of the extensive nature of the field, we will comment only briefly on the general classes of fluoroorganometallic reagents that can be prepared and the general behavior expected. Specific examples, selected particularly from recent literature to illustrate synthetic utility, are listed in Table 6-4.

The organometallic compounds as represented by the general formula $R_nM(X)$ vary in chemical behavior with the nature of the substituent R.

1. Alkyl or aryl organometallics with fluorine remote from the carbon metal bond react as normal alkyl or aryl organometallic compounds.

2. Perfluoroalkyl organometallics are usually stable only at low temperature (−80°C) and tend to give α and β elimination; synthetic utility is limited by stability.

3. Perfluoroalkenyl and alkynyl organometallics are stable, useful derivatives for introducing a perfluoroalkenyl (particularly trifluorovinyl) or alkynyl group.

4. Perfluoroaryl organometallics are readily prepared for a large variety of metals and give reactions typical of aryl organometallic intermediates, but they can decompose readily by β elimination to give benzyne and are susceptible to side reactions of nucleophilic substitution.

For synthetic utility, the most common metal (M) is Li from Group I and Mg from Group II. The common reactions are additions to a carbonyl function to give alcohols (or to related unsaturated center C=X)

$$R_fM(X) + R\overset{\overset{\displaystyle O}{\|}}{C}R' \longrightarrow \left[R_f\overset{\overset{\displaystyle R}{|}}{\underset{\underset{\displaystyle R'}{|}}{C}}\text{—}O^- \; M^+(X) \right] \xrightarrow{H_3O^+} R_f\overset{\overset{\displaystyle R}{|}}{\underset{\underset{\displaystyle R'}{|}}{C}}\text{—OH}$$

or coupling with a reactive halide, usually on a heteroatom Z such as P or Si.

$$R_fM(X) + YZR \longrightarrow R_fZR + M(X)Y$$

[316] (a) J. J. Lagowski, *Quart. Rev. (London)*, **13**, 233 (1959). (b) P. M. Treichel and F. G. A. Stone, *Advan. Organometal. Chem.*, **1**, 143 (1964).

[317] (a) D. Seyferth, *Progr. Inorg. Chem.*, **3**, 129 (1962). (b) C. Tamborski, *Trans. N.Y. Acad. Sci.*, **28**, 601 (1966). (c) M. D. Rausch, *ibid.*, 611. (d) R. D. Chambers and T. Chivers, *Organometal. Chem. Rev.*, **1**, 279 (1966).

Table 6-4. Fluorinated Organometallic

Reagent	Synthesis
A. ALKYL AND ARYL ORGANOMETALLIC REAGENTS WITH FLUORINE SUBSTITUTION REMOTE FROM CARBON-METAL BOND	
$F(CH_2)_nMgX$ $n = 6$ to 10	$F(CH_2)_nCl$ + BuMgCl in ether at room temp. or $F(CH_2)_nX$ + Mg in ether (where X is Cl or Br)
$CF_3CH_2CH_2MgCl$	No details given
$CF_3SO_2CH_2MgX$ $CF_3SO_2CH_2Li$	$CF_3SO_2CH_3$ + RMgX in ether (or RLi)
p-FC_6H_4MgI	p-FC_6H_4I + Mg in ether
2,4-$F_2C_6H_3M$, where M is Li or MgBr	$C_6H_3F_2X$ + Mg in ether (reflux) or BuLi exchange (−78°). X is I or Br
BrMg–C_6H_4–$(CF_2)_3$–C_6H_4–MgBr	From dibromide with Mg in THF
$(F_5S)C_6H_4MgBr$	$F_5SC_6H_4Br$ + Mg (+ CH_3I) in ether (reflux)
FC_6H_4MgBr	C_6H_4FBr + Mg in ether (reflux)

Reagents in Synthesis

Reactions	Comments
$+ CO_2 \rightarrow F(CH_2)_nCO_2H$ (63-73%), by-product $HO_2C(CH_2)_nCO_2H$ (12-18%)	Optimum conditions were determined and effects of halogen and chain length on the course of the reactions were studied. Shorter chain fluoroalkyl chlorides did not work. Ref. 318a
$+ (C_2H_5O)_2POH \rightarrow (CF_3CH_2CH_2)_2POMgCl \xrightarrow{S,\ HCl} (CF_3CH_2CH_2)_2PSOH$	Studying effects of fluorine substitution on properties of phosphorus derivatives. Ref. 318b
$+ CO_2 \rightarrow CF_3SO_2CH_2CO_2H$ (37%), $+ RCH{=}O \rightarrow CF_3SO_2CHCH(OH)R$ (30 to 40%)	The strong electron withdrawing effect of the CF_3 group makes the α-H more acidic than for normal sulfones. Ref. 318c
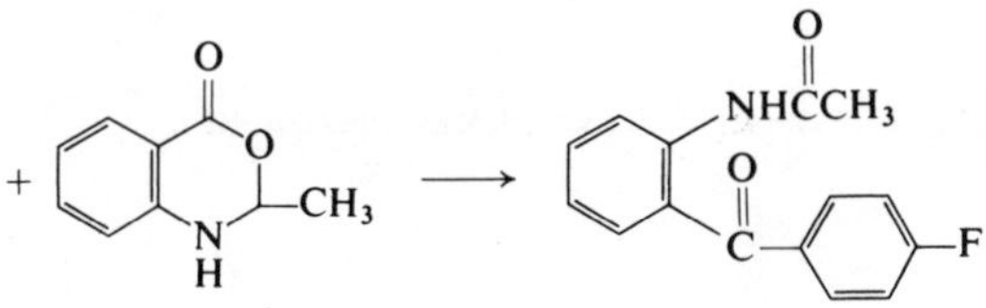 + 2-methyl-1,2-dihydro-4H-3,1-benzoxazin-4-one $\rightarrow$ $2\text{-}(CH_3C(O)NH)C_6H_4C(O)C_6H_4\text{-}F$ (57%)	Used in synthesis of fluorinated derivatives of fluorene for cancer studies. Ref. 318d
+ Acetaldehyde $\rightarrow F_2C_6H_3CH(OH)CH_3$ (39-97%)	Used in synthesis of fluorinated styrenes. Ref. 318e
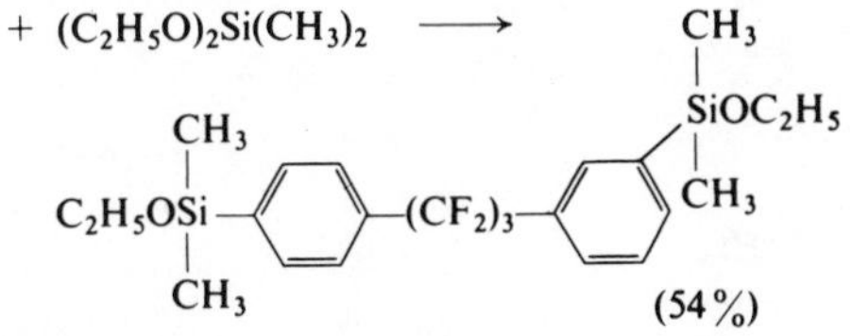 $+ (C_2H_5O)_2Si(CH_3)_2 \rightarrow C_2H_5OSi(CH_3)_2\text{-}C_6H_4\text{-}(CF_2)_3\text{-}C_6H_4\text{-}Si(CH_3)_2OC_2H_5$ (54%)	Used in synthesis of unusual *meta-*, *para-* cyclophanes. Ref. 318f
$+ CO_2 \rightarrow F_5SC_6H_4CO_2H$ (55%) + Acetaldehyde $\rightarrow F_5SC_6H_4CH(OH)CH_3$ (44%)	Noted Grignard formed only when CH_3I used. Ref. 318g
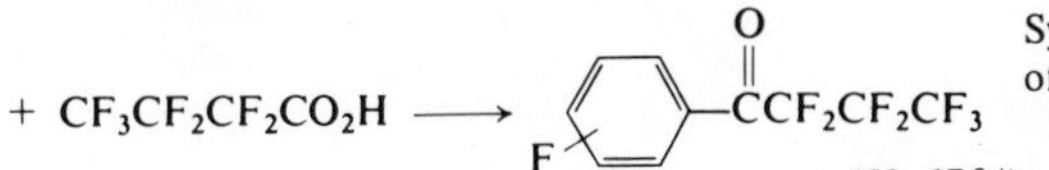 $+ CF_3CF_2CF_2CO_2H \rightarrow F\text{-}C_6H_4\text{-}C(O)CF_2CF_2CF_3$ (50–67%)	Synthesis used in study of electronic properties of fluoroalkyl groups. Ref. 318h

Table 6-4

Reagent	Synthesis
$(CF_3)_2CF-C_6H_4-MgBr$	$(CF_3)_2CF-C_6H_4-Br$ + Mg in ether (reflux)
$CX_2{=}CXC_6H_4Li$ X = Cl or F	$CX_2{=}CXC_6H_4Br + BuLi \xrightarrow{-70^\circ}$
$CF_3C_6H_4Cu$ FC_6H_4Cu	$ArMgX + CuBr \xrightarrow{\text{ether}}$ (add dioxane to precipitate magnesium salts)
B. PERFLUOROALKYL ORGANOMETALLIC REAGENTS	
CF_3MgI	$CF_3I + Mg \xrightarrow[-30\text{ to }-10^\circ]{\text{ethers}}$
$CF_3CF_2CF_2MgX$	$CF_3CF_2CF_2I + Mg \xrightarrow[<0^\circ]{\text{ether}}$ (or C_6H_5MgX, ether 0 to 10°)
R_fMgX R_f is $CF_3CF_2CF_2$ or CF_3 X is I or Br	$R_fX + Mg \xrightarrow[-20\text{ to }25^\circ]{\text{Hg}}$
$(CF_3)_2CFM$ M is Li, MgX, ZnX	$(CF_3)_2CFI \xrightarrow{\text{RMgX}}$ $-78^\circ \longrightarrow$

(cont.)

Reactions	Comments
$+ CO_2 \longrightarrow (CF_3)_2CF\text{-}C_6H_4\text{-}CO_2H$ (61–70%) 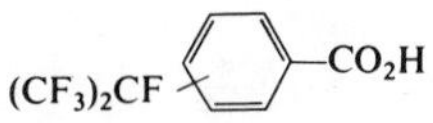	Ref. 318h
$+ CO_2 \rightarrow CX_2{=}CXC_6H_4CO_2H$, $+ CH_3\overset{O}{\overset{\parallel}{C}}H \xrightarrow{(H+)?} CX_2{=}CXC_6H_4CH{=}CH_2$, $+ (C_2H_5)_3SnCl \rightarrow CX_2{=}CXC_6H_4Sn(C_2H_5)_3$	If organolithium reagent was not kept below −70°, vinyl halogens were attacked and poly-condensation products formed. Ref. 318i
$+ R\overset{O}{\overset{\parallel}{C}}X \rightarrow Ar\overset{O}{\overset{\parallel}{C}}R$, $+ RX \rightarrow ArR$	Fluorinated substituents made aryl copper more stable and soluble. Ref. 318j
$+ CO_2 \rightarrow CF_3CO_2H$ (39%) $+ CH_3CN \rightarrow CH_3\overset{O}{\overset{\parallel}{C}}CF_3$ (38%), $+ R\overset{O}{\overset{\parallel}{C}}Cl \rightarrow R\overset{O}{\overset{\parallel}{C}}CF_3$ (15-59%), + Ethylene oxide $\rightarrow CF_3CH_2CH_2OH$ (57%)	Optimum conditions explored for synthesis and reactions of Grignard reagent. Ref. 319a
$+ CO_2 \rightarrow CF_3CF_2CF_2CO_2H$ (75-80%), + aldehydes or ketones to give corresponding alcohols such as 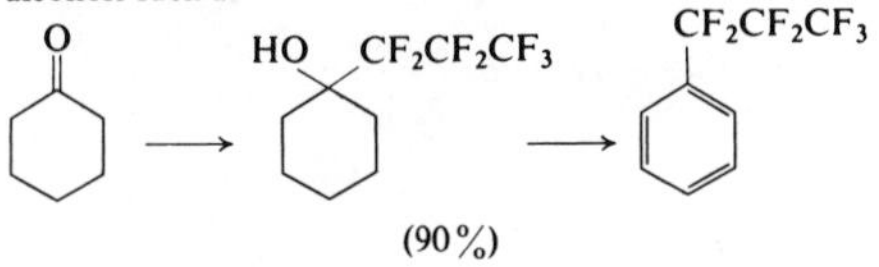 cyclohexanone $\longrightarrow$ 1-$CF_2CF_2CF_3$-cyclohexanol (HO, $CF_2CF_2CF_3$) (90%) $\longrightarrow$ 1-$CF_2CF_2CF_3$-cyclohexene	Conditions of preparation and reactions studied to find optimum. Noted facile decomposition to $CF_3CF_2CF_2H$ or $CF_3CF{=}CF_2$. Ref. 319b, c, d
$+ CH_3\overset{O}{\overset{\parallel}{C}}CH_3 \rightarrow R_f\overset{OH}{\overset{\vert}{C}}(CH_3)_2$ (7-37%)	Varied conditions to get optimum. Noted reaction of CF_3I in presence of cyclohexene gave 1-iodo-2(trifluoromethyl)cyclohexane in a radical type addition instead of 7,7-difluoronorcarane. Ref. 319e
$\xrightarrow{\text{room temp.}} CF_3CF{=}CF_2$ (70-74%), $\xrightarrow{R\overset{O}{\overset{\parallel}{C}}R'} (CF_3)_2CF\overset{OH}{\overset{\vert}{C}}R$ with R′ on the carbinol carbon (13-53%), $\xrightarrow[-50°]{(CH_3)_3SiCl} (CF_3)_2CFSi(CH_3)_3$ (18%)	Reagents did not hydrolyze to $(CF_3)_2CFH$. Ref. 319f

Table 6-4

Reagent	Synthesis
$CF_3CF_2CF_2Li$, (CF_3Li)	$CF_3CF_2CF_2I + RLi \xrightarrow[\text{ether}]{-50^\circ}$
$CF_3CF_2CF_2Li$	$CF_3CF_2CF_2I$ + Li (2% Na), ether, -40°
$LiCF_2CF_2CF_2CF_2Li$	$I(CF_2)_4I$ + *n*-BuLi $\xrightarrow[\text{ether}]{-80^\circ}$
$(CF_3)_3CCu$	$(CF_3)_3CBr + ArCu \rightarrow$ characterized as dioxane complex
$C_7F_{15}Cu$?, R_fCu	$C_7F_{15}I$ + copper bronze $\xrightarrow{\text{pyridine}}$
Li, F_2, F_2, F–F, F_2, F_2, F	H, F_2, F_2, F–F, F_2, F_2, F $+ CH_3Li \xrightarrow[-40^\circ]{\text{ether}}$

(cont.)

Reactions	Comments
$+ RCHO \rightarrow RCH(OH)R_f$ + by-products (up to 50%)	CF_3I reacted with RLi but could not obtain any reaction products of CF_3Li. R_fLi shown to be less reactive than RLi. By-products $CF_3CF_2CF_2H$ and $CF_3CF{=}CF_2$ are also formed. Scope and limitations of the reactions are discussed, particularly reduction of perfluorinated carbonyl compounds. Ref. 319g, h
+ Benzaldehyde $\rightarrow$ $CF_3CF_2CF_2CH(OH)C_6H_5$ (–%), $+ (CH_3)_2SiCl_2 \rightarrow (CF_3CF_2CF_2)_2Si(CH_3)_2$ (20%), + (catechol)BCl $\xrightarrow{-50°}$ (catechol)$BCF_2CF_2CF_3$ (30%)	Reactant present when R_fLi formed; THF and glyme are poorer solvents than ether because of H abstraction. Polymer products and $CF_3CF{=}CF_2$ also noted as by-products with amount dependent on temperature. Ref. 319i, j, k
+ Acetaldehyde $\rightarrow$ $[CH_3CH(OH)CF_2CF_2]_2$ (low yield)	Stability limited at –80° but use of simultaneous or alternating addition techniques found best method. Ref. 3191
$+ RX \rightarrow RF + CuX + [CF_2{=}C(CF_3)_2\,?]$	Part of general study on organocopper chemistry. Ref. 318j
$+ ICH{=}CHCl \rightarrow C_7F_{15}CH{=}CHCl$ (50-96%), series of RX where R is alkenyl or activated aryl, + ArI or ArBr $\rightarrow R_fAr$	Fluoroalkyl copper compound proposed as intermediate but not isolated. Ref. 319m, n
$+ CH_3COCH_3 \rightarrow R_fC(OH)(CH_3)_2$ (44%), $+ CH_3Br \rightarrow R_fCH_3$ (50%), + furan $\xrightarrow{\text{warm}}$ (furan adduct of perfluorobicyclo[2.2.2]octene), $+ D_2O \longrightarrow R_fD$ (60%), R_f is perfluorobicyclo[2.2.2]octyl (bridgehead F)	Loss of LiF to give perfluorobicyclo[2.2.2]oct-2-ene occurs rapidly at room temperature. This olefin was trapped with furan or added LiX which in turn lost LiF: (X-substituted bicyclic carbanion, Li^+) $\longrightarrow$ (X-substituted perfluorobicyclo[2.2.2]octene) (X = Br or I). Ref. 319o

Table 6-4

Reagent	Synthesis
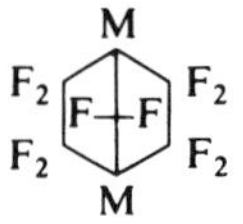 Octafluorocyclohexane ring (F_2 at four positions, F–F bridge) with M at positions 1 and 4	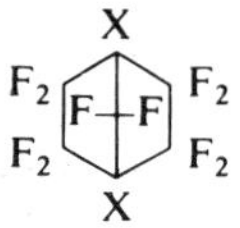 Same ring with X at positions 1 and 4: (a) X = H, CH_3Li, –55°, ether; (b) X = Br or I, Mg
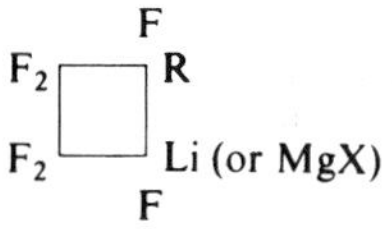 RCF_2CF_2Li	Fluoroolefins + RLi (where R is alkyl or aryl) or RMgX at –80° to 0° in ether
Cyclobutane ring (F_2, F_2) with F, R on one carbon and F, Li (or MgX) on the other (and related compounds)	
$R_f^-K^+$ or $R_f^-Cs^+$	Perfluoroolefin + KF or CsF

C. PERFLUOROVINYL AND ACETYLENIC ORGANOMETALLIC REAGENTS (for review of early work, see Ref. 317a)

Reagent	Synthesis
$CF_2{=}CFLi$	(1) $(CF_2{=}CF)_3SnC_6H_5 \xrightarrow[Et_2O,\ -40^\circ]{3C_6H_5Li}$ (2) $CF_2{=}CFBr + RLi \xrightarrow[-78^\circ]{ether}$ (3) $CF_2{=}CFH + C_4H_9Li \xrightarrow{ether}$
$CF_2{=}CXLi$, X = H, Cl, Br	$CF_2{=}CXBr + BuLi \rightarrow$ or $CF_2{=}CXH$
$CF_3C(Li){=}CH_2$	$CF_3C(Br){=}CH_2 + BuLi \xrightarrow[ether]{-90^\circ}$

(cont.)

Reactions	Comments
(a) + Br_2 or I_2 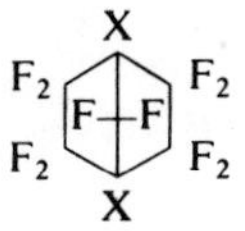 ⟶ [bicyclic structure: F_2, F_2, F_2, F_2, F–F, X at bridgeheads] X = I or Br (b) H_3O^+ ⟶ X = H	Some other bridgehead organometallic reagents also reported. Ref. 319p
$RCF{=}CF_2$ (20–60%) 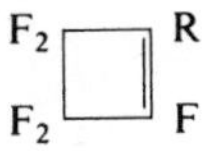 [cyclobutene: F_2, F_2, R, F]	R_fLi (or R_fMgX) postulated as intermediate but has not been trapped nor utilized in synthesis other than to eliminate LiF. Ref. 319q, 347
Nucleophilic substitution or addition reactions of R_f^-	Intermediate R_f^- proposed to explain chemistry but organometallic agent never isolated. Ref. 336 to 342
+ CO_2 → $CF_2{=}CFCO_2H$ (37%), + $(CH_3)_3SnBr$ → $(CH_3)_3SnCF{=}CF_2$ (64%), + $(CH_3)_3SiCl$ → $CF_2{=}CFSi(CH_3)_3$ (65%), 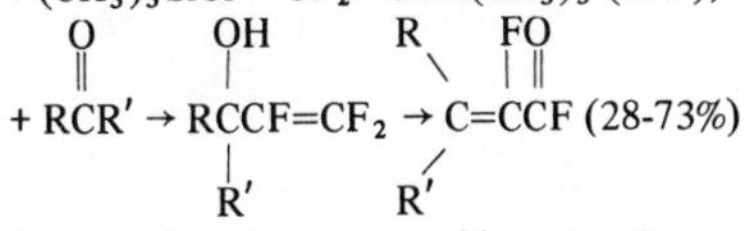 + RC(=O)R′ → RC(OH)(R′)CF=CF₂ → (R)(R′)C=CF–C(=O)F (28-73%) (rearrangement occurs readily unless R is electronegative group)	Reagent decomposes at 0° or above. Suggested decomposition gives FC≡CF but no positive evidence. The proton-exchange method (3) has the advantage over the halogen exchange that the process is not reversible and the starting olefins are usually more readily available. A large additional number of reactions are given in references. Ref. 320a-e
Reaction with series of carbonyl compounds described. Results similar to those obtained for corresponding Grignard reagents	Initial alcohol product rearranges readily unless contains electronegative substituents. Ref. 320d
$\xrightarrow[\text{ether}]{-78^\circ}$ $CH_2{=}C{=}CF_2$ + LiF 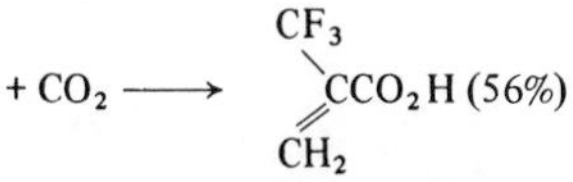 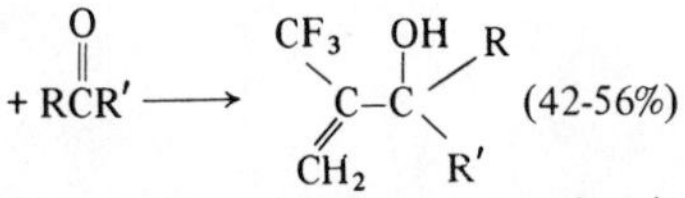 + CO_2 ⟶ $CH_2{=}C(CF_3)CO_2H$ (56%) + RC(=O)R′ ⟶ $CH_2{=}C(CF_3)$–C(OH)(R)(R′) (42-56%) (dehydrates with P_2O_5 to a butadiene)	Side reactions and some unsuccessful reactions also described. Ref. 320f

Table 6-4

Reagent	Synthesis
F_2 Li / F_2 F	F_2 H / F_2 F $+ CH_3Li \xrightarrow[-70^\circ]{\text{ether}}$
F_2 / F_2 Li / F_2 F / F_2	F_2 / F_2 H / F_2 F / F_2 $+ CH_3Li \xrightarrow[\text{ether}]{-70^\circ}$
$CF_2{=}CXMgI$, X = Cl or Br	$CF_2{=}CXI + Mg \xrightarrow[\text{reflux}]{\text{ether}}$
F_2 MgBr / F_2 X	F_2 Br / F_2 X $+ C_2H_5MgBr \xrightarrow{\text{ether}}$
$CF_2{=}CFMgX$, X = Br or I	$CF_2{=}CFX + Mg \xrightarrow[\text{or THF, } -20^\circ]{\text{ether, } 0^\circ}$

(cont.)

Reactions	Comments
+ $CH_3CHO \xrightarrow{-70°}$ 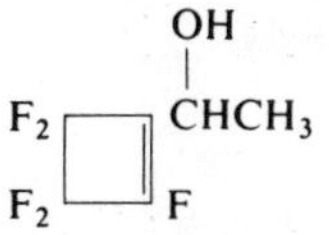 [cyclobutene: F_2, F_2, $CH(OH)CH_3$, F] (23%)	Lithium reagent decomposes between −10 and 15° to give LiF and an intractable polymer. Ref. 320g
+ CO_2 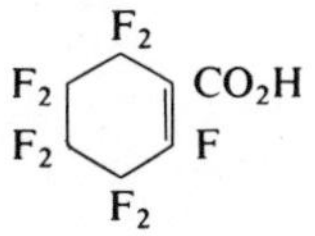 ⟶ [cyclohexene: F_2, F_2, F_2, F_2, CO_2H, F] (77%)	By-products (polymer) result from further loss of F^- at higher temperatures. Olefin intermediate in decomposition was trapped by furan
+ I_2 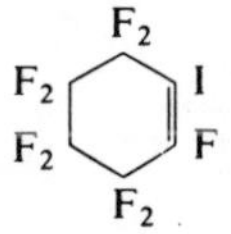 ⟶ [cyclohexene: F_2, F_2, F_2, F_2, I, F] (59%)	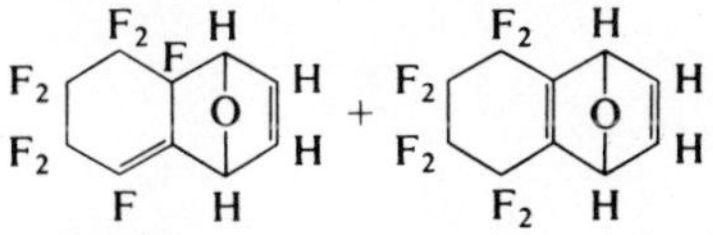 [furan adducts: F_2, F_2, F_2, F, F, H, H, H, H, O] + [F_2, F_2, F_2, F_2, H, H, H, H, O] Ref. 320g
+ $H_2O \rightarrow CF_2{=}CXH$, + $CH_2O \rightarrow CF_2{=}CXCH_2OH$ (56%)	No products were obtained from reaction with CO_2 and other reagents. Ref. 320h
+ H_2O 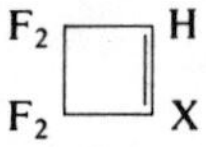 ⟶ [cyclobutene: F_2, F_2, H, X] (∼ 70%) (for X = F obtained 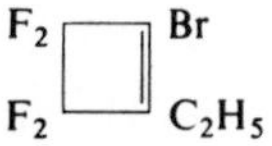 [cyclobutene: F_2, F_2, Br, C_2H_5] as by-product)	Ref. 320i
+ $H_2O \rightarrow CF_2{=}CFH$ (40-69%), + $CO_2 \rightarrow CF_2{=}CFCO_2H$ (38%), 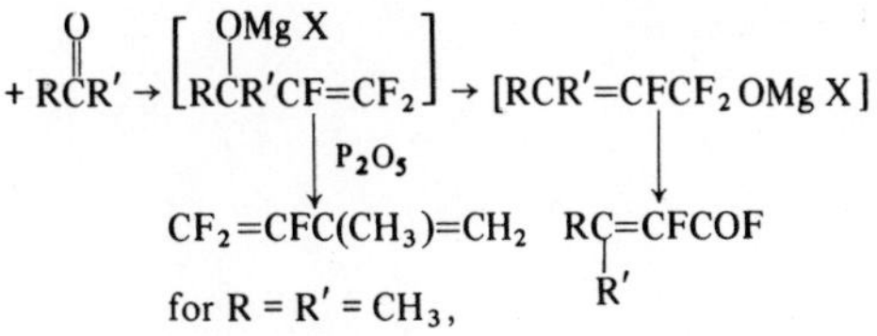 + $RC(O)R' \rightarrow [RC(OMgX)R'CF{=}CF_2] \rightarrow [RCR'{=}CFCF_2OMgX]$ $[RC(OMgX)R'CF{=}CF_2] \xrightarrow{P_2O_5} CF_2{=}CFC(CH_3){=}CH_2$ for R = R′ = CH_3, $[RCR'{=}CFCF_2OMgX] \rightarrow RC(R'){=}CFCOF$ + $(CH_3)_2SiCl_2 \rightarrow (CH_3)_2Si(CF{=}CF_2)_2$ (65%), + $R_3SiCl \rightarrow R_3Si(CF{=}CF_2)$, + $MX_3 \rightarrow (CF_2{=}CF)_3M$ (M = P, As, Sb), + 19-nor-4-androstene-3,17-dione (as 3-dioxolane), → 19-nor-7-hydroxy-17-perfluorovinyl-4-androstene-3-one (–%),	$CF_2{=}CFBr$ preferred because more available. Noted unusual properties of perfluorovinyl metal derivative vs. perfluoroalkyl or vinyl. A large selection of perfluorovinyl organometallics also prepared by this method. Ref. 320b, c, d, j, k, l, m, n, o

Table 6-4

Reagent	Synthesis
$(CF_2{=}CF)_2Hg$	$CF_2{=}CFMgI + HgCl_2 \xrightarrow[\text{ether}]{-10°}$
$CF_3C{\equiv}CMgBr$	$CF_3C{\equiv}CH + C_2H_5MgBr \xrightarrow{\text{THF}}$
$CF_3C{\equiv}CLi$	$CF_3C{\equiv}CH + BuLi \xrightarrow[\text{ether}]{-78 \text{ to } -40°}$

D. PERFLUOROARYL ORGANOMETALLIC REAGENTS (See ref. 317c and d; see also Sections 6-2B and 6-2C)

Reagent	Synthesis
C_6F_5Li	$C_6F_5Br + n\text{-BuLi} \xrightarrow[-78°]{\text{ether}}$ $C_6F_5Br + \text{Li (amalgam)} \xrightarrow[0°]{\text{ether}}$ $C_6F_5H + n\text{-BuLi} \xrightarrow[-70°]{\text{ether}}$
C_6F_5–C_6F_4Li (ring substituents: F F, F, Li F)	C_6F_5Li + C_6F_4 (tetrafluorobenzyne) (decomposition of C_6F_5Li)

(cont.)

Reactions	Comments
$+ I_2 \rightarrow CF_2{=}CFI$ (100%) $+ AlH_3 \cdot N(CH_3)_3 \rightarrow (CF_2{=}CF)_3Al \cdot N(CH_3)_3 + Hg + H_2$	Ref. 320g, p, q
+ 1,4-dihydroestrone-3-methyl ether 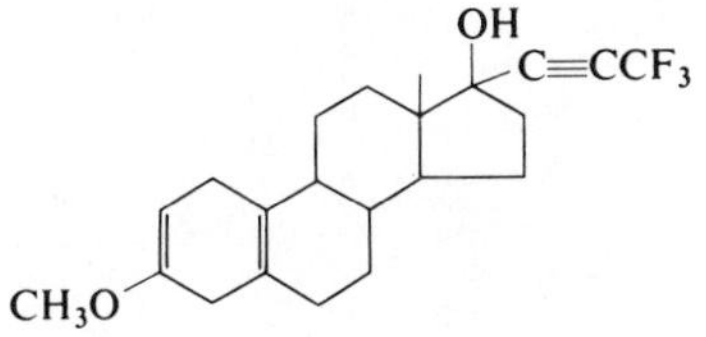 → (structure: OH, $C{\equiv}CCF_3$, CH_3O)	Ref. 320o
$+ RC(O)R' \rightarrow RC(OH)(R')C{\equiv}CCF_3$ (25-70%), $+ CH_3C(O)Cl \rightarrow CH_3C(O)C{\equiv}CCF_3 \rightarrow CH_3C(OH)(C{\equiv}CCF_3)_2$, (55%) $+ (CH_3CH_2)_3SiCl \rightarrow (CH_3CH_2)_3SiC{\equiv}CCF_3$, (81%)	Attempted reaction with hexafluorobenzene, benzonitrile, and fluoroolefins was unsuccessful; $CF_3C{\equiv}CLi$ stable in solution to 0°. Ref. 320f
$+ CO_2 \rightarrow C_6F_5CO_2H$ (91%), $+ RC(O)H \rightarrow C_6F_5CH(OH)R$ $+ BCl_3 \rightarrow (C_6F_5)_3B$, $+ GeCl_4 \rightarrow (C_6F_5)_4Ge$, $+ C_6H_5SiCl_3 \rightarrow C_6H_5Si(C_6F_5)_3$ (50%)	Typical reactions of an aryl lithium reagent, but decomposes over 0° to give tetrafluorobenzyne. Nucleophilic displacements by $C_6F_5{}^-$ ion also observed. Many additional examples given in references and reviews. Note rate of preparation and reaction highly dependent on solvent. Ref. 321a
$+ C_6F_5X \rightarrow C_6F_5Li + 2\text{-}XC_6F_4C_6F_5$, $+ Cl_2Sn(CH_3)_2 \rightarrow (2\text{-}C_6F_4C_6F_5)_2Sn(CH_3)_2$	Also can lose LiF to give the benzyne 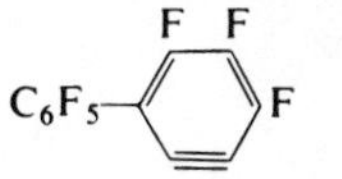 (structure: C_6F_5, F, F, F) Ref. 321b

Table 6-4

Reagent	Synthesis
 F, F, Li, F, Br, F (benzene ring)	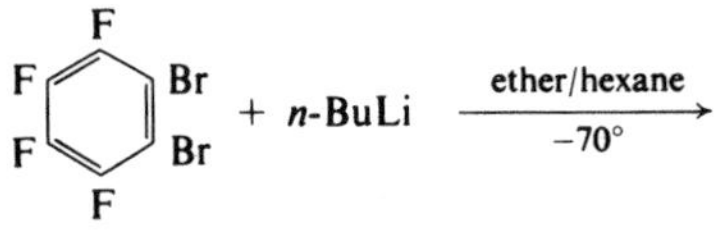 F, F, Br, F, Br, F (benzene ring) + n-BuLi $\xrightarrow[-70^\circ]{\text{ether/hexane}}$
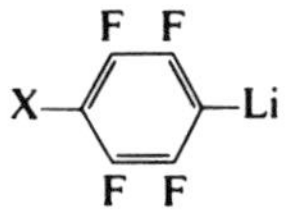 F F, X—(ring)—Li, F F X = F, CF_3, CH_3, H, p-C_6H_4, Br, CO_2H, OH, NH_2	XC_6F_4H + n-BuLi → (used extra equivalents of BuLi if active H present)
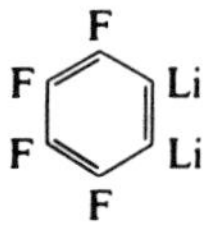 F, F, Li, F, Li, F (benzene ring)	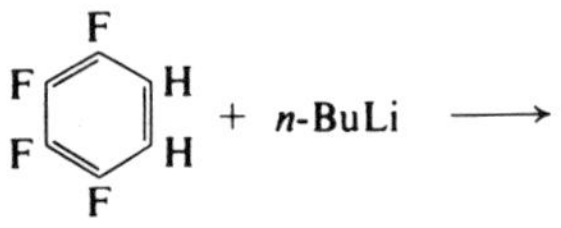 F, F, H, F, H, F (benzene ring) + n-BuLi ⟶
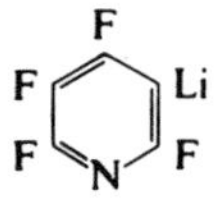 F, F, Li, F, N, F (pyridine ring)	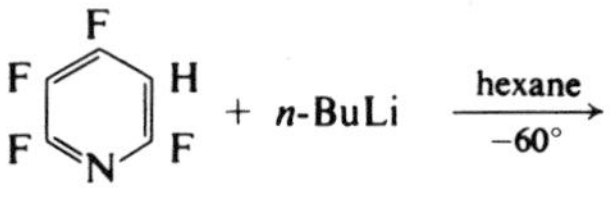 F, F, H, F, N, F (pyridine ring) + n-BuLi $\xrightarrow[-60^\circ]{\text{hexane}}$
 M, F, F, F, N, F (pyridine ring) M = Li, MgX	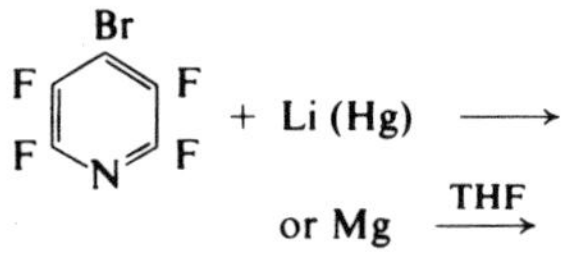 Br, F, F, F, N, F (pyridine ring) + Li (Hg) ⟶ or Mg $\xrightarrow{\text{THF}}$

(cont.)

Reactions	Comments
+ H_3O^+ ⟶ (F, F, F, F, H, Br benzene)	Showed only mono-Li compound formed. Carbonation gave a mixture of acids, including diacid, but concluded mixture arose from secondary reactions. Benzyne formation studied, only product from LiF elimination trapped by furan but reaction complex because of competing LiX addition. Ref. 321c
+ $(CH_3)_3SiCl$ ⟶ (F, F, F, F, $Si(CH_3)_3$, Br benzene) (99%)	
+ CO_2 ⟶ X–C_6F_4–CO_2H,	Showed lithium reagent gave much better yields than Grignard (because of greater nucleophilicity). If X = H could metalate in second step to make disubstituted product. Ref. 321d
+ $(CF_3)_2CO$ ⟶ X–C_6F_4–$C(CF_3)_2OH$ (61–91%)	
+ CO_2 ⟶ (F, F, F, F, CO_2H, CO_2H benzene) (9%)	Monolithio reagent also made and carboxylated. THF was shown to be prepared over ether as solvent. Bis-magnesium reagent was not obtained. Ref. 321e
+ CO_2 ⟶ (F, F, F, F, CO_2H pyridine) (62%)	Some trifluorinated and dicarboxylic acid analogs of the pyridine also prepared. Ref. 321f
+ CO_2 ⟶ N(C_5F_4)–CO_2H (59%)	Decomposed in furan, no evidence for trifluoropyridyne. Ref. 321g
+ RC(=O)R′ ⟶ N(C_5F_4)–C(R)(R′)OH	
+ $HgCl_2$ ⟶ (N(C_5F_4)–$)_2$Hg (72%)	

Table 6-4

Reagent	Synthesis
C_6F_5MgX, X = Cl, Br, I	$C_6F_5I + Mg \xrightarrow[\text{or THF}]{\text{ether}}$ or C_6F_5Br, $C_6F_5Cl + Mg \xrightarrow[\text{ether}]{BrCH_2CH_2Br}$
BrMg–C_6F_4–MgBr (2,3,5,6-tetrafluoro-1,4-phenylene)	H–C_6F_4–H + $3C_2H_5MgBr \xrightarrow{\text{THF}}$
2,2′-dilithio-3,3′,4,4′,5,5′,6,6′-octafluorobiphenyl	2,2′-dihalo-octafluorobiphenyl (F, X) + BuLi ⟶ X = Br or I
C_6F_5Cu	$C_6F_5MgX + CuBr \rightarrow$
$C_6F_5HgCH_3$	$CH_3HgX + C_6F_5MgBr \rightarrow$

(cont.)

Reactions	Comments
$+ CO_2 \rightarrow C_6F_5CO_2H$ $+ RC(=O)R' \rightarrow C_6F_5C(OH)(R)R'$ $+ SnCl_4 \rightarrow (C_6F_5)_4Sn$, $+ PCl_3 \rightarrow (C_6F_5)_nPCl_{3-n}$ (n is 1,2 or 3)	Grignard forms very readily at low temperature in THF but decomposes on warming in contrast to ether solution. Solvent effects in reaction are very pronounced, particularly ether vs. THF. A large number of reactions, typical of ArMgX have been reported for C_6F_5MgX. Decomposition to give polyphenyl derivatives also reported. Ref. 321h
$+ CO_2 \longrightarrow HO_2C{-}C_6F_4{-}CO_2H$ (24%) 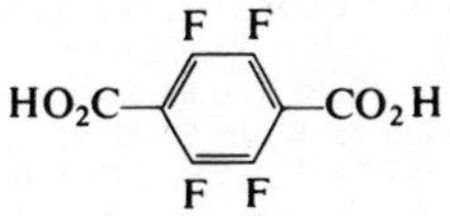	Ref. 321e
$+ H_3O^+ \longrightarrow$ 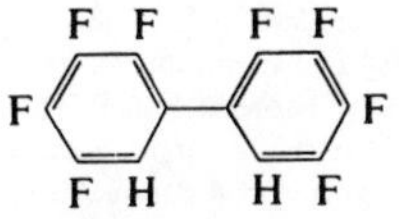 F F F F / F–(biphenyl)–F / F H H F	Starting 2,2′-dibromooctafluorobiphenyl easily prepared by thermal disproportionation of substituted polyphenyl derivative of titanium. Ref. 321i
$+ MCl_4 \longrightarrow$ 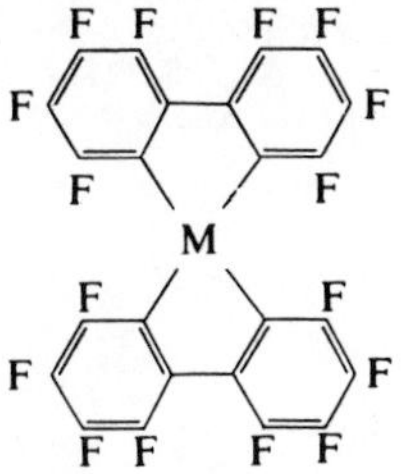 spiro M bis(octafluorobiphenyl-2,2′-diyl)	
$+ CO_2 \longrightarrow$ F F F F / F–(biphenyl)–F / F CO_2H HO_2C F 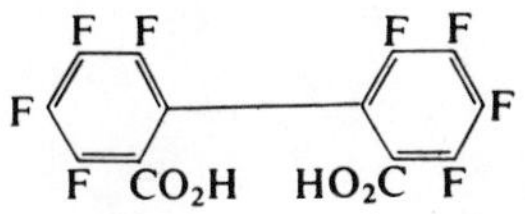	
$+ RC(=O)X \rightarrow C_6F_5C(=O)R$, $+ RX \rightarrow C_6F_5R$, $+ [O] \rightarrow (C_6F_5{-})_2$ or heat	C_6F_5Cu has high solubility in organic solvents and good stability for aryl copper compounds (stable to 200°). Ref. 318j
$+ AlBr_3 \xrightarrow[\text{ether}]{\text{petroleum}} C_6F_5AlBr_2$ or $(C_6F_5)_2AlBr$	Ref. 321j

Table 6-4

[318](a) F. L. M. Pattison and W. C. Howell, *J. Org. Chem.*, **21**, 879 (1956). (b) M. A. Larionova, A. L. Klebanskii, and V. A. Bartashev, *J. Gen. Chem. U.S.S.R. (English Transl.)*, **33**, 257 (1963). (c) L. M. Yagupol'skii, A. G. Panteleimonov, and V. V. Orda, *J. Gen. Chem. U.S.S.R. (English Transl.)*, **34**, 3498 (1964). (d) K. Suzuki, E. K. Weisburger, and J. H. Weisburger, *J. Org. Chem.*, **26**, 2239 (1961). (e) M. M. Nad', T. V. Talalaeva, G. V. Kazennikova, and K. A. Kocheshkov, *Bull. Acad. Sci. U.S.S.R., Div. Chem. Sci. (English Transl.)*, 58 (1959). (f) S. A. Fuqua and R. M. Silverstein, *J. Org. Chem.*, **29**, 395 (1964). (g) W. A. Sheppard, *J. Am. Chem. Soc.*, **84**, 3064 (1962). (h) W. A. Sheppard, *J. Am. Chem. Soc.*, **87**, 2410 (1965). (i) K. A. Kocheskov, E. M. Panov, and R. S. Sorokina, *Bull. Acad. Sci. U.S.S.R., Div. Chem. Sci. (English Transl.)*, 494 (1961). (j) A. Cairncross and W. A. Sheppard, *J. Am. Chem. Soc.*, **90**, 2186 (1968).

[319](a) R. N. Haszeldine, *J. Chem. Soc.*, 1273 (1954). (b) R. N. Haszeldine, *J. Chem. Soc.*, 3423 (1952). (c) O. R. Pierce, A. F. Meiners, and E. T. McBee, *J. Am. Chem. Soc.*, **75**, 2516 (1953). (d) E. T. McBee, C. W. Roberts, and A. F. Meiners, *J. Am. Chem. Soc.*, **79**, 335 (1957). (e) E. T. McBee, R. D. Battershell, and H. P. Braendlin, *J. Org. Chem.*, **28**, 1131 (1963). (f) R. D. Chambers, W. K. R. Musgrave, and J. Savory, *J. Chem. Soc.*, 1993 (1962). (g) O. R. Pierce, E. T. McBee, and G. F. Judd, *J. Am. Chem. Soc.*, **76**, 474 (1954). (h) E. T. McBee, C. W. Roberts, and S. G. Curtis, *J. Am. Chem. Soc.*, **77**, 6387 (1955). (i) J. A. Beel, H. C. Clark, and D. Whyman, *J. Chem. Soc.*, 4423 (1962). (j) H. C. Clark, J. T. Kwon, and D. Whyman, *Can. J. Chem.*, **41**, 2628 (1963). (k) T. Chivers, *Chem. Commun.*, 157 (1967). (l) P. Johncock, *J. Organometal. Chem.*, **6**, 433 (1966). (m) J. Burdon, P. L. Coe, C. R. Marsh, and J. C. Tatlow, *Chem. Commun.*, 1259 (1967). (n) I. M. White, J. Thrower, and V. C. R. McLoughlin, *4th Intern. Fluorine Symp., Estes Park, Colorado*, 1967. (o) S. F. Campbell, R. Stephens, and J. C. Tatlow, *Tetrahedron*, **21**, 2997 (1965). (p) S. F. Campbell, J. M. Leach, R. Stephens, and J. C. Tatlow, *Tetrahedron Letters* (No. 43), 4269 (1967). (q) J. D. Park and R. Fontanelli, *J. Org. Chem.*, **28**, 258 (1963).

[320](a) D. Seyferth, D. E. Welch, and G. Raab, *J. Am. Chem. Soc.*, **84**, 4266 (1962). (b) P. Tarrant, P. Johncock, and J. Savory, *J. Org. Chem.*, **28**, 839 (1963). (c) P. Tarrant and W. H. Oliver, *J. Org. Chem.*, **31**, 1143 (1966). (d) F. G. Drakesmith, R. D. Richardson, O. J. Stewart, and P. Tarrant, *J. Org. Chem.*, **33**, 286 (1968). (e) F. G. Drakesmith, O. J. Stewart, and P. Tarrant, *J. Org. Chem.*, **33**, 472 (1968). (f) F. G. Drakesmith, O. J. Stewart, and P. Tarrant, *ibid.*, 280. (g) S. F. Campbell, R. Stephens, and J. C. Tatlow, *Chem. Commun.*, 151 (1967). (h) J. D. Park, J. Abramo, M. Hein, D. N. Gray, and J. R. Lacher, *J. Org. Chem.*, **23**, 1661 (1958). (i) R. Sullivan, J. R. Lacher, and J. D. Park, *J. Org. Chem.*, **29**, 3664 (1964). (j) J. D. Park, R. J. Seffl, and J. R. Lacher, *J. Am. Chem. Soc.*, **78**, 59 (1956). (k) I. L. Knunyants, R. N. Sterlin, R. D. Iatsenko, and L. N. Pinkina, *Bull. Acad. Sci. U.S.S.R., Div. Chem. Sci. (English Transl.)*,

(cont.)

1297 (1958). (l) R. N. Sterlin, R. D. Yatsenko, L. N. Pinkina, and I. L. Knunyants, *Bull. Acad. Sci. U.S.S.R., Div. Chem. Sci.* (*English Transl.*), 1851 (1960). (m) H. D. Kaesz, S. L. Stafford, and F. G. A. Stone, *J. Am. Chem Soc.*, **81**, 6336 (1959). (n) D. Seyferth, K. Brandle, and G. Raab, *Angew. Chem.*, **72**, 77 (1960). (o) J. H. Fried, T. S. Bry, A. E. Oberster, R. E. Beyler, T. B. Windholz, J. Hannah, L. H. Sarett, and S. L. Steelman, *J. Am. Chem. Soc.*, **83**, 4663 (1961). (p) R. N. Sterlin, Li Wei-kang, and I. L. Knunyants, *Bull. Acad. Sci. U.S.S.R., Div. Chem. Sci.* (*English Transl.*), 1459 (1959). (q) B. Bartocha and A. J. Bilbo, *J. Am. Chem. Soc.*, **83**, 2202 (1961).

[321] (a) (1) D. E. Fenton, A. J. Park, D. Shaw, and A. G. Massey, *J. Organometal. Chem.*, **2**, 437 (1964); (2) P. L. Coe, R. Stephens, and J. C. Tatlow, *J. Chem. Soc.*, 3227 (1962); (3) R. J. Harper, Jr., E. J. Soloski, and C. Tamborski, *J. Org. Chem.*, **29**, 2385 (1964); (4) D. D. Callander, P. L. Coe, and J. C. Tatlow, *Chem. Commun.*, 143 (1966); (5) D. D. Callander, P. L. Coe, and J. C. Tatlow, *Tetrahedron*, **22**, 419 (1966); (6) A. G. Massey and A. J. Park, *J. Organometal. Chem.*, **2**, 245 (1964); (7) D. E. Fenton, A. G. Massey, and D. S. Urch, *J. Organometal. Chem.*, **6**, 352 (1966); (8) F. W. G. Fearon and H. Gilman, *J. Organometal. Chem.*, **10**, 409 (1967). (b) (1) D. E. Fenton and A. G. Massey, *Tetrahedron*, **21**, 3009 (1965); (2) A. J. Tomlinson and A. G. Massey, *J. Organometal. Chem.*, **8**, 321 (1967). (c) (1) C. Tamborski and E. J. Soloski, *J. Organometal. Chem.*, **10**, 385 (1967); (2) S. C. Cohen, D. E. Fenton, D. Shaw, and A. G. Massey, *J. Organometal. Chem.*, **8**, 1 (1967); (3) S. C. Cohen, D. E. Fenton, A. J. Tomlinson, and A. G. Massey, *J. Organometal. Chem.*, **6**, 301 (1966). (d) (1) C. Tamborski, W. H. Burton, and L. W. Breed, *J. Org. Chem.*, **31**, 4229 (1966); (2) C. Tamborski and E. J. Soloski, *ibid.*, 743 and 746. (e) R. J. Harper, Jr., E. J. Soloski, and C. Tamborski, *J. Org. Chem.*, **29**, 2385 (1964). (f) R. D. Chambers, F. G. Drakesmith, and W. K. R. Musgrave, *J. Chem. Soc.*, 5045 (1965). (g) (1) R. D. Chambers, J. Hutchinson, and W. K. R. Musgrave, *J. Chem. Soc.*, 5040 (1965); (2) R. D. Chambers, F. G. Drakesmith, J. Hutchinson, and W. K. R. Musgrave, *Tetrahedron Letters*, 1705 (1967); (3) R. E. Banks, R. N. Haszeldine, E. Phillips, and I. M. Young, *J. Chem. Soc.* (C), 2091 (1967). (h) (1) E. Nield, R. Stephens, and J. C. Tatlow, *J. Chem. Soc.*, 166 (1959); (2) G. M. Brooke and W. K. R. Musgrave, *J. Chem. Soc.*, 1864 (1965); (3) G. M. Brooke, R. D. Chambers, J. Heyes, and W. K. R. Musgrave, *J. Chem Soc.*, 729 (1964); (4) G. Fuller and D. A. Warwick, *Chem. Ind.* (*London*), 651 (1965); (5) C. Tamborski, E. J. Soloski, and J. P. Ward, *J. Org. Chem.* **31**, 4230 (1966); (6) J. M. Holmes, R. D. Peacock, and J. C. Tatlow, *J. Chem. Soc.* (A), 150 (1966); (7) M. G. Barlow, M. Green, R. N. Haszeldine, and H. G. Higson, *J. Chem. Soc.* (C), 1592 (1966); (8) N. N. Vorozhtsov, Jr., V. A. Barkhash, N. G. Ivanova, S. A. Anickina, and O. I. Andreevskaya, *Proc. Acad. Sci. U.S.S.R., Chem. Sect.* (*English Transl.*), **159**, 1135 (1964). (i) S. C. Cohen and A. G. Massey, *J. Organometal. Chem.*, **10**, 471 (1967). (j) (1) R. D. Chambers and T. Chivers, *J. Chem. Soc.*, 4782 (1964); (2) R. D. Chambers and J. Cunningham, *Tetrahedron Letters*, 2389 (1965).

The methods of preparation vary with the type of fluorinated group. The most commonly used are

1. direct reaction of organohalide with metal

$$R_fX + M \longrightarrow R_fM(X)$$

2. metathetical exchange

$$R_fX + RM(Y) \longrightarrow R_fM(Y) + RX$$

3. addition of metal halide to a fluoroolefin

$$R_fCF{=}CF_2 + MX_n \longrightarrow (R_f\underset{}{\overset{CF_2X}{\overset{|}{C}}}F)_nM$$

c. *Addition of bases to fluoroolefins* is an effective way of generating useful carbanion intermediates and has been studied extensively. Depending on conditions, almost all the general types of synthetic reactions are found. This subject has been comprehensively reviewed in discussion of ionic reactions of fluoroolefins.[322, 323] The overall scheme for nucleophilic attack on a fluoroolefin was summarized in Chart 3-1 (see also discussion in Section 5-2D). Depending on the structure of the olefin and conditions, three general courses are found: (1) addition of nucleophilic reagent; (2) substitution (never S_N2) of vinyl halogen; (3) S_N2' substitution of allylic halogen with rearrangement.

Process (1), representing addition of reagent AX, is a very common reaction. Among the reagents that add to fluoroolefins are alcohols,[324–330] mercaptans,[330, 331] phenols,[329, 330, 332] Grignard reagents[333] (and other organometallics[330]), ketoximes,[330, 334] primary and secondary amines,[325, 330, 335]

[322] R. D. Chambers and R. H. Mobbs, *Advan. Fluorine Chem.*, **4**, 50 (1965).
[323] Ref. 6, Houben-Weyl, pp. 280-306.
[324] W. T. Miller, Jr., E. W. Fager, and P. H. Griswold, *J. Am. Chem. Soc.*, **70**, 431 (1948).
[325] D. D. Coffman, M. S. Raasch, G. W. Rigby, P. L. Barrick, and W. E. Hanford, *J. Org. Chem.*, **14**, 747 (1949).
[326] P. Tarrant and H. C. Brown, *J. Am. Chem. Soc.*, **73**, 1781 (1951).
[327] J. D. Park, D. K. Vail, K. R. Lea, and J. R. Lacher, *J. Am. Chem. Soc.*, **70**, 1550 (1948).
[328] J. D. Park, M. L. Sharrah, and J. R. Lacher, *J. Am. Chem. Soc.*, **71**, 2337 (1949).
[329] J. T. Barr, K. E. Rapp, R. L. Pruett, C. T. Bahner, J. D. Gibson, and R. H. Lafferty, Jr., *J. Am. Chem. Soc.*, **72**, 4480 (1950).
[330] (a) D. C. England, L. R. Melby, M. A. Dietrich, and R. V. Lindsey, Jr., *J. Am. Chem. Soc.*, **82**, 5116 (1960). (b) J. W. C. Crawford, *J. Chem. Soc.*, (C), 2395 (1967).
[331] K. E. Rapp, R. L. Pruett, J. T. Barr, C. T. Bahner, J. D. Gibson, and R. H. Lafferty, Jr., *J. Am. Chem. Soc.*, **72**, 3642 (1950).
[332] P. Tarrant and H. C. Brown, *J. Am. Chem. Soc.*, **73**, 5831 (1951).
[333] P. Tarrant and D. A. Warner, *J. Am. Chem. Soc.*, **76**, 1624 (1954).
[334] A. P. Stefani, J. R. Lacher, and J. D. Park, *J. Org. Chem.*, **25**, 676 (1960).
[335] R. L. Pruett, J. T. Barr, K. E. Rapp, C. T. Bahner, J. D. Gibson, and R. H. Lafferty, Jr., *J. Am. Chem. Soc.*, **72**, 3646 (1950).

carbonyl fluoride,[336] halogen fluorides,[337] and activated aryl halides[338–341] as shown by the examples (see Chart 6-10).

Fluoride ion addition to fluoroolefins (as utilized in carbonyl fluoride addition[336] or in nucleophilic aromatic substitution[338–340]) was originally studied by Miller and co-workers[342a] and has developed into a valuable synthetic method for introducing perfluoroalkyl groups.[342b] (See discussion in Section 5-2D.)

Paths (2) and (3) (see Chart 3-1) are clearly illustrated in nucleophilic reactions of cyclic fluoroolefins which have been under extensive study for approximately 20 years. Park and co-workers initially reported on fluorinated cyclobutenes, and continue to publish actively in this field.[343] Considerable interest in this area has been shown by other groups.[329, 331, 344]

A controversy over the mechanism has developed. Park favors carbanion intermediates,[343] but good evidence was presented for an addition–elimination process in reactions leading to ethers or sulfides.[329, 344a] However, the addition part of the process could be concerted or involve a carbanion intermediate. Also, an S_N2' concerted displacement with allylic rearrangement is proposed for some reactions.[342, 344i] Recently, Park and co-workers[343c] summarized the work on mechanism of nucleophilic substitution in halogenated cyclobutenes and argued for a carbanion intermediate. They presented a set of rules on carbanion stabilization which accommodates all results and should be useful in predicting

[336] (a) F. S. Fawcett, C. W. Tullock, and D. D. Coffman, *J. Am. Chem. Soc.*, **84**, 4275 (1962). (b) R. D. Smith, F. S. Fawcett, and D. D. Coffman, *ibid.*, 4285.

[337] C. G. Krespan, *J. Org. Chem.*, **27**, 1813 (1962).

[338] R. D. Chambers, R. A. Storey, and W. K. R. Musgrave, *Chem. Commun.*, 384 (1966).

[339] R. D. Dresdner, F. N. Tlumac, and J. A. Young, *J. Org. Chem.*, **30**, 3524 (1965).

[340] French Patent 1,481,405 (1967) assigned to ICI.

[341] (a) R. L. Dressler and J. A. Young, *J. Org. Chem.*, **32**, 2004 (1967). (b) S. Temple, *J. Org. Chem.*, **33**, 344 (1968).

[342] (a) W. T. Miller, Jr., J. H. Fried, and H. Goldwhite, *J. Am. Chem. Soc.*, **82**, 3091 (1960). (b) J. A. Young, *Fluorine Chem. Rev.*, **1**, 359 (1967).

[343] (a) J. D. Park, C. M. Snow, and J. R. Lacher, *J. Am. Chem. Soc.*, **73**, 2342 (1951). (b) J. D. Park, J. R. Dick, and J. H. Adams, *J. Org. Chem.*, **30**, 400 (1965). (c) J. D. Park, J. R. Lacher, and J. R. Dick, *J. Org. Chem.*, **31**, 1116 (1966). (d) J. D. Park, G. Groppelli, and J. H. Adams, *Tetrahedron Letters*, 103 (1967). (e) J. D. Park, R. Sullivan, and R. J. McMurtry, *ibid.*, 173. (f) J. D. Park and R. J. McMurtry, *ibid.*, 1301. (g) J. D. Park, R. J. McMurtry, and R. Sullivan, *J. Org. Chem.*, **33**, 33 (1968).

[344] The following is a selection of leading references; earlier work from these groups can be found from references listed in these articles:

(a) R. A. Sheppard, H. Lessoff, J. D. Domijan, D. B. Hilton, and T. F. Finnegan, *J. Org. Chem.*, **23**, 2011 (1958). (b) A. B. Clayton, J. Roylance, D. R. Sayers, R. Stephens, and J. C. Tatlow, *J. Chem. Soc.*, 7358 (1965). (c) E. T. McBee, J. J. Turner, C. J. Morton, and A. P. Stefani, *J. Org. Chem.*, **30**, 3698 (1965). (d) D. J. Burton and R. L. Johnson, *Tetrahedron Letters*, 2681 (1966). (e) W. R. Cullen and G. E. Styan, *J. Organometal. Chem.*, **6**, 633 (1966). (f) W. R. Cullen and P. S. Dhaliwal, *Can. J. Chem.*, **45**, 719 (1967); (see also W. R. Cullen, D. S. Dawson, and P. S. Dhaliwal, *ibid.*, 683). (g) P. W. Jolly and F. G. A. Stone, *Chem. Commun.*, 85 (1965). (h) A. W. Frank, *J. Org. Chem.*, **31**, 1917 (1966). (i) S. E. Ellzey, Jr., and W. A. Guice, *ibid.*, 1300. (j) C. M. Sharts and J. D. Roberts, *J. Am. Chem. Soc.*, **83**, 871 (1961).

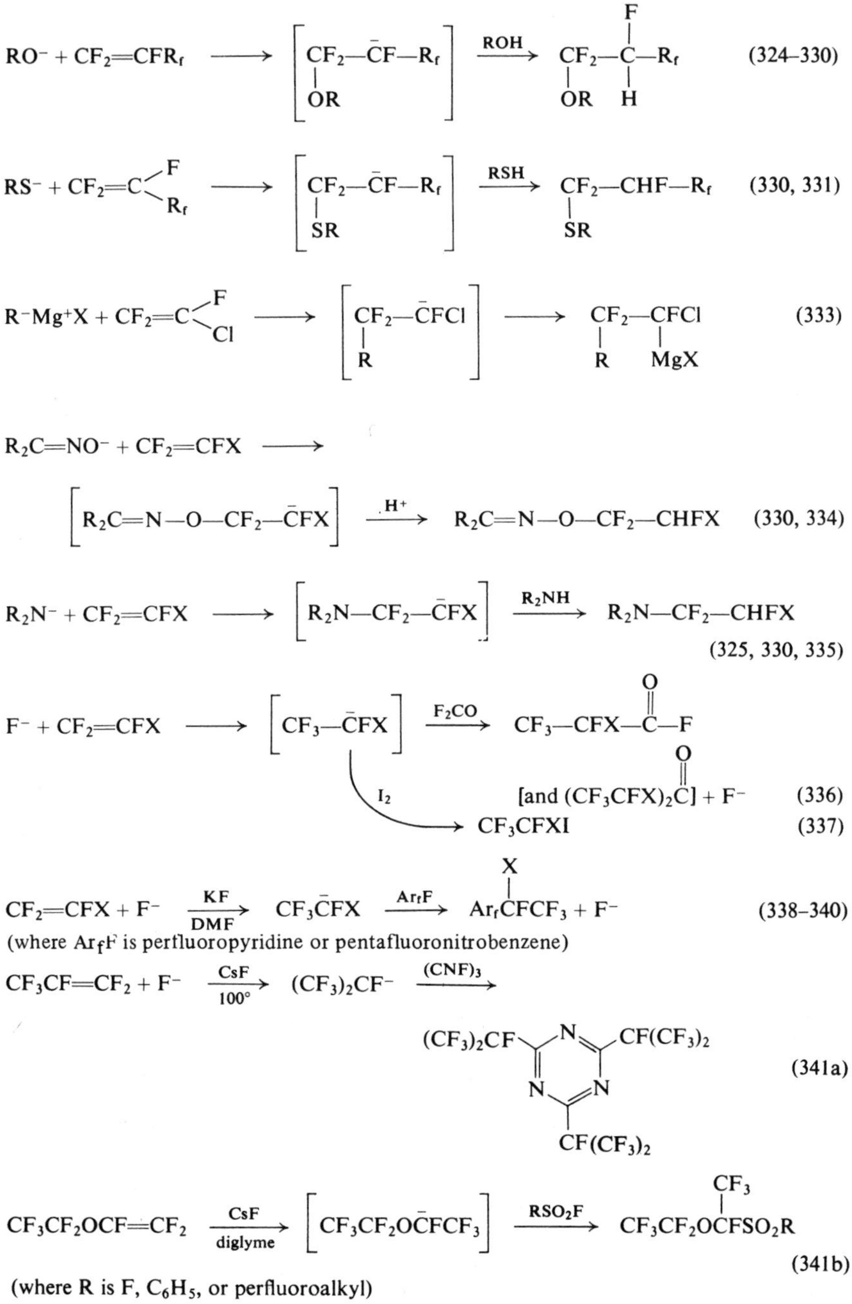

$$RO^- + CF_2{=}CFR_f \longrightarrow \left[\begin{array}{l}CF_2{-}\bar{C}F{-}R_f\\ |\\ OR\end{array}\right] \xrightarrow{ROH} \begin{array}{l}\quad\;\; F\\ \quad\;\; |\\ CF_2{-}C{-}R_f\\ |\qquad |\\ OR\quad H\end{array} \qquad (324\text{–}330)$$

$$RS^- + CF_2{=}C\begin{array}{l}\diagup F\\ \diagdown R_f\end{array} \longrightarrow \left[\begin{array}{l}CF_2{-}\bar{C}F{-}R_f\\ |\\ SR\end{array}\right] \xrightarrow{RSH} \begin{array}{l}CF_2{-}CHF{-}R_f\\ |\\ SR\end{array} \qquad (330, 331)$$

$$R^-Mg^+X + CF_2{=}C\begin{array}{l}\diagup F\\ \diagdown Cl\end{array} \longrightarrow \left[\begin{array}{l}CF_2{-}\bar{C}FCl\\ |\\ R\end{array}\right] \longrightarrow \begin{array}{l}CF_2{-}CFCl\\ |\qquad |\\ R\quad MgX\end{array} \qquad (333)$$

$$R_2C{=}NO^- + CF_2{=}CFX \longrightarrow$$

$$\left[R_2C{=}N{-}O{-}CF_2{-}\bar{C}FX\right] \xrightarrow{H^+} R_2C{=}N{-}O{-}CF_2{-}CHFX \qquad (330, 334)$$

$$R_2N^- + CF_2{=}CFX \longrightarrow \left[R_2N{-}CF_2{-}\bar{C}FX\right] \xrightarrow{R_2NH} R_2N{-}CF_2{-}CHFX$$

(325, 330, 335)

$$F^- + CF_2{=}CFX \longrightarrow \left[CF_3{-}\bar{C}FX\right] \xrightarrow{F_2CO} CF_3{-}CFX{-}\overset{\overset{\displaystyle O}{\|}}{C}{-}F$$

$$\left[\text{and } (CF_3CFX)_2\overset{\overset{\displaystyle O}{\|}}{C}\right] + F^- \qquad (336)$$

$$\left[CF_3{-}\bar{C}FX\right] \xrightarrow{I_2} CF_3CFXI \qquad (337)$$

$$CF_2{=}CFX + F^- \xrightarrow[DMF]{KF} CF_3\bar{C}FX \xrightarrow{Ar_fF} Ar_f\overset{\overset{\displaystyle X}{|}}{C}FCF_3 + F^- \qquad (338\text{–}340)$$

(where Ar_fF is perfluoropyridine or pentafluoronitrobenzene)

$$CF_3CF{=}CF_2 + F^- \xrightarrow[100°]{CsF} (CF_3)_2CF^- \xrightarrow{(CNF)_3}$$

2,4,6-tris[$CF(CF_3)_2$]-1,3,5-triazine: $(CF_3)_2CF$, $CF(CF_3)_2$ and $CF(CF_3)_2$ on the three carbons of the N/N/N triazine ring (341a)

$$CF_3CF_2OCF{=}CF_2 \xrightarrow[diglyme]{CsF} \left[CF_3CF_2O\bar{C}FCF_3\right] \xrightarrow{RSO_2F} CF_3CF_2O\overset{\overset{\displaystyle CF_3}{|}}{C}FSO_2R$$

(341b)

(where R is F, C_6H_5, or perfluoroalkyl)

CHART 6-10

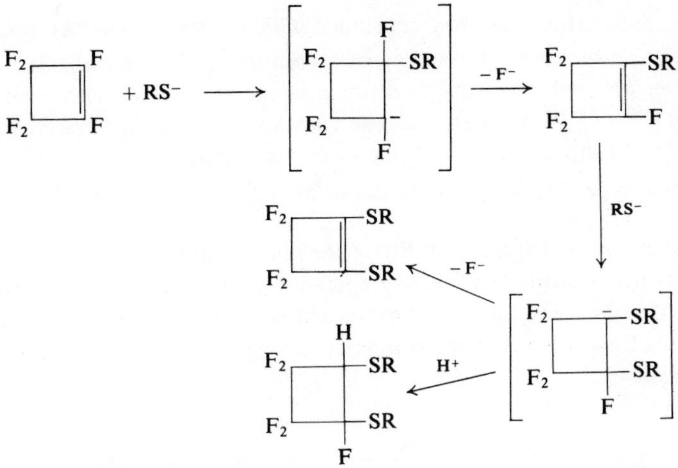

(331)

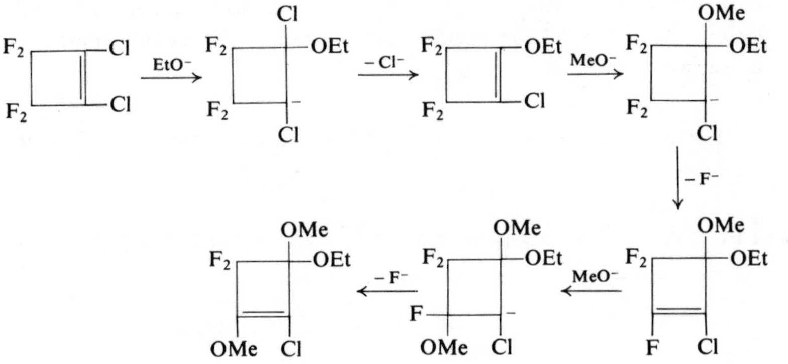

(343c)

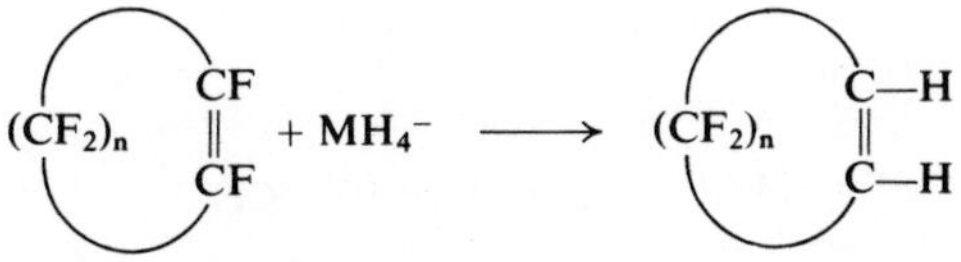

$$(CF_2)_n(CF{=}CF) + MH_4^- \longrightarrow (CF_2)_n(CH{=}CH) \qquad (344d)$$

(n = 2 or 3, M = B or Al)

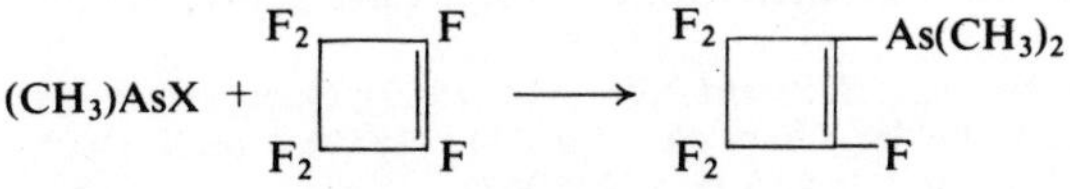

$$(CH_3)AsX + \text{perfluorocyclobutene} \longrightarrow \text{1-}As(CH_3)_2\text{-2-fluoro-3,3,4,4-tetrafluorocyclobutene} \qquad (344e)$$

(where X = H, $As(CH_3)_2$, MgBr)

CHART 6-11

reaction course. However, they concluded with the statement; "An unequivocal differentiation requires the trapping of carbanion. . . ." The carbanion mechanism does not look attractive in reaction of metal hydrides with organic metallics.[344e, f, g] In other studies, the question of electronic vs steric control and effect of ring size[343d, e, 344d] has been examined.

Some typical nucleophilic reactions of cyclic fluorinated olefins are given in Chart 6-11.

An important synthetic utility of the nucleophilic substitution reaction of fluoroolefins in aprotic media is in preparation of fluorovinyl derivatives, particularly ethers and sulfides from alkoxides or sulfides, and alkyl- or aryl-substituted fluoroolefins from organometallic reagents (see Refs. 322 or 323 for general review).

$$RO^- + CF_2{=}CF_2 \xrightarrow[\text{dioxane}]{\text{glyme or}} ROCF{=}CF_2 + F^- \quad (20\text{–}70\%) \qquad (345)$$

The trifluorovinyl ethers are useful monomers.[345a] Presumably the organometallic reagent adds to the olefin but the intermediate organometallic adduct cannot be trapped by carbonyl reagents.[349]

$$C_6H_5MgX + CF_2{=}CCl_2 \longrightarrow C_6H_5CF{=}CCl_2 \quad (64\%) \qquad (346)$$

$$C_6H_5Li + CF_2{=}CF_2 \xrightarrow{-80^\circ} C_6H_5CF{=}CF_2 \xrightarrow{C_6H_5Li} C_6H_5CF{=}CFC_6H_5$$

$$C_6H_5CF{=}CF_2 \xrightarrow[25^\circ]{C_6H_5Li} (C_6H_5)_2C{=}C(C_6H_5)_2 \qquad (347, 348)$$

$$C_6H_5Li + CF_2{=}CFCl \longrightarrow C_6H_5CF{=}CFCl \quad (91\%) \qquad (349)$$

The following series of reactions that must involve carbanion intermediates have been described recently (some of these have good potential for synthesis):

[345] (a) S. Dixon, U.S. Patent 2,917,548 (1959); *Chem. Abstr.*, **54**, 5474e (1960). (b) R. Meier and F. Bohler, *Chem. Ber.*, **90**, 2342 (1957). (c) K. Okuhara, H. Baba, and R. Kojima, *Bull. Chem. Soc. Japan*, **35**, 532 (1962).

[346] P. Tarrant and D. A. Warner, *J. Am. Chem. Soc.*, **76**, 1624 (1954).

[347] S. Dixon, *J. Org. Chem.*, **21**, 400 (1956).

[348] T. F. McGrath and R. Levine, *J. Am. Chem. Soc.*, **77**, 4168 (1955).

[349] R. Meier and F. Bohler, *Chem. Ber.*, **90**, 2344 and 2350 (1957).

$$(CF_3)_2C{=}C{=}O \underset{}{\overset{F^-}{\rightleftharpoons}} \left[\begin{array}{c} (CF_3)_2C{=}\overset{\overset{O^-}{|}}{C}F \\ \updownarrow \\ (CF_3)_2\bar{C}\overset{\overset{O}{\|}}{C}F \end{array} \right] \overset{-F^-}{\rightleftharpoons} CF_2{=}\overset{\overset{CF_3}{|}}{C}{-}\overset{\overset{O}{\|}}{C}F \qquad (350)$$

$$\left[\ \ \right] \xrightarrow{H^+} (CF_3)_2CH\overset{\overset{O}{\|}}{C}F$$

The ketene is readily prepared starting from a variety of fluorocarbons.[350]

$$CF_2{=}CF_2,\ CF_2{=}CFCF_3,\ \text{cyclic-}C_4F_8,\ \text{etc.} \xrightarrow{\text{approx. }800^\circ} (CF_3)_2C{=}CF_2$$

$$\xrightarrow{H_2O} (CF_3)_2CHCO_2H \xrightarrow{P_2O_5} (CF_3)_2C{=}C{=}O$$

Additional chemistry of the resonance stabilized carbanion is reported.[350]

$$F^- + CF_2{=}CF_2 \xrightarrow[\text{diglyme}]{CsF} CF_3CF_2^- \xrightarrow{CF_2=CF_2} CF_3CF_2CF_2\bar{C}F_2 \qquad (351)$$

$$CF_3CF_2CF_2\bar{C}F_2 \longrightarrow CF_3CF_2(CF_2CF_2)_nCF_2\bar{C}F_2 \longrightarrow \text{olefin}$$

$$CF_3CF_2CF_2\bar{C}F_2 \xrightarrow{-F^-} CF_3CF_2CF{=}CF_2 \xrightarrow{F^-} CF_3CF_2\bar{C}FCF_3$$

$$CF_3CF_2\bar{C}FCF_3 \xrightarrow{CF_2=CF_2} CF_3\underset{\underset{CF_2\bar{C}F_2}{|}}{C}F_2CF_2CF_3 \longrightarrow \text{olefin}$$

Yields of products (%) are C_8F_{16} (3-11), $C_{10}F_{20}$ (21-80), $C_{12}F_{24}$ (12-26), $C_{14}F_{28}$ (2-40), and some higher fractions.

[350] D. C. England and C. G. Krespan, *J. Am. Chem. Soc.*, 88, 5582 (1966).
[351] D. P. Graham, *J. Org. Chem.*, 31, 955 (1966).

$$CF_3CF{=}CF_2 \xrightarrow[200^\circ]{\text{solid CsF,}} (CF_3)_2\bar{C}F \longrightarrow (CF_3)_2CFCF_2\bar{C}FCF_3$$

$$\longrightarrow (CF_3)_2CFCF{=}CFCF_3 \xrightarrow{F^-} (CF_3)_2CF\bar{C}FCF_2CF_3 \longrightarrow (CF_3)_2C{=}CFCF_2CF_3 \qquad (339)$$

(Total yield of cis-trans mixture: 83%)

A stabilized fluorocarbanion has been isolated.

$$\pi\text{-}C_6F_8Fe(CO)_3 + CsF \xrightarrow[\text{or THF}]{\text{DMF}} \left[C_6F_8Fe(CO)_3 \right]^- \qquad (352)$$

This could be a useful method for working with carbanions but no synthetic chemistry was reported.[352]

Addition of alkoxides to tetrafluoroethylene in the presence of a suitable carbonyl compound leads to β-alkoxypolyfluoroethyl esters, ketones, and aldehydes, depending on carbonyl compound and conditions[353] (Chart 6-12).

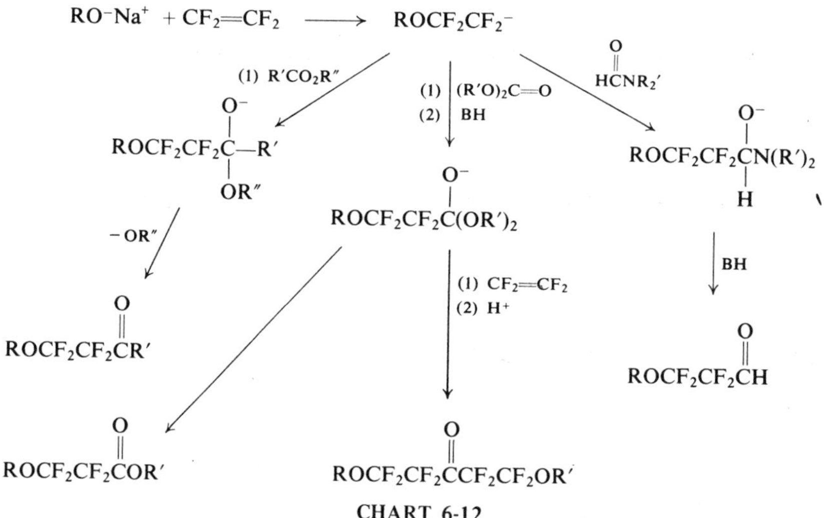

CHART 6-12

[352] G. W. Parshall and G. Wilkinson, *J. Chem. Soc.*, 1132 (1962).
[353] D. W. Wiley, U.S. Patent 2,988,537 (1961); *Chem. Abstr.*, **56**, 330g (1962).

The β-alkoxytetrafluoroethyl carbanion is proposed to be the intermediate. The reaction conditions are simple; tetrafluoroethylene is readily absorbed in a slurry of sodium methoxide in dimethyl carbonate at room temperature and atmospheric pressure. Upon acidification, the clear solution gives β-methoxytetrafluoropropionate and 1,5-dimethoxyperfluoropentan-3-one, indicating incorporation of 1 and 2 moles, respectively, of tetrafluoroethylene. The first mole is absorbed more slowly at room temperature. By taking advantage of the difference in reaction rates, either product can be obtained preferentially, with a combined product yield of 85 to 90%.

Addition of azide ion to fluoroolefins provides perfluoroalkyl or perfluoroalkenyl azides that are valuable intermediates to cyclic nitrogen derivatives[354] (but probably not via nitrenes).

$$CF_3CF_2{=}CF_2 + N_3^- \longrightarrow [CF_3\bar{C}FCF_2N_3]$$

$$[CF_3\bar{C}FCF_2N_3] \xrightarrow{BH^+} CF_3CHFCF_2N_3$$

$$[CF_3\bar{C}FCF_2N_3] \longrightarrow CF_3CF{=}CFN_3 + F^-$$

$$CF_3CF{=}CFN_3 \xrightarrow{\Delta, -N_2} F_3C{-}CF{-}CF \text{ (ring with N)}$$

Cyanide ion on a cyclic fluorinated olefin gives extensive substitution and rearrangement leading to cyanocarbon anions which still contain fluorine.[355]

$$NaCN + F_2\text{-cyclopentene (Cl, Cl; } F_2, F_2, F_2) \longrightarrow (NC)_2C{=}C(CN){-}CF{=}N{-}\bar{C}(CN)_2 + C_{11-12}N_6F^-$$

red (30%) magenta (1–2%)

No intermediates could be isolated even when the olefin was in large excess.

Sulfide ion attacks fluoroolefins that contain two replaceable halogens to give products that must arise from fluorocarbanion intermediates[356] (Chart 6-13).

[354](a) I. L. Knunyants and E. G. Bykhovskaya, *Proc. Acad. Sci. U.S.S.R., Chem. Sect. (English Transl.)*, **131**, 411 (1960). (b) C. S. Cleaver and C. G. Krespan, *J. Am. Chem. Soc.*, **87**, 3716 (1965). (c) R. E. Banks and G. J. Moore, *J. Chem. Soc.* (C), 2304 (1966).

[355]W. R. Carpenter and G. J. Palenik, *J. Org. Chem.*, **32**, 1219 (1967).

[356]C. G. Krespan and D. C. England, *J. Org. Chem.*, **33**, 1850 (1968).

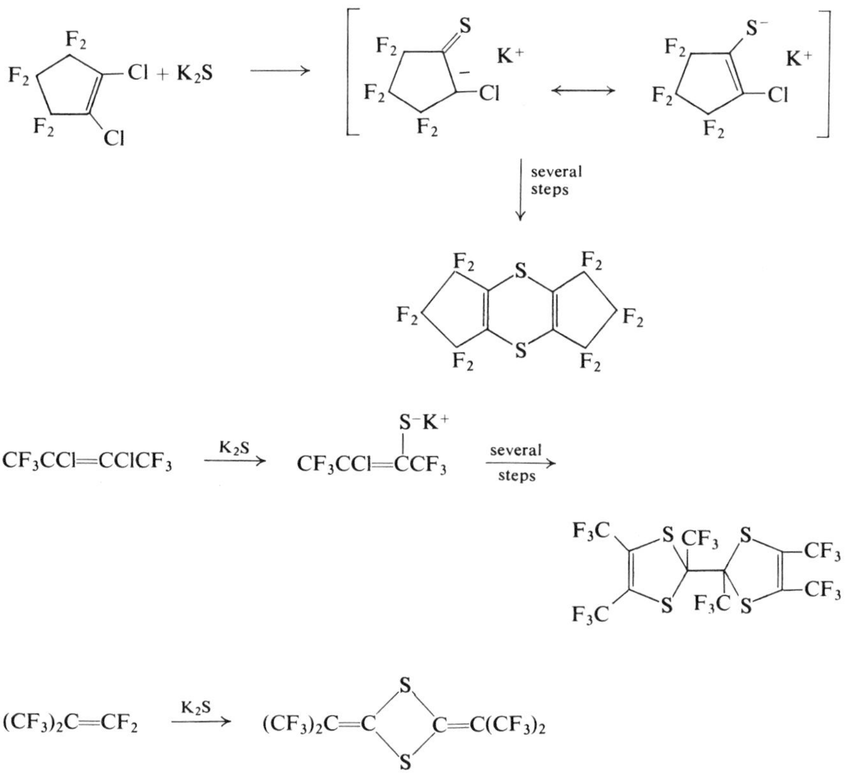

$$\text{1,2-dichlorohexafluorocyclopentene} + K_2S \longrightarrow \left[\text{(thioketone carbanion)}\ K^+ \longleftrightarrow \text{(thiolate)}\ K^+ \right] \xrightarrow{\text{several steps}} \text{bis(hexafluorocyclopenteno)-1,4-dithiin}$$

$$CF_3CCl{=}CClCF_3 \xrightarrow{K_2S} CF_3CCl{=}C(S^-K^+)CF_3 \xrightarrow[\text{steps}]{\text{several}} \text{bis[tetrakis(trifluoromethyl)-1,3-dithiole]}$$

$$(CF_3)_2C{=}CF_2 \xrightarrow{K_2S} (CF_3)_2C{=}C\langle S_2 \rangle C{=}C(CF_3)_2$$

CHART 6-13

Nucleophilic substitution of perfluorinated aromatic compounds definitely involves perfluorinated anionic intermediates.[357–361] Recent publications in this area are extensive, no doubt because of the increased availability and interest in the chemistry of perfluoroaromatic compounds. Burdon and co-workers[358] have elaborated the course of a large number of these nucleophilic reactions and proposed a rational scheme to explain variations in rates of attack of nucleophile

[357] For a review of work in this area, see *Ann. Rept. Progr. Chem. (Chem. Soc. London)*, **62**, 259 (1965). For leading references, see Refs. 358–361; for earlier work, see references contained therein.

[358] (a) J. Burdon, *Tetrahedron*, **21**, 3373 (1965). (b) J. Burdon, P. L. Coe, C. R. Marsh, and J. C. Tatlow, *Tetrahedron*, **22**, 1183 (1966). (c) J. Burdon, D. R. King, and J. C. Tatlow, *ibid.*, 2541.

[359] J. M. Birchall, M. Green, R. N. Haszeldine, and A. D. Pitts, *Chem. Commun.*, 338 (1967).

[360] M. Bellas, D. Price, and H. Suschitzky, *J. Chem. Soc.* (C), 1249 (1967).

[361] (a) G. M. Brooke and Md. A. Quasem, *ibid.*, 865. (b) R. D. Chambers, D. Lomas, W. K. R. Musgrave, *J. Chem. Soc.* (C), 625 (1968).

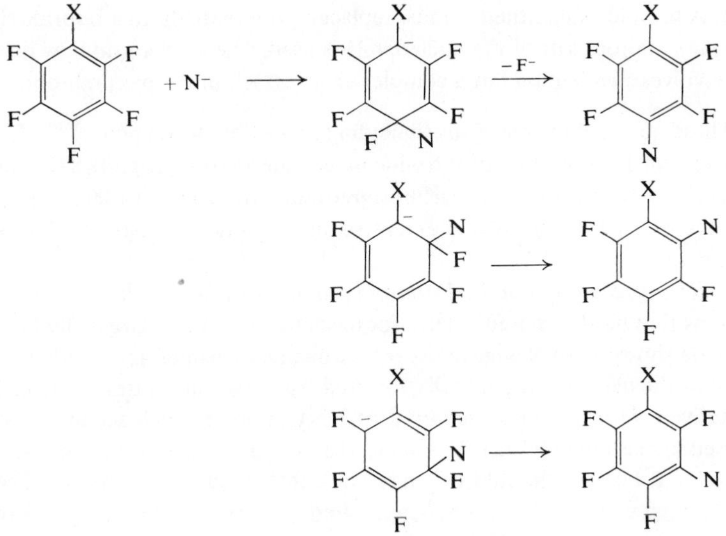

CHART 6-14. Nucleophilic substitution of pentafluorophenyl derivatives by nucleophile N^-.

N^- and orientation to substituent X (see Chart 6-14). The arguments to explain orientation are based on the substituent's ability to stabilize the carbanion intermediate in the transition state, whereas reactivity may be influenced by ground-state energy which is lowered by strong electron-donating substituents. However, other factors can also come into play, as summarized below:

1. If substituent X can stabilize a negative charge, then displacement of the *p* (or *o*)-fluorine is enhanced. Some strong electron-donating groups such as NH_2 or O^- slow the rate and give meta orientation.

2. Ortho substitution becomes major if hydrogen bonding of the nucleophile to X is possible.

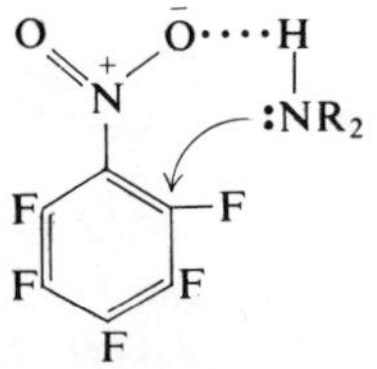

3. Halogen substituents have an $I\pi$ repulsion (orbital penetration effect) and orient substitution to the para position.

4. Steric effect of X decreases ortho substitution.

5. A bromide substituent can be replaced preferentially to a fluoride (in C_6F_5Br) if a cuprous salt of the nucleophile is used. The mechanism now no longer involves a carbanion but a complex from attack of copper on bromine.[362]

These nucleophilic reactions (including carbanion nucleophiles)[361, 363] provide an excellent method of introducing perfluoroaromatic groups (including heterocyclic derivatives such as perfluoropyridine) into a large variety of systems and should parallel the useful and extensive nucleophilic chemistry of fluoroolefins.

d. *Base attack on monohydrofluorocarbons* is a route to fluorinated carbanions that has been used chiefly for mechanistic studies. Originally haloforms were shown to exchange hydrogen through a carbanion intermediate. The stability of the halocarbanion, $^{-}CX_3$, inferred from exchange rates,[364] agreed quantitatively with that measured by decarboxylation of trihaloacetate salts (discussed in Section 6-1C.1a). The halocarbanions are a good source of carbenes (see Section 6-1B.1a). The chloro- and bromo-substituted carbanions can be captured, usually in high yield, with carbonyl compounds, but the fluorinated haloform carbanions are so unstable that the capture reactions are not useful for synthesis.[365]

Andreades[366] studied the acidities of monohydrofluorocarbons bearing hydrogen in primary, secondary, and tertiary positions by carrying out exchange reactions in sodium methoxide–methanol-*O-d* solution. The mechanism was

Relative rate of carbanion formation

CF_3H	$CF_3(CF_2)_5CF_2H$	CF_3CHFCF_3	$(CF_3)_3CH$ (CF_3CHCF_3 with CF_3 on central C)
(1.0)	(6)	(2×10^5)	(10^9)

shown to involve the fluorocarbanion; the stability of the anions, tertiary > secondary > primary, was inferred from the rate of exchange and explained by greater β-fluorine stabilization compared to much smaller α-fluorine stabilization. This enhanced β-fluorine stabilization was rationalized by hyperconjugative resonance (participation: 9 forms for tertiary, 6 for secondary, and 2 for primary).

$$CF_3-\bar{C}(CF_3)-CF_3 \longleftrightarrow CF_3-C(CF_3){=}CF_2\ \ F^- \longrightarrow \text{etc.}$$

[362] L. J. Belf, M. W. Buxton, and G. Fuller, *J. Chem. Soc.*, 3372 (1965).
[363] J. M. Birchall and R. N. Haszeldine, *J. Chem. Soc.*, 3719 (1961).
[364] (a) J. Hine, N. W. Burske, M. Hine, and P. B. Langford, *J. Am. Chem. Soc.*, **79**, 1406 (1957). (b) J. Hine, R. Wiesboeck, and O. B. Ramsay, *J. Am. Chem. Soc.*, **83**, 1222 (1961).
[365] H. G. Viehe and P. Valange, *Chem. Ber.*, **96**, 420 (1963).
[366] S. Andreades, *J. Am. Chem. Soc.*, **86**, 2003 (1964).

Recently, a bridgehead perfluorocarbanion was prepared.[367] Streitwieser and Holtz[368] showed by base-catalyzed proton exchange studies that this bridge-

$$\text{bicyclic } C_8F_{13}H \xrightarrow{\text{base}} \text{bridgehead carbanion } C_8F_{13}^-$$

head carbanion was more stable than tris(trifluoromethyl)methyl anion, and concluded that fluoride ion hyperconjugation was not the explanation for the relative stabilities of anions, but, rather, the inductive field effect of the trifluoromethyl group is a much better stabilizing influence than α-fluorine. Presumably the strong inductive stabilizing effect of an α-fluorine is partly cancelled by a destabilizing interaction of the unshared electron pairs of fluorine with the electrons in the carbanion orbital. Additional evidence in support of this argument comes from measurement of the normal inductive field effects of a trifluoromethyl group in stabilizing 9-trifluoromethylfluorenyl anion.[369] (See Section 3-1A.)

Stabilizing effects of a trifluoromethyl group on carbanions[370] and the effect of α-fluorines on the acidities of substituted nitromethanes[371] have also been studied.

Russian workers have recently reported synthesis using carbanions generated by base attack.[372] If the carbanion is stabilized by resonance with a carbonyl group, stable salts can be isolated and used for further reaction.[372a]

$$(CF_3)_2CHC(=O)X + R_3N \longrightarrow [(CF_3)_2\bar{C}C(=O)X \quad \overset{+}{N}R_3H]$$

stable salt

(X = F or OR)

$$\xrightarrow{RI} (CF_3)_2C(R)C(=O)X \qquad \xrightarrow{Br_2} (CF_3)_2C(Br)C(=O)X$$

[367] (a) S. F. Campbell, R. Stephens, and J. C. Tatlow, *Chem. Commun.*, 134 (1965). (b) S. F. Campbell, R. Stephens, and J. C. Tatlow, *Tetrahedron*, **21**, 2997 (1965).

[368] A. Streitwieser, Jr., and D. Holtz, *J. Am. Chem. Soc.*, **89**, 692 (1967).

[369] A. Streitwieser, Jr., A. P. Marchand, A. H. Pudjaatmaka, *J. Am. Chem. Soc.*, **89**, 693 (1967).

[370] D. J. Cram and A. S. Wingrove, *J. Am. Chem. Soc.*, **86**, 5490 (1964).

[371] H. G. Adolph and M. J. Kamlet, *J. Am. Chem. Soc.*, **88**, 4761 (1966).

[372] (a) Yu. A. Cheburkov, M. D. Bargamova, and I. L. Knunyants, *Bull. Acad. Sci. U.S.S.R., Div. Chem. Sci. (English Transl.)*, 339 (1964). (b) S. T. Kocharyan, E. M. Rokhlin, and I. L. Knunyants, *4th Intern. Symp. Fluorine Chemistry, Estes Park, Colorado, 1967*, Abstracts, p. 60.

Dehydrofluorination is the normal course for the reaction of the 1H-nonafluoro-*t*-butane with base, but perfluorinated groups can be introduced through the carbanion[372b] under certain conditions.

$$R_3N + (CF_3)_3CH \rightleftharpoons (CF_3)_3\bar{C}\ \ R_3\overset{+}{N}H \rightleftharpoons (CF_3)_2C{=}CF_2 + R_3NHF$$

$$(CF_3)_3\bar{C}\ R_3\overset{+}{N}H \xrightarrow{CH_3OH} (CF_3)_2CHCF_2OCH_3$$

$$(CF_3)_3\bar{C}\ R_3\overset{+}{N}H \xrightarrow{H_2O} (CF_3)_2CHCO_2H \xrightarrow{-CO_2} (CF_3)_2CH_2$$

$$(CF_3)_3\bar{C}\ R_3\overset{+}{N}H \xrightarrow{C_6H_5\overset{O}{\overset{\|}{C}}Cl} C_6H_5\overset{O}{\overset{\|}{C}}F + R_3NHCl + (CF_3)_2C{=}CF_2$$

The dehydrochlorination of 1H,2-chlorodecafluorocyclohexanes and related dichloro derivatives with aqueous base have been studied.[373] The stereochemical consequences of this reaction are best explained by a carbanion intermediate. However, little is known about the conformational properties of polyfluorocarbanions of this type.

$$\text{1-H,1-X,2-Cl,2-F-octafluorocyclohexane} \longrightarrow [\text{1-X carbanion}] \longrightarrow \text{1-X-2-F-octafluorocyclohexene}$$

(X = F or Cl)

e. *Base-promoted elimination from esters of perfluorinated acids* has been used to prepare monohydrofluorocarbons.[374] Ethyl trifluoroacetate gave some tetrafluoroethylene, presumably by decomposition of $\bar{C}F_3$ to difluorocarbene.

$$R_fCO_2Et + EtO^-Na^+ \overset{EtOH}{\rightleftharpoons} R_f\overset{O^-}{\overset{|}{\underset{\underset{OEt}{|}}{C}}}{-}OEt \longrightarrow R_f^- + EtO\overset{O}{\overset{\|}{C}}OEt$$

$$R_f^- \xrightarrow{H^+} R_fH + EtO^- \quad (\sim 70\%)$$

The general synthetic utility of this reaction was developed by Wiley[375a] and is the best synthetic route to perfluorinated ketones.[375b]

[373] S. F. Campbell, F. Lancashire, R. Stephens, and J. C. Tatlow, *Tetrahedron*, **23**, 4435 (1967).

[374] E. Bergman, *J. Org. Chem.*, **23**, 476 (1958).

[375] (a) D. W. Wiley, U.S. Patent 3,091,643 (1963); *Chem. Abstr.*, **59**, 11266e (1963). (b) D. W. Wiley, private communications.

$$R_fCOEt\,(C{=}O) + NaOEt \xrightarrow{\text{ether}} R_fC(O^-Na^+)(OEt)-OEt$$

$$\downarrow$$

$$R_fC(O^-)(OEt)R_f \xleftarrow{R_fC(O)OEt} R_f^- + (EtO)_2C{=}O$$

$$R_fC(O^-)(OEt)R_f \xrightarrow{H^+} EtO^- + R_fC(O)R_f \xrightarrow{R_f^-} (R_f)_3COH \qquad R_f^- + (EtO)_2C{=}O \rightarrow \text{olefin}$$

(For $R_f = CF_3CF_2CF_2$, yield of ketone is 90%)

The reaction can also be adapted to prepare the tertiary alcohol or the olefin.

f. *Base attack on perfluoroalkyl iodides* leads to the corresponding monohydroperfluorocarbon[376] with some other by-products. Partially fluorinated iodides such as $CHFI_2$ and CHFClI decompose to halide and formate ester under

$$R_fI \xrightarrow[\text{EtOH}]{\text{KOH}} R_fH \quad (+ R_fOEt)$$

these conditions. For the perfluorinated iodides, a nucleophilic displacement on carbon is considered unlikely. Nucleophilic attack on iodine is suggested to lead to a carbanion intermediate, but mechanism studies[377] have not given a conclusive answer.

The reaction of chlorofluoroalkanes with triethyl phosphite is also proposed to involve a fluorocarbanion intermediate by attack on chlorine[378] and gives three products.

$$(C_2H_5O)_3P + CCl_3(CF_2)_nCl \longrightarrow [(C_2H_5O)_3PCl]^+[CCl_2(CF_2)_nCl]^-$$

$$\swarrow \qquad \swarrow \qquad \searrow$$

$$CH_3CH_2CCl_2(CF_2)_nCl \qquad CCl_2{=}CF(CF_2)_{n-1}Cl \qquad (C_2H_5O)_2P(O)CCl{=}CF(CF_2)_{n-1}Cl$$

$$+ (C_2H_5O)_2P(O)Cl \qquad [+ (C_2H_5O)_3PClF]$$

not isolated

376 R. N. Haszeldine, *J. Chem. Soc.*, 4259 (1952).

377 (a) J. Banus, H. J. Emeleus, and R. N. Haszeldine, *J. Chem. Soc.*, 60 (1951). (b) J. Mason, *J. Chem. Soc.*, 4695 (1960).

378 A. E. Platt and B. Tittle, *J. Chem. Soc.* (C), 1150 (1967).

g. *Negative ion mass spectrometry* is particularly valuable for fluorocarbons because of the marked stability of the fluorinated carbanions.[379] However, synthetic utilization of carbanions generated in this way is limited because of the conditions.

2 Fluorinated Anions (on Heteroatom)

A number of fluorinated anions, F_nX^- or $(R_f)_nX^-$ where X is a heteroatom, have been postulated or isolated as intermediates in synthetic schemes. The ones of interest to organic chemists are

BF_4^-, $R_fBF_3^-$
SiF_6^{-2}, SiF_5^-
NF_2^-, $(R_f)_2N^-$, $(R_f)_2P^-$, PF_6^-, SbF_6^-
FO^-, R_fO^-, SF_5^-, R_fS^-, $R_fSO_2^-$

Some of these anions are used directly to introduce fluorinated heteroatom groups, but others have indirect application to synthetic organic chemistry and mechanism studies.

a. *Boron.* Fluoroborate, BF_4^-, is an extremely stable, generally unreactive anion. Fluoroboric acid and its alkaline metal salts are readily available. The main use in organic synthesis is in the Schiemann reaction where aryldiazonium fluoroborates are isolated as intermediates to aryl fluorides.[380]

$$ArNH_2 \xrightarrow[HBF_4]{HNO_2} Ar\overset{+}{N_2}\overset{-}{BF_4} \xrightarrow{\Delta} ArF + N_2 + BF_3$$

Salts of trifluoromethylfluoroboric acid, $CF_3BF_2^-M^+$, have been prepared[280b] by an unusual CF_3^- transfer. The trimethyltin cation is easily replaced by other

$$(CH_3)_3SnCF_3 + BF_3 \xrightarrow{CCl_4} (CH_3)_3Sn^+CF_3BF_3^-$$

cations such as K^+ or NH_4^+. These salts have been pyrolyzed to eliminate difluorocarbene (see p. 269), but no other synthetic utility has been reported. The anions $C_6F_5BF_3^-$, $CH_2{=}CHBF_3^-$, and $CH_3BF_3^-$ are formed by addition of fluoride ion to the corresponding boron fluoride derivative,[381a, b] and the anion $(C_6F_5)_4B^-$ is obtained from C_6F_5Li and $(C_6F_5)_3B$,[381c] but again these anions have not been used in synthesis.

[379] (a) J. von Hoene and W. M. Hickam, *J. Chem. Phys.*, **32**, 876 (1960). (b) M. M. Bibby and G. Carter, *Trans. Faraday Soc.*, **59**, 2455 (1963). (c) W. B. Askew, Central Research Department, E. I. du Pont de Nemours and Co., Wilmington, Delaware, private communication.

[380] For a recent review, see H. Suschitzky, *Advan. Fluorine Chem.*, **4**, 1 (1965). For typical experimental procedure see D. T. Flood, *Org. Syn.* (Coll. Vol.) **2**, 295 (1943).

[381] (a) R. D. Chambers, T. Chivers, and D. A. Pyke, *J. Chem. Soc.*, 5144 (1965). (b) S. L. Stafford, *Can. J. Chem.*, **41**, 807 (1963). (c) A. G. Massey, A. J. Park, and F. G. A. Stone, *Proc. Chem. Soc.*, 212 (1963).

b. *Silicon.* Silicon hexafluoride dianion, SiF_6^{-2}, also forms stable salts with a variety of inorganic cations[382a] but has not been utilized in organic chemistry. Recently SiF_5^- was reported as a stable species.[382b]

$$(C_6H_5)_4AsCl + SiO_2 + HF \longrightarrow (C_6H_5)_4As^+SiF_5^-$$

c. *Nitrogen.* Difluoroamide anion, NF_2^-, has been suggested to form[383a] from HNF_2 and in hydrolysis of tetrafluoroformamidine and pentafluoroguanidine.[383b] The only synthetic utilization was oxidation. Kinetic studies were made on

$$HNF_2 \xrightarrow{Fe^{3+}} F_2NNF_2$$

reactions of anions with difluoroamine; strong Bronsted bases gave difluorodiazine, but oxidizable anions (for example I^-) gave only NH_3 and F^-, and weak bases (OAc^-) only N_2O and F^-.[384] Alkylation of HNF_2 with carbonium ions[385]

$$2HNF_2 + 2OH^- \longrightarrow N_2F_2 + 2H_2O + 2F^-$$

is a route to a variety of RNF_2 species. This reaction extends the scope of the free radical reactions of $\cdot NF_2$ based on N_2F_2.

$$(C_6H_5)_3C^+ + HNF_2 \xrightarrow[-25^\circ]{SO_2} (C_6H_5)_3CNF_2$$

$$(CH_3)_2C{=}CH_2 + HNF_2 \xrightarrow[\text{or } BF_3\text{—}H_3PO_4]{96\%\ H_2SO_4} (CH_3)_3CNF_2$$

Electrophilic attack of the carbonium ion on the weakly nucleophilic nitrogen is proposed so *that free* NF_2^- *is probably not formed*; however an incipient negative charge must develop on nitrogen during the course of the reaction.

The bis(trifluoromethyl)amide anion, $(CF_3)_2N^-$, is suggested as the intermediate in fluoride ion-catalyzed dimerization of perfluoro-2-azapropene[339] and in COF_2 addition.[336b]

382 (a) D. H. Brown, K. R. Dixon, C. M. Livingston, R. H. Nuttall, and D. W. A. Sharp, *J. Chem. Soc.* (A), 100 (1967). (b) H. C. Clark and K. R. Dixon, *Chem. Commun.*, 717 (1967).

383 (a) K. J. Martin, *J. Am. Chem. Soc.*, 87, 394 (1965). (b) R. L. Rebertus and B. W. Nippoldt, *J. Org. Chem.*, **32**, 4044 (1967).

384 W. T. Yap, A. D. Craig, and G. A. Ward, *J. Am. Chem. Soc.*, **89**, 3442 (1967).

385 W. H. Graham and J. P. Freeman, *J. Am. Chem. Soc.*, **89**, 716 (1967).

$$CF_3N{=}CF_2 \xrightarrow{F^-} [(CF_3)_2N^-] \xrightarrow{CF_3N=CF} (CF_3)_2NCF_2\bar{N}CF_3$$

$$[(CF_3)_2N^-] \xrightarrow{COF_2} (CF_3)_2NC(=O)F \qquad (CF_3)_2NCF_2\bar{N}CF_3 \xrightarrow{-F^-} (CF_3)_2NCF{=}NCF_3$$

Dimerization of *N*-aryldifluoroazomethines also probably involves an anionic intermediate.[386]

$$2ArN{=}CF_2 \xrightarrow{F^-?} ArN(CF_3){-}CF{=}NAr$$

A variety of metal fluorides react with perfluoro-2-azapropene; the product composition depends on whether reaction is ionic or free radical.[144a, 387] However, the adduct **1** from mercuric difluoride is probably formed by an ionic mechanism

$$2CF_3N{=}CF_2 + HgF_2 \longrightarrow \underset{\mathbf{1}}{[(CF_3)_2N]_2Hg}$$

although a four-center addition mechanism is also possible. Ionic or four-center mechanisms are most probable for reactions of $(CF_3)_2NX$ with mercury salts or for mercury adduct **1** with halogens or acylhalides.[388]

$$\underset{\mathbf{1}}{[(CF_3)_2N]_2Hg} + 2RC(=O)Cl \longrightarrow \underset{(60\text{–}80\%)}{RC(=O)N(CF_3)_2} + HgCl_2$$

$$(CF_3)_2NX + (CH_3)_2Hg \longrightarrow (CF_3)_2NHgCH_3 + CH_3X$$

The isomerization (and hydrolysis) of perfluoro-α,ω-bisazomethines is postulated to go by an S_N2' mechanism.[389a] However, a cyclic by-product [in

$$CF_2{=}NCF_2N{=}CF_2 \xrightarrow{F^-} CF_3N{=}C{=}NCF_3$$

[386] W. A. Sheppard, *J. Am. Chem. Soc.*, **87**, 4338 (1965).

[387] J. A. Young, W. S. Durrell, and R. D. Dresdner, *J. Am. Chem. Soc.*, **81**, 1587 (1959).

[388] (a) R. C. Dobbie and H. J. Emeleus, *J. Chem. Soc.* (A), 367 (1966). (b) J. A. Young, S. N. Tsoukalas, and R. D. Dresdner, *J. Am. Chem. Soc.*, **80**, 3604 (1958). (c) J. A. Young, W. S. Durrell, and R. D. Dresdner, *J. Am. Chem. Soc.*, **84**, 2105 (1962).

[389] (a) P. H. Ogden and R. A. Mitsch, *J. Am. Chem. Soc.*, **89**, 5007 (1967); P. H. Ogden, *J. Chem. Soc.* (C), 2302 (1967). (b) R. C. Dobbie and H. J. Emeleus, *J. Chem. Soc.* (A), 933 (1966).

30% yield from isomerization of perfluoro(3,4-dimethyl-2,5-diazahexa-2,4-diene)] is rationalized by a cyclization of an initially formed nitrogen anion, whereas the normal product arises from the typical S_N2' mechanism.

$$CF_2{=}NCF(CF_3){-}CF(CF_3)N{=}CF_2 \xrightarrow{F^-} \left[CF_3N^-{-}CF(CF_3){-}CF(CF_3){-}N{=}CF_2 \right] \longrightarrow \text{cyclic } CF_3N{-}CF(CF_3){-}C(CF_3){=}N{-}CF_2$$

$$CF_2{=}NCF(CF_3){-}CF(CF_3)N{=}CF_2 \xrightarrow{F^-} CF_3N{=}C(CF_3){-}C(CF_3){=}NCF_3$$

A fluorinated anion is also proposed in fluorination of carbylamine chloride 2.[389b]

$$(CF_3)_2NCl + ClCN \xrightarrow[\text{(radical addition?)}]{h\nu} \underset{\mathbf{2}}{(CF_3)_2N{-}N{=}CCl_2}$$

$$\mathbf{2} \xrightarrow{NaF} [(CF_3)_2N{-}N{=}CF_2] \xrightarrow{F^-} [(CF_3)_2N{-}\bar{N}CF_3]$$

$$[(CF_3)_2N{-}\bar{N}CF_3] \xrightarrow{\text{addition to azomethine and } F^- \text{ elimination}} (CF_3)_2NN{=}CFN(CF_3){-}N(CF_3)_2 \quad (77\% \text{ based on } \mathbf{2})$$

Anionic intermediates are probably involved in attack of certain nucleophiles on perfluoroalkylnitriles.[390]

$$R_fC{\equiv}N + N_3^- \longrightarrow \left[R_fC({=}N^-)N_3 \right] \longrightarrow R_fC\text{(tetrazolide anion, }N{-}N{-}N{-}N\text{ ring)}$$

The nitrile group is highly susceptible to nucleophilic reagents because of the strong electron-withdrawing effect of the perfluoroalkyl group.

$$R_f \leftarrow \overset{\delta+}{C}{\equiv}\overset{\delta-}{N}$$

Alcohols, mercaptans, amines, hydroxylamine, and other basic materials including active methylene compounds of type BH have been added to a variety of

[390] H. C. Brown and R. J. Kassal, *J. Org. Chem.*, **32**, 1871 (1967).

fluorinated nitriles. Base catalysis is usually required suggesting an ionic intermediate such as **3**, but no detailed mechanism studies are reported.[391] Many of

$$R_fC{\equiv}N + BH \longrightarrow [R_fC(B){=}N^-] \longrightarrow R_fC(B){=}NH$$

3

where BH is ROH, RSH, RNH_2, R_2NOH (R = H or alkyl)

the adducts are useful intermediates to a variety of heterocycles with perfluoroalkyl substituents, for example,

$$R_fCN + NH_2OH \longrightarrow R_fC({=}NOH)NH_2 \xrightarrow{R_fC(O)Cl} R_fC(NH_2){=}NOC(O)R_f \xrightarrow{P_2O_5} \text{(1,2,4-oxadiazole with } R_f \text{ at C-3 and C-5)}$$

d. *Phosphorus (and Antimony).* Hexafluorophosphate ions, PF_6^-, are extremely stable and inert to chemical attack. Their main value is in preparation of aryl fluorides by thermal degradation of aryldiazonium hexafluorophosphates[392a] (this reaction also provides a controlled source of PF_5 from readily stored derivatives).

$$ArN_2^+PF_6^- \xrightarrow{\Delta} ArF + N_2 + PF_5$$

Fluoroalkylphosphite anions are suggested as intermediates in the reaction of fluoroalkylphosphines with bases.[392b] They give, by loss of fluoride ion, a

$$(CF_3)_2PH \xrightarrow{\text{base}} [(CF_3)_2P^-] \xrightarrow{-F^-} [CF_3P{=}CF_2]$$

$$[CF_3P{=}CF_2] \xrightarrow{CH_3OH} CF_3P(OCH_3)CHF_2; \quad \xrightarrow{NH_3} CF_3P(NH_2)CHF_2; \quad \longrightarrow \text{polymer}$$

phospha-alkene which reacts immediately with base or polymerizes. Several secondary and primary phosphines have been studied with a variety of bases.

[391] For leading references on these addition reactions, see H. C. Brown and C. R. Wetzel, *J. Org. Chem.*, **30**, 3734 (1965); H. C. Brown, P. D. Shuman, and J. Turnbull, *J. Org. Chem.*, **32**, 231 (1967); and H. C. Brown and R. J. Kassal, *ibid.*, 1871.

[392] (a) K. G. Rutherford, W. Redmond, and J. Rigamonti, *J. Org. Chem.*, **26**, 5149 (1961); K. G. Rutherford and W. Redmond, *Org. Syn.*, **43**, 12 (1963). (b) H. Goldwhite, R. N. Haszeldine, and D. G. Rowsell, *J. Chem. Soc.*, 6875 (1965); M. Green, R. N. Haszeldine, B. R. Iles, and D. G. Roswell, *ibid.*, 6879.

Hexafluoroantimonates are also not directly used to introduce fluorinated antimony units but serve as stabilizing counterions for studying carbonium ions[393] (see discussion in Section 6-1D). They may also be of importance in fluorination with antimony fluorides, but their role has not been clearly defined (see Section 4-5).

e. *Oxygen.* For oxygen anions, the species FO^- has been suggested only as an intermediate in fluorinations with elementary fluorine in aqueous solution (see Section 5-3A). However, perfluoroalkoxides, R_fO^-, are stable anions prepared by fluoride ion addition to fluorinated ketones or acid fluorides. They can be alkylated, thus providing a good method whereby R_fO groups are

$$R_fC(=O)R_F \xrightarrow[\text{or KF}]{\text{CsF}} R_fC(O^-)(F)R_F \xrightarrow{\text{RX}} R_fC(OR)(F)R_F$$

(R_f = perfluoroalkyl, R_F = perfluoroalkyl or F;
RX is an alkyl, acyl, sulfonyl halide)

introduced. Hexafluoroacetone has been studied most extensively,[394] but recently other perfluorinated ketones and acid fluorides were used.[395a] An interesting alternative route to perfluoroalkoxides is by fluoride ion-promoted decomposition of fluorosulfates[395]; the alkoxide intermediate readily loses fluoride ion to give acid fluorides, but several perfluorocycloalkyl alkoxides are stable in solution and have been alkylated.[395a]

$$R_fCF(R_f')OSO_2F + F^- \longrightarrow R_fCF(R_f')O^- + SO_2F_2$$

$$R_fCF(R_f')O^- \xrightarrow{-F^-} R_fC(=O)R_f' \qquad R_fCF(R_f')O^- \xrightarrow{\text{RX}} R_fCF(R_f')OR$$

(where R_f is F or perfluoroalkyl)

[393] For leading references, see G. A. Olah and M. B. Comisarow, *J. Am. Chem. Soc.*, 89, 2694 (1967).

[394] (a) A. G. Pittman and D. L. Sharp, *J. Org. Chem.*, **31**, 2316 (1966). (b) A. G. Pittman and D. L. Sharp, *J. Polymer Sci.*, Part B, **3**, 379 (1965); A. G. Pittman, D. L. Sharp, and R. E. Lundin, *J. Polymer Sci.*, Part A, **4**, 2637 (1966). (c) A. G. Pittman and D. L. Sharp, *Textile Res. J.*, **35**, 190 (1965). (d) A. G. Pittman, D. L. Sharp, and B. A. Ludwig, *Div. Polymer Chem. Am. Chem. Soc. Meeting, New York City, September, 1966,* Polymer Preprints, Vol. 7, p. 1093. (e) D. P. Graham and V. Weinmayer, *J. Org. Chem.*, **31**, 957 (1966).

[395] (a) R. W. Anderson, N. L. Madison, and C. I. Merrill, *4th Intern. Symp. Fluorine Chem., Estes Park, Colorado, July, 1967,* Abstracts, p. 64. (b) M. Lustig and J. K. Ruff, *Inorg. Chem.*, **3**, 287 (1964); *Inorg. Chem.*, **4**, 1441 (1965).

The adduct of cesium or potassium fluoride with COF_2 is stable but has not been alkylated.[396] However CF_3O^- must be an intermediate in the reaction of COF_2 with epoxide[396b] or in the base-catalyzed dimerization of carbonyl fluoride.[396c]

$$COF_2 \xrightarrow[\text{(pyridine)}]{\text{base}} \left[\text{base}\cdot F^- F\overset{O}{\overset{\|}{C}}{}^+\right] \xrightarrow{COF_2} [CF_3O^-] \xrightarrow{\overset{\diagup O \diagdown}{CH_2-CH_2}}$$

$$[CF_3OCH_2CH_2O^-] \xrightarrow{COF_2} CF_3OCH_2CH_2O\overset{O}{\overset{\|}{C}}F$$

$$COF_2 + F^- \longrightarrow [CF_3O^-] \xrightarrow{COF_2} CF_3O\overset{O}{\overset{\|}{C}}F + CF_3O\overset{O}{\overset{\|}{C}}OCF_3$$

Addition of bases or nucleophiles, including organometallic reagents, to fluorinated ketones must involve alkoxide intermediates.

$$B^- + R_f-\overset{O}{\overset{\|}{C}}-R \longrightarrow R_f-\underset{B}{\underset{|}{\overset{O^-}{\overset{|}{C}}}}-R \xrightarrow{BY} R_f-\underset{B}{\underset{|}{\overset{OY}{\overset{|}{C}}}}-R$$

This is an extensive area of synthetic chemistry which has already in part been reviewed.[397] As an example, the reaction of cyanide ion with HFA leads to some unusual heterocyclic compounds.

$$CF_3\overset{O}{\overset{\|}{C}}CF_3 + NaCN \longrightarrow (CF_3)_2\overset{CN}{\overset{|}{C}}ONa \xrightarrow{HFA} (CF_3)_2\overset{CN}{\overset{|}{C}}-O\overset{ONa}{\overset{|}{C}}(CF_3)_2 \qquad (398)$$

$$\downarrow$$

$$\underset{\text{ring: } (CF_3)_2C-C(=O)-NNa-C(CF_3)_2-O-}{(CF_3)_2C\overset{\overset{O}{\|}}{C}NNa \;/\; O\text{——}C(CF_3)_2} \longleftarrow \underset{\text{ring: } (CF_3)_2C-C(=NNa)-O-C(CF_3)_2-O-}{(CF_3)_2C\overset{\overset{NNa}{\|}}{C}O \;/\; O\text{——}C(CF_3)_2}$$

[396] (a) D. C. Bradley, M. E. Redwood, and C. J. Willis, *Proc. Chem. Soc.*, 416 (1964). (b) F. S. Fawcett, unpublished results; see P. E. Aldrich and W. A. Sheppard, *J. Org. Chem.*, **29**, 11 (1964), Table II, footnote b; see also C. G. Fritz, E. P. Moore, Jr., and S. Selman, U.S. Patent 3,114,778 (1963). (c) B. C. Anderson and G. R. Morlock, U.S. Patent 3,226,418 (1965); *Chem. Abstr.*, **64**, 9598 (1966).

[397] C. G. Krespan and W. J. Middleton, *Fluorine Chem. Rev.*, **1**, 145 (1967) cover this subject in a review of the chemistry of hexafluoroacetone.

[398] W. J. Middleton and C. G. Krespan, *J. Org. Chem.*, **32**, 951 (1967).

Addition of organometallic reagents to fluorinated aldehydes or ketones is similar to additions to aliphatic counterparts. However, reduction can be a major side reaction, and if R = $(CH_3)_2CH$, the only product is $(CF_3)_2CHOH$.[399]

$$(CF_3)_2C{=}O \xrightarrow[(2)\ BH]{(1)\ RMgX} (CF_3)_2C(R)OH + (CF_3)_2CHOH$$

Perfluorinated epoxides polymerize or isomerize to acid fluorides when treated with fluoride ion.[283b, 400, 401] A perfluoroalkoxide anion must be the intermediate.

$$R_fCF\overset{O}{—}CF_2 \xrightarrow{F^-} [R_fCF_2CF_2O^-] \xrightarrow{-F^-} R_fCF_2CF{=}O$$

$$[R_fCF_2CF_2O^-] \xrightarrow{nR_fCF\overset{O}{—}CF_2} R_fCF_2CF_2O\left(\underset{R_f}{CF}CH_2O\right)_n\underset{R_f}{CF}CF{=}O$$

Perfluorinated alcohols are not usually stable, other than perfluorocyclobutanol[402] or tertiary alcohols. Perfluoro-*t*-butyl alcohol is highly acidic (pK_a 5.4 compared to 20 for *t*-butanol) and may prove to be a useful reagent since a convenient synthesis (involving anionic intermediates) has now been devised.[403]

$$(CF_3)_2C{=}O + CCl_3Li \xrightarrow[-100^\circ]{THF} \left[CF_3\underset{CCl_3}{\overset{CF_3}{C}}—OH\cdot O\bigcirc\right] \xrightarrow[(2)\ SbF_3]{(1)\ H_2SO_4} (CF_3)_3COH$$

[399] (a) I. L. Knunyants, T-Y. Chen, and N. B. Gambaryan, *Bull. Acad. Sci. U.S.S.R., Div. Chem. Sci. (English Transl.)*, 647 (1960). (b) I. L. Knunyants, N. B. Gambaryan, T-Y. Chen, and E. M. Rokhlin, *Bull. Acad. Sci. U.S.S.R., Div. Chem. Sci. (English Transl.)*, 633 (1962). (c) E. T. McBee, S. Resconich, L. R. Belohlav, and H. P. Braendlin, *J. Org. Chem.*, **28**, 3579 (1963).

[400] (a) D. Sianesi, A. Pasetti, and F. Tarli, *J. Org. Chem.*, **31**, 2312 (1966). (b) I. L. Knunyants, V. V. Sohkina, and V. V. Tyuleneva, *Proc. Acad. Sci. U.S.S.R., Chem. Sect. (English Transl.)*, **169**, 722 (1966). (c) R. E. leBleu and J. H. Fassnacht, U.S. Patent 3,293,306 (1966); *Chem. Abstr.*, **66**, 104686g (1967).

[401] (a) D. P. Carlson and A. S. Milian, *4th Intern. Symp. Fluorine Chem., Estes Park, Colorado, July, 1967*, Abstracts, p. 146. (b) W. H. Gumprech, *ibid.*, p. 148. (c) J. L. Warnell, *ibid.*, p. 142.

[402] S. Andreades and D. C. England, *J. Am. Chem. Soc.*, **83**, 4670 (1961).

[403] R. Filler and R. M. Schure, *J. Org. Chem.*, **32**, 1217 (1967).

Fluorinated alcohols of the type R_fCH_2OH and $(R_f)_2CHOH$ are readily available; they are easily converted to alkoxides which are useful in nucleophilic reactions to introduce these partially fluorinated alkoxide groups.[404]

$$\text{cyclo-}[N{=}PCl_2]_3 \xrightarrow{6NaOCH_2R_f} \text{cyclo-}[N{=}P(OCH_2R_f)_2]_3 \quad (81\%)$$

Bis (trifluoromethyl)nitroxy compounds are readily prepared from the sodium salt of bis(trifluoromethyl)hydroxylamine.[405] The ionic reactions of this salt (Chart 6-15) greatly expand the radical-type reactions of $(CF_3)_2NO\cdot$ for introducing the $(CF_3)_2NO$ group into organic compounds.[173e]

Synthesis:

$$CF_3NO \xrightarrow[(2)\ HCl/H_2O/Hg]{(1)\ h\nu} (CF_3)_2NOH \xrightarrow[THF]{NaOH} (CF_3)_2NO^-Na^+$$

Reactions:

$$(CF_3)_2NO^-Na^+ + CH_3I \longrightarrow (CF_3)_2NOCH_3 \quad (28\%)$$

$$+ (CH_3)_3SiCl \longrightarrow (CF_3)_2NOSi(CH_3)_3 \quad (47\%)$$

$$+ C_6F_5N \longrightarrow 4\text{-}[(CF_3)_2NO]C_5F_4N \quad (28\%)$$

$$+ CF_3\overset{O}{\overset{\|}{C}}Cl \longrightarrow (CF_3)_2NO\overset{O}{\overset{\|}{C}}CF_3\ (88\%) \xrightarrow{\Delta} (CF_3)_3N$$

$$+ COF_2 \longrightarrow (CF_3)_2NO\overset{O}{\overset{\|}{C}}F\ (60\%) + [(CF_3)_2NO]_2\overset{O}{\overset{\|}{C}}\ (40\%)$$

CHART 6-15

f. *Sulfur.* The sulfur fluorine anion, SF_5^-, has been postulated in hydrolysis[119] of SF_5Cl and in the Friedel–Crafts reaction of SF_5Cl with benzene (chlorobenzene is only product), although the majority of SF_5Cl chemistry is explained by

[404] M. V. Lenton and B. Lewis, *J. Chem. Soc.* (A), 665 (1966).
[405] R. E. Banks, R. N. Haszeldine, and D. L. Hyde, *Chem. Commun.*, 413 (1967).

homolytic dissociation[182–186] (see Section 6-1A.2). The adduct of sulfur tetrafluoride with cesium fluoride is definitely an intermediate in synthesis of SF_5Cl,[406]

$$CsF + SF_4 \rightleftharpoons CsSF_5 \xrightarrow{Cl_2} SF_5Cl$$

and a stable salt of the SF_5 anion was also isolated by Tunder and Siegel.[407]

$$(CH_3)_4NF + SF_4 \longrightarrow (CH_3)_4\overset{+}{N}SF_5^-$$

However, the SF_5 anion is of sufficiently low stability that most of its salts cannot be isolated and, other than chlorination, it has not been successfully used for synthetic reactions.

The silver and mercury salts of trifluoromethyl thiol are useful intermediates for introducing the CF_3S group.[408]

$$\underset{(\text{or } CF_3SAg)}{RX + (CF_3S)_2Hg} \longrightarrow \underset{(\text{or } AgX)}{RSCF_3 + HgX_2}$$

This reaction is particularly convenient because the salts are easily prepared.

$$2CS_2 + 3HgF_2 \longrightarrow (CF_3S)_2Hg + 2HgS$$

$$CS_2 + 3AgF \longrightarrow CF_3SAg + Ag_2S$$

Although the reactions with the silver salts are probably ionic, the mercury salt appears to be covalent and may react with halides by a four-center concerted reaction. The displacement goes sufficiently easily so that polysubstituted CF_3S compounds are readily prepared.[408c]

$$CBr_2Cl_2 + (CF_3S)_2Hg \xrightarrow[2\text{ hr}]{100°} \underset{(67\%)}{C(SCF_3)_4} \xrightarrow[\text{(radical mechanism)}]{\Delta} (CF_3S)_2C{=}C(SCF_3)_2 + CF_3SSCF_3$$

$$BrCH_2CH_2Br + (CF_3S)_2Hg \xrightarrow[16\text{ hr}]{130°} \underset{(61\%)}{CF_3SCH_2CH_2SCF_3}$$

[406] C. W. Tullock, D. D. Coffman, and E. L. Muetterties, *J. Am. Chem. Soc.*, **86**, 357 (1964).

[407] R. Tunder and B. Siegel, *J. Inorg. Nucl. Chem.*, **25**, 1097 (1963).

[408] (a) E. H. Man, D. D. Coffman, and E. L. Muetterties, *J. Am. Chem. Soc.*, **81**, 3575 (1959). (b) H. J. Emeleus and D. E. MacDuffie, *J. Chem. Soc.*, 2597 (1961). (c) J. F. Harris, Jr., *J. Org. Chem.*, **32**, 2063 (1967).

These reactions (which we tentatively classify as ionic) are a useful complement to the radical-type reactions[72, 73, 188] of CF_3SH and CF_3SCl for introducing a CF_3S group.

Thiocarbonyl fluoride does not add cesium or potassium fluoride to give the expected salt, $MSCF_3$, but instead condenses.[409] Without solvent at −78°C, only a polymer of CSF_2 is formed, but at 0°C, the product is the thiocarbonate.

$$\underset{\text{FCF}}{\overset{\text{S}}{\|}} + \text{CsF} \xrightarrow[-44^\circ]{CH_3CN} F_3CS\overset{\text{S}}{\overset{\|}{C}}F + \underset{\text{chiefly}}{F_3CS\overset{\text{S}}{\overset{\|}{C}}SCF_3}$$

Presumably CF_3S^- is formed as a moderately free anion which initiates polymerization of CF_2S.

Trifluoromethylsulfonate anion is extremely stable and a good leaving group. Thus, perfluoroalkanesulfonate esters are potent alkylating agents.[410]

$$CF_3SO_2OCH_2CF_3 + 2Et_2HN \longrightarrow CF_3CH_2NEt_2 + CF_3SO_2^- H_2\overset{+}{N}Et_2$$

Trifluoromethylsulfinate salts are known, and the free acid adds to benzoquinone.[411]

$$CF_3SO_2Cl \xrightarrow[H_2O]{KSO_3} \underset{(68\%)}{CF_3SO_2K} \xrightarrow[\text{benzoquinone}]{H^+,\ H_2O} \text{2-}(SO_2CF_3)\text{-1,4-}C_6H_3(OH)_2$$

$$(CF_3SO_2)_2Zn + CF_3SCl \longrightarrow CF_3\overset{O}{\overset{\|}{\underset{\underset{O}{\|}}{S}}}SCF_3$$

The zinc salt reacts with trifluoromethylsulfenyl chloride[411b]; this result suggests that a CF_3SO_2 group may be introduced by reaction of trifluoromethylsulfinate salt with active organic halides.

[409] A. Haas and W. Klug, *Angew. Chem. Intern. Ed. Engl.*, **5**, 845 (1966).

[410] (a) R. L. Hansen, *J. Org. Chem.*, **30**, 4322 (1965). (b) J. Burdon and V. C. R. McLaughlin, *Tetrahedron*, **21**, 1 (1965).

[411] (a) R. M. Scribner, *J. Org. Chem.*, **31**, 3671 (1966). (b) R. N. Haszeldine and J. M. Kidd, *J. Chem. Soc.*, 2901 (1955).

6-1D. FLUOROCARBONIUM IONS

Highly electronegative atoms such as fluorine are expected to be poor substituents for carbonium ion species. However, several fluorinated carbonium ions have recently been detected using spectral techniques and are stable in solution.[412]

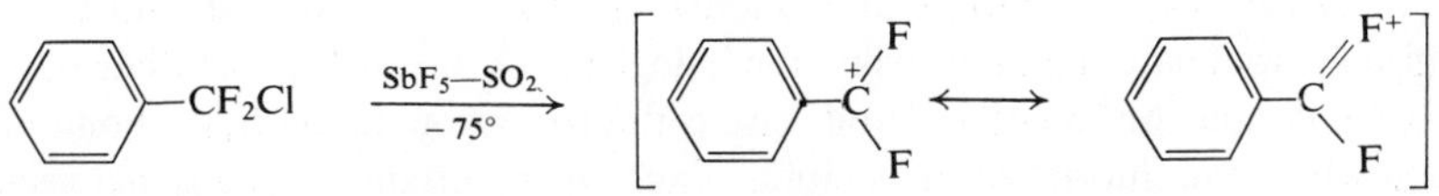

$$C_6H_5CF_2Cl \xrightarrow[-75°]{SbF_5—SO_2} \left[C_6H_5\overset{+}{C}F_2 \longleftrightarrow C_6H_5C(F)=F^+ \right]$$

The α-fluorines were suggested to stabilize the carbonium ion by conjugation of their unshared electron pairs into the vacant p-orbital of the positive carbon, offsetting the inductive destabilizing effect. For phenyldifluorocarbonium ion, the F^{19} nmr signal is a triplet shifted 75 ppm downfield from that of benzotrifluoride. The size of this downfield shift is interpreted as evidence for a much lower electron density on fluorine than normal. The ions that have been generated for spectral study are

$$C_6H_5\overset{+}{C}F_2,\quad (C_6H_5)_2\overset{+}{C}F,\quad (C_6H_5)_2\overset{+}{C}CF_3,\quad (C_6H_5)_2\overset{+}{C}(R)CF_3 \ \text{(R is } CH_3 \text{ or cyclopropyl)},$$

$$(CH_3)_2\overset{+}{C}F,\quad C_6H_5\overset{+}{C}(CH_3)F,\quad (C_6F_5)_3\overset{+}{C},\quad (p\text{-}FC_6H_4)_3\overset{+}{C},\quad C_6F_5\overset{+}{C}H_2$$

The inductive destabilizing effect of CF_3 and F are similar, but the CF_3 cannot contribute a stabilizing resonance effect. Consequently, the ion $(CF_3)_3C^+$ should be much less stable than F_3C^+. In support of this argument, Olah and Pittman[412b] found that several alcohols with only CF_3 groups do not form carbonium ions but are instead protonated on alcohol groups. Fluorinated ketones with more than three α-fluorines are abnormal and do not protonate.

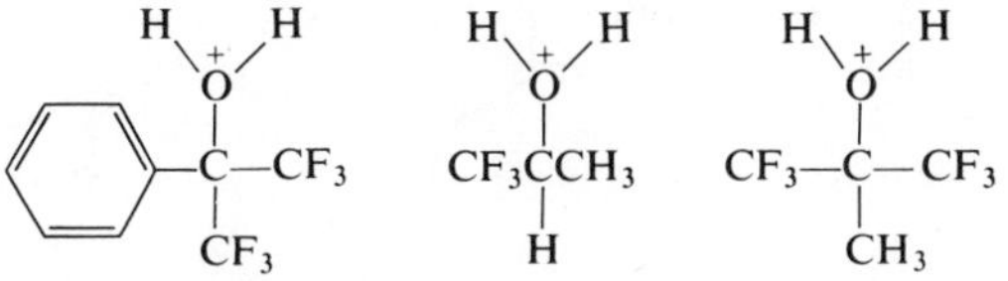

$$C_6H_5C(CF_3)_2\overset{+}{O}H_2 \qquad CF_3CH(CH_3)\overset{+}{O}H_2 \qquad CF_3—C(CH_3)(\overset{+}{O}H_2)—CF_3$$

[412](a) G. A. Olah, C. A. Cupas, and M. B. Comisarow, *J. Am. Chem. Soc.*, **88**, 362 (1966). (b) G. A. Olah and C. U. Pittman, Jr., *ibid.*, 3310. (c) G. A. Olah, R. D. Chambers, and M. B. Comisarow, *J. Am. Chem. Soc.*, **89**, 1268 (1967). (d) G. A. Olah and M. B. Comisarow, *ibid.*, 1027. (e) R. Filler, C.-S. Wang, M. A. McKinney, and F. N. Miller, *ibid.*, 1026. (f) G. A. Olah and J. M. Bollinger, *ibid.*, 4744. (g) G. A. Olah and T. E. Kiovsky, *ibid.*, 5692.

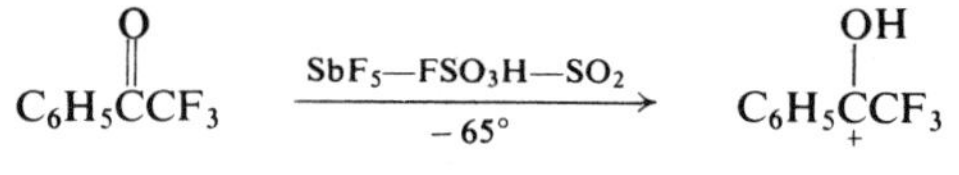

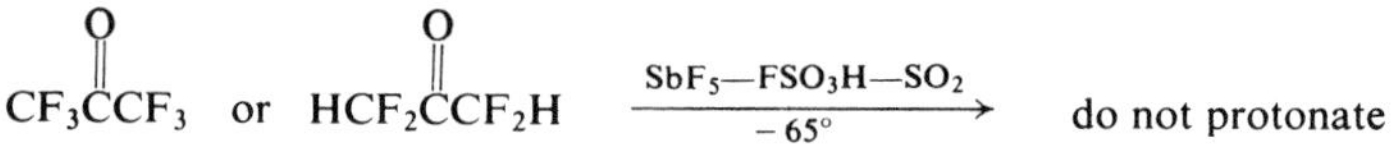

$$C_6H_5\overset{\overset{O}{\|}}{C}CF_3 \xrightarrow[-65^\circ]{SbF_5-FSO_3H-SO_2} C_6H_5\underset{+}{\overset{\overset{OH}{|}}{C}}CF_3$$

$$CF_3\overset{\overset{O}{\|}}{C}CF_3 \text{ or } HCF_2\overset{\overset{O}{\|}}{C}CF_2H \xrightarrow[-65^\circ]{SbF_5-FSO_3H-SO_2} \text{do not protonate}$$

2,3-Dihalo-2,3-dimethylbutanes ionize in SF_5–SO_2 solution at $-60°C$ to give bridged halonium ions when the halogen is Cl, Br, or I. 2,3-Difluoro-2,3-dimethylbutane, on the other hand, gave α-fluoroisopropyldimethylcarbonium ion in which the fluorine atom is rapidly exchanging intramolecularily between the two equivalent sites.[412f]

The tris(pentafluorophenyl)carbonium ion is much less stable than trityl carbonium ion[412d, e] suggesting that resonance-stabilizing effects into ortho and para positions do not completely offset the destabilizing inductive effect from

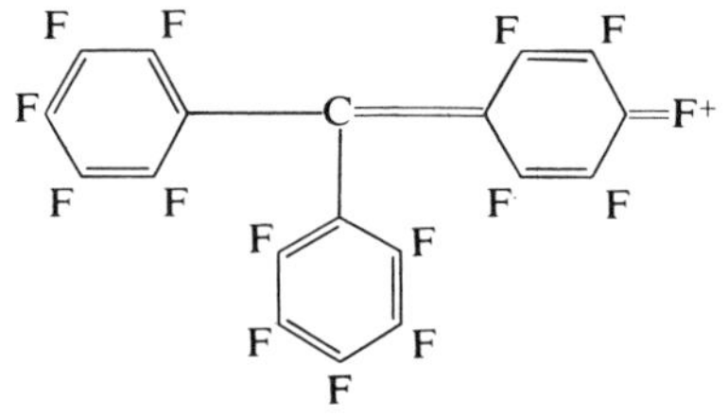

the fifteen fluorines. The resonance contribution must be large since the *o*- and *p*-fluorines show very large downfield F^{19} shifts, indicating large deshielding effects.[412d]

Protonation of mono-, di-, tri-, and tetrafluorobenzenes in fluorosulfonic acid–antimony pentafluoride solution gives solutions of stable fluorobenzenonium ions (fluorocyclohexadienyl cations).[412g] When the solutes were diluted with SO_2, sulfonylation usually occurred. The structures of these ions were inferred from nmr studies.

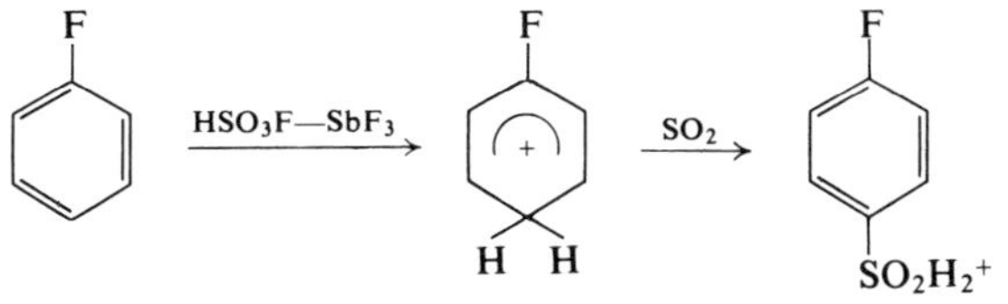

$$C_6H_5F \xrightarrow{HSO_3F-SbF_3} [FC_6H_5\text{H}]^+ \xrightarrow{SO_2} p\text{-}FC_6H_4SO_2H_2^+$$

The synthetic utility of fluorocarbonium ions is much more limited than for radicals or carbanions. Perfluorinated acylcarbonium ions have been isolated and used to acylate benzene.[413]

[413] E. Lidner and H. Kranz, *Chem. Ber.*, 99, 3800 (1966).

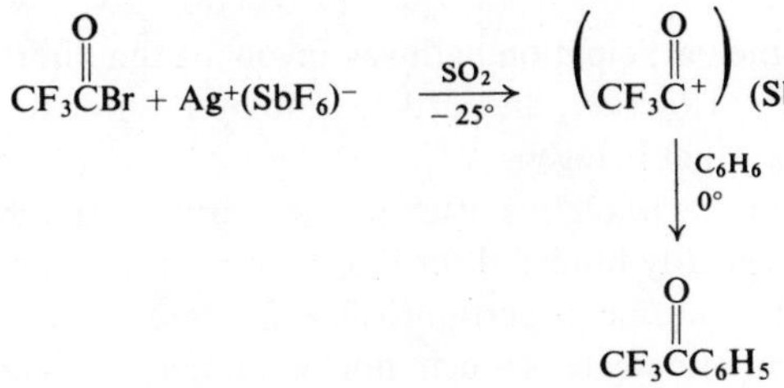

$$CF_3\overset{O}{\overset{\|}{C}}Br + Ag^+(SbF_6)^- \xrightarrow[-25^\circ]{SO_2} \left(CF_3\overset{O}{\overset{\|}{C}}{}^+\right)(SbF_6)^- \xrightarrow[0^\circ]{C_6H_6} CF_3\overset{O}{\overset{\|}{C}}C_6H_5$$

Fluoronium ions are also postulated as intermediates in reaction of trifluoroacetic acid with 5-fluoro-1-pentyl and 5-fluoro-2-pentyltosylate.[414a] Arguments based on rate data and product studies (see Chart 6-16) support a

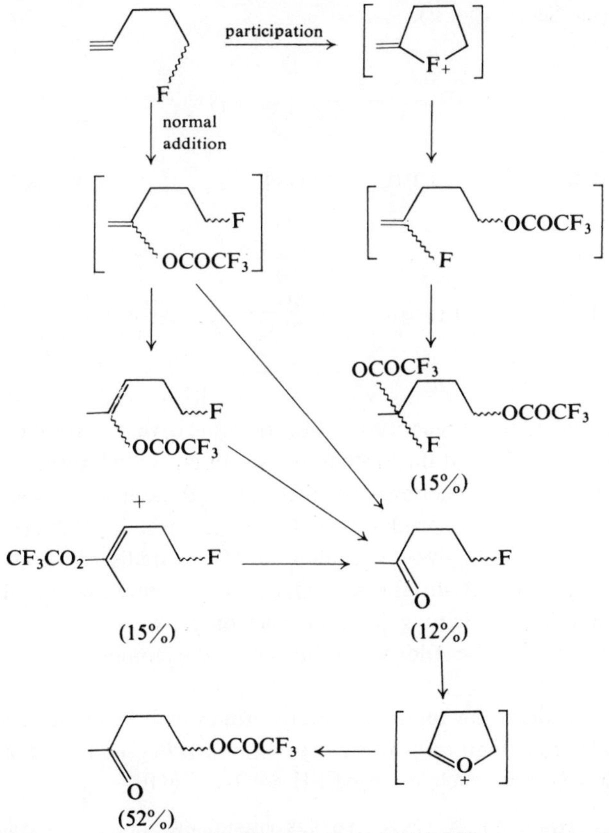

participation

F

normal addition

F_+

F

$OCOCF_3$

$OCOCF_3$

F

$OCOCF_3$

$OCOCF_3$

F

F

$OCOCF_3$

F

(15%)

+

CF_3CO_2

F

F

O

(15%)

(12%)

$OCOCF_3$

O

O^+

(52%)

CHART 6-16. Reaction of 5-fluoro-1-pentyne with trifluoroacetic acid.

[414] (a) P. E. Peterson and R. J. Bopp, *J. Am. Chem. Soc.*, **89**, 1283 (1967). (b) Electron impact-induced dehydrohalogenation studies on alkyl halides show that fluorine behaves differently than chlorine; 1-chlorohexane gives a five-ring chloronium ion as the most abundant ion in the mass spectra [W. Carpenter, A. M. Duffield, and C. Djerassi, *Chem. Commun.*, 1022 (1967).]

competition between the participation pathway involving the fluorocarbonium ion and normal addition. However, the participation by fluorine does not appear to be as effective as for other halogens.[414b]

Electrophilic additions to olefins often require carbonium ion intermediates. This type of addition is greatly hindered for fluoroolefins because of the electronegative character of the fluorine or perfluoroalkyl groups; nucleophilic attack leading to a carbanion intermediate is much more common (see Section 6-1C). However, several types of strong electrophilic reagents add to certain fluoroolefins such as vinyl ethers or amines (the carbonium ion can be stabilized by resonance) or partially hydrogenated species. For example, nitrosyl fluoride can add by either a nucleophilic or electrophilic mechanism depending on olefin and conditions[415] (see Section 5-2E).

$$NOF \rightleftharpoons F^- + \overset{+}{N}O$$

$$\overset{\delta-}{C}F_2{=}\overset{\delta+}{C}F{-}\ddot{\underset{..}{O}}R \xrightarrow{\overset{+}{N}O} [ONCF_2{-}\overset{+}{C}FOR] \xrightarrow[\text{(or NOF)}]{F^-} ONCF_2CF_2OR$$

$$(\overset{\delta-}{F_3C})_2C{=}\overset{\delta+}{C}F_2 \xrightarrow{F^-} [(CF_3)_3C^-] \xrightarrow[\text{(or NOF)}]{\overset{+}{N}O} (CF_3)_3CNO$$

Electrophilic addition reactions of fluoroolefins have recently been reviewed critically,[322, 416] and the mechanism was discussed relative to nucleophilic and radical addition reactions. The mechanism of addition is often inferred from reaction conditions and product structure (orientation in adduct). The reagents that are claimed to give electrophilic additions are listed below. The olefins are usually only partially fluorinated, often substituted with hydrogen or a heteroatom group containing unshared electron pairs.

Types of Electrophilic Additions (with selected examples):

1. Halogens add at low temperature, sometimes with Lewis acid catalysis (radical addition under light or heat is more common): Br_2, I_2, IF, ICl, and Cl_2 with $FeCl_3$ add to olefins such as $CF_2{=}CFH$ or $CF_2{=}CFOR$.

[415] (a) B. L. Dyatkin, R. A. Bekker, Yu. S. Konstantinov, and I. L. Knunyants, *Proc. Acad. Sci. U.S.S.R., Chem. Sect. (English Transl.)*, **165**, 1200 (1965). (b) B. L. Dyatkin, E. P. Mochalina, R. A. Bekker, and I. L. Knunyants, *Bull. Acad. Sci. U.S.S.R., Div. Chem. Sci. (English Transl.)*, 564 (1966). (c) B. L. Dyatkin, E. P. Mochalina, R. A. Bekker, S. R. Sterlin, and I. L. Knunyants, *Tetrahedron*, **23**, 4291 (1967).

[416] B. L. Dyatkin, E. P. Mochalina, and I. L. Knunyants, *Russian Chem. Rev. (Uspekhi Khim., English Transl.)*, **35**, 417 (1966).

$$CF_2{=}CHF \xrightarrow{ICl} \begin{cases} \xrightarrow[\text{via } I^+ \text{ attack}]{\text{liq. phase, } 25^\circ} \overset{+}{C}F_2CHFI \xrightarrow{Cl^-} ClCF_2CHFI \ (90\%) \\ \xrightarrow[\text{via } Cl\cdot \text{ attack}]{h\nu \text{ in gas phase}} \cdot CF_2CHFCl \xrightarrow[\text{or } ICl]{I\cdot} ICF_2CHFCl \ (26\text{–}36\%) \end{cases} \quad (417)$$

2. Hydrogen halides usually require Lewis acid catalysis (HF with BF_3) or olefins with higher hydrogen content (see Section 4-5).

3. Inorganic acids, such as $HOSO_2F$, H_2SO_4, or SO_3, add easily to partially fluorinated olefins and, under drastic conditions such as 200°C, to perfluoroolefins.

$$CF_2{=}CF_2 + SO_3 \longrightarrow \begin{array}{c} CF_2{-}CF_2 \\ | \quad\quad | \\ SO_2{-}O \end{array} \quad (418)$$

4. Mercury(II) fluoride is highly electrophilic and is the only reagent that gives electrophilic addition to perfluoroisobutylene.

$$2CF_3CF{=}CF_2 + HgF_2 \xrightarrow[85^\circ]{HF} (CF_3)_2CFHgCF(CF_3)_2 \ (68\%) \quad (419)$$

5. Chlorocarbons add in presence of aluminum chloride. Acyl halides also add under influence of aluminum chloride.[420]

Reaction	% Yield
$CF_2{=}CF_2 + CFCl_3 \xrightarrow{AlCl_3} CCl_2FCF_2CF_2Cl$	55
$CF_2{=}CF_2 + CCl_4 \xrightarrow{AlCl_3} CCl_3CF_2CF_2Cl$	88
$CF_2{=}CF_2 + CHCl_2F \xrightarrow{AlCl_3} CHClFCF_2CF_2Cl$	50
$CF_2{=}CF_2 + CHCl_3 \xrightarrow{AlCl_3} CHCl_2CF_2CF_2Cl$	83

6. Formaldehyde and water add by acid catalysis; a hydrolysis product, the hydroxy acid, is isolated.

$$CF_2{=}CHF + CH_2O + H_2O \xrightarrow[130^\circ]{90\%\ H_2SO_4} HOCH_2CHFCO_2H \ (25\%) \quad (421)$$

[417] E. R. Bissel, *J. Org. Chem.*, **29**, 252 (1964).

[418] D. C. England, M. A. Dietrich, and R. V. Lindsey, Jr., *J. Am. Chem. Soc.*, **82**, 6181 (1960).

[419] W. T. Miller, Jr., and M. B. Freedman, *J. Am. Chem. Soc.*, **85**, 180 (1963).

[420] D. D. Coffman, R. Cramer, and G. W. Rigby, *J. Am. Chem. Soc.*, **71**, 979 (1949).

[421] B. L. Dyatkin, E. P. Mochalina, and I. L. Knunyants, *Proc. Acad. Sci. U.S.S.R., Div. Chem. Sci. (English Transl.)*, **139**, 646 (1961).

7. Nitration, using nitric acid and HF to generate NO_2^+ (F–NO_2 addition).[422]

Reaction	% Yield
$CF_2{=}CF_2 + HNO_3$ in HF $\rightarrow$ $CF_3CF_2NO_2$	93
$CF_2{=}CH_2 + HNO_3$ in HF $\rightarrow$ $CF_3CH_2NO_2$	58
$CF_2{=}CFCl + HNO_3$ in HF $\rightarrow$ $CF_3CFClNO_2$	88

8. Nitrosation to add X–NO is usually a radical reaction but can be ionic under certain conditions[415] (see examples in above discussion).

Ionization and dissociation energies of the fluorides of the first-row elements have been measured and discussed.[423] The most important application of ionization of fluorocarbons is mass spectral studies for structure elucidation. Very little discussion of this technique has appeared in reviews and texts on mass spectrometry.[424] Recently, several mass spectral studies on fluorinated derivatives have been reported[425] and suggested to be useful in structure determination. Fragmentation involving loss of difluorocarbene was very common. Relative stability of fluorinated carbonium ions in the gas phase could be studied by this technique.

N-Fluoroammonium and iminonium ions have recently been reported.[426]

The perfluoroammonium cation has been prepared by several routes[426] and shown by nmr measurements to be tetrahedral with equivalent fluorines.

$$NF_3 + F_2 + SbF_5 \xrightarrow[80^\circ - 200^\circ]{3000\ \text{psi}} NF_4^+SbF_6^-$$

$$NF_3 + F_2 + AsF_5 \xrightarrow[\text{discharge}]{\text{electrical}} NF_4^+AsF_6^-$$

$$N_2F_4 + AsF_5 \xrightarrow{-78^\circ} N_2F_5^+AsF_6^- \xrightarrow[115^\circ]{F_2} NF_4^+AsF_6^-(\sim 5\%) + NF_3 + AsF_5$$

Cycloalkyldifluoroamines react with fluorosulfonic acid or sulfuric acid to give *N*-fluorimonium ions (resulting from ring expansion with fluoride leaving).

422 (a) A. I. Titov, *Proc. Acad. Sci. U.S.S.R., Chem. Sect. (English Transl.)*, **149**, 222 (1963). (b) I. L. Knunyants, L. S. German, and I. N. Rozhkov, *Bull. Acad. Sci. U.S.S.R., Div. Chem. Sci. (English Transl.)*, 1794 (1963).

423 W. C. Price, T. R. Passmore, and D. M. Roessler, *Discussions Faraday Soc.*, **35**, 201. (1963).

424 (a) J. R. Majer, *Advan. Fluorine Chem.*, **2**, 55 (1961). (b) F. W. McLafferty, *Interpretation of Mass Spectra*, Benjamin, New York, 1966.

425 (a) D. M. Gale, *Tetrahedron*, **24**, 1811 (1968). (b) J. L. Cotter, *J. Chem. Soc.* (B), 1162 (1967). (c) J. M. Miller, *J. Chem. Soc.* (A), 828 (1967). (d) R. C. Dobbie and R. G. Cavell, *Inorg. Chem.*, **6**, 1450 (1967). (e) R. G. Cavell and R. C. Dobbie, *Inorg. Chem.*, **7**, 101 (1968).

426 (a) W. E. Tolberg, R. T. Rewick, D. S. Stringham, and M. E. Hill, *Inorg. Chem.*, **6**, 1156 (1967). (b) K. O. Christe, J. P. Guertin, A. E. Pavlath, and W. Sawodny, *Inorg. Chem.*, **6**, 533 (1967). (c) *Chem. Eng. News*, **45**, 80 (June 12, 1967). (d) K. Baum, *J. Org. Chem.*, **32**, 3648 (1967).

The structure of these ions was based on nmr studies only.

$$\text{N,N-difluorocyclohexylamine} \xrightarrow[\text{or } FSO_3H]{H_2SO_4} \text{F–}\overset{+}{N}\text{=cyclohexylidene}$$

Alkyldiazonium cations are usually unstable; however, 2,2,2-trifluoroethyldiazonium ion is the first such species to be isolated (as determined by spectral measurements) and converted to a fluorosulfate.[427]

$$CF_3CHN_2 \xrightarrow[-60]{FSO_3H} CF_3CH_2N_2^+SO_3F^- \xrightarrow[-N_2]{\text{warming}} CF_3CH_2OSO_2F$$

6-2 FROM STABLE FLUORINATED MOLECULES

Stable fluorinated compounds are often used to introduce fluorinated units into organic compounds. In Section 6-1 we have already reviewed many synthetic aspects, since reactive intermediates must be generated from stable compounds. The objectives in this section are

1. to evaluate and summarize modern synthetic methods from the view of available classes of simple fluorinated compounds,
2. to cover in detail synthetic reactions that have not yet been discussed.

6-2A. FLUOROOLEFINS

Fluoroolefins are a particularly versatile and useful class for introducing fluorinated units. Radical and ionic reactions to generate reactive intermediates were discussed in Sections 6-1A, C, and D and good reviews are available.[1-6, 322, 416, 428] Tables at the end of the reviews by Walling and Huyser[1] and by Stacey and Harris[2] give good summaries of radical reactions. Similar summarizing tables are found for ionic reactions in the review by Chambers and Mobbs.[322] Regardless of mechanism, the addition is effectively represented by

$$RY + CF_2{=}CFX \rightarrow RCF_2CFXY \text{ or } RCFXCF_2Y$$

Thus, a fluorinated unit of two carbons (or more with side chains) is added to reagent R–Y. For radical-type reactions, telomerizations can lead to introduction of multiple units $R(CFXCF_2)_nY$. Substitution often occurs with carbanion reagents, so that a vinyl unit, such as $-CF{=}CF_2$, can also be introduced.

These additions are conveniently classified by mechanism. However,

[427] J. R. Mohrig and K. Keegstra, *J. Am. Chem. Soc.*, **89**, 5492 (1967).
[428] Ref. 6, Houben-Weyl, pp. 247-505.

2 + 1 *Addition:*

$$CCl_3SiF_3 \xrightarrow{140^\circ} [:CCl_2] \xrightarrow{CF_2{=}CFCl} F_2C\text{—}C(Cl_2)\text{—}C(Cl)(F) \ (87\%) \quad (228)$$

$$\xrightarrow{160^\circ\text{–}200^\circ} [:CF_2] + CCl_2{=}CFCl$$

A good difluorocarbene source under neutral conditions

$$F_2C{=}CFCF_3 + \text{alkaline hydrogen peroxide} \xrightarrow[\text{temp.}]{\text{low}} F_2C\text{—}CFCF_3 \ (\text{epoxide, }O) \xrightarrow{\Delta} [:CF_2] + CF_3\overset{O}{\overset{\|}{C}}F \quad (401, 429)$$

$$\xrightarrow{F^-} CF_3CF_2CF_2O(\underset{CF_3}{\underset{|}{C}}FCF_2O)_n\overset{O}{\overset{\|}{C}}F$$

The epoxides are reactive intermediates particularly for preparing difluorocarbenes and polymers.

2 + 2 *Addition:*

$$2CF_2{=}CF_2 \xrightarrow[\substack{(CO_2 \text{ or } (CH_3O)_2SO \\ \text{as inhibitor})}]{150^\circ\text{–}170^\circ} \begin{array}{l} CF_2\text{—}CF_2 \\ | \quad\quad | \\ CF_2\text{—}CF_2 \end{array} \quad \text{Freon C318, propellant gas} \quad (430)$$

$$CF_2{=}CF_2 + CF_2{=}CFOR \longrightarrow \begin{array}{l} CF_2\text{—}CFOR \\ | \quad\quad | \\ CF_2\text{—}CF_2 \end{array} \ (65\%) \quad (431)$$

$$\xrightarrow[175^\circ]{95\% \ H_2SO_4} \begin{array}{l} CF_2\text{—}C{=}O \\ | \quad\quad | \\ CF_2\text{—}CF_2 \end{array}$$

Good preparative method for reactive cyclobutanones

2 + 3 *Addition:*

$$CF_2{=}CFX + M + CH_2{=}CR_2 \longrightarrow \text{FX}, F_2 \ \text{—M—} \ R_2 \ (\text{five-membered ring}) \quad (432)$$

(where M is S or Se) (11–44%)

[429] H. S. Eleuterio and R. W. Meschke, French Patent 1,262,507 (1961); see British Patent 904,877; *Chem. Abstr.*, **58**, 12513b (1963).

[430] (a) E. E. Lewis and M. A. Naylor, *J. Am. Chem. Soc.*, **69**, 1968 (1947). (b) J. C. Tatlow and P. L. Coe, British Patent 991,016 (1965); see British Patent 616,246; *Chem. Abstr.*, 57, 14964d (1962). (c) I. L. Knunyants, G. A. Sokol'skii, and M. A. Belaventsev, Russian Patent 173,733 (1965); *Chem. Abstr.*, **64**, 1979d (1966).

[431] D. C. England, *J. Am. Chem. Soc.*, **83**, 2205 (1961).

[432] C. G. Krespan, *J. Org. Chem.*, **27**, 3588 (1962).

2 + 4 *Addition:*

+ $CF_3CF{=}CF_2$ ⟶ F_2 $F(CF_3)$ (433)

Series of fluoroolefins examined in Diels–Alder reaction to determine stereochemistry and develop evidence for the mechanism.

+ $CF_2{=}CF_2$ $\xrightarrow{475^\circ}$ F_2 F_2 (~ 25%) + F_2 F_2 (~ 50%) (114, 434)

700°–750°/5 mm

O OH ⟵ F_2 F_2 F_2 F_2

(overall yield 20%)
Best tropolone synthesis

RNH_2

H NR NR ⟶ [R N Ni N R]$_2$

Paramagnetic metal chelates useful for studies of spin density.

discrete intermediates are not always involved; in some cases a concerted process must operate. In this regard an important area of olefin chemistry is cycloaddition and Diels–Alder reactions to give cyclic products with three-, four-, five-, and six-membered rings depending on the reagents (see examples above). The products from these cycloadditions are often valuable precursors for further synthetic and mechanistic studies.

Cycloaddition reactions of fluoroolefins have been reviewed[435, 436]: 2 + 2

[433] (a) H. P. Braendlin, G. A. Grindahl, Y. S. Kim, and E. T. McBee, *J. Am. Chem. Soc.*, **84**, 2112 (1962). (b) H. P. Braendlin, A. Z. Zielinski, and E. T. McBee, *ibid.*, 2109.

[434] (a) J. J. Drysdale, W. W. Gilbert, H. K. Sinclair, and W. H. Sharkey, *J. Am. Chem. Soc.*, **80**, 3672 (1958). (b) W. R. Brasen, H. E. Holmquist, and R. E. Benson, *J. Am. Chem. Soc.*, **82**, 995 (1960). (c) D. R. Eaton, A. D. Josey, W. D. Phillips, and R. E. Benson, *J. Chem. Phys.*, **37**, 347 (1962).

[435] (a) J. D. Roberts and C. M. Sharts, *Org. Reactions*, **12**, 1 (1962). (b) W. H. Sharkey, *Fluorine Chem. Rev.*, **2**, in press. (c) D. R. A. Perry, *Fluorine Chem. Rev.*, **1**, 253 (1967). (d) Ref. 6, pp. 253 to 275.

[436] For a general discussion of 2 + 4 cycloaddition reactions, particularly mechanism, see A. Wasserman, *Diels-Alder Reaction*, Elsevier, Amsterdam and New York, 1965.

by Roberts and Sharts[435a] and Sharkey[435b]; 2 + 4 by Perry[435c]; and both 2 + 2 and 2 + 4 by Houben-Weyl.[435d] The summarizing tables in these reviews[435] should be consulted as an entry to the extensive primary literature in this area.

The mechanism of the cycloaddition reaction has long been of interest and is still only partially understood for selected specific cases. On the basis of quantum mechanical calculations, Woodward and Hoffmann[437] predicted that a 2 + 2 cycloaddition cannot go by a concerted mechanism. For simple fluoroolefins, Bartlett and co-workers[74, 77] have shown that the addition must involve a diradical intermediate. They have accumulated evidence based on product structure (orientation and stereochemistry) and kinetic and reactivity studies. For example, the addition of 1,1-dichlorodifluoroethylene to isoprene gives only one product. The most stable intermediate diradical is with α-Cl stabilization on

$CF_2{=}CCl_2$ + $CH_2{=}C(CH_3){-}CH{=}CH_2$ ⟶ [$CF_2{-}\dot{C}Cl_2$ / $H_2C{-}C(CH_3)\cdots\dot{C}H\cdots CH_2$] ⟶ $CF_2{-}CCl_2$ / $CH_2{-}C(CH_3){-}CH{=}CH_2$

the diene half so that only the predicted product is formed and no 3,4- or reverse 1,2-addition is found. Also no 1,4-addition (normal Diels–Alder reaction) nor cycloaddition of fluoroolefin to itself occurs under the reaction conditions. In general, fluoroolefins with geminal fluorine substituents cycloadd 1,2 to dienes to the complete exclusion of the normal 1,4-addition found for dienophiles such as ethylene, maleic anhydride, and tetracyanoethylene. No satisfactory explanation has been offered, but Roberts and Sharts[435a] have rationalized that relief of F–F repulsions would be much greater in formation of a four-member ring than of a six-member ring. The electron-deficient character of the π system promotes reaction of fluoroolefins with an electron-rich system, such as a vinyl ether, but the fluoroolefin will add to itself if no electron-rich system is available.

$CF_3(NC)C{=}C(CN)CF_3$ + $XHC{=}CH_2$ ⟶ zwitterionic intermediates [$CF_3(NC)C{-}C^-(CN)CF_3$ / $H_2C{-}C^+HX$] ⟶ cyclobutanes with F_3C, CN, CN, CF_3, X substituents (two stereoisomers)

[X = RO, RS, p-(RO)C_6H_4]

[437] R. Hoffmann and R. B. Woodward, *J. Am. Chem. Soc.*, **87**, 2046 (1965).

Olefins with very strong electron-withdrawing groups, such as *cis*- and *trans*-1,2-bis(trifluoromethyl)-1,2-dicyanoethylene, cycloadd very readily to electron-rich alkenes.[438] This type of cycloaddition must involve a dipolar intermediate, since the rate of cycloaddition is strongly influenced by solvent. Many of the additions are highly stereospecific; however, with certain substituted electron-rich olefins the stereochemistry of the fluorinated olefin is lost. All the experimental observations are best accommodated by the dipolar intermediate of the geometry shown where the dipoles interact to some extent.

Photosensitized cycloaddition reactions of fluoroolefins with 1,3-dienes usually give four-membered ring products although six-membered ring formation (Diels–Alder reaction) is competitive on occasion.[439] The results are consistent with the attack of the triplet diene on haloethylenes to form a biradical which rapidly cyclizes.

The 2 + 4 cycloadditions (Diels–Alder reaction) are predicted to be concerted,[437] and all available evidence indicates that they, indeed, are,[435c] but no significant mechanism studies have been done with fluorinated olefins (however, see Ref. 74f, 77).

Another important area of fluoroolefin chemistry is formation of adducts with transition metals and in preparation of fluoroorganometallic compounds. Work in this area is extensive and has been reviewed.[317] A typical example is the

$$Fe(CO)_5 + CF_2{=}CF_2 \xrightarrow{h\nu} \underset{\pi\text{ complex?}}{C_2F_4 \cdot Fe(CO)_4} \xrightarrow[\text{irradiation}]{\text{prolonged}} \text{(CO)}_4\text{Fe}\langle CF_2CF_2CF_2CF_2 \rangle \qquad (440)$$

reaction of tetrafluoroethylene with iron carbonyl. This area may have considerable opportunity for use in synthesis, since olefin reactivity should be greatly modified. An example of a synthesis using a transition metal with a fluorinated acetylene will be discussed in the next section.

Summary

The major synthetic uses of fluoroolefins are to

1. add two-carbon fluorinated units, such as $-CF_2CFX-$, by radical or ionic mechanisms;
2. introduce fluorinated vinyl group, such as $-CF{=}CF_2$, by substitution;
3. prepare polymers or telomers;
4. cycloadd for preparation of three-, four-, five-, and six-membered ring compounds that are valuable intermediates in a wide range of studies.

[438] S. Proskow, H. E. Simmons, and T. L. Cairns, *J. Am. Chem. Soc.*, **88**, 5254 (1966).
[439] N. J. Turro and P. D. Bartlett, *J. Org. Chem.*, **30**, 1849 (1965).
[440] R. Fields, M. M. Germain, R. N. Haszeldine, and P. W. Wiggans, *Chem. Commun.*, 243 (1967).

6-2B. FLUOROACETYLENES

Haloacetylenes,[441] $XC{\equiv}CX$ and $XC{\equiv}CH$, where X is halogen, lack stability and often decompose explosively. Fluoroacetylenes are no exception; however, perfluoroalkylacetylenes are easily handled and are of value synthetically for introducing fluorinated units.

Hexafluorobutyne, available as a laboratory reagent, is particularly effective as a highly electrophilic dienophile but also gives other useful adducts. The following examples illustrate the potential of this reagent in synthesis.

1. 4 + 2 Addition (Diels–Alder):

$$CF_3C{\equiv}CCF_3 + \text{durene (1,2,4,5-tetramethylbenzene)} \xrightarrow{200^\circ} \text{adduct } (40\%) \qquad (442a)$$

Hexafluorobutyne is one of the few reagents to give a Diels–Alder addition to durene; the adduct with benzene is unstable and gives further reaction.

2. Cycloaddition to inorganic reagents:

$$CF_3C{\equiv}CCF_3 + S \xrightarrow[\text{flow system}]{445^\circ} CF_3C{=}CCF_3 \text{ (with S—S bridge) } (80\%) \qquad (442b,c)$$

1,2-Dithietene is the only stable example of the unusual ring system and is a reactive reagent.[442c]

3. Radical addition:

$$H_2S + CF_3C{\equiv}CCF_3 \xrightarrow{h\nu} CF_3CH{=}C(SH)CF_3 \quad (60\%) \qquad (443)$$

When substituted by trifluoromethyl groups these novel vinyl thiols are stable.

4. Transition metal-promoted cycloadditions:

$$Rh(CO)_2Cl + CO + CF_3C{\equiv}CCF_3 \xrightarrow{150^\circ} \text{tetrakis(trifluoromethyl)cyclopentadienone} \qquad (444)$$

[441] K. M. Smirnov, A. P. Tomilov, and A. I. Shchekotikhin, *Russian Chem. Rev. (English Transl.)*, **36**, 326 (1967).

[442] (a) C. G. Krespan, B. C. McKusick, and T. L. Cairns, *J. Am. Chem. Soc.*, **83**, 3428 (1961). (b) C. G. Krespan, *J. Am. Chem. Soc.*, **83**, 3434 (1961). (c) C. G. Krespan and B. C. McKusick, *ibid.*, 3438.

[443] F. W. Stacey and J. F. Harris, Jr., *J. Am. Chem. Soc.*, **85**, 963 (1963).

[444] R. S. Dickson and G. Wilkinson, *J. Chem. Soc.*, 2699 (1964).

An adduct of the dienone is isolated with certain other transition metals. This dienone should be a valuable reagent for a variety of chemical programs.

5. Nucleophilic additions:

$$\text{CH}_2\text{=CH-aziridine (N-H)} + CF_3C{\equiv}CCF_3 \xrightarrow{-76^\circ} \text{1,2-divinylaziridine (N-C(CF}_3\text{)=CH(CF}_3\text{))} \xrightarrow{\text{room temp. overnight}} \text{azepine (CF}_3\text{, CF}_3\text{)} \quad (445)$$

The 1,2-divinylaziridine was used for mechanism studies of isomerization to azepine.

A selection of examples has been chosen for Table 6-5 to show the methods for preparing various fluorinated acetylenes and their uses in synthesis for introducing fluorinated units.

Fluorinated benzynes are a special class of reactive acetylenic intermediates that can be effectively used to introduce a tetrafluorophenyl or phenylene unit. Since an early report in 1962,[(449)] a series of papers has described the utility of this highly reactive intermediate in synthesis and compared it to benzyne. Pentafluorophenyllithium (and pentafluorophenylmagnesium halides) are prepared at low temperatures and are useful reagents for preparing a variety of pentafluorophenyl derivatives by normal organometallic reactions.[449, 450] (See Sections 6-1C.1b and 6-2C.) However, when warmed to over 0°C, lithium fluoride (or magnesium fluorohalide) splits out to give tetrafluorobenzyne which can be trapped with a variety of reagents.[451] Chart 6-17 includes a selection of reactions of perfluorobenzyne that illustrates the potential utility of the reagent in synthesis. It parallels benzyne in its reactions but is clearly more electrophilic and as a consequence gives high yields of Diels–Alder adducts particularly with

[445] E. L. Stogryn and S. J. Brois, *J. Am. Chem. Soc.*, **89**, 605 (1967).

[449] P. L. Coe, R. Stephens, and J. C. Tatlow, *J. Chem. Soc.*, 3227 (1962).

[450] D. D. Callander, P. L. Coe, and J. C. Tatlow, *Tetrahedron,* **22**, 419 (1966).

[451] (a) D. E. Fenton and A. G. Massey, *Tetrahedron,* **21**, 3009 (1965). (b) N. N. Varozhtsov, V. A. Barkhasch, N. G. Ivanova and A. K. Petrov, *Tetrahedron Letters,* 3575 (1964). (c) A. J. Tomlinson and A. G. Massey, *J. Organometal. Chem.*, **8**, 321 (1967). (d) J. P. N. Brewer and H. Heaney, *Tetrahedron Letters,* 4709 (1965). (e) D. D. Callander, P. L. Coe, and J. C. Tatlow, *Chem. Commun.*, 143 (1966). (f) H. Heaney and J. M. Jablonski, *Tetrahedron Letters,* 2733 (1967). (g) J. P. N. Brewer and H. Heaney, *Chem. Commun.*, 811 (1967). (h) J. P. N. Brewer, I. F. Eckard, R. Harrison, H. Heaney, J. M. Jaldonski, B. A. Marples, and K. G. Mason, *4th Intern. Symp. Fluorine Chemistry, Estes Park, Colorado, 1967,* Abstracts, p. 90. (i) J. P. N. Brewer, I. F. Eckhard, H. Heaney, and B. A. Marples, *J. Chem. Soc.* (C), 664 (1968). (j) S. C. Cohen, A. J. Tomlinson, M. R. Wiles, and A. G. Massey, *J. Organometal. Chem.*, **11**, 385 (1968).

Table 6-5

Fluorinated Acetylenes–Selected Information on Synthesis and Chemistry

Acetylene	Preparation	Typical reactions (% yield)	Comments
$HC{\equiv}CF$	Fluoromaleic anhydride $\xrightarrow[5\text{-}7\ mm]{650^\circ}$ (100%) $FBrC{=}CHBr + Mg \xrightarrow{THF}$ (80%)	Trimerizes to 1,2,4-Trifluorobenzene and polymerizes; metal salts readily formed. Forms salts with ϕ_3P and tertiary amines. Fluorine replaced in reaction with BuLi or sodium amide	Treacherously explosive in liquid state. Ref. 446a Properties and chemistry compared to other haloacetylenes. Ref. 446b
$FC{\equiv}CF$ $FC{\equiv}CC({=}O)F$	Difluoromaleic anhydride $\xrightarrow[1000^\circ]{500\ to}$ mixture of both acetylenes	$FC{\equiv}CC({=}O)F$ polymerizes	Ref. 446c
$(CH_3)_3CC{\equiv}CF$	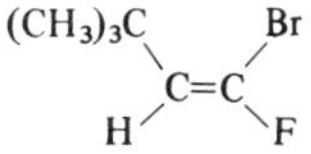 $(CH_3)_3C(H)C{=}C(Br)F \xrightarrow[120^\circ]{solid\ KOH}$	Unstable above 0°, forms oligomers, 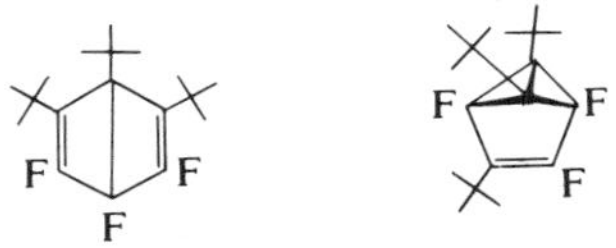 Two trimers have been definitely identified	A suggested prismane structure for another isomeric product has not been verified. Ref. 446d
$HC{\equiv}CCF_3$	$CF_3CCl{=}CCl_2 + 2Zn \xrightarrow[ZnCl_2]{solvent}$ $(CF_3C{\equiv}C)_2Zn \xrightarrow{H_2O} CF_3C{\equiv}CH$ (10%)		Claimed to be best synthesis; others are dehydrohalogenations of $CF_3CH{=}CHX$ and $CF_3CX{=}CH_2$, dehalogenations of $CF_3CX{=}CXH$, $SF_4 + HC{\equiv}CCO_2H$ (see ref. contained in 447a). Ref. 447a.

Reaction	Remarks
$+ H_2S \longrightarrow CF_3C(H){=}C(H)SH$ (73) $+ CF_3CH(SH)CH_2SH$ (~20) $+ (CF_3CH{=}CH)_2S$ (~5),	Adducts for $CF_3C{\equiv}CCF_3$ also. Ref. 443
$+ R_2EH \longrightarrow R_2ECH{=}CHCF_3$ where E = N or As,	Adducts for $CF_3C{\equiv}CCF_3$ also. Ref. 447b
$+ (CH_3)_2C{=}CH_2 \xrightarrow[145\text{–}220^\circ]{\Delta}$ $CH_2{=}C(CH_3){-}CH_2{-}CH{=}C(H)CF_3$ (15%)	Part of a study of allylic addition of olefins to activated acetylenes. Ref. 447c
$+ Fe(CO)_5 \longrightarrow$ 2,5-bis(CF_3)cyclopentadienone–$Fe(CO)_3$	Series of transition-metal acetylides, $CF_3C{\equiv}C{-}M$, also prepared. Ref. 447d
$CF_3C{\equiv}CZnX + CuCl_2 \longrightarrow (CF_3C{\equiv}C{\text{—}})_2$	$CF_3C{\equiv}CCl$ and other acetylenic by-products also formed. Ref. 447a
$CF_3C{\equiv}CLi$ and $CF_3C{\equiv}CMgX$ added to carbonyl reagents or coupled to RX to introduce $CF_3C{\equiv}C{-}$ grouping	See Table 6-4C. Ref. 320f,o

Table 6-5 (Cont.)

Acetylene	Preparation	Typical reactions (% yield)	Comments
$HC{\equiv}CC_6F_5$	$C_6F_5CHBrCH_2Br \xrightarrow[150°]{KOH}$ (low yield) $C_6F_5CClH{-}CCl_3 \xrightarrow{AlCl_3}$ $C_6F_5CCl{=}CCl_2 \xrightarrow[THF]{Mg}$ (68%)	Silver and copper acetylides prepared; chlorine added; did not hydrate readily $+ Fe(CO)_5 \xrightarrow{120°}$ (2,5-bis(C_6F_5)cyclopentadienone, H, H)–$Fe(CO)_3$	$C_6F_5C{\equiv}CCl$ also prepared by similar route. Ref. 447e Series of complexes prepared but free ketone not isolated (see Ref. 444). Ref. 447f
$HC{\equiv}CSF_5$	$SF_5Cl + HC{\equiv}CH \longrightarrow SF_5CH{=}CHCl, \xrightarrow{Br_2} \xrightarrow{Base} \xrightarrow{Zn} HC{\equiv}CSF_5$	Adds methanol to give vinyl ether; diazomethane to give pyrazoles butadiene $+ \xrightarrow{120°}$ (SF_5-cyclohexadiene) $\xrightarrow{Pt}$ $C_6H_5SF_5$	Ref. 447g
$HC{\equiv}C(CH_2)_nF$ (n is 3 to 8)	$F(CH_2)_nX + LiC{\equiv}CH \longrightarrow$ (up to 92%)	Normal acetylene reactions (ω-fluorine not attacked); variety of reactions reported, $\xrightarrow{C_2H_5MgCl} F(CH_2)_nC{\equiv}CMgCl \xrightarrow{CO_2} F(CH_2)_nC{\equiv}CCO_2H$	Part of general study on toxic fluorine compounds; these ω-fluoroacetylenes useful in synthesis; $F(CH_2)_nC{\equiv}CCl$ also prepared. Ref. 447h
$HC{\equiv}CC(OH)(R)CF_3$ (R is alkyl or aryl)	$RC({=}O)CF_3 + HC{\equiv}CCa \longrightarrow$ (up to 7%)	$\xrightarrow{\phi NCO} HC{\equiv}CC(R)(CF_3)OC({=}O)NH\phi$	Diadduct $RC(OH)(CF_3)C{\equiv}CC(OH)(CF_3)R$ also isolated in low yields. Ref. 447i

$HC{\equiv}C\underset{\mid}{C}H{-}\underset{\mid}{C}H_2$ with $CF_2{-}CF_2$ bridging (ring)	$HC{\equiv}CCH{=}CH_2 + CF_2{=}CF_2 \xrightarrow{\Delta} HC{\equiv}CCH{=}CF_2$ 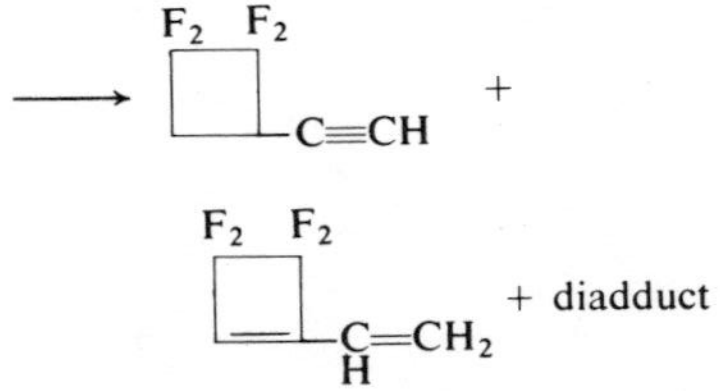 $\longrightarrow$ F_2F_2-cyclobutyl$-C{\equiv}CH$ + F_2F_2-cyclobutenyl$-\underset{H}{C}{=}CH_2$ + diadduct		Ref. 447j
$HC{\equiv}CSCF_3$	$HC{\equiv}CMgX + ClSCF_3$		Reactions similar to other fluoralkyl acetylenes and $CF_3SC{\equiv}CSCF_3$ are expected. Ref. 447k
$HC{\equiv}CN(CF_3)_2$	$ClCH{=}CHN(CF_3)_2 \xrightarrow{KOH}$ (60%)	Polymerized, cycloadded to tetrafluoroethylene	$(CF_3CF_2CF_2)_2NC{\equiv}CH$ also reported. Ref. 447l
$CF_3C{\equiv}CCF_3$ $R_fC{\equiv}CR_f$	(1) dehydrohalogenations of $CF_3\underset{Cl}{C}{=}\underset{Cl}{C}CF_3$ (2) $HO_2CC{\equiv}CCO_2H + SF_4 \longrightarrow$ (3) n-$CF_2{=}CF_2 + CCl_4 \longrightarrow$ $Cl(CF_2CF_2)_nCCl_3 \longrightarrow$ $[Cl(CF_2CF_2)_nCCl{=}]_2 \longrightarrow$	See examples in text of typical reactions of $CF_3C{\equiv}CCF_3$ (Ref. 441 to 445)	Details for preparations 1 and 3 given in Ref. 448a; see references contained therein for information on other preparations. Ref. 448a

Table 6-5 (Cont.)

Acetylene	Preparation	Typical reactions (% yield)	Comments
		+ $(PCF_3)_4 \xrightarrow{170^\circ}$ 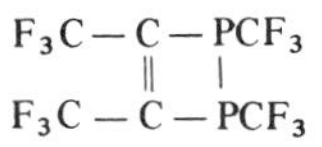 $F_3C-C(-PCF_3)=C(-PCF_3)-CF_3$ (four-membered ring: $F_3C-C-PCF_3$ / $F_3C-C-PCF_3$) (55)	Also for $(PCF_3)_5$ to give corresponding 5 member ring product. Ref. 448b
		+ P $\xrightarrow{I_2}$  [bicyclic diphosphorus cage product bearing six CF_3 groups] (43)	Similar reaction also for arsenic. Ref. 448c
		$\xrightarrow[275^\circ]{CH_3I \text{ or } I_2}$ hexakis(trifluoromethyl) benzene (75)	Hydrolyzed by base to pentakis(trifluoromethyl)benzoic acid. Ref. 448d
		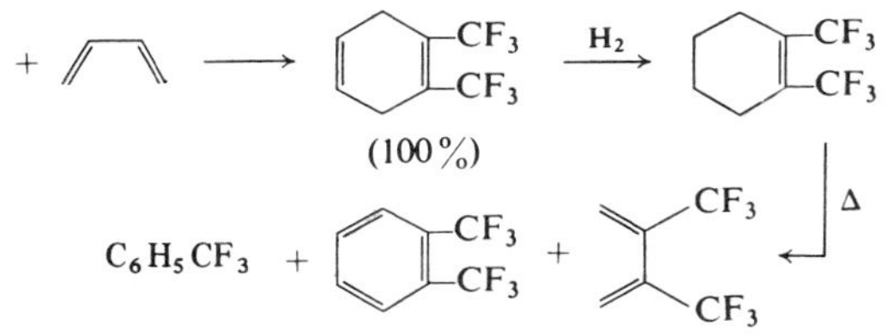 + butadiene ⟶ 1,2-bis(CF_3)cyclohexa-1,4-diene (100%) $\xrightarrow{H_2}$ 1,2-bis(CF_3)cyclohexene $\xrightarrow{\Delta}$ 2,3-bis(CF_3)butadiene + 1,2-bis(CF_3)benzene + $C_6H_5CF_3$	Diene product used for polymerization studies. Ref. 448e
		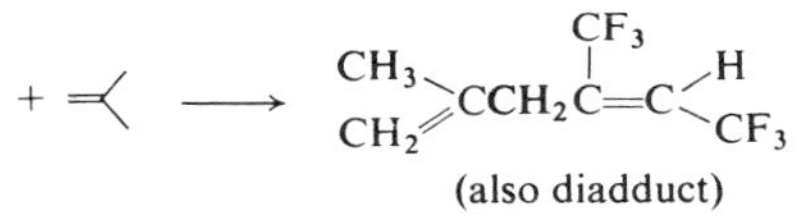 + isobutylene ⟶ $CH_2{=}C(CH_3)CH_2C(CF_3){=}CH(CF_3)$ (also diadduct)	Part of general study of allylic additions to acetylenes (also $CF_3C{\equiv}CH$). Ref. 447c

		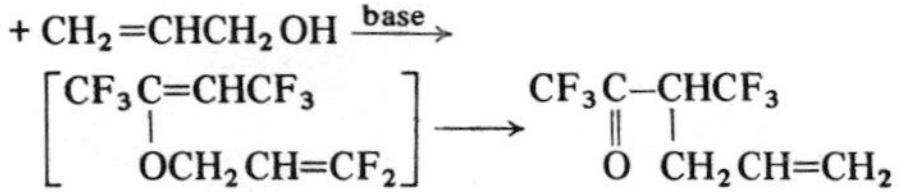 $+ CH_2{=}CHCH_2OH \xrightarrow{\text{base}}$ $[CF_3C(OCH_2CH{=}CF_2){=}CHCF_3] \longrightarrow CF_3C(=O){-}CH(CH_2CH{=}CH_2)CF_3$	An example of Claisen rearrangement under extremely mild conditions. Ref. 448f
		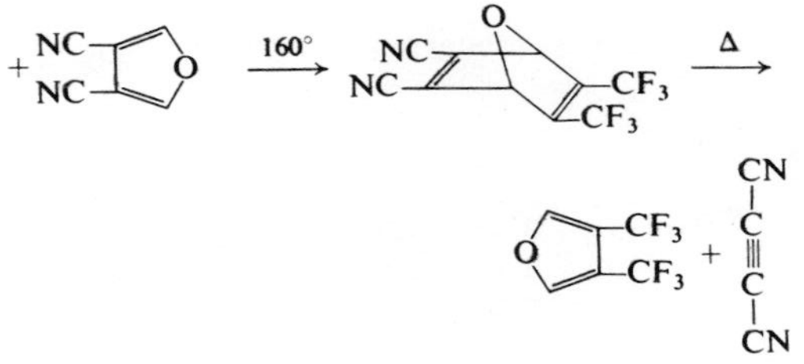 + 3,4-dicyanofuran $\xrightarrow{160^\circ}$ NC, NC, CF$_3$, CF$_3$ oxabicyclic adduct $\xrightarrow{\Delta}$ 3,4-bis(trifluoromethyl)furan + $NC{-}C{\equiv}C{-}CN$	Hexafluorobutyne also added to furan. Ref. 448g
		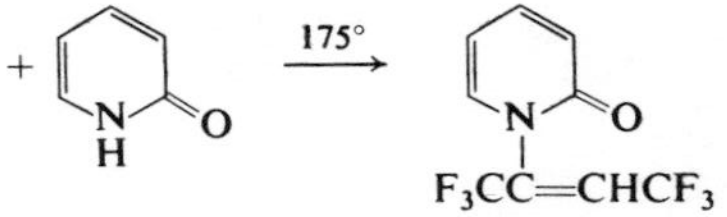 + 2-pyridone $\xrightarrow{175^\circ}$ N-substituted 2-pyridone, $N{-}C(CF_3){=}CHCF_3$ (61)	Trying to get a diene reaction of pyridone. Ref. 448h
		 $+ CF_2{=}C{=}CF_2 \longrightarrow$ cyclobutene with F_3C, CF_3, F_2, $=CF_2$ (50)	Perfluoroallene dimers also formed. Ref. 448i
$C_6H_5C{\equiv}CCF_3$	$C_6H_5CHBrCHBrCF_3 \xrightarrow{NEt_3}$ $\xrightarrow[230^\circ]{KOH}$ (76%)	$+Br_2 \longrightarrow C_6H_5CBr{=}CBrCF_3$, $+Cl_2 \longrightarrow C_6H_5CCl_2{-}CCl_2CF_3$	Ref. 448j
$RC{\equiv}CC_6F_5$ (R is alkyl or aryl)	$RC{\equiv}CLi + C_6F_6 \longrightarrow C_6F_5C{\equiv}CR$, $[+(RC{\equiv}C)_2C_6F_4]$		Ref. 448k

Table 6-5 (Cont.)

Acetylene	Preparation	Typical reactions (% yield)	Comments
$C_6F_5C{\equiv}CC_6F_5$	$2C_6F_5MgBr + IC{\equiv}CI \xrightarrow{CoCl_2} (56\%)$	Triple bond readily hydrogenated, brominated and oxidatively cleaved; does not hydrate or carbonylate. 4,4′-positions easily substituted by nucleophiles 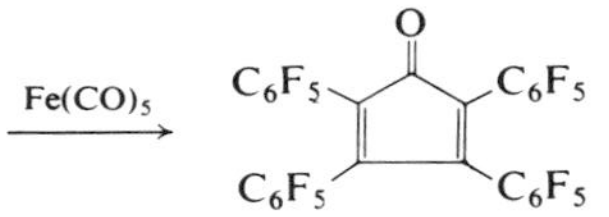 $\xrightarrow{Fe(CO)_5}$ tetrakis(C_6F_5)cyclopentadienone (C_6F_5, C_6F_5, C_6F_5, C_6F_5; C=O) (and metal complexes)	Ref. 448l
	$Ar_3P + C_6F_5CH_2Br \xrightarrow{BuLi}$ $Ar_3\overset{+}{P}\overset{-}{C}HC_6F_5 \xrightarrow{C_6F_5COCl} Ar_3\overset{+}{P}CC_6F_5$ ($=C(-O^-)-C_6F_5$) $\xrightarrow[\text{at 10 mm}]{310^\circ}$ (68% overall)	Excellent dienophiles 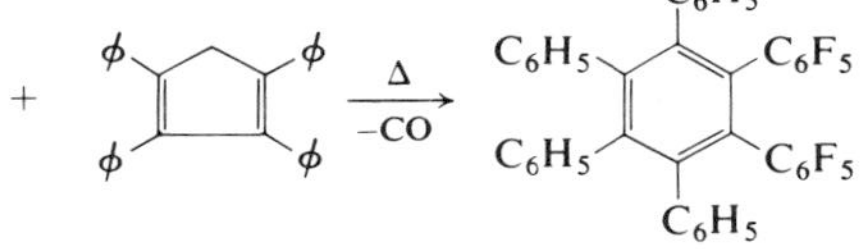 + tetraphenylcyclopentadiene (ϕ, ϕ, ϕ, ϕ) $\xrightarrow[-CO]{\Delta}$ C_6H_5, C_6H_5, C_6H_5, C_6H_5, C_6F_5, C_6F_5-substituted benzene	Also prepared $C_6F_5C{\equiv}CC_6H_5$ by same route or by $C_6H_5C{\equiv}CCu + C_6F_5I \xrightarrow{DMF}$ Ref. 448m
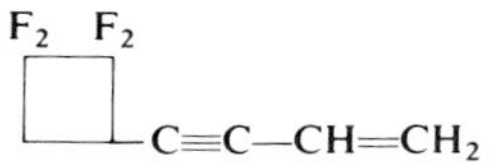 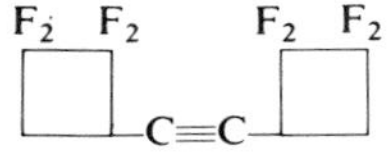 $F_2\ F_2$ cyclobutyl$-C{\equiv}C-CH{=}CH_2$ $F_2\ F_2$ cyclobutyl$-C{\equiv}C-$cyclobutyl $F_2\ F_2$	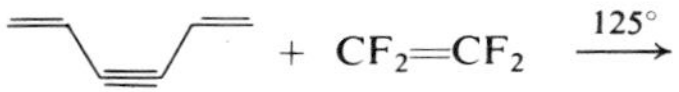 $CH_2{=}CH-C{\equiv}C-CH{=}CH_2 + CF_2{=}CF_2 \xrightarrow{125^\circ}$ mono or diadduct (no attack on acetylene bond)	Cyclobutene derivatives formed by base attack	Adducts using $CF_2{=}CFCl$ also prepared. Ref 448n

$CF_3SC{\equiv}CSCF_3$	From $CF_3SCH_2CH_2SCF_3$ by bromination dehydrohalogenation	+ Diene $\longrightarrow$ 1,2-dialkyl-4,5-bis(trifluoromethylthio)cyclohexa-1,4-diene (R, R; SCF_3, SCF_3) adducts with active hydrogen compounds: $+ CH_3OH \longrightarrow (CF_3S)(CH_3O)C{=}CHSCF_3$ and by radical addition	Ref. 408c
$C_6H_5C{\equiv}CSCF_3$	$CF_3SCl + C_6H_5C{\equiv}CMgBr \rightarrow$		Ref. 408c

---

[446](a) W. J. Middleton and W. H. Sharkey, *J. Am. Chem. Soc.*, **81**, 803 (1959). (b) H. G. Viehe and E. Franchimont, *Chem. Ber.*, **95**, 319 (1962). (c) W. J. Middleton, U.S. Patent 2,831,835 (1958); *Chem. Abstr.*, **52**, 14658f (1958). (d) H. G. Viehe, R. Merenyi, J. F. M. Oth, and P. Valange, *Angew. Chem. Internat. Edit. Engl.*, **3**, 746 (1964); H. G. Viehe, R. Merenyi, J. F. M. Oth, J. R. Senders, and P. Valange, *ibid.*, 755.

[447](a) W. G. Finnegan and W. P. Norris, *J. Org. Chem.*, **28**, 1139 (1963); *J. Org. Chem.*, **31**, 3292 (1966). (b) W. R. Cullen, D. S. Dawson, and G. E. Styan, *Can. J. Chem.*, **43**, 3392 (1965). (c) J. C. Sauer and G. N. Sausen, *J. Org. Chem.*, **27**, 2730 (1962). (d) R. S. Dickson and D. B. W. Yawney, *Australian J. Chem.*, **20**, 77 (1967); M. I. Bruce, D. A. Hawthorne, F. Waugh, and F. G. A. Stone, *J. Chem. Soc.* (A), 356 (1968). (e) P. L. Coe, R. G. Plevey, and J. C. Tatlow, *J. Chem. Soc.* (C), 597 (1966). (f) R. S. Dickson and D. B. W. Yawney, *Inorg. Nucl. Chem. Letters*, **3**, 209 (1967). (g) F. W. Hoover and D. D. Coffman, *J. Org. Chem.*, **29**, 3567 (1964). (h) F. L. M. Pattison and R. E. A. Dear, *Can. J. Chem.*, **41**, 2600 (1963). (i) M. Hudlicky, *Collection Czechoslov. Chem. Commun.*, **26**, 3140 (1961). (j) D. D. Coffman, P. L. Barrick, R. D. Cramer, and M. S. Raasch, *J. Am. Chem. Soc.*, **71**, 490 (1949); J. J. Drysdale, German Patent 1,116,649 (1957). (k) J. F. Harris, Jr., and R. M. Joyce, Jr., U.S. Patent 3,062,893 (1962); *Chem. Abstr.*, **58**, 7869c (1963). (l) F. S. Fawcett, U.S. Patents 3,311,599 and 3,359,319 (1967).

[448](a) C. G. Krespan, R. J. Harder, and J. J. Drysdale, *J. Am. Chem. Soc.*, **83**, 3424 (1961). (b) W. Mahler, *J. Am. Chem. Soc.*, **86**, 2306 (1964). (c) C. G. Krespan, *J. Am. Chem. Soc.*, **83**, 3432 (1961). (d) J. F. Harris, Jr., R. J. Harder, and G. N. Sausen, *J. Org. Chem.*, **25**, 633 (1960). (e) R. E. Putnam, R. J. Harder, and J. E. Castle, *J. Am. Chem. Soc.*, **83**, 391 (1961). (f) C. G. Krespan, *Tetrahedron*, **23**, 4243 (1967). (g) C. D. Weis, *J. Org. Chem.*, **27**, 3695 (1962). (h) L. A. Paquette, *J. Org. Chem.*, **30**, 2107 (1965). (i) R. E. Banks, W. R. Deem, R. N. Haszeldine, and D. R. Taylor, *J. Chem. Soc.* (C), 2051 (1966). (j) L. M. Yagupol'skii and Yu. A. Fialkov, *J. Gen. Chem. U.S.S.R. (English Transl.)*, **30**, 1318 (1960). (k) M. R. Wiles and A. G. Massey, *Chem. Ind. (London)*, 663 (1967). (l) J. M. Birchall, F. L. Bowden, R. N. Haszeldine, and A. B. P. Lever, *J. Chem. Soc.* (A), 747 (1967). (m) R. Filler and E. W. Heffern, *J. Org. Chem.*, **32**, 3249 (1967). (n) C. T. Handy and R. E. Benson, *J. Org. Chem.*, **27**, 39 (1962).

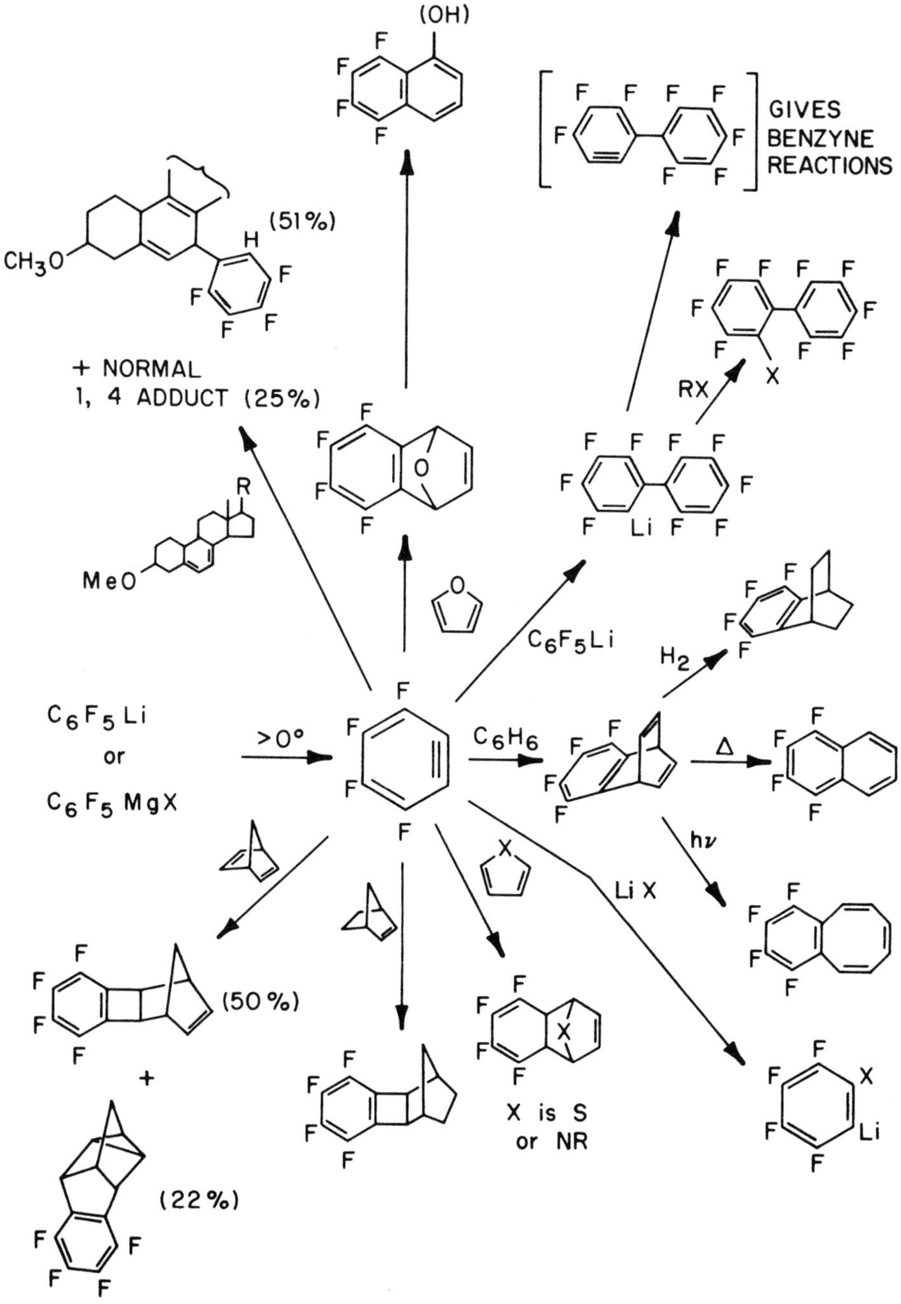

(OH)

GIVES BENZYNE REACTIONS

(51%)

+ NORMAL 1, 4 ADDUCT (25%)

RX

MeO

C_6F_5Li

H_2

C_6F_5Li or C_6F_5MgX

>0°

C_6H_6

Δ

hν

LiX

(50%)

+

(22%)

X is S or NR

CHART 6-17

aromatic substrates. These adducts are useful precursors to a variety of derivatives. However, 1,2-addition (insertion in a C–H bond) can compete under certain conditions.

A study of benzyne generation from bromopentafluorophenyllithium has shown that lithium fluoride eliminates preferentially to lithium bromide.[452] The

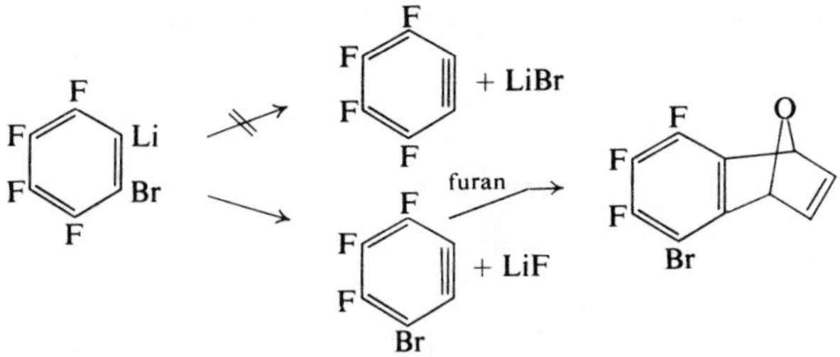

bromotetrafluorobenzyne gave the benzyne reaction of aryl lithium addition, but the reaction mixture was complex because of different modes of addition complicated by bromine–lithium exchanges.

An attempt to generate trifluoropyridyne was not successful.[453]

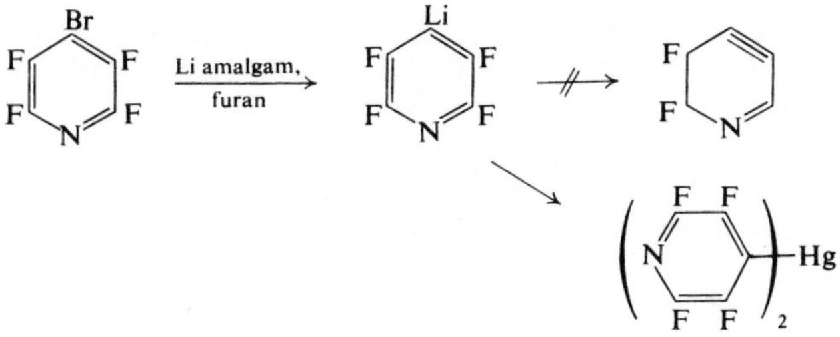

6-2C. FLUORINATED AROMATIC COMPOUNDS

Fluorinated aromatic compounds are conveniently classified into three groups:

1. fluoroaromatic—particularly monofluoro,

[452](a) S. C. Cohen, D. E. Fenton, D. Shaw, and A. G. Massey, *J. Organometal. Chem.*, 8, 1 (1967). (b) C. Tamborski and E. J. Soloski, *J. Organometal. Chem.*, **10**, 385 (1967).

[453]R. D. Chambers, F. G. Drakesmith, J. Hutchinson, and W. K. R. Musgrave, *Tetrahedron Letters*, 1705 (1967).

2. aromatic systems with fluorinated substituents–fluoroalkyl or fluorinated heteroatom groups,

$-C_6H_4-R_f$, $-C_6H_4-XF_n$ or $-C_6H_4-X(R_f)_n$

3. perfluoroaromatic, particularly pentafluorophenyl,

$-C_6F_5$

(This discussion concentrates on benzene derivatives but is applicable to other aromatic systems such as naphthalene or pyridine; however, the majority of published literature deals with benzene derivatives.)

Aromatic units of the first two classifications are easily introduced by typical aromatic reactions, such as:

1. organometallic reagents, usually Grignard or lithium reagents, which add to unsaturated systems, such as C=X (carbonyl), or couple with active halides;

2. electrophilic substitution of aromatic systems–particularly Friedel–Crafts reactions,

$$ArF + R^+ \rightarrow RArF + H^+$$

3. nucleophilic substitution–particularly useful for more highly fluorinated systems;

4. condensation on substituents, for example, fluoroaniline with aldehyde to give a Shiff base or phenolate substitution on an active halide.

Side reactions which can be troublesome when using the fluorinated aromatic derivatives in synthesis are

1. loss of fluorine by nucleophilic substitution, particularly if the fluorine is activated to attack by strong electron-withdrawing groups,[357]

$$O_2N-C_6H_4-F + B^- \longrightarrow O_2N-C_6H_4-B + F^-$$

2. loss of fluorine as metal fluoride giving benzyne when the fluorine is ortho to the metal,[454]

[454] H. Heaney, *Chem. Rev.*, **62**, 81 (1962).

$$\text{o-FC}_6\text{H}_4\text{Li} \xrightarrow{\text{warm}} \text{LiF} + [\text{C}_6\text{H}_4]$$

3. attack on fluorinated substituent,

$$\text{HO}-\text{C}_6\text{H}_4-\text{CF}_3 \xrightarrow{\text{base}} [\text{O}=\text{C}_6\text{H}_4=\text{CF}_2] \longrightarrow \text{polymer} \qquad (455)$$

Also, groups such as SF_3, PF_2 or SiF_3 are easily hydrolyzed.

(1) The classical-type chemistry of *fluorinated aromatic groups* is reviewed through 1961 in an ACS monograph[456a]; a particularly valuable feature of this review is the extensive tables (with a formula index) which are very complete for the partially fluorinated aromatic derivatives.

Aryl fluorine is a valuable chemical label and probe to study reaction mechanisms,[456b] molecular structure and bonding,[456c–e] solvent interaction and complexing,[456f–k] and substituent effects and mechanism of transmission of electronic effects.[456c, d, l–p] In many of these studies F^{19} nmr chemical shift measurements, as developed by R. W. Taft, are particularly valuable because the chemical shift is exceptionally sensitive to changes in charge density in the π system (see discussion in Section 3-1B and note substituent effects in Table 6-6). Also, the F^{19} method is often the only technique to study an unstable or highly reactive substituent.

(2) *Aromatic compounds with fluorinated substituents* are also reviewed to some extent.[456a] We have prepared Table 6-6 to summarize recent work and provide leading references to the more important aromatic systems with fluorinated substituents which have become of interest for mechanistic and biological studies and in potential commercial applications such as dyes, drugs, and stable materials.

[455] R. G. Jones, *J. Am. Chem. Soc.*, **69**, 2346 (1947).

[456] (a) A. E. Pavlath and A. J. Leffler, *Aromatic Fluorine Compounds*, Am. Chem. Soc. Monograph No. 155, Reinhold, New York, 1962. (b) H. Suschitzky, *Angew. Chem. Intern. Ed. Engl.*, **6**, 596 (1967). (c) J. C. Kauer and W. A. Sheppard, *J. Org. Chem.*, **32**, 3580 (1967). (d) W. A. Sheppard, *Trans. N.Y. Acad. Sci.*, Ser. II, **29**, 700 (1967). (e) G. W. Parshall, *J. Am. Chem. Soc.*, **88**, 704 (1966). (f) R. W. Taft, G. B. Klingensmith, and S. Ehrenson, *J. Am. Chem. Soc.*, **87**, 3620 (1965). (g) R. W. Taft and J. W. Carten, *J. Am. Chem. Soc.*, **86**, 4199 (1964). (h) D. Gurka, R. W. Taft, L. Joris, and P. R. Schleyer, *J. Am. Chem. Soc.*, **89**, 5957 (1967). (i) R. G. Pews, Y. Tsuno, and R. W. Taft, *ibid.*, 2391. (j) C. S. Giam and R. W. Taft, *ibid.*, 2397. (k) W. A. Sheppard and R. M. Henderson, *ibid.*, 4446. (l) R. W. Taft, *J. Phys. Chem.*, **64**, 1805 (1960). (m) R. W. Taft, E. Price, I. R. Fox, I. C. Lewis, K. K. Anderson, and G. T. Davis, *J. Am. Chem. Soc.*, **85**, 709, 3146 (1963). (n) M. J. S. Dewar and A. P. Marchand, *J. Am. Chem. Soc.*, **88**, 3318 (1966). (o) W. Adcock and M. J. S. Dewar, *J. Am. Chem. Soc.*, **89**, 379 (1967). (p) R. G. Pews, *ibid.*, 5605.

Table 6-6
Aromatic Compounds with Perfluorinated Substituents

Fluorinated[a] substituent X of ArX	Synthesis	Properties, reactions, use	Method[b]	σ_m	σ_p	σ_I	σ_R	Comments
			Substituent parameter					
A. GROUP III								
$-BF_2$	$ArBCl_2 + SbF_3 \xrightarrow{\text{exothermic}} (66\%)$, $ArMgX + BF_3 \xrightarrow[\text{ether}]{-78^\circ}$ very poor yield, $Ar_2Hg + B_2F_4 \rightarrow$	$C_6H_5BF_2$, bp 94-96°, $ArBF_2 + Et_2O \rightarrow ArB(F)OEt$ $+ (LiCH_2CH_2)_2 \rightarrow$ no reaction	F	.32	.48	.16	.32	$ArBF_2$ compounds shown more stable to solvolysis than corresponding $ArBCl_2$. Ref. 461 a-d
B: GROUP IV								
$-CF_3$	$ArCCl_3 + SbF_3 \rightarrow$ $ArCO_2H + SF_4 \rightarrow$ $ArH + CF_3I \rightarrow$ See Section 7-1 for more extensive coverage	$C_6H_5CF_3$, bp 103°. Common inert substituent on aromatic ring; employed in a wide variety of mechanism studies and in analogs of dyes and in biologically active compounds (particularly drugs). Orients electrophilic substitution of ring to meta position. CF_3 group hydrolyzed only under drastic conditions i.e., 100% H_2SO_4 at 100°	F B A R	.44 .42 .49 .53	.49 .53 .62 .48	.39 .33 .44 .51	.10 .18 .18 –.09	Generally inert, strongly electron-withdrawing substituent. See review (Ref. 456a) for general background. *p*-Phenol derivative decomposes in base to give polymer of  $O{=}C_6H_4{=}CF_2$ Ref. 461e, f, 456a
$-CF_2CF_3$	$ArC(=O)CF_3 + SF_4 \rightarrow$	$C_6H_5CF_2CF_3$ bp 115–116°. Chemically inert like CF_3 group; typical aromatic chemistry shown	F A	.47 .52	.52 .69	.41 –	.11 –	Ref. 461e
$-CF(CF_3)_2$	$ArC(OH)(CF_3)_2 + SF_4 \rightarrow$	$C_6H_5CF(CF_3)_2$, bp 125°. Chemically inert like CF_3 group; typical aromatic chemistry shown	F B A	.50 .37 .52	.52 .53 .68	.48 .25 .48	.04 .26 .17	Group noted to enhance solubility of aromatic compounds in organic solvents. Ref. 461e
$-CF_2CF_2CF_2CF_3$	$ArC(=O)CF_2CF_2CF_3 + SF_4 \rightarrow$	$C_6H_5(CF_2)_3CF_3$, bp 150°; Same comments as for other R_f groups above	F A	.47 .55	.52 .73	.39 –	.11 –	Ref. 461e
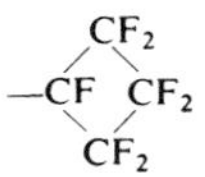 $-CF$ (ring with CF_2, CF_2, CF_2)	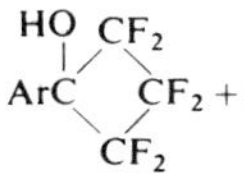 $ArC(HO)$ (ring with CF_2, CF_2, CF_2) $+ SF_4 \rightarrow$	$C_6H_5C_4F_9$, bp 67–68° (25 mm). Group stable and inert, typical of perfluoroalkyl substituents	F	.48	.53	.45	.08	Ref. 461g

$-CF_2CF_2-$ (bis)	$ArCCl_2CCl_2Ar + SbF_3 \xrightarrow[200^\circ]{\text{cat.}(Br_2 \text{ or } SbCl_5)}$	mp 125-126°. Fluorinated group inert to a variety of reactions on the aromatic ring	F (for $FC_6H_4CF_2CF_2-$)	.34	.39	.29	.10	Study carried out to prepare intermediates for preparation of dyes with good light fastness and bright colors. Ref. 461h-k
$-CF_2H$	$ArC(=O)H + SF_4$	$C_6H_5CF_2H$, bp 134°. Hydrolyzes easily; decomposes on storage in glass						Ref. 461l
$-CF_2-O-CF_2-$ (1, 2)	X-substituted phthalan ($C_6H_4(CH_2)_2O$) $\xrightarrow{PCl_5}$ $\xrightarrow{SbF_3}$	o-$C_6H_4(CF_2)_2O$, bp 129-130°. Stable to conversions on aromatic nucleus	B	–	.81	–	–	Noted electronically similar to nitro group. Studied as a dye substituent. Ref. 461m, n
$-CH(CF_3)_2$	$ArC(OH)(CF_3)_2 + HBr \rightarrow$ $ArH + (CF_3)_2C{=}N_2 \rightarrow$	$C_6H_5CH(CF_3)_2$ bp 137.5°. No chemical reactions reported						Ref. 461o
$-C(CF_3)_2(OH)$	$ArH + CF_3C(=O)CF_3 \xrightarrow{AlCl_3}$ $ArMgX + CF_3C(=O)CF_3 \rightarrow$	$C_6H_5C(OH)(CF_3)_2$, bp 163°. Very acidic OH, replaced by F or Cl only under strenuous conditions. Group stable to many transformations on aromatic ring	F A	.29 .35	.30 .38	.38 .31	.02 .15	Large number of aromatic compounds reported substituted by $C(OH)(CF_3)_2$ group. Ref. 461e, p
$-C(CF_3)_2(NH_2)$	$ArH + (CF_3)_2C{=}NH \xrightarrow[\text{catalyst}]{\text{Lewis acid}}$	$C_6H_5C(NH_2)(CF_3)_2$, bp 95° (60 mm). The hexafluoroisopropylamino group is extremely stable and inert to transformation on the aromatic ring						Noted that $C(NH_2)(CF_3)_2$ group is slightly less electron withdrawing than the $C(OH)(CF_3)_2$ group (from F_p^{19} shift for p-fluorine substituent). Ref. 461q
$-C(OH)$ (ring: CF_2, CF_2, CF_2)	$ArMgX$ + hexafluorocyclobutanone ($C(=O)$, CF_2, CF_2, CF_2) $\longrightarrow$ or $ArH(AlCl_3)$	$C_6H_5C(OH)(CF_2)_3$, bp 134° (130 mm). Very acidic OH group; behaves like $C(OH)(CF_3)_2$ group	F	.36	.37	.35	.02	Ref. 461r, g
$-CH_2F$	$ArCH_2Br + HgF_2 \rightarrow$	$C_6H_5CH_2F$, bp 50° (27 mm). Susceptible to nucleophilic attack typical of benzylic halides. Polymerizes easily in presence of catalyst						Variety of other procedures summarized in review (Ref. 456a). Ref. 456a, 461s
$-CH_2CF_3$	$ArCH_2CO_2H + SF_4 \rightarrow$	$C_6H_5CH_2CF_3$, bp 133.5°. Substituent inert to normal chemical conversions on aromatic ring	F A	.12 .16	.09 .14	.14	−.05	Ref. 461g

Table 6-6 (Cont.)

Fluorinated[a] substituent X of ArX	Synthesis	Properties, reactions, use	Substituent parameter Method[b]	σ_m	σ_p	σ_I	σ_R	Comments
$-CH_2SCF_3$	$ArCH_2I + AgSCF_3 \xrightarrow[\text{4 days}]{\text{acetone}}$ $ArCH_2Cl + (CF_3S)_2Hg \rightarrow$	$C_6H_5CH_2SCF_3$, bp 76-77° (30 mm). Prepared a series of benzyl trifluoromethyl sulfides for substituent effect studies	F	.16	.21	.19	–.07	The corresponding acetic acids prepared to measure pKa and determine σ_I value of SCF_3. Ref. 462a, b
$-CH_2\overset{O}{\overset{\|}{S}}CF_3$	$ArCH_2SCF_3 + H_2O_2 \xrightarrow{\text{HOAc}}$	$C_6H_5CH_2\overset{O}{\overset{\|}{S}}CF_3$, mp 78-79°. Prepared a series of benzyl trifluoromethyl sulfoxides for substituent effect studies	F	.25	.24	.26	–.02	The corresponding acetic acid prepared to measure pKa and determine σ_I value of $SOCF_3$. Ref. 462a
$-CH_2SO_2CF_3$	$ArCH_2SCF_3 + H_2O_2$ (excess) $\xrightarrow[\text{heat}]{\text{HOAc}}$	$C_6H_5CH_2SO_2CF_3$, mp 74°. Prepared series of benzyl trifluoromethyl sulfones for substituent effect studies	F	.30	.31	.28	.03	The corresponding acetic acid prepared to measure pKa and determine σ_I value of SO_2CF_3. Ref. 462a
$-CF{=}CF_2$	$ArLi + CF_2{=}CF_2 \rightarrow$ plus di- (and higher) substitution products	$C_6H_5CF{=}CF_2$, bp 68° (75 mm). Studied as monomer						Large number of examples and multiple substitution products (two to four halogens replaced by ArLi) reported for aryllithium reagents with a variety of fluoro(chloro)-olefins. Ref. 462c
$-CF{=}CFCl$	$ArLi + CF_2{=}CFCl \rightarrow$	$C_6H_5CF{=}CFCl$, bp 174°						
$-CF{=}CFCF_3$ (trans)			F	.39	.46	.32	.14	Only electronic properties of group are reported. Ref. 461j, k
$-CH{=}CHCF_3$	$ArCH{=}CHCCl_3 \xrightarrow[\text{dioxane}]{SbF_3}$	$C_6H_5CH{=}CHCF_3$, bp 60-62° (2 mm), + conc. H_2SO_4 → polystyrene,	F	.24	.27	.20 (.21)	.07 (.07)	σ values from Ref. 461g; values in parenthesis from Ref. 461j.
(trans)	$ArCH{=}CHCO_2H \xrightarrow{SF_4}$	+ Br_2 (or Cl_2) → adds to give mixture of two diastereomers	A	.20 (.22)	.34 (.20)	.17 (.23)	.16 (.03)	Ref. 461g, j, k, m, 462d, e
(cis)	Sensitized $h\nu$ irradiation of trans	+ diene → no reaction	F	.15	.17	.12	.05	
			A	.16	.29	.14	.14	
$-C{\equiv}CCF_3$	$ArCHBr{-}CHBr{-}CF_3 \xrightarrow{\text{1. }Et_3N}$	$C_6H_5C{\equiv}CCF_3$, bp 144-145°,	F	.42	.50	.34	.16	Ref. 461j, k, m, 462d, e
	$ArCH{=}CBrCF_3 \xrightarrow{\text{2. KOH}}$	+ $Br_2 \xrightarrow{h\nu}$ dibromoethylene adduct, + $2Cl_2 \xrightarrow{h\nu}$ adduct (tetrachloro derivative)	B	.43	.52	.35	.17	

$-HC{=}CHSO_2CF_3$ (trans)	Not given		F			.34	.25	Listed in a paper on study of substituted effects. Ref. 461j, k
$-C(=O)F$	$ArC(=O)Cl + NaF \xrightarrow[\text{up to } 250°]{TMS}$ $ArCO_2H + SF_4 \xrightarrow{\text{room temp.}}$	C_6H_5COF, bp 157-159°	F	.55	.70	.39	.31	Ref. 461f, l, 462f
$-C(=O)R_f$	$ArMgX + R_fC(=O)X$ $ArH + R_fC(=O)X \xrightarrow{AlCl_3}$	$C_6H_5C(=O)CF_3$, bp 152°. Reactions typical of aromatic ketone, also $+ SF_4 \rightarrow ArCF_2R_f$	F	.63	.80	.47	.33	Ref. 461e, f
$-SiF_3$	$ArSiCl_3 + HF \rightarrow$ (anhydrous or aqueous)	$C_6H_5SiF_3$, bp 101-102°, mp 18-19°. Fluorines easily displaced by nucleophilic agents	F	.54	.67	.42	.25	Ref. 461d, 462g, 462h
C. GROUP V								
$N(CF_3)_2$	$ArN(C(=O)F)_2 + SF_4 \xrightarrow{HF}$ low yield or $ArN(CF_3)C(=O)F$	$C_6H_5N(CF_3)_2$, bp 118°. Inert substituent; directs electrophilic attack on ring to para position	F B A	.49 .40 .47	.50 .53 .53	.49 .29 .44	.01 .23 .06	Nitrogen not basic and compounds will not form salts with mineral acid. Unexpected apparent + R effect explained by p-π interaction mechanism. Ref. 463a
$N(C(=O)F)(CF_3)$	$ArN{=}CF_2 + COF_2 \xrightarrow{F^-}$ Best overall synthesis from $ArNCS + HgF_2 + COF_2 + CsF \xrightarrow[\text{autoclave}]{200°}$	$C_6H_5N(COF)(CF_3)$, bp 134° (180 mm). COF group reactive to nucleophilic reagents typical of carbamyl halide. Fluorinated with SF_4 to $ArN(CF_3)_2$	F	.50	.50	.50	0	Ref. 463a, b
$N(C(=O)F)_2$	$ArNCO + COF_2 \xrightarrow{CsF}$	$C_6H_5N(COF)_2$, mp 93.5-95.5°. Fluorinated with SF_4 to give $ArN(COF)(CF_3)$ and $ArN(CF_3)_2$	F	.58	.57	.58	−.01	Ref. 463a, c
$N{=}CF_2$	$ArNCS + HgF_2 \rightarrow$ $ArN{=}CCl_2 + MF_n \rightarrow$ $ArN{=}CCl_2 + HF \rightarrow ArNHCF_3 \xrightarrow[KF]{\Delta}$	$C_6H_5N{=}CF_2$, bp 58° (53 mm). Highly reactive: $+ HF \rightarrow ArNHCF_3$ $+ F^- \rightarrow ArN(CF_3){-}CF{=}NAr$						Trapping with COF_2 during preparation gave high yields of $ArN(CF_3)(COF)$. Ref. 463b

Table 6-6 (Cont.)

Fluorinated[a] Substituent X of ArX	Synthesis	Properties, reactions, use	Substituent parameter Method[b]	σ_m	σ_p	σ_I	σ_R	Comments
$N{=}C(CF_3)_2$	$ArNCO + CF_3C(=O)CF_3 \xrightarrow[200°,\ 17\ hr]{\phi_3 P{=}O}$	$C_6H_5N{=}C(CF_3)_2$, bp 72-73° (73 mm)	F	.29	.23	+.35	−.12	A variety of HFA anils can be prepared by this method. Ref. 461g, 463d
$NHCF_3$	$ArN{=}CCl_2 + HF \rightarrow$	$C_6H_5NHCF_3$, bp 50-51° (10 mm) $+ KF \xrightarrow{heat} ArN{=}CF_2$. Fluorides easily hydrolyzed or replaced by nucleophiles						Ref. 463e
PF_2	$ArPCl_2 + NaF \xrightarrow{TMS}$	$C_6H_5PF_2$, bp 31° (11 mm). Highly reactive, disproportionates readily: $ArPF_2 \rightarrow ArPF_4 + (ArP)_4$	F	.49	.59	.38	.21	Nmr studies on P and F. Substituent parameter data used as evidence for large π(p-d) interaction involving *d*-orbitals of P. Ref. 463f, g, 461g
$P(CF_3)_2$	$(ArP)_4 + CF_3I \rightarrow$ $ArMgX + IP(CF_3)_2$	$C_6H_5P(CF_3)_2$, bp 148-150° $+$ aqueous base $\rightarrow CF_3H + C_6H_5P(=O)ONa$ $+ Br_2 \rightarrow C_6H_5P(CF_3)_2(Br_2) \xrightarrow{H_2O} C_6H_5P(=O)(CF_3)OH$	F	.60	.69	.50	.19	Substituent parameter data used as evidence for large π(*p-d*) interaction involving *d*-orbitals of P. Ref. 461g, 463g, h
PF_4	$ArPCl_2 + SbF_3 \rightarrow$	$C_6H_5PF_4$, bp 134.5-136°. Not a useful fluorinating agent but gave low yield of $C_6H_5CF_2H$ with C_6H_5CHO, $+$ Siloxane $\rightarrow SiF + ArP(=O)F_2$	F	.63	.80	.45	.35	Substituent parameter data used as evidence for large π(*p-d*) interaction involving *d*-orbitals of P. Other aryl pentavalent phosphorus fluorides such as Ar_2PF_3 are also reported prepared. Ref. 463i, g, 461g
$P(=X)F_2$ (X = O or S)	$ArP(=X)Cl_2 + NaF \xrightarrow{TMS}$	$C_6H_5P(=O)F_2$, bp 44° (2.5 mm); $C_6H_5P(=S)F_2$, bp 47-49° (3 mm)						Ref. 462f

$As(CF_3)_2$	$C_6H_5AsI_2 + CF_3I \rightarrow$	$C_6H_5As(CF_3)_2$, bp 160°. Reacts with Br_2 (product probably $C_6H_5Ar(Br)CF_3$)						Prepared a series of $ArAsR_{f2}$ and Ar_2AsR_f. Ref. 463j
D. GROUP VI								
OCF_3	$ArOCH_3 \xrightarrow[h\nu]{Cl_2} ArOCCl_3 \xrightarrow[SbF_3]{HF\ or}$ $ArOH + COF_2 \rightarrow ArOC(=O)F \xrightarrow[HF]{SF_4}$	$C_6H_5OCF_3$, bp 106°. OCF_3 group inert to all reported chemical transformations on aromatic ring	F B A P	.53 .38 .47 .40	.51 .35 .27 .26	.55 .39 .50 .39	−.04 −.04 −.23 −.15	OCF_3 group para orienting to electrophilic substitution of aromatic rings. OCF_3 shown to enhance biological activity in certain systems (464c). Studied as a substituent for dyes and drugs. (Ref. 464d, e, f). Ref. 464a, b
OCF_2CF_3	$ArOC(=O)CF_3 + SF_4 \xrightarrow{HF}$	$C_6H_5OCF_2CF_3$, bp 115°. Behavior same as for OCF_3	A P	.48 .43	.28 .27	.52 .42	−.25 −.17	Ref. 464a, b
OCF_2CF_2H	$ArOH + CF_2{=}CF_2 \xrightarrow[catalyzed]{base}$	$C_6H_5OCF_2CF_2H$, bp 147°. Behavior appears same as for OCF_3 group	B A	.34 .43	.25 .21	.39 .45	−.14 −.25	Ref. 464a, b, g
OCF_2CFClH	$ArOH + CF_2{=}CFCl \xrightarrow[catalyzed]{base}$? (no information given in reference)		B	.35	.28	.42	−.12	Noted similar electronic effect to OCF_2CF_2H. Ref. 464h
$-O-CF_2-O-$ (1,2)	X-substituted 1,3-benzodioxole ($O-CH_2-O$) $\xrightarrow{PCl_5} \xrightarrow{SbF_3}$	Stable to conversions on aromatic nucleus	B		.36			Studied as a dye substituent. Ref. 462d, 461n, 464e
OCF_2H	$ArOH + CClF_2H \xrightarrow{base}$ (via difluorocarbene)	$C_6H_5OCF_2H$, bp 37° (13 mm). These ethers are stable under basic conditions but hydrolyze rapidly in acid	B	.31	.18	.44	.24	Ref. 464i, j
OSF_5	$ArH + SF_5OOSF_5 \xrightarrow[15\ hr]{150°}$	$C_6H_5OSF_5$, bp 139°. OSF_5 easily hydrolyzed by base and reduced. Stable to nitration and sulfonation of ring (para orientation)	B		.44			Ring substitution proposed to go by radical mechanism ($SF_5O\cdot$) but substituent effects suggest ionic mechanism. Ref. 464k
SF_3	$(ArS)_2 + AgF_2 \xrightarrow{CCl_2FCClF_2}$	$C_6H_5SF_3$, bp 47° (2.6 mm). $+ H_2O \rightarrow HF + [ArSO_2H]$ $+ RC(=O)H \rightarrow RCF_2H + ArSOF$	F	.70	.80	.60	.20	Phenylsulfur trifluoride is a useful fluorinating agent for carbonyl groups (like SF_4) but has the advantage of not requiring special pressure equipment for fluorination. Ref. 465a

Table 6-6 (Cont.)

Fluorinated[a] Substituent X of ArX	Synthesis	Properties, reactions, use	Method[b]	σ_m	σ_p	σ_I	σ_R	Comments
			Substituent parameter					
SF_5	$ArSF_3 + AgF_2 \xrightarrow{120^\circ}$ butadiene + $HC{\equiv}C{-}SF_5$ $\longrightarrow$ SF_5-cyclohexadiene $\xrightarrow{[O_2]}$ $ArH + S_2F_{10} \longrightarrow$ (very low yield)	$C_6H_5SF_5$, bp 149°, $+ H_2SO_4 \xrightarrow{100^\circ} C_6H_5SO_2F + HF$, Stable to all other normal chemical transformations on aromatic ring	F B A P	.59 .61 .63 .61	.67 .68 .86 .70	.55 .55 .56 .53	.07 .11 .27 .12	Meta directing to electrophilic aromatic substitution. Similar in stability to CF_3 group. Imparts high solubility of aromatic systems in organic solvents and consequently found to be a useful activating group in certain biological systems (465bb). Ref. 465b, 186, 447g
SCF_3	$ArSCH_3 \xrightarrow{Cl_2} \xrightarrow[\text{or } SbF_3]{HF}$	$C_6H_5SCF_3$, bp 141-142°. $+ CrO_3 \xrightarrow{H_2SO_4} ArSO_2CF_3$ SCF_3 group stable to majority of chemical transformations carried out on aromatic ring	F B A P	.45 .40 .46 .43	.48 .50 .64 .57	.42 .31 .40 .39	.06 .17 .22 .14	Ortho para directing to electrophilic aromatic substitution; shown to be a stable substituent for use in dyes (Ref. 461n, 465e, f) and drugs (465g). The question of *d*-orbital participation (compared to OCF_3, $S(=O)CF_3$ and SO_2CF_3) has been studied by spectral methods (Ref. 465h, i and dipole moment measurements. Ref. 465j. Ref. 464b, 465c-g
SCF_2CF_2H	$ArSH + CF_2{=}CF_2 \xrightarrow[\text{catalysis}]{\text{base}}$	$C_6H_5SCF_2CF_2H$, bp 104-107° (72-76 mm). Not studied extensively but appears similar to SCF_3	B A	.38 .42	.47 .61	.29 .37	.20 .21	Electronically similar to SCF_3 but with slightly less inductive effect due to replacement of F by CF_2H. Ref. 464b, g
SCF_2H	$ArSH + HCF_2Cl \xrightarrow{\text{base}}$ (via difluorocarbene)	Transformations on aromatic ring carried out by avoiding acidic conditions	B	.33	.36	.30	.27	Ref. 464i, j
$S(=O)F$	$ArSF_3 + {-}C(=O){-} \rightarrow$ (by-product in fluorination of carboxy reagents)	$C_6H_5S(=O)F$, bp 60° (25 mm). Unstable to storage at room temperature						Ref. 465a

$\overset{O}{\parallel}$ SCF_3	$ArSCF_3 \xrightarrow{HNO_3}$	$C_6H_5\overset{O}{\overset{\parallel}{S}}CF_3$, bp 82° (10 mm) $Ar\overset{O}{\overset{\parallel}{S}}CF_3 + PCl_5 \rightarrow POCl_3 + ArSCF_3$. Many normal chemical transformations can be carried out on aromatic rings	F B A	.80 .63 .76	.74 .69 1.07	.67 .60 .70	.13 .11 .32	Orients attacking electrophilic reagent to meta position. Effect on dipole moments, spectral properties and dyes studied. Ref. 461g, j, 465k, h, i
SO_2F	$ArSO_2Cl + KF{\cdot}SO_2 \rightarrow$	Not as reactive to nucleophilic displacement of F^- as the corresponding chloride	F B A	.88 .98 .99	1.01 1.08 1.54	.75 .88 .87	.26 .21 .59	Ref. 461d, 465l
SO_2CF_3	$ArSCF_3 \xrightarrow[H_2SO_4]{CrO_3}$	$C_6H_5SO_2CF_3$, bp 205°. Is stable to many chemical conversions of aromatic ring	F B A P	.90 .76 1.00 .92	1.06 .96 1.65 1.36	.75 .69 .84 .84	.31 .22 .73 .49	One of the strongest uncharged electron-withdrawing groups reported (Refs. 461m, 465f). Effect in dyes, drugs, and other chemicals have been studied. Ref. 461j, k, m, 464b, 465f, g, h, i, j, m
SO_2CF_2H	$ArSCF_2H \xrightarrow{H_2O_2}$	Tolyl derivative oxidized with alkaline permanganate to benzoic acid	B	.75	.86	.65	.19	Ref. 464j
$SeCF_3$	$ArH + CF_3SeCl \rightarrow$	$C_6H_5SeCF_3$, bp 71-72° (44 mm). Group stable to chemical conversions of aromatic ring, such as diazotization and substitution of aniline	F B	.44 .32	.45 .38	.42 .28	.04 .10	Compared electronic effects with other fluorinated substituents. Ref. 465n, o
E. GROUP VII								
IF_2	$ArICl_2 + 2HF + HgO \rightarrow$ or $ArIO + HF$	Used for addition of elements of F_2 to olefins. Also reported to add as ArIF. . .F to steroids						Very reactive reagent. Some of the work on addition to olefin needs reinvestigation. Ref. 466a, b

[a] Ar is benzene or a substituted benzene.

[b] Letter denotes method of measurement.

F F^{19} chemical shift.
B Benzoic acid pK_a.
A pK_a of aniline ions.
P pK_a of phenols.
R Rate measurement.

For a general discussion of methods and definition of parameters, see H. H. Jaffee, *Chem. Rev.*, **53**, 191 (1953) and R. W. Taft, *J. Phys. Chem.*, **64**, 1805 (1960). Also, R. W. Taft, Jr., "Separation of Polar, Steric and Resonance Effects in Reactions," in M. S. Newman, Editor, *Steric Effects in Organic Chemistry* (M. S. Newman, ed.), Wiley, New York, 1956.

Table 6-6 (Cont.)

[461](a) P. A. McCusker and H. S. Makowski, *J. Am. Chem. Soc.*, **79**, 5185 (1957). (b) A. K. Holliday and F. B. Taylor, *J. Chem. Soc.*, 2731 (1964). (c) N. N. Greenwood and J. C. Wright, *J. Chem. Soc.*, 448 (1965). (d) R. W. Taft, private communication. (e) W. A. Sheppard, *J. Am. Chem. Soc.*, **87**, 2410 (1965). (f) R. W. Taft, E. Price, I. R. Fox, I. C. Lewis, K. K. Anderson, and G. T. Davis, *J. Am. Chem. Soc.*, **85**, 709, 3146 (1963). (g) W. A. Sheppard, *Trans. N.Y. Acad. Sci.*, Ser. II, 29, 700 (1967). (h) L. M. Yagupol'skii and V. I. Troitskaya, *J. Gen. Chem. U.S.S.R. (English Transl.)*, **35**, 1616 (1965). (i) L. M. Yagupol'skii, V. I. Troitskaya, B. E. Gruz, and N. V. Kondratenko, *ibid*, 1645. (j) V. F. Bystrov, L. M. Yagupol'skii, A. U. Stepanyants, and Yu. A. Fialkov, *Proc. Acad. Sci. U.S.S.R., Chem. Sect. (English Transl.)* **153**, 1019 (1963). (k) L. M. Yagupol'skii, V. F. Bystrov, A. U. Stepanyants, and Yu. A. Fialkov, *Proc. Acad. Sci. U.S.S.R., Phys. Sci. (English Transl.)* **135**, 3731 (1960). (l) W. R. Hasek, W. C. Smith, and V. A. Engelhardt, *J. Am. Chem. Soc.*, **82**, 543 (1960). (m) L. M. Yagupol'skii and L. N. Yagupol'skaya, *Proc. Acad. Sci. U.S.S.R., Chem. Sect. (English Transl.)*, **134**, 1207 (1960). (n) L. M. Yagupol'skii, G. I. Klyushnik, and V. I. Troitskaya, *J. Gen. Chem. U.S.S.R. (English Transl.)*, **34**, 304 (1964). (o) D. M. Gale, W. J. Middleton, and C. G. Krespan, *J. Am. Chem. Soc.*, **88**, 3617 (1966). (p) For a review of these reactions, see C. G. Krespan and W. J. Middleton, *Fluorine Chem. Rev.*, **1**, 145 (1967). (q) D. M. Gale and C. G. Krespan, *J. Org. Chem.*, **33**, 1002 (1968). (r) D. C. England, *J. Am. Chem. Soc.*, **83**, 2205 (1961). (s) J. Bernstein, J. S. Roth, and W. T. Miller, Jr., *J. Am. Chem. Soc.*, **70**, 2310 (1948).

[462](a) V. V. Orda, L. M. Yagupol'skii, V. F. Bystrov, and A. U. Stepanyants, *J. Gen. Chem. U.S.S.R. (English Transl.)*, **35**, 1631 (1965). (b) E. H. Man, D. D. Coffman, and E. L. Muetterties, *J. Am. Chem. Soc.*, **81**, 3575 (1959). (c) S. Dixon, *J. Org. Chem.*, **21**, 400 (1956). (d) L. M. Yagupol'skii, V. F. Bystrov, and E. Z. Utyanskaya, *J. Gen. Chem. U.S.S.R. (English Transl.)*, **34**, 1059 (1964). (e) L. M. Yagupol'skii and Yu. A. Fialkov, *J. Gen. Chem. U.S.S.R. (English Transl.)*, **30**, 1318 (1960). (f) C. W. Tullock and D. D. Coffman, *J. Org. Chem.*, **25**, 2016 (1960). (g) W. H. Pearlson, T. J. Brice, and J. H. Simons, *J. Am. Chem. Soc.*, **67**, 1769 (1945). (h) L. Spialter, *Tetrahedron Letters*, 11 (1960).

[463](a) F. S. Fawcett and W. A. Sheppard, *J. Am. Chem. Soc.*, **87**, 4341 (1965). (b) W. A. Sheppard, *ibid.*, 4338. (c) F. S. Fawcett, C. W. Tullock, and D. D. Coffman, *J. Am. Chem. Soc.*, **84**, 4275 (1962). (d) Yu. V. Zeifman, N. P. Gambaryan and I. L. Knunyants, *Proc. Acad. Sci. U.S.S.R., Chem. Sect. (English Transl.)*, **153**, 1032 (1963). (e) K. A. Petrov and A. A. Neimysheva, *J. Gen. Chem. U.S.S.R. (English Transl.)*, **29**, 2135, 2662 (1960). (f) R. Schmultzler, *Chem. Ber.*, **98**, 552 (1965). (g) J. W. Rakshys, R. W. Taft, and W. A. Sheppard, *J. Am. Chem. Soc.*, **90**, 5236 (1968). (h) M. A. A. Beg and H. C. Clark, *Can. J. Chem.*, **39**, 564 (1961). (i) R. Schmutzler, *Inorg. Chem.*, **3**, 410 (1964). (j) W. R. Cullen, *Can. J. Chem.*, **38**, 445 (1960) and **39**, 2486 (1961).

[464](a) W. A. Sheppard, *J. Org. Chem.*, **29**, 1 (1964). (b) W. A. Sheppard, *J. Am. Chem. Soc.*, **85**, 1314 (1963). (c) P. G. Heytler and W. W. Prichard, *Biochem. Biophys. Res. Commun.*, **7**, 272 (1962); W. L. Miller and J. J. Krake, *Proc. Soc. Exptl. Biol. Med.*, **113**, 449 (1963). (d) L. M. Yagupol'skii and M. S. Marenets, *J. Gen. Chem. U.S.S.R. (English Transl.)*, **27**, 1477 (1957). (e) L. M. Yagupol'skii and V. I. Troitskaya, *J. Gen. Chem. U.S.S.R. (English Transl.)*, **30**, 3102 (1960); *J. Gen. Chem. U.S.S.R. (English Transl.)*, **31**, 845 (1961). (f) French Patent 1,245,552 (1960). (g) D. C. England, L. R. Melby, M. A. Dietrich, and R. V. Lindsey, Jr., *J. Am. Chem. Soc.*, **82**, 5116 (1960). (h) R. Van Poucke, R. Pollet, and A. de Cat, *Bull. Soc. Chim. Belges*, **75**, 573 (1966). (i) T. G. Miller and J. W. Thanassi, *J. Org. Chem.*, **25**, 2009 (1960). (j) R. Van Poucke, R. Pollet and A. De Cat, *Tetrahedron Letters*, 403 (1965); *Bull. Soc. Chim. Belges*, **75**, 40 (1966). (k) J. R. Case, R. Price, N. H. Ray, H. L. Roberts, and J. Wright, *J. Chem. Soc.*, 2107 (1962).

[465](a) W. A. Sheppard, *J. Am. Chem. Soc.*, **84**, 3058 (1962); *Org. Syn.*, **44**, 39 and 82 (1964). (b) W. A. Sheppard, *J. Am. Chem. Soc.*, **84**, 3064 and 3072 (1962). (bb) R. L. Metcalf and T. R. Fukuto, *J. Econ. Entomol.* **55**, 340 (1962). (c) W. A. Sheppard, *J. Org. Chem.*, **29**, 895 (1964). (d) S. Andreades, J. F. Harris, Jr., and W. A. Sheppard, *ibid.*, 898. (e) L. M. Yagupol'skii and M. S. Marenets, *J. Gen. Chem. U.S.S.R. (English Transl.)*, **25**, 1725 (1955). (f) L. M. Yagupol'skii and B. E. Gruz, *J. Gen. Chem. U.S.S.R. (English Transl.)*, **31**, 1219 (1961); L. M. Yagupol'skii, V. I. Troitskaya, I. I. Levkoev, E. B. Lifshits, P. A. Yuta, and N. S. Barvyn, *J. Gen. Chem. U.S.S.R. (English Transl.)*, **37**, 174 (1967). (g) E. A. Nodiff, S. Lipschutz, P. N. Craig, and M. Gordon, *J. Org. Chem.*, **25**, 60 (1960). (h) V. F. Kulik, Y. P. Egorov, M. S. Marenets, and L. M. Yagupol'skii, *J. Struct. Chem. U.S.S.R. (English Transl.)*, **4**, 495 (1963). (i) A. E. Lutskii, L. M. Yagupol'skii, and S. A. Volchenok, *J. Gen. Chem. U.S.S.R. (English Transl.)*, **34**, 2749 (1964). (j) A. E. Lutskii, L. M. Yagupol'skii, and E. M. Obukhova, *ibid.*, 2663. (k) L. M. Yagupol'skii, M. S. Marenets, and N. V. Kondratenko, *J. Gen. Chem. U.S.S.R. (English Transl.)*, **35**, 377 (1965). (l) F. Seel, H. Jonas, L. Riehl, and J. Langer, *Angew. Chem.*, **67**, 32 (1955). (m) S. M. Shein, *J. Appl. Chem. U.S.S.R. (English Transl.)*, **35**, 2482 (1962). (n) L. M. Yagupol'skii and V. G. Voloshchuk, *J. Gen. Chem. U.S.S.R. (English Transl.)*, **36**, 173 (1966). (o) V. G. Voloshchuk, L. M. Yagupol'skii, G. P. Syrova, and V. F. Bystrov, *J. Gen. Chem. U.S.S.R. (English Transl.)*, **37**, 105 (1967).

[466](a) W. Carpenter, *J. Org. Chem.*, **31**, 2688 (1966). (b) P. G. Holton, A. D. Cross, and A. Bowers, *Steroids*, **2**, 71 (1963).

(3) *Perfluoroaromatic chemistry* has developed during the past few years into a very active and flourishing field. Over a hundred papers and a large number of patents have appeared since the late 1950's when synthetic routes were developed to make perfluoroaromatics readily available. Major contributions to this field have been made by the Imperial Smelting Company (which provides the basic commercial supply of a variety of perfluoroaromatic derivatives), government laboratories (particularly McLoughlin in England and Tamborski and Wall in the United States), and academic groups (particularly those headed by Tatlow at Birmingham, Musgrave at Durham, and Filler at Illinois Institute of Technology). Unfortunately no comprehensive review of this important field is available (most of this work is too recent for Pavalath and Leffler[456a]; see, however, reviews[317b, d] on metal and metalloid derivatives of perfluoroaromatic derivatives).

Perfluoroaromatic compounds are usually prepared by one of the following routes:

1. oxidative fluorination of highly chlorinated aromatics followed by high-temperature dehalogenation over metal gauze,

$$C_6Cl_6 \longrightarrow \underset{\text{average composition of mixture}}{C_6F_6Cl_6} \xrightarrow[500^\circ]{Fe} C_6F_6 \qquad (457)$$

2. nucleophilic displacement of fluoride ion on polychloroaromatic compounds (see discussion in Section 5-4A.1),

$$C_6Cl_6 + 5KF \xrightarrow[230^\circ,\ 48\ hr]{\text{sulfolane}} 5KCl + C_6F_5Cl \qquad (458)$$

This looks like the most economical commercial route.

3. pyrolysis of monofluorohalomethanes (see discussion in Section 6-1B.1b, page 254).

$$CHFX_2 \xrightarrow{\Delta} C_6F_6 + 3X_2 + 6HX \qquad (207\text{–}218)$$

[457] G. M. Brooke, R. D. Chambers, J. Heyes, and W. K. R. Musgrave, *J. Chem. Soc.*, 729 (1964).

[458] G. Fuller, *J. Chem. Soc.*, 6264 (1965).

Perfluoroaromatic units are introduced into organic compounds mainly by two basic reactions: (*1*) organometallic reagents–particularly lithium or magnesium and (*2*) nucleophilic aromatic substitution. Both of these approaches involve fluorocarbanion intermediates and, as such, were reviewed in Section 6-1C. We have examples below of typical synthesis where perfluoroaromatic units were introduced[459, 460] (Chart 6-18), but first, further amplification is needed on the organometallic route, which can be employed in three entirely different ways:

1. normal addition to unsaturated systems such as carbonyl groups;
2. coupling to active halides, particularly metal halides which has led to a large selection of pentafluorophenyl organometallics (see discussion below);
3. elimination of LiF to give perfluorobenzyne as a reactive dienophile (see discussion and examples in Section 6-2B).

The pentafluorophenyl group often exerts unusual stabilizing influence on organometallic reagents[317d]; the strong inductive electron-withdrawing effect of the fluorines must withdraw electron density from the metal through the σ bond making it less susceptible to electrophilic attack, but at the same time the high p-electron density in the pentafluorophenyl ring must stimulate the π system to interact with empty p- or d-orbitals on the metal so that nucleophilic attack is discouraged. Fluorinated aromatic compounds appear to have better solubility in organic solvents than many other substituted aromatic derivatives.

6-2D. FLUORINATED CARBONYL REAGENTS

This section is divided into three parts: (*1*) fluorinated aldehydes and ketones where the fluorinated unit usually is introduced by reaction on the carbonyl function; (*2*) fluoroketenes which can be considered as a special class of ketones; and (*3*) fluorinated acids, acid halides, anhydrides, and esters where the carbonyl functionality is often retained.

A review by Braendlin and McBee[467] on the "Effect of adjacent perfluoroalkyl groups on carbonyl reactivity" provides a general description of the chemistry of fluorinated carbonyl reagents. Several synthetic reactions are outlined and mechanistic implications discussed.

1. Fluorinated Ketones and Aldehydes

The most readily available and useful compound in this class is hexafluoroacetone (HFA). Krespan and Middleton[468, 469] have provided an up-to-date

[459] R. D. Chambers and T. Chivers, *J. Chem. Soc.*, 4782 (1964).
[460] R. Filler and H. H. Kang, *Chem. Commun.*, 626 (1965).
[467] H. P. Braendlin and E. T. McBee, *Advan. Fluorine Chem.*, **3**, 1 (1963).
[468] C. G. Krespan and W. J. Middleton, *Fluorine Chem. Rev.*, **1**, 145 (1967).
[469] The reactions of HFA are also reviewed by N. P. Gambaryan, E. M. Rokhlin, Yu. V. Zeifman, C. Ching-Yun, and I. L. Knunyants, *Angew. Chem. Intern. Ed. Engl.*, **5**, 947 (1966).

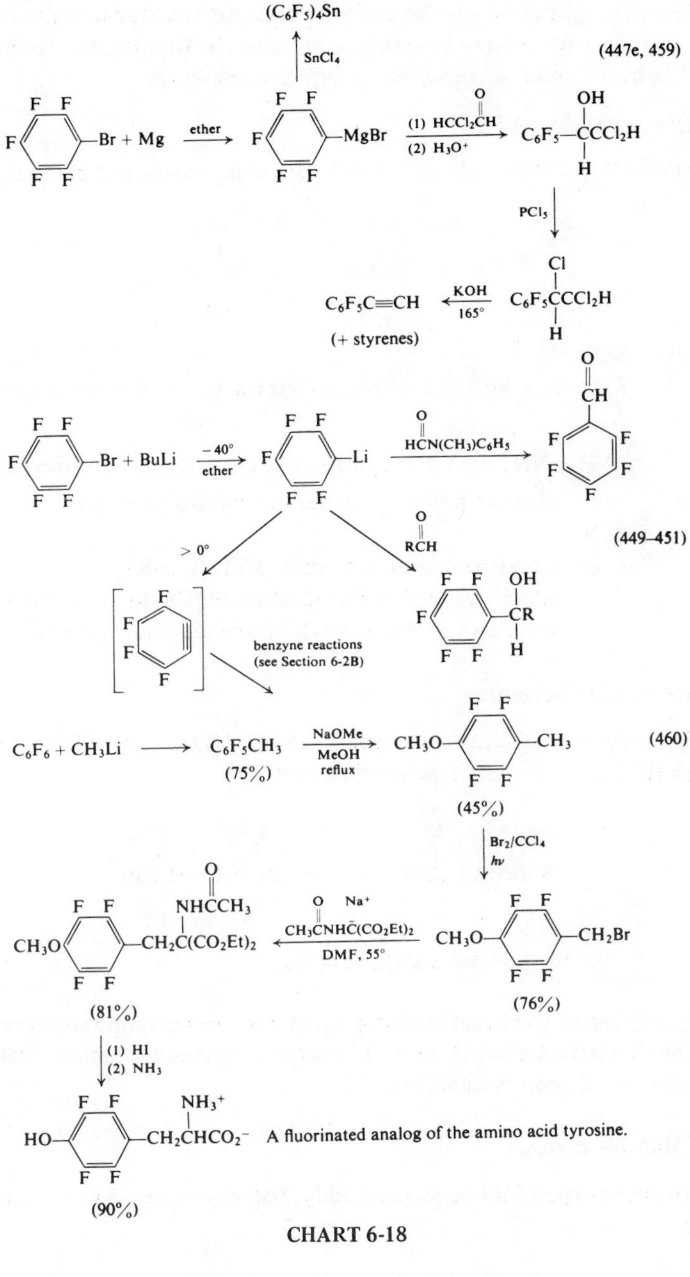

$(C_6F_5)_4Sn$

$SnCl_4$

(447e, 459)

$C_6F_5Br + Mg \xrightarrow{\text{ether}} C_6F_5MgBr \xrightarrow[(2)\ H_3O^+]{(1)\ HC(=O)CCl_2H} C_6F_5-CH(OH)CCl_2H$

PCl_5

$C_6F_5CHClCCl_2H \xrightarrow[165^\circ]{KOH} C_6F_5C{\equiv}CH$

(+ styrenes)

$C_6F_5Br + BuLi \xrightarrow[\text{ether}]{-40^\circ} C_6F_5Li \xrightarrow{HC(=O)N(CH_3)C_6H_5} C_6F_5CHO$

$C_6F_5Li \xrightarrow{RC(=O)H} C_6F_5CH(OH)R$

(449–451)

$C_6F_5Li \xrightarrow{> 0^\circ}$ [tetrafluorobenzyne] → benzyne reactions (see Section 6-2B)

$C_6F_6 + CH_3Li \longrightarrow C_6F_5CH_3$ (75%) $\xrightarrow[\text{MeOH, reflux}]{NaOMe}$ $CH_3O-C_6F_4-CH_3$ (45%) (460)

Br_2/CCl_4, $h\nu$

$CH_3O-C_6F_4-CH_2Br$ (76%) $\xrightarrow[\text{DMF, } 55^\circ]{CH_3C(=O)NH\bar{C}(CO_2Et)_2\ Na^+}$ $CH_3O-C_6F_4-CH_2C(NHCOCH_3)(CO_2Et)_2$ (81%)

(1) HI
(2) NH_3

$HO-C_6F_4-CH_2CH(NH_3^+)CO_2^-$ (90%) A fluorinated analog of the amino acid tyrosine.

CHART 6-18

picture of the chemistry of this useful reagent. Consequently, we will outline with illustrative examples only the main methods for introducing a fluorinated ketone unit. We have followed the organization of the Krespan and Middleton review[468] which should be consulted for primary references.

a. *Addition of nucleophiles.*

Active methylene compounds, active hydrogen compounds, and bases add as shown.

$$BH + CF_3\overset{\overset{\displaystyle O}{\|}}{C}CF_3 \longrightarrow B\underset{\underset{\displaystyle CF_3}{|}}{\overset{\overset{\displaystyle CF_3}{|}}{C}}OH$$

where BH is ROH,

RSH (can lead to cyclized product with difunctional reagent),

RNH_2 (NH_3 used in preparation of $CF_3\overset{\overset{\displaystyle NH}{\|}}{C}CF_3$ and a variety of nitrogen heterocyclic derivatives also prepared),

HCN,

active methylene compounds such as $CH_2(CO_2R)_2$,

F^- which can lead to introduction of $(CF_3)_2CFO-$ unit (see discussion in Section 6-1C.2 and see also page 117).

b. *Organometallic Reagents.*

This reaction was discussed in Section 6-1C. Addition usually goes normally, although HFA is much more reactive than acetone.

$$RMe + CF_3\overset{\overset{\displaystyle O}{\|}}{C}CF_3 \longrightarrow R\underset{\underset{\displaystyle CF_3}{|}}{\overset{\overset{\displaystyle CF_3}{|}}{C}}OH + H\underset{\underset{\displaystyle CF_3}{|}}{\overset{\overset{\displaystyle CF_3}{|}}{C}}OH$$

(where Me is MgX, Li or Na)

Reduction to give hexafluoroisopropanol is a troublesome side reaction; with a sterically hindered Grignard, such as isopropylmagnesium bromide, reduction occurs to the exclusion of addition.

c. *C–C Multiple Bonds.*

An allylic-type addition occurs readily, but vinylic addition, such as with

$$(CF_3)_2C{=}O + CH_3CH{=}CH_2 \longrightarrow (CF_3)_2\overset{\overset{\displaystyle OH}{|}}{C}CH_2CH{=}CH_2$$

ethylene, can be made using a Lewis acid catalyst. Cycloaddition, or Diels–Alder reactions with activated olefins, acetylenes, or dienes will be discussed later.

d. *Aromatics.*

Electrophylic substitution of aromatic systems to introduce hexafluoroisopropyl groups occurs readily under typical Friedel–Crafts conditions.

$$XC_6H_5 + CF_3\overset{\overset{\displaystyle O}{\|}}{C}CF_3 \xrightarrow{\text{(Lewis acid catalyst)}} XC_6H_4\underset{\underset{\displaystyle CF_3}{|}}{\overset{\overset{\displaystyle CF_3}{|}}{C}}{-}OH \xrightarrow{SF_4} XC_6H_4\underset{\underset{\displaystyle CF_3}{|}}{\overset{\overset{\displaystyle CF_3}{|}}{C}}F \qquad (470)$$

Orientation is normal: ortho and para substitution for activating, and meta for deactivating substituents. The hexafluoroisopropanol group is a very stable electron-attracting group that survives most chemical transformations on the aromatic residue and can be readily converted with SF_4 to the useful heptafluoroisopropyl substituent.[470]

e. *Inorganic Reagents.*

Phosphorus reagents give a variety of reactions: the carbonyl can be replaced by chlorines using PCl_5, triphenylphosphine imines give the corresponding HFA imines,

$$(CF_3)_2C{=}O + \phi_3P{=}NR \rightarrow (CF_3)_2C{=}NR$$

and Wittig reagents give olefins.

$$(CF_3)_2C{=}O + \phi_3P{=}CH\overset{\overset{\displaystyle O}{\|}}{C}CH_3 \longrightarrow (CF_3)_2C{=}CH\overset{\overset{\displaystyle O}{\|}}{C}CH_3$$

Phosphite esters give complex reactions, usually as a result of secondary reaction, but primary products can be isolated.

$$(CF_3)_2C{=}O + P(OC_2H_5)_3 \longrightarrow \begin{array}{c}(CF_3)_2C{-}{-}C(CF_3)_2\\ |\qquad\quad | \\ O\diagdown_{P}\diagup O \\ (OC_2H_5)_2\end{array}$$

$$\xrightarrow{\Delta} \text{several products including } (CF_3)_2C\overset{O}{\frown}C(CF_3)_2 \text{ (epoxide)}$$

$$\xrightarrow{H^+/H_2O} \begin{array}{c}OH\quad OH\\ |\qquad | \\ (CF_3)_2C{-}{-}C(CF_3)_2\end{array}$$

(the perfluoropinacol is also prepared by bimolecular reduction)

[470] W. A. Sheppard, *J. Am. Chem. Soc.*, 87, 2410 (1965).

Arsenic hydrides add as expected to HFA, but silicon, germanium, and tin hydride add so that the metal bonds to oxygen of the HFA.

$$R_3MeH + HFA \rightarrow R_3MeOCH(CF_3)_2$$

f. *Heterocyclic Derivatives.*

These are easily formed in a number of condensations of HFA with reagents such as nitriles, isonitriles, or orthoformates, so that a variety of unusual gem(trifluoromethyl)-substituted five- and six-membered ring compounds can be prepared.

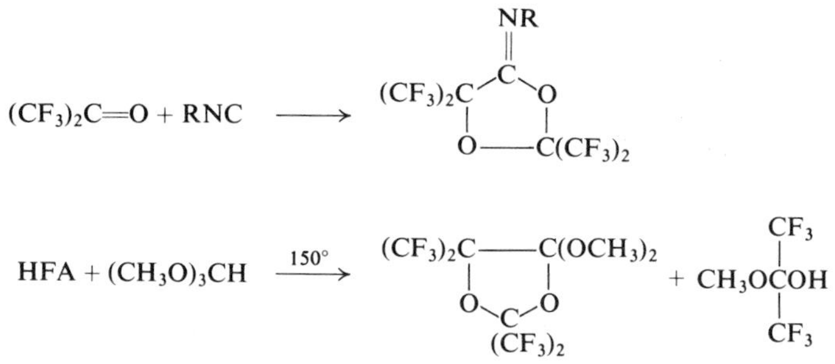

$$(CF_3)_2C{=}O + RNC \longrightarrow \text{(five-membered ring: } (CF_3)_2C{-}C({=}NR){-}O{-}C(CF_3)_2{-}O{-}\text{)}$$

$$HFA + (CH_3O)_3CH \xrightarrow{150^\circ} \text{(ring: } (CF_3)_2C{-}C(OCH_3)_2{-}O{-}C(CF_3)_2{-}O{-}\text{)} + CH_3OC(CF_3)_2OH$$

g. *Radical-Type Additions*

Radical-Type Additions to HFA can give some unusual products (see pages 196 and 232) since radical attack can be either on carbon or oxygen of the carbonyl. These radical reactions have been applied to modifying polymer with HFA and in copolymerization with olefins.

h. *Cycloaddition and Diels–Alder reactions*

Cycloaddition and Diels–Alder reactions of fluorinated ketones and aldehydes lead to heterocyclic four- and six-membered ring products. Since this area has been reviewed,[435b, c, 471] we cite only some typical examples[472] in Chart 6-19.

i. *Miscellaneous.*

Decomposition of HFA either thermally or photolytically provides a source of $CF_3\cdot$ or CF_2: as discussed in Sections 6-1A or B.

[471] J. Hamer and J. A. Turner in 1,4-*Cycloaddition Reaction.* (J. Hamer, ed.), Academic, New York, 1967, Chapter 8.

[472] (a) J. F. Harris, Jr., and D. D. Coffman, *J. Am. Chem. Soc.*, **84**, 1553 (1962). (b) E. W. Cook and B. F. Landrum, *J. Heterocyclic Chem.*, **2**, 327 (1965). (c) W. J. Linn, *J. Org. Chem.*, **29**, 3111 (1964).

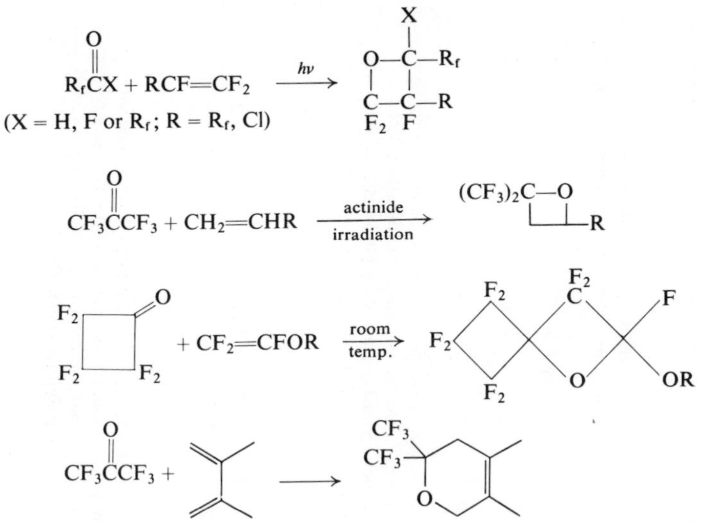

$$R_fCX(=O) + RCF{=}CF_2 \xrightarrow{h\nu} \text{[O—C(X)(R}_f\text{)—C(F)(R)—CF}_2\text{ ring]} \qquad (472a)$$

($X = H$, F or R_f; $R = R_f$, Cl)

$$CF_3C(=O)CF_3 + CH_2{=}CHR \xrightarrow[\text{irradiation}]{\text{actinide}} (CF_3)_2C\text{—O (oxetane ring bearing R)} \qquad (472b)$$

$$\text{hexafluorocyclobutanone } (F_2, F_2, F_2) + CF_2{=}CFOR \xrightarrow{\text{room temp.}} \text{spiro product } (F_2, F_2, F_2, C F_2, O, F, OR) \qquad (431)$$

$$CF_3C(=O)CF_3 + \text{2,3-dimethylbutadiene} \longrightarrow \text{2,2-bis(CF}_3\text{)-4,5-dimethyl-3,6-dihydro-2H-pyran} \qquad (472c)$$

CHART 6-19

Hexafluoroacetone has also been used to prepare bis(trifluoromethyl)-diazomethane or bis(trifluoromethyl)diazirine, which are good sources of bis(trifluoromethyl)carbene $(CF_3)_2C$: (see Section 6-1B).

$$(CF_3)_2CN_2 \qquad (CF_3)_2C\langle{}^{N}_{N}\|$$

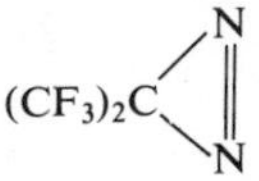

Many fluorinated ketones and aldehydes have been studied; in general, they can be used to introduce a fluorinated unit by the methods outlined for hexafluoroacetone. The more important ones are summarized in Table 6-7. Points to consider in use of these reagents are noted under properties. In particular, the less highly fluorinated compounds condense with esters (Claisen condensation) at an active methylene group.

Finally, the haloform reaction of fluorinated aldehydes and ketones should be noted both as a useful and by-product reaction in synthesis.

$$RC(=O)CF_3 \xrightarrow{100\%\ KOH} RCO_2H + CHF_3 \qquad (473)$$

[473] (a) H. Shechter and F. Conrad, *J. Am. Chem. Soc.*, **72**, 3371 (1950). (b) J. H. Simons and E. O. Ramler, *J. Am. Chem. Soc.*, **65**, 389 (1943). (c) A. Sykes, J. C. Tatlow, and C. R. Thomas, *J. Chem. Soc.*, 835 (1956).

Table 6-7

Fluorinated Aldehydes and Ketones – Availability for Synthesis

Aldehyde or ketone	Preparation	Properties and use in synthesis
$R_f\overset{O}{\overset{\|}{C}}H$	Prepared by a variety of methods (see Ref. 5, p. 143). Best general synthesis is controlled reduction (with $LiAlH_4$) of corresponding acid, anhydride or ester	Gives most of the typical aldehyde reactions; Irradiation with fluoroolefins gives oxetanes: $\overset{H}{\text{O—C—}R_f}$ / $F_2\text{C—C—R}$ Ref. 474a
$H(CF_2)_n\overset{O}{\overset{\|}{C}}H$	$H(CF_2)_nCH_2OH + Cl_2 \xrightarrow[10\text{-}40^\circ]{h\nu}$ recovered from hemiacetal by distillation $+NO_2 \xrightarrow[300\text{-}400^\circ]{air} H(CF_2)_2CH(OH)_2 \xrightarrow{dehydrate}$	Used as reactive reagent for the preparation of stable hydrates, carbonyl derivatives and decarbonylated products. Noted condensation products such as oximes are very stable. Ref. 474b,c
$C_6F_5\overset{O}{\overset{\|}{C}}H$	$C_6F_5MgX + HC(OC_2H_5)_3$ (excess) *or* $C_6H_5N(CH_3)\overset{O}{\overset{\|}{C}}H \rightarrow$	mp 20°, bp 168-170°. Typical reactions of benzaldehyde. Ref. 474d
$C_5NF_4\overset{O}{\overset{\|}{C}}H$	$C_5NF_5 + CH_3CH{=}CHLi \rightarrow C_5NF_4CH{=}CHCH_3 \xrightarrow{O_3}$ $+ NaCN \rightarrow C_5NF_4CN \xrightarrow{reduction}$	bp 52-54° (20 mm). Typical aldehyde reactions with hydroxylamine, arylhydrazine, aniline, Grignard reagent and oxygen but was decarbonylated by aqueous KOH (typical for polyfluoroaldehyde). Ref. 474e
$CF_3\overset{O}{\overset{\|}{C}}CF_3$ (HFA)	$CCl_3\overset{O}{\overset{\|}{C}}CCl_3 \xrightarrow[300^\circ]{HF/Cr^{+3}}$ Commercial method. Other routes such as oxidation of olefins were reported earlier	bp – 27°; commercially available. *See text* for chemical reactions. A new potentially useful reaction of HFA is condensation with α-amino acids to give oxazolidones. Useful for gc identification of amino acids and in peptide synthesis. Ref. 468,475a

$R_f\overset{\overset{O}{\|}}{C}R_f$	$R_f\overset{\overset{O}{\|}}{C}OEt + C_2H_5O^- \xrightarrow[\text{hydrolysis}]{\text{after}} R_f\overset{\overset{O}{\|}}{C}R_f$ $COF_2 + CF_2{=}CFR_f \xrightarrow{F^-}$ *or* $R_f'\overset{\overset{O}{\|}}{C}F + CF_2{=}CFR_f \xrightarrow{F^-}$	Similar to HFA in limited studies reported. Ref. 336, 375
Series of chlorofluoroacetones $CF_3\overset{\overset{O}{\|}}{C}CF_2Cl$ to $CCl_3\overset{\overset{O}{\|}}{C}CFCl_2$	Controlled condition for HFA synthesis so that Cl is only partially replaced in hexachloroacetone	$CF_3\overset{\overset{O}{\|}}{C}CF_2Cl$, bp 45°; $ClCF_2\overset{\overset{O}{\|}}{C}CF_2Cl$, bp 8°. Most of series is commercially available. Similar to HFA in most reactions, but because of steric effect of Cl, the more highly chlorinated ketones appear less reactive. A different course is noted for reaction with phosphite esters (Ref. 475b). Ref. 468, 475b
Series of fluorinated acetones: $CF_3\overset{\overset{O}{\|}}{C}CF_2H$ to $CH_3\overset{\overset{O}{\|}}{C}CH_2F$	Best general method $CF_3\overset{\overset{O}{\|}}{C}OEt + F_2HCCO_2Et \xrightarrow{NaH} CF_3\overset{\overset{O}{\|}}{C}CF_2CO_2Et \xrightarrow[\text{reflux}]{H_2SO_4(aq.)} CF_3\overset{\overset{O}{\|}}{C}CF_2H$ Other routes are $CH_3MgX + R_fCN \rightarrow CH_3\overset{\overset{O}{\|}}{C}R_f$. Replacement of Cl by F using reagents such as KHF_2 (for less highly fluorinated acetones)	$CH_3\overset{\overset{O}{\|}}{C}CFH_2$, bp 77-79° Typical enolizable ketone condensations given for series (see Ref. 483). Other normal ketone reactions also found. With less highly fluorinated derivatives, often found side reaction of replacement of F by bases. Ref. 475c, d, e
$R_f\overset{\overset{O}{\|}}{C}R$	$RMgX + R_fCN \rightarrow$ (claimed as general method) $R'CH_2CO_2C_2H_5 + R_fCO_2C_2H_5 \xrightarrow{Na} \xrightarrow{H_3O^+}$	Semicarbazones and 2,4-dinitrophenylhydrazone derivatives prepared. Trifluoromethyl ketones also source of α,β-unsaturated ketones $CF_3\overset{\overset{O}{\|}}{C}R + \phi_3P{=}CH\overset{\overset{O}{\|}}{C}R' \rightarrow CF_3\overset{\overset{R}{\|}}{C}{=}CH\overset{\overset{O}{\|}}{C}R' + \phi_3PO$ Ref. 475f, g, h

Table 6-7 (Cont.)

Aldehyde or ketone	Preparation	Properties and use in synthesis
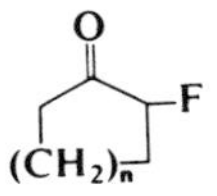 2-fluorocycloalkanone, $(CH_2)_n$ ring	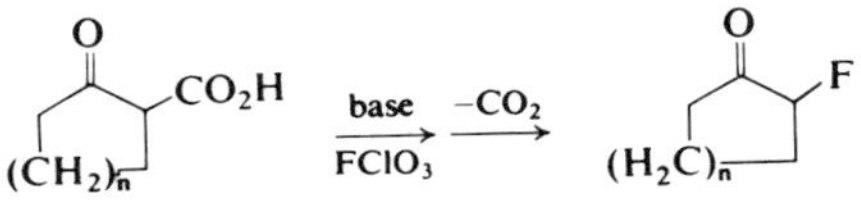 2-oxocycloalkanecarboxylic acid (CO_2H), $(CH_2)_n$ ring $\xrightarrow[FClO_3]{base}$ $\xrightarrow{-CO_2}$ 2-fluorocycloalkanone, $(H_2C)_n$ ring. Also some bicyclic ketones prepared	By-product of α-fluorodiester also obtained. Mechanism of formation discussed. Ref. 476a
$RCF_2\overset{O}{\overset{\Vert}{C}}CH_3$ R = steroid residue	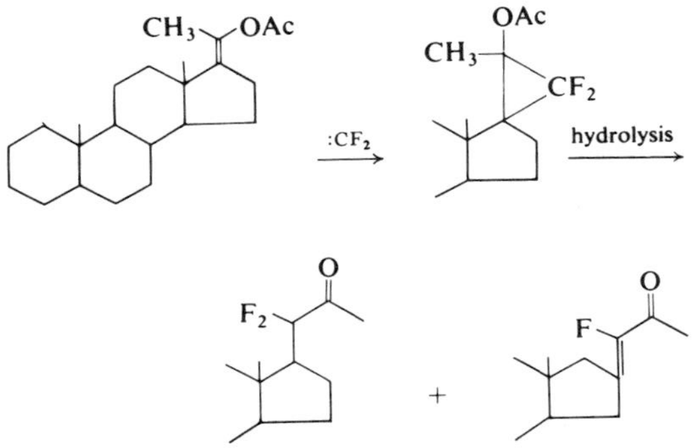 Steroid enol acetate (CH_3, OAc) $\xrightarrow{:CF_2}$ difluorocyclopropane adduct (CH_3, OAc, CF_2) $\xrightarrow{hydrolysis}$ F_2 methyl ketone + F vinyl methyl ketone	Several derivatives of ketones prepared. Ref. 476b
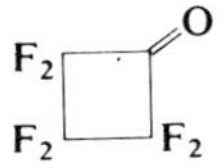 Hexafluorocyclobutanone (F_2, F_2, F_2; O)	$CF_2=CF_2 + CF_2=CFOR \rightarrow$ alkoxyheptafluorocyclobutane $\xrightarrow[H_2SO_4,\ 175°]{95\%}$	bp 1°. Thermally stable to over 300°; chemically more reactive than HFA; gives 2+2 and 2+4 cycloadditions; adds nucleophilic reagents; gives Friedel-Crafts substitution of aromatic compounds. Ref. 431
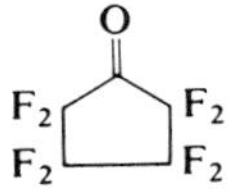 Octafluorocyclopentanone (F_2, F_2, F_2, F_2; O)	Cyclopentanone + F_2 $\xrightarrow{CoF_3}$	bp 24°. Formed a semicarbazone. Ref. 476c

Compound	Preparation	Properties and reactions
$ArC(O)CF_3$ (Also $ArC(O)R_f$)	$ArMgX + CF_3CO_2H \rightarrow$ $ArH + CF_3C(O)Cl$ (or $CF_3C(O)OC(O)CF_3$) $\rightarrow$	$C_6H_5C(O)CF_3$, bp 66-67° (33 mm). Reduced to alcohol, fluorinated with SF_4 to $ArCF_2CF_3$. Reacted with phosphines to give olefins $+ (n\text{-}C_4H_9)_3P \rightarrow (Ar)(CF_3)C{=}C(H)(CH_2)_2CH_3$ (cis-trans mixture) Ref. 476d, e, f, g
Tetrakis(trifluoromethyl)cyclopentadienone ($(CF_3)_4C_4C{=}O$)	$CF_3C{\equiv}CCF_3 + RhCl(CO)_2 + CO \xrightarrow[100\ atm]{150°}$ See Table 6-5 for preparation of related cyclopentadienones	mp 44-45°. Forms complexes with transition metal compounds. Ref. 444
$CF_3C(O){-}C(O)CF_3$	$CF_3CCl{=}CClCF_3 + CrO_3 + H_2SO_4 \rightarrow$	bp 20°, hydrates readily giving both mono and diadducts. $+ o\text{-}C_6H_4(NH_2)_2 \longrightarrow$ 2,3-bis(trifluoromethyl)quinoxaline $+$ strong base $\rightarrow CF_3H$ (haloform reaction). Ref. 477a
$R_fC(O){-}C(O)R_f$	For $R_f = CF(CF_3)_2$: $FC(O){-}C(O)F + CF_2{=}CFCF_3 \xrightarrow{F^-}$ For $R_f = CF_2CF_2CF_3$ or C_4F_9: $R_fC(O)Cl + Ni(CO)_5 \rightarrow \xrightarrow{heat}$	$[(CF_3)_2CFC(O)]_2$ bp 91-93°; $[CF_3CF_2CF_2C(O)]_2$ bp 95° Characterized by condensation with *o*-phenylene diamine (see reaction of hexafluorobiacetyl above). Ref. 336,55
Tetrafluorocyclobutane-1,2-dione (F_2, F_2)	$2CF_2{=}CFOCH_3 \rightarrow$ 1,2-dimethoxyhexafluorocyclobutane $\xrightarrow{H_2SO_4}$	bp 35°, deep blue in gaseous, liquid and solid state; polymerizes readily, extremely reactive at carbonyl group similar to hexafluorocyclobutanone. Ref. 431

Table 6-7 (Cont.)

Aldehyde or ketone	Preparation	Properties and use in synthesis
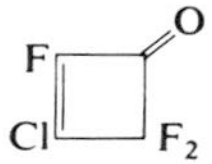 2-chloro-3-fluoro-4,4-difluorocyclobut-2-enone (F, Cl, F_2, =O)	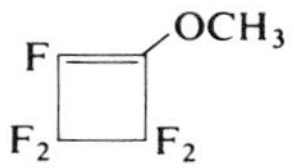 1-methoxy-2-fluoro-3,3,4,4-tetrafluorocyclobutene (F, OCH_3, F_2, F_2) $\xrightarrow{SnCl_4}$ (Note Lewis acid replacement of F by halogens)	More extensive replacement of F noted with certain other Lewis acids. Ref. 477b
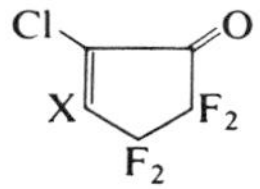 2-chloro-3-X-4,4,5,5-tetrafluorocyclopent-2-enone (Cl, X, F_2, F_2, =O); X is Cl or F	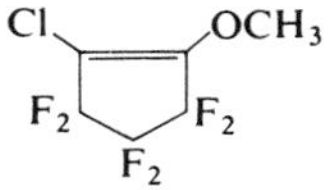 1-methoxy-2-chloro-3,3,4,4,5,5-hexafluorocyclopentene (Cl, OCH_3, F_2, F_2, F_2) $\xrightarrow[\text{(b) } SnCl_4]{\text{(a) } CaF_2/FeCl_3}$ followed by BF_3. (a gives X = F), (b gives X = Cl)	Ref. 477b
R_fCCH_2CR (with C=O at both carbonyl carbons)	$R_fCO_2CH_3 + CH_3\overset{O}{\overset{\Vert}{C}}R' \xrightarrow{NaOCH_3}$ *or* $RCO_2CH_3 + CH_3\overset{O}{\overset{\Vert}{C}}R_f \xrightarrow{NaOCH_3}$ (Cyclic by-products also obtained)	$CF_3\overset{O}{\overset{\Vert}{C}}CH_2\overset{O}{\overset{\Vert}{C}}R + NCCH_2\overset{O}{\overset{\Vert}{C}}NH_2 \xrightarrow{NaOEt}$ 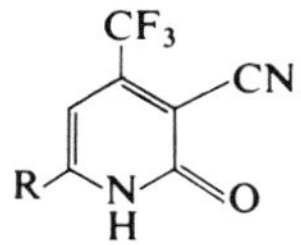 6-R-4-CF_3-3-CN-2-pyridone (R, CF_3, CN, N–H, =O); + metals → coordination compounds. Ref. 477c,d,e,f

[474](a) J. F. Harris and D. D. Coffman, *J. Am. Chem. Soc.*, **84**, 1553 (1962). (b) N. O. Brace, *J. Org. Chem.*, **26**, 4005 (1961). (c) R. M. Scribner, *J. Org. Chem.*, **29**, 279 (1964). (d) A. K. Barbour, M. W. Buxton, P. L. Coe, R. Stephens, and J. C. Tatlow, *J. Chem. Soc.*, 808 (1961). (e) R. E. Banks, R. N. Haszeldine, and I. M. Young, *J. Chem. Soc. (C)*, 2089 (1967).

[475](a) F. Weygand, K. Burger, and K. Engelhardt, *Chem. Ber.*, **99**, 1461 (1966). (b) D. W. Wiley and H. E. Simmons, *J. Org. Chem.*, **29**, 1876 (1964). (c) E. T. McBee, O. R. Pierce, H. W. Kilbourne, and E. R. Wilson, *J. Am. Chem. Soc.*, **75**, 3152 (1953). (d) E. D. Bergman, S. Cohen, E. Hoffman, and Z. Rand-Meir, *J. Chem. Soc.*, 3452 (1961). (e) E. Cherbuliez, A. de Picciotto, and J. Rabinowitz, *Helv. Chim. Acta*, **43**, 1143 (1960). (f) J. Burdon and V. C. R. McLoughlin, *Tetrahedron*, **20**, 2163 (1964). (g) E. T. McBee, O. R. Pierce, and D. D. Meyer, *J. Am. Chem. Soc.*, **77**, 917 (1955). (h) D. L. Dull, I. Baxter, and H. S. Mosher, *J. Org. Chem.*, **32**, 1622 (1967).

[476](a) H. Machleidt and V. Hartman, *Ann.*, **679**, 9 (1964). (b) P. Crabbe, H. Carpio, A. Cervantes, J. Iriarte, and L. Tökes, *Chem. Commun.*, 79 (1968). (c) F. F. Holub and L. A. Bigelow, *J. Am. Chem. Soc.*, **72**, 4879 (1950). (d) K. T. Dishart and R. Levine, *J. Am. Chem. Soc.*, **78**, 2268 (1956). (e) R. Fuchs and G. J. Park, *J. Org. Chem.*, **22**, 993 (1957). (f) W. A. Sheppard, *J. Am. Chem. Soc.*, **87**, 2410 (1965). (g) D. J. Burton, F. E. Herkes, and K. J. Klabunde, *J. Am. Chem. Soc.*, **88**, 5042 (1966).

[477](a) L. O. Moore and J. W. Clark, *J. Org. Chem.*, **30**, 2472 (1965). (b) O. Scherer, G. Hörlein, and H. Millauer, *Chem. Ber.*, **99**, 1966 (1967). (c) R. A. Moore and R. Levine, *J. Org. Chem.*, **29**, 1439 and 1883 (1964). (d) S. Portnoy, *J. Org. Chem.*, **30**, 3377 (1965). (e) R. C. Fay and T. S. Piper, *J. Am. Chem. Soc.*, **85**, 500 (1963). (f) R. D. Gillard and G. Wilkinson, *J. Chem. Soc.*, 5885 (1963).

2. Fluorinated Ketenes

The fluoroketenes have only been known since 1963, but have already been reviewed by one group of the major contributors to the area.[478] Bis(trifluoromethyl)ketene is readily prepared by a variety of routes and has been studied most extensively. The most practical synthesis was described by England and Krespan[479] and is based on perfluoroisobutylene. Bis(trifluoromethyl)ketene is in equilibrium with perfluoromethacroyl fluoride in the vapor phase over anionic

$$(CF_3)_2C{=}CF_2 \xrightarrow{RO^-} (CF_3)_2C{=}CFOR \xrightarrow{H_3O^+} (CF_3)_2CHCO_2H \xrightarrow{P_2O_5} (CF_3)_2C{=}C{=}O$$

catalysts such as sodium fluoride and is very reactive to a variety of reagents as summarized in Chart 6-20.

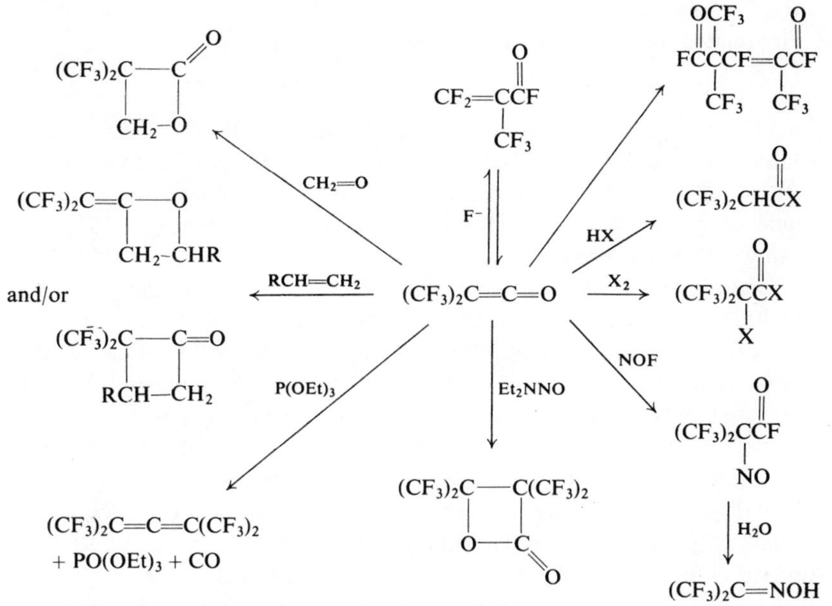

CHART 6-20

The products from these reactions are often useful intermediates so that bis(trifluoromethyl)ketene should become a valuable common reagent for introducing the hexafluoroisopropylidene unit.

Some other fluorinated ketenes are known.[478a] The most interesting is the parent difluoroketene which is a highly reactive and unstable compound.[480] It has

[478] (a) Yu. A. Cheburkov and I. L. Knunyants, *Fluorine Chem. Rev.*, **1**, 107 (1967). (b) For the most recent publication, see Yu. A. Cheburkov, N. Makhamadaliev, and I. L. Knunyants, *Tetrahedron*, **24**, 1341 (1968).

[479] D. C. England and C. G. Krespan, *J. Am. Chem. Soc.*, **87**, 4019 (1965); *J. Am. Chem. Soc.*, **88**, 5582 (1966).

[480] D. C. England and C. G. Krespan, *J. Org. Chem.*, **33**, 816 (1968).

been detected by dissociation at 35°C to carbon monoxide and tetrafluoroethylene and by trapping reactions.

$$CF_2{=}CF_2 \xrightarrow{NaOCH_3} CF_2{=}CFOCH_3 \xrightarrow{Br_2} BrCF_2CFBrOCH_3 \xrightarrow{ClSO_3H}$$

$$BrCF_2\overset{O}{\overset{\|}{C}}X \xrightarrow{Zn} [CF_2{=}C{=}O] \xrightarrow{BrCF_2COX} CF_2{=}CXO\overset{O}{\overset{\|}{C}}CF_2Br + \text{other products}$$

$$[CF_2{=}C{=}O] \xrightarrow{\text{acetone}} \begin{array}{l} CF_2{-}C{=}O \\ \;|\quad\;\;\, | \\ (CH_3)_2C{-}O \end{array}$$

$$[CF_2{=}C{=}O] \xrightarrow{35^\circ} CO + [{:}CF_2] \longrightarrow CF_2{=}CF_2$$

3. Fluorinated Acids, Acid Halides, Anhydrides, and Esters

These reagents are useful for introducing $R_f\overset{O}{\overset{\|}{C}}{-}$ or $R_f\overset{O}{\overset{\|}{C}}O{-}$ units. Most published work is with trifluoroacetyl derivatives (R_f is CF_3), but longer-chain perfluorinated compounds are also available and appear to behave in a similar manner. A special class is fluoroformyl groups (R_f is F, particularly COF_2) which will also be reviewed.

Certain of these fluorinated compounds, such as trifluoroacetic acid and anhydride, are useful reagents as solvents, catalysts, or the like; often they participate in organic reactions but do not appear in the final product. Since our primary concern in this chapter is synthesis of fluorinated compounds by adding fluorinated units, we refer the reader to a good review by Hudlicky[481] on use of fluorinated compounds as chemical reagents.

Useful background material on condensation reactions of various fluorinated carbonyl reagents with active methylene compounds is given in reviews by Braendlin and McBee,[467] Lovelace, Rausch, and Postelnek,[482] and Hudlicky.[483] Also, many reactions of fluorinated carbonyl compounds are discussed in Houben-Weyl,[6] although the examples are unfortunately scattered through the text. Use of trifluoroacetic acid and anhydride in synthesis has also been reviewed.[484]

These reagents are best reviewed as the general class R_fCOX, where X is OH, OR, O_2CR, halogen (usually F or Cl), or NR_2, because they can often be used interchangeably to introduce a perfluoroacyl group, R_fCO. The general types of reactions are listed below and specific examples from the recent literature are given in Table 6-8. More examples can be obtained from the reviews.[5, 6, 481–484]

[481] Ref. 5, Chapter 7, p. 278.

[482] A. M. Lovelace, D. A. Rausch, and W. Postelnek, *Aliphatic Fluorine Compounds*, Reinhold, New York, 1958, Chapters 7, 8, and 9.

[483] Ref. 5, pp. 227-233.

[484] J. M. Tedder, *Chem. Rev.*, **55**, 787 (1955).

a. *Acylation of Active Hydrogen Compounds*

$$\mathrm{ROH} + \mathrm{R_f\overset{O}{\overset{\|}{C}}X} \longrightarrow \mathrm{RO\overset{O}{\overset{\|}{C}}R_f} \xrightarrow{H_2O} \mathrm{ROH} + \mathrm{R_f\overset{O}{\overset{\|}{C}}OH}$$

$$\text{or} \qquad\qquad \text{or} \qquad\qquad \text{or}$$

$$\mathrm{RNH_2} \qquad\qquad \mathrm{RNH\overset{O}{\overset{\|}{C}}R_f} \qquad\qquad \mathrm{RNH_2}$$

Trifluoroacetic acid, anhydride,[484] and acid chloride are used as acylating agents, particularly in providing temporary blocking of OH or NH_2 in carbohydrates[485a] and peptides,[485b] when synthetic operations are done at other sites in the molecule. They are easily removed by hydrolysis.

b. *Friedel–Crafts Acylation.*

Acid halides or anhydrides acylate aromatic rings using normal Friedel–Crafts conditions to give aryl perfluoroalkyl ketones.

$$\mathrm{ArH} + \mathrm{R_f\overset{O}{\overset{\|}{C}}X} \xrightarrow[\text{or HF}]{\mathrm{AlCl_3}} \mathrm{Ar\overset{O}{\overset{\|}{C}}R_f}$$

c. *Reaction with Organometallic Reagents.*

Organometallic reagents react with acids, acid halides, or esters to give ketones. Use of excess organometallic reagent can lead to tertiary alcohol.

$$\mathrm{RMgX} + \mathrm{R_f\overset{O}{\overset{\|}{C}}X} \longrightarrow \mathrm{R\overset{O}{\overset{\|}{C}}R_f} \xrightarrow{\mathrm{RMgX}} \mathrm{R\overset{OH}{\overset{|}{\underset{R}{\underset{|}{C}}}}R_f}$$

Perfluorinated ketones are prepared from fluorinated esters by reaction with alkoxides,[375] as described in Section 6-1C (see pages 310-311).

d. *Active Methylene Condensations.*

Condensations at active methylene groups is an area of classical organic chemistry already reviewed.[467, 482, 483] The condensations of the Claisen type can

[485] (a) T. G. Bonner, *Advan. Carbohydrate Chem.*, **16**, 59 (1961). (b) F. Weygand, *Bull. Soc. Chim. Biol.*, **43**, 1269 (1961).

Table 6-8

Synthetic Routes to Introduce Fluorinated Units by Use of $R_fC(=O)X$ Reagents

Reactants	Product(s) (% yield)	Comments
A. ACYLATION OF ACTIVE HYDROGEN COMPOUNDS		
ArOH + trifluoroacetic anhydride $\xrightarrow[\text{benzene}]{\text{reflux}}$	ArO_2CCF_3 (quantitative)	Series of phenols thiophenols and *N*-hydroxy imide derivatives prepared and used to synthesize active esters of acylamino acids by ester interchange (for peptide synthesis). Fluorinated with SF_4 to $ArOCF_2CF_3$. Ref. 487a, 490a,b
Imidizole + trifluoroacetic anhydride $\xrightarrow[\circ]{\text{THF}}$	1-Trifluoroacetylimidizole (80)	Active reagent for trifluoroacetylating ROH and RNH_2. Ref. 490c
$ArOH + CF_3C(=O)OC(=O)CH_3 \xrightarrow{CCl_4}$	ArO_2CCF_3 (quantitative)	Mechanism study. Acetyltrifluoroacetate prepared by reaction of trifluoroacetic anhydride with acetic anhydride. Ref. 490d
$RCH(NH_2)CO_2H$ + trifluoroacetic anhydride $\xrightarrow{CF_3CO_2H}$	$RCH(NHC(=O)CF_3)CO_2H$ (70-95) Series of α-amino acids studied. Special techniques needed for some of more sensitive ones	Studied in peptide synthesis. Derivatives also useful for gas chromatographic analysis of amino acids. Ref. 491

Reactants	Product (yield, %)	Remarks
$(RO)_2C_6H_3CH_2NHCH(CO_2H)C_6H_5$ + trifluoroacetic anhydride	$(RO)_2C_6H_3CH_2N(COCF_3)CH(CO_2H)C_6H_5$	Cyclized product to [structure: $(RO)_2$-substituted tetrahydroisoquinoline-fused oxazolidinone bearing F_3C and C_6H_5] (or isomer). Ref. 492a
$HONH_2\cdot HCl$ + trifluoroacetic anhydride $\xrightarrow{reflux}$	$CF_3C(=O)ONHC(=O)CF_3$ (80)	Useful reagent to convert aldehydes to nitriles. Also prepared $CF_3C(=O)NHOH$ but less effective in same reaction. Ref. 492b
$ArOH + R_fCOCl \xrightarrow[ether]{pyridine}$	ArO_2CR_f (76)	$ArO_2CR_f + SF_4 \xrightarrow{HF} ArOCF_2R_f$. Ref. 487a
$C_6H_5C(=NOH)NH_2 + ClC(=O)(CF_2)_3C(=O)Cl$	$[C_6H_5C(NH_2)=N-O-C(=O)]_2(CF_2)_3$ (60–65) With excess benzamidoxime: $C_6H_5C\langle N{=}C(-O-N{=})\rangle C(CF_2)_2CO_2^-\ C(=NOH)C_6H_5\ (H_3N^+)$ (21) plus $[C_6H_5C{=}N{-}O{-}C{=}N]_2(CF_2)_3$ (1,2,4-oxadiazole) (10)	Preparation of 1,2,4-oxadiazoles from perfluoromonoacyl chlorides reported earlier. Ref. 492c

Table 6-8 (Cont.)

Reactants	Product(s) (% yield)	Comments
Polyvinyl alcohol (PVA) + $R_fCOCl \xrightarrow[\text{benzene}]{\text{pyridine}}$	Fluorinated esters of PVA	Studied solubility, heat stability, and oil repellency. Ref. 492d
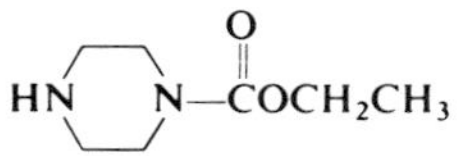 $HN(CH_2CH_2)_2N\text{—}COCH_2CH_3$ (C=O) + $CF_3COC_2H_5$ (C=O)	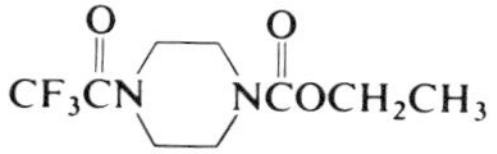 $CF_3CN(CH_2CH_2)_2NCOCH_2CH_3$ (both C=O) (–)	Reduced with borane to 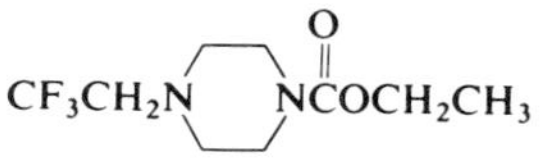 $CF_3CH_2N(CH_2CH_2)_2NCOCH_2CH_3$ (C=O). Ref. 492e
B. FRIEDEL–CRAFTS ACYLATION		
$(CF_3C)_2O + C_6H_5N(CH_3)_2$, $AlCl_3$, Et_3N (C=O)	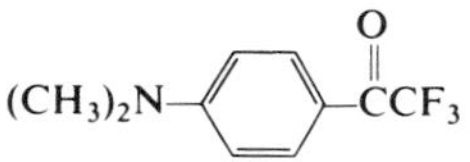 $(CH_3)_2N\text{—}C_6H_4\text{—}CCF_3$ (C=O, para) (39)	Other aryl perfluoroalkyl ketones also prepared. Ref. 493a
$ClCCF_2CCl + C_6H_6$, $AlCl_3$ (both C=O)	$(C_6H_5C)_2CF_2$ (C=O) (68)	Ref. 493b
$C_6H_5CNCH_2CO_2H$ (C=O, N–R) $+ (CF_3C)_2O$ (C=O)	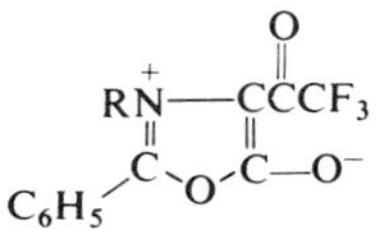 RN^+—$CCCF_3$ (C=O), ring: $RN^+{=}C(C_6H_5)\text{—}O\text{—}C(O^-){=}C(COCF_3)$ (almost quantitative)	Proposed mechanism of formation: 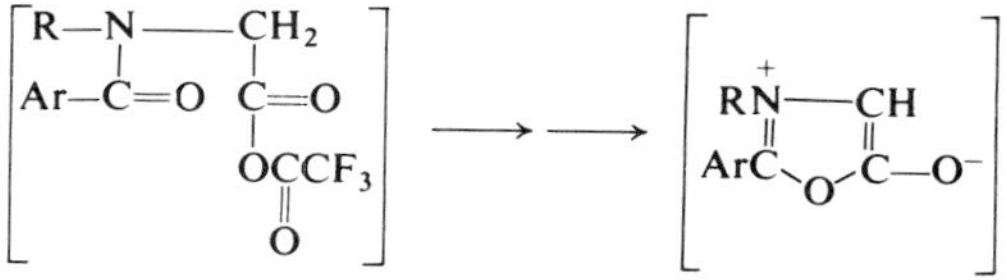 [R—N(—C(=O)Ar)—CH$_2$—C(=O)—OC(=O)CF$_3$] ⟶ ⟶ [RN^+—CH ring: ArC—O—C—O^-]. Noted trifluoroacetyl group stabilized mesoionic system. Ref. 493c

C. REACTION WITH ORGANOMETALLIC REAGENTS

Reactants	Product (yield, %)	Remarks
$ArMgX + R_fCO_2H$	$ArC(=O)R_f$ (50-67)	General reaction for preparation of fluorinated ketones (see Table 6-7). Ref. 493a
$CH_3MgBr + HO_2CCF_2CF_2CO_2H \xrightarrow{ether}$	$CH_3C(=O)CF_2CF_2C(=O)CH_3$ (45)	A series of Grignards with perfluorinated acids and anhydrides were run as part of a study of fluorinated paraffins. Only by-products such as alcohols were formed in some reactions. Ref. 494a
2-Thienylmagnesium bromide $+ C_3F_7COCl \xrightarrow{ether}$	2-Heptafluorobutyrothiophene (32)	Product converted by series of steps to $C_3F_7(CH_2)_4CH_3$. Ref. 494b

D. ACTIVE METHYLENE CONDENSATIONS (see also Table 6-7)

Reactants	Product (yield, %)	Remarks
$CH_3C(=O)OC_2H_5 + CF_3C(=O)OC_2H_5 \xrightarrow{Na}$ (Claisen ester condensation conditions)	$CF_3C(=O)CH_2C(=O)OC_2H_5$ (61)	Acid hydrolysis of product gives $CF_3C(=O)CH_3$. Series of fluorine containing β-ketoesters prepared from $RCH_2CO_2C_2H_5$ and $R_fCO_2C_2H_5$. Ref. 495a
$RR'C{=}O + FCH_2CO_2C_2H_5 \xrightarrow{C_2H_5O^-}$	$RR'C(OH)CHFCO_2C_2H_5$ (17-60) R, R′ is alkyl, aryl or H	Noted as useful synthesis of α-F, β-OH acids. Ref. 495b

Table 6-8 (Cont.)

Reactants	Product(s) (% yield)	Comments
$CH_3C(=O)CH_2CH_3 + CH_2FC(=O)OC_2H_5 \xrightarrow[\text{ether}]{\text{NaH}}$	$FCH_2C(=O)CH_2C(=O)CH_2CH_3$ (20) + $CH_3CH_2C(OH)(CH_3)CHFC(=O)OC_2H_5$ (14.5)	Part of a general study on use of ethyl **fluoroacetate in** synthesis; discusses methylene reactivity. A variety of other condensations of ethyl fluoroacetate given in earlier papers referenced in this article. Ref. 495c
$C_2H_5OC(=O)Cl + NaCHFC(=O)OC_2H_5$	$HFC(CO_2C_2H_5)_2$ (20-25)	Condensed product with urea to give a fluorobarbituric acid. Other fluoropyrimidines also prepared. Ref. 495d
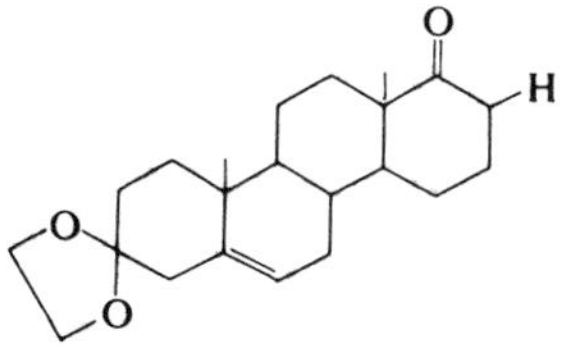 H + $CF_3CO_2C_2H_5 \xrightarrow[\text{benzene}]{\text{NaH}}$	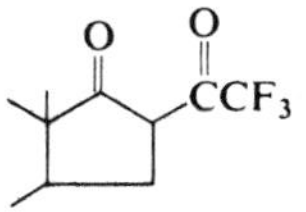 CCF₃ (–)	Series of steroids trifluoroacetylated at carbon 16. Ref. 496a
Aminoacids + excess trifluoroacetic anhydride	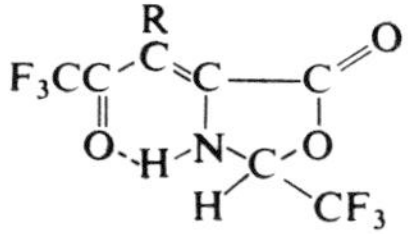 F_3CC R CF_3 For R = H, CH_3, C_6H_5 Other products formed depending on conditions and nature of R	Heat with base 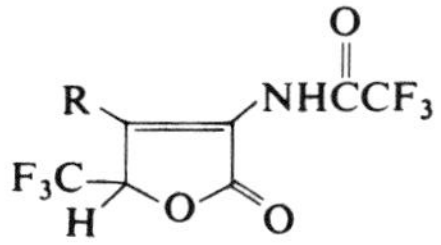 → R $NHCCF_3$ F_3C H Ref. 496b

Reaction	Products (yield %)	Comments
$RCHO + CF_3COCH_2CO_2C_2H_5$ $\xrightarrow{\text{piperidine}}$ (usual Knoevenagel conditions); R = alkyl or aryl	Cyclic product: $C_2H_5O_2C$, R, H, $CO_2C_2H_5$; H–C–C(R)(H)–C–H; HO–C(CF$_3$)–O–C(CF$_3$)–OH	Products hydrolyzed by base to 3-substituted glutaric acids. Ref. 496c
$(EtO)_2P(O)CH_2CO_2Et + CH_2FCOCHFCO_2Et$ (Wittig reaction conditions)	$CH_2FC(=CHCO_2Et)CHFCO_2Et$	Recent paper in series of 39 by Bergmann and co-workers on organic fluorine compounds. This series provides extensive examples of a large variety of condensation reactions which have been used to introduce mono- (and di-) fluoro substituted units, particularly for making analogs of biologically active materials (note other papers in series cited above). Ref. 496d
o-Phenylenediamine + $CF_3COCH_2COOC_2H_5$ (a) alkaline conditions	Benzodiazepinone: N(H)–C(=O)–CH$_2$–C(CF$_3$)=N (80)	Some chemical reactions of products described. Ref.496e
(b) acidic conditions	2-(CH_2COCF_3)benzimidazole, CCH_2CCF_3 (O) (15)	
$RC_6H_4NH_2 + CF_3COCH_2COOC_2H_5$ $\xrightarrow[\text{polyphosphoric acid}]{150°}$	R-4-hydroxy-2-CF_3-quinoline + R-2-hydroxy-4-CF_3-quinoline (Yields varied from 100 to 9 depending on R)	Products converted to other quinoline derivatives. Ref. 497a

Table 6-8 (Cont.)

Reactants	Product(s) (% yield)	Comments
E. ADDITION REACTIONS		
Cyclopentene + CF_3CO_2H	Trifluoroacetoxycyclopentane. A series of cyclic olefins and hexenes studied	Mechanism studied; shown to involve carbonium ion intermediate which can lead to rearranged products. Noted remarkable effect of remote substituent. Ref. 497b
$H_2C{=}C_6H_4{=}CH_2 + CF_3CO_2H$	$CF_3CO_2CH_2C_6H_4CH_3$ and $CF_3CO_2(CH_2C_6H_4CH_2)_nH$	Ref. 497c
$CH_2{=}C(NHO_2CCH_3)CO_2C_2H_5 + (C_2H_5O_2C)_2CHF$ (Michael condition of Na in alcohol)	$(C_2H_5O_2C)_2CFCH_2CH(NHO_2CCH_3)CO_2C_2H_5$ (–)	Hydrolyzed without isolation to γ-fluoroglutamic acid $HO_2CCHFCH_2CH(NH_2)CO_2H$ (56% yield overall). Ref.497d
$CH_2{=}CHCN + FCH(CONH_2)_2 \xrightarrow{NaOCH_3}$	$NCCH_2CH_2CF(CONH_2)_2$ (90)	Hydrolyzed and decarboxylated to γ-fluoroglutamic acid. Ref. 498a
$R_fC(O)X + CH_2N_2 \xrightarrow{ether}$	$R_fC(O)CH_2X$ (28) (X is Cl or Br)	Noted Arndt-Eistert reaction does not work on perfluorinated acids, only get α-haloketone product. Ref. 498b,c Also, $RC(O)F + CH_2N_2 \rightarrow RC(O)CH_2F$ (see Ref. 499f)

$(C_6H_5O)_3P + CF_3CO_2H$	$C_6H_5O_2CCF_3$ (55-60)	Claimed to be easiest and cheapest route to phenyltrifluoroacetate needed for peptide synthesis (see Part A). Ref. 498d

F. CARBONYL FLUORIDE AND FORMYL FLUORIDE IN SYNTHESIS

$ROH + COF_2$ or $ClC(=O)F$ (base may be used to remove HX but not necessary) $RNH_2 + COF_2$ (R is alkyl or aryl)	$ROC(=O)F$ (high yields) $RNHC(=O)F$ (including some vicinal difunctional derivatives)	$ROC(=O)F \xrightarrow{\text{amine catalysis}} RF + CO_2$ $ROC(=O)F + SF_4 \xrightarrow{HF} ROCF_3$ Ref. 487, 489, 499a
Ethylene oxide + $COF_2 \xrightarrow{F^-}$	$CF_3OCH_2CH_2OC(=O)F$	Formed via $F^- + COF_2 \rightarrow CF_3O^-$ which attacks epoxide. Ref. 396b, 487b
$(CH_2)_x$–O (cyclic ether) + COFCl $\xrightarrow[\text{(tertiary amine or glycol)}]{\text{cat.}}$	$Cl(CH_2)_xOC(=O)F$ (47-100)	$\xrightarrow{BF_3\text{-}Et_2O} Cl(CH_2)_xF \xrightarrow[\text{oxide}]{\text{metal}} CH_2{=}CH(CH_2)_{x-2}F.$ Ref. 499b
COF_2 anionic catalyst, heat	$CF_3OC(=O)F$ or $CF_3OC(=O)OCF_3$	Ref. 499c
COF_2 + MF (M is Cs, K or Rb)	$MOCF_3$	Decomposed readily on attempted reactions or solution. Ref. 499d
$COF_2 + RC(=O)R'$	$RC(F)(OC(=O)F)R'$ (78) Usually loses CO_2 to give RCF_2R'	Ref. 336a

Table 6-8 (Cont.)

Reactants	Product(s) (% yield)	Comments
$COF_2 + R_fCF{=}CF_2$	$R_fC(CF_3)C(=O)F$ (13-80) (or $R_fC(=O)R_f$ 39%)	
$+ RNCO$	$RN(C(=O)F)_2$ (95)	
$+ CF_3N{=}CF_2$	$(CF_3)_2NC(=O)F$ (56)	
$+ R_fCN$	R_fCF_2NCO (18)	
Most additions required a catalyst, such as CsF or tertiary amine		
$COF_2 + HCN + MF$	$FC(=O)CN$	Ref. 499e
$C_2H_5OC(=O)F + C_2H_5CHN_2 \xrightarrow[\text{ether}]{HF}$	$C_2H_5CHFC(=O)OC_2H_5$ (68)	Reaction can also be used for preparation of α-fluoroketones by starting with an acyl fluoride. Ref. 499f
$HC(=O)F + ArH$	$RC(=O)H$	$HC(=O)F$ prepared by reaction $HC(=O)OH + KHF_2 + C_6H_5C(=O)Cl \xrightarrow{\Delta}$
$+ R_2NH$	$R_2NC(=O)H$	Excellent reagent for C-, O-, S- and N- formulations but F not retained in product. Ref. 499g

[490](a) S. Sakakibara and N. Inukai, *Bull. Chem. Soc. Japan*, **38**, 1979 (1965). (b) F. Weygand and A. Röpsch, *Chem. Ber.*, **92**, 2095 (1959). (c) H. A. Staab and G. Walther, *Angew. Chem.*, 72, 35 (1960) and *Chem. Ber.*, **95**, 2070 (1962). (d) T. G. Bonner and E. G. Gabb, *J. Chem. Soc.*, 3291 (1963).

[491](a) F. Weygand and R. Geiger, *Chem. Ber.*, **89**, 647 (1956). (b) F. Weygand and H. Rinno, *ibid.*, **92**, 517 (1959). (c) F. Weygand, A. Prox, M. A. Tilak, D. Hoffter, and H. Fritz, *ibid.*, **97**, 1024 (1964). (d) P. A. Cruickshank and J. C. Sheehan, *Anal. Chem.*, **36**, 1191 (1964). (e) A. Darbre and K. Blau, *Biochem. Biophys. Acta.*, **100**, 298 (1965).

[492](a) J. Gardent, *Compt. Rend.*, **257**, 3621 (1963). (b) J. H. Pomeroy and C. A. Craig, *J. Am. Chem. Soc.*, **81**, 6340 (1959). (c) J. P. Critchley, E. J. P. Fear, and J. S. Pippett, *Chem. Ind. (London)*, 806 (1964). (d) H. Maramatsu and K. Inukai, *J. Chem. Soc. Japan, Ind. Chem. Sec.*, **64**, 1512 (1961). (e) W. V. Curran and R. B. Angier, *J. Org. Chem.*, **31**, 3867 (1966).

[493](a) W. A. Sheppard, *J. Am. Chem. Soc.*, **87**, 2410 (1965). (b) E. J. P. Fear, J. Thrower, and J. Veitch, *J. Chem. Soc.*, 3199 (1956). (c) G. Singh and S. Singh, *Tetrahedron Letters*, 3789 (1964).

[494](a) R. H. Groth, *J. Org. Chem.*, **24**, 1709 (1959). (b) S. Portnoy and H. Gisser, *J. Org. Chem.*, **27**, 3331 (1962).

[495](a) J. Burdon and V. C. R. McLoughlin, *Tetrahedron*, **20**, 2163 (1964). (b) V. F. Martynov and M. I. Titov, *J. Gen. Chem. U.S.S.R., (English Transl.)*, **30**, 4072 (1960), *ibid.*, **32**, 716 (1962). (c) E. D. Bergmann and S. Cohen, *J. Chem. Soc.*, 3537 (1961). (d) E. D. Bergmann, S. Cohen and I. Shahak, *J. Chem. Soc.*, 3286 (1959).

[496](a) M. Harnik, E. Hürzeler and E. V. Jensen, *Tetrahedron*, **23**, 335 (1967). (b) W. Steglich and V. Austel, *Angew. Chem. Int. Ed.*, **6**, 184 (1967). (c) A. S. Dey and M. M. Joullié, *J. Org. Chem.*, **30**, 3237 (1965). (d) E. D. Bergmann, I. Shahak, and I. Gruenwald, *J. Chem. Soc.* (C), 2206 (1967). (e) F. B. Wigton and M. M. Joullié, *J. Am. Chem. Soc.*, **81**, 5212 (1959).

[497](a) A. S. Dey and M. M. Joullié, *J. Heterocyclic. Chem.*, **2**, 113 (1965). (b) P. E. Peterson and G. Allen, *J. Org. Chem.*, **27**, 1505 (1961) and **28**, 2290 (1962). (c) H. R. Davis, L. A. Errede, and B. F. Landrum, U. S. Patent 3,055,931 (1962); *Chem. Abstr.*, **58**, 4468e (1963). (d) R. L. Buchanan, F. H. Dean, and F. L. M. Pattison, *Can. J. Chem.*, **40**, 1571 (1962).

[498](a) E. D. Bergmann, S. Cohen, and A. Shani, *Israel J. Chem.*, **1**, 79 (1963). (b) J. D. Park, E. R. Larsen, H. V. Holler, and J. R. Lacher, *J. Org. Chem.*, **23**, 1166 (1958). (c) G. Olah and S. Kuhn, *Chem. Ber.*, **89**, 864 (1956). (d) L. Benoiton, H. N. Rydon, and J. E. Willet, *Chem. Ind. (London)*, 1060 (1960).

[499](a) H. J. Emeleus and J. F. Wood, *J. Chem. Soc.*, 2183 (1948). (b) K. O. Christe and A. E. Pavlath, *J. Org. Chem.*, **30**, 1639 (1965). (c) B. C. Anderson and G. R. Morlock, U. S. Patent 3,226, 418 (1965) and *Chem. Abstr.*, **64**, 9598h (1966). (d) M. E. Redwood and C. J. Willis, *Can. J. Chem.*, **43**, 1893 (1965). (e) C. W. Tullock, U. S. Patent 2,816,131 (1957); *Chem. Abstr.*, **52**, 7346h (1958). (f) E. D. Bergmann and I. Shahak, *Israel J. Chem.*, **3**, 73 (1965). (g) G. Olah and S. Kuhn, *Chem. Ber.*, **89**, 2211 (1956) and *J. Am. Chem. Soc.*, **82**, 2380 (1960).

either lead to introduction of $R_f\overset{O}{\overset{\|}{C}}-$ using a perfluorinated ester or $RO_2C\underset{H}{\overset{R_f}{C}}-$ with an active methylene ester of type $R_fCH_2\overset{O}{\overset{\|}{C}}OR$. This area has been explored

$$R_fCO_2R + CH_3CO_2R \xrightarrow{\text{base}} R_f\overset{O}{\overset{\|}{C}}CH_2CO_2R$$

$$(Et\overset{O}{\overset{\|}{OC}})_2 + CH_2FCO_2Et \xrightarrow{\text{NaOEt}} EtO_2C\overset{O}{\overset{\|}{C}}CHFCO_2Et$$

extensively by Bergmann in Israel for synthesis of mono- and difluorinated compounds of interest in biological studies.[486]

e. *Addition Reactions.*

Perfluorinated acids add readily to olefins and acetylenes

$$R_f\overset{O}{\overset{\|}{C}}OH + RCH{=}CH_2 \longrightarrow R_fCO_2\overset{R}{\overset{|}{C}}HCH_3$$

$$+ HC{\equiv}CH \xrightarrow[\text{P}_2\text{O}_5,\ \text{or anhydride of acid}]{\text{HgO}} R_fCO_2CH{=}CH_2$$

Fluorinated acid fluorides (and carbonyl fluoride, see below) add to a variety of unsaturated fluorinated systems with fluoride ion catalysis, providing a useful route to perfluorinated ketones.[336b] This is an important synthetic advance that involves a fluorocarbanion intermediate as discussed in Section 6-1C.1.

$$R_f\overset{O}{\overset{\|}{C}}F + R_f'CF{=}CF_2 \xrightarrow{F^-} R_f\underset{CF_3}{\underset{|}{C}}F\overset{O}{\overset{\|}{C}}R_f$$

4. Carbonyl Fluoride

Carbonyl fluoride is a special reagent that falls into the category of fluorinated carbonyl compounds. It can be used in direct fluorination[336a] (see Section 5-4B.2), but is frequently used to introduce a COF unit. Some particularly valuable applications in synthesis are

$$ROH + COF_2 \longrightarrow FCO_2R \xrightarrow[\text{HF}]{\text{SF}_4} ROCF_3 \qquad (487)$$

(R is aryl or negatively substituted aliphatic)

[486] See Ref. 496d in Table 6-8.

[487] (a) W. A. Sheppard, *J. Org. Chem.*, 29, 1 (1964). (b) P. E. Aldrich and W. A. Sheppard, *ibid.*, 11.

$$RN{=}CF_2 + COF_2 \xrightarrow{F^-} RN(COF)(CF_3) \xrightarrow[HF]{SF_4} RN(CF_3)_2 \qquad (488)$$

$$COF_2 + R_fCF{=}CF_2 \xrightarrow{F^-} R_fCF(CF_3)C(=O)F \text{ and/or } R_fCF(F_3C)C(=O)CF(CF_3)R_f \qquad (336b)$$

$$ROH \xrightarrow[\text{or } COClF]{COF_2} ROC(=O)F \xrightarrow[\text{pyridine or } BF_3]{\Delta} RF + CO_2 \qquad (487, 489)$$

The preparation of aryl fluorides from phenols and carbonyl chlorofluoride is accomplished using a high-temperature gas phase decomposition[489] of the intermediate aryl fluoroformate (see Section 5-4A.3).

6-2E. THIOCARBONYL AND OTHER FLUORINATED SULFUR REAGENTS

Although thiocarbonyl derivatives are much less stable than carbonyl analogs, a number of fluorinated derivatives have been prepared in recent years and used in synthesis. In many respects they parallel carbonyl reagents. Other fluorinated sulfur units such as $-SF_3$, $-SF_5$, $-SOF$, $-SO_2F$, $-SCF_3$, $-SO_2CF_3$ which can be introduced from available reagents, will also be reviewed briefly.

1. Fluorinated Thiocarbonyl Compounds

The most extensive work on perfluorinated thiocarbonyl compounds was described in a series of papers from du Pont.[500] (See also Table 6-9.)

The three compounds which were studied most extensively and are representative of the series are hexafluorothioacetone, trifluoroacetyl fluoride, and thiocarbonyl fluoride, respectively,

$$CF_3C(=S)CF_3 \qquad CF_3C(=S)F \qquad FC(=S)F$$

[488](a) W. A. Sheppard, *J. Am. Chem. Soc.*, **87**, 4338 (1965). (b) F. S. Fawcett and W. A. Sheppard, *ibid.*, 4341.

[489]K. O. Christe and A. E. Pavlath, *J. Org. Chem.*, **30**, 3170 and 4104 (1965); *J. Org. Chem.*, **31**, 559 (1966).

[500](a) W. J. Middleton, E. G. Howard, and W. H. Sharkey, *J. Am. Chem. Soc.*, **83**, 2589 (1961). (b) W. J. Middleton, E. G. Howard, and W. H. Sharkey, *J. Org. Chem.*, **30**. 1375 (1965). (c) W. J. Middleton and W. H. Sharkey, *ibid.*, 1384. (d) W. J. Middleton, *J. Org. Chem.*, **30**, 1390 (1965). (e) *ibid.*, **30**, 1395 (1965). (f) W. J. Middleton, H. W. Jacobson, R. E. Putnam, H. C. Walter, D. G. Pye, and W. H. Sharkey, *J. Polymer Sci.*, **A3**, 4115 (1965). (g) A. L. Barney, J. M. Bruce, Jr., J. N. Coker, H. W. Jacobson, and W. H. Sharkey, *J. Polymer Sci.* (A-1), **4**, 2617 (1966). (h) *Chem. Eng. News*, **41**, 46 (Sept. 23, 1963). (i) W. H. Sharkey, in *The Polymer Chemistry of Synthetic Elastomers* (J. P. Kennedy and E. Tornquist, eds.), Interscience, New York, 1968. (j) W. H. Sharkey, in *Polyaldehydes* (O. Vogel, ed.), Marcel Dekker, New York, 1967, pp. 91-100.

Several synthetic routes have been reported but the ones of choice are[500 a,b]

$$CF_3CF{=}CF_2 + HgF_2 \longrightarrow (CF_3)_2CF{-}Hg{-}CF(CF_3)_2 \xrightarrow[S]{\text{boiling}} CF_3C(=S)CF_3$$

$$(CF_3CClF)_2Hg \xrightarrow[S]{\text{boiling}} CF_3C(=S)F$$

$$Cl_2C\langle S_2 \rangle CCl_2 \xrightarrow[\text{tetramethylene sulfone}]{SbF_3,} F_2C\langle S_2 \rangle CF_2 \xrightarrow[500^\circ]{475^\circ \text{ to}} FC(=S)F$$

thiophosgene dimer

These reagents are all extremely reactive monomers and a series of homo- and copolymers was prepared.[500f-j] Some of these polymers have unusual elastomeric properties. The polymerization is best initiated by fluoride ion, which is one indication of the extreme reactivity of the thiocarbonyl compounds to nucleophiles. A series of products from reaction of hexafluorothioacetone with nucleophiles has been described by Middleton and Sharkey[500c]; an interesting aspect is that the nucleophiles can attack at sulfur as well as at carbon, suggesting resonance contributions from **1** and **2** are both important. The direction of

$$\underset{\text{HFTA}}{CF_3C(=S)CF_3} \longleftrightarrow \underset{\mathbf{1}}{CF_3\overset{+}{C}(S^-){-}CF_3} \longleftrightarrow \underset{\mathbf{2}}{CF_3\overset{-}{C}(S^+){-}CF_3}$$

$$\text{HFTA} + {}^-SO_3H \longrightarrow HC(CF_3)_2SSO_3^-$$

$$\text{HFTA} + RSH \longrightarrow HSC(CF_3)_2{-}SR + HC(CF_3)_2{-}S{\leftarrow}\text{HFTA}{\rightarrow}_n SR$$

(R is ethyl or methyl) (*n* is **1** or **2**)

addition can be completely reversed (derived from **2**) or normal (derived from **1**) depending on acidity of the thiol or mechanism of reaction.

Phosphites react by attack on sulfur, but an unusual product is formed with trimethyl phosphite,

$$\text{HFTA} + 2\text{P(OR)}_3 \longrightarrow (CF_3)_2C{=}P(OR)_3 + S{=}P(OR)_3$$

$$\xrightarrow[\text{for R = CH}_3]{\Delta} CF_3{-}C(CH_3)(CF_3){-}P(=O)(OCH_3){-}OCH_3$$

The exceptional reactivity of this thioketone is demonstrated by reaction with halogens, halogen halides, water, diazo compounds, hydrogen sulfide, and sulfur dioxide.[500c]

The thiocarbonyl derivatives are extremely active dienophiles in Diels–Alder addition. For example, hexafluorothioacetone reacts instantaneously with butadiene at $-78°C$; the adduct can be converted to a thiopyrane. To illustrate this extensive reactivity, twenty-four Diels–Alder reactions of thiocarbonyl compounds are reported. These Diels–Alder reactions have been reviewed[435c, 471] although one review[471] is of questionable value.

HFTA + butadiene ⟶ 2,2-bis(CF$_3$)-3,6-dihydro-2H-thiopyran (90%) $\xrightarrow{Br_2}$ $\xrightarrow{KOH}$ 2,2-bis(CF$_3$)-2H-thiopyran

Cycloaddition of hexafluorothioacetone to electron-rich olefins is also reported.[435b, 500e]

$$\text{HFTA} + CH_3OCH{=}CH_2 \longrightarrow \begin{array}{c} CH_3OCH{-}CH_2 \\ | \quad\quad | \\ S{-}C(CF_3)_2 \end{array}$$

Allylic addition of a variety of olefins to hexafluorothioacetone to give allyl sulfides is also reported to occur under mild conditions.[500e]

$$(CH_3)_2C{=}C(CH_3){-}CH_3 + S{=}C(CF_3)_2 \longrightarrow H_2C{=}C(CH_3){-}C(CH_3)_2{-}S{-}CH(CF_3)_2$$

2. Bis(trifluoromethyl)thioketene

Bis(trifluoromethyl)thioketene is the first thioketene isolable under normal handling conditions and is an extremely reactive reagent for introducing a hexafluoroisopropylidene grouping.[501] Interestingly, this reagent reacts either on C=C or C=S functionality depending on reagents.

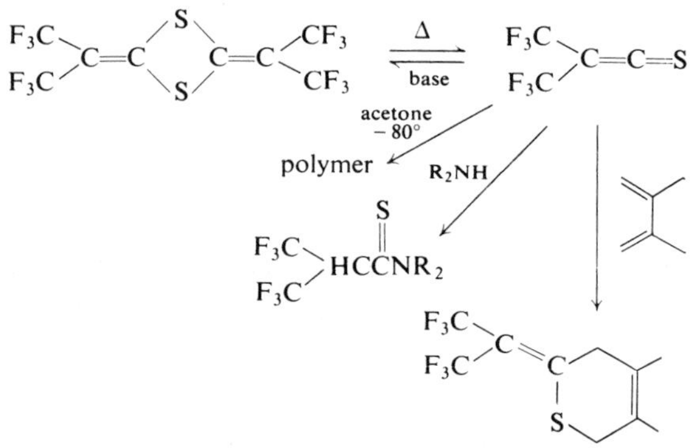

3. **Other Fluorinated Sulfur Units**

Fluorinated sulfur groups that can be added as a unit are listed in Table 6-9. Only selected examples and references are given. Many of these reagents have not been studied in reactions with organic reagents but we feel that they have good potential for synthetic utilization.

Review articles that are useful in this area are available for S–F compounds[119] and SCF_3 derivatives.[502]

6-2F. NITROGEN FLUORIDES, FLUORINATED NITRILES, AND OTHER NITROGEN COMPOUNDS

Nitrogen fluorides, particularly NF_3 and N_2F_4, have been extensively studied because of their potential in high-energy fuel sources and as a source of NF_2 units. The major synthetic chemistry for introducing N–F-containing units is by radical mechanisms, as reviewed in Section 6-1A.2a,[116–118, 120] although anionic N–F species can also be used (see Section 6-1C.2). One of the most important classes of fluoroalkyl nitrogen compounds that can be used to add a fluorinated unit is nitriles, particularly CF_3CN; anionic additions to nitriles are reviewed briefly in Section 6-1C.2. A number of other fluorinated nitrogen species are also reviewed in Sections 6-1A.2a and 6-1C.2 in discussion of radical and carbanion chemistry. Of particular potential for synthesis are imines (or

[501] M. S. Raasch, *Chem. Commun.*, 577 (1966).

[502] (a) A. Senning and S. Kaae, *Quart. Rept. Sulfur Chem.*, **2**, 1 (1967). (b) H. J. Emeleus, *Angew. Chem.*, **74**, 189 (1962). (c) E. Kuhle, E. Klauke, and F. Grewe, *Angew. Chem.*, **76**, 807 (1964). (d) A. Senning, *Chem. Rev.*, **65**, 385 (1965). (e) N. Kharasch, Z. S. Ariyan, and A. J. Havlik, *Quart. Rept. Sulfur Chem.*, **1**, 93-186 (1966). (f) R. E. Banks and R. N. Haszeldine, *Org. Sulfur Compounds*, **2**, 137 (1966).

azomethines), $R_2C{=}NR'$, where one or both R and R′ are perfluoroalkyl or fluorines. The two most important compounds are

1. hexafluoroacetone imine, $(CF_3)_2C{=}NH$; and
2. perfluoroazomethine, $CF_3N{=}CF_2$.

The latter compound is particularly useful for introducing $(CF_3)_2N$ units.

Diazo and aziridine derivatives are useful to introduce a fluorinated unit; however, nitrogen is usually lost to give a reactive carbene, as reviewed in Section 6-1B.1f. They are not discussed further in this section.

Perfluoronitroso compounds are also highly reactive and have potential economic importance in rubbers.[506]

The use of fluorinated nitrogen units in synthesis is summarized in Table 6-10 by using typical examples with leading references.

PROBLEMS

1. When a perfluoroalkyl iodide is added to the C=C double bond of a substituted ethylene, frequently only one of the two possible products of the addition is formed in a significant yield. For example, heptafluoropropyl iodide adds to vinyl acetate to form 1-iodo-3,3,4,4,5,5,5-heptafluoropentyl acetate. Use appropriate mechanisms and explain in detail why only one product is frequently formed. Under what conditions are two products expected by addition in both orientations? In your explanation consider both steric and electronic effects.

2. Apply the principles deduced in Problem 1 and predict the product expected to predominate when perfluoropropyl iodide is added to the alkenes listed below. Where two products might be expected to form in *roughly* comparable amounts, indicate this fact and give structures for both products. If the reaction can take an unusual course because of some special structure of the alkene, indicate this possibility (telomerization? cyclization of radical intermediates?). (Consult Refs. 6-1 and 6-2 and Section 6-2 of Ref. 6-3.)

(a) 1-Propene
(b) 3,3,3-Trifluoro-1-propene
(c) 1-Propyne
(d) Acrylonitrile
(e) Methyl acrylate
(f) Methylenecyclopropane
(g) Vinyl fluoride
(h) 2-Hydroperfluoropropene
(i) Chlorotrifluoroethylene
(j) Norbornadiene
(k) 1,5-Hexadiene

(continued on page 397)

[506] M. C. Henry, C. B. Griffis, and E. C. Stump, *Fluorine Chem. Rev.*, **1**, 1 (1967).

Table 6-9
Synthetic Methods to Introduce Fluorinated Sulfur Units

Fluorinated sulfur reagent	Preparation and physical properties	Synthetic reaction to introduce fluorinated sulfur unit	Comments
A. DIVALENT SULFUR			
FSSF	S + AgF	Rearranges to $S{=}SF_2$ at room temp.	Identified by nmr. Mechanism of rearrangement not discussed; could it be F· + S∸S–F? Ref. 503a
CX_3SF (X = Cl or F)	$CCl_3SCl + KSO_2F$ 150° gas phase	Warm to room temp. → rearrange to sulfenyl chloride	Characterized by nmr at low temperature in solution. Ref. 503b
$(CF_3)_2CFSF$	$[(CF_3)_2C]_2SF_2 \xrightarrow[200°]{\Delta}$	Highly reactive and unstable	Possible mechanism involves $(CF_3)_2C{\div}SF$ F· Ref. 503c
CF_3SCl (Also CF_3SBr and CF_3SI)	Best route is $NaF + CCl_3SCl \xrightarrow{TMS}$ (TMS = tetramethylene sulfone) (bp–1°)	$+ R_2NH \rightarrow R_2NSCF_3$ $+ R'_2C{=}CR_2 \xrightarrow{h\nu} R'_2C(Cl){-}CR_2(SCF_3)$ (R′ = H, alkyl, halogen, etc.) $+ RH \xrightarrow{h\nu} RSCF_3$ $+ ArH \xrightarrow{Lewis\ Acid} ArSCF_3$ $+ ArMgX \rightarrow ArSCF_3$	Extremely useful reagent. A variety of biologically active derivatives formed from CF_3SCl, particularly of the CF_3SNR_2 type (Ref. 502c). Extensive synthetic and mechanism studies on radical addition of $CF_3S\cdot$ generated from CF_3SCl. Ref. 72, 188, 503d, e, f
CF_2ClSCl	$CSF_2 + Cl_2 \rightarrow$ $CCl_3SCl + NaF \xrightarrow{TMS}$ (bp 52°)	$+ R_2NH \rightarrow R_2NSCF_2Cl$ and related derivatives for biological study	Should behave like CF_3SCl in reaction; bromo analogs also known. Ref. 502c, 503g
$CFCl_2SCl$	$CCl_3SCl + HgF_2$ (bp 97-98°)	$+ C_6H_5OC({=}S)NHC_6H_5 \rightarrow C_6H_5NHC({=}O)SSCFCl_2$ + cyclohexene → 1-Cl-2-$(SCFCl_2)$–cyclohexane $+ KI \rightarrow (FCl_2CS)_2$	Structure initially reported incorrectly as CCl_3SF. Should be similar to CF_3SCl in reactions. Bromoanalogs also reported. Ref. 502c, 503h

CF_3SH	$(CF_3S)_2Hg + HCl$ or $CF_2{=}S + HF \rightarrow$ (bp $-37°$)	$+ RCH{=}CH_2 \xrightarrow{h\nu} CF_3SCH_2CH_2R$ hydrolyzed by base	Mechanism study of $CF_3S\cdot$ generated from CF_3SH compared to that from CF_3SCl. Ref. 73, 408c, 503e
CF_3SSCF_3	$NaF + CCl_3SCl \xrightarrow{TMS}$ or $CSCl_2 + NaF \xrightarrow{TMS}$ or $CS_2 + HgF_2$ or IF_5 (bp $34°$)	$+ Cl_2 \rightarrow CF_3SCl$ $+ CF_3CF{=}CF_2 \xrightarrow[\text{or heat}]{h\nu} CF_3S[CF(CF_3)CF_2]_nSCF_3$	Ref. 189, 190, 503e, 503d
CF_3SAg	$CS_2 + 3AgF \xrightarrow{140°}$ Stable in air; decomposes over 80° *in vacuo*	Reactions similar to those for $(CF_3S)_2Hg$ given below	Advantage over mercury compound in giving insoluble silver halide when reacted with active halogen compound. Ref. 408b
$(CF_3S)_2Hg$	$HgF_2 + CS_2 \xrightarrow{250°}$ (mp 37-38°)	$+ RX \rightarrow RSCF_3$ $+ RC(=O)Cl \rightarrow RC(=O)SSF_3$ $+ HCl \rightarrow CF_3SH$ $+ Cl_2 \rightarrow CF_3SCl$	Useful reagent for introducing CF_3S groups. Ref. 408
$\overset{S}{CF_2{-}CF_2}$ (cyclic)	(a) $CF_2{=}S$ + perfluoropropylene oxide $\xrightarrow{175°}$ (b) $CF_2{=}CF_2$ + refluxing S (bp $-10.5°$)	$\xrightarrow[CF_3SSCF_3]{h\nu} CF_3S(SCF_2CF_2)_nSCF_3$ (other initiators also used) $+ CF_2{=}CF_2 \xrightarrow{R\cdot}$ copolymer + (perfluorotetrahydrothiophene ring: F_2, F_2, F_2, F_2, S) $+ R_2NH \rightarrow R_2NCF_2C(=S)NR_2$ $+ AlCl_3 \rightarrow CF_3C(=S)F \rightarrow [CF_3C(=S)SCF_2CF_3]_2$	A number of copolymers reported. Ref. 190
$S{-}S$ / $CF_2{-}CF_2$ (cyclic)	$\text{-}(SCF_2CF_2{-}S)_n\text{-} \overset{\Delta}{\rightleftharpoons}$ (Also $CF_2{-}CF_2$ bridged by $(S)_{3,4}$)	Heat $\rightarrow$ $F_2C{-}S{-}CF_2$ / $F_2C{-}S{-}CF_2$ (ring) $+ S_x$ Base $\rightarrow$ polymer	Interconversions in this series (depending on conditions and amount of sulfur as well as ring size of cyclic sulfides) are discussed. Copolymers prepared. Ref. 503i

Table 6-9 (Cont.)

Fluorinated sulfur reagent	Preparation and physical properties	Synthetic reaction to introduce fluorinated sulfur unit	Comments
$CF_3C(S{-}S)=CCF_3$ (S–S ring)	$CF_3C{\equiv}CCF_3 + S \xrightarrow[(I_2)]{445^\circ}$ (bp 93-94°)	$\rightleftarrows$ dimer (1,2,5,6-tetrathiocin: CF_3, CF_3 / S–S, S–S / CF_3, CF_3) → tricyclic dimer with six CF_3 groups and four S	Other perfluorinated dithietenes, $R_fC(S{-}S)=CR_f$ also prepared. Ref. 442
		+ $>C{=}C<$ ⟶ 2,3-bis(CF_3)-1,4-dithiene ring (S, S)	
		+ $HC{\equiv}CH \xrightarrow{70^\circ}$ 2,3-bis(CF_3)-1,4-dithiin (H, H) $\xrightarrow{125^\circ}$ 3,4-bis(CF_3)thiophene (H, H)	
		+ Hg ⟶ $(-HgS-C(CF_3){=}C(CF_3)-S-)_x$ $\xrightarrow{CH_3I}$ $CF_3(CH_3S)C{=}C(CF_3)SCH_3$	
$FC(=S)Cl$	$CFCl_2SCl \xrightarrow{Sn}$ (bp 9°)	+ Br_2 → CFClBrSBr + ROH → $ROC(=S)F$ Similar to reactions of CF_2S	Ref. 503j

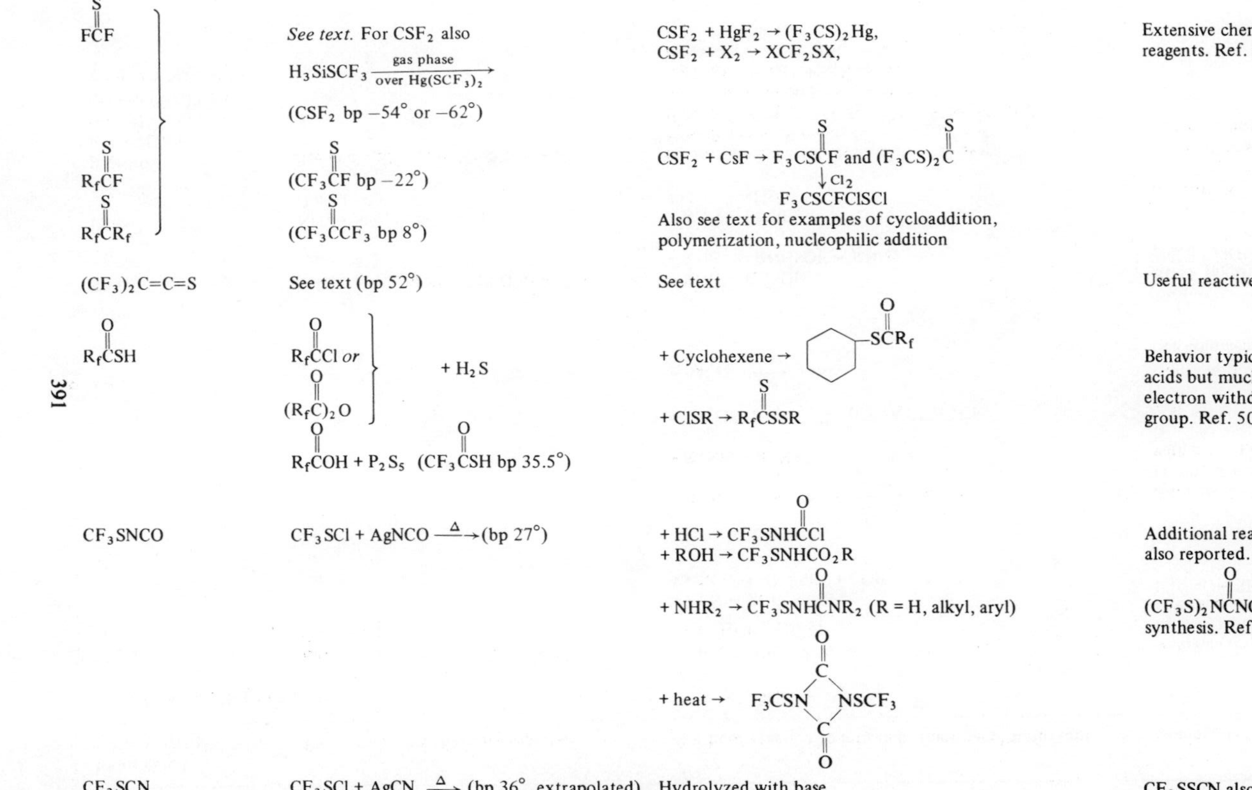

$\mathrm{FC(=S)F}$, $\mathrm{R_fC(=S)F}$, $\mathrm{R_fC(=S)R_f}$	*See text.* For $\mathrm{CSF_2}$ also $\mathrm{H_3SiSCF_3 \xrightarrow[\text{over } Hg(SCF_3)_2]{\text{gas phase}}}$ ($\mathrm{CSF_2}$ bp −54° or −62°) ($\mathrm{CF_3C(=S)F}$ bp −22°) ($\mathrm{CF_3C(=S)CF_3}$ bp 8°)	$\mathrm{CSF_2 + HgF_2 \rightarrow (F_3CS)_2Hg}$, $\mathrm{CSF_2 + X_2 \rightarrow XCF_2SX}$, $\mathrm{CSF_2 + CsF \rightarrow F_3CSC(=S)F}$ and $\mathrm{(F_3CS)_2C{=}S}$; $\mathrm{F_3CSC(=S)F \xrightarrow{Cl_2} F_3CSCFClSCl}$ Also see text for examples of cycloaddition, polymerization, nucleophilic addition	Extensive chemistry reported for these reagents. Ref. 500, 503j, k, l
$\mathrm{(CF_3)_2C{=}C{=}S}$	See text (bp 52°)	See text	Useful reactive intermediate. Ref. 501
$\mathrm{R_fC(=O)SH}$	$\mathrm{R_fC(=O)Cl}$ *or* $\mathrm{(R_fC(=O))_2O}$ $\mathrm{+ H_2S}$ $\mathrm{R_fC(=O)OH + P_2S_5}$ ($\mathrm{CF_3C(=O)SH}$ bp 35.5°)	+ Cyclohexene → $\mathrm{C_6H_{11}SC(=O)R_f}$ $\mathrm{+ ClSR \rightarrow R_fC(=S)SSR}$	Behavior typical of aliphatic thiolcarboxylic acids but much stronger acid because of electron withdrawing effect of fluoroalkyl group. Ref. 503m
$\mathrm{CF_3SNCO}$	$\mathrm{CF_3SCl + AgNCO \xrightarrow{\Delta}}$ (bp 27°)	$\mathrm{+ HCl \rightarrow CF_3SNHC(=O)Cl}$ $\mathrm{+ ROH \rightarrow CF_3SNHCO_2R}$ $\mathrm{+ NHR_2 \rightarrow CF_3SNHC(=O)NR_2}$ (R = H, alkyl, aryl) + heat → $\mathrm{F_3CSN(C{=}O)_2NSCF_3}$ (cyclic dimer)	Additional reactions typical of isocyanates also reported. Also isolated linear dimer, $\mathrm{(CF_3S)_2NC(=O)NCO}$, as by-product in synthesis. Ref. 503n, o
$\mathrm{CF_3SCN}$	$\mathrm{CF_3SCl + AgCN \xrightarrow{\Delta}}$ (bp 36°, extrapolated)	Hydrolyzed with base	$\mathrm{CF_3SSCN}$ also prepared. Ref. 503n

Table 6-9 (Cont.)

Fluorinated sulfur reagent	Preparation and physical properties	Synthetic reaction to introduce fluorinated sulfur unit	Comments
B. TETRAVALENT SULFUR			
SF_4	Best synthesis: $SCl_2 + NaF \xrightarrow{CH_3CN}$ (bp $-40°$)	$+ CF_3CF{=}CF_2 \xrightarrow{CsF} (CF_3)_2CFSF_3$ and $[(CF_3)_2CF]_2SF_2$, $+ RCN \rightarrow RCF_2N{=}SF_2$, $+ RNCO \rightarrow RN{=}SF_2 + COF_2$,	Valuable reagent for fluorination of carbonyl groups which probably involves intermediate adduct $R_2C{=}O + SF_4 \rightarrow [R_2C(F)OSF_3] \rightarrow R_2CF_2 + SOF_2$. Ref. 503c, 504a, b, c
		$+ RNH_2 \rightarrow RN{=}SF_2$ ($\xrightarrow[\text{amine}]{\text{excess}}$ $RN{=}S{=}NR$), $+ R_2NH \rightarrow R_2NSF_3$	Adducts $RN(SF_3)(COF)$ or $RCF{=}NSF_3$ proposed as intermediates in reaction with nitriles or isocyanates.
CF_3SF_3	$CS_2 + CoF_3$ or AgF_2 (bp $-20°$)	$+ CF_3CF{=}CF_2 \xrightarrow{CsF} (CF_3)_2CFSF_2CF_3$,	Ref. 503c
$ArSF_3$	$(ArS)_2 + AgF_2$ ($C_6H_5SF_3$ bp 47° at 2.6 mm)	$+ AgF_2 \xrightarrow{120°} ArSF_5$,	Useful in fluorination of carbonyl groups (see comments on SF_4 above). Ref. 504d
SOF_2	$SOCl_2 + NaF$ or $SF_4 + H_2O$ or R_2CO (bp $-44°$)	$+ ROH \rightarrow ROS(=O)OR$ $+ RNH_2 \rightarrow RNHSOF \rightarrow RNSO$	$ROS(=O)F$ prepared by $ROS(=O)Cl + MF$, decomposed to $RF + SO_2$. Ref. 504e, f, g
$FS{\equiv}N$	$NH_3 + SF_4$, fluorination of S_4N_4, etc. (bp + 0.4°)	$+ AgF_2 \rightarrow N{\equiv}SF_3$ $+ H_2O \rightarrow NHSO$	No report on reaction with organic reagents. Ref. 504h
$RN{=}SF_2$ R = aryl or R_f (For SF_5 containing iminosulfur difluorides see under Section C)	See under SF_4. ($CF_3N{=}SF_2$ bp $-6°$)	Hydrolysis and alcoholysis leads to cleavage of S–F and S–N bonds. $R_fN{=}SF_2 + CF_3CF{=}CF_2 \xrightarrow{CsF} R_fN{=}SF(CF(CF_3)_2)$ plus some unusual by-products. $R_fN{=}SF_2 + AgO \rightarrow R_fN{=}S{=}O$ $R_fN{=}SF_2 + AlCl_3 \rightarrow R_fN{=}SCl_2$	These iminosulfur difluorides are very stable thermally. Ref. 504a, b, i, j
$FC(=O)N{=}SF_2$	$SF_4 + M(NCO)_n$ (bp 49°, extrapolated) M = P, Si, or RSO_2	Hydrolyzed readily in aqueous base. $+ PCl_5 \rightarrow FOCN{=}SCl_2$	No report on reaction with organic reagents. Ref. 504k

$SF_2{=}N{-}CN$	$H_2NCN + SF_4 \xrightarrow{NaF}$ (mp −33°, estimated bp 104°)	$\xrightarrow[170^\circ]{heat}$ $NSF + (FCN)_3 + CF_3N{=}SF_2 + N_2$	By-product in synthesis is $SF_2{=}NCF_2N{=}CF_2$ (also $CF_3N{=}SF_2$ and $N{\equiv}SF$ under certain conditions). Ref. 5041
$SF_2{=}NSO_2F$	$FSO_2NH_2 + SF_4 \xrightarrow{NaF}$ (mp 77-78°, bp 105.5°)	Stated useful intermediate to other N,S compounds but no details reported	Ref. 504m
$S{=}SF_2$	$AgF + S$; $S_2Cl_2 + KSO_2F$ (bp −10°)	Isomerized by BF_3 to $S + SF_4$; $+ HX \rightarrow$ other disulfur halides	No report on reaction with organic reagents. Structure proven by microwave. FSSF also prepared but thermodynamically unstable relative to $S{=}SF_2$ (see Part A). Ref. 504n, o, 503a
C. HEXAVALENT SULFUR			
SF_5Cl	$SF_4 + Cl_2 + CsF$ or $S + ClF_3$ (bp −21°)	$+ RCH{=}CH_2 \xrightarrow[peroxide]{heat,\ light} RCHClCH_2SF_5$ $+ HC{\equiv}CH \rightarrow ClHC{=}CHSF_5$ + diene $+ C_2F_4 \rightarrow Cl(C_2F_4)_nSF_5$ $+ CF_3CN \rightarrow SF_5N{=}CClCF_3$ + benzene $\rightarrow C_6H_5Cl$ only, no $C_6H_5SF_5$ $+ RNH_2 \rightarrow SF_4$ + solid residue	Initial adducts converted to a variety of other products. SF_5 group is stable under many of the reaction conditions but is often hydrolyzed with basic reagents, e.g., $CH_3OCH{=}CHSF_5 \xrightarrow{H_3O^+} HC(=O)CH_2SF_5 \xrightarrow[H_3O^+]{NaMnO_4} HOC(=O)CH_2SF_5$ (dissociation constant 3.9×10^{-3} intermediate between FCH_2CO_2H and F_2CHCO_2H). Ref. 119, 182, 183, 184, 186, 505a
F_5SSF_5	$ClSF_5 + H_2 \xrightarrow{h\nu}$ (bp 29°)	$+ CF_3CF{=}CF_2 \rightarrow CF_3CF(SF_5)CF_2SF_5$ $+ C_6H_6 \xrightarrow{heat} C_6H_5SF_5$ (low yield) $+ RCH{=}CH_2 \xrightarrow{125\text{-}140^\circ} RCH(SF_5){-}CH_2F$ (R = Cl, F, Br) $+ C_2H_5C{\equiv}CH \rightarrow C_2H_5CF_2CHFSF_5$ (structure proof based on mass spec)	Noted reagent is a powerful oxidizing agent and degradatively oxidizes many organic reagents. Ref. 184-186.
SF_5OOSF_5	$SF_5Cl + O_2 \xrightarrow{h\nu}$ (bp 49°)	$+ CF_3CF{=}CF_2 \xrightarrow{150^\circ} F_5SO[C_3F_6]_nOSF_5$ (n = 2 to 4) $+ ArH \xrightarrow{150^\circ} ArOSF_5 + SOF_4 + HF$ (where Ar = C_6H_5, $C_6H_4CH_3$ and C_6H_4Cl)	Too strong an oxidizing agent to give clean reactions with most organic systems, even $CF_2{=}CF_2$. $ArOSF_5$ + Na/EtOH $\rightarrow$ ArOH, $C_6H_5OSF_5 + HNO_3 \rightarrow p\text{-}O_2N{-}C_6H_4{-}OSF_5$, reduced to aniline. Ref. 175, 176

Table 6-9 (Cont.)

Fluorinated sulfur reagent	Preparation and physical properties	Synthetic reaction to introduce fluorinated sulfur unit	Comments
SF_5OF	$SOF_2 + F_2 \xrightarrow{AgF_2}$ (bp −35°)	$+ C_2H_4 \rightarrow FCH_2CH_2OSF_5$, $+ C_2F_4 \rightarrow F_3CCF_2OSF_5$, $+ ClCH{=}CH_2 \rightarrow ClFCHCH_2OSF_5$	Reaction run in gas phase at room temperature; not practical synthetic method because of powerful oxidizing character of SF_5OF. Ref. 165a, b, 505b
SF_5NH_2 (Note SF_5NHCF_3 listed below under reactions of $SF_5N{=}CF_2$)	$NSF_3 + 2HF$ (mp ca. 43°, sublimation temperature by extrapolation 61°)	Dissociates to NSF_3 + HF slowly at room temperature; easily hydrolyzed, soluble in dry ether	Should be studied with selected organic reagents. Ref. 505c
$SF_5N{=}CX_2$ (X = Cl or F or R_f)	$SF_5Cl + XCN \xrightarrow{h\nu}$ $SF_5N{=}CCl_2 + NaF \rightarrow$ ($SF_5N{=}CCl_2$, bp 86-88°) ($SF_5N{=}CF_2$, bp 11-13°)	Degraded by aqueous alkali: $SF_5N{=}CF_2 \xrightarrow[KF]{\Delta} SF_4{=}NCF_3$ $+ HgF_2 \longrightarrow \left(F_5SN(CF_3)\right)_2Hg \xrightarrow{RC(=O)Cl} RC(=O)N(CF_3)SF_5$ $+ HF \longrightarrow SF_5{-}NHCF_3 \xrightarrow[ArCOSH]{NaF} SF_5N{=}C{=}S$ $SF_5{-}NHCF_3 \xrightarrow[ArCO_2H]{NaF} SF_5NCO \xrightarrow{ROH} SF_5NHC(=O)OR$ $+ AgF_2 \longrightarrow \left((SF_5)(CF_3)N{-}\right)_2 \xrightarrow{Cl_2} (SF_5)(CF_3)NCl$	Similar conversions for fluoroalkyl azomethines also reported. Ref. 184a
$SF_5N{=}SF_2$	$SF_4 + NSF_3 \xrightarrow{BF_3}$ (bp 38°, extrapolated)	Hydrolyzed by aqueous base $+ H_2O \rightarrow [SF_5NSO]$?	No reactions of this material with organic reagents are reported. Expected to be similar to $R_fN{=}SF_2$. Ref. 505d

SOF_4	$SF_4 + O_2 \xrightarrow{cat.}$ (bp −49°)	$+ NH_2R \rightarrow RN{=}SOF_2 \xrightarrow{R'_2NH} RN{=}SOF(NR'_2)$, $+ ArOH \rightarrow ArF + SO_2F_2 + HF$, $+ NH_3 \rightarrow HN{=}SO_2F \xrightarrow{-HF}$ rubbery polymer	Ref. 505e, f, g, h
$F_3S{\equiv}N$	$NH_3 + S + AgF_2 \xrightarrow{CCl_4}$ $NH_3 + S_2F_{10} \rightarrow$ (bp 27°) For other methods, see review (Ref. 504h)	$+ NaOH \rightarrow$ hydrolyzed $+ 2HF \rightarrow SF_5NH_2$ $+ (C_2H_5)_2NH \rightarrow N{\equiv}SF_2N(C_2H_5)_2$	Reactions with organic reagents not reported. Ref. 504h
SO_2F_2	$SO_2Cl_2 + 2NaF \rightarrow$ (bp −55°)	$+ ArOH \rightarrow ArOSO_2F$ (or $(ArO)_2SO_2$) $+ NH_3 \rightarrow NH_2SO_2NH_2$	Ref. 341b, 503d, 505g
		$+ R_fOCF{=}CF_2 \xrightarrow[glyme]{CsF} \left(R_fOCF(CF_3)\right)_2SO_2$	Added SO_2F_2 (and RSO_2F) to several fluoroolefins to give sulfones or sulfonyl fluorides
SO_2ClF	$SO_2Cl_2 + NaF \rightarrow$ (bp 7°)	$+ ArOH \rightarrow ArOSO_2F$	Ref. 503d, 505g
FSO_2NH_2	$NH_2SO_2Cl + KF \xrightarrow{CH_3CN}$ $S_2O_5F_2 + 2NH_3 \rightarrow$ (mp 8°, bp 88° at 12 mm)	$+ H_2O \rightarrow H_2NSO_3H$	R_2NSO_2F also prepared. Ref. 505i, j
FSO_2NCO	$OCNSO_2Cl + Na(H)F \rightarrow$	+ $CH_2{=}CHR$ $\longrightarrow$ 4-R-1-(SO_2F)-azetidin-2-one $+ RCH{=}CHC(=O)NHSO_2F$	Paper deals chiefly with reactions of $ClSO_2NCO$. Reactions of FSO_2NCO stated similar to those of $ClSO_2NCO$ but fluoride noted more stable to heat and hydrolysis. Ref. 505k, l
CF_3SO_2X (X = Cl or F)	$CF_3SCl + 2Cl_2 + 2H_2O$ (CF_3SO_2Cl bp 32°)	$CF_3SO_2Cl \xrightarrow{RH} RCF_3 + SO_2 + HCl$ $ROH + CF_3SO_2F \rightarrow ROSO_2CF_3$	$ROSO_2CF_3$ is a powerful alkylating agent. [$C_6F_5SO_3R$ also noted useful for alkylations (Ref. 505m).] Ref. 410, 411b, 505m
CF_3SO_2M	$CF_3SO_2Cl + K_2SO_3$ or $Zn(H_2O)$, $CF_3SO_2K \xrightarrow{H_3O^+} (CF_3SO_2H)$	$(CF_3SO_2)_2Zn + CF_3SCl \rightarrow CF_3SO_2SCF_3$ CF_3SO_2H + *p*-benzoquinone $\xrightarrow{Ag_2O}$ 2-(trifluoromethylsulfonyl)-*p*-benzoquinone	No extensive studies with organic reagents; however $CF_3SO_2^-$ is a good leaving group in nucleophilic reactions. Trifluoromethyl sulfonylquinones better prepared by oxidation of trifluoromethylsulfides. Ref. 411

Table 6-9 (Cont.)

[503](a) F. Seel, R. Budenz, and D. Werner, *Chem. Ber.*, **97**, 1369 (1964). (b) F. Seel, W. Gombler, and R. Budenz, *Angew. Chem. Intern. Ed. Engl.*, **6**, 707 (1967). (c) R. M. Rosenberg and E. L. Muetterties, *Inorg. Chem.*, **1**, 756 (1962). (d) C. W. Tullock and D. D. Coffman, *J. Org. Chem.*, **25**, 2016 (1960). (e) R. N. Haszeldine and J. M. Kidd, *J. Chem. Soc.*, 3219 (1953). (f) W. A. Sheppard, *J. Org. Chem.*, **29**, 895 (1964); S. Andreades, J. F. Harris, Jr., and W. A. Sheppard, *ibid.*, 898. (g) C. W. Tullock, U.S. Patent 2,884,453 (1959); *Chem. Abstr.*, **53**, 16963b (1959). (h) W. A. Sheppard and J. F. Harris, Jr., *J. Am. Chem. Soc.*, **82**, 5106 (1960). (i) C. G. Krespan and W. R. Brasen, *J. Org. Chem.*, **27**, 3995 (1962). (j) N. N. Yarovenko and A. S. Vasileva, *J. Gen. Chem. U.S.S.R. (English Transl.)*, **29**, 3754 (1959). (k) A. Haas and W. Klug, *Angew. Chem. Intern. Ed. Engl.*, **5**, 845 (1966). (l) A. J. Downs, *J. Chem. Soc.*, 4361 (1962). (m) W. A. Sheppard and E. L. Muetterties, *J. Org. Chem.*, **25**, 180 (1960). (n) H. J. Emeleus and A. Haas, *J. Chem. Soc.*, 1272 (1963). (o) A. Haas, *Chem. Ber.*, **98**, 111 (1965) and **97**, 2189 (1964).

[504](a) W. C. Smith, C. W. Tullock, R. D. Smith, and V. A. Engelhardt, *J. Am. Chem. Soc.*, **82**, 551 (1960); F. S. Fawcett and C. W. Tullock, *Inorg. Syn.*, **7**, 119 (1963). (b) R. D. Cramer, *J. Org. Chem.*, **26**, 3476 (1961); see also R. D. Peacock and I. N. Rozhkov, *J. Chem. Soc.* (A), 107 (1968). (c) B. Cohen and A. G. MacDiarmid, *Angew. Chem. Intern. Ed. Engl.*, **2**, 151 (1963). (d) W. A. Sheppard, *J. Am. Chem. Soc.*, **84**, 3058 (1962); *Org. Syn.*, **44**, 39, 82 (1964). (e) K. Wiechert and R. Hoffmeister, *J. Prakt. Chem.*, **410**, 290 (1960). (f) A. Zappel, *Chem. Ber.*, **94**, 873 (1961). (g) M. Goehring and G. Voigt, *Chem. Ber.*, **89**, 1050 (1956). (h) O. Glemser and M. Field, *Halogen Chem.*, **2**, 1 (1967). (i) R. D. Dresdner, J. S. Johar, J. Merritt, and C. S. Patterson, *Inorg. Chem.*, **4**, 678 (1965). (j) M. Lustig, *Inorg. Chem.*, **5**, 1317 (1966). (k) A. F. Clifford and C. S. Kobayashi, *ibid.*, **4**, 571 (1965); see also H. W. Roesky and R. Mews, *Angew Chem. Intern. Ed. Engl.*, 7, 217 (1968). (l) O. Glemser and U. Biermann, *Inorg. Nucl. Chem. Letters*, **3**, 223 (1967). (m) O. Glemser, H. W. Roesky, and P. R. Heinze, *Angew. Chem. Intern. Ed. Engl.*, **6**, 179 (1967). (n) V. F. Seel and H. D. Gölitz, *Z. Anorg. Allg. Chem.*, **327**, 32 (1964). (o) E. B. Wilson, Jr., and R. L. Kuczkowski, *J. Am. Chem. Soc.*, **85**, 2028 (1963).

[505](a) N. H. Ray, *J. Chem. Soc.*, 1440 (1963). (b) F. B. Dudley, G. H. Cady, and D. F. Eggers, Jr., *J. Am. Chem. Soc.*, **78**, 1553 (1956). (c) A. F. Clifford and L. C. Duncan, *Inorg, Chem.*, **5**, 692 (1966). (d) A. F. Clifford and J. W. Thompson, *ibid.*, 1424. (e) R. Cramer and D. D. Coffman, *J. Org. Chem.*, **26**, 4010 (1961). (f) G. W. Parshall, R. Cramer, and R. E. Foster, *Inorg. Chem.*, **1**, 677 (1962). (g) R. Cramer and D. D. Coffman, *J. Org. Chem.*, **26**, 4164 (1961). (h) H. E. Holmquist and R. E. Benson, *J. Am. Chem. Soc.*, **84**, 4720 (1962). (i) R. Appel and G. Eisenhauer, *Z. Anorg. Allg. Chem.*, **310**, 90 (1961). (j) H. Jonas and D. Voigt, *Angew Chem.*, **70**, 572 (1958). (k) R. Graf, *Ann.*, **661**, 111 (1963). (l) R. Graf, *Angew Chem. Intern. Ed. Engl.*, **7**, 172 (1968); see also p. 181. (m) J. E. Connett, *Chem. Ind.* (*London*), 1695 (1965).

3. In Section 6-1A several methods are given for generating perfluoroalkyl radicals for use in synthesis of organic fluorine compounds. List these methods in order of decreasing utility for practical synthesis. State succinctly the reasons for your ordering and briefly state scope and limitations for each method.

4. Thoroughly compare physical properties and chemical reactions of molecular bromine (Br—Br) with tetrafluorohydrazine ($F_2N{-}NF_2$). What are the similarities and what are the differences? Consult appropriate references in Section 6-1A.

5. (a) Consult review articles and appropriate leading references in Section 6-1A.2 in order to compile a summary of the properties and chemical reactions of difluoroamine (HNF_2). Compare HNF_2 with known weak acids that react with organic compounds and speculate on the reactions that HNF_2 might have with organic compounds.

(b) Consult *Chemical Abstracts* Subject Index from 1968 on and look up the reactions of HNF_2 with organic compounds. Compare your speculations in Problem 5(a) with data reported subsequent to publication of this book.

6. (a) Geminally substituted dichlorocyclopropanes are most easily formed by reacting an alkene with chloroform and a metallic alkoxide. Dichlorocarbene ($:CCl_2$) is the intermediate. Why isn't this method useful in geminal difluorocyclopropanes?

(b) Evaluate the methods for synthesizing geminal difluorocyclopropanes. List them in order of generally decreasing synthetic utility. Justify your ordering.

7. After consulting the appropriate references explain in detail the results given in Table 6-3. Account for the difference in reactions of bis(trifluoromethyl)-carbene formed from bis(trifluoromethyl)diazomethane and bis(trifluoromethyl)-diazirine.

8. Select appropriate commercially available starting materials and present in detail syntheses of the following compounds using methods presented in Sections 6-1A and 6-1B:

(a) 7,7-Difluoronorcarane
(b) 7-Fluoro-7-chloronorcarane
(c) 1-Trifluoromethylcyclohexene
(d) Bis(perfluoropropyl)disulfide
(e) Phenyl difluoromethyl ether
(f) Perfluoromethylamine.
(g) 1-Chloro-1-fluoro-2,2,3,3-tetramethylcyclopropane
(h) 1,1,2,2-Tetrafluoroethanethiol
(i) 2,2′-Difluoro-2,2′-dichlorodicyclopropyl
(j) 1,2-Bis(difluoroamino)propane.
(k) 6α-Trifluoromethyl-17α-acetoxyprogesterone
(l) 1,3-Dibromo-1,1,3,3-tetrafluoropropane
(m) Perfluoro-2-heptyl-3-methyloxetane

(*continued on page 406*)

Table 6-10

Introduction of Fluorinated Nitrogen Units by Synthesis

(Other than NOF, no NF compounds are included in this table; for use of these compounds in synthesis, see pages 207–227, 313 or Reviews 116–120)

Fluorinated nitrogen reagent	Synthesis and physical properties	Synthetic reaction	Comments
A. AMINES			
$R_fCH_2NH_2$ $(R_fCH_2)_2NH$	Reduction of $R_fC(=O)NH_2$, R_fCN and other nitrogen species such as oximes or nitro compounds; addition of ammonia to fluoro-olefins; substitutions of ammonia on R_fCH_2X. ($CF_3CH_2NH_2$ bp 37°)	Typical amine reactions, e.g., Maleic anhydride + $CF_3CH_2NH_2 \rightarrow$ *N*-(2,2,2-trifluoroethyl)monohydrogenmaleamide	Amine group less basic than normal because of electronegativity of fluoroalkyl groups. For general review, see Ref. 482, Chapters X and XI. Ref. 507
$(CF_3)_2NH$	$ClCN + HF \xrightarrow{250°}$ $CF_3N{=}CF_2 + HF \longrightarrow$ (bp –8 to –6°)	Unstable to loss of HF, easily hydrolyzed. $+ Cl_2 \rightarrow (CF_3)_2NCl$	Not a useful reagent because of instability. Ref. 508, 388b
$(CF_3)_2CFNH_2$	$(CF_3)_2C{=}NH + HF \longrightarrow$	Reacted rapidly with glass	Too unstable to study. Ref. 509
$RC(R_f)_2NH_2$ (R = aryl or alkyl)	$ArH + (R_f)_2C{=}NH \xrightarrow{catalyst}$ $R{-}H + (R_f)_2C{=}NH \longrightarrow$ (activated H, such as allylic) (bp 90°)	$C_6H_5C(CF_3)_2NH_2$ Amino group very stable, resists chromic acid oxidation, acetylated only by way of Na salt (from NaH reaction); aromatic ring nitrates in meta position	A large number of analogs and derivatives prepared. Ref. 510

B. IMINES

$CF_3N{=}CF_2$ and $R_fN{=}CF_2$	$ClCN + HF \xrightarrow{250°} (CF_3)_2NH \xrightarrow[200°]{KF}$ $(CF_3)_2NCF{=}NCF_3 \xrightarrow[NaF]{525°}$ ($CF_3N{=}CF_2$, bp −28 to −25°)	Easily hydrolyzed to R_fCO_2H or R_fCN, etc., depending on conditions: + $F^- \rightarrow (CF_3)_2NCF{=}NCF_3$ (dimer), + $HgF_2 \rightarrow [(CF_3)_2N]_2Hg$, + $COF_2 \xrightarrow{CsF} (CF_3)_2NC({=}O)F$, + $AgF_2 \rightarrow (CF_3)_2NN(CF_3)_2$ + dimer. Other metal fluorides give hydrazine, dimer, or NF compounds, often mixtures. + $H_2O \rightarrow CF_3NCO$ (or SiO_2), +ROH $\rightarrow CF_3NHCO_2R \rightarrow RNCO$, + Cyanuric fluoride $\xrightarrow{CsF} F_{3-n}C_3N_3[N(CF_3)_2]_n$ (where n is 1,2, or 3)	Very useful intermediate for introducing a variety of $R_{f2}N$- and $-R_fN(R)-$ groups. For example, mercury adduct can be acylated to prepare $RC({=}O)N(CF_3)_2$ (Ref. 388) Ref. 336a, 387, 508a, 511
$(R_f)_2C{=}NR$ $(CF_3)_2C{=}NH$, $ArN{=}C(CF_3)_2$ $RN{=}C(CF_3)_2$	Best route: $(CF_3)_2C{=}O + NH_2R \rightarrow (CF_3)_2C(OH)NHR \xrightarrow{Base} (CF_3)_2C(O^-)NHR \xrightarrow{POCl_3} (CF_3)_2C{=}NR$ Also: $ArNCO + (R_f)_2C{=}O \xrightarrow{\phi_3P}$ $(CF_3)_2C(NHR)(NHAr) \xrightarrow{HCl} (CF_3)_2C{=}NR$ Ar is aryl; R is H or alkyl [$(CF_3)_2C{=}NH$, bp 16-16.5°]	+ $NH_3 \rightarrow (CF_3)_2C(NH_2)_2$ stable geminal diamine + HX $\rightarrow (CF_3)_2C(X)\text{-}NH_2$ (X = $NHNH_2$, NCO, N_3, OCH_3, F, CN, etc.) + butadiene $\longrightarrow$ 2,2-bis(trifluoromethyl)-1,2,3,6-tetrahydropyridine [ring NH, $(CF_3)_2$] + isobutylene $\longrightarrow (CF_3)_2C(NH_2){-}CH_2{-}C(CH_3){=}CH_2$ + $LiAlH_4 \rightarrow RNH{-}CH(R_f)_2$	$(CF_3)_2C{=}NH$ is an extremely valuable reagent. Products have been used to prepare a variety of other nitrogen reagents, e.g., $(CF_3)_2C(N_3){-}NH_2 \xrightarrow{h\nu} (CF_3)_2C\langle NH{-}NH \rangle \xrightarrow{[O]} (CF_3)_2C\langle N{=}N \rangle$ Electrophilic reagents like ketene do not react or react poorly with imine. NH of imine is acidic and forms stable salts $(CF_3)_2C{=}NH \xrightarrow{CH_3Li} (CF_3)_2C{=}NLi \xrightarrow{Br_2} (CF_3)_2C{=}NBr \xrightarrow{h\nu} (CF_3)_2C{=}N{-}N{=}C(CF_3)_2$ Ref. 509, 512

Table 6-10 (Cont.)

Fluorinated nitrogen reagent	Synthesis and physical properties	Synthetic reaction	Comments
C. AMINE DERIVATIVES (N-chloro, hydrazines, isocyanates)			
CF_3NCO R_fNCO	$CF_3N{=}CF_2 \xrightarrow[\text{hydrolysis}]{\text{controlled}}$ (SiO_2 works well) $R_fCN + COF_2 \xrightarrow[300°]{CsF}$ $R_fC(=O)Cl + NaN_3$ (activated) $\longrightarrow$ $R_fN{=}NR_f + CO \longrightarrow$ (CF_3NCO, bp–36°)	Normal reactions of isocyanates: $R_fNCO \xrightarrow{ROH} R_fNHCO_2R$ $\xrightarrow{RSH} R_fNHC(=O)SR$ $\xrightarrow{HX} R_fNHC(=O)X$	Noted that $-CF_2-NH-$ system is easily hydrolyzed and could lead to decomposition of product under certain conditions. Large number of examples of reactions reported. Ref. 387, 511, 513, 514
$(R_f)_2NCl$ $(R_f)_2NBr$	$(R_f)_2NH + Cl_2 \longrightarrow$ $[(R_f)_2N]_2Hg + X_2 \longrightarrow$ (X is Cl, Br)	$+ RCH{=}CH_2 \xrightarrow[\text{or light}]{\text{heat}} (R_f)_2NCH_2-CHR$, (with X on CHR) $+ HC{\equiv}CH \xrightarrow[\text{or light}]{\text{heat}} (R_f)_2NCH{=}CHX$, $+$ heat or light $\rightarrow (R_f)_2NN(R_f)_2$	Most of work reported with $(CF_3)_2NCl$; series of adducts with olefins dehydrohalogenated to vinyl derivatives. Reactions shown to involve $(CF_3)_2N\cdot$ Ref. 143, 144, 388b
$CF_3N{=}NCF_3$ $R_fN{=}NR_f$, $R_fN{=}NR$	Variety of methods; best appears to be: $2ClCN + 2Cl_2 + 6NaF \xrightarrow{225°} CF_3N{=}NCF_3$ (bp –32°) $R_fCN + AgF + Cl_2 \rightarrow R_fCF_2N{=}NCF_2R_f$ $R_fNO + RNH_2 \rightarrow R_fN{=}NR$	$R_fN{=}NR_f$ + reducing agent such as HI, H_2S $\xrightarrow[CH_3OH]{\text{cold}} R_fNHNHR_f$ $CF_3N{=}NCF_3 + HX \rightarrow X_2C{=}N-N(CF_3)-X$ or $X_2C{=}N-N{=}CX_2$ where $X = NR_2$, OR, SR $+ CH_2N_2 \rightarrow$ 1,2-bis(trifluoromethyl)diazirdine $+ RMgI \rightarrow \xrightarrow{H_2O} CF_3N(R)NHC(=O)F$	$R_fN{=}NR_f$ is a useful source of R_f radicals for synthesis, but particularly mechanism studies (see Section 6-1A and Table 6-1B) Ref. 515, 516, 56

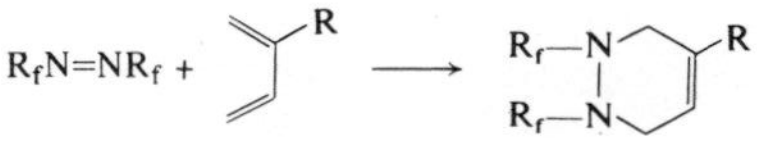

		$R_fN{=}NR_f$ + (2-R-butadiene) $\longrightarrow$ (1,2-di-R_f-4-R-tetrahydropyridazine) $+ CO \rightarrow R_fNCO$ $(R_fCF_2N\!\!\rightarrow\!\!)_2 + M(CO)_x \rightarrow R_fCF{=}N{-}N{=}CFR_f + MF_x + CO$	
D. NITRILES			
FCN	Cyanuric chloride $\xrightarrow{NaF}$ cyanuric fluoride $\xrightarrow{1300^\circ}$ (bp –46°)	Polymerized at room temp. Forms explosive mixture with air	Does not appear to be a useful reagent for organic reactions. Ref. 517
CF_3CN R_fCN	Best method: $R_fC(=O)NH_2 + P_2O_5 \rightarrow$ Other useful methods are: $R_fC(=S)F + NaN_3 \rightarrow$ fluoroolefins + NH_3 $\rightarrow$ allylic halide + NaCN $\rightarrow$ (CF_3CN, bp –64°)	$R_fCN + HX \rightarrow R_fC(=NH)X$ (X = OR, SR, NH_2, NH_2NH_2, etc.) Usually stable adducts which are useful intermediates to variety of R_f substituted products. For example, 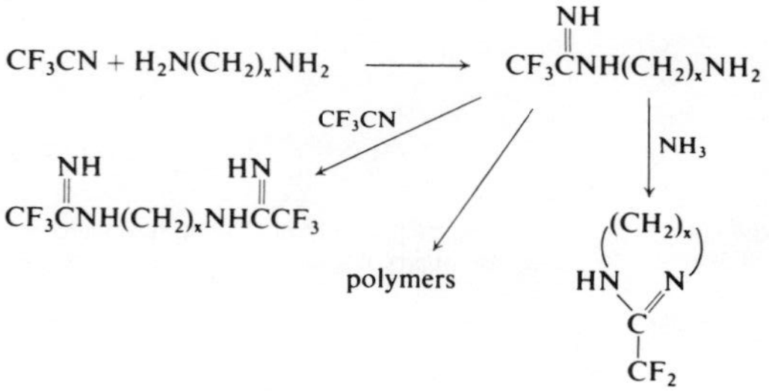 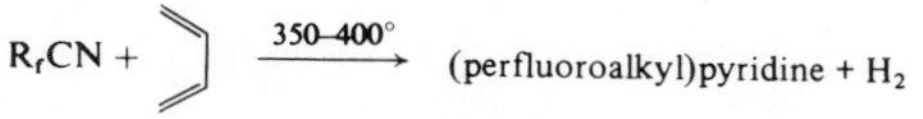 $CF_3CN + H_2N(CH_2)_xNH_2 \longrightarrow CF_3C(=NH)NH(CH_2)_xNH_2$ $\xrightarrow{CF_3CN}$ $CF_3C(=NH)NH(CH_2)_xNHC(=NH)CF_3$ $\rightarrow$ polymers $\xrightarrow{NH_3}$ (cyclic amidine: HN–C(CF₂)=N bridged by $(CH_2)_x$) R_fCN + (butadiene) $\xrightarrow{350\text{–}400^\circ}$ (perfluoroalkyl)pyridine + H_2 $\xrightarrow[\text{pressure}]{\text{heat}}$ 2,4,6-tris(perfluoroalkyl)-1,3,5-triazine	Addition of bases and nucleophilic reagents to perfluorinated nitriles has been studied extensively (see Ref. 390, 391). Nitrile highly susceptible to nucleophilic reagents because of strong electron withdrawing effect of the perfluoralkyl group. The R_fCN are considered similar to NCCN in chemical behavior. For general review see Ref. 482, Chapters X and XI. Ref. 390, 391, 514a, 518, 519

Table 6-10 (Cont.)

Fluorinated nitrogen reagent	Synthesis and physical properties	Synthetic reaction	Comments
E. AMIDES AND AMIDINES			
$R_fC(=O)NH_2$	Most general are $R_fC(=O)X + NH_3 \rightarrow$ (X = OR, Cl, $OC(=O)R_f$) Addition of ammonia to fluoroolefins Hydrolysis of nitriles $R_fC(=O)R_f + NH_3 \rightarrow$	Typical amide reactions. Dehydrated to nitrile. Forms imides, stable *N*-bromoamides $CF_3C(=O)NHCH_3 \xrightarrow{PCl_5} CF_3C(Cl)=NCH_3$	For general review, see Ref. 482 Chapters X and XI. Ref. 520
$R_fC(=O)NHNHC(=O)R_f$	$R_fC(=O)X + NH_2NH_2 \rightarrow$ (X = Cl or $OC(=O)R_f$)	$+ P_2O_5 \xrightarrow[300°]{200-}$ 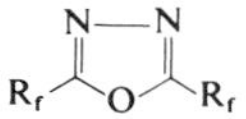 2,5-di-R_f-1,3,4-oxadiazole $+ P_2S_5 \xrightarrow[300°]{250-}$ 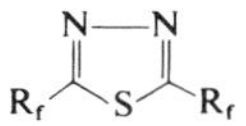 2,5-di-R_f-1,3,4-thiadiazole	Oxadiazoles used as intermediates to fluorinated tetrazines. Ref. 521, 522, 523
$R_fC(=NH)-NH_2$	$R_fCN + NH_3 \rightarrow$	$\xrightarrow[NH_3]{heat}$ 2,4,6-tris(perfluoroalkyl)-1,3,5-triazine $\xrightarrow{heat} R_fC(=NH)N=C(NH_2)R_f \xrightarrow{M^{2+}} [R_fC(=NH)-N=C(R_f)-NH]_2M$ (chelate) 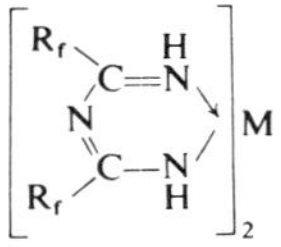	Mechanism of condensation studied. Ref. 524

F. NITROSO AND NITROXY COMPOUNDS

FNO	Variety of methods (see review*) but preferred procedure is $HF + BF_3 + NO_2 \xrightarrow{\leq 0^\circ} NOBF_4 \xrightarrow[\Delta]{NaF} NOF$ (bp–60°)	Adds to halogenated olefins by either electrophilic or nucleophilic mechanism. For highly fluorinated olefins (except $CF_2{=}CF_2$) fluoride ion catalyzed. $(CF_3)_2C{=}CF_2 \xrightarrow{F^-} [(CF_3)_3C^-] \xrightarrow{FNO} (CF_3)_3CNO$, $CF_2{=}CFOR \xrightarrow{+NO} [ONCF_2\overset{+}{C}FOR] \xrightarrow{NOF}$ $ONCF_2CF_2OR$. Variety of other products produced under radical conditions: $R_fC(=O)R_f + NOF \rightleftarrows R_fC(ONO)(F)R_f$ rearrange under certain conditions. Cholesteryl acetate $\xrightarrow[CCl_4,\ 0^\circ]{NOF}$ X, F, N–NO_2 $\xrightarrow{H_2O}$ X, F, O	Reactions often give complex mixtures but useful reagent for synthesis under controlled conditions with certain reagents. Ref. 415, 525, 526
R_fNO R_f = fluoroalkyl	Review reports on 15 methods (see Ref. 506); most useful or practical are $R_fI + NO$	Polymerizes and copolymerizes with fluoroolefins and other monomers, e.g., $CF_3NO + CF_2{=}CF_2 \xrightarrow{\leq 0^\circ} \text{-(}N(CF_3)\text{-}OCF_2CF_2\text{)}_n$	Copolymers have attractive properties as rubbers. Ref. 506, 527, 528, 529

* C. Woolf, *Advan. Fluorine Chem.*, **5**, 1 (1965).

Table 6-10 (Cont.)

Fluorinated nitrogen reagent	Synthesis and physical properties	Synthetic reaction	Comments
	R_fCO_2Ag or $(R_fCO)_2O$ + ClNO $\rightarrow R_fC(=O)ONO \xrightarrow[\Delta]{h\nu \text{ or}}$ addition of nitrogen oxides or FNO to fluoroolefins, Tabulation of synthesis and properties of known R_fNO given in review. (CF_3NO, bp $-84°$)	but above 0° $CF_3NO + CF_2{=}CF_2 \longrightarrow$ 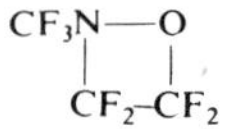 CF_3N—O / CF_2—CF_2 (four-membered ring) Cycloadds to a variety of unsaturated reagents. Highly reactive in the Diels-Alder addition are   CF_3NO + butadiene $\xrightarrow{-78°}$ CF_3—N (3,6-dihydro-1,2-oxazine ring, N—O) isobutylene $\longrightarrow$ $CH_2{=}C(CH_3)$—CH_2—$N(OH)CF_3$ $CF_3NO + F_2 \xrightarrow{AgF_2} (CF_3)_2NOCF_3$ $\xrightarrow{h\nu} (CF_3)_2NONO$	
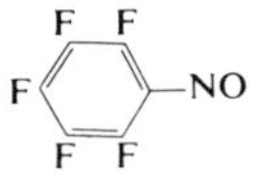 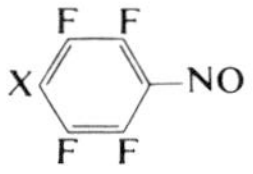 C_6F_5NO (pentafluoronitrosobenzene) X-C_6F_4-NO (X para to NO) X = Br, CO_2H	$C_6F_5NH_2 \xrightarrow[CH_2Cl_2]{\text{performic acid}}$ (C_6F_5NO mp 42-45°, blue crystals)	Monomeric. Reacts rapidly with nucleophiles; 1,3 - cyclohexadiene + $XC_6F_4NO \longrightarrow$ 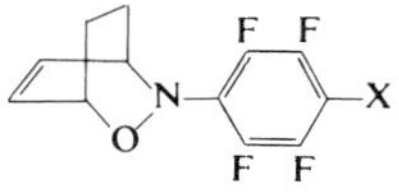 bicyclic N—O adduct bearing N-(C_6F_4-X)	Noted C_6F_5NO was monomeric in crystalline state or in solution but XC_6F_4NO were dimeric in crystalline state but formed monomer on heating or solution in polar solvents. Ref. 530, 531
$(CF_3)_2NO\cdot$	$(CF_3)_2NOH \xrightarrow[KMnO_4]{Ag_2O \text{ or}} (CF_3)_2NO\cdot$ (from $CF_3NO\cdot$ see below), [$(CF_3)_2NO\cdot$ purple gas, bp $-20°$]	See Chart 6-4 for reactions (page 233).	Useful reagent for introducing $(CF_3)_2NO$ group. Ref. 173

$(CF_3)_2NO^-Na^+$	$CF_3NO \xrightarrow{h\nu} (CF_3)_2NONO$ (bp 10°) $\xrightarrow{\text{aq. HCl}}$ $(CF_3)_2NOH$ (bp 35°) $\xrightarrow[\text{THF}]{\text{NaOH}}$	See Chart 6-15 for reactions (page 320)	Useful method to introduce $(CF_3)_2NO$ groups. Ref. 405

[507] T. V. Sheremeteva, Z. V. Borisova, and V. V. Kudryavtsev, *Bull. Acad. Sci. U.S.S.R., Div. Chem. Sci. (English Transl.)*, 2093 (1961).

[508] (a) F. S. Fawcett, C. W. Tullock, and D. D. Coffman, *J. Chem. Eng. Data*, **10**, 398 (1965). (b) D. A. Barr and R. N. Haszeldine, *J. Chem. Soc.*, 2532 (1955).

[509] W. J. Middleton and C. G. Krespan, *J. Org. Chem.*, **30**, 1398 (1965).

[510] D. M. Gale and C. G. Krespan, *J. Org. Chem.*, **33**, 1002 (1968).

[511] D. A. Barr and R. N. Haszeldine, *J. Chem. Soc.*, 3428 (1956).

[512] (a) Yu. V. Zeifman, N. P. Gambaryan, and I. L. Knunyants, *Proc. Acad. Sci. U.S.S.R., Chem. Sect. (English Transl.)*, **153**, 1032 (1963). (b) Yu. V. Zeifman, N. P. Gambaryan, and I. L. Knunyants, *Bull. Acad. Sci. U.S.S.R., Div. Chem. Sci. (English Transl.)*, 1431 (1965).

[513] W. J. Chambers, C. W. Tullock, and D. D. Coffman, *J. Am. Chem. Soc.*, **84**, 2337 (1962).

[514] (a) N. N. Yarovenko, S. P. Motornyi, L. I. Kirenskaya, and A. S. Vasilyeva, *J. Gen. Chem. U.S.S.R. (English Transl.)*, **27**, 2301 (1957). (b) S. P. Motornyi, L. I. Kirenskaya and N. N. Yarovenko, *J. Gen. Chem. U.S.S.R. (English Transl.)*, **29**, 2122 (1959).

[515] G. O. Pritchard, H. O. Pritchard, H. I. Scheff, and A. F. Trotman-Dickenson, *Trans. Faraday Soc.*, **52**, 849 (1956).

[516] (a) S. P. Makarov, A. Ya. Yakubovich, V. A. Ginsburg, A. S. Filatov, M. A. Englin, N. F. Privezentseva, and T. Ya. Nikoforova, *Proc. Acad. Sci. U.S.S.R., Chem. Sect. (English Transl.)*, **141**, 1130 (1961). (b) V. A. Ginsburg, A. Ya. Yakubovich, A. S. Filatov, V. A. Shpanskii, E. S. Vlasova, G. E. Zelenin, L. F. Sergienko, L. L. Martynova, and S. P. Makarov, *Proc. Acad. Sci. U.S.S.R., Chem. Sect. (English Transl.)*, **142**, 4 (1962). (c) V. A. Ginsburg, A. Ya. Yakubovich, A. S. Filatov, G. E. Zelenin, S. P. Makarov, V. A. Shpanskii, G. P. Kotel'nikova, L. F. Sergienko, and L. L. Martynova, *ibid.*, 34.

[517] F. S. Fawcett and R. D. Lipscomb, *J. Am. Chem. Soc.*, **86**, 2576 (1964).

[518] R. N. Johnson and H. M. Woodburn, *J. Org. Chem.*, **27**, 3958 (1962).

[519] (a) G. J. Janz and A. R. Monahan, *J. Org. Chem.*, **29**, 569 (1964). (b) G. J. Janz and A. R. Monahan, *J. Org. Chem.*, **30**, 1249 (1965). (c) A. R. Monahan and G. J. Janz, *J. Phys. Chem.*, **69**, 1070 (1965).

[520] W. P. Norris and H. B. Jonassen, *J. Org. Chem.*, **27**, 1449 (1962).

[521] H. C. Brown, M. T. Cheng, L. J. Parcell, and D. Pilipovich, *J. Org. Chem.*, **26**, 4407 (1961).

[522] W. J. Chambers, and D. D. Coffman, *ibid.*, 4410.

[523] H. C. Brown, H. J. Gisler, and M. T. Cheng, *J. Org. Chem.*, **31**, 781 (1966).

[524] (a) W. L. Reilley, and H. C. Brown, *J. Org. Chem.*, **22**, 698 (1957). (b) H. C. Brown, and P. D. Schuman, *J. Org. Chem.*, **28**, 1122 (1963).

[525] S. Andreades, *J. Org. Chem.*, **27**, 4157, 4163 (1962).

[526] G. A. Boswell, Jr., *Chem. Ind. (London)*, 1929 (1965).

[527] (a) D. A. Barr, and R. N. Haszeldine, *J. Chem. Soc.*, 1881 (1955). (b) R. E. Banks, M. G. Barlow, and R. N. Haszeldine, *J. Chem. Soc.*, 4714, 6149 (1965).

[528] (a) S. P. Makarov, V. A. Shpanskii, V. A. Ginsburg, A. I. Shchekotikhin, A. S. Filatov, L. L. Martynova, I. V. Pavlovskaya, A. F. Golovaneva, and A. Ya. Yakubovich, *Proc. Acad. Sci. U.S.S.R., Chem. Sect. (English Transl.)*, **142**, 62 (1962). (b) V. A. Ginsburg, S. S. Dubov, A. N. Medvedev, L. L. Martynova, B. I. Tetel'baum, M. N. Vasil'eva, and A. Ya. Yakubovich, *Proc. Acad. Sci. U.S.S.R., Chem. Sect. (English Transl.)*, **152**, 796 (1963).

[529] J. M. Shreeve, and D. P. Babb, *J. Inorg. Nucl. Chem.*, **29**, 1815 (1967).

[530] G. M. Brooke, J. Burdon, and J. C. Tatlow, *Chem. Ind. (London)*, 832 (1961).

[531] J. A. Castellano, J. Green, and J. M. Kauffman, *J. Org. Chem.*, **31**, 821 (1966).

9. Select appropriate commercially available starting materials and present in detail syntheses of the following compounds *using methods presented in Section 6-1C:*

(a) $C_4H_9OCF_2CHFCl$

(b) $C_4H_9SCF_2CHCl_2$

(c) $(C_2H_5)_2NCF_2CF_2H$

(d) $CF_3CF_2C(=O){-}F$

(e) $CF_3CF_2CF_2C(=O)CF_2CF_2CF_3$

(f) 2,4-Difluorostyrene

(g) *p*-$F_5SC_6H_4COOH$

(h) $CF_3CF_2CF_2CH_2CH_2OH$

(i) $CF_3CF_2CF_2CH(OH)CH_2CH_3$

(j) $CF_2{=}CFCOOH$

(k) OH, CHCH$_3$, F$_2$, F$_2$, F

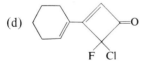

(l) $(CH_3)_2C(OH)C{\equiv}CCF_3$

(m) C_6F_5COOH

(n) C_6F_5Cu

10. Synthesize the compounds listed below from fluoroethylenes and other unfluorinated commercially available starting materials using the methods in Sections 6-2A and 6-2B:

(a) $C_6H_5CH{-}CF_2$, $CH_2{-}CF_2$

(b) $CF_2{-}C{=}O$, $CF_2{-}CF_2$

(c) $CF_2{-}CF$, $CF_2{-}CF$

(d) F Cl, O

(e) CF_2, CF_2

(f) F_2, F_2

(g) CF_2, CF_2

(h) CF_3

11. (a) Compare methods for preparing the following pairs of acetylenes. What are the similarities and the differences?

$HC{\equiv}CH$ and $HC{\equiv}CF$

$CH_3C{\equiv}CH$ and $CF_3C{\equiv}CH$

$CH_3C{\equiv}CCH_3$ and $CF_3C{\equiv}CCF_3$

(b) Compare the chemical reactions of the three pairs of acetylenes given in Problem 11(a). What are the similarities and differences for each pair?

(c) Propose a synthesis of difluoroacetylene (FC≡CF). Speculate on its chemical reactions and speculatively compare it with the acetylenes in Problem 11(a).

(d) Compare the chemical reactions and the synthetic utility of tetrafluorobenzyne to benzyne. Explain the apparent differences in chemical reactivity. (For a review of benzyne chemistry, see R. W. Hoffman, *Dehydrobenzene and Cycloalkynes,* Academic, New York, 1967.)

12. Starting with commercially available hexafluorobenzene (C_6F_6) or chloropentafluorobenzene (C_6ClF_5) give details for synthesis of the structures given below. Use techniques from any part of the book as required by the problem.

(a) $C_6F_5NH_2$
(b) C_6F_5OH
(c) $C_6F_5CH_3$
(d) C_6F_5D
(e) C_6F_5COOH
(f) $C_6F_5CF_2CF_3$
(g) 1-Bromoperfluorobiphenyl
(h) 1-Bromo-2,3,4-trifluoronaphthalene
(i) 5,6,7,8-Tetrafluoro-1-naphthol
(j) Pentafluorophenylacetylene

13. Synthesize the compounds listed below from commercially available starting materials. Base your answers on information in Tables 6-6 to 6-10.

(a) $C_6H_5CF_2CF_2CF_3$

(b) $C_6H_5C(OH)(CF_3)_2$

(c) $C_3F_7C(=O)C_3F_7$

(d) $CF_3C(=O)CF_2H$

(e) $C_6H_5OC(=O)CF_3$

(f) $C_6H_5C(=O)CF_2C(=O)C_6H_5$

(g) Trifluoroacetoxycyclopentane

(h) α-Fluoroglutamic acid

(i) 1,3-Bis(trifluoromethoxy)benzene

(j) 2,3,4,5-Tetrakis(trifluoromethyl) cyclopentadienone

(k) Perfluorobiacetyl

(l) Tetrafluorocyclobutane-1,2-dione

(m) $CF_3CF_2\overset{\overset{\displaystyle O}{\|}}{C}CH_2\overset{\overset{\displaystyle O}{\|}}{C}CH_3$

(n) $CF_3SCH_2CH_2CH_3$

(o) $CF_3S(SCF_2CF_2)_nSCF_3$ polymer

(p) $C_6H_{11}\text{—}\overset{\overset{\displaystyle O}{\|}}{S}CF_3$ (cyclohexyl)

(q) $C_6H_5SF_3$

(r) $C_6H_5SF_5$

(s) $CH_3CHClCH_2SF_5$

(t) $C_6H_5(CF_3)_2CNH_2$

(u) $(CF_3)_2CFNH_2$

(v) $(CF_3)_3N$ (by 2 methods)

(w) $C_3F_7\overset{\overset{\displaystyle O}{\|}}{C}N(CF_3)_2$

(x) $C_2F_5OCF{=}CF_2$ (by 2 methods)

(y) $CF_3CF_2CH_2SCH{=}CH_2$

Chapter Seven

Synthesis and Chemistry of Fluorinated Functional Groups

In preceding chapters we have reviewed synthesis of organic fluorine compounds with the emphasis and organization based on methods and reagents. However, the practicing organic chemist usually wants information on the synthesis and chemistry of specific types of fluorinated compounds. This chapter is designed to summarize routes to the major classes of fluorinated functional groups (particularly referenced to discussions in the preceding three chapters) and to outline the chemical reactivity of these groups so that handling and subsequent synthetic reactions of the fluorinated organic compound can be easily accomplished.

The organization by functional groups is

1. trifluoromethyl, $-CF_3$, bonded to aliphatic, aromatic, and heteroatoms (groups such as OCF_3 and SCF_3);

2. difluoromethylene, $>CF_2$, as in $CX_2CF_2CX_2$, where X is H or halogen;

3. monofluorides in aliphatic compounds, $\geq CF$, as in R_3CF, RCH_2F, or $(RCX_2)_2CFR$, where X and R can be hydrogen or halogen;

4. perfluoroalkyl, R_f, in various bonding situations (similar to trifluoromethyl);

5. arylfluorides, ArF, ranging from monofluoro to perfluoro aromatic derivatives;

6. vinylic fluorides, $=C{<}^{F}_{X}$, where X is flourine or some other group;

7. heteroatom fluorides, X–F, where fluorine is bonded to O, N, S, P, and other heteroatoms in an organic compound.

The emphasis for discussion of chemistry is on stability to oxidation, reduction, hydrolysis, substitution, and elimination. Synthetic methods are summarized in tables emphasizing preferred methods with critical comments on limitations. Methods of limited utility can be found in the text of Chapters 4, 5, and 6.

7-1 THE TRIFLUOROMETHYL GROUP (CF_3)

A trifluoromethyl group is inert to most chemical reagents under typical reaction conditions. It is not attacked by oxidizing or reducing agents and does not undergo substitution, elimination, or hydrolysis unless activated by some functional group. For example, β,β,β-trifluoroisobutyric acid is hydrolyzed by

$$CH_3CH(CF_3)COOH \xrightarrow[(2)\ H_3O^+]{(1)\ 2N\ NaOH,\ 100^\circ} CH_3{-}CH(COOH)COOH \quad (\text{via } CH_3C(CO_2H){=}CF_2) \tag{1}$$

hot caustic solution; no doubt the highly acidic α-hydrogen is easily attacked by base so that the logical sequence is dehydrofluorination, hydration of olefin, and subsequent hydrolysis.

Trifluoromethyl groups attached to aromatic rings are stable to base but are activated to hydrolysis by strong acids (via electrophilic attack on fluorine).

$$C_6H_5{-}CF_3 \xrightarrow[100\%\ H_2SO_4,\ \text{heat}]{HF,\ H_2O,\ 100^\circ} C_6H_5{-}CO_2H\ (94\text{–}100\%) \tag{2}$$

Both Hudlicky[3] and Houben-Weyl[4] cite numerous examples that show the excellent stability of a trifluoromethyl group; the following sequence of reactions carried out by Swarts[5] is a good illustration.

[1] M. W. Buxton, M. Stacey, and J. C. Tatlow, *J. Chem. Soc.*, 366 (1954).

[2] (a) J. H. Simons and R. E. McArthur, *Ind. Eng. Chem.*, **39**, 364 (1947). (b) G. M. Le Fave, *J. Am. Chem. Soc.*, **71**, 4148 (1949). (c) E. T. BcBee and M. R. Frederick, *ibid.*, 1490.

[3] M. Hudlicky, *Chemistry of Organic Fluorine Compounds,* Macmillan, New York, 1962, pp. 205-207.

[4] *Methoden der organishen Chemie (Houben-Weyl),* Vol. 5, Part 3, "Halogen Verbindungen–Fluor und Chlor" (E. Müller, ed.), Thieme, Stuttgart, Germany, 1962, pp. 439-440, 442-450, 473-482, 493-501.

[5] (a) F. Swarts, *Bull. Acad. Roy. Belg.*, 389 (1920); *Chem. Abstr.*, **16**, 2316 (1922). (b) F. Swarts, *Bull. Sci. Acad. Roy. Belg.*, **8**, 343 (1922); *Chem. Abstr.*, **17**, 769 (1923). (c) F. Swarts, *Bull. Sci. Acad. Roy. Belg.* (5) **12**, 679 (1926); *Chem. Abstr.*, **21**, 2120 (1927). (d) F. Swarts, *Bull. Sci. Acad. Roy. Belg.*, **13**, 175 (1927); *Chem. Abstr.*, **22**, 58 (1928).

$$C_6H_5CF_3 \xrightarrow[0^\circ]{HNO_3} m\text{-}O_2NC_6H_4CF_3 \xrightarrow{H_2/Pt} m\text{-}H_2NC_6H_4CF_3 \xrightarrow[Na_2Cr_2O_7]{H_2SO_4} CF_3COOH$$

$$CF_3COOH \xrightarrow[H_2SO_4]{EtOH} CF_3COOC_2H_5$$

$$CF_3COOC_2H_5 \xrightarrow[EtO_2CCH_3,\ 16\ hr]{NaOC_2H_5} CF_3C(=O)CH_2C(=O)OC_2H_5$$

$$CF_3C(=O)CH_2C(=O)OC_2H_5 \xrightarrow[70^\circ]{10\%\ H_2SO_4} CF_3C(=O)CH_3 \xrightarrow{NaOH} CF_3H$$

Trifluoromethyl groups are usually introduced into organic compounds by one of four general methods:

1. fluoride exchange on a trihalomethyl group;
2. reaction of sulfur tetrafluoride with a carboxylic acid group or derivative;
3. addition of CF_3X to unsaturation;
4. condensation reactions with CF_3– containing compounds.

A detailed summary of synthetic methods is given in Table 7-1.

7-2 THE DIFLUOROMETHYLENE GROUP ($>CF_2$)

The difluoromethylene group is comparable in stability to a trifluoromethyl group. As discussed in Section 3-1, the C–F bond in $-CF_2-$ is expected to be marginally weaker than in $-CF_3$, but differences in chemical stability are not apparent in many fluorinated compounds. With perfluoroalkyl substituents, $R_f-CF_2-R_f$, the difluoromethylene group is essentially inert, for example, Teflon, polytetrafluoroethylene, is only attacked by highly reactive metal systems such as sodium in liquid ammonia.

Even when substituted by methylene groups, $-CH_2-CF_2-CH_2-$, the difluoromethylene group is not attacked readily by oxidizing or reducing agents or by nucleophiles or electrophiles unless the α-hydrogen is activated by substituents. Substituents on the CH_2 group that enhance the acidity of the hydrogen facilitate elimination of hydrogen fluoride; the initial reaction must be base-catalyzed elimination of a proton so that subsequent loss of fluoride ion and hydrolysis to a carbonyl group should be rapid. As discussed in Section 2-2, the heat of formation of hydrogen fluoride is so large that the elimination is expected whenever a favorable mechanism with low activation energy is available.

Table 7-1
Summary of Synthetic Routes for Trifluoromethyl ($-CF_3$) Groups

Type	Method	Scope and limitations	Comments	Book section	Leading references or reviews[a]
Alkyl–CF_3 Aryl–CF_3 Acyl–CF_3 O–CF_3 S–CF_3	Fluoride replacement of halogen; common reagents are HF and/or SbF_3 or SbF_5	Unsaturated groups often add HF; reactive halogens at other sites in molecule also replaced; CO_2H, SO_3H, ether and nitro groups usually not attacked	Usually CCl_3 used rather than CBr_3 or CI_3 because of availability. $RCCl_3$ easily prepared by chlorination of RCH_3, or $\cdot CCl_3$ addition to olefin. Classical method of major industrial value	4-5 (Tables 4-7 and 4-8)	4-95 (R), 4-96 (R), 4-97 (R)
Alkyl–CF_3 Aryl–CF_3 Vinyl–CF_3 Acetylenic–CF_3 O–CF_3 N–CF_3	SF_4 or RSF_3 on RCO_2H; HF good catalyst; RCO_2Et, $(RC(=O)F)_2O$, $RC(=O)F$ can also be used	More reactive carbonyl functional groups also fluorinate but selectivity can be attained. Major limitation is SF_4 reactivity with functional groups such as amide, quinone, imino, nitrile, halide, amine and some alcohol groups although side reaction with these other functional groups can often be controlled; C=C and C≡C functionality are not attacked	Best general laboratory method	5-4B.1 (Table 5-6)	5-205 (R), 5-209a
Alkyl–CF_3 Alkenyl–CF_3	CF_3X addition to olefin (and other unsaturated materials); CF_3I most common	Addition initiated by heat, light or peroxide, involves $CF_3\cdot$ intermediate; radical type by-products. Reaction broadly applicable; may get mixture of products with unsymmetrical olefins	HX easily removed to give alkenyl derivative, or used as reactive function for further synthesis	6-1A (Table 6-1A)	6-1 to 6-6 (R)
Ar–CF_3	CF_3I high-temperature substitution	Radical reaction; substituted aromatic compounds give mixture of isomers	Method has unexplored potential	6-1A	6-34 to 6-37
Alkyl–CF_3 $RC(=O)-CF_3$ $R_2C(X)-CF_3$	Condensation and addition reactions of active methylene compounds and various CF_3 substituted carbonyl and thiocarbonyl compounds and CF_3CN. Addition of organometallic reagents and other nucleophiles to fluorinated aldehydes, ketones, acids, esters, nitriles	Much of chemistry parallels classical condensation reaction in organic chemistry (typical Claisen-ester, Knoevenagel, aldol condensation)	Broad area involving many available reagents, such as $CF_3C(=O)H$, $CF_3C(=O)CF_3$, $CF_3C(=O)X$, CF_3CN, that can be exploited to much greater extent	6-1C.2 6-2C, D, E, F. (Tables 6-7, 6-8, 6-10)	6-481 (R), 6-482 (R), 6-483 (R), 6-484 (R)
Alicyclic (4 or 6-ring)–CF_3 Aryl–CF_3	Cycloaddition and Diels-Alder reactions of olefins and acetylenes of type C=C–CF_3 or C≡CCF_3	2 + 2 cycloaddition reactions often favored over 2 + 4 Diels-Alder. Reactions best with electron-rich dienes or dienophiles	Products useful for further conversion. Also heteroatom cyclic compounds from fluorinated carbonyl and thiocarbonyl derivatives	6-2A, B	6-435 (R)

[a] Chapter-reference number; R indicates a review.

Many examples of reactions of difluoromethylene groups and conditions required to achieve these reactions are given in Houben-Weyl.[6] A few representative examples that illustrate not only typical reactions but the high stability of the CF_2 group (and some other fluorinated groups) are given in Chart 7-1.

$$CH_3CHF_2 \xrightarrow[500^\circ]{CaF_2} CH_2{=}CHF \quad (98\%) \tag{7}$$

$$HO_2CCH_2{-}CF_2CO_2H \xrightarrow[100^\circ,\ 16\ hr]{NaOH/H_2O} HO_2C{-}CF{=}CHCO_2H \ (77\%) \tag{8}$$

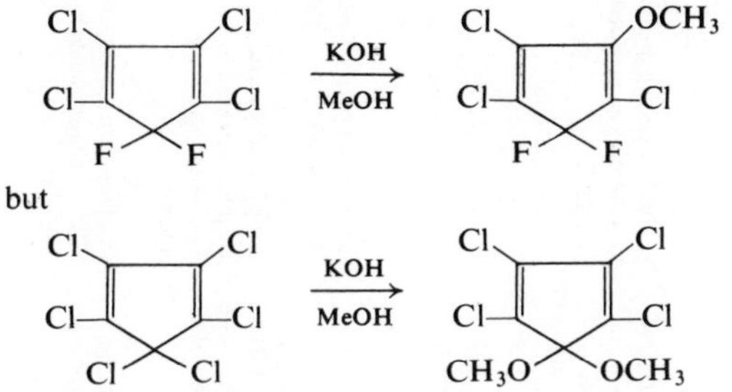

$$\text{1,2,3,4-tetrachloro-5,5-difluorocyclopentadiene} \xrightarrow[MeOH]{KOH} \text{1-methoxy-2,3,4-trichloro-5,5-difluorocyclopentadiene} \tag{9}$$

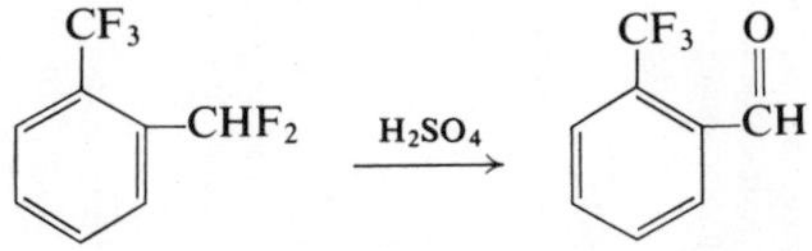

but

$$\text{hexachlorocyclopentadiene} \xrightarrow[MeOH]{KOH} \text{1,2,3,4-tetrachloro-5,5-dimethoxycyclopentadiene}$$

$$o\text{-}CF_3C_6H_4CHF_2 \xrightarrow{H_2SO_4} o\text{-}CF_3C_6H_4CHO \tag{10}$$

$$\text{1-trifluoromethyl-2-fluoro-3,3,4,4,5,5,6,6-octafluorocyclohexene} \xrightarrow[(2)\ KOH]{(1)\ KMnO_4} HO_2C(CF_2)_4CO_2H + CHF_3 \tag{11}$$

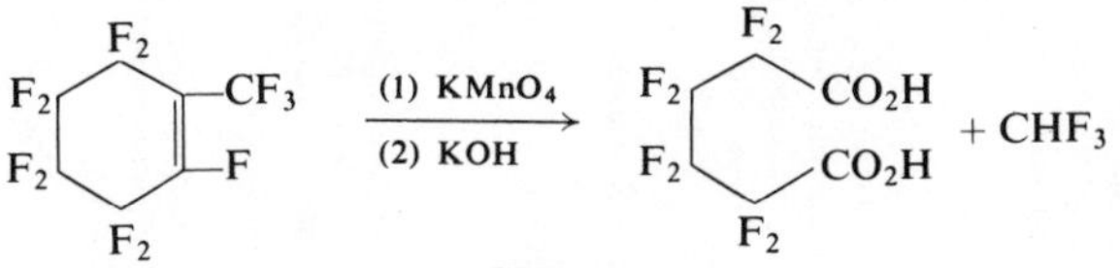

CHART 7-1

At least six methods are important for introducing difluoromethylene groups into organic compounds:

1. reaction of sulfur tetrafluoride with ketones and aldehydes;
2. addition of hydrogen fluoride to acetylenes;
3. fluoride ion exchange of dihalomethylene group;
4. condensation reactions of $-CF_2-$ containing compounds;
5. perchloryl fluoride with active methylene compounds;
6. addition of difluorocarbene to unsaturation.

The scope, limitations, and utility of each of these methods are summarized in Table 7-2.

[6] See Ref. 4; most examples are in sections contained between pp. 435 and 501.
[7] J. Harmon, U.S. Patent 2,599,631 (1952); *Chem. Abstr.*, **47**, 1725d (1953).
[8] M. S. Raasch, R. E. Miegel, and J. E. Castle, *J. Am. Chem. Soc.*, **81**, 2678 (1959).
[9] E. T. McBee, D. K. Smith, and H. E. Ungnade, *J. Am. Chem. Soc.*, **77**, 387 (1955).
[10] O. Scherer, U.S. Patent 2,180,772 (1936); *Chem. Abstr.*, **34**, 16892 (1940).
[11] R. N. Haszeldine and J. E. Osborne, *J. Chem. Soc.*, 61 (1956).

Table 7-2

Summary of Synthetic Routes for Difluoromethylene ($-CF_2-$) Groups

Type	Method	Scope and limitations	Comments	Book sections	Leading reference or reviews
Alkyl$-CHF_2$	Fluoride replacement of halogen on $RCHX_2$ with HgF_2 or by addition of HF and substitution on RCH=CHX	Very limited utility. HgF_2 usually prepared in situ from HgO and HF. $-CHBr_2$ preferred. Addition substitution can go well. Other halogens or double bonds may react	Better with $-CHBr_2$. $RHCl_2$ can be prepared by chlorination of RCH_3	4-3, 4-6A (Tables 4-4 and 4-9)	4-95 (R), 4-96 (R), 4-140 (R)
Alkyl$-CHF_2$, Aryl$-CHF_2$, Alkyl$-CF_2-$Alkyl, Aryl$-CF_2-$Aryl, Alkenyl$-CF_2-$R, Alkynyl$-CF_2-$R	SF_4 or RSF_3 on $R\overset{O}{\overset{\Vert}{C}}H$ or $R-\overset{O}{\overset{\Vert}{C}}-R$, HF is good catalyst. COF_2 also has potential for aliphatic ketones	Other carbonyl groups may fluorinate. C=C and C≡C not attacked. SF_4 will react with other functional groups such as amide, quinone, imino, nitrile, amine, halide, some alcohols, etc.	Method of choice in the laboratory. Good yields usually obtained	5-4B.1 (Table 5-6)	5-180, 5-209, 5-205 (R)
Alkyl$-CF_2-$Alkyl, Aryl$-CF_2-$Aryl	$SbF_3-Sb(V)$ on RCX_2R, HgO$-$HF on RCX_2R	$-CX_2-$ usually $-CCl_2-$ because of synthesis. Cannot have neighboring $-CX_3$ or $-CX_2-$ groups. Yields are not high	Not generally useful as $-CCl_2-$ normally prepared from $-\overset{O}{\overset{\Vert}{C}}-$. SF_4 on ketones preferred	4-5B, 4-6A (Tables 4-8 and 4-9)	4-95 (R), 4-96 (R), 4-140 (R)
Alkyl$-CF_2-$Alkyl	Addition of 2 HF to $-C\equiv C-$	Generally high yields with alkyl substituted acetylenes. Phenyl acetylene gives 18% yield. With unsymmetrical dialkyl acetylenes, two products are formed	Not generally useful because of the methods used to synthesize acetylenes. SF_4 on ketones preferred	4-3B (Table 4-5)	4-61 (R), 4-62, 4-63
$R'\overset{O}{\overset{\Vert}{C}}CF_2R$, $R\overset{O}{\overset{\Vert}{C}}CF_2H$	Claisen condensation and related condensation reactions	Unknown scope. Many scattered examples but no systematic study. Typical example is: $HCF_2\overset{O}{\overset{\Vert}{C}}OR \xrightarrow{\text{base}} CHF_2\overset{O}{\overset{\Vert}{C}}CF_2\overset{O}{\overset{\Vert}{C}}OCH_3$ (81%) $\rightarrow HCF_2\overset{O}{\overset{\Vert}{C}}CHF_2$ (62%). Condensation product readily cleaved providing a useful ketone synthesis	Condensations to introduce $-CF_2-$ groups needs systematic research	6-2D (Tables 6-7 and 6-8)	7-3 (pp. 230-231), 7-4 (pp. 435-501)

$A-CF_2-A$, $A-CF_2-R$, A = activating group	$FClO_3$ on anion of $A-CH_2-A$ or $A-CH_2-R$	Generally useful where A is $-\overset{O}{\overset{\parallel}{C}}-$, $-CN$, $-COOR$ or similar type function. Requirement is an activated $-CH_2-$ group. Basic conditions may cause unwanted side reactions such as elimination or condensation.	Explosion hazard exists with some systems (see ref. 5-112). This is a generally useful synthetic method	5-3B (Table 5-3)	5-111, 5-113
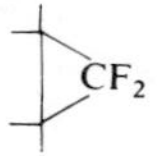 (cyclopropane ring with CF_2)	$:CF_2 + \rangle C{=}C\langle$ $:CF_2$ generated by methods described in Section 6-1B.1.b	General reaction. Convenient sources of $:CF_2$ are $ClCF_2CO_2Na$, $(CF_3)_3PF_2$, CF_3CF-CF_2 (epoxide, O bridging), and CX_2-CX_2 (ring with CF_2)	Best general method for synthesizing specific gem-difluorocyclopropanes	6-1B.1 (Table 6-2)	6-193 (R), 6-194 (R)
Alicyclic (4 or 6 ring) containing $-CF_2-$ in ring	Cycloaddition and Diels-Alder reactions of olefins and acetylenes with $CF_2{=}C\langle$ and similar compounds	2 + 2 cycloaddition reactions usually favored over 2 + 4 Diels-Alder	Fluorocyclobutanes most conveniently synthesized by this method. Compounds useful as intermediates	6-2A, 6-2B	6-435 (R)
Alkyl$-OCHF_2$, Aryl$-OCHF_2$, Alkyl$-SCHF_2$, Aryl$-SCHF_2$	$:CF_2$ + ROH or RSH. Generate $:CF_2$ by dehydrohalogenation of $HXCF_2$ in an aprotic solvent	Useful for aliphatic and aromatic compounds. Products must be stable to strong base. Better yields for aromatic derivatives	Most convenient synthesis of difluoromethyl ethers and thioethers	6-1B.1.b (Table 6-2)	6-197 to 6-200
$(A)(A)(R)C-CHF_2$	$:CF_2$ + compound containing active methylene hydrogen. Generate $:CF_2$ from dehydrohalogenation of $HXCF_2$	Scope not known. May be general for activated methylene compounds	Research may be in order	6-1B.1.b (Table 6-2)	6-201

7-3 MONOFLUOROCARBON GROUPS IN ALIPHATIC COMPOUNDS (⋗CF)

Compounds containing monofluorinated carbon functionality have a wide range of chemical stability—from relatively inert to highly labile fluorine. Although the C—F bond is stronger than C—C and C—H bonds (see Section 3-1), the driving force for elimination of hydrogen fluoride from monofluoroaliphatic compounds is very high and a low activation energy process is often available. A practical problem is that hydrogen fluoride elimination is autocatalytic but can be inhibited by storing the sample with an added inert solid base such as calcium oxide. The addition of hydrogen fluoride to alkenes (see discussion in Sections 2-2 and 4-3A) is another aspect of this problem. Selected portions of Houben-Weyl[12] and the review on monofluorosteroids by Taylor[13] give useful information and examples.

Seven methods that we consider important for introducing saturated monofluorinated carbon groups into organic compounds are

1. alkali-metal fluoride substitution of halogen or oxygen-bonded function in a polar solvent;
2. addition of hydrogen fluoride to alkenes;
3. addition of hydrogen fluoride to epoxides and other functions;
4. addition of XF reagents to alkenes and other unsaturated functions, particularly BrF or IF;
5. sulfur tetrafluoride fluorination of alcohols and halides;
6. perchloryl fluoride with active methylene compounds;
7. condensations with monofluorocarbonyl reagents.

These methods are reviewed in more detail in Table 7-3, and some of them are evaluated in the practical synthetic problem of making a series of monofluorocarboxylic acids.[14]

7-4 PERFLUOROALKYL GROUPS

Perfluoroalkyl groups are generally unreactive and unlikely to undergo oxidation, reduction, substitution, elimination, or hydrolysis. They are quite similar to the trifluoromethyl group in stability toward chemical reagents. The

[12] See Ref. 4, pp. 421-426, 462-463, 493-501.
[13] N. F. Taylor and P. W. Kent, *Advan. Fluorine Chem.*, **4**, 113-141 (1965).
[14] F. L. M. Pattison, R. L. Buchanan, and F. H. Dean, *Can. J. Chem.*, **43**, 1700 (1965).

highest point of lability is where the fluoroalkyl group is joined to other atoms, particularly to atoms bearing hydrogen. Elimination of hydrogen fluoride, as already discussed for CF_3 or CF_2 groups, is the most likely mode of reaction.

The $X(CF_2CF_2)_n$ group should also be included in this class, since it is readily available by addition (n is 1) or telomerization (n is 2 to 6) reactions of tetrafluoroethylene[16] and has been used in synthetic studies.[17] The ω-hydrogen (X = H) is extremely inert so that this type of CF_2H group is comparable in stability to a CF_3.

The perfluoroalkyl group can be introduced by a stepwise buildup, but direct introduction of a complete fluorinated unit is usually more practical.

Methods for preparing compounds containing perfluoroalkyl groups, including two-step procedures, are summarized in Table 7-4; many of these methods closely parallel those discussed already for a trifluoromethyl group.

7-5 FLUORINE ON AN AROMATIC RING

Isolated fluorine atoms on an aromatic ring (not activated by other groups) are extremely stable to chemical reagents that are normally used for oxidation, reduction, substitution, elimination, or hydrolysis. Thus the aromatic ring usually reacts before the C–F bond is ruptured. On the other hand, isolated fluorines which are activated by electron-withdrawing substituents in ortho and para positions are labile to nucleophilic attack. This type of reaction has been discussed (see Section 6-1C) and reviewed in detail by Parker.[18, 19]

Fluorine as an aromatic substituent orients electrophilic attack on the aromatic ring to the ortho and para positions and is less deactivating than chlorine[20] (see discussion in Section 3-1B).

Perfluoroaromatic compounds are stable to oxidation and electrophilic attack but are extremely reactive to nucleophilic attack. The chemistry of perfluoroaromatic compounds was discussed in Sections 6-1C and 6-2C.

Synthesis of mono-, di-, and sometimes trifluorinated compounds is usually by the Balz–Schiemann reaction (or one of its variations) which has been

[16] (a) R. M. Joyce, Jr., U.S. Patent 2,559,628 (1951); *Chem. Abstr.*, **46**, 3063b (1952). (b) K. L. Berry, U.S. Patent 2,559,629 (1951); *Chem. Abstr.*, **46**, 3063f (1952). (c) D. C. England, L. R. Melby, M. A. Dietrich, and R. V. Lindsey, Jr., *J. Am. Chem. Soc.*, **82**, 5116 (1960). (d) Y. K. Kim, *J. Org. Chem.*, **32**, 3673 (1967).

[17] (a) P. D. Faurote and J. G. O'Rear, *J. Am. Chem. Soc.*, **78**, 4999 (1956). (b) N. O. Brace, *J. Org. Chem.*, **26**, 3197 and 4005 (1961).

[18] R. E. Parker, *Advan. Fluorine Chem.*, **3**, 63 (1963).

[19] See also Ref. 4, pp. 433-435.

[20] L. M. Stock, and H. C. Brown, *Advan. Phys. Org. Chem.*, **1**, 35 (1963).

Table 7-3

Summary of Synthetic Routes for Monofluorocarbon $\left(\gt C-F\right)$ Groups

Type R = alkyl or aryl	Method	Scope and limitations	Comments	Book sections	Leading references or reviews
$R-CH_2F$	Fluoride replacement of halogen. Common reagents are anhydrous KF, AgF or HgF_2, usually in a polar solvent of moderate to high dielectric constant. $R-X + KF \rightarrow R-F$	KF usually used in high boiling polar solvent such as *N*-methylpyrrolidone, most commonly on more readily available chlorides. Iodides and bromides react more easily. AgF usually used only on bromides or iodides. KF on secondary or tertiary halides usually gives alkenes	Together with following method, this is most useful and general synthesis of primary fluorides	4-6, 5-4A.1 (Tables 4-9, 4-10, 5-4)	A-8 (R), 5-149, 5-151, 5-152
$R-CH_2F$	Fluoride replacement of oxygen function, $ROSO_2Ar(OCH_3)$, derived from alcohol and $ArSO_2Cl$ or CH_3OSO_2Cl. Anhydrous KF, either dry or in a solvent, is usually used. $ROSO_2Ar + KF \rightarrow R-F + KOSO_2Ar$	Most normal primary alcohols can be converted by this route to corresponding primary fluoride. Secondary fluorides also work well. Better yields usually obtained if a polar solvent used	This method and that described above are the most useful and general routes to primary fluorides	4-6, 5-4A.1 (Table 4-10)	A-8 (R)
$R-CH_2F$	Decomposition of a fluoroformate ester or fluorosulfinate ester formed from an alcohol (usually catalyzed by BF_3 or amine). $ROH + COF_2 \rightarrow R-O\overset{O}{\overset{\Vert}{C}}F \rightarrow RF + CO_2$	Limited to R groups that are unreactive to COF_2, COFCl, SOF_2	Good method where applicable. Decomposition of fluoroformates used more widely. Mechanism needs study	5-4A.3	5-186, 5-187, 5-190
$R-CHF-R'$	Fluoride replacement of halogen. AgF usual reagent on bromide or iodide. $-\overset{Br}{\overset{\vert}{C}}H- + AgF \rightarrow -\overset{F}{\overset{\vert}{C}}H- + AgBr$ Usually used with a solvent	Best on iodides and bromides. In general KF leads to elimination forming alkenes	Method widely used to introduce fluorine into steroids	4-6A (Table 4-9)	4-144 to 149
$R-CHF-R'$	Fluoride replacement of oxygen function (see above). $ROSO_2Ar + KF \rightarrow$ (or $ROSO_2OCH_3$)	See discussion above for RCH_2F	Method widely used to synthesize fluorosteroids	4-6B (Table 4-10)	4-168, 4-170, 4-171, 4-186 to 4-189
$R-CHF-R'$ $RR'R''C-F$	Addition of HF to an alkene. Temperature $< 0°$ usually used. Anhydrous conditions required	Best for mono and 1,2-dialkyl ethylenes. Yields poor with trisubstituted ethylenes with concomitant polymer formation. When tertiary carbonium ion formed Wagner-Meerwein rearrangement is possible	Good method for derivatives of 1-alkenes and 1,2-dialkyl ethylenes. Carbonium ion mechanism	4-3A (Tables 4-3, 4-4)	A-2 (R) (pp. 99-108)

RCHXCHFR′ RCHXCFR′R″	Addition of XF (IF or BrF) to an alkene. *N*-bromosuccinimide, *N*-bromoacetamide, *N*-iodosuccinimide or similar reagent used in presence of liquid anhydrous HF at temperature $< 0°C$. May use free I_2. Ether frequently used as solvent. $RC(=O)N(R)Br + {>}C{=}C{<} \xrightarrow{HF} -C(Br)-C(F)-$	Useful for any alkene which is not easily polymerized by HF. Elimination of HX may occur to give fluoroalkene	Broad applications. Only simple equipment required. Reaction has been widely applied to synthesis of fluorinated steroids. May use to get fluoroalkenes	5-2G (Table 5-2)	5-90 to 5-99
$RCHFCH_2OH$, RCHFCH(OH)R, RR′CFCH(OH)R	Addition of HF to epoxide 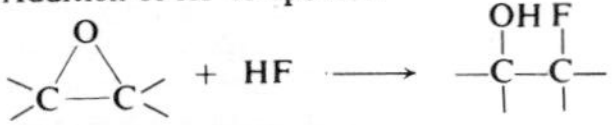 ${>}C(-O-)C{<} + HF \longrightarrow -C(OH)-C(F)-$ May use anhydrous HF, aqueous HF or other reaction media	Widely applicable to most epoxides	Reaction widely used in synthesis of fluorinated steroids	5-2A (Table 5-1)	5-39 to 5-41
R–CHF–R′	Substitution of an alcohol –OH by –F using α-fluorinated amines 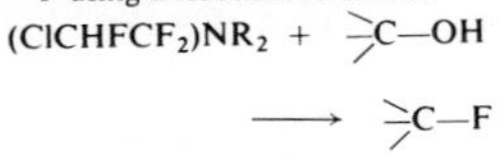 $(ClCHFCF_2)NR_2 + {\geq}C{-}OH \longrightarrow {\geq}C{-}F$	No limitations apparent. Can be used on carboxylic acids to get acid fluorides	Reaction used mainly to convert steroid–OH to steroid–F. Appears to be a general reaction	5-4A-2 (Table 5-5)	5-174
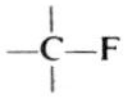 $-\overset{\vert}{\underset{\vert}{C}}-F$	Substitution of OH by –F using SF_4. Substitution of halogen by SF_4	Applicable to specialized cases where carbon bearing OH is substituted by two or more activating groups	Highly acidic alcohols usually react cleanly but normal aliphatic alcohols are decomposed	5-4A-2 (Table 5-5)	5-182 7-15
$RCHFCO_2R$ RCHF–A A–CHF–A (A = activating group)	Reaction of anion with $FClO_3$. A strong base required in an appropriate solvent 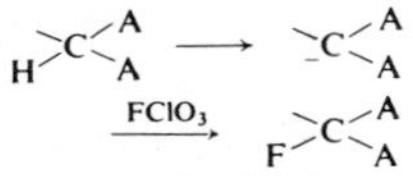 $H-C(A)_2 \longrightarrow {}^{-}C(A)_2 \xrightarrow{FClO_3} F-C(A)_2$	Reactants and products must be stable to strong bases. A may be ester, cyano, nitro, ketone or similar groups	General method with compounds containing an active methylene group	5-3B (Table 5-3)	5-111, 5-113, 7-14
$-C(=O)-CHF-C(=O)-$ $-\overset{\vert}{\underset{\vert}{C}}-CHF-C(=O)-$	Claisen ester condensation or similar condensation reactions	Unknown scope. Exploited extensively by E. D. Bergmann. Typical example 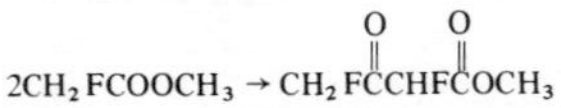 $2CH_2FCOOCH_3 \rightarrow CH_2FC(=O)CHFC(=O)OCH_3$	The $-C(=O)CHF-$ group introduced into a variety of compounds for biological studies. Good opportunities for further research	6-2D (Tables 6-7, 6-8)	7-3 (pp. 230-231), 7-4 (pp. 435-501), 7-14

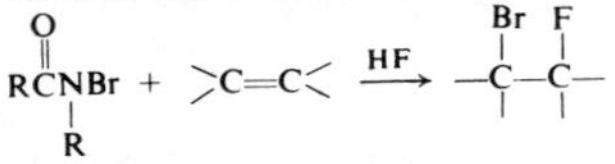

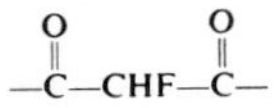

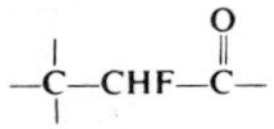

[15] R. E. A. Dear and F. F. Gilbert, *J. Org. Chem.*, 33, 819 (1968).

Table 7-4
Summary of Synthetic Routes for Perfluoroalkyl (C_nF_{2n+1}) Groups

Type $R_f = C_nF_{2n+1}$	Method	Scope and limitations	Comments	Book section	Leading references or reviews
$R_f{-}\overset{\vert}{\underset{\vert}{C}}{-}\overset{\vert}{\underset{\vert}{C}}{-}X$	Addition of perfluoroalkyl halides to alkenes. $R_fX + \rangle C{=}C\langle \longrightarrow R_f{-}\overset{\vert}{\underset{\vert}{C}}{-}\overset{\vert}{\underset{\vert}{C}}{-}X$ Radical mechanism initiated by $h\nu$, heat, or initiator such as dibenzoylperoxide, azobisisobutyronitrile, etc.	General reaction for perfluoroalkyl halides and hydrocarbon alkenes or substituted alkenes. Perfluoroiodides react best. Perfluoroalkenes or polyfluorohaloalkenes frequently undergo polymerization or telomerization	Excellent reaction for introducing perfluoroalkyl groups. Probably the most useful method. Halide X is useful functionally to attach other groups, can be eliminated (to alkene) or reduced (to alkane)	6-1A.1.a (Table 6-1A)	6-1(R), 6-19 to 6-27
$R_f{-}\overset{A}{\underset{F}{C}}{-}\overset{B}{\underset{F}{C}}{-}R'_f$ where R'_f is F or Perfluoroalkyl	Anionic addition of reagent A–B to perfluorinated alkenes or polyfluoropolyhaloalkenes. $R_fCF{=}CFR'_f + A{-}B \rightarrow R_f\overset{A}{C}F{-}\overset{B}{C}FR'_f$ Reactions are catalyzed by fluoride ion (when A is F)	Almost any perfluoroalkene is satisfactory. A–B may vary widely: RO–H, RNH–H, COF_2, NOF, HF–$H_2C{=}O$, R–MgX, ArO–H, $R_2C{=}NOH$, Ar_fF	Perfluoroalkenes are attached preferentially at a terminal vinyl CF_2 group. Yields are frequently high	6-1C.1c	A7-2(4) (R) A6-4(3) (R), 6-324 to 6-341
$A{+}CF_2CF_2{+}_nB$ $A{+}CF_2CX_2{+}_nB$ X = halogen, F	Telomerization of fluoroalkenes, especially $CF_2{=}CF_2$ and $CF_2{=}CFCl$. Radical mechanism with initiators such as $\cdot CH_2OH$ from CH_3OH. For $A{-}B + n(CF_2{=}CF_2)_x \rightarrow A{+}CF_2CF_2{+}_xB$. A is often H	Telomers are most frequently formed from $CF_2{=}CF_2$ and $CF_2{=}CFCl$	Telomer products have industrial and consumer utility	6-1A.1a	6-3 (R), 6-4 (R)
$R_f\overset{O}{\overset{\Vert}{C}}CH_2\overset{O}{\overset{\Vert}{C}}{-}OR$ $R_f\overset{O}{\overset{\Vert}{C}}CH_2\overset{O}{\overset{\Vert}{C}}{-}R$ $R_f\overset{O}{\overset{\Vert}{C}}R$	Condensation of perfluoroalkyl carbonyl derivative with ester, ketone or other compound containing an active methylene group. $R_f\overset{O}{\overset{\Vert}{C}}OCH_3 + CH_3\overset{O}{\overset{\Vert}{C}}CH_3 \rightarrow R_f\overset{O}{\overset{\Vert}{C}}CH_2\overset{O}{\overset{\Vert}{C}}CH_3$	Wide range of perfluorocarbon esters, R_fCO_2R condensed with aliphatic or aromatic compounds containing active methylene groups (customarily used in enolate ion condensations). Yields are generally good. Products are often used as synthetic intermediates (to ketones by cleavage or for other condensation reactions)	Convenient reaction because of ready availability of perfluorocarbon acids and esters.	6-2D (Tables 6-7, 6-8)	6-467 (R), 6-482 (R) to 6-484 (R)

$R_f{-}C(OH){<}$, $R_f{-}COOH$, $R_f{-}I$, $R_f{-}H$, $R_f{-}A$	Grignard synthesis with R_fMgX and compounds generally reactive with a Grignard reagent. For some reactions R_fLi is acceptable. A can be a variety of groups derived from reaction of Compound B (such as CF_3CN) with a Grignard reagent, and M is a metal or metalloid	R_fMgX will undergo expected Grignard reactions with $H_2C{=}O$, RCHO, $RC(=O)R$, H_2O, RX, CO_2, I_2, MX_n	Extensive area of high synthetic importance	6-1C.1b (Table 6-4)	6-316 (R), 6-317 (R)
$Ar{-}CF_2{-}R_f$, $R_fCF_2CH_2CF_2R_f$, $R_fCF_2CH_2CF_2R$, $ArOR_f$, ROR_f	Reaction of SF_4 with carbonyl groups of compounds formed by condensation reactions, such as Friedel-Crafts and Claisen condensations	$ArH + R_fC(=O)X \longrightarrow ArC(=O){-}R_f \longrightarrow ArCF_2R_f$ $R_fC(=O)CH_3 + CH_3C(=O)OC_2H_5 \longrightarrow R_fC(=O)CH_2C(=O){-}OC_2H_5 \xrightarrow{SF_4} R_fCF_2CH_2CF_3$ General for carbonyl compounds. Only limited by which carbonyl compounds may be synthesized	Used to form perfluoroalkyl benzene derivatives for further studies	5-4B.1 (Tables 5-6, 6-6)	5-180, 5-205 (R), 5-209, 6-461e
$R_f{-}R_f$, R_fCl, $R_f{-}(C_nF_{2n}){-}Cl$, $R_fOCF(R_f)_2$	Photolysis of $R_fC(=O)Cl$ or $R_fC(=O){-}F$. $R_f\cdot$ from $R_fC(=O)Cl$ may couple, add Cl, or add to unsaturated compound $R_f\cdot$ from $R_fC(=O){-}F$ may couple, or *O*-alkylate $R_fC(=O){-}F$ to $R_fO\dot{C}FR_f \xrightarrow{R_f} R_fOCF(R_f)_2$	Products formed depend on nature of R_f and whether $R_fC(=O){-}F$ or $R_fC(=O){-}Cl$. $R_fC(=O){-}F$ preferentially coupled to $R_f{-}R_f$, while $R_fC(=O){-}Cl$ gave primarily R_fCl	A useful synthetic method that has not been widely exploited. Starting materials are readily available	6-1A.1a (Table 6-1A)	6-53

Table 7-4 (cont.)

Type $R_f = C_nF_{2n+1}$	Method	Scope and limitations	Comments	Book section	Leading references or reviews
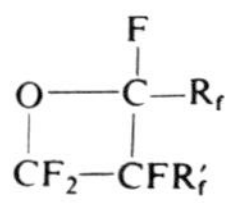 Perfluorooxetane: O—C(F)(R$_f$) ring with CF_2—CFR'_f	Photolysis of mixture of $R_fC(=O)-F$ and $CF_2{=}CFR'_f$	A general method of perfluorooxetane formation		6-1A.1 (Table 6-1)	6-54
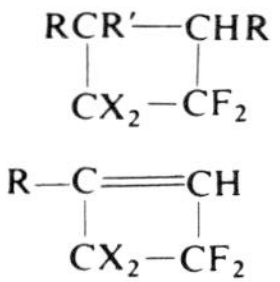 RCR′—CHR ring with CX_2—CF_2; R—C═CH ring with CX_2—CF_2 (X is F or halogen)	Thermal cycloaddition of tetrafluoroethylene or related compound to an alkene or an alkyne 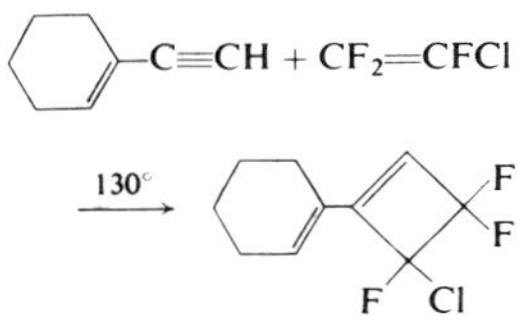 1-cyclohexenyl—C≡CH + $CF_2{=}CFCl$ $\xrightarrow{130^\circ}$ 1-(1-cyclohexenyl)-3-chloro-3,4,4-trifluorocyclobutene	Almost any alkene or alkyne with $CF_2{=}CF_2$, $CF_2{=}CFCl$, $CF_2{=}CCl_2$ or $CF_2{=}CXY$. Yields vary depending on kind and number of substituents on the alkene or alkyne	Widely used reaction to give a dazzling array of new cyclobutanes and cyclobutenes. Six-membered ring adducts (2 + 4 cycloaddition) can be obtained with certain reagents	6-2A	6-435 (R)

extensively reviewed[21-23] and discussed earlier in Sections 4-7 and 5-4A.4. Other methods, some of which are valuable for perfluoroaromatic derivatives, are

1. replacement of halogens or other groups (not NH_2 or N_2^+) by fluoride ion;

2. addition of F_2 and simultaneous substitution of hydrogen or halogens in haloaromatics by elementary fluorine or a high-valence metal fluoride (oxidative fluorination), followed by dehalogenation or dehydrohalogenation;

3. pyrolysis of halofluorocarbons to give reactive species that condense.

These methods are summarized in detail in Table 7-5. The methods for introducing a fluoroaromatic or perfluoroaromatic unit into organic compounds are reviewed in Section 6-2C.

7-6 FLUORINE BONDED TO VINYL CARBONS

Fluorine bonded to vinyl carbons is much more reactive to displacement than fluorine on aliphatic carbon, but less reactive than a fluorine of a perfluoroaromatic compound. The majority of displacements occur by addition of a nucleophile to give a carbanion intermediate which is stabilized by the electronegative fluoride:

$$B^- + CF_2{=}CFR \longrightarrow B\underset{\underset{R}{|}}{\overset{\overset{F}{|}}{C}}{-}\bar{C}F_2 \xrightarrow{-\bar{F}} B\underset{\underset{R}{|}}{C}{=}CF_2$$

$$\text{or} \quad R\bar{C}F{-}CF_2B \longrightarrow RCF{=}CFB$$

These nucleophilic reactions were already discussed in Sections 6-1C and 6-2A and are important and useful synthetic reactions of fluoroolefins.

The perturbing influence of the electronegative fluorines on an olefinic bond is also shown by the high reactivity of olefins to radical attack (see Sections 3-1B and 6-1A) and cycloaddition reactions (Section 6-2A).

A simple explanation of high reactivity is that the unshared electrons in the filled 2-p orbitals of fluorine strongly repel the 2p-π electrons of the double bond. As a result the normal π system is distorted and orbital bonding energies changed, so that the fluoroolefins are chemically very reactive, particularly to any nucleophilic or radical species (see discussion in Section 3-1B).

[21] H. Suschitzky, *Advan. Fluorine Chem.*, **4**, 1 (1965).

[22] A. E. Pavlath and A. J. Leffler, *Aromatic Fluorine Compounds*, Reinhold, New York, in particular pp. 12-16 and 42-45.

[23] See Ref. 4, pp. 213-247.

Table 7-5

Summary of Synthetic Routes to Fluoroaromatic Compounds

Type	Method	Scope and limitations	Comments	Book sections	Leading references or reviews
Ar—F Mono- and difluoro-benzenes, naphthalenes, etc.	$ArN_2^+BF_4^-$ or $ArN_2^+PF_6^-$ decomposition	General method from $ArNH_2$. Intermediate diazonium salt must be anhydrous or phenols will form. Easier to isolate anhydrous diazonium hexafluorophosphate with less hazard from explosive decomposition	Best general method. Typical utility is preparation of fluorobenzenes and fluoronaphthalenes for study of substituent effects[a]	4-7, 5-4A.4	5-199, 5-198 (R), 7-21 (R) to 7-23 (R)
	A—Ar·X + MF_n where X is halogen (or some other good leaving group) and A is an ortho- or para-activating group(s)	Group X must be activated to nucleophilic displacement by electron-withdrawing groups in ortho- or para- position of Ar	A highly polar solvent, such as dimethyl sulfoxide or *N*-methylpyrrolidone, with KF is probably best system.	4-6B, 5-4A.1 (Tables 4-10, 5-4)	4-165 (R)
	ArOCOF pyrolysis in gas phase	Limited to aromatic substituents that contain no active hydrogen and are thermally stable	ArOCOF prepared easily from COF_2 or COClF; potentially cheap route	5-4A-3	5-187b
Perfluoroaromatic C_6F_5X, C_5NF_4X, and fused ring aromatics	$C_6X_6 + F_2 (MF_n) \rightarrow C_6F_nX_{12-n} \xrightarrow{\text{Fe, heat}} C_6F_6$ or C_6F_5X	Adapted for large-scale preparation of highly fluorinated aromatic compounds including heterocyclic and fused-ring aromatics	Has been employed extensively by Birmingham group	4-4 6-2C	4-67 (R), 4-85, 6-457
	C_6X_6 + KF in polar solvent or high temperature vapor phase	Useful for a variety of aromatic systems	The best potential for a commercial synthesis	5-4A-1 (Table 5-4)	5-150, 5-156 to 5-159, 6-458
	Pyrolysis of $CFCl_2H$	Claimed as useful preparation of hexafluorobenzene	Limited to hexafluorobenzene	6-1B-1b-2	6-207 to 6-218
ArF_nY	Condensation or substitution reactions: Friedel-Crafts (for n = 1 to 3), Organometallic (n = 1 to 5), Nucleophilic substitution (n is > 3)	Fluorinated aryl unit introduced. Because of stability of aryl fluorides, a wide variety of chemical reactions can be carried out on other groups attached to aryl ring. Aromatic compounds highly fluorinated or substituted with strongly electronegative groups are susceptible to loss of fluorine by nucleophilic displacement	Extensive chemistry using aryl fluoride derivatives. Numerous reports on chemistry of perfluoroaromatic compounds in recent literature	6-2C.1	6-456a (R), 6-317d, 6-357 to 6-361
	Perfluorobenzyne additions to dienes and allylic hydrogen compounds	Perfluorobenzynes prepared by warming perfluoroaryl lithium or magnesium bromide over 0° in presence of trapping reagent	Adducts used to prepare a variety of other perfluoroaryl derivatives	6-2B, 6-2C (See Chart 6-17)	6-449 to 6-453

[a] W. Adcock and M. J. S. Dewar, *J. Am. Chem. Soc.*, **89**, 386 (1967) report the synthesis of 55 α and *p*-fluoronaphthalenes for substituent effect studies. The Schiemann reaction was used extensively in these syntheses.

The common synthetic routes to vinyl fluoride derivatives are

1. dehalogenation of saturated halofluorocarbons;
2. dehydrohalogenation of saturated halofluorocarbons;
3. nucleophilic substitution reactions, particularly with organometallic reagents or nucleophiles on fluoroolefins in aprotic medium.
4. Fluorocarbenes with phosphines on aldehydes and ketones

These synthetic routes are discussed fully in Houben-Weyl,[24] and the nucleophilic substitution reactions are also reviewed in Sections 6-1C and 6-2A. The carbene reaction is discussed in Section 6-1B.1c.

7-7 FLUORINE BONDED TO HETEROATOMS

This topic, even when restricted to organic compounds, requires a volume of its own to cover it exhaustively. We can only outline in Table 7-6 some of the major classes of interest to an organic chemist,

B–F
Si–F
N–F, P–F, As–F
O–F, S–F, Se–F
I–F

and indicate preferred synthetic routes, stability, utility, and importance. Review articles, if available, are listed and should be consulted for more details.

PROBLEMS

1. Select appropriate commercially available starting materials and outline appropriate syntheses of the compounds listed. Give an alternative synthesis when possible.

(a) 3,3-Difluorooctane
(b) 2,2,3,3-Tetrafluorooctane
(c) 2,2,4,4-Tetrafluorooctane
(d) 2,2,3,3,4,4-Hexafluorooctane
(e) 1,2-Difluorooctane
(f) 1,3-Difluorooctane
(g) 1,8-Difluorooctane
(h) 1,1,1,8,8,8-Hexafluorooctane
(i) 2,7-Diiodo-1,1,1,8,8,8-Hexafluorooctane
(j) 1,1,1,2,2,3,3,4,4-Nonafluorooctene-7
(k) 1,1,1,2,2-Pentafluoro-3-octanone
(l) 2-Fluorooctanal
(m) Perfluorooctane
(n) Perfluorooctene-4
(o) Perfluorocyclooctane

[24] Ref. 4, pp. 377-395 and 426-429.

Table 7-6

Summary of Synthetic Routes to Fluoro-Heteroatom Organic Compounds ($R{-}XF_n$)

Type	Method	Scope and limitations	Comments	Book section	Leading references or reviews
RBF_2 R_2BF	Halogen exchange on boron halide (usually B–Cl); organometallic on BF_3; Friedel-Crafts substitution of ArH by BF_3	Alkyl and aryl (including perfluorinated) and vinyl boron difluorides prepared	RBF_3^- salts prepared and shown stable	6-1C.2a	6-381a, b 6-461a-d
$RSiF_3$ R_2SiF_2 R_3SiF	Halogen exchange on silicon chlorides $RSiCl_3 + HF$; $R_2Si(OR')_2 + BF_3$ ·etherate → R_2SiF_2, $RSi(OR')_3 + HSO_3F \rightarrow RSiF_3$, $RSiF_3 + RMgX \rightarrow R_3SiF$	General for alkyl and aryl silanes (mono-, di-, and trifluoro)	Noted $RSiF_3$ is chemically highly reactive but R_3SiF is relatively inert (even to RMgX)	6-2C (Table 6-6)	6-462g, h[a]
RNF_2	$RNH_2 + F_2 \xrightarrow[\text{aqueous solution}]{\text{buffered}} RNF_2$ on $RCONHR' \rightarrow R'NF_2$	Yields are often low and organic group often fluorinated or oxidized product often results from rearrangement	*N*-Fluoroanilines as yet unknown *N*-fluoro derivatives are high energy compounds and should be handled with proper precautions against explosion hazards	5-3A	5-101 to 5-106
	also on related nitrogen derivatives, $\left.\begin{matrix}R_2C{=}NR\\RC{\equiv}N\end{matrix}\right\} + F_2 \rightarrow RNF_2$	Generally limited to where R is perfluorinated		5-1A	
	$N_2F_4 \rightleftharpoons \cdot NF_2 \rightarrow$ addition to unsaturated system, → coupling with R·, → substitution for R–H,	Extensive area of chemistry discussed in reviews		6-1A.2.a (Table 6-1C)	6-116 to 6-118 (R)
	$(C_6H_5)_3C^+ + HNF_2 \rightarrow (C_6H_5)_3CNF_2$, $(CH_3)_2C{=}CH_2 + HNF_2 \xrightarrow{H_2SO_4}$ $(CH_3)_3CNF_2$	Limited to stable carbonium ions and certain alkenes	Not generally useful. HNF_2 is exceedingly dangerous	6-1C.2	6-385
$R_2C{=}NF$	$R_2CH{-}NF_2 \xrightarrow[\text{or base}]{\text{heat}}$	RCH_2NF_2 often converted to $RC{\equiv}N$ *N*-fluoroimines often formed as by-products in preparation of RNF_2		6-1A.2.a	5-105, 6-128 to 6-135

RPF_2 R_2PF (R is alkyl or aryl)	$RPCl_2 + NaF \xrightarrow{TMS}$	Used to prepare a variety of alkyl, aryl and fluoroalkyl phosphine fluorides	Disproportionates readily $RPF_2 \rightarrow (RP)_4 + RPF_5$ Also extensive work on phosphonitrilic fluorides	6-1A.2b, 6-1C.2d, 6-2C.2 (Table 6-6)	[b]
RPF_4 R_2PF_3 $RF_2P{=}X$	$RPCl_2 + SbF_3 \rightarrow$ $RCl_2P{=}X + NaF \xrightarrow{TMS}$ (X = S or O)	A very large variety of alkyl and aryl phosphoric (V) fluorides has been prepared	Extensive studies of structure (by nmr) and chemical and physical properties		
$RAsF_2$ $RAsF_4$	$RAsCl_2 + MF \rightarrow RAsF_2$ $ArAsCl_2 + SF_4 \rightarrow ArAsF_4$	Halide exchange used to prepare both alkyl and aryl arsine (III) fluorides. Only aryl arsenic (V) fluoride prepared (also Ar_3AsF_2)		5-1B	5-37b[c]
ROF	Elementary fluorine addition to carbonyl group or substitution on OH as in $(CF_3)_2C(OH)_2$	Limited to R as perfluoroalkyl. Yields often low	High-energy compounds very hazardous to handle; explosions on contact with many organic reagents	5-1A 5-3A	5-5, 5-6
RSF RSF_3 RSF_5 RSO_2F	$RSSR \xrightarrow{AgF_2} RSF_3$ or RSF_5 Addition of SF_4 or SO_2F_2 to fluoroolefins Addition of SF_5Cl to olefins or acetylenes $RSO_2X \xrightarrow{MF}$	Major methods summarized in Table 6-9; See also Table 6-6 for aryl derivatives	Other than for RSF, R can be alkyl, aryl or perfluoroalkyl. Only perhalosulfenyl fluorides (RSF) are known and these are unstable to storage at room temperature	6-2E (Tables 6-6, 6-9)	6-119 (R)
$ArIF_2$	$ArICl_2 + HgO + HF$ or $ArIO + HF$	Only aryliododifluorides are known	$ArIF_2$ used as a source of F_2 for addition to unsaturated systems	5-1B	5-12, 5-27 to 5-30

[a] H. Harry Szmant, G. W. Miller, J. Maklouf, and K. C. Schreiber, *J. Org. Chem.*, **27**, 261 (1962); L. W. Breed and M. E. Whitehead, *J. Org. Chem.*, **27**, 632 (1962).
[b] For a complete review of the fluorides of phosphorus (which include organophosphorus derivatives) see R. Schmutzler, *Advan. Fluorine Chem.*, **5**, 31 (1965) or *Halogen Chemistry*, 2, 31 (1967). [c] L. H. Long, H. J. Emeleus, and H. V. A. Briscoe, *J. Chem. Soc.*, 1123 (1946).

2. Select appropriate commercially available starting materials and outline syntheses of the compounds listed. Give an alternative synthesis when possible.

(a) Perfluoroisobutyl iodide
(b) Perfluorocyclopentene
(c) Perfluoroterephthaloyl fluoride
(d) Perfluoropyridine
(e) Perfluoroocta-1,7-diene
(f) Perfluorotetradecane
(g) Heptafluorobutyric acid
(h) Perfluoroethanol
(i) Bromoperfluorocyclohexane

3. Select appropriate commercially available starting materials and outline syntheses of the compounds listed. Give an alternative synthesis when possible.

(a) 2-Fluorobutanoic acid
(b) 3-Fluorocamphor
(c) 2-Fluorocyclohepta-2,4,6-trien-1-one
(d) Limonene hydrofluoride
(e) 4-(Perfluoroheptyl)butanoic acid
(f) 1-(Perfluoropropyl)cyclohexene
(g) (Difluoromethylene)cyclohexane
(h) 1,1-Bis(trifluoromethyl)cyclohepta-2,4,6-triene
(i) 9,9-Bis(trifluoromethyl)bicyclo [6.1.0] nonene-4
(j) Heptafluoropropylbenzene

4. Select appropriate commercially available starting materials and outline appropriate syntheses of the following compounds. Give an alternative synthesis when possible.

(a) 1,3,5-Trifluorobenzene
(b) 3-Amino-1-fluoronaphthalene
(c) 2,4-Difluorotrifluoromethoxybenzene
(d) 3-Pentafluoroethylaniline
(e) 1-Phenyl-3,3,4,4-tetrafluorocyclobutene-1
(f) 1,2-Diphenyl-3,3,4,4-tetrafluorocyclobutene
(g) 1,4-Dideuterotetrafluorobenzene
(h) 4-(Trifluoromethyl)nitrobenzene
(i) 2-Bromo-1-fluoro-1-phenylethane
(j) Methyl 2-bromo-3-fluoropropionate
(k) Ethyl 2-fluoro-2-bromoacetate
(l) 1,2-Bis(trifluoromethyl)benzene
(m) 1,1,3,3-Tetramethyl-2,2,4,4-tetrafluorocyclobutane
(n) 7,7-Difluoronorcarane

(o) 1,1-Dimethyl-2-fluoro-2-chlorocyclopropane
(p) 1,2-Dimethyl-4,5-bis(trifluoromethyl)-4,5-diazacyclohex-1-ene
(q) 1,2-Bis(trifluoromethyl)-4,5-dimethyl-1,4-cyclohexadiene
(r) 1,2-Dimethylene-3,3,4,4-tetrafluorocyclobutane
(s) 3,3-Difluoro-4,4-dichloro-1-phenylcyclobut-1-ene
(t) 1,2-Bis(ethylmercapto)-3,3,4,4-tetrafluorocyclobut-1-ene
(u) Perfluoro(2-*n*-propyl-1,2-oxazetidine)

5. Select appropriate commerically available starting materials and indicate appropriate syntheses of the compounds listed.

(a) $(CF_3)_2CFSF_3$
(b) $FCH_2CHClSF_5$
(c) *N,N,N',N'*-Tetrafluorohexamethylenediamine
(d) *N,N*-Difluoroacetamide
(e) 1,2-Bis(difluoroamino)octane
(f) $(CF_3)_2NCH_2CH_2Br$
(g) Bis(trifluoromethyl)cyclohexylcarbinol
(h) Trifluoromethylcyclohexyl sulfide
(i) 4-Nitrophenyldifluoromethyl ether
(j) 1,2-Diphenyl-1,2-difluoroethylene
(k) (Heptafluoropropyl)phenylcarbinol
(l) $CF_3OCH_2CH_2OC(O)F$
(m) 6-Methyl-2-trifluoromethylpyridine
(n) 2-Trifluoromethylpyridine
(o) 2,2,3,4,4,4-Hexafluorobutanol-1
(p) 1,1,1,2,3,3-Hexafluoroheptanone-4
(q) Perfluoroisopropyl acrylate
(r) 2,2,3-Trifluoro-3-chloropropionamide
(s) 1,1,1,4,4,4-Hexafluoro-2-acetoxybutene-2
(t) (1,1,2-Trifluoro-2,2-dichloroethyl)-*t*-butyl ether
(u) Vinyl perfluorobutyrate

6. Indicate the experimental details required to accomplish the desired conversion. Indicate alternative methods.

(a) 5α-Pregnen-3β,17α-diol-20-one diacetate → 5α-fluoropregnan-17α-ol-3,20-dione acetate
(b) 16β-Bromo-17α,20:20,21-bis(methylenedioxy)-Δ^4-pregnene-3-one → 16β-fluoro-17α,20:20,21-bis(methylenedioxy)-Δ^4-pregnene-3-one
(c) α-1-Bromo-2,3,4,6-tetraacetyl-*d*-galactose → β-1-fluoro-2,3,4,6-tetraacetyl-*d*-galactose
(d) Hydrocortisone → 21,9α-difluoro-Δ^4-pregnene-11β,17α-diol-3,20-dione

(e) 5-Pregnen-3β-ol-20-one acetate → 5,6-difluoropregnan-3β-ol-20-one
(f) 5-Pregnen-3β-ol-20-one → 20-cyano-17α-fluoro-5-pregnen-3β,20-diol
(g) Cholesterol acetate → 5-fluorocholestanol-6-one
(h) 3β-Hydroxypregn-5-en-20-one → 3β-fluoropregn-5-en-20-one
(i) 19-Nor-4-androstene-3,17-dione → 19-nor-4-androstene-17-ol-17-trifluorovinyl-3-one

Chapter Eight Comparative Reactivity of Fluorinated Organic Compounds to Hydrocarbon Analogs

The general chemical properties of various classes of organic compounds are well known to organic chemists. The purpose of this chapter is to compare systematically chemical properties of hydrocarbons to their fluorinated analogs, so that fluorinated compounds can be better used in synthetic programs and for practical applications. Background information and comments on chemical properties can be obtained from standard reference works,[1-4] but no comprehensive comparison is available. This chapter supplements the discussion on the effect of fluorine on chemical reactivity in Sections 3-1 and 3-3.

8-1 ELECTRONIC AND STERIC EFFECTS

Fluorine, as the most electronegative element, is inductively strongly electron-withdrawing. However, it does have six unshared electrons in filled 2p-orbitals which interact with an adjacent carbon π system. The mechanism of this interaction is a subject of some controversy (see Section 3-1B), but the effects on chemical properties are to make a benzene system susceptible to electrophilic attack in the ortho–para position and to enhance the chemical reactivity of olefins. A donation of charge from the p-electrons of fluorine to the π system is the generally accepted mechanism; however, an alternative proposal that the π electrons are repelled by the concentrate of charge density on fluorine

[1] M. Hudlicky, *Chemistry of Organic Fluorine Compounds,* Macmillan, New York, 1964, Chapters 6 and 8.

[2] *Methoden der organischen Chemie (Houben-Weyl),* Vol. 5, Part 3, "Halogen Verbindungen–Fluor und Chlor" (E. Müller, ed.), 4th Ed., Thieme, Stuttgart, Germany, 1962.

[3] A. M. Lovelace, D. A. Rausch, and W. Postelnek, *Aliphatic Fluorine Compounds,* Am. Chem. Soc. Monograph No. 138, Reinhold, New York, 1958.

[4] A. E. Pavlath and A. J. Leffler, *Aromatic Fluorine Compounds,* Am. Chem. Soc. Monograph No. 155, Reinhold, New York, 1962.

can be argued from reactivity of fluoroolefins and the preference for para vs ortho substitution in aromatic electrophilic substitution, particularly in comparison to the effects of other halogens. Perfluoroalkyl groups show a strong electron-withdrawing inductive effect which is slightly enhanced by a small resonance interaction with a π system. The origin of this resonance effect is a controversial question (discussed in Section 3-1B). In contrast, alkyl groups appear slightly electron-donating relative to hydrogen, both inductively and in

Table 8-1
Electronic Effects of Fluorine and Fluoroalkyl Groups

		Parameters			
Substituent[a]	Method	σ_m	σ_p	σ_I	σ_R
H	–[b]	0	0	0	0
F	B[c]	0.34	0.06	0.45	−0.40
	F[c]	0.35	0.18	0.51	−0.33
Cl	B	0.37	0.23	0.42	−0.25
	F	0.35	0.26	0.44	−0.18
CH_3	B	−0.07	−0.17	−0.05	−0.10
	F	−0.15	−0.23	−0.08	−0.15
CF_3[d]	A[c]	0.49	0.65	0.44	0.18
	F	0.44	0.49	0.39	0.10
CCl_3[d]	F	0.32	0.33	0.31	0.02
$CF(CF_3)_2$[d]	A	0.52	0.68	0.48	0.17
	F	0.50	0.52	0.48	0.04

[a] Unless indicated otherwise, data from Refs. 5 and 6.
[b] Zero by definition for all methods.
[c] Sigma parameters calculated from B, benzoic acid pK_a; F, F^{19} chemical shift; and A, pK_a of anilinium ions.
[d] Data from W. A. Sheppard, *J. Am. Chem. Soc.*, 87, 2410 (1965); *Trans. N.Y. Acad. Sci.*, Ser. II, 29, 700 (1967).

resonance interaction with a π system. Quantitative comparison of electronic effects are conveniently made using σ parameters determined from pK_a or reactivity rate measurements[5] or by F^{19} chemical shift studies.[6] Sigma parameters for fluorine and fluoroalkyl groups, as well as for other halogens and the hydrocarbon groups, are given in Table 8-1. The larger the positive value of σ, the

[5] (a) C. D. Ritchie and W. F. Sager, *Prog. Phys. Org. Chem.*, **2**, 323 (1964). (b) R. W. Taft, Jr., and I. G. Lewis, *J. Am. Chem. Soc.*, **81**, 5343 (1959). (c) R. W. Taft, Jr., *J. Phys. Chem.*, **64**, 1805 (1960).

[6] R. W. Taft, E. Price, I. R. Fox, I. C. Lewis, K. K. Andersen, and G. T. Davis, *J. Am. Chem. Soc.*, **85**, 709, 3146 (1963).

stronger the electron-withdrawing effect of the group; the σ_I and σ_R parameters are a qualitative measure of the inductive and resonance effects, respectively.[5]

A final point in this discussion is steric size of fluorine relative to hydrogen. The van der Waals radii are hydrogen, 1.2 Å, and fluorine, 1.35 Å. A steric bulk of a CF_3 group can be estimated from the effective radius of 2.73 Å (C–F bond distance of 1.38 plus van der Waals radius of fluorine) compared to 2.3 Å for CH_3. This difference appears small but may in practice be much larger because of effective electronic repulsion of the fluorines.

8-2 ALKANES

Perfluoroalkanes are significantly less reactive to all chemical reagents except alkali metals than are the corresponding alkanes. Perfluoroalkanes also have greater thermal stability. Polytetrafluoroethylene is unique in its thermal unzippering depolymerization to form tetrafluoroethylene. The article by Reed on "Physical chemistry of fluorocarbons"[7] superbly reviews the properties of fluorocarbons. Some typical chemical properties are compared in Table 8-2.

Table 8-2
Comparison of Chemistry of Alkanes and Perfluoroalkanes

Reagent and conditions	Typical reaction of alkane	Typical reaction of perfluoroalkane
Oxygen–elevated temp.	Nonselective oxidation or total combustion	No reaction
Chlorine–$h\nu$ or elevated temp.	Free-radical chlorination $RH + Cl_2 \xrightarrow[h\nu]{25^\circ} RCl + HCl$	No reaction
Nitric acid–elevated temp.	Nitration to nitroalkane $RH + HNO_3 \xrightarrow{425^\circ} RNO_2 + H_2O$	No reaction
Alkali metals	No reaction	Etching of Teflon, potential violent reaction

8-3 ALKENES

The unusual reactivity of perfluoroalkenes, particularly as compared with alkenes, was discussed in Section 3-3 and further in Sections 6-1C and 6-2A. The more important reactions for the two classes of olefins are compared in Table 8-3. Two major differences for perfluoroalkenes are susceptibility to nucleophilic attack and tendency to cycloadd.

[7] T. M. Reed, III, *Advan. Fluorine Chem.*, **5**, 133 (1964).

Table 8-3

Comparison of Chemistry of Alkenes and Perfluoroalkenes

Reagents and conditions	Typical reaction of alkene	Typical reaction of fluoroalkene
Hydrogen, catalyst	Hydrogenation $R_2C{=}CR_2 \longrightarrow R_2CH{-}CHR_2$	Hydrogenation $F_2C{=}CF_2 \longrightarrow F_2CH{-}CHF_2$
Bromine, low temp.	Addition $R_2C{=}CR_2 \longrightarrow R_2CBr{-}CBrR_2$	Addition $R_fCF{=}CF_2 \longrightarrow R_fCFBr{-}CBrF_2$
Hydrogen halide, low temp.	Addition $R_2C{=}CR_2 \xrightarrow{HX} R_2CH{-}CXR_2$	Reluctant Addition $CF_2{=}CF_2 \xrightarrow{HX} F_2CH{-}CXF_2$
Alcohol, sodium alcoholate–low temp.	No reaction	Addition $F_2C{=}CF_2 \xrightarrow[NaOC_2H_5]{C_2H_5OH} C_2H_5O{-}CF_2{-}CF_2{-}H$
Primary or secondary amine	No reaction	Addition $F_2C{=}CF_2 \xrightarrow{R_2NH} R_2N{-}CF_2{-}CF_2{-}H$
Nucleophile, X^-, aprotic medium	No reaction	Substitution $F_2C{=}CF_2 \longrightarrow F_2C{=}CFX$

Heat alone	May dimerize, polymerize, or other reaction. No cyclobutane formation	Cycloaddition to form cyclobutane 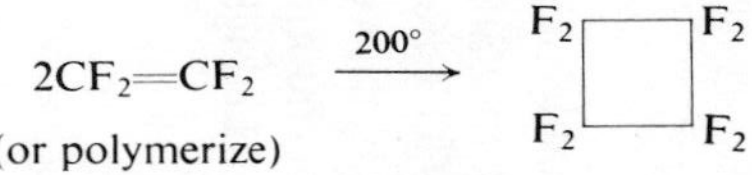 $2CF_2{=}CF_2 \xrightarrow{200°}$ [F_2 cyclobutane, F_2 on each corner] (or polymerize)
Heat with diene	Diels-Alder adduct, 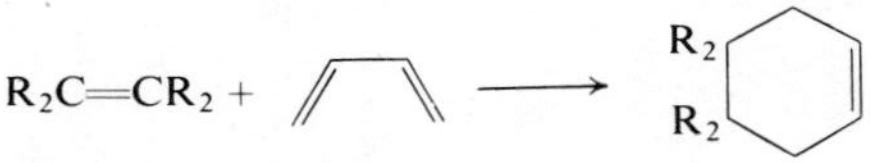 $R_2C{=}CR_2 +$ [butadiene] $\longrightarrow$ [cyclohexene with R_2, R_2]	Cycloadduct and possibly Diels-Alder adduct depending on fluoroolefin 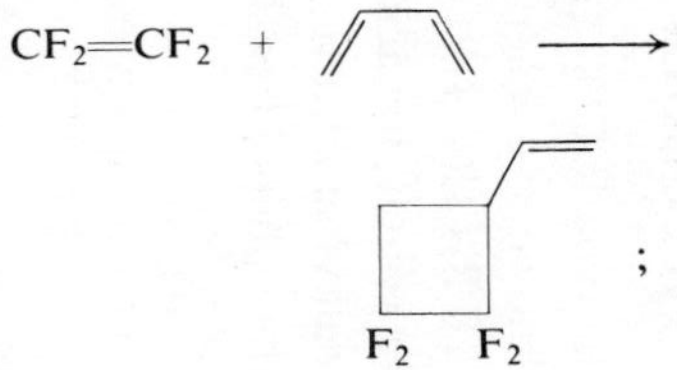 $CF_2{=}CF_2 +$ [butadiene] $\longrightarrow$ [vinylcyclobutane with F_2, F_2]; no [cyclohexene with F_2, F_2] 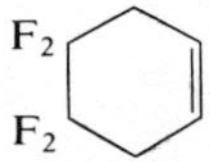
Alkene treated with appropriate polymerization catalyst	Anionic, cationic, or free radical polymerization	Usually free radical polymerization
Lewis acid	Ready complexing with π electrons	Much more difficult to form complex with π electrons
Lewis base	No ready reaction	Easy attack on CF_2 of double bond.

8-4 ALCOHOLS

Perfluorinated alcohols of the type R_fCF_2–OH are unstable to elimination of hydrogen fluoride and are not known as a general class of compounds. Alcohols of the structure R_fCH_2OH, $R_fCH(OH)R_f$, and $(R_f)_3COH$ are known.[3, 8] They are more acidic than normal alcohols; the acidity of the hydroxyl group increases with increasing perfluoroalkyl substitution so that perfluorinated tertiary alcohols are more acidic than phenol (see Table 8-4). The highly

Table 8-4
Acidities of Fluorinated Alcohols[a]

Alcohol	pK_a
CH_3OH	15.5
$(CH_3)_3COH$	20.0
CF_3CH_2OH	11.3
CCl_3CH_2OH	11.8[b]
$(CF_3)_2CHOH$	9.30
$(CF_3)_3COH$	5.4
$(CF_3)_2C(OH)_2$	6.58
C_6H_5–OH	9.9

[a] Data from Refs. 3 and 8 unless indicated otherwise.
[b] J. M. Birchall and R. N. Haszeldine, *J. Chem. Soc.*, 3653 (1959).

fluorinated alcohols are extremely strong hydrogen-bonding donors, strongly enhancing their solvent properties and ability to form complexes. In general when acidity and hydrogen-bonding differences are taken into consideration, the chemistry of normal alcohols and perfluoroalkyl-substituted carbinols is quite similar. One significant difference is a much greater stability of perfluoroalkyl carbinols to dehydration. (Note in particular adducts of perfluoroketones in condensations, see Table 8-5.)

8-5 ETHERS

Di(perfluoroalkyl) ethers have good thermal stability and generally are inert to chemical reagents. They are much less basic than their hydrocarbon analogs. Aryl perfluoroalkyl ethers, $Ar(OCF_3)_n$, are also extremely stable and are reported to be stable fluid candidates.[9] Ethers that contain the function $-CF_2-O-CH_2-$ (such as an alkyl perfluoroalkyl ether) are easily hydrolyzed or reacted with alcohols to give esters or orthoesters, respectively.

[8] (a) W. J. Middleton and R. V. Lindsey, Jr., *J. Am. Chem. Soc.*, **86**, 4948 (1964). (b) R. Filler and R. M. Schure, *J. Org. Chem.*, **32**, 1217 (1967).

[9] W. A. Sheppard, *J. Org. Chem.*, **29**, 1 (1964); U.S. Patent 3,265,741 (1966); *Chem. Abstr.*, **65**, 13610a (1966).

$$CH_3OCF_2H \xrightarrow{H_2SO_4} CH_3O\overset{\overset{\displaystyle O}{\|}}{C}H \qquad (10)$$

Aryl difluoromethyl ethers are stable in base but rapidly hydrolyze in contact with acid.[11]

8-6 ALDEHYDES AND KETONES

The effects of adjacent perfluoroalkyl groups on carbonyl reactivity were reviewed by Braendlin and McBee.[12]

The most striking difference between alkyl aldehydes and ketones and their perfluorinated analogs is the great reactivity of the carbonyl group to add active hydrogen compounds. For example, perfluorocyclobutanone reacts rapidly with water to form an extraordinarily stable hydrate which can be dehydrated only with a powerful reagent such as phosphorus pentoxide.[13]

The discussion in Section 6-2D.1 on fluorinated aldehydes and ketones as synthetic intermediates gives a good outline of the chemistry. Hexafluoroacetone, as representative of the class, has been comprehensively reviewed by Krespan and Middleton.[14]

The unusual reactivity of the carbonyl group in perfluorinated ketones is rationalized by the abnormal polarization effect derived from the strong electron-withdrawing character of the fluoroalkyl group. This abnormal polarization is shown by spectral measurements.[12]

$$R\overset{\overset{\displaystyle O}{\|}}{C}R \longleftrightarrow \underset{\text{normal}}{R\overset{\overset{\displaystyle O^-}{|}}{\underset{+}{C}}R} \longleftrightarrow \underset{\text{abnormal}}{R\overset{\overset{\displaystyle O^+}{|}}{\underset{-}{C}}R}$$

Ketones in which alkyl and perfluoroalkyl groups are substituted on the carbonyl group R_fCOR are more easily reacted at the α-methylene group than are normal dialkyl ketones. For example, 1,1,1-trifluoroacetone is easily condensed with itself in the presence of a suitable base to form 1,1,1,5,5,5-hexafluoro-2-methylpentan-2-ol-4-one.

$$CF_3\overset{\overset{\displaystyle O}{\|}}{C}CH_3 \xrightarrow[CHCl_3,\ 10^\circ]{NaNH_2} CF_3\overset{\overset{\displaystyle O}{\|}}{C}CH_2\overset{\overset{\displaystyle OH}{|}}{\underset{\underset{\displaystyle CH_3}{|}}{C}}CF_3 \qquad (15)$$

In Table 8-5 the reactions of acetone and hexafluoroacetone with a number of reagents are compared.

[10] J. Hine and J. J. Porter, *J. Am. Chem. Soc.*, **79**, 5493 (1957).

[11] (a) T. G. Miller and J. W. Thanassi, *J. Org. Chem.*, **25**, 2009 (1960). (b) R. Van Poucke, R. Pollet and A. de Cat, *Tetrahedron Letters*, 403 (1965).

[12] H. P. Braendlin and E. T. McBee, *Advan. Fluorine Chem.*, **3**, 1 (1963).

[13] D. C. England, *J. Am. Chem. Soc.*, **83**, 2205 (1961).

[14] C. G. Krespan and W. J. Middleton, *Fluorine Chem. Rev.*, **1**, 145 (1967).

[15] E. T. McBee, D. H. Campbell, R. J. Kennedy, and C. W. Roberts, *J. Am. Chem. Soc.*, **78**, 4597 (1956).

Table 8-5
Comparison of Reactions of Acetone and Hexafluoroacetone

Reagent and conditions	Reactions of acetone	Reactions of hexafluoroacetone
Water, room temp.	No reaction	Forms stable 1,1-diol $(CF_3)_2C(OH)_2$
Ethanol, room temp.	No reaction	Forms stable hemiacetal $(CF_3)_2C(OH)(OC_2H_5)$
Ethanol, dry HCl, reflux	Acetal formation $(CH_3)_2C(OC_2H_5)_2$	Hemiacetal only $(CF_3)_2C(OH)(OC_2H_5)$
Ammonia, room temp.	No stable adduct	Stable adduct $(CF_3)_2C(NH_2)(OH)$
Semicarbazide	$(CH_3)_2C{=}NNHC(=O)NH_2$	$(CH_3)_2C(OH)(NHNHC(=O)NH_2)$ Adduct will not dehydrate to semicarbazone
Hydrogen cyanide	$(CF_3)_2C(OH)(CN)$	$(CF_3)_2C(OH)(CN)$
Diethyl malonate	$(CH_3)_2C{=}C(CO_2C_2H_5)_2$	$(CF_3)_2C(OH){-}CH(CO_2C_2H_5)_2$ Adduct not readily dehydrated
Grignard-RMgX	$(CH_3)_2C(OH)R$	$(CF_3)_2C(OH)R$
Isobutylene, 100°	No reaction	Allylic addition $(CF_3)_2C(OH){-}CH{=}C(CH_3)CH_2C(OH)(CF_3)_2$
Methylvinyl ether	No reaction	Cycloaddition $(CF_3)_2C{-}O{-}CH(OCH_3){-}CH_2$ (four-membered ring: $(CF_3)_2C$—O, CH_2—$CHOCH_3$)
Benzene, $AlCl_3$, < 20°	No reaction (unless forced)	Electrophilic substitution $C_6H_5C(CF_3)_2OH$

8-7 ACIDS AND DERIVATIVES

Perfluoroalkyl carboxylic acids are significantly more acidic than their corresponding alkyl analogs; for example, trifluoroacetic acid has a $pK_a = 0.23$ (comparable to a mineral acid) compared to $pK_a = 4.74$ for acetic acid.[3]

In spite of the large acidity difference, most of the chemistry of perfluoroalkyl carboxylic acids is quite similar to that of the corresponding aliphatic acids.

Trifluoroacetic acid has found wide use as a reaction solvent because of its strong acidity, great solvent power toward polar compounds, and ability to catalyze esterification and polymerization reactions. Trifluoroacetic anhydride is a powerful agent for esterifications and catalyzes esterification of difficultly esterified alcohols and acids. Leading references can be obtained from Hudlicky[16] and Section 6-2D.3.

The reactions of perfluoroaliphatic acids are compared with aliphatic acids in Table 8-6.

8-8 AMINES

Perfluoroalkylamines are not chemically similar to their alkyl analogs and, consequently, are of much less importance. On one hand, the tertiary perfluoroalkylamines, such as $(CF_3)_3N$, are not basic (will not form a salt with strong mineral acids), are stable to oxidation, and do not pyrolyze until heated over 500°C. But the primary and secondary amines are relatively unstable and are dehydrofluorinated to imino compounds by weak bases (see page 398).

$$(CF_3)_2NH \xrightarrow{KF} CF_3N{=}CF_2$$

8-9 AROMATIC DERIVATIVES

Perfluoroaromatic compounds are highly susceptible to nucleophilic reactions, whereas the hydrocarbon analogs undergo facile electrophilic substitution unless highly substituted by electronegative groups. The chemistry of perfluoroaromatic compounds (pertaining to use in synthesis) was discussed in Section 6-2C.3; the nucleophilic reactions were discussed in Section 6-1C.1c on fluorinated carbanions.

The electronic character of an aromatic ring is grossly affected by the replacement of hydrogen by fluorine. As pointed out previously (see discussion in Section 3-1B), the highly electronegative fluorine strongly withdraws electrons

[16] Ref. 1, pp. 227-232 and Chapter 7.

Table 8-6

Comparison of the Reactions of Perfluorocarboxylic Acids with Aliphatic Carboxylic Acid

Reagents and conditions	Typical Reactions of aliphatic acid or derivative	Typical reaction of perfluoro acid or derivative
PCl_5	Acid chloride	Perfluorinated acid chloride
Alcohol, proper catalyst	Ester formed. Difficult to esterify more acidic alcohols such as phenols	Ester formed more readily. Even phenols easily acylated particularly by trifluoroacetic anhydride
Hunsdieker on silver salt	$RCOOAg \xrightarrow{X_2} RX$	$R_fCOOAg \xrightarrow{X_2} RX$
Heat to high temp.	$RCOOH \xrightarrow{\Delta}$ various decomposition reactions	$R_fCF_2COOH \xrightarrow{550^\circ} R_f{=}CF_2 + HF + CO_2$
Salt heated to high temp.	$RCOONa \xrightarrow[CaO]{NaOH} RH + CO_2$ (poor yield)	$CF_3CF_2CF_2COOAg \xrightarrow[(90\%)]{260-270^\circ} CF_3CF_2CF_2CF_2CF_2CF_3$ $CF_3CF_2CF_2CF_2COOK \xrightarrow{165-200^\circ} CF_3CF_2CF{=}CF_2 + CF_3CF{=}CFCF_3$ $CF_3CF_2CF_2CF_2COOK \xrightarrow[200^\circ]{HOCH_2CH_2OH} CF_3CF_2CF_2CHF_2$
$LiAlH_4$, ether, -5° on free acid	$RCOOH \rightarrow RCH_2OH$	$R_fCOOH \rightarrow R_fCH_2OH + R_fCH(OH)_2$ (poorer yields)
$LiAlH_4$, ether, -5° on ester	$RCOOR \rightarrow RCH_2OH$	$R_fCOOR \xrightarrow{eq.\ LiAlH_4} R_fCHO$ $R_fCOOR \xrightarrow{excess\ LiAlH_4} R_fCH_2OH$
$LiAlH_4$, ether, -5° on amide	$RCONH_2 \rightarrow RCH_2NH_2$	$R_fCONH_2 \rightarrow R_fCH_2NH_2$
Claisen ester condensation with acetone	$CH_3\overset{O}{\overset{\parallel}{C}}{-}OC_2H_5 \xrightarrow{acetone} CH_3\overset{O}{\overset{\parallel}{C}}{-}CH_2\overset{O}{\overset{\parallel}{C}}{-}CH_3$ Major side reaction of self-condensation: $2CH_3\overset{O}{\overset{\parallel}{C}}{-}OC_2H_5 \rightarrow CH_3\overset{O}{\overset{\parallel}{C}}{-}CH_2\overset{O}{\overset{\parallel}{C}}OC_2H_5$	$R_f\overset{O}{\overset{\parallel}{C}}{-}OC_2H_5 \xrightarrow{acetone} R_f\overset{O}{\overset{\parallel}{C}}{-}CH_2\overset{O}{\overset{\parallel}{C}}{-}CH_3$ Very clean reaction under mild conditions

from the σ framework; an accumulation of five fluorines in the pentafluorophenyl ring obviously causes a major distortion of electron density. A chemical consequence is that nucleophilic attack is strongly promoted and carbanion intermediates are stabilized. This subject was discussed in Section 6-1C.1c and has been summarized in recent publications,[17, 18] particularly in the series by Burdon of which leading references[17] are given.

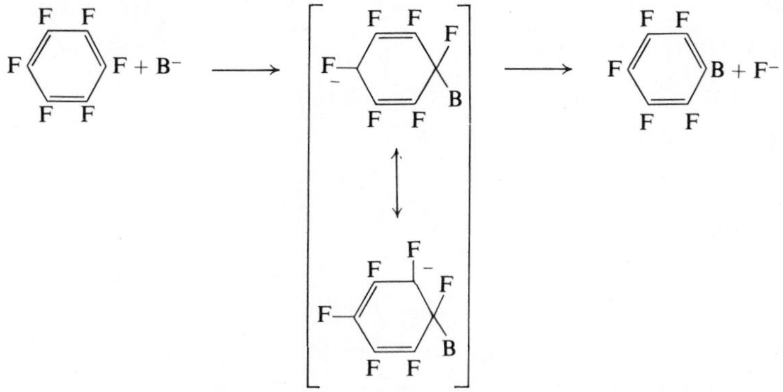

However, the six unshared electrons in 2p-orbitals of each fluorine will interact with the p-π system of the aromatic ring more effectively than for other halogens because of the short C–F bond distance (1.30 Å) and similar size of p-orbitals. As a result the π system appears to have a higher than normal density of electrons so that it strongly interacts with an electron-deficient substituent to feed charge density to the substituent.

The ionization constants of some perfluorinated phenols and aromatic acids[19] are given in Table 8-7 and compared to hydrocarbon analogs. Clearly, the pentafluorophenyl group is strongly electron-withdrawing since the fluorinated phenol and acids are much more acidic than their hydrocarbon counterparts. However, pentachlorophenol is more acidic than pentafluorophenol, although inductively a perfluorinated group is more strongly electron-withdrawing than the corresponding perchlorinated group (in Table 8-4 note acidity of trichloroethanol compared to trifluoroethanol). This apparent abnormality in acidity was explained by back coordination with electron release from fluorine which,

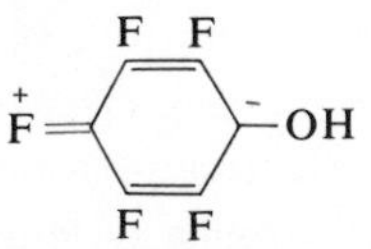

[17](a) J. Burdon, *Tetrahedron,* **21,** 3373 (1965). (b) J. Burdon, D. R. King, and J. C. Tatlow, *Tetrahedron,* **22,** 2541 (1966).

[18]R. J. de Pasquale and C. Tamborski, *J. Org. Chem.,* **32,** 3163 (1967).

[19](a) J. M. Birchall and R. N. Haszeldine, *J. Chem. Soc.,* 3653 (1959). (b) R. D. Chambers, F. G. Drakesmith, and W. K. R. Musgrave, *J. Chem. Soc.,* 5045 (1965).

Table 8-7
Ionization Constants of Some Perfluorinated Aromatic Acids and Phenols

Phenols[a]	pK_a	Acids[b]	pK_a
C_6H_5OH	9.9	$C_6H_5CO_2H$	4.21
C_6F_5OH	5.5	$C_6F_5CO_2H$	3.38[c]
C_6Cl_5OH	5.2	3-pyridinecarboxylic acid	4.82
		tetrafluoro-3-pyridinecarboxylic acid	3.45
		4-pyridinecarboxylic acid	4.84
		tetrafluoro-4-pyridinecarboxylic acid	3.21

[a] Data from Ref. 19a.
[b] Data from Ref. 19b.
[c] A much more acidic value for pentafluorobenzoic acid reported earlier was noted as incorrect.

as pointed out above, is better than from chlorine, and partly offsets the strong inductive effect.[19a]

Nuclear magnetic resonance F^{19} chemical shift measurements on pentafluorophenyl derivatives[20] (Table 8-8) show that strong electron-withdrawing substituents or groups which can accept electron density from the π system (such as BX_2) effectively deshield the *o*- and *p*-fluorine; thus, p-π resonance forms of the type shown above, where the ring π system interacts with the empty p-orbitals of the substituent, must contribute significantly.

Quantitative measurements of the inductive and resonance effects of a

[20] (a) I. J. Lawrenson, *J. Chem. Soc.*, 1117 (1965). (b) R. D. Chambers and T. Chivers, *ibid.*, 3933.

pentafluorophenyl ring can only be obtained by pK_a, reactivity rate measurements, or F^{19} chemical shift studies on the appropriate pentafluorobiphenyl derivatives. However, a value for σ_m of +0.22 was inferred from F^{19} nmr chemical shift correlations on a series of pentafluorophenyl phosphorus compounds, C_6F_5PXY.[21] The σ_p was estimated to be +0.4 from a correlation of rates of reaction of substituted pentafluorobenzenes with sodium pentafluorophenolate.[18] These results confirm qualitatively the inductive electron-withdrawing character of the pentafluorophenyl ring, but the larger value of σ_p over σ_m is unexpected if an electron-donating resonance effect operates. The σ_m value

Table 8-8
Fluorine-19 Nuclear Magnetic Resonance Chemical Shift Data on Pentafluorophenyl Derivatives

X of C_6F_5X	δ, in ppm, relative to C_6F_6 ortho	meta	para
H	− 23.8	− 0.03	− 8.9
CH_3	− 18.9	+ 1.4	− 3.6
CF_3	− 22.9	− 2.3	− 15.0
Cl	− 22.0	− 1.3	− 6.6
Br	− 30.2	− 2.0	− 8.0
I	− 43.6	− 3.0	− 10.1
OCH_3	− 4.4	+ 2.0	+ 2.0
NH_2	+ 0.7	+ 2.8	+ 11.2
CN	− 30.4	− 3.7	− 19.4
BF_2	− 34.1	− 2.5	− 19.9
$BF_2{\cdot}NC_5H_5$	− 30.0	+ 0.2	− 7.3
PF_2	− 22.7	− 2.1	− 14.1

appears much too small, but further discussion is not warranted until proper data are obtained.

An overall result of the strong inductive withdrawal in the σ- framework coupled with π electron interaction is that the pentafluorophenyl group greatly stabilizes certain organometallic species. For example, organocopper compounds are highly unstable and usually decompose to give copper metal and coupling products on attempted isolation at room temperature, but pentafluorophenylcopper is stable to 200°C.[22] Also pentafluorophenylboron[20b] and aluminum[23] derivatives and pentafluorophenyl derivatives of transition metals such as titanium,[24] zirconium,[25] nickel, palladium, and platinum[26] have been shown

[21] M. G. Barlow, M. Green, R. N. Haszeldine, and H. G. Higson, *J. Chem. Soc.* (B), 1025 (1966).

[22] A. Cairncross and W. A. Sheppard, *J. Am. Chem. Soc.*, **90**, 2186 (1968).

[23] R. D. Chambers and J. A. Cunningham, *J. Chem. Soc.* (C), 2185 (1967).

[24] M. A. Chaudhari, P. M. Treichel, and F. G. A. Stone, *J. Organometal. Chem.*, **2**, 206 (1964).

[25] M. A. Chaudhari and F. G. A. Stone, *J. Chem. Soc.* (A), 838 (1966).

[26] (a) F. J. Hopton, A. J. Rest, D. T. Rosevear, and F. G. A. Stone, *J. Chem. Soc.* (A), 1326 (1966). (b) *Chem. Eng. News*, **42**, 40 (Feb. 10, 1964).

Table 8-9
Comparison of Chemistry of Aromatic and Perfluoroaromatic Compounds

Reagent and conditions	Typical reaction of aromatic derivatives	Typical reaction of perfluoroaromatic derivatives	Comments and Refs.
HNO_3, H_2SO_4	Electrophilic substitution to give $C_6H_5NO_2$	C_6F_6 no reaction C_6F_5H nitrated to $C_6F_5NO_2$ with difficulty using sulfuric acid-fuming nitric acid mixtures	33
Br_2, $FeBr_3$	Electrophilic substitution to give C_6H_5Br	No reaction	–
RBr, $AlCl_3$ 80°	Friedel–Crafts substitution to give C_6H_5R	C_6F_6 no reaction but $C_6F_5H + CHCl_3 \xrightarrow[150°]{AlCl_3} (C_6F_5)_3CH$	34
30% aq. NH_3	No reaction	Substitution to give $C_6F_5NH_2$	35
NaOR, Roh	No reaction	Substitution to give C_6F_5OR	17,18
$LiAlH_4$	No reaction	Replacement of fluorine to C_6F_5H	36 Benzene hydrogenates with ruthenium catalyst but no comparable experiment for C_6F_6
RLi + ArH	No reaction unless activated, such as [anisole with OCH_3 → anisole with OCH_3 and Li on the ring next to it]	$C_6F_6 + RLi \rightarrow C_6F_5R$	37
RLI + ArBr	ArLi	$C_6F_5Br + RLi \xrightarrow{-78°} C_6F_5Li$	Section 6-2C.3, C_6F_5Li or C_6F_5MgX decomposes, on warming over 0°, to tetrafluorobenzyne

Cl_2, $h\nu$	Addition to give $C_6H_6Cl_6$	No reaction?	
$(C_6H_5CO_2)_2$, heat	Radical substitution to give $(C_6H_5)_2$ and $C_6H_5CO_2C_6H_5$	$C_6H_5C_6F_5$	38 Mechanism of radical attack on C_6F_6 proposed to be different from that on C_6H_5X
ArH, $h\nu$	Low yields of  [structure]; if highly substituted with sterically large groups get bicyclo [2.2.0]-hexa-2-5-dienes	C_6F_6 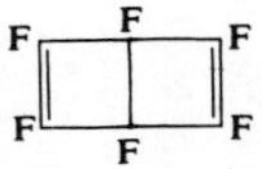 ⟶ [hexafluorobicyclo[2.2.0]hexa-2,5-diene structure]	39
C_6H_6, O_2, V_2O_5 400°	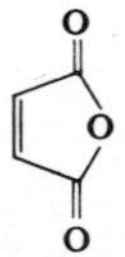 [maleic anhydride structure]	No reaction?	

[33]P. L. Coe, A. E. Jukes, and J. C. Tatlow, *J. Chem. Soc.* (C), 2323 (1966).
[34]W. F. Beckert and J. U. Lowe, Jr., *J. Org. Chem.*, **32**, 582 (1967).
[35]G. G. Yakobson, V. D. Shteingartz, G. G. Furin, and N. N. Vorozhtsov, Jr., *J. Gen. Chem. U.S.S.R. (English Transl.)* **34**, 3560 (1964).
[36]G. M. Brooke, J. Burdon, and J. C. Tatlow, *J. Chem. Soc.*, 3253 (1962).
[37]J. M. Birchall and R. N. Haszeldine, *J. Chem. Soc.*, 3719 (1961).
[38]P. A. Claret, J. Coulson, and G. H. Williams, *Chem. Ind. (London)*, 228 (1965).
[39](a) G. Camaggi, F. Gozzo, and G. Cevidalli, *Chem. Commun.*, 313 (1966). (b) I. Haller, *J. Am. Chem. Soc.*, **88**, 2070 (1966).

to have much better thermal and chemical stability than their hydrocarbon counterparts. However, pentafluorophenyltin derivatives hydrolyze much more easily than their phenyl analogs,[27] but without detailed mechanism studies any explanation would be speculative.

The pentafluorophenyl system has been used as a probe to determine substituent effects[20a, 28] and in studies on structure of organometallic reagents.[29] In regard to substituent effect studies, correlation equations have been proposed[20a] but, unfortunately, are not nearly as precise as for the *m*- and *p*-fluorobenzenes—no doubt because of complex interdependency resulting from the extensive p-π interactions, particularly of paramagnetic interactions of *o*-fluorines with substituents such as bromine or iodine passed onto the *m*- and *p*-fluorines.[28d, 28e]

The thermal stability of perfluoroaromatic derivatives (with the exception of Grignard or lithium reagents which give perfluorobenzyne) is excellent. Decafluorobiphenyl is slightly more stable than biphenyl[30]; some other perfluoroaromatic derivatives have stability comparable to hydrocarbon analogs. However, in certain practical applications, such as high temperature materials, the perfluoroaromatic compounds have been found less stable than their hydrocarbon analogs.[31]

Perfluoroaromatic derivatives and perfluoroaromatic analogs of biologically active compounds have been prepared but so far have not shown biological activity of special interest or practical value.[32]

The more important chemical properties of the aromatic compounds are compared in Table 8-9. Unfortunately, direct comparisons for several common reactions are not available but the gross differences in electrophilic and nucleophilic reactivity are apparent.

[27] C. Eaborn, J. A. Treverton, and D. R. M. Walton, *J. Organometal. Chem.*, **9**, 259 (1967).

[28] (a) R. J. Abraham, D. B. MacDonald, and E. S. Pepper, *Chem. Commun.*, 542 (1966). (b) J. Homer and L. F. Thomas, *J. Chem. Soc.* (B), 141 (1966). (c) P. Bladon, D. W. A. Sharp, and J. M. Winfield, *Spectrochim. Acta*, **20**, 1033 (1964). (d) A. J. R. Bourn, D. G. Gillies, and W. E. Randall, *Proc. Chem. Soc.*, 200 (1963). (e) N. Boden, J. W. Emsley, J. Feeney, and L. H. Sutcliffe, *Mod. Phys.*, **8**, 133 (1964).

[29] D. F. Evans and M. S. Khan, *J. Chem. Soc.* (A), 1643 (1967).

[30] L. A. Wall, R. E. Donadio, and W. J. Pummer, *J. Am. Chem. Soc.*, **82**, 4846 (1960).

[31] *Chem. Week*, **92**, 60 (April 27, 1963).

[32] (a) E. H. P. Young, *4th Intern. Symp. Fluorine Chem., Estes Park, Colorado, July 1967*, Abstracts, p. 86. (b) Belgian Patents 601,586 and 606,483 (1961); 622,852 and 622,853 (1963) to National Smelting Co., Ltd.; *Chem. Abstr.*, **59**, 8661h and 8661g (1963).

Chapter Nine

Safety, Toxicity, and Biological Importance of Organofluorine Compounds

The purpose of this chapter is threefold: (1) to emphasize safety practices necessary in working with fluorinating agents (leading references that should be consulted before doing any fluorinations are given) and to discuss hazards ranging from explosions to burns, particularly pointing out insidious problems that occur in synthesis and workup; (2) to point out toxicity of common fluorinated reagents and products to provide a general guide in handling; (3) to outline applications of organic fluorine chemistry to biological problems. This rapidly growing and fascinating field is described briefly both to reemphasize the problems of toxicity in Section 9-2 and to show the great potential of fluorine chemistry in medicinal chemistry and biological research.

The message of this chapter is that organic fluorine chemistry is potentially very hazardous, however, no more so than many other fields of chemistry, particularly, if the chemist is aware of the hazards and takes proper safety precautions.

9-1 SAFETY IN HANDLING FLUORINATING AGENTS

9-1A. FLUORINE

Fluorine, the most electronegative element, is extremely hazardous, primarily because the weak F–F bond in molecular fluorine (37.6 kcal/mole) leads to extraordinary reactivity. Fluorine is most conveniently handled from commercially available cylinders typically containing 0.5 or 6 lb. of fluorine at 400 psi. Special reducing valves are commercially available for the cylinders. Fluorine cylinders are normally kept behind steel barricades and the valve opened remotely. The reason for this is that a metal valve in contact with high-pressure fluorine may ignite if the metal-fluoride coating on the valve is broken.

Stainless steel or Monel® tubing (properly passified) is recommended for transfer of the gas. Monel valves or high-nickel steel valves with Teflon® packing, marketed under the trade names like Hoke, Ermeto, or Whitey, are highly recommended. Data sheets distributed by commercial vendors provide the detailed information needed to work with fluorine under pressure. These data sheets and a review by Cady are highly recommended as starting points for assessing the problems of working with fluorine.[1-5]

Needless to say, fluorine can severely damage any animal tissue with which it comes in contact and, hence, it is exceedingly toxic when inhaled at the level of parts per million.[6, 7]

In areas where cylinders of fluorine are not available, electrolytic generation of fluorine from potassium hydrogen fluoride is common. Because of the lack of a pressure head; electrolytically generated fluorine is less convenient for laboratory work, but it is safer. An electric switch controls the flow of fluorine so that the hazards of a burning metal valve and a fluorine fire are eliminated.

9-1B. HALOGEN FLUORIDES

In one sense, halogen fluorides (ClF_3, BrF_3, IF_5) are less hazardous than fluorine because they are less reactive and less volatile—in the same sense, a rattlesnake is less hazardous than a cobra. We recommend that halogen fluorides be handled with the same careful techniques that are used with fluorine and that commercial data sheets[8, 9] and reviews[10] be carefully studied.

Chlorine trifluoride presents an unusual explosion hazard not likely to be encountered with fluorine. It has a boiling point above ice-bath temperature (bp 12°C) and may inadvertently be condensed with organic material. One of the authors experienced an unexpected explosion when chlorine trifluoride was condensed into a trap containing chloroform.

[1] *Fluorine,* Product Data Sheet TA-85412 or TA-85413 or subsequent revisions, Allied Chemical Co. Gen. Chem. Div.

[2] *Fluorine Handling and Safety,* New Product Data Sheet, Pennsalt Chemicals.

[3] *Matheson Fluorine Data Sheet,* reprinted from 3rd Ed. of *Matheson Gas Data Book,* The Matheson Co., Inc.

[4] *The Handling and Storage of Liquid Propellants,* Office of the Director of Defense Research and Engineering, January 1963. A revision is planned for 1968 according to *J. Chem. Ed.,* **44,** A-1057, (1967).

[5] G. H. Cady in *Fluorine Chemistry* (J. H. Simons, ed.), Vol. 1, Academic, New York, 1950, pp. 293-318.

[6] H. C. Hodge and F. A. Smith, in *Fluorine Chemistry* (J. H. Simons, ed.), Vol. 4, Academic, New York, 1963, pp. 38, 197, 231, 280.

[7] B. C. Saunders, *Advan. Fluorine Chem.,* **2,** 184-5 (1961).

[8] *Chlorine Trifluoride and Other Halogen Fluorides,* Tech. Bull. TA-8532-2 or subsequent revision, Gen. Chem. Div., Allied Chemical Co.

[9] "Chlorine Trifluoride", Safety in the Chemical Laboratory, *J. Chem. Ed.,* **44,** A-1057 (1967). This review is reprinted from pp. 71-78 of Ref. 4.

[10] W. K. R. Musgrave, *Advan. Fluorine Chem.,* **1,** 1-28 (1960).

9-1C. HYDROGEN FLUORIDE

Hydrogen fluoride is the fluorinating agent that most organic chemists will use at some point in their careers. It is dangerous because of the extreme severity of its burns and because of high inhalation toxicity and accompanying damage to the lungs (pulmonary edema). We do not have the space to describe properly the destructiveness of hydrogen fluoride to animal tissue but urge the reader to consult the leading references.[11-15] In Ref. 11, plate 2 (opposite p. 3) should rapidly convince the skeptic of the damage potential of hydrogen fluoride toward human tissue. The proper treatment of hydrogen fluoride burns is given on pp. 35-38 of Ref. 6.

Hydrogen fluoride can be safely handled using proper techniques[11-15]; developments in plastic equipment have simplified the problems. Polypropylene or polyethylene flasks, bottles, tubing, and stopcocks permit easy, safe transfer of liquid or gaseous hydrogen fluoride. Copper or mild steel tubing and valves permit easy handling of the gas under low pressure. But proper protective clothing, safety devices, and good ventilation are a must in working with hydrogen fluoride.[11-14]

9-1D. GASEOUS FLUORINATING AGENTS

The most common reagents are sulfur tetrafluoride and carbonyl fluoride. These agents are particularly dangerous because of ease of hydrolysis to form hydrogen fluoride *in vivo* at sites where it does immediate damage.

$$SF_4 + 2H_2O = SO_2 + 4HF$$

Experiments with these reagents under pressure must be carried out with adequate safety precautions, particularly with regard to venting of gases in case of an uncontrolled reaction leading to rupture of blow-out discs. Appropriate manufacturers' data sheets should be consulted.

9-1E. OXIDIZING INORGANIC FLUORIDES

Reagents such as CoF_3, AgF_2, PbF_4, BiF_5, and UF_6 may react explosively with organic compounds, often after an induction period.

$$AgF_2 + R{-}H = R{-}F + HF + \text{HEAT} + \text{subsequent chain reactions}$$

Small-scale trial experiments are good practice when using these fluorinating

[11] A. J. Rudge, *The Manufacture and Use of Fluorine and Its Compounds,* Oxford Univ. Press, London and New York, 1962, pp. 2-3, 10-17, 78.

[12] *Hydrofluoric Acid,* Chemical Safety Data Sheet SD-25, Manufacturing Chemists' Assoc., Inc., 1957.

[13] M. Hudlicky, *Chemistry of Organic Fluorine Compounds,* Macmillan, New York, 1962, pp. 24-25.

[14] J. H. Simons (ed.), *Fluorine Chemistry,* Vol. 1, Academic, New York, 1950, pp. 225-268, particularly 262-268.

[15] See Ref. 6, pp. 35-38, 197-198, 209-210, 232, 280-281, 365.

agents, but watch a scaled-up reaction where heat is not as readily dissipated. Use of fluorocarbon diluents for dissipation of heat is a good technique.

9-1F. INORGANIC FLUORIDES

These reagents present only the normal dangers of toxicity of inorganic fluorides. The toxicity of inorganic fluorides is discussed in detail in Ref. 6. But be aware of a major hazard of inorganic fluorides—fluoride plus a strong inorganic acid forms hydrogen fluoride!

9-1G. GENERAL COMMENT

Adequate information is available to permit safe handling of inorganic fluorinating agents. This accumulated knowledge must be used by the chemist to protect himself. A safety conscience is as important as safety equipment.

As in other areas of chemistry, compatibility of reagents and solvents must be considered in any chemical reaction, and the chemist must be prepared for the unexpected. For example, the normally stable fluorocarbons are often commonly used as very inert solvents, but they react violently with sodium, potassium, or barium.[16]

The prudent chemist should take one additional step before working with fluorinating agents and organic fluorine compounds. He should contact his local health service at his laboratory and alert the staff as to the nature of the chemicals in use. Possible injuries should be discussed and provision made for proper treatment.

9-2 TOXICITY

A major problem in any research laboratory is toxicity of chemical reagents, particularly by inhalation, absorption through the skin, or eye contact, since ingestion is avoided by extremely simple precautions. Fluorinated compounds are unusual in that some fluorocarbons are completely inert and can be safely inhaled in place of nitrogen in air while others are among the most lethal materials known. Often only relatively minor changes in structure produce astonishing changes in toxicity; for example, octafluorocyclobutane, $\begin{array}{l}CF_2\text{—}CF_2\\ |\quad\ \ \ |\\ CF_2\text{—}CF_2\end{array}$, is so inert that it is safely used as a propellant for food stuffs in aerosol cans and as an 80% mixture with 20% oxygen no effects on rats are detected. (At 10% concentration, there are no effects from chronic inhalation). However, hexafluorocyclobutene, $\begin{array}{l}CF_2\text{—}CF\\ |\quad\ \ \ \|\\ CF_2\text{—}CF\end{array}$, is almost as toxic as phosgene with $L(ct)_{50}$ of 630 to

[16] *Chem. Eng. News,* **46,** 39 (Jan. 15, 1968).

1300 ppm/minute and should be handled only in a well ventilated area. Unfortunately, the close similarity in names can lead to confusion.[17]

Numerous reviews[18–22] that discuss toxicity problems of certain classes of fluorinated compounds in great depth are available. In particular, the highly toxic phosphofluoridate derivatives[21] and the ω-fluoro acids (and related derivatives that degrade *in vivo* to monofluorocitric or monofluoroacetic acid)[20] are discussed extensively; these are special classes of fluorine compounds that will probably not be encountered by the average fluorine chemist but the reviews[18–23] give background material needed to develop a good safety attitude.

A number of fluorine compounds, both common and uncommon, are listed with toxicity data in Table 9-1 and compared to a few well-known laboratory reagents. The toxicity of most fluorine chemicals used in the laboratory has not been evaluated and cannot be safely estimated from comparison to related materials.[24] Thus, we strongly advise: *Treat all fluorine compounds as if they were highly toxic* (another nerve gas or phosgene) until properly examined by a competent toxicologist. Some materials are insidious in that they have no appreciable odor and no effects are felt until hours later; perfluoroisobutylene is an example of such a hazardous chemical that is odorless and appears innocuous until edema of the lungs develops several hours after exposure.

In Section 9-1 the hazards of skin burns from fluorine chemicals such as hydrogen fluoride were noted, but should be reemphasized here. Also, absorption of toxic fluorine chemicals through the skin can occur easily (facilitated by the high solubility of organic fluorine compounds in lipids) so that proper covering of hands and arms when handling fluorine chemicals is a good rule. Another good general rule is that any compound that readily hydrolyzes to produce HF will have a high inhalation toxicity (note toxicity of HF in Table 9-1).

[17] *Chem. Eng. News,* **45,** 44 (Oct. 16, 1967); *ibid.,* p. 8 Nov. 20 and p. 8 Dec. 4.

[18] H. C. Hodge, F. A. Smith, and P. S. Chen, in *Fluorine Chemistry* (J. H. Simons, ed.), Vol. 3, Academic, New York, 1963.

[19] H. C. Hodge and F. A. Smith, in *Fluorine Chemistry* (J. H. Simons, ed.), Vol. 4, Academic, New York, 1965.

[20] F. L. M. Pattison, *Toxic Aliphatic Fluorine Compounds,* Elsevier, Amsterdam and New York, 1959.

[21] B. C. Saunders, *Some Aspects of the Chemistry and Toxic Action of Organic Compounds Containing Phosphorus and Fluorine,* Cambridge Univ. Press, London and New York, 1957.

[22] "Pharmacology of Fluorides," *Handbook of Experimental Pharmacology,* Vol. 20, Part 1, Springer, Berlin and New York, 1966. A collection of ten chapters contributed by experts.

[23] J. W. Clayton, Jr., *Fluorine Chem. Rev.,* **1,** 197 (1967).

[24] Dr. E. C. Stump, Peninsular Chemresearch, Inc., Box 14318, Industrial Research Campus, Gainesville, Florida 32601, and Dr. J. Wesley Clayton, Jr., Haskell Laboratory for Toxicology and Industrial Medicine, E. I. du Pont de Nemours and Co., Wilmington, Delaware 19898, are compiling data on toxicity effects of fluorinated compounds and ask to be informed of any toxicity studies or observations of unusual toxic effects.

Table 9-1

Toxicity of Fluorinated Compounds

Compound	Inhalation toxicity [a]	Comments	Ref.
Alkanes			
CF_4	TLV (1000)	Fluorocarbons uniformly show low toxicity, whereas chlorocarbons are usually highly toxic. $CBrF_3$ is a useful fire-extinguishing agent. Fluorocarbons have been safely used for many years as refrigerants or food propellants	23,25
$CBrF_3$	ALC 83.2% vol. (15 min)		
CCl_3F (F-11)	TLV 1000		
CCl_4	TLV 10 ALC 2.9% vol. (15 min)		
CCl_2FCClF_2 (F-113)	ALC 10% vol.		
$CClF_2CClF_2$ (F-114)	ALC $>$20% vol.		
$CClF_2CF_3$ (F-115)	ALC $>$80% vol.		
CF_3CF_3 (F-116)	ALC $>$80% vol.		
Octofluorocyclobutane (Freon C-318)	ALC $>$80% vol.		
Alkenes			
$CF_2{=}CF_2$	ALC 40,000 ppm	Note unusually high toxicity of perfluoroisobutylene (approximately 10 times more than phosgene). Toxicity of alkenes generally increases with number of chlorines	23,25
$CF_2{=}CFCl$	LC_{50} 1000 ppm		
$CCl_2{=}CCl_2$	ALC 4000 ppm		
$CF_3CF{=}CF_2$	LC_{50} 3000 ppm		
$(CF_3)_2C{=}CF_2$ (PFIB)	ALC 0.5 ppm LCT_{50} 3 ppm/hr $L(ct)_{50}$ 180 ppm/min		(see also Ref. 17)
hexafluorocyclobutene	$L(ct)_{50}$ 630-1300 ppm/min	More toxic than phosgene	17
$CF_3CCl{=}CClCF_3$	ALC 100 ppm		23
$CF_3CCl{=}CHCF_3$	LC_{50} $\sim$3 ppm	Almost as toxic as PFIB	27
Aromatic			
C_6H_5F	Not highly toxic	Oral toxicity studies show toxicity like benzene. Partly metabolized, partly eliminated in expired air	25
C_6H_6	TLV 25 ppm		
C_6F_6	None reported		
Oxygen derivatives			
COF_2	ALC 100 ppm	Hydrolyzed to HF (toxicity of HF)	25
$COCl_2$	LCT_{50} 25-40 ppm/hr TLV 1 ppm		25,26
CF_3CO_2H	TLV 2.5 mg/m^3 air	Listed under general fluoride toxicity	26
CH_2FCO_2H	Very high	Extensive studies of oral toxicity (LD_{50} 6.6 mg/kg)	20
$(CF_3)_2C{=}O$ (or hydrate)	LC_{50} 300 ppm	Caused permanent eye damage to rabbit	28
$(CF_3)_2CHOH$	ALC 3200 ppm	Caused permanent eye damage to rabbit	28
$H(CF_2CF_2)_nCH_2OH$	ALC 2000-2500 ppm		25
CF_3OOCF_3	$L(ct)_{50}$ 3200-5000 ppm/min	Estimated from limited study	29
Sulfur-containing fluorine compounds			
CF_3SCl	$L(ct)_{50}$ 440-880 ppm/min		17
CF_3SSCF_3	$L(ct)_{50}$ 200 ppm/min	Almost as toxic as PFIB; causes death by breaking down lung tissue	17
SF_4	Approximately same as phosgene	Hydrolyzes instantly; toxicity of HF product	30

$C_6H_5SF_3$	Less toxic than SF_4 or phosgene	Hydrolyzed to HF (HF toxicity)	31
SF_6	TLV 1000 ppm in air ALC > 80% vol.	Replaced nitrogen in air for extended periods with no apparent effect on test animals. Suggested to be nontoxic because of extreme stability and inertness to hydrolysis.	26,32
S_2F_{10}	TLV 0.01 ppm LCT 1 ppm/18 hours	Approximately 100 times more toxic than phosgene; considered as a potential war gas	32
Fluorinating agents			
HF	ALC 200 ppm, TLV 3 ppm	Also causes extremely bad skin burns	22
F_2	TLV 0.1 ppm in air	Highly dangerous because reactive with all organic materials	25
OF_2	$L(ct)_{50}$ 100 ppm/min	Similar to ozone but more toxic; more toxic than F_2 or HF	33
$FClO_3$	LD_{50} 2000-4000 ppm	Technical information claims no unusual toxicity problems	34
NF_3	ALC 2500 ppm	Toxicity effects similar to nitrogen oxides (but not as toxic). Noted to be insidious because no odor detectable at dangerous concentrations	
BF_3	TLV 0.3 ppm	Produces severe pulmonary edema	36

[a] Direct comparative data are not available for series. The toxicity (in parts per million, unless indicated otherwise) is quoted with an indication of time. Toxicity data are usually from studies on mice or rats.

TLV–Threshold limit value (in parts per million, level at which workers can be exposed day after day without harmful effects.

ALC–Approximate lethal concentration for 4-hr exposure unless indicated otherwise.

LC_{50}–Lethal concentration for 50% of animals, 4-hr exposure.

LCT_{50}–Lethal concentration for 50% of animals per time unit.

$L(ct)_{50}$–A lethal toxicity index, the product of the concentration of toxin in parts per million times the number of minutes required to cause death in 50% of the laboratory animals tested.

LD_{50}–Lethal dose (usually oral or injected) to kill 50% of animals.

---

[25] Ref. 22, Chapter 9.

[26] N. I. Sax, *Dangerous Properties of Industrial Materials,* 2nd ed., Reinhold, New York, 1963.

[27] *Chem. Eng. News,* **44,** 6, 8 (Nov. 28, 1966).

[28] C. G. Krespan and W. J. Middleton, *Fluorine Chem. Rev.,* **1,** 145 (1967) (see p. 191).

[29] Private communication from Dr. David Lester, Center of Alcohol Studies, Rutgers, The State University, New Brunswick, New Jersey 08903. The toxicology studies were carried out on compound supplied by and under grant-in-aid from the Industrial Chemicals Division of Allied Chemical Corporation.

[30] C. W. Tullock, F. S. Fawcett, W. C. Smith, and D. D. Coffman, *J. Am. Chem. Soc.,* **82,** 539 (1960).

[31] W. A. Sheppard, *J. Am. Chem. Soc.,* **84,** 3058 (1962).

[32] (a) D. Lester and L. A. Greenberg, *Arch. Ind. Hyg. Occup. Med.,* **2,** 348 (1950). (b) L. A. Greenberg and D. Lester, *ibid.,* 350. (c) G. Kimmerle, *Arch. Tox.,* **18,** 140 (1960).

[33] D. Lester and W. R. Adams, *Am. Ind. Hyg. Assn. J.,* **26,** 562 (1965).

[34] *Perchloryl Fluoride,* dsA-1819 and dsF-1819, Pennsalt Chemicals Corp., Philadelphia, Pennsylvania, 19157.

[35] (a) R. L. Jarry and H. C. Miller, *J. Phys. Chem.,* **60,** 1412 (1956). (b) T. R. Torkelson, F. Oyen, S. E. Sadek, and V. K. Rowe, *Tox. Appl. Pharmacol.,* **4,** 770 (1962).

[36] T. R. Torkelson, S. E. Sadek, and V. K. Rowe, *Am. Ind. Hyg. Assoc. J.,* **22,** 263 (1961).

9-3 BIOLOGICAL APPLICATIONS

Fluorinated compounds have become extremely important in a variety of biological applications: medicinal compounds (particularly steroids); agricultural chemicals; fluorinated derivatives of natural products (which are useful as tracers or inhibitors in biochemical studies and include fluorinated nucleic bases, carbohydrates, amino acids, and fatty acids); and miscellaneous uses (such as catalysts and temporary blocking groups in synthesis, or analytical purposes such as in amino acid separation).

Since our review (Chapters 5 to 7) of synthetic methods for introducing fluorine should be particularly valuable to chemists interested in using fluorinated compounds in biological problems, we want to summarize briefly the current status of this area. Again this is a topic which, to be treated exhaustively, requires at least a volume of its own, so that we rely heavily on available reviews and only select examples to illustrate major areas of interest with summarizing tables to cover recent significant studies.

9-3A. MEDICINAL CHEMICALS

The major impact of fluorine chemistry on drugs is in the area of steroids. A total of seventeen fluorinated steroids were commercially available or under active clinical investigation in 1964, chiefly as progestogen and antifertility agents, and corticoids (trade names and principal producers also given).[37] Eighteen fluorinated steroids (listed in Table 9-2) are currently (January 1968) marketed in the United States.[38] Methods for introducing fluorine into steroids were reviewed,[39] and the pharmacological aspects of the fluorine substituents in cortical hormones are discussed by Fried and Borman.[40] A recent review summarizes both synthetic approaches and physiological activity of monofluorosteroids.[41]

Buu-Hoi[42] has provided a good comprehensive review of fluorinated compounds that have been of interest in pharmacological studies under the topics: antimetabolites; diuretics; anesthetics; convulsants; anticonvulsants, and muscle relaxants; steroids; central nervous system drugs; antihistamine and analgesics; anthelmintic, bactericides, antiviral, and fungicides; and miscellaneous. The effects of halogenation on pharmacological properties and mechanism of

[37]N. Applewig, *Steroid Drugs,* **2,** 444 (1964).

[38]We thank Dr. G. A. Boswell of the du Pont Central Research Department for providing us with this information.

[39](a) J. W. Chamberlain, in *Steroid Reactions* (C. Djerassi, ed.), Holden-Day, San Francisco, 1963, Chapter 3. (b) A. A. Akhren, I. G. Reshetova, and Yu. A. Titov, *Russian Chem. Rev. (English Transl. Uspekhi Khim.),* **34,** 926 (1965).

[40]J. Fried and A. Borman, *Vitamins and Hormones,* **16,** 303 (1958).

[41]N. F. Taylor and P. W. Kent, *Advan. Fluorine Chem.,* **4,** 113 (1965).

[42]N. P. Buu-Hoi, *Drug Res.,* **3,** 9 (1961).

action have been reviewed (with a special section on fluorine).[43] Fluorinated anesthetics were definitively reviewed by Krantz and Rude.[44] One further important use to be noted is fluoride ion in preventing tooth decay.[45]

Because of the extensive coverage by reviews in this area, no further tabulation of medicinal fluorinated compounds will be made. Numerous patents continue to appear claiming useful biological activity of fluorinated compounds. Specific and selective introduction of fluorine or fluorine substituents should continue to be of interest particularly because of beneficial changes in properties of molecules:

1. higher fat solubility giving different *in vivo* absorption and transport rate;
2. altered electronic effects;
3. improved stability;
4. approximately same steric size as hydrogen.

On the other hand, some fluorinated compounds are relatively reactive so that fluoride or hydrogen fluoride could be liberated *in vivo* making the compound highly toxic.

9-3B. AGRICULTURAL CHEMICALS

The major use of fluorinated compounds in agriculture is in the area of killing pests or controlling disease. Fluoroacetic acid (and derivatives) is a common rat poison.[20] The effects of fluoride on plants and animals have been reviewed.[46] Inorganic fluorides have long been used as insecticides and probably have as a common mechanism of action the inhibitory effect of fluoride ion on magnesium-containing enzymes. A large variety of organofluorochemicals have been studied and used as pesticides[47] including fluoroacetic acid derivatives, fluorophosphorus derivatives, sulfur fluorine compounds,[47b] and DDT analogs. These compounds are best classified by mode of action.

1. Blocking of vital enzyme reaction: fluorine is substituted for a hydrogen in a molecule where the hydrogen is involved in enzymatic transfer and irreversibly complexes and blocks the enzyme from normal function; fluoroacetic acid and derivatives that degrade to fluoroacetic acid *in vivo* are the best examples.

2. Reactive organic fluorides, which couple by a fluoride displacement to functional groups at the active site of the enzyme, block vital function. The fluorophosphorus derivatives are the best example in this category.

[43] M. B. Chenoweth and L. P. McCarty, *Pharmacol. Rev.*, **15**, 673 (1963).

[44] Ref. 22, Chapter 10, p. 501.

[45] (a) Ref. 22, Chapter 4, p. 173. (b) Ref. 6, Chapter 2.

[46] Ref. 22, Chapters 5 and 6, pp. 231 and 301.

[47] (a) Ref. 22, Chapter 7, p. 355. (b) E. Kuhle, E. Klawke, and F. Grewe, *Angew. Chem.*, **76**, 807 (1964) review use of trihalomethanesulfenyl derivatives as pesticides.

Table 9-2

Fluorinated Steroids (Glucocorticoids) Marketed in the United States

Structure	Generic name	Chemical name	Trade name	Company
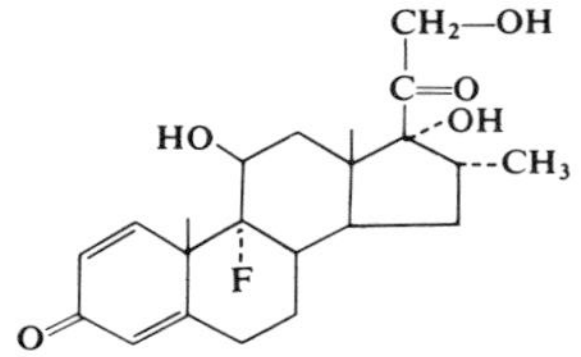	Dexamethasone	9α-Fluoro-16α-methyl-11β, 17α, 21-trihydroxypregna-1,4-diene-3,20-dione	Decardon Deronil	Merck Schering
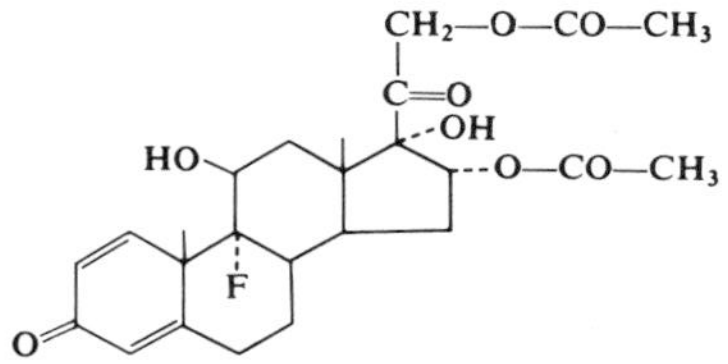	Triamcinolone diacetate	9α-Fluoro-11β,16α,17α,21-tetrahydroxypregna-1,4-diene-3,20-dione 16α,21-diacetate	Aristocort Kenacort	Lederle Squibb
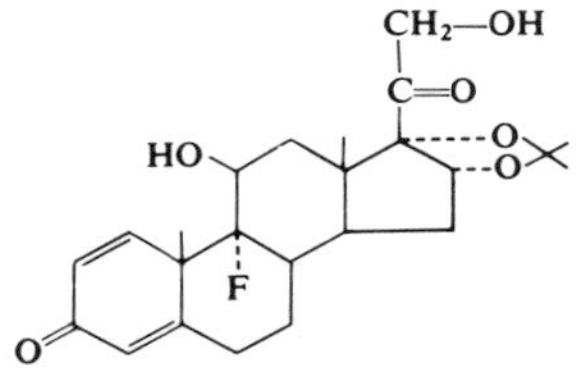	Triamcinolone acetonide	9α-Fluoro-11β,16α,17α, 21-tetrahydroxypregna-1,4-diene-3,20-dione 16,17-acetonide	Aristoderm Kenalog	Lederle Squibb
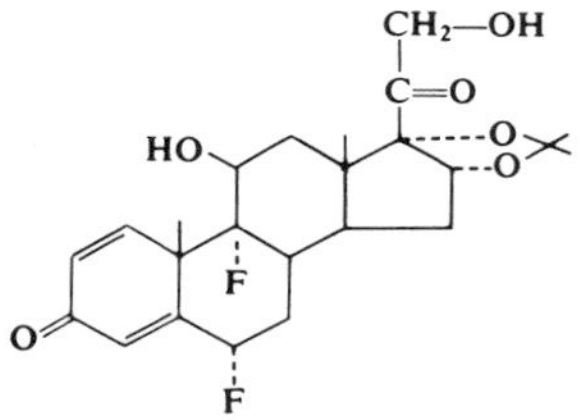	Fluocinolone acetonide	6α,9α-Difluoro-11β,16α, 17α,21-tetrahydroxypregna-1,4-diene-3,20-dione 16,17-acetonide	Synalar	Syntex

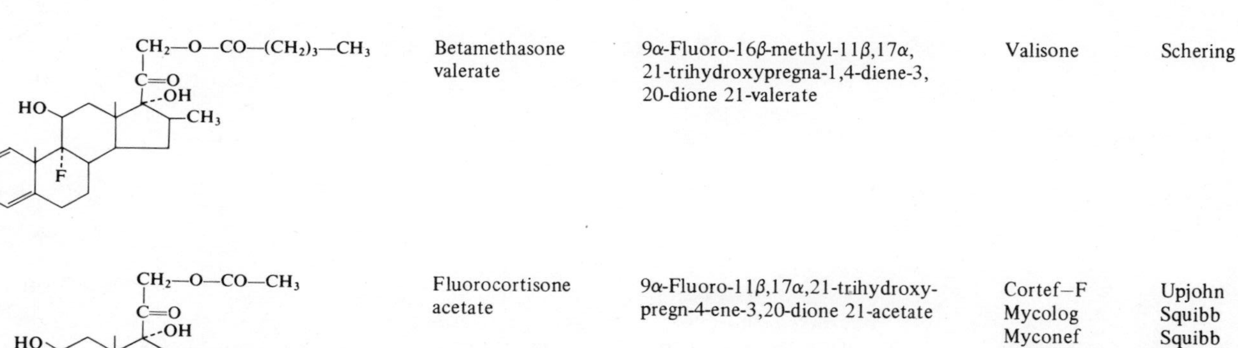

Name	Chemical name	Trade name	Manufacturer
Flurandrenolone acetonide	6α-Fluoro-11β,16α,17α,21-tetrahydroxypregn-4-ene-3,20-dione 16,17-acetonide	Cordran	Lilly
Betamethasone	9α-Fluoro-16β-methyl-11β,17α,21-trihydroxypregna-1,4-diene-3,20-dione	Celestone	Schering
Betamethasone valerate	9α-Fluoro-16β-methyl-11β,17α,21-trihydroxypregna-1,4-diene-3,20-dione 21-valerate	Valisone	Schering
Fluorocortisone acetate	9α-Fluoro-11β,17α,21-trihydroxypregn-4-ene-3,20-dione 21-acetate	Cortef–F Mycolog Myconef Florinef	Upjohn Squibb Squibb Squibb

Table 9-2 (cont.)

Structure	Generic name	Chemical name	Trade name	Company
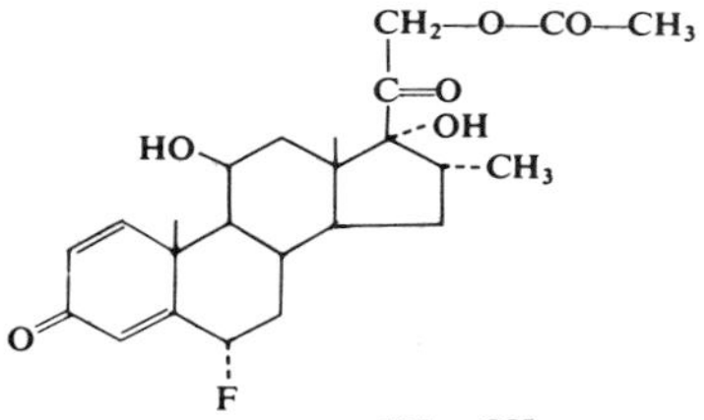 $CH_2—O—CO—CH_3$; $C=O$; OH; CH_3; HO; O; F	Paramethasone acetate	6α-Fluoro-16α-methyl-11β, 17α,21-trihydroxypregna-1,4-diene-3,20-dione 21-acetate	Haldrone	Lilly
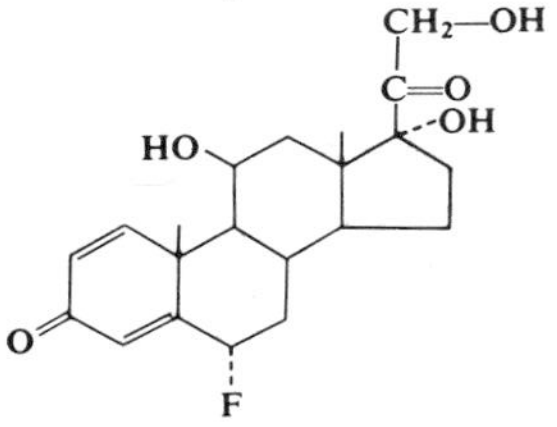 $CH_2—OH$; $C=O$; OH; HO; O; F	Fluprednisolone	6α-Fluoro-11β,17α,21-trihydroxy-pregna-1,4-diene-3,20-dione	Alphadrol	Upjohn
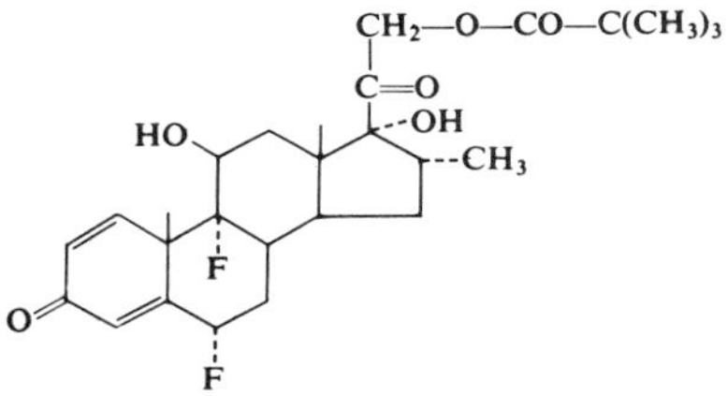 $CH_2—O—CO—C(CH_3)_3$; $C=O$; OH; CH_3; HO; F; O; F	Flumethasone pivalate	6α,9α-Difluoro-16α-methyl-11β, 17α,21-trihydroxypregna-1,4-diene-3,20-dione 21-pivalate	Locacorten	Ciba
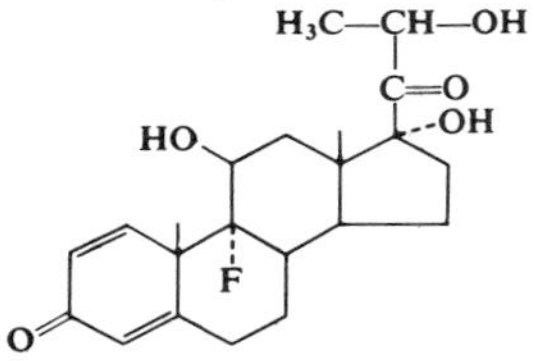 $H_3C—CH—OH$; $C=O$; OH; HO; F; O	Fluperolone acetate	9α-Fluoro-21-methyl-11β,17α, 21-trihydroxypregna-1,4-diene-3,20-dione	Methral	Pfizer

Name	Chemical name	Trade name	Company
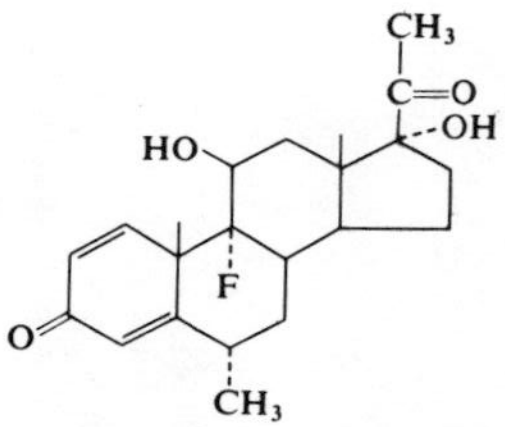 Fluorometholone	6α-Methyl-9α-fluoro-11β,17α-dihydroxypregna-1,4-diene-3,20-dione	Oxylone	Upjohn
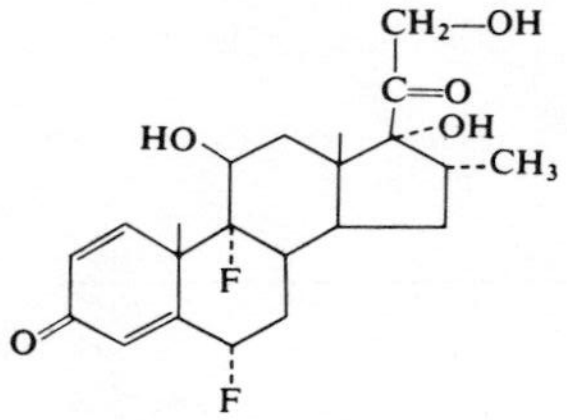 Flumethasone	6α,9α-Difluoro-16α-methyl-11β,17α,21-trihydroxypregna-1,4-diene-3,20-dione		Upjohn Syntex
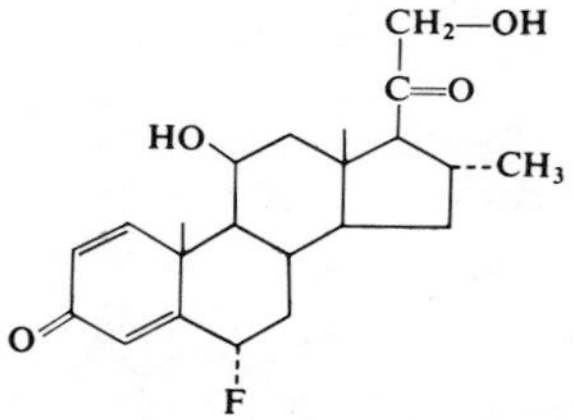 Fluocortolone	6α-Fluoro-11β,21-dihydroxy-16α-methylpregna-1,4-diene-3,20-dione		Berlin
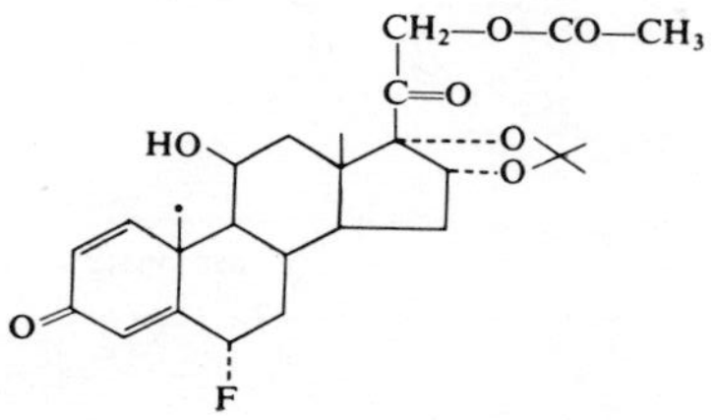 Fluoxolonate	6α-Fluoro-11β,16α,17α,21-tetrahydroxypregna-1,4-diene-3,20-dione 16,17-acetonide 21-acetate		Lilly

Table 9-2 (cont.)

Structure	Generic name	Chemical name	Trade name	Company
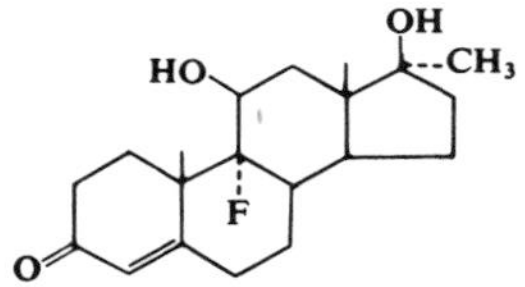	Fluoxymesterone	9α-Fluoro-11β-hydroxy-17α-methyltestosterone	Halotestin Ora-Testryl Ultandren	Upjohn Squibb Ciba
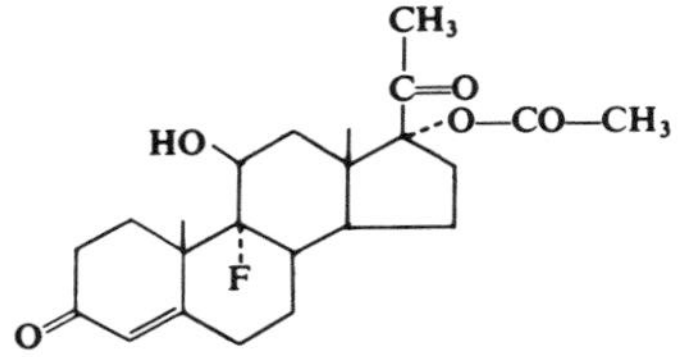	Fluorogestone acetate	9α-Fluoro-11β,17α-dihydroxypregn-4-ene-3,20-dione 17-acetate	Syncro-mate	Searle

3. Fat-solubilizing function to get an aromatic or other toxic group into the system; fluorinated DDT's are in this category.

A major contribution of fluorinated compounds has been in studies of insect biochemistry and mechanism of detoxification and of development of resistance to insecticides.

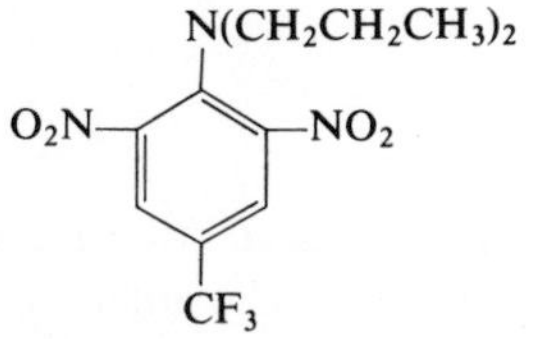

$N(CH_2CH_2CH_3)_2$

O_2N — — NO_2

CF_3

1

A benzotrifluoride derivative (**1**), trade name Treflan, is sold by Elanco Products Company, a Division of Eli Lilly and Company, for herbal uses.[48] Also, hexafluoroacetone hydrate is marketed as an herbicide.[28]

9-3C. FLUORINATED DERIVATIVES OF NATURAL PRODUCTS

This area is conveniently organized into normal biological classifications of amino acids, sugars, fatty acids, and nucleic bases. These fluorinated analogs are often antimetabolites and as such are also medicinal chemicals.[42] Steroids, another class of natural products, are reviewed under medicinal chemicals because the major interest for fluorinated analogs is as drugs.[39–42]

1. Amino Acids

A number of fluorinated analogs of amino acids have been prepared and are reviewed.[42, 49] Most of these have low value in biological systems.[50] However, often they have only been studied superficially by observing external growth effects on a few classes of microorganisms. But we conclude that even very minor structural changes, which result from replacement of hydrogen by fluorine of only slightly larger steric size in a part of the molecule remote from the reactive functional groups, are sufficient to limit greatly the ability of the fluorinated amino acids to function in place of natural ones at molecular level in biochemical

[48] *Weed Society of America Herbicide Handbook,* Ed. 1, W. F. Humphrey Press, Geneva, New York, 1967, pp. 281-284.

[49] W. Shive and C. G. Skinner, in *Metabolic Inhibitors* (R. M. Hochster and J. H. Quastel, eds.), Vol. 1, Academic, New York, 1963, pp. 2-73.

[50] (a) J. Lazar and W. A. Sheppard, *J. Med. Chem.,* **11**, 138 (1968). (b) M. Hudlicky, *154th Meeting Am. Chem. Soc., Chicago, Illinois, September 1967,* Abstracts of papers, K003. (c) R. Filler and H. H. Kang, *Chem. Commun.,* 626 (1965). (d) H. Lettré and U. Wölcke, *Ann.,* **708**, 75 (1967).

processes. Actual incorporation of fluorinated amino acid analogs into proteins is limited to *p*-fluorophenylalanine[51] and (4-trifluoromethyl)-2-aminopentanoic acid (trifluoroeucine).[52] Substitution of fluorinated analogs of amino acids for natural ones in enzymatically active proteins is intriguing because the fluorinated analog could alter activity but of more importance would be as a useful marker and probe in elucidation of the structure of protein and in studying the mechanism of enzyme action.

2. Sugars

Fluorine can be easily introduced into sugars particularly by hydrogen fluoride ring opening–addition to cyclic ethers and F_2 or XF addition to unsaturated systems (see Sections 5-1 and 5-2). The synthesis and chemistry of several classes of fluorinated sugars have been reviewed.[53] However, the fluorinated carbohydrates do not appear to have any unusual or interesting biological activity.

3. Fatty Acids

The area of monofluoroaliphatic compounds, which primarily involves monofluorocarboxylic acid derivatives, is exhaustively reviewed by Pattison.[20, 54] Extensive substitution of fluorine completely changes the properties of carboxylic acids so that they cannot replace naturally occurring materials in biological systems. However, monofluoro substitution produces remarkable effects. Fluoroacetic acid and all ω-fluorocarboxylic acids $F(CH_2)_nCO_2H$, where n is an odd number, are extremely toxic, but acids, where n is even, are relatively inocuous. Fluoroacetic acid (or derivatives) is converted *in vivo* to fluorocitric acid which blocks the fundamental citric acid and tricarboxylic acid cycles by irreversibly complexing with certain enzymes that carry out hydrogen transfer. Any monofluoroaliphatic compounds that metabolize to fluoroacetic or fluorocitric acids are also highly toxic.

Some toxic ω-fluorocarboxylic acids occur in nature in a species of South African and Australian plants[20, 54]; needless to say, the biological origin of these fluorinated materials is of considerable interest.

4. Nucleic Bases

Fluorinated nitrogen heterocyclic derivatives are of major importance in cancer treatment and research. As antimetabolites in nucleic acid synthesis, they

[51] (a) R. Munier and G. N. Cohen, *Biochim. Biophys. Acta,* **21**, 347, 378 (1959). (b) W. L. Fangman and F. C. Neidhardt, *J. Biol. Chem.*, **239**, 1839 and 1844 (1964).

[52] O. M. Rennert and H. S. Anker, *Biochemistry,* **2**, 471 (1963).

[53] (a) F. Michel and A. Klemer, *Adv. Carbohydrate Chem.*, **16**, 85 (1961). (b) J. E. G. Barnett, *Adv. Carbohydrate Chem.*, **22**, 177 (1967).

[54] Ref. 22, Chapter 8, p. 387.

provide the basis for a wide variety of biological and biochemical studies. Fluorinated pyrimidines have been reviewed by Heidelberger,[55] a leading worker in the field; this review briefly covers synthesis but mainly discusses biological aspects (see also Section 5-4B.1 for use of SF_4 in synthesis of fluorinated purines and pyrimidines). The compound of major interest is 5-fluorouracil which can replace uracil in nucleic acids and grossly alter the biochemical operation of living cells.

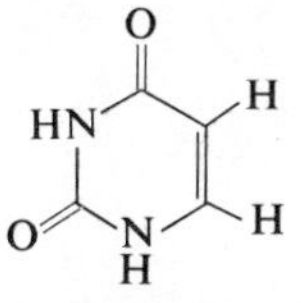

uracil

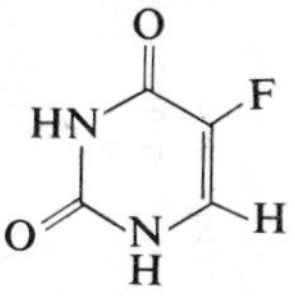

5-fluorouracil

9-3D. MISCELLANEOUS APPLICATIONS

A variety of other applications of fluorine chemistry or fluorinated reagents to biological problems have been reported. One important area is use of trifluoroacetic acid or anhydride as solvents, catalysts, or reagents in syntheses where fluorine does not appear in the final product.[56–59] (For a discussion of use in peptide synthesis and amino acid analysis, see Section 6-2D.3 and Table 6-8.)

Several recent examples of use of fluorinated reagents are:

1. cleavage of peptides with anhydrous HF or trifluoroacetic acid,[60]

2. removal of peptide protecting groups with HF,[61]

3. use of trifluoroacetic acid and hydrates of fluoroketones as solvents for proteins (powerful hydrogen-bonding properties of solvent breaks helix), particularly for nmr studies,[62]

4. trifluoroacetyl as a protecting group during reactions of sugars,[63]

5. 2,4-dinitro-1-fluorobenzene, phenylmethanesulfonyl fluoride and alkyl phosphorofluoridates react with enzymes to provide blocking agents for studies of enzyme action or in degradation of proteins.[21, 64]

[55] C. Heidelberger, *Progr. Nucleic Acid Res. Mol. Biol.*, **4**, 1 (1965).

[56] T. G. Bonner, *Advan. Carbohydrate Chem.*, **16**, 59 (1961).

[57] F. Weygand, *Bull. Soc. Chim. Biol.*, **43**, 1269 (1961).

[58] M. Hudlicky, *Chemistry of Organic Fluorine Compounds*, Macmillan, New York, 1962, Chapter 7.

[59] J. M. Tedder, *Chem. Rev.*, **55**, 787 (1955).

[60] (a) S. Sakakibara, K. H. Shin, and G. P. Hess, *J. Am. Chem. Soc.*, **84**, 4921 (1962). (b) K. D. Kopple and E. Bachli, *J. Org. Chem.*, **24**, 2053 (1959).

[61] (a) S. Sakakibara, Y. Shimonishi, Y. Kishida, M. Okada, and H. Sugihara, *Bull. Chem. Soc. Japan (English Transl.)*, **40**, 2164 (1967). (b) J. Lenard, *J. Org. Chem.*, **32**, 250 (1967). (c) J. Lenard and A. B. Robinson, *J. Am. Chem. Soc.*, **89**, 181 (1967).

[62] (a) R. Longworth, *Nature*, **203**, 295 (1964). (b) J. A. Ferretti, *Chem. Commun.*, 1030 (1967).

[63] H. Newman, *J. Org. Chem.*, **30**, 1287 (1965).

[64] (a) A. M. Gold and D. Fahrney, *Biochemistry*, **3**, 783-91 (1964). (b) B. R. Baker, *Design of Active-Site-Directed Irreversible Enzyme Inhibitors*, Wiley, New York, 1967.

Appendix A

Recommended General References on Fluorine Chemistry

Ref. no.	Title of book, author, publisher	Chapter title, author	No. of pages (no. of ref.)	Comments
A1	*Chemistry of Organic Fluorine Compounds,* M. Hudlicky, Macmillan, New York, 1962.		536 (1105)	Best general text for the organic chemist desiring an introduction to organic fluorine chemistry. Has become outdated with references only through 1959. Inclusion of experimental procedures enhances the usefulness for teaching. The final one-third of the book contains a list of references (1105), author index, subject index, and an extremely useful 98 page formula index
		1. Introduction	6	
		2. Apparatus and Material	11	
		3. Fluorinating Reagents	27	
		4. Methods for Introducing Fluorine Into Organic Compounds	58	
		5. Preparation of Organic Compounds of Fluorine	36	
		6. Reactions of Organic Fluorine Compounds	119	
		7. Fluorinated Compounds as Chemical Reagents	9	
		8. Properties of Organic Fluorine Compounds	26	
		9. Analysis of Organic Fluorides	17	
		10. Practical Applications of Organic Fluorine Compounds	24	
A2	*Methoden der Organischen Chemie,* Vol. 5, Part 3 (Houben-Weyl), Halogenverbindungen-Fluor und Chlor, Georg Thieme Verlag, Stuttgart, 1962		502	Outstanding volume in German. This is the authoritative single most useful treatise on organic fluorine chemistry. The volume is comprehensive and completely referenced through 1960. The high price ($70.00) discourages individual ownership. For the organic chemist expecting to do more than casual work in fluorine chemistry, familiarization with Houben-Weyl is a must
		Part A. Preparation of Fluorine Compounds	397	
		Part B. Reactivity and Conversions of Fluorine Compounds	105	

A3-1	*Fluorine Chemistry*, Vol. I, J. H. Simons Ed., Academic, New York, 1950		615 (1787)	Taken with Volume II provides a complete review of fluorine chemistry through 1950. Although some chapters are obsolete, many chapters are still exceedingly valuable, mainly because of the comprehensive tables provided
		1. Nonvolatile Inorganic Fluorides, H. J. Emeléus	76 (219)	Eighteen valuable tables. Chapter retains usefulness as an initial reference source.
		2. Volatile Inorganic Fluorides, Anton B. Burg	47 (242)	Complements Chapter 1. Same general organization
		3. The Chemistry of the Fluoro Acids of Fourth, Fifth and Sixth Group Elements, Willy Lange	64 (183)	Description of organic esters of inorganic acids is of interest to organic chemists.
		4. The Halogen Fluorides, H. S. Booth and J. T. Pinkston, Jr.	12 (35)	Made obsolete by later reviews. Early references of value
		5. Boron Trifluoride, H. S. Booth and D. R. Martin	24 (108)	Made obsolete by later reviews. Early references of value
		6. Hydrogen Fluoride, J. H. Simons	36 (137)	Nine valuable tables. Tables of solubilities of salts in HF are of particular value
		7. Hydrogen Fluoride Catalysis, J. H. Simons	32 (84)	Good introduction to field. Should be read with Chapter 6
		8. Preparation of Fluorine, G. H. Cady	22 (36)	Better reviews are now available (see B7a-1)
		9. Physical Properties of Fluorine, G.H. Cady and L. L. Burger	4 (9)	Three tables summarize properties of fluorine. Very convenient reference
		10. The Theoretical Aspects of Fluorine Chemistry, G. Glockler	54 (166)	Extremely valuable chapter. Must reading for all chemists. Valuable tables
		11. The Action of Elementary Fluorine upon Organic Compounds, L. A. Bigelow	18 (41)	Now obsolete. A table of compounds of value for mp's and bp's
		12. Fluorocarbons and Their Production, J. H. Simons	22 (28)	Historical interest. Valuable tables on properties of perfluorocarbons

Appendix A (cont.)

Ref. no.	Title of book, author, publisher	Chapter, title, author	No. of pages (no. of ref.)	Comments
		13. Fluorocarbons – Their Properties and Wartime Development, T. J. Brice	40 (98)	Interesting background discussion. Nine pages of valuable tables giving bp, mp, d_4^{20}, n_D^{20}, method of preparation and references for 77 perfluorinated compounds
		14. Fluorocarbon Derivatives, W. H. Pearlson	60 (222)	Made obsolete by later reviews. Three valuable tables list properties of 455 fluorinated compounds
		15. Aliphatic Chlorofluoro Compounds, J. D. Park	30 (141)	Made obsolete by later reviews
		16. Fluorine Compounds in Glass Technology and Ceramics, W. A. Weyl	22 (38)	Unique topic not of interest to organic chemists
A3-2	*Fluorine Chemistry,* Vol. II, J. H. Simons, Ed., Academic, New York, 1954.		565 (2652)	Supplements Volume I. Literature covered through 1952
		1. Fluorine Containing Complex Salts and Acids, A. G. Sharpe	38 (182)	Supplements Chapters 1-3 of Volume I. General background material for an organic chemist
		2. Halogen Fluorides – Recent Advances, H. J. Emeléus	12 (55)	Supplements Chapter 5 of Volume I. Later reviews are available (Ref. B6j-2). Some valuable discussion provided which is not duplicated by later authors
		3. Analytical Chemistry of Fluorine and Fluorine-Containing Compounds, P. J. Elving, C. A. Horton, and H. H. Willard	162 (1369)	Obsolete techniques for modern organic fluorine analysis. Authoritative article on classical methods
		4. Organic Compounds Containing Fluorine, P. Tarrant	108 (556)	Still valuable because of discussion and 37 pages of tables listing approximately 1460 fluorinated organic compounds

		5. Metallic Compounds Containing Fluorocarbon Radicals and Organometallic Compounds Containing Fluorine, H. J. Emeléus	12 (38)	Obsolete. Provides state-of-the-art through 1953 (See Refs. B51-1 to 4)
		6. Fluorocarbon Chemistry, J. H. Simons	116 (445)	When taken with Chapters 13-15 of Volume I, a complete review of perfluorocarbon chemistry through August 1952 is provided. Physical properties of 1000 compounds are given in 16 pages of tables
		7. The Infrared Spectra of Fluorocarbons and related Compounds, D. G. Weiblen	55 (163)	Actual spectra of 150 fluorinated compounds gives this chapter lasting value. This chapter and Ref. A6-4-6 nicely complement each other
A3-3	*Fluorine Chemistry,* Vol. III, J. H. Simons, Ed., Academic, New York, 1963	A single long chapter entitled Biological Effects of Organic Fluorides (54 pages) followed by Tables (130 pages), H. C. Hodge, F. A. Smith, and P. S. Chen	240 (1623)	Authoritative discussions on fluoroacetates and phosphofluoridates followed by discussions of the effects of organofluorine compounds in animals. The tables are extensive and complete. Organic chemists should read this chapter to increase their safety consciousness
A3-4	*Fluorine Chemistry,* Vol. IV, J. H. Simons, Ed., Academic, New York, 1965		786 (3568)	Highly specialized text, of limited use to organic chemists
		1. Biological Properties of Inorganic Fluorides, H. C. Hodge and F. A. Smith	376 (-)	About 100 pages of text and 276 pages of tables. The organic chemist should skim read the material as a safety precaution. Data is useful for those embroiled in water fluoridation campaigns
		2. Effects of Fluorides on Bones and Teeth, H. C. Hodge and F. A. Smith	410 (-)	Of little interest to organic chemists. Extremely valuable material to assist in political water-fluoridation campaigns

Appendix A (cont.)

Ref. no.	Title of book, author, publisher	Chapter, title, author	No. of pages (no. of ref.)	Comments
A3-5	*Fluorine Chemistry,* Vol. V, J. H. Simons, Ed., Academic, New York, 1964		505 (1271)	Supplements Volumes I and II. Same general style
		1. General Chemistry of Fluorine-Containing Compounds, J. H. Simons	131 (819)	Supplements the general chapters of Volumes I and II covering the literature for 1952-1962. The organic chemist will not find too much of specific interest but is advised to skim read the chapter for familiarity. Pages 29-42 on carbon are of special interest because of interpretations by the author
		2. Physical Chemistry of Fluorocarbons, T. M. Reed, III	106 (183)	Outstanding chapter which all chemists should read. Of special value are the comparisons between properties of fluorocarbons and hydrocarbons
		3. Radiochemistry and Radiation Chemistry of Fluorine, J. A. Wethington, Jr.	58 (231)	Production and use of fluorine isotopes are covered. Fluorine-18 appears to be the only practical isotope, $t_{1/2}$ = 107-112 min
		4. Industrial and Utilitarian Aspects of Fluorine Chemistry, H. G. Bryce	211 (190)	Of particular value to industrial chemists. Economical and historical sections are followed by applications of fluorine chemistry to 15 areas such as leather, textiles, lubricants, refrigerants, etc.
A4	*Aliphatic Fluorine Compounds,* A. C. S. Monograph No. 138, A. M. Lovelace, D. A. Rausch, and W. Postelnek, Reinhold, New York, 1958	1. Fluorination; 2. Alkanes; 3. Alkenes and Alkynes: 4. Alcohols; 5. Ethers; 6. Ketones, Aldehydes and Acetals; 7. Carboxylic Acids; 8. Acyl Halides and Anhydrides; 9. Esters; 10. Nitrogen compounds I;	370 (1161)	Valuable text which should be owned. Complete coverage through 1955 with some 1956 and 1957 citations. The initial chapter is followed by chapters covering various functional groups as specified. Each chapter is divided into

<table>
<tr><td></td><td></td><td>11. Nitrogen Compounds II;
12. Organometallic and Organometalloidal Compounds; 13. Sulfur Compounds</td><td></td><td>three parts: Method of Synthesis, Table of Compounds, and References. The extensive and complete tables include empirical formula, structural formula, coded method of synthesis, yield, bp, mp, refractive index, density, and reference. This volume is probably the best starting point to find out how to synthesize a specific aliphatic fluorine compound. A total of 166 pages of tables</td></tr>
<tr><td>A5</td><td>Aromatic Fluorine Compounds, A. C. S. Monograph, A. E. Pavlath and A. J. Leffler, Reinhold, New York, 1962</td><td>I. Preparation of Aromatic Fluorine Compounds
II. Theoretical Consideration of Properties and Reactivity
III. Aromatic Fluorohydrocarbons
IV. Halogenated Aromatic Fluorohydrocarbons
V. Oxygen Containing Aromatic Fluorine Compounds
VI. Nitrogen Containing Aromatic Fluorine Compounds
VII. Sulfur, Silicon and other Metalloorganic Aromatic Compounds
VIII. Heterocyclic Compounds
IX. Applications of Fluoroaromatic Compounds</td><td>820 (2433)</td><td>Complements the Lovelace, Rausch, and Postelnek Monograph but is organized differently: not by functional group but by elemental substitution on fluorinated aromatic ring which makes it difficult to use. Chapters I and II present broad general discussions. The next six chapters describe preparation of the type compound followed by extensive tables (531 pages total of tables plus a 35 page formula index for specific use with tables). Chapter IX nicely supplements Chapter IV of Simons Volume V. This book is an excellent initial source for a literature search and belongs on the organic fluorine chemist's bookshelf. Literature covered through 1958</td></tr>
<tr><td>A6-1</td><td>Advances in Fluorine Chemistry, Vol. 1, M. Stacey, J. C. Tatlow and A. G. Sharpe, Ed., Academic, New York, 1960</td><td>1. The Halogen Fluorides – Their Preparation and Uses in Organic Chemistry, W. K. R. Musgrave</td><td>203 (970)
28 (85)</td><td>Supplements earlier reviews in Simons Volumes I and II (Refs. A-3-1 and A-3-2). Describes only the preparation of halogen fluorides and their reactions with organic compounds</td></tr>
</table>

Ref. no.	Title of book, author, publisher	Chapter, title, author	No. of pages (no. of ref.)	Comments
		2. Transition Metal Fluorides and Their Complexes, A. G. Sharpe	39 (251)	Good introduction to the subject
		3. Fluoroboric Acids and Their Derivatives, D. W. A. Sharp	61 (412)	The Balz-Schiemann reaction and other reactions catalyzed by BF_3 or HBF_4 make this chapter of value to the organic chemist
		4. The Electrochemical Process for the Synthesis of Fluoro-Organic Compounds, J. Burdon, and J. C. Tatlow	37 (85)	A Valuable and unique chapter. Information is not conveniently available elsewhere. Updates Simons articles (Refs. A3-1 and 2)
		5. Exhaustive Fluorinations of Organic Compounds with High Valency Metallic Fluorides, M. Stacey, and J. C. Tatlow	33 (137)	Excellent chapter presenting work originally developed for the Manhattan Project and recently used by the authors. Methods probably won't be used by average fluorine chemist, but has industrial importance (Refs. B1-b-2, B1-b-3).
A6-2	*Advances in Fluorine Chemistry*, Vol. 2; M. Stacey, J. C. Tatlow, and A. G. Sharpe, Ed., Butterworths, Washington, D. C., 1961		220 (675)	
		1. The Thermochemistry of Organic Fluorine Compounds, C. R. Patrick	34 (174)	Excellent chapter. Contains the basic material for understanding the difference between organic fluorine chemistry and organic chemistry. Thermodynamic and structural data summarized in 12 useful tables
		2. Fluorine Resources and Fluorine Utilization, G. C. Finger	20 (42)	Of little value to the research chemist. Useful background material for an economic report
		3. Mass Spectrometry of Fluorine Compounds, J. R. Majer	49 (61)	Good introduction to subject. Comparison of spectral and peculiarities of fluorine compounds are presented, 33 tables

		4.	The Fluorination of Organic Compounds using Elementary Fluorine, J. M. Tedder	34 (75)	This chapter taken with Chapter 1 gives the organic chemist and excellent perspective for comparing fluorine with the halogens
		5.	The Fluorides of the Actinide Elements, N. Hodge	45 (231)	Has little of interest to organic chemists
		6.	The Physiological Action of Organic Compounds Containing Fluorine, B. C. Saunders	29 (88)	Too brief. Subject treated better in Refs. A8 to A10
A6-3	*Advances in Fluorine Chemistry*, Vol. 3; M. Stacey, J. C. Tatlow, and A. G. Sharpe, Ed., Butterworths, Washington, D. C., 1963			281 (1315)	For the organic chemist, probably most valuable volume of the series
		1.	Effects of Adjacent Perfluoroalkyl Groups on Carbonyl Reactivity, H. P. Braendlin and E. T. McBee	18 (94)	A good, although narrow, dissertation on how and why a perfluoroalkyl group differs from an alkyl group. Describes effect of perfluoroalkyl group on remainder of molecule
		2.	Perfluoralkyl Derivatives of the Elements, H. C. Clark	44 (214)	Excellent survey of synthesis, properties, and reactions of subject compounds
		3.	Mechanisms of Fluorine Displacement, R. E. Parker	55 (66)	Very valuable contribution but unfortunately hard to read. Rate data excellently tabulated and analyzed. Must reading for the organic chemist
		4.	Nitrogen Fluorides and Their Inorganic Derivatives, C. B. Colburn	20 (66)	Short interesting review. More comprehensive reviews available. (See Refs. B5e-5, B6g-2)
		5.	The Organic Fluorochemicals Industry, J. M. Hamilton, Jr.	64 (173)	Title misleading. Mainly discusses fluorochloromethanes and ethanes (Freons) and fluorinated olefins. Compounds discussed comprise the bulk of manufactured organic fluorine compounds. Well worth reading
		6.	The Preparation of Organic Fluorine Compounds by Halogen Exchange, A. K. Barbour, L. J. Belf, and M. W. Buxton	90 (672)	Excellent discussion of the classical method for synthesizing organic fluorine compounds. Complete and as applicable today as when it was written. Must reading for the organic chemist

Ref. no.	Title of book, author, publisher		Chapter, title, author	No. of pages (no. of ref.)	Comments
A6-4	*Advances in Fluorine Chemistry*, Vol. 4, M. Stacey, J. C. Tatlow, and A. G. Sharpe, Ed., Butterworths, Washington, D. C., 1965			319 (2283)	
		1.	The Balz-Schiemann Reaction, H. Suschitzky	32 (139)	Supplements Pavlath and Leffler (Ref. A5). Sections on experimental procedure and scope and limitations are particularly valuable
		2.	Some Techniques and Methods of Inorganic Fluorine Chemistry, R. D. Peacock	18 (107)	Presents one point of view on preparing inorganic fluorides
		3.	Ionic Reactions of Fluoroolefins, R. D. Chambers and R. H. Mobbs	63 (266)	Good summarizing review of subject with excellent discussion of mechanism. Outstanding 10-page table summarizing data
		4.	Structural Aspects of Monofluorosteroids, N. F. Taylor and P. W. Kent.	29 (160)	Good discussion of selective methods for introducing fluorine into sensitive molecules. Too much attention on features of steroid chemistry. Limited discussion and inadequate tables greatly limit value of the chapter
		5.	Fluorides of the Main Group Elements, R. D. W. Kemmitt and D. W. A. Sharp	111 (1183)	Excellent tabulation of properties of subject compounds. Physical, thermodynamic, structural and spectral data provided. Complements Chapters in Ref. A3-1 and A3-5
		6.	The Vibrational Spectra of Organic Fluorine Compounds, J. K. Brown, and K. J. Morgan	61 (424)	Well-referenced correlation of infrared data on organic fluorine compounds. Nicely complements Chapter 7 of A3-2
A6-5	*Advances in Fluorine Chemistry*, Vol. 5, M. Stacey, J. C. Tatlow, and A. G. Sharpe, Ed.,			288 (906)	
		1.	Oxyfluorides of Nitrogen, C. Woolf	30 (135)	Preparation, properties and reactions of the compounds are described

	Butterworths, Washington D. C., 1965	2.	Fluorides of Phosphorus, R. Schmutzler	255 (771)	Encyclopedic review covering preparation and properties of organic and inorganic phosphorus-fluorine compounds. 69 comprehensive tables
A7	*Fluorine Chemistry Reviews,* P. Tarrant, Ed. (Review journal published by Marcel Dekker, New York)				A valuable new review journal to be published semiannually. This journal should provide up-to-date reviews of timely topics. Also available in book edition
A7-1	Vol. 1, Number 1, 1967	1.	Synthesis, Compounding and Properties of Nitroso Rubbers, M. C. Henry, C. B. Griffis and E. C. Stump, p. 1	76 (88)	Thorough review of both chemistry and technical aspects of fluorinated nitroso rubbers
		2.	Electrochemical Fluorination, S. Nagase, p. 77	30 (125)	A brief survey which complements earlier reviews (A3-1, Chapter 12, A3-2, Chapter 6, and A6-1, Chapter 4) by covering more recent work and emphasizing the need for mechanistic studies
		3.	The Fluoroketenes, Yu. A. Cheburkov and I. H. Knunyants, p. 107	37 (286)	Good survey of this recently developed area of fluorine chemistry by the major contributors
		4.	Hexafluoroacetone, C. G. Krespan and W. J. Middleton, p. 145	52 (157)	A thorough survey of particular importance because of value of hexafluoroacetone as both a reactive laboratory and commercial intermediate
A7-2	Vol. 1, Number 2, 1967	1.	Fluorocarbon Toxicity and Biological Action, J. Wesley Clayton, p. 197	55 (57)	An important review of toxicity of common commercial organic fluorochemicals, specifically fluoroalkanes, fluoralkenes and fluoro polymers
		2.	Diels-Alder Reactions of Organic Fluorine Compounds, D. R. A. Perry, p. 253	60 (76)	A comprehensive review with good summarizing tables of Diels-Alder addition of fluorinated compounds, both alkene and hetero atom unsaturated systems as dienophiles and dienes

Appendix A (cont.)

Ref. no.	Title of book, author, publisher	Chapter, title, author	No. of pages (no. of ref.)	Comments
		3. Methods for the Introduction of Hydrogen into Fluorinated Compounds, F. J. Mettille and D. J. Burton, p. 315	43 (103)	A thorough compilation which primarily covers catalytic hydrogenation reactions and reactions with metal hydrides
		4. Reactions Involving Fluoride Ion and Polyfluoroalkyl Anions, J. A. Young, p. 359	48 (72)	A timely review by one of the pioneers of an increasingly important field of synthetic organic fluorine chemistry
A7-3	Vol. 2, Number 1, 1968	1. The Cycloaddition Reaction of Fluoroolefins, W. H. Sharkey	53 (56)	
		2. The Reaction of Halogenated Cycloalkenes with Nucleophiles, J. D. Park, R. J. Murtry, and J. H. Adams	20 (41)	
		3. Ionization Potentials and Molecule-Ion Dissociation Energies for Diatomic Metal Halides, J. W. Hastie and J. L. Margrave	33 (60)	
		4. Nuclear Magnetic Resonance Spectra of N-F Compounds, W. L. Brey and J. L. Hynes	48 (98)	
		5. The F^{19} Chemical Shifts and Coupling Constants of Fluoroxy Compounds, C. J. Hoffman	11 (13)	
A7-4	Vol. 3, Number 2, 1968 (in press)	1. Fluorine Compounds in Anesthesiology, E. R. Larsen		

		2. Reactions of Fluoroolefins with Electrophilic Reagents, B. L. Dyatkin, E. P. Mochalina, and I. L. Knunyants 3. Phosphorous, Arsenic and Antimony Pentafluorophenyl Compounds, M. Fild and O. Glemser 4. Fluoroalicyclic Derivatives of Metals and Metalloids, W. R. Cullen		
A8	*Toxic Aliphatic Fluorine Compounds,* F. L. M. Pattison, Elsevier, New York, 1959	1. Introduction and General Survey 2. Fluoroacetates 3. ω-Fluorocarboxylic Acids and Derivatives 4. Other ω-Fluoro Compounds 5. Potential Uses and Applications 6. Summary and Final Remarks Appendices I. An Introduction to the Chemistry of Monofluoro Compounds II. Some Representative Preparative Procedures III. First Aid and Hospital Treatment	227 (335)	This well-written informative monograph should be read by any chemist planning to do experimental work with organic fluorine compounds so that he may be aware of the potential danger that these compounds present. Only one class of toxic fluorine compounds is discussed; fluoroacetic acid and aliphatic compounds that can be metabolically converted to fluoroacetic acid. The point is quite well made that ω-substituted aliphatic compounds with an even number of carbons in the chain are toxic The tables in the book are a valuable reference source on ω-fluorine aliphatic compounds. The chemistry discussed is worth knowing. The information on the Krebs cycle and how fluoroacetate leads to the killing fluorocitrate is good general background knowledge

Ref. no.	Title of book, author, publisher	Chapter, title, author	No. of pages (no. of ref.)	Comments
A9	*Some Aspects of the Chemistry and Toxic Action of Organic Compounds Containing Phosphorus and Fluorine,* B. C. Saunders, Cambridge University Press, 1957	1. Introduction and General Survey 2. Nomenclature of Esters Containing Phosphorus 3. Notes on the Mammalian Nervous System 4. The Phosphorofluoridates 5. I. Phosphorodiamic Fluorides. II. Tabun and Sarin 6. Selected Reactions of Esters Containing Phosphorus 7. The Fluoroacetates 8. Other Compounds Containing the C-F Link 9. Insecticides 10. Esterase Activity and Medical Aspects Appendices A. Determination of Fluorine in Organic Compounds B. Table of Properties of Typical Fluoro Compounds C. First-Aid Treatment for Nerve-gas Poisoning	231 (49)	This is a highly personalized volume that deals mainly with fluorophosphate chemistry developed in World War II and the action of these compounds on the nervous system of animal life (nerve gases). Fluoroacetates are also considered. Most of the material in the book relates directly to the work of the author and his associates. Synthetic chemistry and physiological responses of the compounds synthesized are neatly woven into a scientific narrative that makes this volume enjoyable to read In all likelihood most organic fluorine chemists will not be exposed to fluorophosphates and cannot expect the information of the book to be applicable. On the other hand, fluorine substitution of phosphorus does produce fantastically toxic materials and other series of compounds with other elements could also be dangerous. This volume and the Pattison volume should be read to develop a safety conscience to the hazards in fluorine chemistry

Ref.	Book	No.	Chapter	Pages (Refs.)	Comments
A10	*Handbook of Experimental Pharmacology*, Vol. 20, Part 1. "Pharmacology of Fluorides," O. Eichler, A. Farah, H. Herken and A. D. Welch, Editors, Springer-Verlag, New York, 1966	1.	Fluorine Chemistry, R. E. Banks and H. Goldwhite	52 (105)	An excellent survey of fluorine chemistry including inorganic and organic syntheses, analyses, industrial aspects and a comprehensive bibliography
		2.	Metabolism of Inorganic Fluoride, F. A. Smith	88 (261)	These five chapters duplicate to a great extent material contained in Ref. A3-3 and A3-4. They mainly cover biological aspects and are not of immediate interest to an organic chemist
		3.	Fluoride and the Skeletal and Dental Tissues, J. A. Weatherell	32 (113)	
		4.	The Role of Fluorides in Tooth Chemistry and in the Prevention of Dental Caries, F. Brudevold	58 (295)	
		5.	The Effects of Fluoride on Plants, M. D. Thomas and E. W. Alther	76 (242)	
		6.	The Effects of Fluorides on Livestock, with Particular Reference to Cattle, J. L. Shupe and E. W. Alther	49 (87)	
		7.	Fluorine-Containing Insecticides, R. L. Metcalf	32 (102)	The only comprehensive review of this important area of biological applications of fluorine chemistry
		8.	Monofluoro Aliphatic Compounds, F. L. M. Pattison and R. A. Peters	72 (293)	An up-dated and shortened version of Ref. A8. This chapter includes good reviews of synthesis and properties in addition to biological effects
		9.	The Mammalian Toxicology of Organic Compounds Containing Fluorine, J. W. Clayton, Jr.	38 (95)	An important chapter for all organic chemists; complements and to some extent duplicates similar review by Clayton in Ref. A7-2
		10.	The Fluorinated Anesthetics, J. C. Krantz, Jr. and F. G. Rudo	64 (110)	A comprehensive survey of an important application of organic fluorine chemistry to the medicinal field

Appendix B. Specialized or

Ref. no.	Topic	Title of review	Authors
B1.a.1	General	Fluorine and Its Compounds	R. N. Haszeldine and A. G. Sharpe
B1.a.2	General	Fluorocarbons and Their Derivatives	R. E. Banks
B1.a.3	General	Organic Fluorine Chemistry	C. G. Krespan
B1.a.4	General	New and Varied Paths for Fluorine Chemistry	O. R. Pierce and A. M. Lovelace
B1.a.5	General	Laboratory and Technical Production of Fluorine and Its Compounds	H. R. Leech
B1.a.6	General	Progress in Organic Fluorine Chemistry	W. Funasaka
B1.a.7	General	Text of papers on fluorine chemistry presented at Chicago ACS Meeting in September 1946	E. T. McBee, *et al.*
B1.a.8	General	Preparations, Properties, and Technology of Fluorine and Organic Fluoro Compounds	C. Slesser and S. R. Schram, Ed.
B1.a.9	General	Halogen Chemistry	V. Gutmann, Ed.
B1.a.10	General	Organic Fluorine Chemistry	C. M. Sharts
B1.b.1	General and aliphatic	Aliphatic Fluorides	A. L. Henne

Limited Reviews on Fluorine Chemistry

Reference	No. of pages (no. of ref.)	Comments
Wiley, New York, 1951	153 (101)	Good introduction to field of fluorinated compounds, but now out of date
Oldbourne Press, London, 1964	161 (408)	Synthesis and chemistry of highly fluorinated compounds, both aliphatic and aromatic and of elements. Areas surveyed are well done but limited coverage because of size
Science, **150,** 13 (1965)	6 (15)	Development of organic fluorine chemistry is reviewed. Free-radical and anionic reactions, properties of the products, and new fluorinating agents are described
Chem. and Eng. News, **40,** July 9, 1962, p. 72	9 (4)	Develops thesis that fluorine chemistry is on the move with research pointed to new fluorination methods and new compounds, particularly the fluoroaromatics, cyclic and organometallic derivatives
Quart. Rev. (London), **3,** 22 (1949)	14 (52)	History of production through German and American wartime efforts
J. Chem. Soc. Japan, Ind. Chem. Sect. **65,** 1156 (1962)	6 (31)	In Japanese
Ind. and Eng. Chem., **39,** 236 (1947)	198 (–)	Text of 53 papers presented at Symposium. Reports of major work on fluorine chemistry carried out under Manhattan Project
McGraw-Hill, New York, 1951	868 (–)	Forty chapters by 60 authors reporting on the work carried out under the Manhattan Project, particularly in organic derivatives
Halogen Chemistry, Vols. 1, 2 and 3 Academic, New York, 1967	–	Chapters on various phases of halogen chemistry; many chapters include review on aspects of fluorine chemistry–both organic and inorganic. The chapters containing pertinent material on fluorine chemistry are listed under appropriate section
J. Chem. Ed., **45,** 185 (1968)	8 (85)	Papers presented at the Summer Conference of the California Association of Chemistry Teachers, August, 1967. An introductory outline of fluorine chemistry is presented as an incentive to teachers to include this subject in their normal chemistry courses
Organic Chemistry, An Advanced Treatise, Vol. 1, H. Gilman, Ed., Wiley, New York, 1943, p. 944	21 (106)	A classical work giving a survey of field at an early stage. Methods of preparation, effects on physical properties, chemical properties, applications, and analysis

Appendix B

Ref. no.	Topic	Title of review	Authors
B1.b.2	General and aromatic	The Emergence of Fluorocarbon Chemistry	E. J. Forbes
B1.b.3	General and fluoroaromatic	Progress in Organic Fluorine Chemistry	M. Stacey
B1.b.4	General and hypofluorites	Some Recent Advances in Fluorocarbon Chemistry	G. H. Cady
B2.a.1	Fluorination methods	Organic Fluorine Compounds	W. Bockemüller (translated and revised by C. J. Kibler)
B2.a.2	Fluorination methods and fluorocarbon properties	Fluorinations and Properties of Fluoro Derivatives of Paraffins and Cycloparaffins	E. T. McBee and O. R. Pierce
B2.b.1	Fluorination methods–aliphatic	The Preparation of Aliphatic Fluorine Compounds	A. L. Henne
B2.c.1	Fluorination methods–aromatic	Preparation of Aromatic Fluorine Compounds from Diazonium Fluoroborates–The Schiemann Reaction	A. Roe
B2.c.2	Fluorination methods–aromatic	Aromatic	D. Bryce-Smith
B2.d.1	Fluorination methods–C–F bond formation	Fluorocarbon Chemistry. Part 1. The Fluorination of Organic Compounds	R. Stephens and J. C. Tatlow
B2.d.2	Fluorination methods–C–F bond formation	Introduction to Fluorination of Organic Compounds	L. Motta and E. Conti
B2.e.1	Fluorination methods–electrolytic	Modern Aspects and Development of Fluorine Chemistry	D. Sianesi
B2.e.2	Fluorination methods–electrolytic	Electrochemical Fluorination	M. Schmeisser and P. Sartori

cont.)

ℓeference	No. of pages (no. of ref.)	Comments
Science Prog., **51**, 353 (1963)	11 (18)	A historical review of the emergence of fluorocarbon chemistry especially of the aromatic fluorocarbons
Royal Institute of Chem. J., **4**, 11 (1960)	4 (1)	Review of types of preparations with emphasis on perfluorobenzene derivatives
roc. Chem. Soc., 133 1960)	6 (28)	Short history of fluorine use and development in synthesis, properties and applications of saturated fluorocarbons; trifluoromethyl hypofluorite chemistry is emphasized
Jewer Methods of Preparative)rganic Chemistry, Vol. 1, nterscience, New York, 948, pp. 229 and 593	23 (130)	Comprehensive review of methods of preparation of organic fluorine compounds up until 1948. Now out of date
he Chemistry of Petroleum Iydrocarbons, Vol. 3, B. T. rooks, C. E. Boord, S. S. urtz, Jr., and L. Schmerling, ds., Reinhold, New York, 1955, . 73	12 (33)	Review of methods of fluorination and properties of products. Particularly emphasizes classical fluorination methods
)rg. Reactions, **2,** 49 (1944)	45 (153)	Methods and experimental procedures for preparation of aliphatic fluorine compounds. Now out of date
)rg. Reactions, **5,** 193 (1949)	36 (142)	Mechanism, preparation, and decomposition of diazonium fluoroborates
nnual Reports, The Chemical ociety, London, **61,** 326 (1964)	20 (192)	Fluorination of various aromatic compounds by cobalt (III) fluoride followed by vapor phase dehydrofluorination
)uart. Rev. (London), **16,** 44 1962)	27 (177)	Review of both selective fluorination at a functional group and exhaustive fluorination
him. Ind. (Milan), **46,** 775 1964)	14 (163)	Shows choice of fluorination process is as important as choice of fluorinating agent; divides into groups and reviews. (In Italian)
him. Ind. (Milan), **46,** 883 1964)	12 (7)	Review of present state of fluorine production and uses; description of electrolytic fluorine generator. (In Italian)
hem.-Ing.-Tech., **36,** 9 (1964)	6 (41)	Review of electrolytic fluorination methods, particularly adaption of Simon's cell to fluorinate gases. (In German)

Appendix B

Ref. no.	Topic	Title of review	Authors
B2.f.1	Fluorination methods–fluorinated steroid preparation	Introduction of Fluorine into the Steroid System	J. W. Chamberlin
B2.g.1	Fluorination methods–hydrogen fluoride	Use of Hydrogen Fluoride in Organic Reactions	K. Wiechert (translated and revised by J. E. Jones)
B2.h.1	Fluorination methods–sulfur tetrafluoride	The Chemistry of Sulfur Tetrafluoride	W. C. Smith
B3.a.1	Chemical properties–mechanism	The Mechanism of Permanganate Oxidation of Fluoroalcohols in Aqueous Solution	R. Stewart and R. van der Linden
B3.b.1	Chemical properties–reactions	The Reactions of Organic Fluorine Compounds	W. K. R. Musgrave
B3.b.2	Chemical properties–reactions	Reactions of Halogen Methyl Radicals	J. M. Tedder and J. C. Walton
B3.b.3	Chemical properties–reactions	Cyclobutane Derivatives from Thermal Cycloaddition Reactions	J. D. Roberts and C. M. Sharts
B3.b.4	Chemical properties–reactions	1,2- and 1,4-Cycloaddition to Conjugated Dienes	P. D. Bartlett
B4.a.1	Physical properties	Physical Chemical Properties of Fluorine	E. Wicke and E. U. Franck
B4.b.1	Physical properties–spectral F^{19} nmr	Nuclear Magnetic Resonance Spectra of Other Nuclei	J. A. Pople, H. J. Bernstein, and W. G. Schneider

:ont.)

.eference	No. of pages (no. of ref.)	Comments
'teroid Reactions, C. Djerassi, :d., Holden-Day, San 'rancisco, 1963, p. 155	24 (58)	A good review up to 1963 composed chiefly of examples as equations with appropriate comments on conditions, scope and limitations
'ewer Methods of Preparative 'rganic Chemistry, Vol. 1, nterscience, New York, 1948, . 315	54 (157)	Extensive review of earlier work on use of hydrogen fluoride as a reactant (for introducing fluorine), as a solvent and as a catalyst
'ngew. Chem. Int. Ed., **1,** 467 1962)	9 (65)	Current knowledge regarding the chemical and physical properties of SF_4 is reviewed. The structural features of the compound are discussed, and its toxicological properties are presented. Particular attention is given to its reactions as a fluorinating agent
'araday Soc. Disc., **29,** 211 1960)	8 (13)	Oxidation of trifluoromethyl carbinols to ketones; isotope effect and ring substituent effect in particular
'uart. Rev. (London), **8,** 331 1954)	24 (166)	Describes reaction of fluorocarbons, fluorohalogen-carbons, organometallic, organometalloid, carboxylic acid and derivatives, ethers, aldehydes, ketones, alcohols and monofluorides
'rogress in Reaction Kinetics, 'ol. 4, G. Porter, Ed., Per-gamon, New York, 1967, 'hapter 2	24 (82)	The review, which includes fluorocarbon radicals, covers sources of halogenomethyl radicals, radical-radical reactions, radical transfer reactions, and radical addition reactions. Considerable emphasis is on the mechanistic aspects
'rg. Reactions, **12,** 1 (1962)	56 (128)	A large portion of this extensive review is devoted to cycloaddition reaction of fluoroolefins (a common reaction of fluorinated olefins)
'cience, **159,** 833 (1968)	6 (19)	An up-to-date complement to B3.b.3 on question of mechanism of cycloaddition reactions. The dependence of products and rates on reactant structure is used to tell something about competing reaction mechanisms. Fluoroolefins are a major reactant for these studies
'ngew Chem., **66,** 701 (1954)	10 (71)	Dissociation energies, thermal properties, electron affinities, etc., are given
High-Resolution Nuclear Magnetic Resonance, McGraw-Hill, New York, 1959, Chapter 12	29 (42)	A good introduction to F^{19} nmr on binary fluorides, halomethanes, fluorobenzenes, fluorinated cyclic compounds, fluorinated olefins, halogen fluorides, perfluoroalkyl sulfur hexafluoride, sulfur tetra-fluoride, fluorosulfonates, fluorosilanes. Now out of date

Ref. no.	Topic	Title of review	Authors
B4.b.2	Physical properties–spectral F^{19} nmr	Nuclear Magnetic Resonance	E. O. Bishop
B4.b.3	Physical properties–spectral F^{19} nmr	Nuclear Magnetic Resonance in Electrolyte Solution	R. A. Craig
B4.b.4	Physical properties–spectral F^{19} nmr	^{19}F Nuclear Magnetic Resonance	J. W. Emsley, J. Feeney, and L. H. Sutcliffe
B4.c.1	Physical properties–structure	Crystallography	H. M. Powell, C. K. Prout, and S. C. Wallwork
B4.d.1	Physical properties–thermo	The Dissociation Energies of Gaseous Alkali Halides	L. Brewer and E. Brackett
B4.d.2	Physical properties–thermo	Thermodynamic Properties of Gaseous Metal Dihalides	L. Brewer, G. R. Somayajulu, and E. Brackett
B4.d.3	Physical properties–thermo	Some Aspects of the Thermochemistry and Thermodynamics of the Fluorocarbons	C. R. Patrick
B4.d.4	Physical properties–thermo	Thermochemistry	C. T. Mortimer
B5.a.1	Fluorocarbons–acetylenic	Haloacetylenes	K. M. Smirnov, A. P. Tomilov, and A. I. Shchekotikhim
B5.b.1	Fluorocarbons–aliphatic	Chemie und Technologie Aliphatischer Fluoroorganischer Verbindungen	D. Osteroth
B5.b.2	Fluorocarbons–aliphatic	The Chemistry of Perchlorocyclopentenes and Cyclopentadienes	H. E. Ungnade and E. T. McBee
B5.c.1	Fluorocarbons–aromatic	Aromatic Fluorocarbons	P. L. Coe

ont.)

eference	No. of pages (no. of ref.)	Comments
nnual Reports, The Chemical ociety, London, **58,** 55 (1962)	24 (228)	Fluorine resonance briefly discussed—chiefly on page 70
nnual Reports, The Chemical ociety, London, **59,** 63 (1962)	11 (54)	F^{19} nmr resonance studies
ligh Resolution Nuclear lagnetic Spectroscopy, Vol. 2, ergamon, New York, 1966, . 871	97 (207)	An up-to-date fairly comprehensive review on F^{19} nmr for both organic and inorganic compounds
nnual Reports, The Chemical ociety, London, **61,** 567 (1964)	49 (401)	Structural determinations of fluorides including IF_7 niobium, tantalum and ruthenium pentafluorides
hem. Rev., **61,** 425 (1961)	8 (56)	Values given for Li, Na, K, Pb, and Cs fluorides
hem. Rev., **63,** 111 (1963)	11 (90)	Thermo values given for Be, Mg, Cu, Si, Ba, Zn and other fluorides
etrahedron, **4,** 26 (1958)	10 (38)	Thermochemistry of some simple fluoro compounds is discussed
nnual Reports, The Chemical ociety, London, **61,** 61 (1964)	19 (192)	Heats of fluorination reviewed
ussian Chem. Rev. (*English Transl.*), **36,** 326 (1967)	13 (167)	Review covers synthesis, properties and chemical reactions of halogenated acetylenes XC≡CH and XC≡CX, where X is I, Br, Cl or F
ollection of Chemical and hemical Technological eviews, New Series 59, R. ummerer, L. Birkofer, and . Goubeau, Eds., Ferdinand nke Verlag, Stuttgart, 1964	195 (572)	A survey of the development of the chemistry and technology of aliphatic fluorine compounds up to the year 1961, covering syntheses of fluorinated hydrocarbons, alcohols, ethers, aldehydes, ketones, carboxylic acids, and compounds containing nitrogen and sulfur as well as with their chemical reactions. Unfortunately, many recent important developments are left out (such as perchloryl fluoride synthesis) and reaction mechanisms are hardly touched upon, but patents and technological aspects are more fully covered. (In German)
hem. Rev., **58,** 249 (1958)	72 (264)	Many fluoro-substituted chlorocyclics are listed with properties, both chemical and physical
hem. Prod., **26,** 17 (1963)	5 (25)	Popular type review emphasizing unusual properties, chemistry and industrial potential

Appendix B

Ref. no.	Topic	Title of review	Authors
B5.c.2	Fluorocarbons–aromatic	Aromatic Compounds	J. M. Tedder
B5.c.3	Fluorocarbons–aromatic	The Aromatic Fluorocarbons and Their Derivatives	J. C. Tatlow
B5	Fluorocarbons–arsenic and antimony derivatives	See B6.h.5 and B6.h.10	
B5.d.1	Fluorocarbons–carbenes	Halocyclopropanes from Halocarbenes	W. E. Parham and E. E. Schweizer
B5.d.2	Fluorocarbons–carbenes	Halocarbenes	W. Kirmse
B5.d.3	Fluorocarbons–carbenes	Dihalomethylenes, Monohalomethylenes	J. Hine
B5.e.1	Fluorocarbons–fluoroalkyl heteroatom derivatives	Compounds of Fluorocarbon Radicals with Metals and Non-Metals	H. J. Emeleus
B5.f.1	Fluorocarbons–α-fluoro ethers	The α-Haloalkyl Ethers	L. Summers
B5.g.1	Fluorocarbons–fluoroolefins	Electrophilic Addition Reactions of Fluoroolefins	B. L. Dyatkin, E. P. Mochalina, and I. L. Knunyants
B5.h.1	Fluorocarbons–free radical	Formation of Carbon-Hetero Atom Bonds by Free Radical Chain Addition to Carbon-Carbon Multiple Bonds	F. W. Stacey and J. F. Harris, Jr.
B5.h.2	Fluorocarbons–free radicals	Free Radical Additions to Olefins to Form Carbon-carbon Bonds	C. Walling and E. S. Huyser

(cont.)

Reference	No. of pages (no. of ref.)	Comments
Annual Report, The Chemical Society, London, **58,** 222 (1961)	16 (120)	Chemistry of hexafluorobenzene given as basis for study of other fluorinated aromatic structures. Nucleophilic reactions of these compounds discussed
Endeavor, **22,** 89 (1963)	7 (17)	Topics discussed are synthetic routes to aromatic fluorides, syntheses of highly fluorinated aromatics, pentafluorophenyl derivatives from pentafluorobenzene, displacement of F from hexafluorobenzene by nucleophilic reagents, displacement of F from pentafluorophenyl derivatives by nucleophilic reagents, reactions of polycyclic aromatic fluorocarbons and physical properties of aromatic fluorocarbons
Org. Reactions, **13,** 55 (1963)	36 (140)	Methods of preparation and reactions of some fluorinated carbenes
Carbene Chemistry, Academic, New York, 1964, Chapter 8	57 (192)	Good critical review of synthetic and mechanistic aspects of dihalocarbenes, which includes fluorocarbenes
Divalent Carbon, Ronald, New York, 1964, Chapters 3 and 4	31 (76)	Includes discussion both of synthesis and mechanistic aspects of di- and monofluorocarbenes
1959 International Congress of Pure and Applied Chemistry, **1,** 82 (1960)	13 (20)	Properties, reactivity, reactions of fluoroalkyl heteroatom derivatives
Chem. Rev., **55,** 301 (1955)	53 (423)	Only a few fluoro-chloro ethers listed with physical properties and references
Russian Chem. Rev. (*English Transl.*), **35,** 417 (1966)	11 (26)	A comprehensive review of additions of halogen, halogen halides, inorganic acids, sulfur trioxide, and mercuration, alkylation, acylation, condensation nitration, nitrosation to fluoroolefins. Mechanism is critically discussed
Org. Reactions, **13,** 150 (1963)	227 (631)	Covers free radical additions to fluoroolefins and radicals from CF_3SH, $ClSO_2F$, CF_3SCl, SF_5Cl and H_2S, RSH, RSO_2Cl, etc.
Org. Reactions, **13,** 91 (1963)	59 (160)	Covers fluorinated olefins in free radical addition of various organic reagents, including fluorocarbon radicals

Appendix B

Ref. no.	Topic	Title of review	Authors
B5.i.1	Fluorocarbons–hypofluorites	Organic and Inorganic Hypofluorites	C. J. Hoffman
B5.j.1	Fluorocarbons–ketones	Reactions of the Carbonyl Group in Fluorinated Ketones	N. P. Gambaryan, E. M. Rokhlin, Yu. V. Zeifman, C. Ching-Yun, and I. L. Knunyants
B5.k.1	Fluorocarbons–nitrogen fluorides	Nitrogen Fluorides and Their Organic Derivatives	C. J. Hoffman and R. G. Neville
B5.k.2	Fluorocarbons–nitrogen fluorides	The Fluorides of Nitrogen	C. B. Colburn
B5.k.3	Fluorocarbons–nitrogen fluorides	Radical Reactions of Tetrafluorohydrazine	J. P. Freeman
B5.k.4	Fluorocarbons–nitrogen fluorides	Derivatives of the Nitrogen Fluorides	J. K. Ruff
B5.l.1	Fluorocarbons–organometallic	Recent Studies on Fluoroalkyls and Related Compounds	H. J. Emeleus
B5.l.2	Fluorocarbons–organometallic	Perfluoroalkyl Derivatives of Metals and Non-metals	J. J. Lagowski
B5.l.3	Fluorocarbons–organometallic	Fluorocarbon Derivatives of Metals	P. M. Treichel and F. G. A. Stone

cont.)

eference	No. of pages (no. of ref.)	Comments
hem. Rev., **64**, 91 (1964)	8 (82)	Properties and limited reactions of perfluoroalkyl hypofluorites and groups V, VI, and VII hypofluorites
ngew. Chem. Int. Ed., **5**, 947 1966)	10 (121)	Reviews chemistry of hexafluoroacetone as representative of fluorinated ketones. Mechanisms are critically discussed
hem. Rev., **62**, 1 (1962)	18 (183)	Review covers binary and ternary inorganic N–F compounds and organic compounds containing N–F bonds. The literature of N–F chemistry was critically reviewed from the date of discovery of each inorganic N–F compound through 1960. The literature of organic compounds containing the N–F bond was reviewed from 1900 through 1960
ndeavor, **24**, 138 (1965)	5 (32)	Author briefly reviews the synthesis of N_2F_4, the discovery of the ease of dissociation of this compound into NF_2 free radicals, and the subsequent preparations of inorganic, organic and polymeric fluorides of nitrogen
ree Radicals in Inorganic hemistry, Advances in hemistry, Series No. 36, merican Chemical Society, Vashington, D. C., 1962, hapter 13	3 (14)	Brief outline of radical reactions of N_2F_4 to prepare organic and inorganic N–F derivatives; F^{19} nmr of derivatives also discussed
hem. Rev., **67**, 665 (1967)	14 (142)	Up-to-date coverage of synthesis of NF compounds, chemistry of NF_2 radical, other nitrogen fluoride functional groups and complexes of NF compounds
ngew. Chem. Int. Ed., **1**, 129 (1962)	5 (35)	Use of fluoroalkyliodides for preparation of new organometallic and organometalloidal compounds containing fluoroalkyl groups. Comparison of properties and behavior is made
uart. Rev. (*London*), **13**, 233 (1959)	32 (153)	Chemistry and properties of the known perfluoroalkyl derivatives of the elements other than C, O, N, and the halogens. Effective electronegativity of perfluoroalkyl groups, preparative methods, and perfluoroalkyl derivatives of element Groups I, II, III and IV are covered
dvances in Organometallic hemistry Vol. 1, F. G. A. tone and R. West, Eds., cademic, New York, 1964, . 143	78 (156)	An extensive review of organometallic derivatives where a fluorocarbon is bonded to a metal; also covers synthesis, chemistry and properties (including spectral)

Appendix B

Ref. no.	Topic	Title of review	Authors
B5.l.4	Fluorocarbons–organometallic	New Chemistry of Fluorine: Organometal and Organometalloid Binding of Fluorine	R. N. Haszeldine
B5.l.5	Fluorocarbons–organometallic	Polyfluoroalkyl Derivatives of Metalloids and Non-metals	R. E. Banks and R. N. Haszeldine
B5.l.6	Fluorocarbons–organometallic	Vinyl Compounds of Metals	D. Seyferth
B5.l.7	Fluorocarbon–organometallic	Fluorocarbon–Transition Metal Chemistry	M. D. Rausch
B5.l.8	Fluorocarbons–organometallic	Perfluoroorganometallic Compounds	C. Tamborski
B5.l.9	Fluorocarbons–organometallic	Pentafluorophenylmetal Compounds	R. D. Chambers and T. Chivers
B5.l.10	Fluorocarbons–organometallic	Organoelement Halides of Germanium, Tin, and Lead	I. Ruidisch, H. Schmidbaur, and H. Schumann
B5.m.1	Fluorocarbons–phosphorus fluorine	Chemistry and Stereochemistry of Fluorophosphoranes	R. Schmutzler
B5.m.2	Fluorocarbons–phosphorus fluorine	Fluorophosphoranes	R. Schmutzler
B5.n.1	Fluorocarbons–silicon	Organosilicon Compounds of Fluorine	G. V. Odabashyan, V. A. Ponomarenko, and A. D. Petrov
B5.o.1	Fluorocarbons–sulfur fluorine	The Chemistry of Compounds Containing Sulfur-fluorine Bonds	H. L. Roberts

(cont.)

Reference	No. of pages (no. of ref.)	Comments
Angew Chem., **66,** 693 (1954)	9 (1)	P, As, Hg, and fluorocarbons are discussed with reactions such as the Grignard
Adv. Inorg. Chem. and Radiochem., **3,** 337 (1961)	97 (210)	Comprehensive review of synthesis and chemistry of perfluoroalkyl derivatives of the elements: mercury, boron, silicon, nitrogen, phosphorus, arsenic, antimony, oxygen, sulfur, and selenium
Progress in Inorganic Chemistry, Vol. 3, F. A. Cotton, Ed., Interscience, New York, 1962	132 (496)	Seven pages are devoted to review of perfluorovinyl metal compounds
Trans. N. Y. Acad. Sci., **28,** 611 (1966)	12 (50)	A good resumé of the fluorocarbon derivatives of transition metals with emphasis on methods of synthesis and comments on properties and bonding
Trans. N. Y. Acad. Sci., **28,** 601 (1966)	10 (30)	A good summary of the organometallic chemistry of perfluorobenzene derivatives
Organometallic Chem. Rev., **1,** 279 (1966)	25 (33)	Perfluoroaromatic organometallic derivatives are systematically reviewed by periodic classification as to synthesis, chemistry, and properties
Halogen Chemistry Vol. 2, V. Gutmann, Ed., Academic, New York, 1967, p. 233	92 (711)	Extensive review that includes both fluorocarbon and fluoride substituted derivatives of germanium, tin, and lead
Angew. Chem. Int. Ed., **4,** 496 (1965), or *Angew. Chem.*, **77,** 530 (1965)	13 (69)	A review on chemical and stereochemical aspects of phosphorus-halogen compounds containing pentacoordinate phosphorus. Compounds of this type are derived from PF_5 by substitution of F atoms with various groups
Halogen Chemistry Vol. 2, V. Gutmann, Ed., Academic New York, 1967, p. 31	83 (143)	A good review of preparation, properties and structure of fluorophosphoranes (chiefly organic derivatives) with extensive table of properties
Russian Chem. Rev. (*English Transl.*), **30,** 407 (1961)	20 (254)	Preparation and properties of fluoroorganosilicon compounds
Quart. Rev. (*London*), **15,** 30 (1961)	26 (105)	The chemistry of known S–F compounds is classified under the following headings: S_2F_2, SF_2 and their derivatives of the type R·SF where there is a univalent element or radical; SF_4 and its derivatives $R{\cdot}SF_3$; SF_6 and its derivatives $R{\cdot}SF_5$ and SF_4R_2; fluorooxyacids and oxyfluorides of S; and S–N–F compounds

Appendix B

Ref. no.	Topic	Title of review	Authors
B5.o.2	Fluorocarbons–sulfur fluorine	Compounds Containing Sulphur and Fluorine	R. N. Haszeldine
B5.o.3	Fluorocarbons–sulfur fluorine	Polyfluoroalkyl Derivatives of Sulfur	R. E. Banks and R. N. Haszeldine
B5.p.1	Fluorocarbons–trifluoromethyl hetero	*N*-, *O*- and *S*-Trihalomethyl Compounds	A. Senning
B5.q.1	Fluorocarbons–trihalomethyl sulfur	Dichlorofluoromethylthio Compounds and Their Application as Pesticides	E. Kühle, E. Klauke, and F. Grewe
B5.q.2	Fluorocarbons–trihalomethyl sulfur	Recent Studies of Trihalo-methylsulfur Compounds	A. Senning and S. Kaae
B6.a.1	Inorganic	Fluor	Gmelin Institut
B6.a.2	Inorganic	Some General Aspects of the Inorganic Chemistry of Fluorine	A. G. Sharpe
B6.a.3	Inorganic	Advances in Inorganic Fluorine Chemistry	S. Yoshizawa
B6.a.4	Inorganic	The Nature of Metal–Halogen Bonds	R. G. Pearson and R. J. Mawby

ont.)

eference	No. of pages (no. of ref.	Comments
hemical Society Symposia, ristol, 1958, Special Publica- on, No. 12, The Chemical ociety, Burlington House, . I., London, p. 317	14 (31)	Short review of di-, tetra- and hexavalent sulfur compounds with S–F or S-fluoroalkyl structure
rg. Sulfur Compounds, **2,** 37 (1966)	50 (131)	Comprehensive review of polyfluoroalkyl derivatives of sulfur under headings: (1) compounds containing bivalent sulfur, (2) sulfonic acids, and (3) derivatives of sulfur hexa- and tetrafluorides. Main review includes references up to 1962. Twelve additional references up to 1964 are covered in an appendix (pp. 417-418). Six comprehensive tables of derivatives listing physical properties are very useful
hem. Rev., **65,** 385 (1965)	28 (413)	Review of the chemistry, biological activity, and uses of title compounds containing identical halogen atoms in the trihalomethyl group (i.e., $-CF_3$, $-CCl_3$, $-CBr_3$, and $-CI_3$). Mixed halomethyl compounds (i.e., $-CCl_2F$, etc.) are treated only as far as they appear as intermediates or by-products
ngew. Chem., **76,** 807 (1964)	10 (44)	The trihalomethyl sulfur derivative, particularly CF_3S- and CF_2ClS-, are reviewed. The emphasis is on method of synthesis of derivatives of biological interest, particularly as agricultural chemicals. Some toxicity data also given
uart. Rept. Sulfur Chem., 1 (1967)	92 (150)	A collection of references with a short summary that includes literature on compounds that contain sulfur and trihalomethyl groups (where halo is usually F or Cl)
nelins Handbuch der organischen Chemie, stem-Nummer 5, Verlag emie, GmbH; Weinheim/ rgstrasse, 1959	258 (in text)	Very detailed review of fluorine and its inorganic compounds, preparations, chemistry, and properties
uart. Rev. (London), **11,** 49 957)	12 (49)	Fluorides of metals and nonmetals and physical properties of fluorine
Chem. Soc. Japan, Ind. Chem. ct., **65,** 1141 (1962)	5 (41)	In Japanese
logen Chemistry, Vol. 3, Gutmann, Ed., Academic, w York, 1967	29 (65)	Review covers bond energies (in tables), various physical and spectral properties related to bonding and theoretical calculations

Appendix B

Ref. no.	Topic	Title of review	Authors
B6.b.1	Inorganic–actinides	The Halogen Chemistry of the Actinides	K. W. Bagnall
B6.b.2	Inorganic–actinides	The Chemistry of Uranium Fluorides	I. V. Tananaev and N. S. Nikolaev
B6.b.3	Inorganic–actinides	The Role of Fluorine in Uranium Chemistry	A. Level
B6.c.1	Inorganic–complexes	Complexes	D. Nicholls
B6.d.1	Inorganic–Group I	The Typical Elements	A. K. Holliday
B6.d.2	Inorganic–Group I	Inorganic Chemistry	C. C. Addison and N. N. Greenwood
B6.d.3	Inorganic–Group I	Higher Oxidation States of Silver	J. A. McMillan
B6.f.1	Inorganic–Group III	Boron Subhalides and Related Compounds with Boron-boron Bonds	A. K. Holliday and A. G. Massey
B6.f.2	Inorganic–Group III	Boron Trifluoride Coordination Compounds	N. N. Greenwood and R. L. Martin
B6.f.3	Inorganic–Group III	The Halides of Boron	A. G. Massey
B6.f.4	Inorganic–Group III	Boron Halides	G. Urry
B6.g.1	Inorganic–Group IV	The Reaction of Fluorosilicates with Some Organic Halides	J. Dahmlos
B6.g.2	Inorganic–Group IV	Inorganic Silicon Halides	E. Hengge

(cont.)

Reference	No. of pages (no. of ref.)	Comments
Halogen Chemistry, Vol. 3, V. Gutmann, Ed., Academic, New York, 1967	79 (483)	Extensive review of preparation, structure and chemistry of fluorides of tri-, tetra-, penta-, and hexavalent actinides
Russian Chem. Rev. (English Transl.), **30**, 654 (1961)	18 (175)	UF_3, UF_4, UF_6, UO_2F_2, double fluorides, and intermediate fluorides are reviewed as to preparation, physical and chemical properties
Energie Nucléaire, **4**, 278 (1962)	9 (–)	After recalling the uranium processing cycle, the author describes the principal methods used in the preparation of uranium tetrafluoride and uranium hexafluoride. He then gives a summary of the different methods used to separate uranium by volatilization of uranium hexafluoride. (In French)
Annual Reports, The Chemical Society, London, **60**, 222 (1963)	22 (198)	The preparation and magnetic properties of complex fluorides having the perovskite structure
Annual Reports, The Chemical Society, London, **60**, 179 (1963)	25 (342)	Reactions of fluorine with alkali-metal halides
Annual Reports, The Chemical Society, London, **55**, 111 (1958)	57 (517)	Properties of alkali halides
Chem. Rev., **62**, 65 (1962)	16 (194)	Review brings up to date the chemistry of Ag(II) and Ag(III) compounds including fluorinated compounds
Chem. Rev., **62**, 303 (1962)	16 (108)	Review of boron structure including diboron tetrafluoride
Quart. Rev. (London), **8**, 1 (1954)	39 (175)	Review of complexes of BF_3 emphasizing questions of structure, stability, and physical properties
Advan. Inorg. Chem. and Radiochem., **10**, 1 (1967)	129 (879)	Comprehensive review of preparation, structure and chemistry of boron halides, which deals mainly with fluorides and halides
The Chemistry of Boron and its Compounds, E. L. Muetterties, Ed., Wiley, New York, 1967, p. 325	51 (204)	Review of preparation, structure, properties, and chemistry of boron halides which includes boron fluorides
Angew. Chem., **71**, 274 (1959)	3 (5)	Uses of fluorosilicates in synthesis
Halogen Chemistry, Vol. 2, V. Gutmann, Ed., Academic, New York, 1967, p. 169	54 (381)	Silicon fluorides are included as part of general review of silicon halides

Ref. no.	Topic	Title of review	Authors
B6.h.1	Inorganic–Group V	Halides of Phosphorus, Arsenic, Antimony, and Bismuth	L. Kolditz
B6.h.2	Inorganic–Group V	The Difluoramino Radical	F. A. Johnson
B6.h.3	Inorganic–Group V	Chemistry of Difluoramine	A. V. Fokin and Yu. M. Kosyrev
B6.h.4	Inorganic–Group V	Trifluorophosphine Complexes of Transition Metals	Th. Kruck
B6.h.5a	Inorganic–Group V	Halides of Arsenic and Antimony	L. Kolditz
B6.h.5b	Inorganic–Group V	Chemistry of Some Inorganic Nitrogen Fluorides	A. V. Pankratov
B6.h.6	Inorganic–Group V	Recent Develpoments in N–F Chemistry	C. B. Colburn
B6.h.7	Inorganic–Group V	The Difluoroamino Free Radical and Its Reactions	C. B. Colburn
B6.h.8	Inorganic–Group V	The Chemistry of Nitrous Acid and Its Derivatives in Liquid Hydrogen Fluoride	F. Seel
B6.h.9	Inorganic–Group V	Phosphonitrilic Derivatives and Related Compounds	N. L. Paddock
B6.h.10	Inorganic–Group V	Addition Compounds of Group V Pentahalides	M. Webster

(cont.)

Reference	No. of pages (no. of ref.)	Comments
Advan. Inorg. Chem. and Radiochem., **7**, 1 (1965)	22 (192)	Preparation, structure and properties of halides of metals in + 3 and + 5 oxidation state are reviewed. The fluorinated derivatives compose a major portion of this class
Free Radicals in Inorganic Chemistry, Advances in Chemistry, Series No. 36, American Chemical Society, Washington, D.C., 1962, Chapter 12	5 (19)	Discussion of synthesis of $\cdot NF_2$ for N_2F_4, equilibrium and reactions with inorganic type reagents
Russian Chem. Rev. (English Transl.), **35**, 791 (1966)	12 (45)	Reviews properties, methods of preparation, chemistry, and explosive properties ofdifluoramine (HNF_2)
Angew. Chem. Int. Ed., **6**, 53 (1967)	15 (94)	Reviews coordination chemistry of trifluorophosphines
Halogen Chemistry Vol. 2, V. Gutmann, Ed., Academic, New York, 1967, p. 115	45 (389)	Review concentrates on + 3 and + 5 halides of arsenic and antimony but includes a short section on compounds with As–C and Sb–C bonds
Russian Chem. Rev. (English Transl.), **32**, 157 (1963)	9 (75)	Review is concerned with the chemistry of nitrogen fluorides including tetrafluorohydrazine, dichloroamine, chlorodifluoroamine, fluorine azide and difluorodiazine isomers. The structure of the nitrogen trifluoride molecule is also discussed
J. Chem. Ed., **38**, 180 (1961)	3 (21)	Development of NF_3, FN_3, N_2F_4, HNF_2, N_2F_2, NF_2Cl properties and uses and preparation
Chemistry in Britain, **2**, 336 (1966)	5 (52)	Text of paper delivered as the centenary lecture to the Chemical Society in February. A good review of the preparation and chemistry (including with organic compounds) of NF_2 radical
Angew. Chem. Int. Ed., **4**, 635 (1965)	7 (31)	The reaction of liquid HF with HNO_2, N_2O_3, N_2O_4, and nitrosyl compounds leads to unexpectedly high-boiling liquids which contain NOF in the form of remarkably stable, electrolytically dissociated HF solvates
Quart. Rev. (London), **18**, 168 (1964)	43 (219)	Preparation, substitution reactions, electronic and molecular structure, thermochemistry, and properties
Chem. Rev., **66**, 87 (1966)	32 (575)	Structure, spectral data, physical properties, thermochemistry and chemistry of P, As and other pentafluorides

Appendix B

Ref. no.	Topic	Title of review	Authors
B6.i.1	Inorganic–Group VI	The Oxygen Fluorides	A. G. Streng
B6.i.2	Inorganic–Group VI	The Typical Elements	A. K. Holliday
B6.i.3	Inorganic–Group VI	Synthesis of Oxyhalides	K. Dehnicke
B6.i.4	Inorganic–Group VI	Fluorosulfonic Acid as a Reaction Medium and Fluorinating Agent	A. Engelbrecht
B6.i.5	Inorganic–Group VI	Fluorine-containing Compounds of Sulfur	G. H. Cady
B6.i.6	Inorganic–Group VI	Sulfur-Nitrogen-Halogen Compounds	O. Glemser
B6.i.7	Inorganic–Group VI	Sulphur-Nitrogen-Halogen Compounds	O. Glemser and N. Field
B6.i.8	Inorganic–Group VI	Halide Chemistry of Chromium, Molybdenum and Tungsten	J. E. Fergusson
B6.j.1	Inorganic–Group VII	Physiochemical Properties of Chlorine Trifluoride	Yu. D. Shishkov and A. A. Opalovskii
B6.j.2	Inorganic–Group VII	Halogen Fluorides and Other Covalent Fluorides	H. C. Clark
B6.j.3	Inorganic–Group VII	The Physical Inorganic Chemistry of the Halogens	A. G. Sharpe
B6.j.4	Inorganic–Group VII	Physical and Chemical Properties of Halogen Fluorides	L. Stein
B6.j.5	Inorganic–Group VII	Polyhalogen Complex Ions	A. I. Popov

(cont.)

Reference	No. of pages (no. of ref.)	Comments
Chem. Rev., **63,** 607 (1963)	18 (109)	Review gives preparation and properties of oxygen difluoride, dioxygen difluoride, and trioxygen difluoride
Annual Reports, The Chemical Society, London, **61,** 115 (1964)	33 (468)	Reviews of inorganic and organic hypofluorites, oxides and oxyfluorides of the halogens are cited
Angew. Chem., **77,** 22 (1965)	8 (127)	Applications, limits and problems in individual methods for the synthesis of oxyhalides are discussed with reference to characteristic examples
Angew. Chem. Int. Ed., **4,** 368 (1965)	1 (5)	Use of FSO_2OH as a solvent, particularly for preparation of inorganic fluorides
Adv. Inorg. Chem. and Radiochem., **2,** 105 (1960)	53 (336)	Many physical properties of fluorine-sulfur compounds including those containing C, S, F, O, halides, and elements other than oxygen
Angew. Chem., **75,** 697 (1963)	10 (77)	Preparations, properties and structure of cyclic and acyclic S–N halides are described including fluorocarbons with S–F
Halogen Chemistry, Vol. 2, V. Gutmann, Ed., Academic, New York, 1967, p. 1	30 (102)	Good review of preparation and properties of both acyclic and cyclic derivatives
Halogen Chemistry, Vol. 3, V. Gutmann, Ed. Academic, New York, 1967	75 (543)	Review concentrates on preparation, structure and physical properties which include fluorides
Russian Chem. Rev. (*English Transl.*), **29,** 357 (1960)	8 (93)	Review considers the physical and chemical properties of ClF_3
Chem. Rev., **58,** 869 (1958)	26 (149)	Physical and chemical properties; isotopic exchange reactions of halogen fluorides
Halogen Chemistry, Vol. 1, V. Gutmann, Ed., Academic, New York, 1967, p. 1	39 (112)	Comprehensive review of the physical properties of the halogens (including fluorine) and some aspects of inorganic chemistry
Halogen Chemistry, Vol. 1, V. Gutmann, Ed., Academic, New York, 1967, p. 133	72 (454)	Extensive review of all halogen fluorides, particularly a comprehensive tabulation of physical properties
Halogen Chemistry, Vol. 1, V. Gutmann, Ed., Academic, New York, 1967, p. 255	35 (166)	Review includes ions where fluorine is involved in polyhalogen complex ions

Appendix B

Ref. no.	Topic	Title of review	Authors
B6.1.1	Inorganic–halogen exchange	Isotopic Halogen Exchange Reactions	M. F. A. Dove and D. B. Sowerby
B6.1.2	Inorganic–hexafluorides	The 25-year Revolution in Hexafluoride Chemistry	B. Weinstock
B6.1.3	Inorganic–lanthanides	Preparation of Anhydrous Lanthanide Halides	M. D. Taylor
B6.1.4	Inorganic–octahedral complexes	Structural Chemistry of Octahedral Fluorocomplexes of the Transition Elements	D. Babel
B6.m.1	Inorganic–rare gas	Noble Gas Compounds	H. H. Hyman, Ed.
B6.m.2	Inorganic–rare gas	The Chemistry of Noble Gas Compounds	H. H. Hyman
B6.m.3	Inorganic–rare gas	The Chemistry of Noble Gases	B. Thomson
B6.m.4	Inorganic–rare gas	Compounds of Elements of the Zero Group	A. B. Neiding
B6.m.5	Inorganic–rare gas	The Valence Compounds of Rare Gases	R. Hoppe
B6.m.6	Inorganic–rare gas	Rare Gas Compounds	K. Rossmanith
B6.m.7	Inorganic–rare gas	Chemical Compounds of the Noble Gases	C. L. Chernick
B6.m.8	Inorganic–rare gas	Xenon and Radon Fluorides	A. B. Neiding

cont.)

Reference	No. of pages (no. of ref.)	Comments
Halogen Chemistry, Vol. 1, V. Gutmann, Ed., Academic, New York, 1967, p. 41	79 (379)	Reviews each type of halide exchange. The section on fluorine is only six pages and very little is reported on organic fluorine compounds
Chem. and Eng. News, **42** (Sept. 21, 1964), p. 86	15 (4)	Metal and non-metal hexafluorides, synthesis and properties, handling, potential, use as fluorinating agents
Chem. Rev., **62,** 503 (1962)	9 (112)	Methods of preparation of lanthanide halides including anhydrous lanthanide fluorides and hydrates
Structure and Bonding, Vol. 3, Springer-Verlag, New York, 1967, p. 1	87 (348)	Extensive review on structure and bonding these complexes
University of Chicago Press, Chicago, Illinois, 1963	404	Review on fluoride of noble gas compounds, chemistry, properties, physical, theoretical, spectral and physiological data. Extensive references listed by sections
Science, **145,** 773 (1964)	11 (71)	The synthesis of simple fluorides, oxyfluorides, oxides, xenates, and perxenates is discussed
Advancement of Science, **22,** 208 (1965)	10 (59)	Review of some reactions and physical properties of noble gas compounds
Russian Chem. Rev. (*English Transl.*), **34,** 403 (1965)	42 (210)	Fluorides, oxyfluorides and oxygenated compounds of xenon, radon, and krypton, their chemical bonds, thermodynamics, and applications are reviewed
Angew. Chem. Int. Ed., **3,** 538 (1964) or *Angew. Chem.,* **76,** 455 (1964)	9 (121)	Review covers older studies; discovery of the first rare gas fluorides; preparation of the rare gas compounds; XeF_2, XeF_4; higher fluorides and oxy-fluorides of xenon; fluorine compounds of krypton; fluorine compounds of higher order; oxygen compounds of xenon, thermochemistry and bonding
Oesterr. Chem.-Ztg., **65,** 82 (1964)	7 (51)	This review covers the xenon fluorides (preparation, structure, thermodynamic stability, etc.) and the nature of the bonds in these compounds. (In German)
Rec. Chem. Progr., **24,** 139 (1963)	17 (123)	Review covers the preparation and properties of Xe–metal hexafluorides, XeF_2' XeF_6, other Xe fluorides, Xe oxyfluorides, Xe trioxide, and compounds of gases other than Xe (Rn, Kr)
Russian Chem. Rev. (*English Transl.*), **32,** 224 (1963)	4 (24)	A review is presented on the separation of the title compounds

Appendix B

Ref. no.	Topic	Title of review	Authors
B6.m.9	Inorganic–rare gas	Rare Gas Fluorides and Other Compounds	J. Serre
B6.m.10	Inorganic–rare gas	The Chemistry of Xenon	J. G. Malm, H. Selig, J. Jortner, and S. A. Rice
B6.n.1	Inorganic–spectral properties	Review of Metal-halogen Vibrational Frequencies	R. J. H. Clark
B6.o.1	Inorganic–transition metals	Pentahalides of the Transition Metals	A. D. Beveridge and H. C. Clark
B6.o.2	Inorganic–transition metals	Electronic Structure and Molecular Orbital Treatment of Halogen and Noble Gas Complexes in Positive, Negative and Undefined Oxidation States	C. K. Jorgensen
B7.a.1	Industrial	The Manufacture and Use of Fluorine and Its Compounds	A. J. Rudge
B7.a.2	Industrial	Industrial Aspects of Fluorine Chemistry	A. K. Barbour
B7.a.3	Industrial	Industrial Organic Fluorine Chemistry	M. Katori and R. Kojima
B7.b.1	Industrial–polymers	Polymeric Fluoro Complexes	L. Kolditz
B7.b.2	Industrial–polymers	Ethylene and Fluoroethylene Polymers	R. C. Voter
B7.b.3	Industrial–polymers	Fluorine Containing Organic Polymers	C. A. Barson and C. R. Patrick
B7.b.4	Industrial–polymers	Polyfluorinated Linear Bifunctional Compounds Containing Like Functions as Potential Monomers	I. L. Knunyants, Li Chih-yuan, and V. V. Shokina

(cont.)

Reference	No. of pages (no. of ref.)	Comments
Bull. Soc. Chim. Fr., **1964**, 671	6 (81)	Preparation and properties of rare gas fluorides
Chem. Rev., **65**, 199 (1965)	38 (183)	Covers xenon fluorides
Halogen Chemistry, Vol. 3, V. Gutmann, Ed., Academic, New York, 1967	36 (154)	First half of chapter devoted to metal fluorine vibrational frequencies
Halogen Chemistry, Vol. 3, V. Gutmann, Ed., Academic, New York, 1967	36 (255)	Reviews preparation and physical and chemical properties of pentafluorides (including mixed fluorohalides) of transition metals of Group V through Group VIII
Halogen Chemistry, Vol. 1, V. Gutmann, Ed., Academic, New York, 1967	119 (824)	Extensive treatment particularly from theoretical point of view
Oxford University Press, London, 1962	87 (14)	A unique volume describing the industrial aspects of fluorine chemistry and particularly HF and elementary fluorine but includes sections on inorganic and organic compounds of fluorine
Chem. Ind. (London), 958 (1961)	15 (82)	Production of HF, organic fluorine compounds and their uses are discussed
J. Chem. Soc. Japan, Ind. Chem. Sect., **65**, 1161 (1962)	4 (40)	In Japanese
Z. Chem. (Leipzig), **2**, 186 (1962)	8 (41)	Review considers compounds which are regarded as polymeric fluoro complexes
High Polymers, Vol. 12, Part I, G. M. Kline, Ed., Interscience, New York, 1959, p. 165	32 (86)	Properties and uses of polyfluoroethylene
Brit. Plastics, **36**, 70 (1963)	6 (130)	F-containing organic polymers developed for use in special purposes because of their unusually high thermal and chemical stabilities and their high resistance to solvents
Russian Chem. Rev. (*English Transl.*), **32**, 461 (1963)	16 (189)	Review surveys the methods of preparation and the properties of polyfluorinated bifunctional compounds containing two like groups, which may be potential monomers for the preparation of new F-containing polycondensation polymers

Appendix B

Ref. no.	Topic	Title of review	Authors
B7.b.5	Industrial–polymers	Fluorine-containing Elastomers	J. C. Montermoso
B7.b.6	Industrial–polymers	Production and Properties of Polymers Containing Fluorine	G. Bier, R. Schaff, and K.-H. Kahrs
B7.b.7	Industrial–polymers	Fluorine-containing Polymers Fluorinated Vinyl Polymers with Functional Group, Condensation Polymers, and Styrene Polymers	W. Postelnek, L. E. Coleman, and A. M. Lovelace
B7.b.8	Industrial–polymers	Fluorocarbon Coatings Resist Corrosion, Heat and Friction	F. West
B7.b.9	Industrial–polymers	Polymers Containing Fluorine	R. E. Banks, J. M. Birchall, and R. N. Haszeldine
B7.b.10	Industrial–polymers	Fluorocarbons	M. A. Rudner
B7.b.11	Industrial–polymers	Fluorine-containing Elastomers	J. C. Cooper
B7.b.12	Industrial–polymers	Poly (Thiocarbonyl Fluoride) and Related Elastomers	W. H. Sharkey
B7.b.13	Industrial–polymers	Polymerization of Fluorothiocarbonyl Compounds	W. H. Sharkey
B8.a.1	Utility	Fluorochemical Properties and Applications	C. G. Klaus

(cont.)

Reference	No. of pages (no. of ref.)	Comments
Rubber Chem. Technol. (Rubber Reviews), **34,** 1521 (1961)	32 (137)	A review is presented concerning the function of F in polymer molecules, research on F-containing elastomers, and properties of fluorinated polybutadienes, vinylidene fluoride copolymers, fluoroacrylate polymers, fluoropolyesters, fluorinated silicone elastomers, fluoropolyamidine, and fluorinated nitroso elastomers
Angew Chem., **66,** 285 (1954)	8 (52)	In German
Fortschritte der Hochpolymeren-forschung, Berlin, **1,** 75 (1958-1960)	39 (113)	Vinyl polymers with functional groups and condensation polymers. (In English)
Mater. in Design Eng. **62,** No. 3, 112 (1965)	4 (–)	Article summarizes the key types of fluorocarbon coatings based on tetrafluoroethylene, chlorotrifluoroethylene, fluorinated ethylene propylene, vinylidene fluoride and fluoroelastomer compounds, their properties and how they are applied to resist corrosion, heat, and friction
Thermal Degradation of Polymers, Soc. Chem. Ind. Monograph No. 13, Macmillan, New York, 1961, p. 270	24 (38)	Survey of chemical and physical properties of newer homo- and copolymers containing fluorine, particularly discussing thermal stability and chemical and solvent resistance
Reinhold, New York, N. Y., 1958	238 (70)	A volume in the "Reinhold Plastics Application Series" which reviews the development and technology of fluorocarbons, resins, and plastics and application in various commercial fields
The Polymer Chemistry of Synthetic Elastomers, J. P. Kennedy and E. Tornquist, Eds, Interscience, New York, 1968	–(53)	Summarizes methods of preparation and properties of elastomers prepared from fluorinated monomer
The Polymer Chemistry of Synthetic Elastomers, J. P. Kennedy and E. Tornquist, Eds., Interscience, New York, 1968	– (7)	The preparation and properties of the polymers from thiocarbonyl fluoride, thioacyl fluorides, and hexafluorothioacetone are reviewed
Polyaldehydes, O. Vogel, Ed., Marcel Dekker, New York, 1967	9 (12)	The preparation and properties of polymers from CF_2S, R_fCSF, and CF_3CSCF_3 are described
Chem. Eng. Progr., **62,** 98 (1966), July	4 (17)	Fluorochemicals have found applications in chromium plating, corrosion control, petroleum fire fighting, textile fabric treatment, paper treating, and leather surface treating

Appendix B

Ref. no.	Topic	Title of review	Authors
B8.a.2	Utility	Examples from Fluorine Chemistry and Possible Industrial Applications	J. Schröder
B8.a.3	Utility	Fluorochemicals and Their Applications	H. J. Emeléus
B8.a.4	Utility	Fluorine-containing Lubricants	A. J. Rudge
B8.b.1	Utility–catalytic	Catalysis of Organic Reactions by Boron Trifluoride	D. Kästner (translated and revised by J. E. Jones)
B8.b.2	Utility–catalytic	Friedel-Crafts and Related Reactions	G. A. Olah, Ed.
B8.b.3	Utility–catalytic	Boron Fluoride and Its Compounds as Catalysts in Organic Chemistry	A. V. Topchiev, S. V. Zavgorodnii, and Ya. M. Paushkin (translated by J. T. Greaves)
B8.c.1	Utility–dyes	Fluorinated Dyes	H. Hopff
B8.d.1	Utility–fluoroaromatics	The Opportunity for Chemists in Aromatic Fluorocarbons	J. C. Tatlow
B8.e.1	Utility–hydrofluoric acid	HF Capacity Hits New High	
B8.f.1	Utility–propellants	Advanced Propellant Chemistry	Symposium Papers
B8.g.1	Utility–synthesis	The Use of Trifluoroacetic Anhydride and Related Compounds in Organic Synthesis	J. M. Tedder

(cont.)

Reference	No. of pages (no. of ref.)	Comments
Philips Tech. Rev., **26**, 111 (1965)	6 (–)	General introduction with discussion of utility in fluorine reactions, which owing to their great reaction energies are suitable for the production of actinic light in "combustion-type flash lamps." The reactivity of fluorine and the stability of fluorides are made use of to carry out chemical transport reactions at very high temperatures in incandescent lamps
ICSU Rev., **4**, 117 (July 1962)	6 (–)	Applications of fluorocarbons and their derivatives as based on physical and chemical properties are discussed
Chem. Ind. (*London*), 452 (1955)	10 (19)	Fluoro and chlorofluoro lubricants production
Newer Methods of Preparative Organic Chemistry, Interscience, New York, 1948, p. 249	65 (152)	Extensive review of use of BF_3 as a catalyst in organic reactions
Interscience, New York, Vols. I to IV, 1963-1965		Comprehensive review of Friedel-Crafts reaction include discussion of use of organic and inorganic fluorides (particularly HF and BF_3) throughout four volumes
Pergamon, New York, 1959	326 (1580)	Comprehensive review of BF_3 as a catalytic reagent in all types of reactions
Chimia, **15**, 193 (1961)	2 (–)	In German
New Scientist, **17**, 236 (1963)	3 (–)	The discovery of methods for making fluorocarbon compounds analogous to the well-known aromatic dyes, drugs, and polymers, opens up a great and promising vista for the organic chemist. Whether any of these compounds will find practical application will depend on their virtues outweighing their cost
Chem. Week, **86**, 101 (March 26, 1960)	2 (4)	A discussion of HF production during recent years and outlook for increased production based upon outlets such as fluorocarbons and use of AlF_3 in Al production. Producers and their capacities are mentioned
Advances in Chemistry Series No. 54, American Chemical Soc., Washington, D. C., 1966	290 (–)	A collection of papers from a symposium held at Detroit ACS Meeting, 1965, of which over half are on fluorine chemicals particularly on nitrogen fluoride derivatives, oxygen difluoride, and halogen fluorides
Chem. Rev., **55**, 787 (1955)	41 (214)	Use of F_3CCOOH in esterification, electrophilic substitution of aromatic compounds, acyladdition to unsaturated compounds, other condensation reactions, and preparation of anhydride and acid

Appendix B

Ref. no.	Topic	Title of review	Authors
B8.h.1	Utility–unusual properties	Fluorine Chemistry	O. R. Pierce
B9.a.1	Commercial availability	Fluoroketone Boosters	
B9.a.2	Commercial availability	Some Newly Available Chemical Compounds	W. Bradley
B9.a.3	Commercial availability	Faster Pace in Fluorochemicals	
B10.a.1	Biological–medicinal	The Discovery of the Anaesthetic Halothane–An Example of Industrial Research	J. Ferguson
B10.a.2	Biological–medicinal	Les Dérivés Organiques du Fluor d'Intérêt Pharmacologique	N. P. Buu-Hoi
B10.a.3	Biological–medicinal	The Trifluoromethyl Group in Medicinal Chemistry	H. L. Yale
B10.b.1	Biological–nucleic bases	Fluorinated Pyrimidines	C. Heidelberger
B10.c.1	Biological–peptides	The Analytical and Synthetic Importance of *N*-trifluoro Acetylated Amino Acids and Peptides	F. Weygand
B10	Biological–pesticides	See B5.q.1	
B10.d.1	Biological–steroids (see also B2.f.1)	Fluorosteroids	A. A. Akhren, I. G. Reshetova, and Yu. A. Titov

(cont.)

Reference	No. of pages (no. of ref.)	Comments
Int. Sci. and Tech. (Nov. 31, 1962)	7 (–)	A description is given of the extraordinary properties of fluorine which have led to a large number of new uses for this chemical element
Chem. Week, **92,** (March 31, 1963)	2 (–)	Allied Chemical Company's production of hexafluoroacetone, fluorochloroacetones, trifluoroacetic and difluorochloroacetic acids is discussed. General application areas and prices are mentioned
Ind. Chemist, **35,** 230 (1959)	7 (45)	General review of fluorine compounds, fluorination of hydrocarbons, isocyanates, hydrazines, lithium compounds, organometallic compounds and peracids, and peroxides
Chem. Week, **90,** 53 (Jan. 6, 1962)	1 (1)	A brief discussion of fluorochemicals recently developed either commercially or to an advanced research stage
Chem. Ind. (London), 818 (1964)	7 (8)	Text of lecture on fluorothane, $CF_3CHClBr$
Progress in Drug Research, **3,** 9 (1961)	66 (172)	An extensive review of organic compounds containing fluorine that are of interest in the drug field. (In French)
J. Med. and Pharm. Chem., **1,** 121 (1959)	13 (37)	Fluorinated hydrocarbons as inhalation anaesthetics, fluorinated phenothiozine derivatives as ataractic and antiemetic agents, others as diuretic agents
Progress in Nucleic Acid Research and Molecular Biology, Vol. 4, J. N. Davidson and W. E. Cohn, Eds., Academic, New York, 1965, p.1	50 (196)	A significant review by a pioneer in this important field of antimetabolites. A short discussion of synthesis and chemical properties is followed by an extensive evaluation of biological effects and medicinal use including discussion of mechanism of action
Bull. Soc. Chim. Biol., **43,** 1269 (1961)	12 (34)	In French. Reviews synthesis of trifluoroacetylated amino acids and use in analysis and synthesis of peptides
Russian Chem. Rev., (English Transl.), **34,** 926 (1965)	17 (326)	Methods of introducing fluorine into the steroid molecule (opening of steroid oxides with fluorine containing reagents, preparation of fluorosteroids from steroid olefins, substitution of fluorine for other groups of halogen atoms) and physiological activity of the fluorosteroids are reviewed

Appendix B

Ref. no.	Topic	Title of review	Authors
B10.d.2	Biological–steroids	Synthetic Derivatives of Cortical Hormones	J. Freid and A. Borman
B10.e.1	Biological–sugars	Application of Trifluoro-acetic Anhydride in Carbohydrate Chemistry	T. G. Bonner
B10.e.2	Biological–sugars	Glycosyl Fluorides and Azides	F. Micheel and A. Klemer
B10.e.3	Biological–sugars	Halogenated Carbohydrates	J. E. G. Barnett
B10.f.1	Biological–toxicity	Toxic Phosphorus and Fluorine Compounds	B. C. Saunders
B10.f.2	Biological–toxicity	Physiological Action of Diisopropyl Phosphorofluor-idate	B. C. Saunders

(cont.)

Reference	No. of pages (no. of ref.)	Comments
Vitamins and Hormones, Vol. 16, R. S. Harris, G. F. Marrian, and K. V. Thimann, Eds., Academic New York, 1958	371 (150)	Review concentrates on halogenated steroids. Synthetic methods are reviewed but main discussion is on the biological effects
Adv. Carbohydrate Chem., **16,** 59 (1961)	26 (68)	Trifluoroacyl group-synthesis and use as a blocking group, other acylations, and ring opening reactions
Adv. Carbohydrate Chem., **16,** 85 (1961)	19 (51)	Preparations and reactions of glycosyl fluorides and properties of their derivatives, aldosyl azides, and ω-fluoro carbohydrates
Adv. Carbohydrate Chem., **22,** 177 (1967)	50 (214)	Review of monosaccharides monohalogenated at positions other than C-1. Synthesis, reactions, and biological activity are covered
Endeavour, **19,** 32 (1960)	7 (26)	Article presents a short review concerning toxic P and F compounds and their uses in clinical medicine and as insecticides. Among the main aspects discussed are the mechanism of fluoroacetate poisoning and the mechanism of phosphorylation and transiodination
Phosphoric Esters and Related Compounds, Chemical Society Special Publication No. 8, London, 1957, p. 165	6 (7)	Effect of diisopropyl phosphorofluoridate on eyes, lacrimal and salivary glands, heart, bronchioles, alimentary canal, urinary, bladder, etc., and therapeutic use of peroxidase in pyrophosphate formation

Appendix C

Commercial Sources of Fluorinating Agents and Organic Fluorine Chemicals (Chiefly for U.S.A.)

Name of company	Address	Type of fluorinated chemical available	Comments (data sheets and other information)
Air Products and Chemicals, Inc., Specialty Gas Dept.	P.O. Box 538, Allentown, Pa. 18105	Hydrogen fluoride, nitrogen trifluoride, difluorodiazine	
Aldrich Chemical Co., Inc.	2371 N. 30th St., Milwaukee, Wis. 53210	Selection of organic reagents which includes a sprinkling of fluorinated materials	Source of a selection of fluorinated organic reagents; should be consulted early
Allied Chemical Corp., Industrial Chemicals Div., Plastics Div.	P.O. Box 70, Morristown, N.J. P.O. Box 365, Morristown, N.J. 07960	FK fluoroketones (series of chlorofluoroketones including hexafluoroacetone), chlorofluoroacetic acids; fluorine and halogen fluorides; SF_6; series of low-boiling fluorocarbons (Genetrons)	Technical data bulletin available on inquiry; halogenating reagents available under trade name of Baker and Adamson reagents
American Potash & Chem. Corp.	3000 W. 6th St., Los Angeles, Calif. 90054	Alkali metal fluorides, particularly cesium fluoride	
Armour Agricultural Chemical Co.	685 DeKalb Industrial Way, Decatur, Ga.	Laboratory and semicommercial production of fluorinated phosphorus derivatives such as PF_5, PF_3, POF_3 and fluorophosphoric acids	Technical data and literature survey available on some products
Columbia Organic Chem.	912 Drake St., Cedar Terrace, Columbia, S.C. 29209 (Mailing address: P.O. Box 5273)	Good selection of highly fluorinated organic reagents	
E. I. du Pont de Nemours & Co., Inc., Product Information Center	1007 Market St., Wilmington, Del. 19898	Variety of fluorocarbons and chlorofluorocarbons (well known	Primarily commercial suppliers; data sheets with extensive back-

		Freons), selected fluorinated olefins and telomers and polymers (esp. polytetrafluoroethylene (Teflon), polyvinylfluoride (Tedlar), hexafluoropropylene, :, hexafluoropropylene oxide (Freon E series), hexafluoroacetone, and hexafluoroisopropanol (commercial solvent))	ground information available on request; originally supplied SF_4 (with data sheet) but not available any longer
Eastman Distillation Products Industries, Eastman Organic Chem. Dept.	Rochester, N.Y. 14603	Selection of organic reagents which includes a sprinkling of fluorinated materials	Source of a selection of fluorinated organic reagents; should be consulted early
Halocarbon Products Corp.	82 Burlews Court, Hackensack, N.J. 07601	Laboratory and bulk supplies of trifluoroacetic acid and anhydride, trifluoroethanol and halocarbons, especially halocarbon lubricants (highly inert)	Technical data available on use and handling of products
Harshaw Chemical Div. of the Kewanee Oil Co.	Barclay Bldg., 1 Belmont Ave., Bala Cynwyd, Pa. 19004	Hydrogen fluoride and variety of inorganic fluorides useful as fluorinating agents; fluorine generator available	Commercial manufacturers and laboratory suppliers. Excellent data sheets available on hydrogen fluoride
Henley & Co., Inc.	202 E. 44 St., New York, N.Y. 10017	Fluorinated aromatic and alicyclic compounds	Offers data on physical properties and suggested uses
Hooker Chem. Corp., Industrial Chem. Div.	1200 47 St., Niagara Falls, N.Y. 14302	Few benzotrifluorides and fluorolubes	Used to supply much larger variety of fluorinated derivatives
Hynes Chemical Research Corp. (Agent, Bodman Chemicals)	308 Bon Air Ave., Durham, N.C. 27704 (106 N. Essex Ave., Narberth, Pa. 19072)	Variety of fluorinated organic intermediates available, particularly those derived from chlorofluoroacetones and hexafluoroacetone	Limited source of fluorinated reagents for laboratory use

Appendix C (cont.)

Name of company	Address	Type of fluorinated chemical available	Comments (data sheets and other information)
Imperial Smelting Corp. Ltd.	6 St. James's Square, London, S.W.1, England	Hexafluorobenzene and a wide variety of perfluoroaromatic compounds (over 70 compounds)	Laboratory and commercial suppliers
K & K Laboratories Inc.	121 Express St., Engineers Hill Plainview, N.Y. 11803	Variety of fluorinated organic reagents including fluorinated heterocyclic compounds such as fluoropyridines	
The Matheson Co.	P.O. Box 85, East Rutherford, N.J.	A wide variety of fluorinated gases, particularly those useful as fluorinating agents (SF_4, COF_2, halogen fluorides) and reactive fluoroorganic intermediates (tetrafluoroethylene, other fluorinated olefins, hexafluoroacetone)	Commercial and laboratory supplier. Data sheets available on many reagents; gas catalog lists products
Minnesota Mining & Manufacturing Co., Product Information Center	2501 Hudson Rd., St. Paul, Minn. 55119	Fluorochemical acids, liquids, plastics, elastomers, oils, waxes, greases	Technical information available from Product Information Center
Moleculon Research Corp.	139 Main St., Cambridge, Mass. 02142	α,β,β-Trifluorostyrene as monomer	
The Nalge Co. Inc., Dept. 27864	75 Pamorama Creek Drive, Rochester, N.Y. 14625	Suppliers of Nalgene laboratory ware made of Teflon FEP-fluorocarbon resins	Chemically inert labware excellent for working with fluorinated materials that attack glass (such as HF)
Ozark-Mahoning Co.	310 W. 6th St., Tulsa, Okla. 74119	Source of a variety of unusual inorganic fluorinating agents	Interested in developing efficient methods of synthesizing inorg-

		(AsF_3, TiF_4, precursors of NOF, PF_5), principally of fluorine with antimony, arsenic, boron, iron, phosphorus, tin, and titanium	anic fluorine compounds of potential usefulness to industry
Peninsular ChemResearch Inc. (subsidiary of Calgon Corp.)	Box 14318, Gainesville, Fla. 32601	A varied selection of fluorinating agents and organofluorine compounds, including many unusual intermediates not available from other sources	Suppliers of both laboratory and bulk quantities; custom synthesis available; a good source of reagents and intermediates for laboratory work
Pennsalt Chem. Corp.	3 Penn Center, Philadelphia, Pa. 19102	Perchloryl fluoride and fluorinated olefins	Product information bulletin available on perchloryl fluoride
Pierce Chemical Co.	P.O.Box 117, Rockford, Ill. 61105	A varied selection of fluorinating agents and organofluorine compounds, particularly aromatic derivatives with fluorinated substituents [such as $ArCF_3$, $ArOCF_3$, $ArSCF_3$, and fluorine (mono to penta) substitution]	Suppliers of both laboratory and bulk quantities; custom synthesis available on request; a good source of reagents and intermediates for laboratory work
Narmco Research and Development Div. of Whittaker Corp.	3540 Aero Court, San Diego, Calif. 92123	Supplies hexafluorobenzene for laboratory use at a reasonable price	

Appendix D.

PROBLEM SOLUTIONS AND INFORMATION

In writing problems it was desired to illustrate thoroughly the various important points of organic fluorine chemistry. Usually the problems were prepared from the text and from the tables. For this reason the answers can usually be found by referring to the appropriate portion of the text or tables. We have provided many of the answers in the form of reference numbers which will lead to a rapid check.

We have not tried to be comprehensive in the answers. Many excellent alternative solutions are not included. We encourage the reader to find satisfactory alternative solutions.

ANSWERS

CHAPTER 4

1. Sections 2-1 and 4-1.

2. (a) Table 4-1, Ref. 4-6.
   (b) Table 4-1, Ref. 4-5.
   (c) Table 4-1, Ref. 4-5.
   (d) Table 4-1, Ref. 4-10.
   (e) Table 4-2, Ref. 4-22.

3. Ref. 4-6 and 4-7.

4. (a) Table 4-3, Ref. 4-29.
   (b) Table 4-3, Ref. 4-30.
   (c) Table 4-3, Ref. 4-32.
   (d) Table 4-3, Ref. 4-34.
   (e) Table 4-3, Ref. 4-38.

5. Section 4-3A.1, Table 4-3.

6. (a) Fluorocyclohexane.
   (b) Table 4-4, Ref. 4-41.
   (c) Table 4-4, Ref. 4-41.
   (d) 1-Chloro-1-fluoroethane.
   (e) 1,1-Dichloro-1-fluoro-(28%) and 1-chloro-1,1-difluorobutane (15%).
   (f) 1,2-Dichloro-2-fluoropropane.

7. Section 4-3B, Table 4-5.

8. (a) Add HF to 1-hexyne, Table 4-5.
   (b) Add HF to 3-hexyne, Table 4-5.
   (c) Diels–Alder reaction of maleic anhydride and butadiene; add HF to product and hydrolyze carefully.
   (d) Diels–Alder reaction as in (c); add F_2 to product with $Pb(OAc)_4$–HF or F_2 at −78°C in $CFCl_3$.

9. (a) Table 4-1, Ref. 4-5.
   (b) Table 4-2, Ref. 4-24.
   (c) $[C_6H_5N(CF_3)S-]_2$, Ref. 4-21.
   (d) Table 4-2, Ref. 4-22.
   (e) 2,2-Difluoropropane, Ref. 4-20.

12. (a) Chlorinate toluene to α,α,α-trichlorotoluene and fluorinate with HF or SbF_3. Table 4-7, Ref. 4-112.
   (b) Chlorinate propane to octachloropropane; dehalogenate with Zn to hexachloropropene and fluorinate with SbF_3. Ref. 4-112a.
   (c) Fluorinate with BrF_3, Ref. 4-20; or convert to 2,2-dichloropropane, then 1-propyne, and add HF; or treat 2,2-dichloropropane with SbF_3–Br_2. Ref. 4-118.
   (d) Add Br_2 and fluorinate with SbF_3, Ref. 4-109.
   (e) Chlorinate to carbon tetrachloride; then HF–Sb(V) at 240°C, 70 atm, Ref. 4-121.
   (f) Fluorinate with SbF_3–Sb(V), 155°C, 2 hours, 6 atm, Ref. 4-131.

13. See article by Barbour, Belf, and Buxton in Ref. A6-3. Also see Hudlicky, Ref. Al, pp. 87–133 and Houben–Weyl, Ref. A2, pp. 145–212.

16. (a) Add HF at elevated temperature (BF_3 catalyst) or add HI and fluorinate with AgF. Convert fluoroalcohol to tosylate and treat with KF in glycol.
   (b) AgF in wet CH_3CN, 30°–40°C; Ref. 4-147.
   (c) HgO–HF, $<$ 0°C.
   (d) Chlorinate and couple to 1,2-diphenyltetrachloroethane. Fluorinate with HgO–HF, $<$ 20°C; Ref. 4-157.
   (e) PBr_3–Br_2 to α-bromoacetic acid; fluorinate with KF or AgF and reduce acid function.
   (f) Oxidize to butyraldehyde. Add HCN, tosylate OH, and fluorinate with KF in diethylene glycol. Ref. 4-171.
   (g) *N*-Bromosuccinimide to 1,4-dibromobutene-2. Add F_2 to double bond with $Pb(OAc)_4$–HF or F_2 at −78°C. Fluorinate with KF in ethylene glycol or with AgF or HgO–HF.

(h) Tosylate and treat with KF in glycol; or Ref. 4-164.
(i) Couple (Grignard) to 1,5-hexadiene and add HBr anti-Markownikoff.
(j) Add HI to open ring. React first with KCN and then with tosyl chloride. Fluorinate with KF in glycol.

CHAPTER 5

2. (a) F_2–H_2O to *N,N*-difluorourea; then H_2SO_4. Refs. 5-101, 5-102.
   (b) F_2–H_2O; Ref. 5-104.
   (c) F_2–H_2O, HCO_3^- to $C_6H_{11}NF_2$; dehydrofluorinate with CsF in CH_3CN; Ref. 5-105.
   (d) Na in toluene, then ClO_3F; Ref. 5-113.
   (e) Product from (d); Na in toluene, then ClO_3F; Ref. 5-113.
   (f) Malonate condensation with butyl iodide. Form enolate of product, add ClO_3F; decarboxylate. Ref. 5-96c.
   (g) F_2 at $-78°C$ in $CFCl_3$; Ref. 5-3a.

3. (a) Peroxidize and add HF; Ref. 5-39a.
   (b) Peroxidize, add HF, react with tosyl chloride, fluorinate with KF in ethylene glycol.
   (c) Dehydrate, peroxidize, add HF; convert alcohol to tosylate and fluorinate with KF.
   (d) Peroxidize and add HF.
   (e) Form adduct of CsF and HFA. React with bromide. Add HF to epoxide.

6. (a) Table 5-4, Ref. 5-149a.
   (b) Table 5-4, Ref. 5-149a.
   (c) Table 5-4, Ref. 5-163.
   (d) Table 5-4, Ref. 5-150b.
   (e) Table 5-4, Ref. 5-150g.
   (f) Table 5-4, Ref. 5-158.
   (g) Table 5-4, Refs. 5-150h, 5-157a and b.

8. (a) Phthalic acid plus SF_4; Ref. 5-180.
   (b) Cycloaddition of $CF_2{=}CF_2$ and $CH_2{=}CHCOOCH_3$. React with SF_4; Ref. 5-180.
   (c) Balz-Shiemann first on nitroaniline, then on 4-fluoroaniline, or COF_2 on hydroquinone and pyrolyze.
   (d) Isobutyric acid plus SF_4; Ref. 5-216.
   (e) React COF_2 and CH_3OH; react product with SF_4. Refs. 5-187a, 5-207c.

(f) React 4–Cl–C_6H_4OH and COF_2; then SF_4. Refs. 5-187a, 5-207c.
(g) React 4-heptanone with SF_4; Ref. 5-180.
(h) React heptanal with SF_4; Ref. 5-180.
(i) React aniline with COF_2; then SF_4. Ref. 5-207c.
(j) React 3β-hydroxy steroid with $(ClCFHCF_2)N(C_2H_5)_2$ in CH_2Cl_2; Ref. 5-174d.
(k) Use KF on *n*-butyl tosylate or halide; or use AgF on butyl iodide.
(l) Phenyl magnesium bromide plus hexafluoroacetone; use SF_4 to replace OH. Ref. 5-182.

9. (a) Refs. 5-5c or 5-6c.
(b) Ref. 5-48.
(c) Ref. 5-67c.
(d) Ref. 5-65.
(e) Ref. 5-83.
(f) Ref. 5-82b.
(g) Ref. 5-82c.
(h) Table 5-2, Ref. 5-90b.
(i) Table 5-2, Ref. 5-90a.
(j) Ref. 5-96d.
(k) Table 5-2, Ref. 5-96a.
(l) Peroxidize, add HF, oxidize. See Ref. 7-15.

10. (a) Tosylate 1-pentanol; react with KF in glycol.
(b) Use KF on ditosylate of 1-pentane-1,5-diol.
(c) Add IF to 1-pentene and react with AgF. Or add HF to epoxide and convert to tosylate; then KF.
(d) React 3-pentanone with SF_4.
(e) React 2,4-pentanedione with SF_4.
(f) Difluorinate 2,4-pentanedione with ClO_3F; then SF_4.
(g) Alkylate diethylmalonate with allyl bromide. Introduce fluorine with ClO_3F. Saponify. Add HF to double bond. Decarboxylate. Fluorinate with SF_4.
(h) Claisen ester condensation to obtain 4-ketopentanoic acid; then SF_4.
(i) React glutaraldehyde with SF_4.
(j) Fluorinate pentane over CoF_3.
(k) Add HF to 1,2-epoxypentane; react product with COF_2 and then SF_4.
(l) React half-ester of glutaric acid with SF_4 and saponify.
(m) Add BrF to 4-pentenoic acid; then Hunsdieker reaction.
(n) Chlorinate product from (f).
(o) Isn't this challenging?

11. (a) Schiemann on aniline.
    (b) Sequential Schiemann via *o*-nitroaniline.
    (c) React SF_4 and quinone; or sequential Schiemann.
    (d) Sequential Schiemann via *m*-nitroaniline.
    (e) Schiemann on 4-aminobenzoic acid; then SF_4.
    (f) Friedel–Crafts on *N*-acetylaniline with trifluoroacetyl chloride; then SF_4.
    (g) React COF_2 with 4-hydroxybenzoic acid; then SF_4.
    (h) Benzene plus COF_3; dehalogenate products with Fe.
    (i) Hexafluorobenzene plus methyl lithium.
    (j) Hexafluorobenzene plus sodium methoxide.
    (k) React 3,6-dichlorophthalic anhydride with KF at 270°C and then with SF_4. Ref. 5-172.
    (l) Table 5-4, Ref. 150d.
    (m) React 1,3-dichloro-4,6-dinitrobenzene with KF.
    (n) Hexafluorobenzene plus one equivalent $LiAlH_4$.

12. (c) Ref. 5-3a.
    (d) Ref. 5-90b.
    (e) Ref. 5-90a.
    (f) Ref. 5-39b.
    (g) Ref. 5-41a.

13. (a) Cycloadd $CF_2{=}CF_2$ and ethyl propynoate; then SF_4.
    (b) Dimerize dimethyl ketene and react product with SF_4.
    (c) Add HF to cyclobutene.
    (d) Add HF to cyclopentene; or KF displacement on tosylate.
    (e) ClO_3F on enolate ion.
    (f) Add BrF to cyclooctene and eliminate HBr.
    (g) React SF_4 with $HOOC{-}C{\equiv}C{-}COOH$.
    (h) React 1,1,1-trichlorobutane with HF–Sb(V).

CHAPTER 6

2. (a) $C_3F_7CH_2CHICH_3$
   (b) $C_3F_7CH_2CHICF_3$
   (c) $C_3F_7CH{=}CICH_3$
   (d) $C_3F_7CH_2CHICN$ (plus telomers).
   (e) $C_3F_7CH_2CHICOOCH_3$ + cyclic lactone; Ref. 6-25.
   (f) $(CH_2)_2CICH_2C_3F_7$
   (g) $C_3F_7CH_2CHFI + C_3F_7CHFCH_2I$ (plus telomers).
   (h) $C_3F_7CH(CF_3)CF_2I$

(i) $C_3F_7CF_2CFClI$ (plus telomers). Recent work with other systems suggests $C_3F_7CFClCF_2I$ should also occur.
(j) 3-(Perfluoropropyl)-5-iodonortricyclene.
(k) $C_3F_7CH_2CHICH_2CH_2CH{=}CH_2$, $C_3F_7CH_2CHICH_2CH_2CHICH_2C_3F_7$, 1-(Iodomethyl)-2-(2,2,3,3,4,4,4-heptafluorobutyl)-cyclopentane.

8. (a) Ref. 6-234, also Refs. 6-250, 6-265, 6-278.
(b) Refs. 6-241, 6-246, 6-243, 6-279.
(c) Add CF_3I to cyclohexene and dehydroiodinate.
(d) Ref. 6-42.
(e) Refs. 6-197, 6-199.
(f) Ref. 6-57.
(g) Ref. 6-243.
(h) Ref. 6-191f.
(i) Ref. 6-192g.
(j) Propene + N_2F_4 at 150°C.
(k) Ref. 6-29.
(l) Ref. 6-191d.
(m) Ref. 6-54.

9. (a) Refs. 6-(324–330).
(b) Refs. 6-330, 6-331.
(c) Refs. 6-325, 6-330, 6-335.
(d) Ref. 6-336.
(e) Refs. 6-374, 6-375.
(f) Ref. 6-318e.
(g) Ref. 6-318g.
(h) Refs. 6-319a and b.
(i) Ref. 6-319e.
(j) Ref. 6-320a.
(k) Ref. 6-320c.
(l) Ref. 6-320b.
(m) Ref. 6-319h.
(n) Ref. 6-318j.

10. Answers in text Sections 6-2A and B or in Refs. 6-435a, b.

CHAPTER 7

2. (a) KF, I_2, CH_3CN on perfluoroisobutylene; Ref. 5-94.
(b) KF in *N*-methylpyrrolidone on C_5Cl_8; Ref. 5-149a.
(c) KF at 230°C on tetrachlorophthaloyl chloride; Ref. 5-168.

(d) KF at 500°C on perchloropyridine; Refs. 5-157a, b.
(e) Refs. 6-15, 6-16.
(f) Ref. 6-53.
(g) Ref. 6-319b.
(h) Ref. 4-22.
(i) Ref. 5-5c.

3. (a) React ClO_3F with enolate of diethyl ethylmalonate and decarboxylate; Ref. 5-96c.
(b) React ClO_3F with 3-carbethoxycamphor and decarboxylate; Ref. 5-129.
(c) React SF_4 and tropolone; Ref. 5-180.
(d) Add HF to α-pinene.
(e) Add $C_7F_{15}I$ to ethyl but-3-enoate; Ref. 6-21.
(f) Add $(CF_3)_2CFI$ to cyclohexene and dehydroiodinate; Refs. 6-10, 6-23.
(g) Use $(C_4H_9)_3P$ and $ClCF_2COONa$ to generate ylid to react with cyclohexanone; Ref. 6-195d.
(h) Ref. 6-254.
(i) Ref. 6-255.
(j) Ref. 6-319b.

4. (i) Use Br_2, AgF in benzene to add BrF to styrene; Ref. 5-90e.
(j) Add BrF from *N*-bromoacetamide and HF to methyl acrylate; Ref. 5-99.
(k) Add BrF to ethyldiazoacetate; Ref. 5-97.
(l) Best is SF_4 on phthalic acid.
(m) Best is SF_4 on dimethylketene dimer.
(n) Ref. 6-234.
(o) Ref. 6-242.
(p) Ref. A7-1, p. 282.
(q) Ref. A7-1, p. 281.
(r) Ref. A2, p. 254.
(s) Ref. A2, p. 257.
(t) Ref. A2, p. 285.
(u) Ref. A2, p. 308.

5. (a) Add SF_4 to $CF_3CF{=}CF_2$; Ref. 5-88.
(b) Ref. 5-89b.
(c) Ref. 5-105.
(d) Ref. 6-121b.
(e) Add N_2F_4 to 1-octene.
(f) Ref. 6-144b.
(g) Ref. 6-52.

(h) Ref. 6-188.
(i) Ref. 6-197.
(j) Refs. 6-347, 6-348.
(k) Ref. 6-319f.
(l) Ref. 5-68.
(m) Ref. A2, p. 307.
(n) Ref. A2, p. 307.
(o) Ref. A2, p. 277.
(p) Ref. A2, p. 278.
(q) Ref. 5-57.
(r) Ref. A2, p. 279.
(s) Ref. A2, p. 282.
(t) Ref. A2, p. 280.
(u) Ref. A2, p. 283.

6. (a) Table 4-3, Ref. 4-37.
(b) Ref. 4-138.
(c) Ref. 4-139.
(d) Use $FClO_3$ to introduce 9α-fluorine. Then form methyl sulfonate at 21-position and treat with KF in dimethyl sulfoxide; Ref. 4-170.
(e) Ref. 5-23.
(f) Ref. 5-39d.
(g) Use NOF in CCl_4; Ref. 5-72.
(h) Ref. 5-174d.
(i) Use $CF_2{=}CFMgX$

Addendum Current Significant References

The cut-off date for inclusions of published literature in the book was March 1968, although a few more recent references were inserted. This addendum lists significant references that have appeared through the remainder of 1968. We have assigned these to the appropriate sections in order of appearance in the book, and cross-referenced where necessary.

CHAPTER 2 THE NATURE OF FLUORINE AND FLUORIDE ION

AD2-1 The energy of the O–F bond in trifluoromethyl hypofluorite. J. Czarnowski, E. Castellano, and H. J. Schumacher, *Chem. Commun.*, 1255 (1968).

Section 2-1 Fluorine and Fluorination by Fluorine

AD2-2 The effective shape of the covalently bound fluorine atom. S. C. Nyburg and J. T. Szymanski, *Chem. Commun.*, 669 (1968).

CHAPTER 3 THE CARBON-FLUORINE BOND

Section 3-1A Bonding characteristics on saturated carbon

AD3-1 Bond strengths in fluorinated compounds: decomposition of chemically activated 1,1,1-trifluoropropane. P. Cadman, D. C. Phillips, and A. F. Trotman-Dickenson, *Chem. Commun.*, 796 (1968).

AD3-2 Structural chemistry of donor-acceptor interactions. See p. 628 and following for discussion of C–F bond length and some other properties resulting from fluorine substitution. H. A. Bent, *Chem. Rev.*, **68**, 587 (1968).

AD3-3 The polar effects on base strength of fluorine and other substituents have been determined for β-substituted ethylamines. P. Love, R. B. Cohen, and R. W. Taft, *J. Am. Chem. Soc.*, **90**, 2455 (1968).

AD3-4 Acidity of hydrocarbons. XXIX. Kinetic acidities of benzal fluoride and 9-fluorofluorene. A pyramidal benzyl anion. The results are generalized with the working hypothesis: a fluorine substituent stabilizes a pyramidal or nonconjugated anion but can destabilize a conjugated anion. A. Streitwieser, Jr., and F. Mares, *J. Am. Chem. Soc.*, **90**, 2444 (1968).

AD3-5 Enhancement of equilibrium acidities of hydrocarbons by polyfluoroaryl substitution. R. Filler and C.-S. Wang, *Chem. Commun.*, 287 (1968).

Section 3-1B Bonding characteristics in unsaturated systems

AD3-6 Thermodynamic properties of fluorine compounds. Heat capacity and entropy of pentafluorochlorobenzene and pentafluorophenol. R. J. L. Andon, J. F. Counsell, J. L. Hales, E. B. Lees, and J. F. Martin, *J. Chem. Soc.* (A), 2357 (1968).

AD3-7 Enthalpies of addition of methanol, pyrrolidine, and 1-butenethiol to $CF_2{=}CCl_2$. Paper reports quantitative data showing addition is strongly exothermic and compares results to ethylene additions. J. Hine and F. E. Rogers, *J. Am. Chem. Soc.*, **90**, 6701 (1968).

AD3-8 (a) Cycloaddition. VI. Competitive 1,2- and 1,4-cycloaddition of 1,1-dichloro-2,2-difluoroethylene to butadiene. J. S. Swenton and P. D. Bartlett, *J. Am. Chem. Soc.*, **90**, 2056 (1968); see also *ibid.*, 6067. (b) V. 2-Alkylbutadienes and 1,1-dichloro-2,2-difluoroethylene. The effect of diene conformation on mode of cycloaddition. P. D. Bartlett, G. E. H. Wallbillich, A. S. Wingrove, J. S. Swenton, L. K. Montgomery, and B. D. Kramer, *ibid.*, 2049. (c) X. Reversibility in the biradical mechanism of cycloaddition. Tetrafluoroethylene and 1,1-dichloro-2,2-difluoroethylene with 2,4-hexadiene. P. D. Bartlett, C. J. Dempster, L. K. Montgomery, K. E. Schueller, and G. E. H. Wallbillich, *J. Am. Chem. Soc.*, **91**, 405 (1969).

AD3-9 The accepted stabilizing effect of fluorine on small strained ring systems such as cyclobutanes does not agree with the results in recent papers (and also with the observation that the FCF angle in Difluoromethylenes is less than expected, p. 24). (a) 11,11-Dihalogeno-1,6-methano-[10] annulenes as dihalocarbene transfer agents. The difluoro is most stable in the open form, whereas the dichloro and dibromo favor the closed form with the halogens on the cyclopropane ring. V. Rautenstrauch, H. J. Scholl, and E. Vogel, *Angew. Chem. Intern. Ed. Engl.*, **7**, 228 (1968). (b) Hexafluorobenzene adds carbenes to give a stable hexafluorotropilidine system, whereas benzene gives the norcaradiene (with a cyclopropane ring). D. M. Gale, *J. Org. Chem.*, **33**, 2536 (1968); M. Jones, Jr., *ibid.*, 2538. (c) A fluorine substituent in bullvalene is more highly favored on the olefinic position than the cyclopropane position in contrast to other halogen substituents. J. F. M. Oth, R. Merényi, H. Röttele, and G. Schröder, *Tetrahedron Letters,* 3941 (1968). The small ring system would be stabilized by fluorine substitution if the predominant factors were increased s-orbital contribution in the C–C bond (less strain energy in small rings) or relief of repulsive interactions between geminal fluorines. However, significant p-p attractive interactions between geminal fluorines would destabilize small rings. The systems in the above references all involve a π system. We believe that repulsive effects between a fluorine and a π system may dominate over all other factors. A study of models of the above systems suggests that p-π repulsion would be at a minimum if the compounds existed in the open form. For other halogens the repulsions would be of less consequence in spite of larger steric bulk because of the longer bond and more polarizable atom. Predictions on any new system are not definitive at this time because of these conflicting effects, but an examination of molecular models could be instructive.

AD3-10 Rotational isomers of the 1-fluoro-2-haloethanes. More refined data on this problem. M. F. El Bermani and N. Jonathan, *J. Chem. Phys.,* **49**, 340 (1968).

AD3-11 Through space interaction is well documented in several recent papers on through space coupling. (a) Nuclear magnetic resonance studies of ^{19}F-^{19}F spin-spin coupling in 1-substituted 4,5-difluoro-8-methylphenanthrene. Evidence of long-range coupling. K. L. Servis and K. N. Fang, *J. Am. Chem. Soc.,* **90**, 6712 (1968). (b) F-F spin-spin coupling in perfluoro-*o*-xylene. J. Jonas, *J. Chem. Phys.,* **47**, 4884 (1967). (c) Dihedral angle and bond angle dependence of vicinal proton-fluorine spin-spin coupling. K. L. Williamson, Y.-F. Li Hsu, F. H. Hall, S. Swager, and M. S. Coulter, *J. Am. Chem. Soc.,* **90**, 6717 (1968). (d) Long-range proton-fluorine spin-spin coupling in 5-fluoropyrimidine nucleosides. R. J. Cushley, I. Wempen, and J. J. Fox, *J. Am. Chem. Soc.,* **90**, 709 (1968). (e) Long-range proton-fluorine coupling in a rigid system. The nmr spectra of 5-substituted difluorotetrachlorobicyclo[2.2.1]heptenes. K. L. Williamson and J. C. Fenstermaker, *J. Am. Chem. Soc.,* **90**, 342 (1968). (f) The angular dependencies of long-range ^{19}F-^{1}H coupling constants. A. B. Foster, R. Hems, L. D. Hall, and J. F. Manville, *Chem. Commun.,* 158 (1968).

AD3-12 Fluorine and proton nmr studies of the conformations of 4,4-difluorocyclohexanol and its benzoate. The gem-difluoro group stabilizes the axial conformation; possible explanations are discussed. R. D. Stolow and T. W. Giants, *Tetrahedron Letters,* 5777 (1968).

AD3-13 Evidence for p-π interaction from signs of spin-spin coupling constants between methyl protons and ring fluorine nuclei in fluorotoluene derivatives. Further evidence for a positive hyperfine interaction in the C–F bond. D. J. Blears, S. S. Danyluk, and T. Schaefer. *J. Chem. Phys.,* **47**, 5037 (1967).

AD3-14 The relationship between charge density and the chemical shifts of fluorine nuclei in aromatic compounds. J. W. Emsley, *J. Chem. Soc.* (A), 2018 (1968).

AD3-15 Fluorine-19 chemical shifts in saturated systems. E. W. Della, *Chem. Commun.,* 1558 (1968).

Section 3-3A Chemical reactivity of saturated fluorocarbons

AD3-16 Boron fluorine-catalyzed alkylation. Ethylation with ethyl fluorine at low temperatures. R. Nakane, R. Oyama, and A. Natsubori, *J. Org. Chem.,* **33**, 275 (1968).

Section 3-3B Chemical reactivity of unsaturated systems

AD3-17 Phenylmagnesium bromide induced S_N2' reaction and migrations of fluorine in 1,1-bis(chlorodifluoromethyl)ethylene. J. G. Shdo, M. H. Kaufman, and D. W. Moore, *J. Org. Chem.,* **33**, 2173 (1968).

AD3-18 Proton exchange rates of fluorobenzene and benzotrifluoride with lithium cyclohexylamide. A. Streitwieser, Jr., and F. Mares, *J. Am. Chem. Soc.,* **90**, 644 (1968).

AD3-19 Hydrogen isotope exchange of polyfluorobenzenes with sodium methoxide in methanol. A. Streitwieser, Jr., J. A. Hudson, and F. Mares, *J. Am. Chem. Soc.,* **90**, 648 (1968).

AD3-20 Kinetic studies on reactions of fluoro- and chloro-2,4-dinitrobenzene. F. Pietra and D. Vitali, *J. Chem. Soc.* (B), 1200 (1968).

CHAPTER 4 CLASSICAL METHODS OF FLUORINATION

Section 4-2 Halogen fluorides with organic compounds

AD4-1 Fluorination of toluene with chlorine trifluoride in the gas phase. G. Schiemann, M. Kühnhold, and B. Cornils, *Ann.* **714**, 62 (1968).

Section 4-5 Hydrogen fluoride and/or antimony fluorides in substitution of halogen or oxygenated functions

AD4-2 Preparation of fluorocyclohexanes using HF on chlorides or 2-chloro-1,1,2-trifluoroethylamine on alcohols (for conformational studies). E. L. Eliel and R. J. L. Martin, *J. Am. Chem. Soc.,* **90**, 682 (1968); *ibid.,* 689.

Section 4-7 The Schiemann Reaction

AD4-3 Explosion hazards of diazonium fluoroborate. G. O. Doak and L. D. Freedman, *Chem. Eng. News.,* **46**, 8 (Dec. 18, 1968).

CHAPTER 5 MODERN SELECTIVE METHODS OF FLUORINATION

Section 5-1B Addition of F_2 from a fluorinating agent to unsaturated centers

AD5-1 On the nature of the lead tetrafluoride fluorinating agent. Lead(IV) diacetate difluoride was isolated and shown to be the active species in F–F addition to alkenes (not PbF_4). J. Bornstein and L. Skarlos, *J. Am. Chem. Soc.*, **90**, 5044 (1968).

AD5-2 Organic reactions of fluoroxy compounds: Electrophilic fluorination of activated olefins. CF_3OF reacts with vinyl ester to ethers and enamines (also dienes) of steroids to give mainly the corresponding α-fluoroketone in addition to adducts from both F–F and CF_3O–F addition. D. H. R. Barton, L. S. Godinho, R. H. Hesse, and M. M. Pechet, *Chem. Commun.*, 804 (1968).

AD5-3 Xenon difluoride as an oxidative fluorinator. Converted SO_2 to SO_2F_2. N. Bartlett and F. O. Sladky, *Chem. Commun.*, 1046 (1968).

AD5-4 Dimethylselenium difluoride, properties and preparation from dimethylselenide and AgF_2. K. J. Wynne and J. Puckett, *Chem. Commun.*, 1532 (1968).

Section 5-2 Addition of XF to unsaturated centers

AD5-5 Hydrofluoric acid with cycloalkene epoxides. Steric and electronic requirements elucidated (Section 5-2A). I. Shahak, Sh. Manor, and E. D. Bergmann, *J. Chem. Soc.* (C), 2129 (1968).

AD5-6 Perfluoroalkylsilver compounds by addition of AgF to fluoroalkenes (Section 5-2B). W. T. Miller, Jr., and R. J. Burnard, *J. Am. Chem. Soc.*, **90**, 7367 (1968).

AD5-7 Addition of CF_3O–F to olefinic bonds in steroids (Section 5-2F). See Ref. AD5-2.

AD5-8a Addition reactions of SF_5OF to alkenes. Preparation of some new pentafluorosulfuroxyalkanes and -alkenes (Section 5-2F). R. D. Place and S. M. Williamson, *J. Am. Chem. Soc.*, **90**, 2550 (1968).

AD5-8b The reaction of dioxygen difluoride and perfluoropropene: preparation of 1-fluoroperoxyperfluoropropane and 2-fluoroperoxyperfluoropropane. I. J. Solomon, A. J. Kacmarek, J. N. Keith, and J. K. Raney, *J. Am. Chem. Soc.*, **90**, 6557 (1968).

AD5-9 Preparation of perfluoroalkylsulfinyl halides, $R_fS(O)X$. SOF_2 was added to tetrafluoroethylene in presence of CsF. Trifluoromethylsulfur trifluoride was prepared from CF_3OF and CF_3SCl and converted to $CF_3S(O)X$, where X is F, Cl, or Br (Section 5-2F). C. T. Ratcliffe and J. M. Shreeve, *J. Am. Chem. Soc.*, **90**, 5403 (1968).

AD5-10 The addition of the elements of "BrF" and of "IF" (from halogen, AgF in benzene) to unsaturated carbohydrates (Section 5-2G). L. D. Hall and J. F. Manville, *Chem. Commun.*, **35**, (1968).

AD5-11 Trifluoromethyl hypochlorite by addition of ClF to fluorinated carbonyl reagents. D. E. Gould, L. R. Anderson, D. E. Young, and W. B. Fox, *Chem. Commun.*, 1564 (1968).

Section 5-3A Replacement of H by F using elementary fluorine in aqueous solution

AD5-12 Fluorodinitroacetonitrile was prepared by aqueous fluorination (with elementary fluorine) of sodium dinitroacetonitrile. R. A. Wiesboeck and J. K. Ruff, *J. Org. Chem.*, **33**, 1257 (1968).

AD5-13 Aqueous fluorination of nitronate salts by elementary fluorine was used to prepare a series α-fluoronitroaliphatics. V. Grakauskas and K. Baum, *J. Org. Chem.*, **33**, 3080 (1968).

AD5-14 Fluoronitroaliphatics—synthetic approaches and general properties. Preparation by aqueous fluorination of the carbanion with F_2 and by use of perchloryl fluoride. Methods compared. M. J. Kamlet and H. G. Adolph, *J. Org. Chem.*, **33**, 3073 (1968); H. G. Adolph, R. E. Oesterling, and M. E. Sitzmann, *ibid.*, 4296.

AD5-15 Fluorination of nitroaromatic amines in liquid hydrogen fluoride and acetonitrile with F_2. First report of an *N,N*-difluoroaniline. C. L. Coon, M. E. Hill, and D. L. Ross, *J. Org. Chem.*, **33**, 1387 (1968).

Section 5-3B Replacement of H by F using perchloryl fluoride

For fluorination of nitroaliphatic aliphatic compounds with perchloryl fluoride, see Ref. AD5-14.

AD5-16 Fluorination of methyl isobutyrate with perchloryl fluoride; preparation of 4-fluoro-3-oxo-2,2,4-trimethylpentanoate and α-fluoroisobutyrate. W. J. Gensler, Q. A. Ahmed, and M. V. Leeding, *J. Org. Chem.*, **33**, 4279 (1968).

AD5-17 Used perchloryl fluoride on α-diketones as early step in sequence to the 3,3-difluoro-2,4-dialkyloxetane system. R. A. Stein and K. B. Haack, *J. Org. Chem.*, **33**, 3784 (1968).

Section 5-4A Replacement of halogen or halogenlike groups, RX to RF

AD5-18 Cyclopropanecarboxylic acid fluoride. An improved synthesis from KF and 4-chlorobutyryl chloride. R. E. A. Dear and E. E. Gilbert, *J. Org. Chem.*, **33**, 1690 (1968).

AD5-19 Preparation by KF substitution and nucleophilic substitution of tetrafluoropyridazine. R. D. Chambers, J. A. H. MacBride, and W. K. R. Musgrave, *J. Chem. Soc.* (C), 2116 (1968).

AD5-20 Triphosphonitrilic and tetraphosphonitrilic fluoride chlorides by NaF fluorination. J. Emsley and N. L. Paddock, *J. Chem. Soc.* (A), 2590 (1968).

AD5-21 Preparation of fluorocyclohexanes using 2-chloro-1,1,2-trifluoroethylamine on cyclohexanols. See Ref. AD4-2.

AD5-22 Use of Et_2NCF_2CHClF to bring about the Beckman rearrangement under very mild conditions. R. L. Autrey and P. W. Scullard, *J. Am. Chem. Soc.*, **90**, 4924 (1968).

AD5-23 Reaction of nitro alcohols with sulfur tetrafluoride (replacement of OH by F). K. Baum, *J. Org. Chem.*, **33**, 1293 (1968).

AD5-24 Conversion of ethers to alkyl fluorides by methyl fluoroformate. Z. Suzuki and K. Morita, *Bull. Chem. Soc. Japan*, **41**, 1724 (1968).

AD5-25 Electrophilic fluorination of aromatic rings (including steroids) using CF_3OF. D. H. R. Barton, A. K. Ganguly, R. H. Hesse, S. N. Loo, and M. M. Pechet, *Chem. Commun.*, 806 (1968).

AD5-26 Aryl fluorides prepared from arenediazonium hexafluoroantimonates and -arsenates. C. Sellers and H. Suschitzky, *J. Chem. Soc.* (C), 2317 (1968).

AD5-27 Explosive hazards of diazonium fluroborate. See Ref. AD4-3.

AD5-28 4,4-Difluorocyclohexanone and 6,6-difluoro-*cis*-decal-2-ones, 5,5-difluoro-*cis*-hydrindan and 9-methyl-5,5-difluoro-*cis*-hydrindan, prepared by SF_4 reactions and used to study conformational equilibria and equilibration by nmr. R. E. Lack, C. Ganter, and J. D. Roberts, *J. Am. Chem. Soc.*, **90**, 7001 (1968); R. E. Lack and J. D. Roberts, *ibid.*, 6997.

CHAPTER 6 ADDITION OF FLUORINATED UNITS

Section 6-1A.1 Fluorocarbon Radicals

AD6-1 The preparation of perfluoroalkylene oxide polymers by iodide coupling reactions–irradiation of $O(CF_2CF_2I)_2$. D. E. Rice, *J. Polymer Sci. B.*, **6**, 335 (1968).

AD6-2 Differences in the reactivities of trifluoromethyl and methyl radicals produced by photodissociation of trifluoromethyl iodide. C. L. Kibby and R. E. Weston, Jr., *J. Am. Chem. Soc.*, **90**, 1084 (1968).

AD6-3 Free radical addition of trifluoroacetonitrile to propylene (gas phase, 400-450°); mechanism study shows trifluoromethyl radicals are intermediates. Products are 1,1-adducts. B. Hardman and G. J. Janz, *J. Am. Chem. Soc.*, **90**, 6272 (1968).

AD6-4 Study of cage reaction of trifluoromethyl radical generated by photolysis of $CF_3N{=}NCF_3$. Reacted with CH_3 radicals. K. Chakravorty, J. M. Pearson, and M. Szwarc, *J. Am. Chem. Soc.*, **90**, 283 (1968); O. Dobis, J. M. Pearson, and M. Szwarc, *ibid.*, 278.

AD6-5 Radical additions of silanes to vinyl fluoride and to 1-chloro-2-fluoroethylene. D. Cooper, R. N. Haszeldine, and M. J. Newlands, *J. Chem. Soc.* (A), 2098 (1967).

AD6-6 Electron spin resonance spectra of fluorinated aromatic free radicals. S. V. Kulkarni and C. Trapp, *J. Am. Chem. Soc.*, **91**, 191 (1969).

AD6-7 Molecular-orbital theory of geometry and hyperfine coupling constants of fluorinated methyl radicals. D. L. Beveridge, P. A. Dobosh, and J. A. Pople, *J. Chem. Phys.*, **48**, 4802 (1968).

Section 6-1A.2 Fluorinated heteroatom radicals

AD6-8 Kinetics of the thermal dissociation of N_2F_4 in shock waves. A. P. Modica and D. F. Hornig, *J. Chem. Phys.*, **49**, 629 (1968).

AD6-9 Radical reactions of tetrafluorohydrazine with allylic halides. S. F. Reed, Jr., *J. Org. Chem.*, **33**, 1861 (1968).

AD6-10 Radical reactions of tetrafluorohydrazine with saturated hydrocarbons and ethers. S. F. Reed, Jr., and R. C. Petry, *Tetrahedron*, **24**, 5089 (1968).

AD6-11 Radical reactions of tetrafluorohydrazine with cyclic olefins. S. F. Reed, Jr., *J. Org. Chem.*, **33**, 2634 (1968).

AD6-12 Photochemical reaction of perfluoroacyl fluorides with tetrafluorohydrazine as a synthesis of perfluoroamines. R. A. Mitsch and E. W. Newvar, *J. Org. Chem.*, **33**, 3675 (1968).

AD6-13 Reaction of NF_3 and N_2F_4 with a variety of CN compounds. U. Biermann, O. Glemser, and J. Knaak, *Chem. Ber.*, **100**, 3789 (1967).

AD6-14 Photolysis of tetrafluorohydrazine with cyclopropane. C. L. Bumgardner and E. Lawton, *Tetrahedron Letters*, 3059 (1968).

AD6-15 Chemical activation during photodifluoroamination. C. L. Bumgardner, E. L. Lawton, and H. Carmichael, *Chem. Commun.*, 1079 (1968).

AD6-16 Reactions of *N*-fluoriminonitriles. T. O. Stevens, *J. Org. Chem.*, **33**, 2660 (1968).

AD6-17 Free radical reactions of perfluoro-*N*-fluoropiperidine by irradiation. R. E. Banks, K. Mullen, and G. E. Williamson, *J. Chem. Soc.* (C), 2608 (1968).

AD6-18 Matrix-isolation study of the infrared spectrum of the free radical F_2CN. M. E. Jacox and D. E. Milligan, *J. Chem. Phys.*, **48**, 4040 (1968).

AD6-19 Free radical addition of trimethylsilane to trifluoronitrosomethane. A. C. Delany, R. N. Haszeldine, and A. E. Tipping, *J. Chem. Soc.* (C), 2537 (1968).

AD6-20 The thermal dissociation of diphosphorus tetrafluoride and the formation of tetraphosphorus hexafluoride. D. Solan and P. L. Timms, *Chem. Commun.*, 1540 (1968).

AD6-21 Bis(fluoroformyl)peroxide was decomposed to give radical fragments for reaction with reagents such as difluorodiazomethane as a route to new fluorinated peroxides. R. L. Talbott, *J. Org. Chem.*, **33**, 2095 (1968); see also, *ibid.*, 2099.

AD6-22 Bis(trifluoromethyl)semidiazoxide from electrolysis of CF_3NO and CF_3NO_2; esr studies. J. L. Gerloch and E. G. Janzen, *J. Am. Chem. Soc.*, **90**, 1652 (1968).

AD6-23 Bis(trifluoromethyl)nitroxide: a novel synthesis and some reactions. H. G. Ang, *Chem. Commun.*, 1320 (1968).

AD6-24 Preparation of mercury (II) bis(trifluoromethyl)nitroxide and its use in synthesis. H. J. Emeleus and P. M. Spaziante, *Chem. Commun.*, 770 (1968).

AD6-25 EPR study of the equilibrium between peroxydisulfuryl difluoride and fluorosulfate free radicals in the gaseous state. R. A. Seward and S. Fujiwara, *J. Chem. Phys.*, **48**, 5524 (1968).

AD6-26 Some new reactions of trifluoromethanesulfenyl ($CF_3S\cdot$) radical. B. W. Tattersall and G. H. Cady, *J. Inorg. Nucl. Chem.*, **29**, 2819 (1967).

Section 6-1B.1 Carbenes: methods of preparation and reactions of fluorinated carbenes

AD6-27 Phenylfluorocarbene generated from α-chloro-α-fluorotoluene and potassium *t*-butoxide and added to olefins. T. Ando, Y. Kotoku, H. Yamanaka, and W. Funasaka. *Tetrahedron Letters*, 2479 (1968).

AD6-28 Monofluorocarbene generated from $CHFBr_2$ and butyl lithium and its syn/anti selectivity in addition to olefins. M. Schlosser and G. Heinz, *Angew. Chem. Intern. Ed. Engl.*, **7**, 820 (1968).

AD6-29 Difluorocarbene from pyrolysis of difluorochloroacetonitrile and reaction with cyclohexene. A. R. Monahan, D. J. Perettie, and G. J. Janz, *Tetrahedron Letters*, 3955 (1968).

AD6-30 Experimental and theoretical kinetics of difluoromethylene from fluorocarbons by shock waves. A. P. Modica and S. J. Sillers, *J. Chem. Phys.*, **48**, 3283 (1968).

AD6-31 Reactions of dihalogenocarbenes with ring B steroid olefins: difluorocarbene from CF_2ClCO_2Na. F. T. Bond and R. H. Cornelia, *Chem. Commun.*, 1189 (1968).

AD6-32 Chemistry of difluorocarbene adducts to sterically hindered acetylenes. Difluorocarbene generated from CF_2ClCO_2Na in diglyme. P. Anderson, P. Crabbe, A. D. Cross, J. H. Fried, L. H. Knox, J. Murphy, E. Velarde, *J. Am. Chem. Soc.*, **90**, 3888 (1968).

AD6-33 Additions of difluorocarbene to an ene-yne system in a steroid molecule. Difluorocarbene from CF_2ClCO_2Na in diglyme. P. Crabbe, P. Anderson, and E. Velarde, *J. Am. Chem. Soc.*, **90**, 2998 (1968).

AD6-34 Insertion reactions of trifluoromethylcarbene generated photochemically from 2,2,2-trifluorodiazomethane (also cycloaddition reactions). J. H. Atherton and R. Fields, *J. Chem. Soc.* (C), 1507 and 2276 (1968).

AD6-35 Photolysis studies on trifluoroacetyl diazoacetic ester. H. Dworschak and F. Weygand, *Chem. Ber.*, **101**, 289, 302, 308 (1968).

AD6-36 Reductive defluorination-cyclization of R_fNFCNF_2-type compounds to form perfluorodiaziridines, and $-C(NF_2)_2$ to fluorodiazirines. R. A. Mitsch, *J. Org. Chem.*, **33**, 1847 (1968).

AD6-37 Anomalous reactions of bis(trifluoromethyl)diazomethane and bis(trifluoromethyl)diazirine with saturated hydrocarbons. W. J. Middleton, D. M. Gale, and C. G. Krespan, *J. Am. Chem. Soc.*, **90**, 6813 (1968).

AD6-38 The abstraction of oxygen by fluorinated carbenes (generated from diazirines). W. Mahler, *J. Am. Chem. Soc.*, **90**, 523 (1968).

AD6-39 Addition of dichloro- and difluorocarbene prepared by the trihalomethyl-metal route to vinyl and allyl compounds of carbon, silicon, and germanium. D. Seyferth and H. Dertouzos, *J. Organometal. Chem.*, **11**, 263 (1968).

AD6-40 The role of chlorofluorocarbene in the photolysis and light-induced oxidation of chlorotrifluoroethylene. W. J. R. Tyerman, *Chem. Commun.*, 392 (1968).

Section 6-1B.2 Structure and properties of fluorinated carbenes

AD6-41 (a) Microwave spectrum of CF_2. F. X. Powell and D. R. Lide, Jr., *J. Chem. Phys.*, **45**, 1067 (1966). (b) 2500-Å absorption spectrum of CF_2. C. W. Mathews, *J. Chem. Phys.*, **45**, 1068 (1966).

AD6-42 Mass spectrometric study of the high-temperature equilibrium $C_2F_4 \rightleftharpoons 2CF_2$ and heat of formation of the CF_2 radical. K. F. Zmbov, O. M. Uy, and J. L. Margrave, *J. Am. Chem. Soc.*, **90**, 5090 (1968).

AD6-43 Molecular orbital calculation on the electronic structure of methylenes, including difluoromethylene and other fluorinated methylenes. R. Hoffmann, G. D. Zeiss, and G. W. Van Dine. *J. Am. Chem. Soc.*, **90**, 1485 (1968).

Section 6-1B.3 Fluorinated heteroatom carbenes

AD6-44 The reaction of boronmonofluoride with acetylenes. P. L. Timms, *J. Am. Chem. Soc.*, **90**, 4585 (1968).

AD6-45 Microwave spectrum of SiF_2. V. M. Rao, R. F. Curl, Jr., P. L. Timms, and J. L. Margrave, *J. Chem. Phys.*, **43**, 2557 (1965).

AD6-46 Electron spin resonance studies of SiF_2. H. P. Hopkins, J. C. Thompson, and J. L. Margrave, *J. Am. Chem. Soc.*, **90**, 901 (1968).

Section 6-1C.1 Fluorinated Carbanions

(See also Refs. AD3-4, -5, -17 to -20.)

AD6-47 The preparation and hydrolysis of tertiary alkyl (perfluoroalkyl)-phosphines using heptafluoropropyllithium. K. Gosling, D. J. Holman, J. D. Smith, and B. N. Ghose, *J. Chem. Soc.* (A), 1909 (1968).

AD6-48 Perfluoroalkylsilver compounds. Ref. AD5-6.

AD6-49 Pentafluorophenylarsenic compounds from pentafluorophenylmagnesium bromide. M. Green and D. Kirkpatrick, *J. Chem. Soc.* (A), 483 (1968).

AD6-50 Reactions of polyfluoroaryllithium with dimethyl carbonate. R. D. Chambers and D. J. Spring, *J. Chem. Soc.* (C), 2394 (1968).

AD6-51 Synthesis and reactions of organolithium and -magnesium compounds of perfluorodiphenyl ethers. R. J. De Pasquale and C. Tamborski, *J. Organometal. Chem.*, **13**, 273 (1968).

AD6-52 Reactions of pentafluorophenyl and ortho-bromotetrafluorophenyl organometallic compounds. R. D. Chambers, J. A. Cunningham, and D. A. Pyke, *Tetrahedron,* **24**, 2783 (1968).

AD6-53 A new intramolecular nucleophilic arylation: synthesis of octafluoro-9-fluorenone using lithium reagents. R. Filler and A. E. Fiebig, *Chem. Commun.*, 606 (1968).

AD6-54 Synthesis of perhalostyrene compounds from pentafluorophenylcopper and perfluorovinyl iodides. C. Tamborski, E. J. Soloski, and R. J. De Pasquale, *J. Organometal. Chem.*, **15**, 494 (1968).

AD6-55 Fluoroolefin addition and substitution. See Refs. AD3-7 and AD3-8.

AD6-56 Reactions of 1,2-dichloroperfluorocycloalkenes and perfluorocycloalkenes with various trivalent phosphines. R. F. Stockel, F. Megson, and M. T. Beachem, *J. Org. Chem.*, **33**, 4395 (1968).

AD6-57 Polyfluoroalkylation of fluorinated aromatic systems. R. D. Chambers, J. A. Jackson, W. K. R. Musgrave, and R. A. Storey, *J. Chem. Soc.* (C), 2221 (1968).

AD6-58 Addition of perfluorobenzoyl fluoride (and related perfluorinated aromatic acid derivatives) to fluoroalkenes to give perfluorinated ketones. R. D. Chambers, C. A. Heaton, and W. K. R. Musgrave, *J. Chem. Soc.* (C), 1933 (1968).

AD6-59 Isomerization of β-substituted perfluoroolefins involving fluorocarbanions. Kinetic vs equilibrium control. D. J. Burton and F. E. Herkes, *J. Org. Chem.*, **33**, 1854 (1968).

AD6-60 Nucleophilic displacement reactions of fluorinated diphenyl ethers. R. J. De Pasquale and C. Tamborski, *Chem. Ind. (London),* 1438 (1968).

Section 6-1C.2 Fluorinated anions (on heteroatom)

AD6-61 The formation of the pentafluorosilicate anion in dehydrofluorination reactions. J. J. Harris and B. Rudner, *J. Am. Chem. Soc.*, **90**, 515 (1968).

AD6-62 The reaction of difluoroamine with carbonyl compounds and acetylenes is suggested to go by an ionic mechanism. K. Baum, *J. Am. Chem. Soc.*, **90**, 7083 and 7089 (1968).

AD6-63 Hexafluoroisopropoxides of metals. P. N. Kapoor, R. N. Kapoor, and R. C. Mehrotra, *Chem. Ind. (London),* 1314 (1968).

AD6-64 The dilithium salt of hexafluoropropane-2,2-diol. Pyrolyzed to tetrafluoroethylene and alkylated. P. H. Ogden and G. C. Nicholson, *Tetrahedron Letters,* 3553 (1968); J. H. Prager and P. H. Ogden, *J. Org. Chem.*, **33**, 2100 (1968).

AD6-65 Fluoride adducts of fluorinated ketones prepared and used in synthesis of fluorinated ethers by addition to olefins. F. W. Evans, M. H. Litt, A-M. Weidler-Kubanek, and F. P. Avonda, *J. Org. Chem.*, **33**, 1837 and 1839 (1968).

Section 6-1D Fluorocarbonium ions

AD6-66 Studies on carbonium ions from nitro compounds, including 2-fluoro-2-nitropropane, in $FSO_3H{-}SbF_5$ solvent to give the 2-fluoroisopropyl carbonium ion. G. A. Olah and T. E. Kiovsky, *J. Am. Chem. Soc.*, **90**, 6461 (1968).

AD6-67 Halonium ions were formed from 2-fluoro-3-halo-2-methylbutanes (including α-fluorocarbonium ions). G. A. Olah and J. M. Bollinger, *J. Am. Chem. Soc.*, **90**, 947 (1968).

AD6-68 Methylfluorobenzenonium ions. G. A. Olah and T. E. Kiovsky, *J. Am. Chem. Soc.*, **90**, 2583 (1968).

AD6-69 Addition of difluoroamine to aldehydes and ketones to give α-difluoroaminocarbinols which are a source of difluoroaminomethyl carbonium ion. J. P. Freeman, W. H. Graham, and C. O. Parker, *J. Am. Chem. Soc.*, **90**, 121 (1968).

AD6-70 The synthesis of fluorammonium salts. V. Grakauskas, A. H. Remanick, and K. Baum, *J. Am. Chem. Soc.*, **90**, 3839 (1968).

AD6-71 Octafluoronaphthalene radical cation. N. M. Bazhin, *Tetrahedron Letters,* 4449 (1968).

Section 6-2A Fluoroolefins

The following references report new fluorinated olefins that have been used or have potential in synthesis.

AD6-72 (a) Perfluorocyclopropene and fluorinated cyclopropanes (cycloadditions). P. B. Sargeant and C. G. Krespan, *J. Am. Chem. Soc.*, **91**, 415 (1969); D. C. F. Law and S. W. Tobey, *J. Am. Chem. Soc.*, **90**, 2376 (1968). (b) Perfluoroalkyl trifluorovinyl ketones. B. C. Anderson, *J. Org. Chem.*, **33**, 1016 (1968). (c) Polyfluorovinylamines, *N,N*-bistrifluoromethylvinylamine, trifluoromethyl trifluorovinyl ether. E. S. Alexander, R. N. Haszeldine, M. J. Newlands, A. E. Tipping, *J. Chem. Soc.* (C), 796 (1968); R. N. Haszeldine and A. E. Tipping, *ibid.*, 398. (d) Perfluoropenta-1,2-diene. R. E. Banks, A.

Braithwaite, R. N. Haszeldine, and D. R. Taylor, *J. Chem. Soc.* (C), 2593 (1968).

Further work on cycloaddition reactions, particularly on the mechanism, is given in Ref. AD3-8 and

AD6-73 (a) 10-Chloro-9-fluorobicyclo[6.2.0]deca-2,4,6,9-tetraene and 3-chloro-2-fluorobicyclo[2.2.0]hexa-2,5-diene from 1,1-dichloro-2,2-difluoroethylene and cyclooctatetrene. G. Schröder and Th. Martini, *Angew. Chem. Intern. Ed. Engl.*, **6**, 806 (1967). (b) Reaction of fluoroalkenes with pyridine N-oxide to give fluoroalkylpyridines. E. A. Mailey and L. R. Ocone, *J. Org. Chem.*, **33**, 3343 (1968). (c) Synthesis of 2-substituted tetrafluoropyridines by a 2 + 4 cycloaddition of perfluoro-1,3-cyclohexadiene with nitriles. L. P. Anderson, W. J. Feast, and W. K. R. Musgrave, *Chem. Commun.*, 1433 (1968).

Section 6-2B Fluorinated acetylenes

AD6-74 Perfluoro(methylacetylene), $CF_3C{\equiv}CF$. R. E. Banks, M. G. Barlow, W. D. Davies, R. N. Haszeldine, K. Mullen, and D. R. Taylor, *Tetrahedron Letters*, 3909 (1968).

AD6-75 Reactions of 3,3,3-trifluoropropyne with metal carbonyls. D. A. Harbourne and F. G. A. Stone, *J. Chem. Soc.* (A), 1765 (1968).

AD6-76 Reaction of hexafluorobutyne-2 with phosphorins to prepare substituted 1-phosphavarrelenes (1-phospha[2.2.2]octa-2,5,7-trienes). G. Markl and F. Lieb, *Angew. Chem. Intern. Ed. Engl.*, **7**, 733 (1968).

AD6-77 Synthesis of some fluoroaromatic acetylenes by nucleophilic substitution by phenylacetylide salts on perfluoroaromatics. P. L. Coe, J. C. Tatlow, and R. C. Terrell, *J. Chem. Soc.* (C), 2626 (1967).

AD6-78 Preparation of *N,N*-bistrifluoromethylethynylamines. J. Freear and A. E. Tipping, *J. Chem. Soc.* (C), 1096 (1968).

AD6-79 The following series records recent developments on fluorinated arynes: (a) S. C. Cohen, M. L. N. Reddy, D. M. Roe, A. J. Tomlinson, and A. G. Massey, *J. Organometal. Chem.*, **14**, 241 (1968). (b) S. C. Cohen, D. Moore, R. Price, and A. G. Massey, *J. Organometal. Chem.*, **12**, P37 (1968). (c) Cleavage of ethers and thioethers by tetrafluorobenzyne. J. P. N. Brewer, H. Heaney, and J. M. Jablonski, *Tetrahedron Letters*, 4455 (1968). (d) Cycloaddition reactions of the isomeric trifluorobenzynes. R. Harrison and H. Heaney, *J. Chem. Soc.* (C), 889 (1968). (e) Cycloaddition reactions of tetrachlorobenzyne (compared to tetrafluorobenzyne). H. Heaney and J. M. Jablonski, *J. Chem. Soc.* (C), 1895 (1968).

Section 6-2C Fluorinated Aromatic Compounds

Section 6-2C.1 Monofluoroaromatics

For nucleophilic substitution studies, see Refs. AD3-18 and -20. For preparation of arylfluorides by electrophilic fluorination, see Ref. AD5-2.

AD6-80 The *p*-fluoro substituent as a label for the study of mass spectral reactions. M. M. Bursey, R. D. Rieke, T. A. Elwood, and L. R. Dusold, *J. Am. Chem. Soc.,* **90**, 1557 (1968).

AD6-81 Nuclear magnetic resonance studies on fluorobenzenes: the effect of substituents on the meta and para F—F coupling constants. R. J. Abraham, D. B. Macdonald, and E. S. Pepper, *J. Am. Chem. Soc.,* **90**, 147 (1968).

Section 6-2C.2 Aromatic Compounds with Fluorinated Substituents

For the first report of *N,N*-difluoroanilines, see Ref. AD5-15.

AD6-82 Substituent parameters (σ_R°) on a series of substituents including fluorinated ones calculated by infrared intensity measurements. R. T. C. Brownlee, R. E. J. Hutchinson, A. R. Katritzky, T. T. Tidwell, and R. D. Topsom, *J. Am. Chem. Soc.,* **90**, 1757 (1968).

AD6-83 7-Trifluoromethoxy and 7-trifluoromethylthio derivatives of 1,4-benzodiazepines. F. J. McEvoy, E. N. Greenblatt, A. C. Osterberg, and G. R. Allen, Jr., *J. Med. Chem.,* **11**, 1248 (1968).

AD6-84 Photohydrolysis of monosubstituted benzotrifluorides. R. Grinter, E. Heilbronner, T. Petrzilka, and P. Seiler, *Tetrahedron Letters,* 3845 (1968).

Section 6-2C.3 Perfluoroaromatics

For perfluoroaryl metal derivatives and perfluorobenzynes, see also references in Sections 6-1C.1 and 6-2B. For nucleophilic substitution, see Refs. AD3-19 and AD5-19. Other references dealing with perfluoroaromatic compounds are AD3-6, AD3-9(b), AD3-11(b).

AD6-85 A large amount of data on chemical shifts and coupling constants in pentafluorophenyl derivatives have been tabulated and analyzed. M. G. Hogben and W. A. G. Graham, *J. Am. Chem. Soc.,* **91**, 283, 291 (1969); R. Fields, J. Lee, and D. J. Mowthorpe, *J. Chem. Soc.* (B), 308 (1968).

AD6-86 Chlorides of polyfluoroaromatic acids in the Friedel-Crafts reaction. N. N. Vorozhtsov, Jr., V. A. Barkhash, and S. A. Anichkina, *J. Org. Chem., U.S.S.R.* (English Transl.), **2**, 1871 (1966).

AD6-87 Preparation and chemistry of perfluorinated aromatic mono- and dicarboxylic acids. P. Sartori, *Angew. Chem. Intern. Ed. Engl.,* **6**, 994 (1967).

AD6-88 Reactions of tetrafluoroisonicotinic acid and pentafluorobenzoic acid. R. D. Chambers, C. A. Heaton, and W. K. R. Musgrave, *J. Chem. Soc.* (C), 1933 (1968).

AD6-89 Reactions of pentafluorobenzaldehyde. E. V. Aroskar, P. J. N. Brown, R. G. Plevey, and R. Stephens, *J. Chem. Soc.* (C), 1569 (1968).

AD6-90 2,3,5,6-Tetrafluoro-4-methylpyridine and related compounds. R. D. Chambers, B. Iddon, W. K. R. Musgrave, and L. Chadwick, *Tetrahedron,* **24**, 877 (1968).

AD6-91 The thermal isomerization of hexafluorobicyclo[2.2.0] hexa-2,5-diene to hexafluorobenzene. E. Ratajczak and A. F. Trotman-Dickenson, *J. Chem. Soc.* (A), 509 (1968).

AD6-92 The relative donor strengths of nitrobenzene and pentafluoronitrobenzene in complexing with boron halides (studied by nmr). E. F. Mooney, M. A. Qaseem, and P. H. Winson, *J. Chem. Soc.* (B), 224 (1968).

AD6-93 Perfluorophenyl silicon compounds. M. F. Lappert and J. Lynch, *Chem. Commun.*, 750 (1968).

AD6-94 Ligands containing fluoroaromatic groups. M. L. N. Reddy, M. R. Wiles, and A. G. Massey, *Nature,* **217**, 740 (1968).

Section 6-2D Fluorinated Carbonyl Reagents

Section 6-2D.1 Fluorinated ketones and aldehydes

For reaction of pentafluorobenzaldehyde, see Ref. AD6-89; for synthesis of polyfluoroaromatic ketones, see Refs. AD6-58, AD6-86; and for perfluoroalkyl trifluorovinyl ketones, see Ref. AD6-72(b). The following reactions of hexafluoroacetone (and/or related polyfluorinated ketones) are reported (see also Ref. AD6-65).

AD6-95 (a) With allylic olefins. R. L. Adelman, *J. Org. Chem.*, **33**, 1400 (1968). (b) With ketenimines. A-M. Weidler-Kubanek and M. H. Litt, *J. Org. Chem.*, **33**, 1844 (1968). (c) With alkylthiosilanes and disilthianes. E. W. Abel, D. J. Walker, and J. N. Wingfield, *J. Chem. Soc.* (A), 1814 (1968). (d) With silanes. R. E. A. Dear, *J. Org. Chem.*, **33**, 3959 (1968). (e) With cyclic silthianes. E. W. Abel, D. J. Walker, and J. N. Wingfield, *J. Chem. Soc.* (A), 2642 (1968). (f) With secondary phosphines. R. F. Stockel, *Chem. Commun.*, 1594 (1968). (g) With tertiary phosphines. F. Ramirez, C. P. Smith, J. F. Pilot, and A. S. Gulati, *J. Org. Chem.*, **33**, 3787 (1968). (h) With a Wittig reagent to give a stable four-ring adduct. G. Chioccola and J. J. Daly, *J. Chem. Soc.* (A), 568 (1968).

AD6-96 Other chemistry based on hexafluoroacetone or other polyfluorinated ketones: (a) Sulfur-containing esters of perfluoropinacol (from hexafluoroacetone). M. Allan, A. F. Janzen, and C. J. Willis, *Chem. Commun.*, 55 (1968). (b) Mechanism of thermal decomposition of fluoroacetone hemiketal esters. P. E. Newallis, P. Lombardo, and E. R. McCarthy, *J. Org. Chem.*, **33**, 4169 (1968). (c) 7,7-Bis(trifluoromethyl)-quinonemethide. W. A. Sheppard, *J. Org. Chem.*, **33**, 3297 (1968); J. J. Murray, *ibid.*, 3306. (d) The preparation and some reactions of perhalocyclopentanones and -hexanones. L. G. Anello, A. K. Price, and R. F. Sweeney, *J. Org. Chem.*, **33**, 2692 (1968). (e) Fluorine-containing polyethers were prepared starting from trifluoroacetone (through epoxide). F. D. Trischler and J. Hollander, *J. Polymer Sci.*, **5**, (Part A-1), 2343 (1967).

Section 6-2D.2 Fluorinated Ketenes

AD6-97 Preparation and chemistry of monohaloketenes including monofluoroketene. W. T. Brady and E. F. Hoff, Jr., *J. Am. Chem. Soc.*, **90**, 6256 (1968).

AD6-98 Addition reactions across the C=C in bis(trifluoromethyl)-ketene with peroxodisulfuryldifluoride tetrafluorohydrazine, difluoroaminofluorosulfate, and fluorine fluorosulfate. D. T. Meshri and J. M. Shreeve, *J. Am. Chem. Soc.*, **90**, 1711 (1968).

Section 6-2D.3 Fluorinated Acids, Acid Halides, Anhydrides, and Esters

AD6-99 A general method for the preparation of β-perfluoroalkylalanines starting from perfluorocarboxylic anhydride. W. Steglich, H.-U. Heininger, H. Dworschak, and F. Weygand, *Angew. Chem. Intern. Ed. Engl.*, **6**, 807 (1967).

AD6-100 Trifluoroacetylation of hydroxy compounds with trifluoroacetic anhydride and with acetyl trifluoroacetate (including sugars). T. G. Bonner, E. G. Gabb, and P. M. McNamara, *J. Chem. Soc.* (B), 72 (1968); T. G. Bonner, P. M. McNamara, and B. Smethurst, *J. Chem. Soc.* (B), 114 (1968).

AD6-101 The reaction of isocyanic acid with trifluoroacetic anhydride. Preparation of trifluoroacetyl isocyanate and 2,2,2,2′,2′,2′,-hexafluorodiacetamide. W. C. Firth, Jr., *J. Org. Chem.*, **33**, 441 (1968).

AD6-102 Synthesis of esters of ω-nitroso perfluorinated carboxylic acids from anhydride. E. C. Stump, W. H. Oliver, and C. D. Padgeti, *J. Org. Chem.*, **33**, 2102 (1968).

AD6-103 Trifluoroacetolysis of neopentyl *p*-toluenesulfonate. W. G. Dauben and J. L. Chitwood, *J. Am. Chem. Soc.*, **90**, 6876 (1968).

AD6-104 Facile N–O bond cleavages of amine oxides by trifluoroacetic anhydride. A. Ahond, A. Cavé, C. Kan-Fan, H. P. Husson, J. de Rostolan, P. Potier, *J. Am. Chem. Soc.*, **90**, 5622 (1968).

AD6-105 Rapid mass spectrometric identification of the *N*-terminal amino acid residue in terminal *N*-thiobenzoyl polypeptides using trifluoroacetic acid. G. C. Barrett and J. R. Chapman, *Chem. Commun.*, 335 (1968).

AD6-106 Reaction of 2-thienyl metals with perfluoroaliphatic acid derivatives to prepare perfluoroacyl thiophens. E. Jones and I. M. Moodie, *J. Chem. Soc.* (C), 1195 (1968).

AD6-107 Condensations of diethyl fluorooxaloacetate with aldehydes. E. D. Bergmann, I. Shahak, E. Sal'i, and Z. Aizenshtat, *J. Chem. Soc.* (C), 1232 (1968).

AD6-108 Inductive and participation effects in the addition of trifluoroacetic acid to cyclopropanes. P. E. Peterson and G. Thompson, *J. Org. Chem.*, **33**, 968 (1968).

AD6-109 *t*-Butyl fluoroformate—a new reagent for peptide syntheses. E. Schnabel, H. Herzog, P. Hoffmann, E. Klauke, and I. Ugi, *Angew. Chem. Intern. Ed. Engl.*, 7, 380 (1968); *Ann.*, **716**, 175 (1968).

AD6-110 Some novel perfluoroalkanedioic acid derivatives and α,ω-diiodoperfluoroalkanes. V. C. R. McLoughlin, *Tetrahedron Letters*, 4761 (1968).

AD6-111 Oxonium ions in trifluoroacetic acid using trimethyloxonium fluoroborate. P. E. Peterson and F. J. Slama, *J. Am. Chem. Soc.*, **90**, 6516 (1968).

AD6-112 Solvent shifts of methoxy group nmr resonances induced by trifluoroacetic acid as an aid to structure elucidation. R. G. Wilson and D. H. Williams, *J. Chem. Soc.* (C), 2477 (1968).

Section 6-2E Thiocarbonyl and other fluorinated sulfur reagents

For perfluoroalkylsulfinyl halides, see Ref. AD5-9. For addition of SF_5O–F to alkenes, see Ref. AD5-8(a).

AD6-113 Derivatives from CSF_2 and CSFCl. A. Haas and W. Klug, *Chem. Ber.*, **101**, 2609 and 2617 (1968).

AD6-114 Fluorothiocarbonyl isothiocyanate. A. Haas and W. Klug, *Angew. Chem. Intern. Ed. Engl.*, **6**, 940 (1967).

AD6-115 Preparation and properties of trifluorodithioacetic acid. E. Lindner and H.-G. Karmann, *Angew. Chem. Intern. Ed. Engl.*, 7, 301 (1968).

AD6-116 Preparation and properties and structure of $RN{=}SF_2$ derivatives, where R is –COF, COCl, and SO_2F. U. Biermann and O. Glemser, *Chem. Ber.*, **100**, 3795 (1967); B. Krebs, E. Meyer-Hussein, O. Glemser, and R. Mews, *Chem. Commun.*, 1578 (1968).

AD6-117 Ethyl trifluoromethanesulfonate—rates and mechanism of solvolysis. A. Streitwieser, Jr., C. L. Wilkins, and E. Kiehlmann, *J. Am. Chem. Soc.*, **90**, 1598 (1968).

AD6-118 Alkylations with methyl and ethyl fluorosulfonates. M. G. Ahmed, R. W. Alder, G. H. James, M. L. Sinnott, and M. C. Whiting, *Chem. Commun.*, 1533 (1968).

AD6-119 Trifluoromethyl thiolsulfonates and their reactions with mercaptans and amines. J. P. Weidner and S. S. Block, *J. Med. Chem.*, **10**, 1167 (1967).

AD6-120 Recent papers describing preparation and chemistry (usually inorganic) of derivatives containing the NSO_2F group. H. W. Roesky and A. Hoff, *Chem. Ber.*, **101**, 162 (1968); H. W. Roesky, *Angew, Chem. Intern. Ed. Engl.*, **7**, 63, 630 (1968); H. W. Roesky and U. Biermann, *Angew. Chem. Intern. Ed. Engl.*, **6**, 882 (1967).

Section 6-2F Nitrogen fluorides, fluorinated nitriles, and other nitrogen compounds.

For N–F compounds, see under Sections 6-1A.2 and 6-1C.2. For introduction of $(CF_3)_2N$ group, see Refs. AD6-23, -24, -72(c), -78. For fluorinated nitriles in cycloaddition, see Ref. AD6-73(c).

AD6-121 Bis(trifluoromethyl)aminotrimethylsilane. H. G. Ang, *J. Chem. Soc.* (A), 2734 (1968).

CHAPTER 8 COMPARATIVE REACTIVITY OF FLUORINATED ORGANIC COMPOUNDS TO HYDROCARBON ANALOGS

For information on electronic effects, see Ref. AD6-81; on aromatic derivatives see Refs. AD6-79, -85; on the photohydrolysis of a CF_3 group, see Ref. AD6-84.

AD8-1 The electrochemical transformation of α,α,α-trifluoroacetophenone into acetophenone; an unusually ready hydrogenolysis of the C–F bond. J. H. Stocker and R. M. Jenevein, *Chem. Commun.*, 934 (1968).

CHAPTER 9 SAFETY, TOXICITY, AND BIOLOGICAL IMPORTANCE OF ORGANOFLUORINE COMPOUNDS

Section 9-3A Medicinal chemicals

For fluorinated steroids, see Refs. AD6-31, -32, -33; for 7-trifluoromethoxy and 7-trifluoromethylthio derivatives of 1,4-benzodiazepines, see Ref. AD6-83.

AD9-1 1-2-[4′-(Trifluoromethyl)-4-biphenyloxyl] ethylpyrrolidine–a potent hyposterolemic agent. S. Gordon and W. P. Cekleniak, *J. Med. Chem.,* **11**, 993 (1968).

AD9-2 Monofluoroisoquinolines–pharmacological activity. J. C. Belsten and S. F. Dyke, *J. Chem. Soc.* (C), 2073 (1968).

Section 9-3C.1 Fluorinated derivatives of amino acids

For a general method for preparation of β-perfluoroalkylalanines, see Ref. AD6-99. Synthesis of 3,3,3-trifluoroalanine and of 3,3,3-trifluoroalanyl peptides. F. Weygand, W. Steglich, W. Oettmeier, A. Maierhofer, and R. S. Loy, *Angew. Chem. Intern. Ed. Engl.*, **5**, 600 (1966).

Section 9-3C.2 Fluorinated derivatives of sugars

See Refs. AD5-10, AD6-100.

Section 9-3C.4 Fluorinated derivatives of nucleic bases

AD9-3 Synthesis of 8-fluoroadenosine. M. Ikehara and S. Yamada, *Chem. Commun.*, 1509 (1968).

Section 9-3D Miscellaneous applications

For peptide structure elucidation, see Ref. AD6-105.

AD9-4 Use of fluorocarbons in place of blood. *New Scientist,* **38**, 245 (1968).

AD9-5 Trifluoroisopropyl cyanoacrylate shows promise as a surgical tissue adhesive. *Chem. Eng. News,* **46**, 35 (July 1, 1968).

AD9-6 2-Fluoropyridine N-oxide and its reactions with amino acid derivatives (blocking reagent in peptide analysis). D. Sarantakis, J. K. Sutherland, C. Tortorella, and V. Tortorella, *J. Chem. Soc.* (C), 72 (1968).

AD9-7 Use of sulfonyl fluorides as irreversible enzyme inhibitors. B. R. Baker and R. B. Meyer, Jr., *J. Med. Chem.,* **11**, 489 (1968).

AD9-8 *t*-Butyl fluoroformate–a new reagent for peptide syntheses. See Ref. AD6-109.

APPENDIX B SPECIALIZED OR LIMITED REVIEWS ON FLUORINE CHEMISTRY

ADB-1 See Ref. AD3-2.

ADB-2 Nucleophilic reactions of metal carbonyl anions with fluorocarbons. M. I. Bruce and F. G. A. Stone. *Angew. Chem. Intern. Ed. Engl.,* **7**, 747 (1968).

ADB-3 Nitrogen oxide fluorides. R. Schmutzler, *Angew. Chem. Intern. Ed. Engl.,* **7**, 440 (1968).

ADB-4 Fluorine-19 Nuclear Magnetic Resonance Spectroscopy. A 68 page review with 182 references. E. F. Mooney and P. H. Winson. *Ann. Rev. NMR Spectroscopy,* **1**, 244 (1968).

Index*

*A designation in parentheses following a page entry refers to the reference number of a review appearing on that page.

Page numbers in boldface type indicate the most significant discussion of the topic.